Proceedings of the 35th International MATADOR Conference

Srichand Hinduja • Kuang-Chao Fan
Editors

Proceedings of the 35th International MATADOR Conference

Formerly The International Machine Tool Design and Research Conference

Springer

Srichand Hinduja, PhD
School of Mechanical, Aerospace and Civil Engineering
The University of Manchester
Sackville Street
Manchester
M60 1QD
UK

Kuang-Chao Fan, PhD
Department of Mechanical Engineering
National Taiwan University
1, Sec. 4, Roosevelt Rd.
Taipei 10627
Taiwan
ROC

British Library Cataloguing in Publication Data
A catalogue record for this book is available from the British Library

Library of Congress Control Number: 2007929078

ISBN-13: 978-1-84628-987-3 e-ISBN-13: 978-1-84628-988-0

Printed on acid-free paper.

9 8 7 6 5 4 3 2 1

Springer Science+Business Media
springer.com

Proceedings of The Thirty-Fifth International

MATADOR

Conference

Organising Committee

Professor S. Hinduja *(Co-Chairman)*
Professor K.-C. Fan *(Co-Chairman)*
Professor Z. Q. Lin *(Co-Chairman)*
Professor L. Li
Professor C. L. Liu
Professor Y. S. Liao
Professor Y. Y. Tsai
Professor L. F. Xi
Dr J. Atkinson
Dr P. Mativenga
Dr S. Mekid
Dr M. A. Sheikh
Dr R. Heinemann

Organising Secretaries

Mrs C. Collins
Ms J. Chen

The University of Manchester,
National Taiwan University
and
Shanghai Jiao Tong University

Foreword by the Conference Chairmen

It gives us pleasure to introduce the Proceedings for the 35th internal MATADOR conference. The Proceedings include 88 refereed papers from several countries on different continents, thus making this Conference truly international.

Since its inception in 1959, the MATADOR conferences have been held at various locations within the United Kingdom and this is the first time that it is being held in Taiwan. This reflects the increasingly important role played by the East Asian countries in manufacturing.

When the MATADOR Conference was established, it dealt with metal working fundamentals in cutting, grinding and forming and the design of metal working machines for this purpose. Since those early days, new areas of research in manufacturing have emerged which therefore have become part of the conference's scope as well. Some of the emerging areas that have been addressed in this conference include:

- Micro- and nano-manufacture;
- Laser materials processing;
- Advanced coatings technology;
- Globalisation of manufacture and its implications;
- Fuel Cell technology;
- Rapid Manufacture, and
- Inspection technology for micro parts.

We would like to express our gratitude to all the members of the organising committee and the companies and organisations that have made this conference possible. Our special thanks go to Ms Christine Collins and Dr Robert Heinemann for their extra effort in organising this conference and in the preparation of these proceedings.

We hope you find the papers interesting and stimulating.

Professors Srichand Hinduja and Kung-Chao Fan
Manchester, UK and Taipei, Taiwan
2007

Contents

4 Design of Optical Components

5 Electro-physical and chemical Processes

6 Forming

7 Fuel Cell Technology

12 Precision Machine Design

13 Process and Machine Tool Control

14 Rapid Prototyping

15 Robotics

16 Sensor Technology

1

Bio- and Nano-technologies

1–1

Preheated Carbon Source for Carbon Nanotube Synthesis

C.Y. Liu[1], N. K. Chang[1], Y. T. Chang[1], J. H. Hsu[1], S. H. Chang[1,2]

[1]Department of Mechanical Engineering National Taiwan University, Taiwan, ROC, d92522010@ntu.edu.tw
[2]Mechanical and Systems Research Laboratories Industrial Technology Research Institute, Taiwan, ROC, shuochang@itri.org.tw

Abstract. In this paper, we synthesized single-walled carbon nanotubes (SWCNTs) using chemical vapor deposition (CVD) process in a furnace with three different heating regions. Ferritin nanoparticles were used as catalyst for the CNTs synthesis through carbon source which are methane and ethylene mixture. The reason we used a three stage furnace is to preheat the carbon source at higher temperatures and to synthesize CNTs at lower temperatures. This method can be effective to grow SWNTs at lower temperature. Furnace temperature, methane flow, productivity of CNTs and the diameter distributions of CNTs were considered in the process of CNTs synthesis. The lowest synthesis temperature of CNTs grown from ferritin nanoparticles was 700°C. In the future, it is possible to apply this method to grow CNTs on electronic devices at the lower temperature.

Keywords: carbon nanotubes, nanoparticles, CVD, synthesis temperature,

1. Introduction

Carbon nanotubes (CNTs) were discovered by Iijima in 1991 [1], carbon nanotubes have been intensively studied in chemistry, physics, and materials science. Numerous methods to grow CNTs have been developed, including arc discharge, chemical vapor deposition, pulsed laser ablation and so on [2-4]. Chemical vapor depositions (CVD) of SWNTs has been proved to be a classic method to produce SWNTs. Compared to the arc discharge and laser ablation methods, CVD could offer more control over the growth temperature, catalyst composition, feeding gas, and other parameters [5-7].

There were already a few reports on carbon nanotube growth utilizing ferritin as a catalyst. Zhang et al. dispersed iron oxide nanoparticles produced from apoferritin on a mesoporous alumina support and synthesized single-walled carbon nanotubes (SWNTs) at 900°C [8]. Diameter-controlled growth of MWNTs has been achieved by CVD with ferrtin by Daiyu et al. [9].

In this paper, we report on the single-wall carbon nanotube synthesis on a silicon oxide substrate using ferritin as a catalyst by chemical vaper deposition process in a furnace with three different heating regions at a temperature as low as 700°C.

2. Experimental

A CVD set-up equipped with a 1 inch quartz tube furnace was used. In the CVD, three heating zones in the furnace can be individually controlled to heat up. The catalyst we used was ferritin consisting of a protein shell and iron particles [10]. Concentration of the ferritin solution was controlled to 0.5 g/L by deionized water dilution. Silicon wafers with 500 nm thermal oxide layer were used as substrates. Then, the substrates were placed on a flat quartz board and inserted into the three positions of a 1 inch diameter quartz tube into the CVD furnace, as show in Figure 1.1(a). The furnace was heated up to 950°C, 850°C, and 750°C in 50 minutes. Growing nanotubes on the substrate took place for 10 minutes under combined flows of 500 sccm of H_2, 20 sccm of C_2H_4 and 62-1000 sccm of CH_4. The region at 950°C was a preheating zone, 850°C was a transition zone and 750°C was a growth zone, as show in Figure 1.1(b).

The diameters of the ferritins, nanoparticles, and nanotubes were mainly determined with an Atomic Force Microscope (AFM) in the tapping mode. High-resolution transmission electron microscopy (HR-TEM; 2100F, JEOL) and scanning electron microscopy (SEM; ERA 8800, Elionix) were used to characterize the carbon nanotubes.

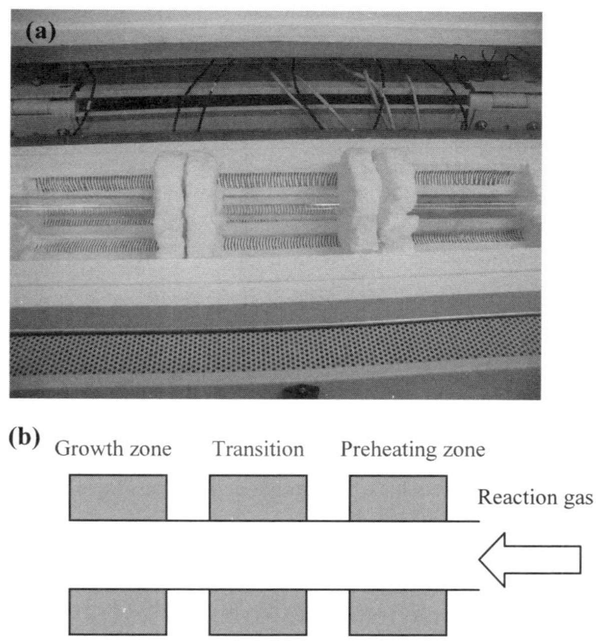

Figure 1.1. (a) The form of the CVD furnace; (b) Sketch of the three-segment CVD.

3. Results and Discussion

An AFM image of ferritin on a silicon substrate is shown in Figure 1.2. The diameter of ferritin is about 12 nm. It can be seen in the image that the ferritin molecules are uniformly distributed on the substrate. After nanotube growth, we scanned 20 different areas (unit area 10 μm × 10 μm) on the substrate by AFM and recorded the counts, lengths and diameters of nanotubes. We defined the productivity of nanotubes as the total length of nanotubes per unit area. The counts of nanotubes synthesized at 950°C and 850°C, under the flow rate of 500 sccm of H_2, 20 sccm of C_2H_4 and 62-1000 sccm of CH_4, are as listed in Table 1.1. The productivity of SWNTs per 100 μm^2 at 950°C and 850°C are shown in Figure 1.3. The optimal productivity of SWNTs was obtained when the 500 sccm of CH_4 at 950°C and 250 sccm of CH_4 at 850°C were conducted, respectively. Figure 1.4 shows the diameter distribution of SWNTs grown at 950°C and 850°C. The SEM images of the diameters of SWNTs were 3.2 nm and 3.8 nm at 950°C and 850°C, respectively, as shown in Figure 1.5.

Few SWNTs were grown at 750°C, which was controlled by the catalyst of ferritin only when the flow rate with CH_4 of 250 sccm, C_2H_4 of 20 sccm and H_2 of 500 sccm. The diameters of SWNTs were 0.85~3.3 nm and the length was about 1~5 μm, as shown in Figure 1.6. The productivity of SWNTs at 700°C was less than that at 750°C only when the flow rate was controlled with CH_4 of 125 sccm, C_2H_4 of 20 sccm and H_2 of 500 sccm. The diameters of SWNTs were 0.9~2.2 nm and the length about 2~5μm, as shown in Figure 1.7.

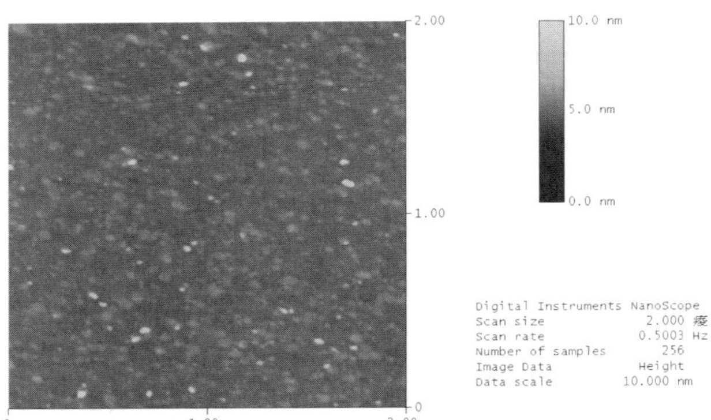

Figure 1.2. AFM image of the concentrations of the ferritin solutions were controlled to 0.5 g/L by deionized water dilution.

Table 1.1. The counts of nanotubes at different flow rate on CH_4 (H_2:500 sccm, C_2H_4:20 sccm)

CH_4 (sccm)	1000	750	500	250	125	62
The counts of CNTs per 100 μm^2 at 950°C	3	1	10	6	2	0
The counts of CNTs per 100 μm^2 at 850°C	1	2	4	8	3	1

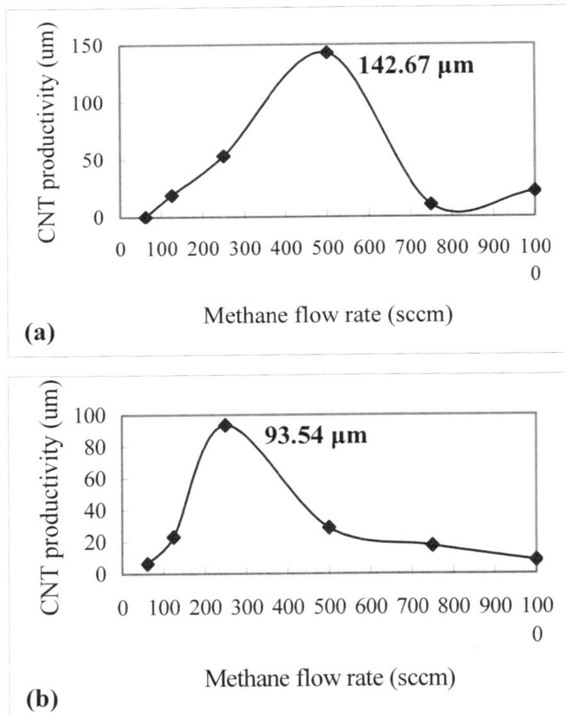

Figure 1.3. The productivity of carbon nanotubes per 100μm^2, (a) 950°C (b) 850°C (H_2:500 sccm, C_2H_4:20 sccm and CH_4:62-1000 sccm)

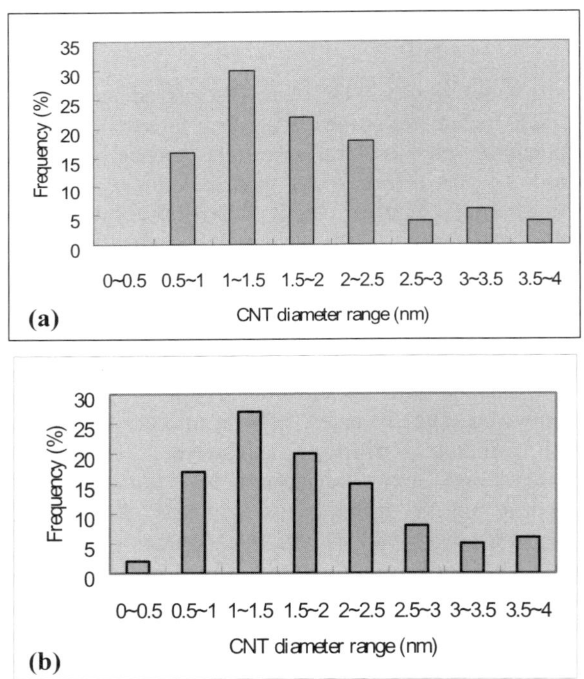

Figure 1.4. Diameter distribution of carbon nanotubes grown at (a) 950°C (b) 850°C.

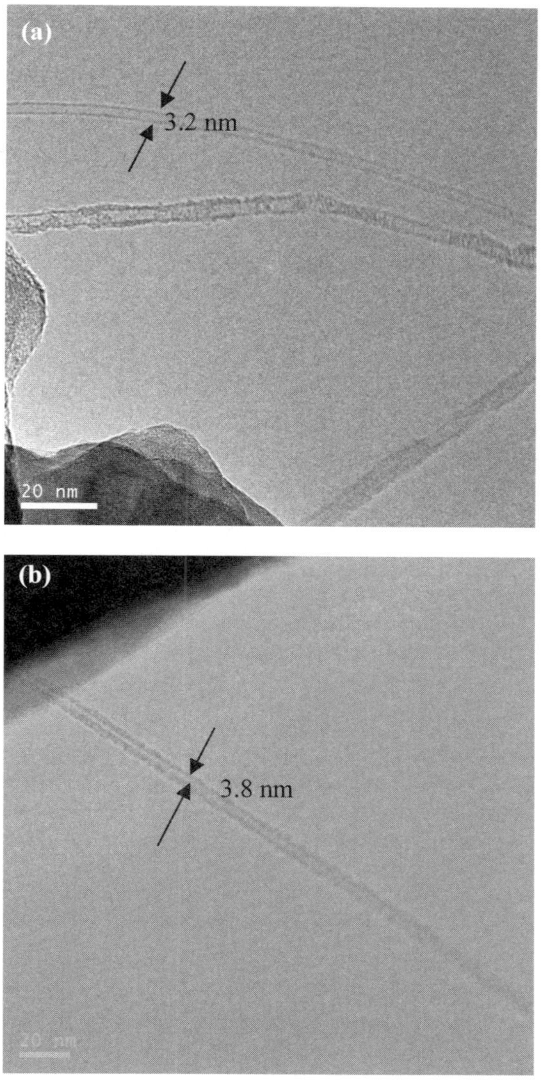

Figure 1.5. TEM images of carbon nanotubes grown at (a) 950°C (b) 850°C

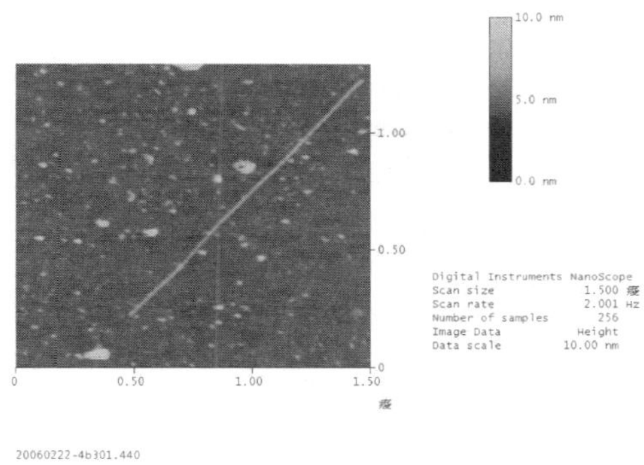

Figure 1.6. The AFM image of SWNT at 750°C.

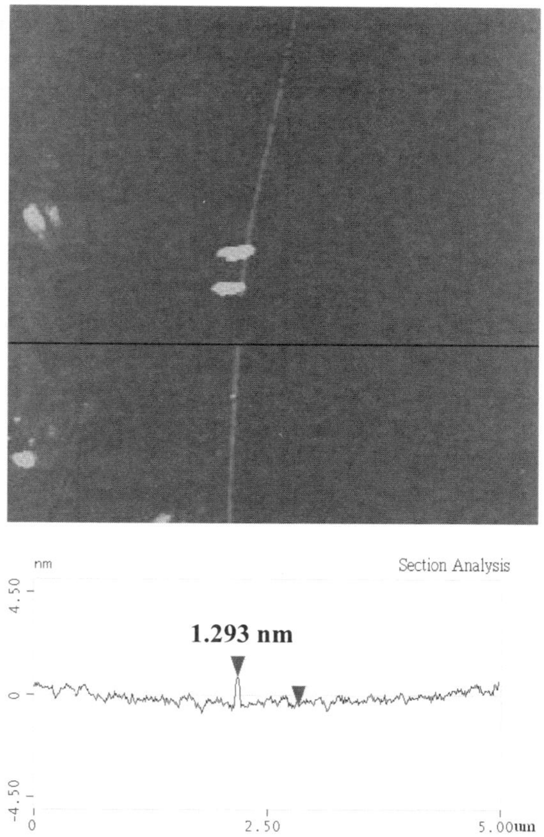

Figure 1.7. The AFM image of SWNT at 700°C.

SWNTs was synthesized by chemical vapor deposition (CVD) process in a furnace with three different heating regions. The optimal parameters of grown SWNTs in our experiments are shown in Table 1.2.

Table 1.2. The optimal parameters of grown SWNTs at different temperatures.

Temperature	Optimal parameters
950°C	CH_4:500 sccm, H_2:500 sccm, C_2H_4:20 sccm
850°C	CH_4:250 sccm, H_2:500 sccm, C_2H_4:20 sccm
750°C	CH_4:250 sccm, H_2:500 sccm, C_2H_4:20 sccm
700°C	CH_4:125 sccm, H_2:500 sccm, C_2H_4:20 sccm

4. Conclusions

In summary, experimental results revealed that the reaction gases showed an optimal flow rate at different growth temperatures. The flow rate of methane is important to synthesize SWNTs at low temperature. A higher flow rate of methane would reduce the catalytic activity by the great amount of carbon decomposition at the suitable growth temperature. The flow rate of methane can be reduced by decreasing the growth temperature. We couldn't synthesize SWNTs from ferritin on the substrate at 675°C. It is possible that the catalyst and carbon source are hard to react at this temperature. In our study, the lowest temperature at which a growth of SWNTs by the catalyst of ferritin was observed was 700°C and the mean diameter of SWNTs was 1~4 nm.

References

[1] S. Iijima, 1991, Nature, 354, P.56.
[2] P.M. Ajayan, S. Iilima, 1993, Nature, 361, P.333.
[3] Z. W. Pan, S. S. Xie, B. H. Chang, L. F. Sun, W. Y. Zhou, G. Wang, 1999, Chem. Phys. Lett, 299, P.97.
[4] B. I. Yakobson, R. E. Smalley, 1997, Am. Sci, 85, P.324.
[5] Su. M, Zheng. B, Liu, 2000, J. Chem. Phys. Lett, 322, P.321.
[6] H. E. Unalan, M. Chhowalla, 2005, Nanotechnology, 16, P.2135.
[7] G. H. Jeong, S. Suzuki, 2005, J. App. Phys, 98, P.124311.
[8] Y. Zhang, Y. Li, W. Kim, D. Wang and H. Dai, 2002, Appl. Phys. A, 74, P.325.
[9] D. Kondo, S. Sato, A. Kawabata, Y. Awano, 2005, J. J. Appl. Phys, 44, P.5292.
[10] Yiming Li, Woong Kim, et al., 2001, J. Phys. Chem. B, 105, P. 11424.

Nano Manufacturing of a Small-tipped Stylus for Measurement of Micro-Machine Elements with Concave Surfaces

S. Kurokawa[1] and S. Imamura[2]

[1]Kyushu University, Fukuoka, Japan, 819-0395, kurobe-@mech.kyushu-u.ac.jp
[2]Graduate School, Kyushu University, Fukuoka, Japan, 819-0395

Abstract. The purpose of this research is to manufacture a new small-tipped stylus for measurement of micro machine elements with concave shapes. It is very difficult to measure surfaces of concave shapes because its overhanging part becomes the obstacle against measurement. We need to develop a small tipped stylus to be able to reach the position to be measured. In machining process, one of the photolithography technique is applied: the combination of the electron-beam exposure and the etching process of a silicon chip. In this paper, a preferable stylus shape is proposed and the trial manufacturing of the stylus has been performed with wet etching and dry deep reactive ion etching (DRIE). Several process conditions of etching and resultant shapes of manufactured styli are investigated and evaluated with SEM observation. The difficulty of achieving adequate etched features is shown especially due to the degree of under-etch phenomena.

Keywords: Micro Machine Element, Measurement, Small-tipped Stylus, Nano Manufacturing

1. Introduction

It is very important to measure machined parts in order to check the manufacturing accuracy. Nowadays, micro machine elements can be manufactured with MEMS [1,2]. Those which have concave parts such as micro gear surfaces [3] should be also measured, but at this moment, it is not possible because their overhanging part becomes the obstacle against measurement. For example, a conventional contact type stylus is too large compared to concave spaces of micro machine parts so that it can not reach deeply inside into a concave space. A conventional detecting unit for AFM also has the same problem. On the other hand, non-contact type probes such as a laser light beam is blocked off or reflects with a very small angle, which can not come back to a detector in consequence.

To overcome those difficulties, the authors tried to manufacture a small tipped stylus in a special shape which is preferable for measurement of concave shapes. In machining process, one of the MEMS techniques is applied, i.e. the photolithography which utilizes the combination of the electron beam exposure and the etching process of a silicon chip.

In this paper, preferable stylus shapes are proposed first. Each of them should have a small tip and has relatively large stiffness compared to a conventional probe of AFM. Styli used for AFM are essentially cantilever beams that bend easily under small forces and have sharp pointed tips. However, the principle of AFM can not be used for measuring micro concave parts. The rigidity of a probe is necessary for measurement and the tip shape should be controlled with certain curvature. Secondly, the trial manufacturing of the styli has been performed with electron beam lithography, wet etching and dry deep reactive ion etching (DRIE). Several process conditions of etching and resultant shapes of manufactured styli are investigated and evaluated with SEM observation. It is discussed the possibility for realization of measuring styli for concave surfaces of micro machine elements and manufacturing conditions are suggested.

2. Design of a small-tipped stylus

The example subject of measuremet is supposed to be a micro gear whose module is about 0.05. This size is about 1/100 times less than that of conventional industrial gears. For surface topography measurement of such micro gears, a small-tipped stylus must be used in contact with very light measuring force. For this purose, the designed stylus should have large rigidity and not have a sharpe pointed tip. The tip of stylus must have certain curvature so that the tip doesn't scratch the surfaces. On the other hand, it is important to consider the attachment to the measuring instrument so as to be handled easily. Therefore, the stylus must have a small tip and also have a quite large root form. It should also have a large volume so as to keep the rigidity of the body. The designed shape that satisfies those conditions is shown in figure 1 for example.

The stylus consists of an extremely small part as well as a relatively large part and it has a monolithic construction. The radius of curvature of the tip is 2 μm, while the root size is about 3.0 mm in this example. Another stylus, whose radius of curvature of the tip is 4 μm, is also designed so as to observe the influence of the tip radius. In order to keep the rigidity, those styli have the adequate thickness. Such characteristics are essentially different from those of conventional AFM styli.

3. Nano Manufacturing Method

The designed stylus has the construction of a combination of a tiny part and relatively large part in one body. The quality of the tip shape, expecially its radius of curvature, will be a key characteristic of the stylus. One of some

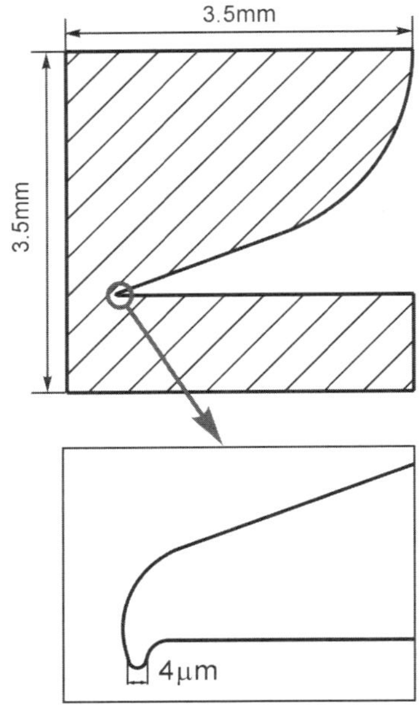

Figure 1.1. Example of a designed stylus

adequate techniques to achieve the quality and to have manucfacturing capability of tiny parts is the lithography technique. There are sevral lithography techniques, for example, photo lithography, electron beam lithography, ion beam lithography, nano-imprint lithography, and their combination technique [4,5]. Most of those techniques are normally used to create small pits or high aspect walls. Trial manufacturing of ultra sharp pointed tips for AFM are performed with the combination of lithography and etching technique [6-8], but the control of tip curvature and the rigidity of the stylus is not considered. In order to manufacture both tiny tips under the controlled radius of curvature and a rigid whole body with a large volume, we utilize here the combination of electron beam lithography and etching technique of silicon substrate.

The configuration of the workpiece of a silicon chip is shown in Table 1.1 with thickness of each layer.

Table 1.1. The configuration of workpiece

Configuration Layer		Thickness
Material substrate	Si	500μm
Si etching mask	SiO_2	850nm
	Cr	250nm
EB resist	ZEP520A	320nm

The material substrate is silicon (Si), on which silicon dioxide (SiO_2) is formed as a Si etching mask. The thickness of SiO_2 is quite large, about 850 nm. However, in the Si etching process, even 850nm-thick mask of SiO_2 is not enough for deep etching. We need to etch the substrate passing through 500 μm thickness to create a rigid stylus. The SiO_2 layer could not bear such a deep etching process. Therefore, the metal layer of chromium (Cr) is created on the SiO_2 layer by sputtering. This metal layer is strong enough for the deep Si etching process. Electron beam resist (EB resist: ZEP520A, ZEON corp.) is pasted on the Cr layer by spin coating. The design pattern shown in Figure 1.1 is drawn by electron beam and the EB resist of the hatched area is exposed. After development, the EB resist of the exposed area is removed. The remained EB resist will act as the mask for Cr and SiO_2 etching.

4. Results of Nano Manufacturing

4.1 Electron Beam Exposure

Electron beam exposure was carried out using a high resolution electron beam lithography system (S-4300SE type SEM from Hitachi Ltd.) at 30kV. Figure 1.2 shows the SEM image of the designed stylus pattern, whose radius of curvature is 2 μm after electron beam drawing and development.

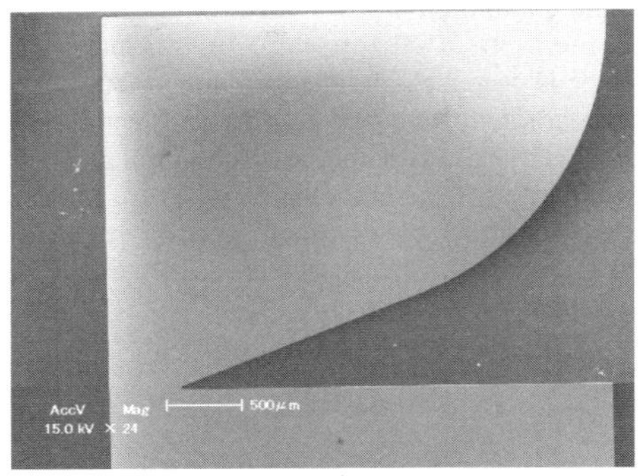

a. The whole pattern

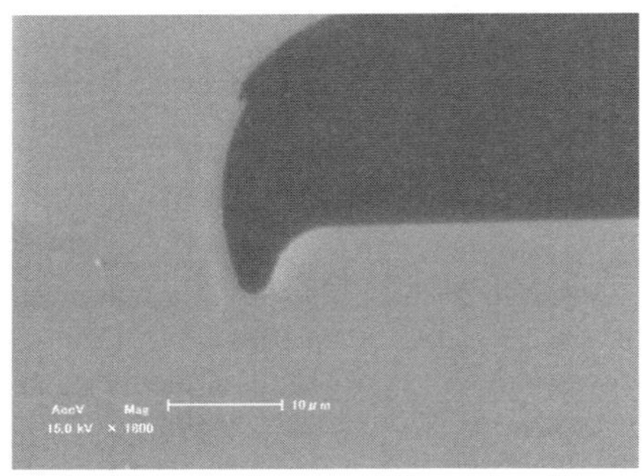

b. Close-up image of the stylus tip

Figure 1.2. SEM images of the designed stylus after EB exposure and development

The developed result shows that it is exposed as designed and the radius of curvature of the stylus tip is also patterned exactly. This EB resist pattern is used as a mask of etching of Cr and SiO_2 layers.

4.2 Etching

Etching of Cr and SiO_2 layers was carried out using the wet etching process. Etchant of the Cr layer is mixed liquid of perchloric acid into cerium ammonium nitrate and etching time is 5 min. Etchant of the SiO_2 layer is buffered hydrogen fluoride and etching time is 20 min. The wet etching has an isotropic characteristic by nature.

This is the disadvantage for the high precision micro machining, because the etched pattern tends to become slightly smaller than the designed one. However, the shrinking phenomenon can be improved to feed the amount of shrinkage to the designed pattern so as to be exposed as expected. On the other hand, the combination of Cr and SiO_2 layers can be expected to have quite long endurance against the etching of Si substrate. It is the essential characteristic in application of deep etching.

Etching of Si substrate was carried out using a dry etcher (SAMCO Model RIE-10iP). In order to etch the substrate passing through 500 μm thickness, deep reactive ion etching (DRIE) was used in this process. In DRIE process, a mixture of SF_6 and Ar gases was used in etching Si substrate, while a mixture of C_4F_8 and Ar gases was used in the passivation process. There are many parameters to be controlled in etching and passivation.

We made a lot of trials and two cases of them are explained here. The parameters for the first case are listed in Table 1.2.

Table 1.2. The etching and passivation parametes in the case 1

Si etching			Passivation		
Gas flow rate (sccm)			Gas flow rate (sccm)		
SF_6	Ar	C_4F_8	SF_6	Ar	C_4F_8
50	20	0.5	0.5	20	50
Bias (W)		50	Bias (W)		20
ICP (W)		700	ICP (W)		550
Pressure (Pa)		4	Pressure (Pa)		4
Time (s)		4	Time (s)		2

Etched results of the stylus pattern, whose radius of curvature is 2 μm, after 200 loops of the etching-passivation step, are shown in Figure 1.3.

Although DRIE process was applied, Si substrate has etched excessively. From Figure 1.3 b and Figure 1.3 c, the sidewall around the tip has completely vanished and the quite large undercut can be observed. The amount of undercut is about 54 μm, while the depth of etched substrate is about 75 μm. This etching seems to be almost isotropic. From this result, even if the DRIE process is applied, the etching is not always anisotropic and it depends strongly on the parameter conditions of DRIE. From Figure 1.3 d, it is found that Cr and SiO_2 layers still exist in enough thickness of about 1μm. It may be interesting to use this thin layer as a cantilever beam for AFM for example, but our purpose is to manufacture a more rigid and thicker stylus that is capable of to be used as a contact type stylus. We have to adjust some conditions to realize weaker etching and stronger passivation than the first case.

The selected parameters for the second case are listed in Table 1.3. Some conditions for Si substrate etching have changed slightly; bias power was decreased from 50 W to 40 W, and the ICP power was decreased from 700 W to 500 W too. Conditions for passivation have changed more drastically. The flow rate of the deposition gas C_4F_8

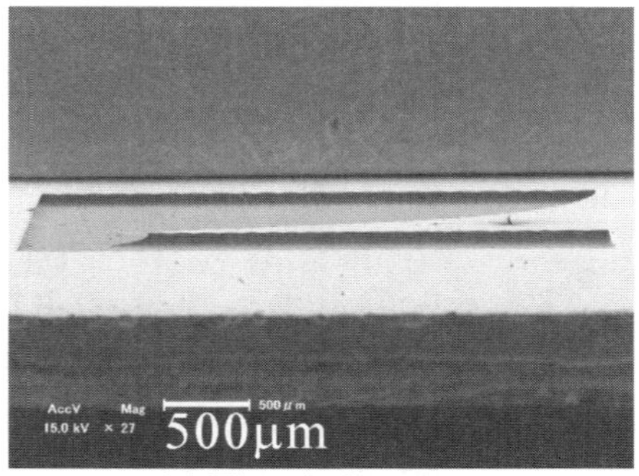

(a) The whole pattern

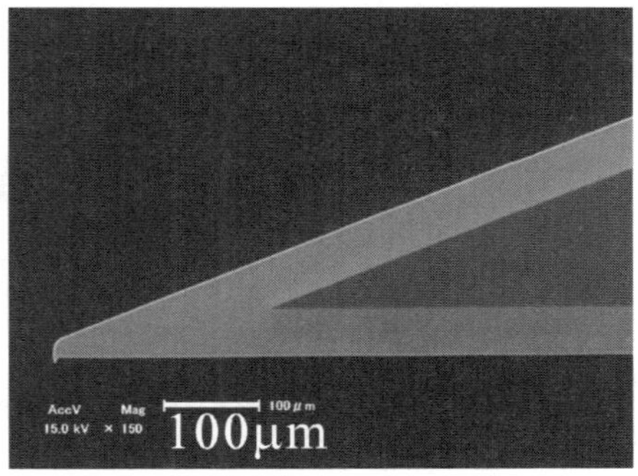

(b) The tip area

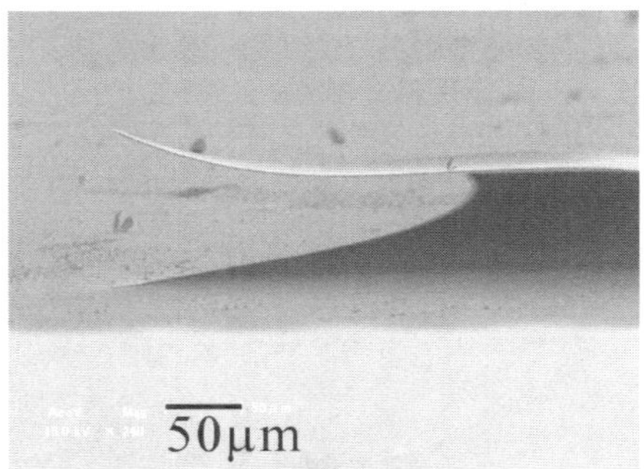

(c) Close-up of the tip

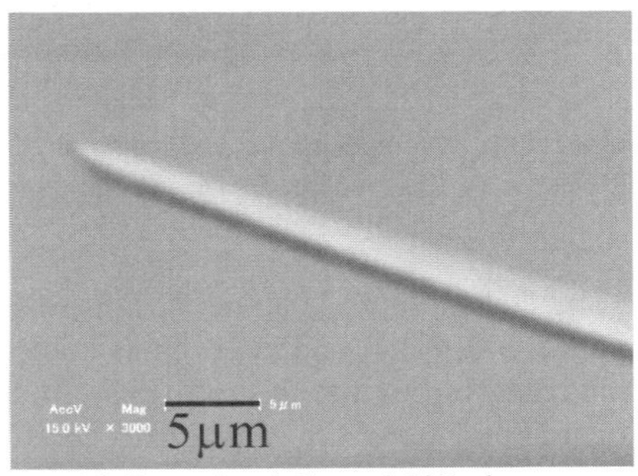

(d) Remained Cr and SiO_2 layers

Figure 1.3. SEM images of the designed stylus after DRIE etching in 200 loops under the first condition

increased from 50 sccm to 70 sccm. ICP power also increased from 550 W to 700 W. Finally, passivation time has doubled from 2s to 4s. Under those conditions, etched results of the stylus pattern, whose radius of curvature is 4μm, after 300 loops of the etching-passivation step, are shown in figure 1.4.

From Figure 1.4, it seems that the passivation process is a little bit strong, but the sidewall remains as desired and the stylus tip keeps the same shape along in the depth

Table 1.3. The reselected etching and passivation parametes in the case 2

Si etching			Passivation		
Gas flow rate (sccm)			Gas flow rate (sccm)		
SF_6 50	Ar 20	C_4F_8 0.5	SF_6 0.5	Ar 20	C_4F_8 70
Bias (W)		40	Bias (W)		20
ICP (W)		500	ICP (W)		700
Pressure (Pa)		4.5	Pressure (Pa)		4.5
Time (s)		4	Time (s)		4

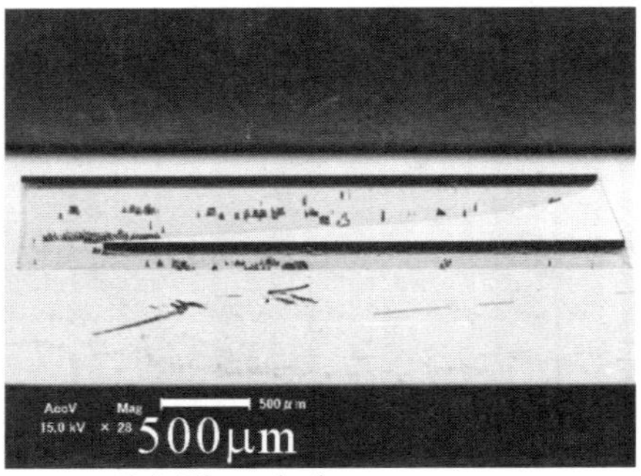

(a) The whole pattern

(b) Close-up of the tip

Figure 1.4. SEM images of the designed stylus after DRIE etching in 300 loops under the second condition

direction. The etched depth is about 72 µm, so the aspect ratio is about 9.0. The undercut phenomenon hardly observed and this is just the desired result. If the DRIE process is continued under the same DRIE conditions, the thick and rigid stylus will be expected.

5. Conclusion

A new small-tipped stylus for measurement of micro machine elements with concave shapes was designed and trial manufacturing was performed using the combination of the electron-beam exposure and the etching process of a silicon chip. By choosing appropriate conditions for the deep reactive ion etching process, the thick and rigid stylus which has the controlled tip shape and the controlled radius of curvature at least to 4 µm can be manufactured.

The next step is to etch the substrate passing through the chip and to complete the stylus. After that, the optimum tip shape should be investigated by applying some etched styli for actual measurement.

Acknowledgement

The electron beam exposure was performed using S-4300SE type SEM drawing system and deep reactive ion etching was performed using SAMCO Model RIE-10iP etcher. The authors would like to thank Prof. Atsushi Takahara at the Institute for Materials Chemistry and Engineering, Kyushu University for his help in using both systems mentioned above at the Collabo-station II, Kyushu University. The wet etching and sputtering were performed at Department of Aeronautics and Astronautics, Faculty of Engineering, Kyushu University under the help of Assoc. Prof. Koji Takahashi and Mr. Tatsuya Ikuta.

References

[1] B. Bhushan ed.: 2004, Handbook of Nanotechnology, Springer.
[2] Robert W. Johnstone, M. Parameswaran: 2004, An Introduction to Surface-Micromachining, Kluwer Academic Publishers.
[3] Toshiyuki Horiuchi, et al.: 2006, Micro-gear fabrication using optical projection lithography on copper-clad plastic substrates and electroplating of nickel, Microelectronic Engineering, Vol.83, P1316-1320.
[4] Xing Cheng, L. Jay Guo: 2004, A combined-nanoimprint-and-photolithograpy patternig technique, Microelectronic Engineering, Vol.71, P277-282.
[5] C. M. Waits, B. Morgan, et al.: 2005, Microfabrication of 3D silicon MEMS structures using gray-scale lithography and deep reactive ion etching, Sensors and Actuators A, Vol.119, P245-253.
[6] Mitsuhiro Shikida, Michiaki Odagaki, et al.: 2004, Non-photolithograhic pattern transfer for fabricating arrayed three-dimensional microstructures by chemical anisotropic etching, Sensors and Actuators A, Vol.116, P264-271.
[7] Jiarui Tao, Yifang Chen, et al.: 2005, A systematic study of dry etch process for profile control of silicon tips, Microelectronic Engineering, Vol.78-79, P147-151.
[8] X. D. Bai, Zhi Xu, et al.: 2005, Aligned 1D silicon nanostructure arrays by plasma etching, Science and Technology of Advanced Materials, Vol.6, P804-808.

Surface Plasmon Resonance Instrument to Detect Biomolecular Interactions

C.L. Chiang[1], H.C. Shih[2], T.K. Chou[3] and C.T. Yang[4]

[1]Industrial Technology Research Institute, Taiwan, 886-3-5741211, C.L.Chiang@itri.org.tw
[2]Industrial Technology Research Institute, Taiwan, 886-3-5741222, hcshih@itri.org.tw
[3]National Central University, Taiwan, 886-3-4227151#34357, 943207024@cc.ncu.edu.tw
[4]Industrial Technology Research Institute, Taiwan, 886-3-5741206, ctyang@itri.org.tw

Abstract. Surface Plasmon Resonance (SPR) is a powerful technique to detect biomolecular interactions such as DNA and DNA, RNA and protein, antigen and antibody interactions in real-time. The change of SPR situation can be measured by angle, wavelength, intensity or phase interrogation. Since a first commercial SPR instrument was presented at 1990 by Biacore life sciences, there were many different types of SPR instruments bulit. We propose an SPR instrument which adds imaging function to intensity interrogation and can be applied in array chip observation. The wavelength interrogation function also integrated in this instrument. This instrument can be used to detect biomolecular interactions in real time and produce sensorgram. For preliminary experiment, we used different concentration rabbit IgG and Anti-IgG to verify our instrument design, and the experimental results are shown as sensorgram. Optical detection limit at intensity interrogation and wavelength interrogation of the instrument has achieved 8×10^{-4} RI/° and 4×10^{-4} RI/nm.

Keywords: Surface Plasmon Resonance, SPR, biomolecular interactions, fluid interactions

1. Introduction

Surface Plasmon Resonance (SPR) is very sensitive to slight changes of thickness and refractive index (RI) of thin films. For this reason, SPR instruments can monitor dynamic molecular interactions happening on bio thin-film surfaces. SPR sensor platform has quite a lot different types, such are Attenuated-Total Reflection (ATR), optical prism coupling, optical fiber coupling, grating coupling and etc. Most commercial products use a prism coupling configuration, because it is comparatively simple to build.

There are four kinds of interrogation modes in prism coupling SPR instruments, which are

- wavelength,
- angular,
- intensity, and
- phase interrogations.

In 1990, Pharmacia Biosensor released the first commercially optical biosensor detection platform called Biacore. It launched the field of label-free real-time molecular interaction analysis. In 1996, Pharmacia spun the biosensor division out as a new company, Biacore AB, and present a new Biacore 2000 platform. This platform allowed user to run one sample over four flow cell surfaces independently or in series [1]. For this reason, new SPR instrument products should consider high throughput capabilities [2].

Since the Human Genome Project has been completed, biomedical research has been shifting from genomics to proteomics. Proteomics research has two main areas, one is the morphology of proteins and the other is protein to protein interaction [3,4]. Biomedical researchers also want to further understand bio-molecule interaction in real-time and to help the genomics development [5]. As above, SPR instruments become very useful detection tool for monitoring bio-molecular interactions.

We proposed an SPR instrument, which adds imaging function to intensity interrogation for observing multiple detection points as once. In addition, we add spectrum meter to get wavelength interrogation at the same time. The experiments in this paper were chose to demonstrate the potential of this instrument in proteomic research.

2. Theory of the instrument

The SPR system is built up in the Kretschmann configuration. A glass slide with a metal thin-film coating on surface and attahch to a prism coupler (either triangular or hemicylindrical prism). Then, when p-polarized light illuminates the interface through the prism under conditions of total internal reflection, at a certain incident angle, the prism provides the coupling condition to generate SPW (Surface Plasmon Wave). As wave vector of SPW (k_{sp}) equals the wave vector of incident TM wave (k_x), the SPR happen, this can be written as

$$k_x = k_0 \cdot n_p \cdot \sin(\theta) = k_0 \sqrt{\frac{\varepsilon_m \cdot n^2}{\varepsilon_m + n^2}} = \left[\frac{2\pi}{\lambda}\right] n_p \sin(\theta) = k_{sp} \qquad (1)$$

where λ is wavelength of incident light, θ is incident angle (resonance angle), ε_m is the dielectric constant of metal thin-film, k_0 is the wave vector in vacuum, n_p is the refractive index of prism, and n is the refractive index of sample. Applying Fresnel's law and multi-reflection theory, we obtain reflective coefficient r_p of p-polarized light,

$$r_p = \frac{r_{12} + r_{23}\exp(2ik_{z2}d_2)}{1 + r_{12}r_{23}\exp(2ik_{z2}d_2)} \quad (2)$$

where

$$r_{12} = (Z_1 - Z_2)/(Z_1 + Z_2)$$

$$r_{23} = (Z_2 - Z_3)/(Z_2 + Z_3)$$

$$Z_i = \varepsilon_i / k_{zi}, k_{zi} = \left[\varepsilon_i (k_0)^2 - k_1\right]^{0.5},$$

$$k_1 = k_0 \sin(\theta)\varepsilon_1^{0.5}, i = 1,2,3$$

where 1, 2, and 3 are defined as prism (ε_1) / metal film (ε_2) / sample layer (ε_3), and k_z is the wave vector in z direction of metal film. r_{12} is the reflective coefficient of a prism and metal film, r_{23} is reflective coefficient of metal film and sample. k_1 is same as the k_x of the Eq.1. The ε_1 and ε_2 will change under different incident wavelength of light, thus can be utilized for angular or wavelength scanning [6]. When SPR occurs, most incident light energy couples into SPW, so output light intensity drops down. The effective reflectance r_p represents the final result of SPR, which can be written as

$$R_p = |r_p|^2 \quad (3)$$

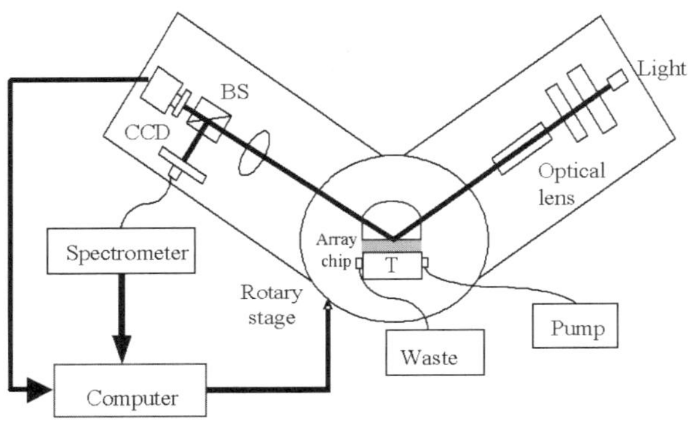

Figure 1.1. System configuration of SPR instrument. T: Sample fluid tank.

3. Experiments and Results

The intrumental setup is as shown in the Figure 1.1. We use BK7 glass as prism coupler and the refractive index is 1.515. Matching oil with same refractivity as the prism was applied between glass array chip and prism. White light of 400~1000 nm was used as a light source. It emitted into a polarizer and converted to p-polarized light, and then passed through an optical assembly into parallel light beam. The parallel light beam incident into the prism and reflected by gold film layer on array chip. A sample fluid tank was placed behind the array chip. The change of the reflective light was collected by a CCD and a spectrum meter, and the change of the image was collected by a CCD. The signal acquisition and data representation is done in PC with a LabVIEW^{TM} program.

There were two kinds of chips, array chip and slide chip, and a 50 nm thick gold film was deposited on glass substrate. In the array chip, the diameter of each point was 1 mm, and the distance between point to point was 1.5 mm. The chips were cleansed by 99.5% ethanol and ultra-pure water first, then immerse in 10 mM α-lipoic acid fluid (>98.0%, Fluka) for 30 minutes to make Self-Assembled Mono-layers (SAMs) on gold surface. After that, we activated the SAMs by 0.1 M EDC (≥97.0%, Fluka)/NHS (98+%, Alfa Aesar) for 30 minutes. Using 10 μL pipette spotting the IgG (from rabbit serum, reagent grade ≥95%, SIGMA) on the gold film for 24 hours. After the IgG immobilized, the chips rinsed with 1x PBS (phosphate buffer saline, pH=7.4) and dried. We prepared the 100 μg/ml, 50 μg/ml Anti-IgG (Anti-Rabbit IgG, antibody produced in goat, SIGMA) by 1x PBS Buffer to be reactant.

The resonance angle should be obtained before real-time monitoring for optimal detection. We scan from 71° to 65°, with 0.2° interval, and we got the resonance angle at 68.4°. Figure 1.2 shows array chip image while scanning.

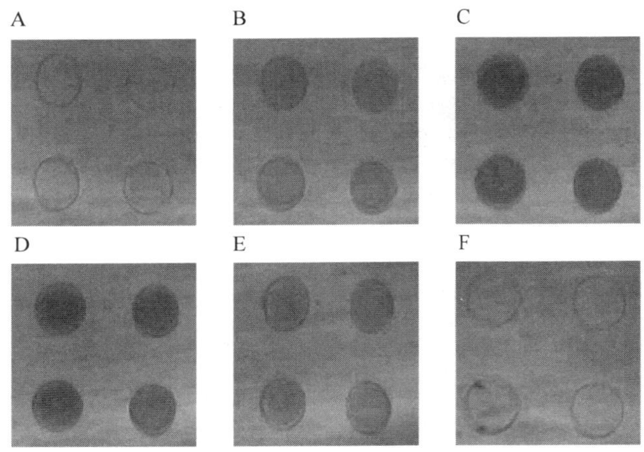

Figure 1.2. Image change at gold film in SPR sensor. Total detection time is 90 secconds. A-F images are catch from real-time images and each period is about 15 sec.

To verify the sensitivity of our instrument, we did an experiment by measuring different concentration of PBS buffer. We injected water and then PBS buffer at different time and record the intensity change of the reflected light. The angle of the platform was set at resonant angle. The PBS buffer's concentration was set at 0.4x, 0.8x, 1.2x, and 1.6x, which correspond to fluid refractive index at 1.3333, 1.3341, 1.3346, and 1.3354. The experiment results was shown in Figure 1.3. Each PBS buffer corresponded to a light intensity, which was 70.0, 71.5, 72.8, and 74.8. The minimum distinguishable RI is about 8×10^{-4}.

To verify wavelength scanning resolution, we scanned resonance wavelength by 0.25 nm from 600~1000 nm. Using ethanol of refractive index at 1.3333, 1.3352, 1.3380, 1.3410, 1.3440, and 1.3469 to test, we measured the resonance wavelength at 676.1 nm, 688.6 nm, 693.2 nm, 701.3 nm, 707.6 nm, and 716.2 nm respectively. The calculated wavelength resolution is 4×10^{-4} RI/nm, see Fig. 1.4.

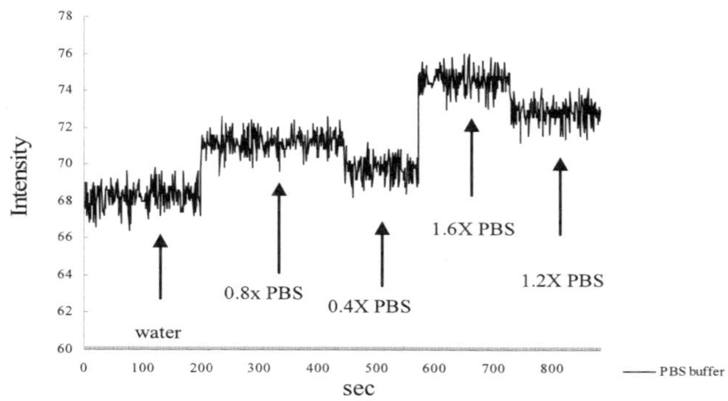

Figure 1.3. Intensity detection of water and different concentration of PBS buffer.

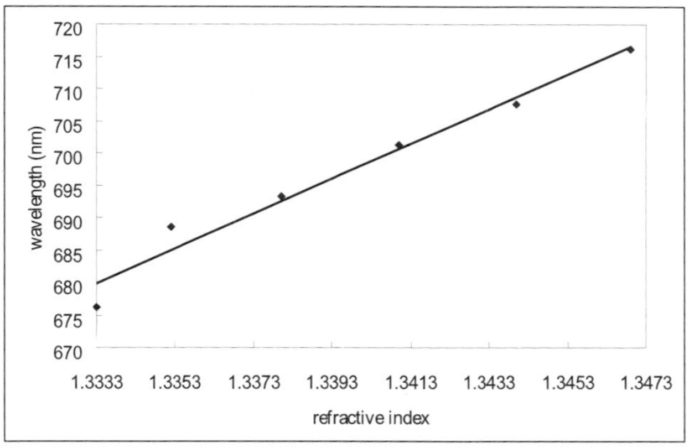

Figure 1.4. Wavelength interrogation experiment was measured using different concentration ethanol.

An experiment was done to evaluate the angle shift resulting from protein binding. The IgG of 100 μg/ml was immobilized on gold film of array chip. The resonance angle was at 68.8°, after binding with Anti-IgG the resonance angle shifted, see Table 1.1. Figure 1.5 shows intensity to angle change of 100 μg/ml IgG interaction with 50 μg/ml Anti-IgG before and after, environment temperature in experiment period was 24°C. Figure 1.6 shows intensity to angle change of 100 μg/ml IgG interaction with 100 ug/ml Anti-IgG before and after, environment temperature in experiment period was 17°C. The cause of the difference was the concentration of Anti-IgG.

Table 1.1. The resonance angle and angle shift after IgG binded with Anti-IgG.

	IgG (100 μg/mL)	After interaction with Anti-IgG (50 μ g/mL)	Angle Shift	IgG (100 μ g/mL)	After interaction with Anti-IgG (100 μ g/mL)	Angle Shift
Spot 1	68.8°	69.6 °	0.8 °	68.8 °	69.8 °	1 °
Spot 2	68.8 °	69.6 °	0.8 °	68.8 °	69.8 °	1 °
Spot 3	68 °	68.8 °	0.8 °	68.8 °	69.8 °	1 °
Spot 4	68 °	68.8 °	0.8 °	68.8 °	69.8 °	1 °

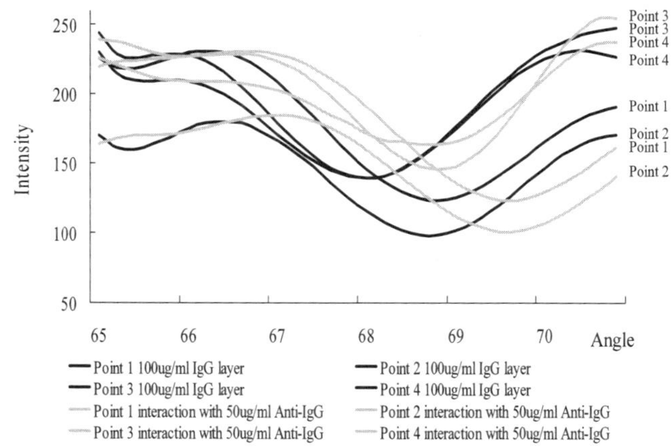

Figure 1.5. The resonance angle and angle shift after 100 ug/ml IgG binded with 50 μg/ml Anti-IgG.

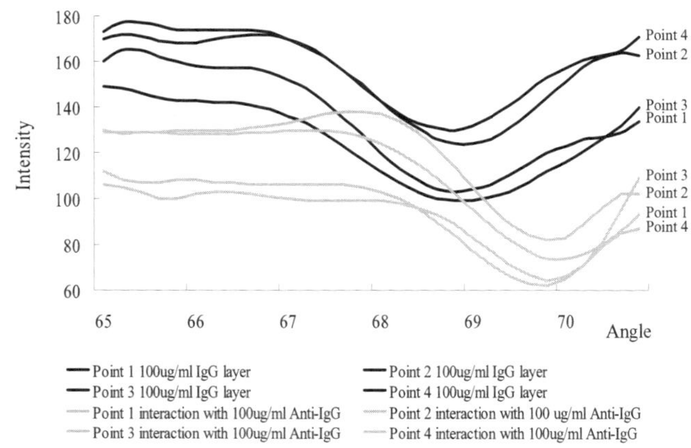

Figure 1.6. The resonance angle and angle shift after 100 ug/ml IgG binded with 100 μg/ml Anti-IgG.

To see the IgG and Anti-IgG binding progress in real-time, another experiments were done. Figure 1.7 shows 100 μg/ml IgG interacted with 100 μg/ml Anti-IgG at 17 °C.

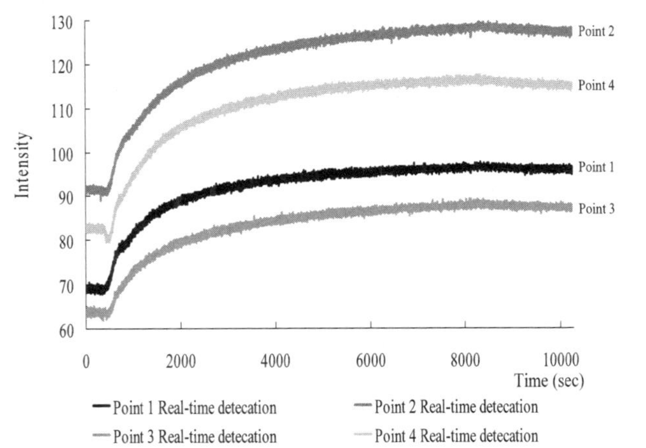

Figure 1.7. Real-time detection of interaction between 100 ug/ml IgG with 100 ug/ml Anti-IgG.

Since light source of the instrument was divergent, many efforts were put to make light source parrallel. The parallelization of light affect the size of field of view. The filed of view in this study was 7 mm x 5 mm, allowing 15 points to detect simultaneously.

4. Conclusion

According to experimental data, the resolution, the image function of detecting multiple test points at same time, and the function of real-time binding process monitoring were successfully verified. The resolution was adequate for doing biological experiments, especially proteomic experiments. The image function allows the user to use array chip to do high-throughput experiments. The wavelength interrogation function was also integrated in the instrument, providing ability of more precise measurment. The construction of the instrument is not finished, more work is needed to be done. The real-time monitoring of binding using wavelength interrogation will be next work to do. The optical configuration of this SPR instrument is simple, and the setup and maintenance cost will be adequate compared with current commercial products.

Reference

[1] Phillippe Guedon, et al., 2000, Characterization and optimization of a real-time, parallel, label-free, polypyrrole-based DNA sensor by surface plasmon resonance imaging, Anal. Chem., Vol. 72, No.24, P6003-6009.

[2] R. L. Rich, D. G. Myszka, 2007, Higher-throughput, label-time molecular interaction analysis, Analytical Biochemistry, Vol.361, P1-6.

[3] Yu Xinglong, et al., 2005, A surface plasmon resonance imaging interferometry for protein micro-array detection, Sensor and Actuators, B108, P765-771.

[4] Rina Y.P. Lue, et al. 2004, Versatile protein biotinylation Strategies for potential High-throughput proteomics, J. Am. Chem. Soc., Vol.126, P1055-1062.

[5] Andrew J. Thiel, et al. 1997, In situ surface plasmon resonance imaging detection of DNA hybridization to oligonucleotide arrays on gold surfaces, Anal. Chem., Vol.69, P4948-4956.

[6] R. M. Corn, et al. 2001, Fourier transform surface plasmon resonance adsorption sensor instrument, US patent no.6330062.

2

Coating Technology

2–1

Plasma-Ion Enhanced Optical Multilayer Thin-Film Coating for Consumer Electronics Products

G.L. Chen[1], C.C. Chang[2], S.C. Chien[3], T.S. Tung[4], H.H. Lee[5], W.C. Ling[6], and H.L. Lee[7]

[1]Hon-Hai Precision Ind. Co., Ltd., Taipei 236, Taiwan, galane.chen@foxconn.com
[2]Hon-Hai Precision Ind. Co., Ltd., Taipei 236, Taiwan, ching-chou.chang@foxconn.com
[3]Hon-Hai Precision Ind. Co., Ltd., Taipei 236, Taiwan, shih-che.chien@foxconn.com
[4]Hon-Hai Precision Ind. Co., Ltd., Taipei 236, Taiwan, tsai-shih.tung@foxconn.com
[5]Hon-Hai Precision Ind. Co., Ltd., Taipei 236, Taiwan, hsin-ho.lee@foxconn.com
[6]Hon-Hai Precision Ind. Co., Ltd., Taipei 236, Taiwan, wei-cheng.ling@foxconn.com
[7]Hon-Hai Precision Ind. Co., Ltd., Taipei 236, Taiwan, han-lung.lee@foxconn.com

Abstract. In this study, the optical multilayer thin-film coatings, started from optical design, are successfully deposited on glass substrates. Thermal evaporation method is used to prepare optical multilayer thin-film coatings, such as band-pass filter, edge filter, anti-reflective coating, for key components of consumer electronics products. Electron beam gun evaporation with and without plasma-ion enhanced source during deposition are compared in our system. The coating products, deposited with plasma-ion enhanced source, have high packing density, high refractive index, good adhesion, in comparison to those of conventional thermal evaporation or electron beam gun evaporation without plasma-enhanced source in coating chamber. In addition, excellent environmental stability, impervious to humidity, good uniformity among the substrate, and low wavelength shift with temperature in coating layers, which are especially important for mass prodution, are sucessfully obtained in this study.

Keywords: coating, consumer electronics, plasma-ion, thin-film

1. Introduction

Thin-film deposition techniques are important and intensively investigated in research and applicable for many commercial products [1]. In the computer, consumer electronics, and car industry, thin-film coatings are widely used for these products. Besides this, multilayer thin-film coatings are especially crucial in optical applications, such as in cameras, cell phones, projectors, liquid crystal displays, and lighting applications. In order to achieve high performance and good quality of these optical products, the key components inside need to be coated with optical thin-films. Optical thin-films, designed as components or substrate coated single layer or multilayer of dielectric films or metal to modify the transmission of optical waves, are developed intensively in research and industry fields.

Optical thin-film coatings can be formed in liquid or vapor type processes. Liquid processes, such as electroplating, chemical reaction, and sol-gel method, are easy to form antireflective films, mirror, or protective films. The latter one can be separated to physical vapor deposition (PVD) and chemical vapor deposition (CVD). In PVD [2], such as thermal evaporation, sputtering, and ion beam sputtering deposition, the coating materials are evaporated in atomic or cluster particles in vacuum and deposited on substrate. In this study, thermal evaporation method is used to form optical multilayer coatings, such as band-pass filter, edge filter, anti-reflective coatings, for key components in our consumer electronics products. Conventional thermal evaporation techniques with resistive-heating have relatively low process stability. Electron beam gun evaporation with a plasma-ion enhanced source in the coating chamber is applied in our system. With a plasma-ion enhanced source in coating process, the optical multilayer coating has high stability, especially for improvement in optical properties. In addition, high packing density, high refractive index, good adhesion of coating, are obtained in comparison to those of conventional thermal evaporation. Impervious to humidity, good uniformity, and low wavelength shift with temperature behavior, are also overcome.

The optical multilayer coatings, started from optical design, are deposited by using electron beam gun evaporation with and without a plasma-ion enhanced source in vacuum. The transmission spectra are compared, and the reliability tests are tested to ensure the environmental stability of these coatings. The thin-film coatings are successfully deposited and applicable for key components in our consumer electronics products.

2. Experimental

The transmission spectra in visible and near IR region according to the product specification are designed by optical simulation software. Using high refractive-index and low refractive-index material, we can design multilayer thin-film coating to meet the spectra specification by parameter optimization.

The glass substrates are cleaned with detergent and dried by ultrasonic cleaning process. Then the glass substrates are loaded on the calotte holder and moved into the coating chamber. After reaching process temperature over the substrates and opening polycold system to remove the moisture inside the coating chamber, the coating processes are started by using electron beam gun thermal evaporation system. The deposited thickness of each layer is well controlled in-situ by optical-monitor system by reflective mode. The main parameter of this study is plasma-ion enhanced or without plasma-ion enhanced.

The transmission spectra of the coating are measured by a spectrophotometer. The performance of the coating is compared by the theoretical prediction and the experimental result, and the characteristic of the spectra is also studied. Finally, the reliability tests, such as thermal shock, static humidity, temperature cycling are tested to identify the environmental stability of these coatings.

3. Results and Discussions

The results of theoretical design of the transmission spectra by optical simulation software are shown in Fig. 1.1. The prediction of the design spectra has a transmission value of larger than 90% in the visible region. The half of the transmission percentage (T_{50}) locates at about 650 nm. Besides, the transmission is below 3% in the range of 700-1100 nm. The spectra show an IR region-cut characteristic, and the thin-film coating can apply for IR-cut filter. The comparison between theoretical design and experimental result of the transmission spectra are shown in Fig 1.2. The experimental results are also measured and compared in Fig 1.2. The thin-film coating shows a T_{50} of about 649 nm, which is close to the theoretically designed value. In addition, the transmission in the range of 700-1150 nm accurately fits the theoretical values, as shown in Fig 1.3. Furthermore the spectra of the experimental results in the visible region has small ripple and has an average value of larger than 90%. That means the coating has a low reflectivity, which is crucial in optical applications. The multilayer thin-film coating is successfully designed and deposited. The deviation of the transmission in visible region may result from neglecting absorption in the coating material and glass substrate in the optical design.

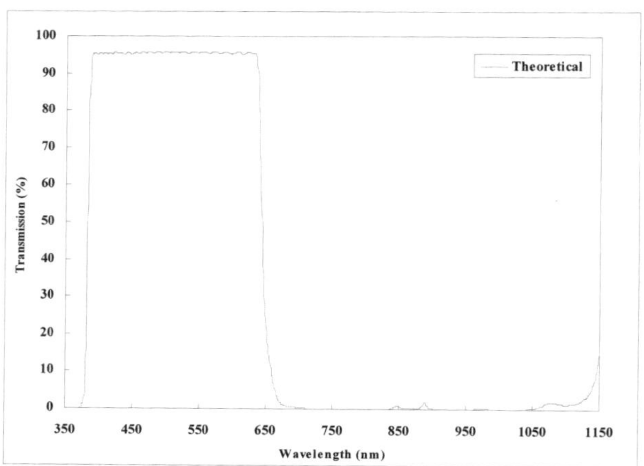

Figure 1.1. The theoretical design of the transmission spectra by optical simulation software

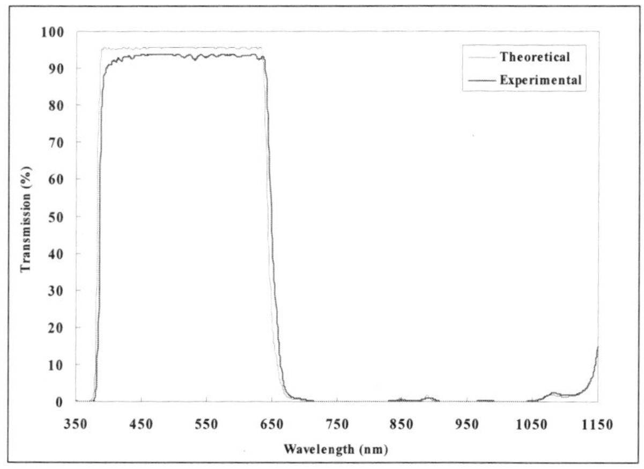

Figure 1.2. The comparison between theoretical design and experimental result of the transmission spectra in the visible and near IR region

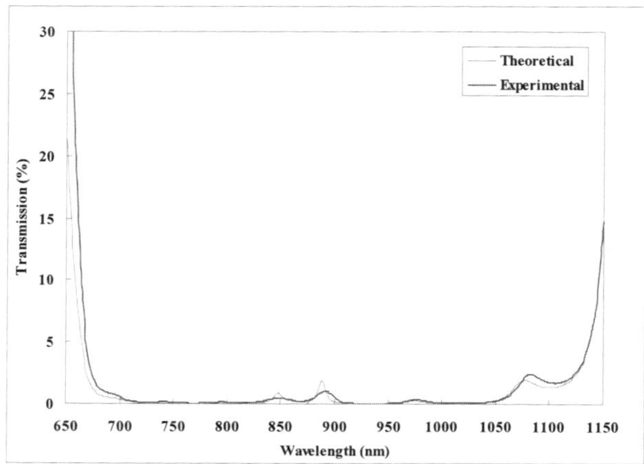

Figure 1.3. The comparison between theoretical design and experimental result of the transmission spectra in the isolation band region

The transmission spectra of plasma-ion and non-plasma-ion enhanced coating processes are compared in Fig 1.4. When the coating process is not plasma-ion enhanced, the characteristic of the spectra is far from the theoretical design. Besides, the ripple in the visible region in non-plasma-ion enhanced process is larger than that of the plasma-ion enhanced process. T_{50} value (locates at 667 nm) deviates the design value about 17 nm in non-plasma-ion enhanced process; in contrast, T_{50} value (locates at 649 nm) well fits the design value in the plasma-ion enhanced process. The transmission in the IR-region is not suppressed and has a value of larger than 3% in the non-plasma-ion enhanced process. Although the average values of the transmission in visible region are close in both coating processes, the ripple of the spectra is relatively large in the non-plasma-ion enhanced coating process. The deviation of the spectra without the plasma-ion source in coating process can be explained below. When the coating process is assisted by the plasma source, argon plasma that bombards the substrate and ionizes the reactive gas can cause high packing density of the deposited thin-film. Thus, the optical properties of the thin-film, such as absorption and scattering losses, homogeneity, are effectively improved.

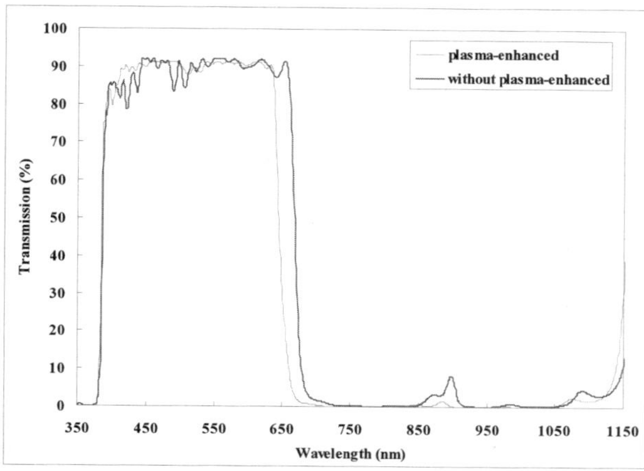

Figure 1.4. The transmission spectra of plasma-enhanced and non-plasma-enhanced coating processes

When considering mass production of the multilayer optical coatings, the uniformity of the optical properties of the coatings at different position on the calotte holder is important. The transmission spectra of plasma-enhanced coating processes at different deposited position, from the inner layer to the outer layer, designed as p1 to p5, are shown in Fig 1.5. The figure, which is rescaled to the characteristic of T_{50} and isolation band region, is shown in Fig. 1.6. From Fig 1.5, it is found that the transmission spectra show good uniformity at different position on the calotte holder. In Fig 1.6, T_{50} values are close in the first three inner layers (T_{50} at p1~p3, 644~645 nm) and shift to long wavelength as the glass substrate posited at outer layer on the calotte holder (p4 = 647 nm, p5 = 651 nm). T_{50} as a function of the calotte holder position are plotted in Fig 1.7. Because the coating process is assisted by the plasma-ion source, the spectra show the uniformity overall the calotte holder is quite good (the T_{50} deviation is controlled in ±5 nm). In addition, the transmission in near IR region is below 1% overall the calotte holder. That means the coatings cut off light in near IR region, so the coatings can reduce low noise fluctuation when the optical products need to receive IR region signal, for example, the coating can be applied for optical products as IR-cut filter.

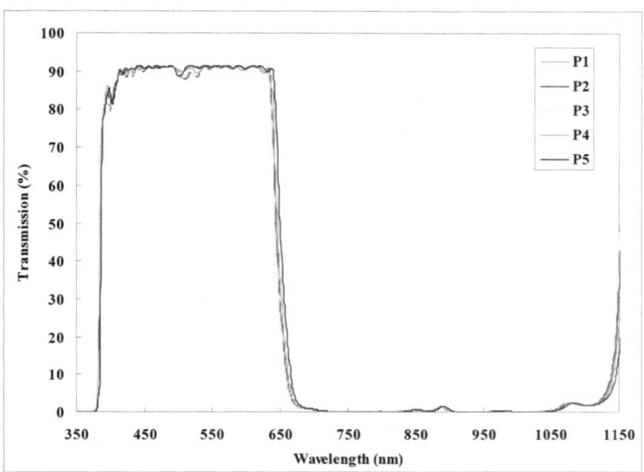

Figure 1.5. The transmission spectra of plasma-enhanced coating processes at different deposited position on the calotte holder

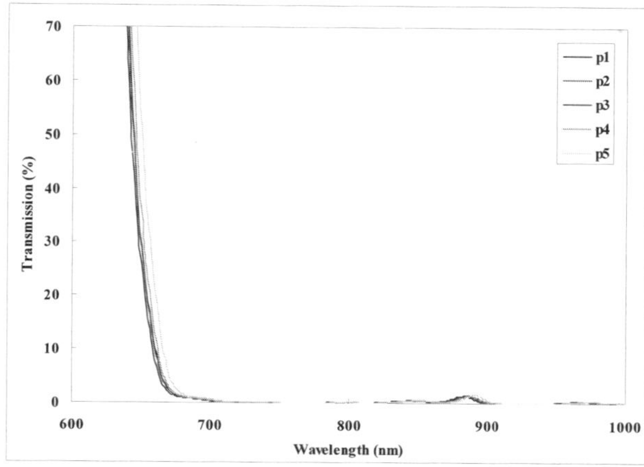

Figure 1.6. The transmission spectra at T_{50} region of plasma-enhanced coating processes at different deposited position on the calotte holder

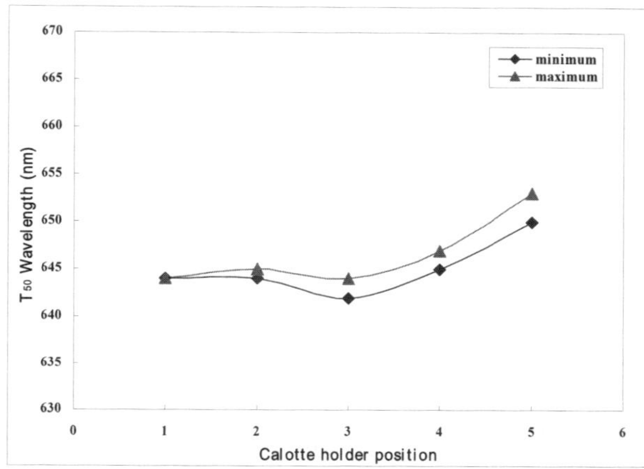

Figure 1.7. T_{50} as a function of the calotte holder position when the coating processes are plasma-ion enhanced

In contrast, the transmission spectra at different deposited position on the calotte substrate when the coating processes are not assisted by plasma-ion source are shown in Fig 1.8 and Fig 1.9 (from the inner layer to the outer layer, designed as np1 to np5). Unlike the spectra in Fig 1.5, the uniformity of the transmission spectra in Fig 1.8 is not good. From the inner layer to the outer layer on the calotte substrate, the transmission spectra vary largely in visible and near IR region. Besides, the ripples are large in near IR region, and that may result in noise fluctuation when the optical products need to receive IR region signal. It means that the ion bombardment is not uniform at different layers when the coating process is not assisted by the plasma-ion source. The transmission spectra in the near IR-region are also not suppressed because of the low packing density of the deposited thin-film. Moreover T_{50} values range from 667 to 631 nm, as shown in Fig 1.10, and that implies that the process result in poor uniformity at different holder position. Obviously we can conclude that it is not suitable for mass production if the coating process is not plasma-ion enhanced.

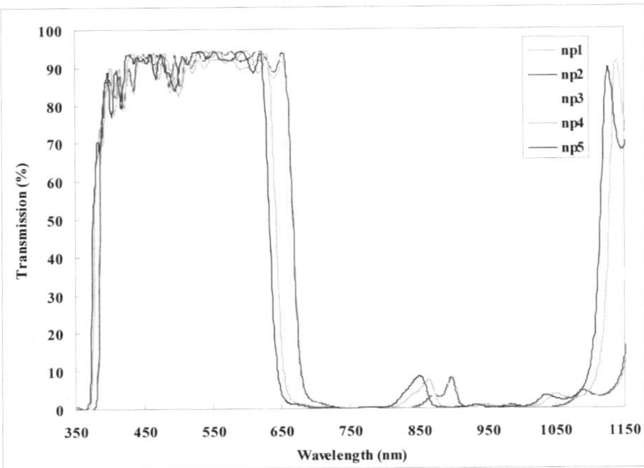

Figure 1.8. The transmission spectra at different deposited position on the calotte holder when the coating processes are not plasma-ion enhanced

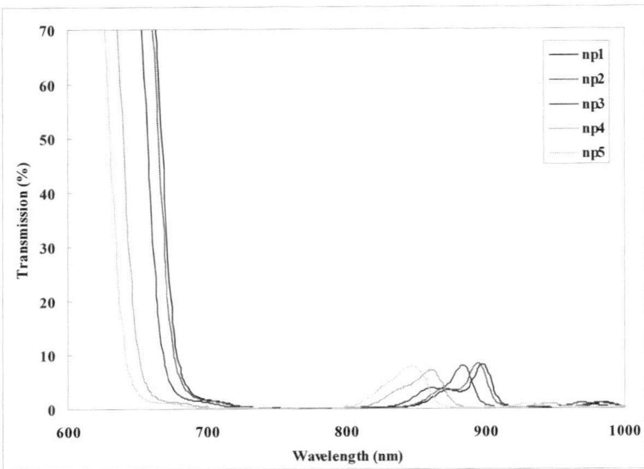

Figure 1.9.. The transmission spectra at T_{50} region at different deposited position on the calotte holder when the coating processes are not plasma-ion enhanced

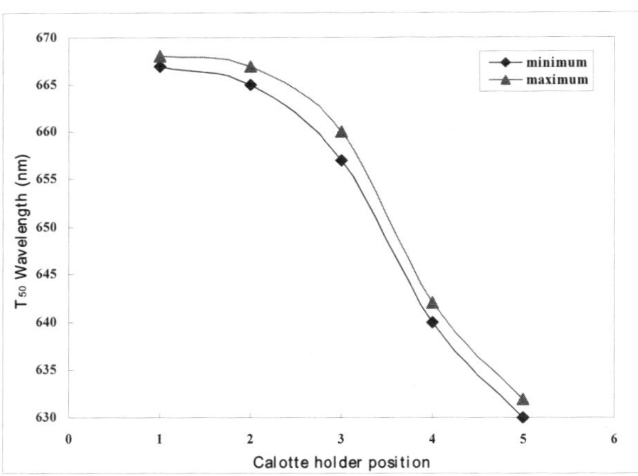

Figure 1.10. T_{50} as a function of the calotte holder position when the coating processes are not plasma-ion enhanced

In order to apply the optical coatings in filters for the commercial products, the environmental stability of the coatings are important. The thin-film coatings, which are coated by plasma-ion enhanced coating processes, are prepared to test the environmental stability. The reliability test, including aging accelerated life test (-40°C keep 45 mins, 85°C keep 45 mins, 20 s transition, 60 cycles), static humidity (65°C, 90% RH, keep 100 hrs), low temperature and high temperature cycling tests (-40°C keep 20 mins, 75°C keep 20 mins, 10 mins transition, 100 cycles), are studied. The test samples are 50 pieces for each test, and the performance of transmission spectra is compared before and after the tests. The transmission spectra in visible region (average transmission in 400~600 nm, and minimum transmission in 400~600 nm) and near IR region (average transmission in 700~1100 nm) and at T_{50} are characterized to see if the coatings are stable after the environmental test. The average transmission in 400~600 nm, minimum transmission in 400~600 nm, and average transmission in 700~1100 nm all meet the specification after the tests. Besides, all the T_{50} derivation are in ±5 nm after the environmental tests. From the reliability tests, it can conclude that the coatings under the plasma-enhanced source are very stable after environmental tests. Thus the optical coatings are suitable used in IR-cut filters for the commercial consumer electronics products in mass production.

4. Conclusion

The optical multilayer thin-film coatings are successfully deposited on glass substrates. Starting from optical design, multilayer thin-film coatings, like IR-cut filter, can be deposited to obtain good optical properties and good uniformity with a plasma-ion enhanced coating process. The coating products, which are deposited with a plasma-ion enhanced source, have high packing density, high refractive index, good adhesion, in comparison to those of conventional thermal evaporation. In addition, high environmental stability (aging accelerated life test, static humidity, low temperature and high temperature cycling tests), impervious to humidity, good uniformity overall the calotte holder, which are especially important for mass production, are successfully obtained in this study. Thus, the optical coatings can be used in IR-cut filters for the commercial optical products in mass production. In conclusion, the thin-film coating techniques can apply for other optical components in consumer electronics products, especially for mass production.

References

[1] Macleod H. A. 2001, Thin Film Optical Filters, 3rd ed., Institute of Physics Publishing. Bristol and Philadelphia.

[2] Smith D., 1995, Thin Film Deposition, MaGraw-Hill, Inc.

2–2

The Structure and Properties of Electroless Ni-P and Ni-Mo-P Deposits

Y. H. Chou[1], C. Y. Bai[2], M. D. Ger[2,a], C. L. Chao[2], S. J. Lee[3], C. Y. Lee[3]

[1] Graduate School of Defense Science, Institute of Technology, National Defense University, Ta-His, Tao-Yuan, Taiwan, ROC.
[2] Department of Applied Chemistry & Materials Science, Institute of Technology, National Defense University, Ta-His, Tao-Yuan, Taiwan, ROC. mdger@ccit.edu.tw
[3] Department of Mechanical Engineering & Fuel Cells Research Center, Yuan Ze University, Tao-Yuan, Taiwan, ROC

Abstract. The structure and properties of electroless Ni-P and Ni-Mo-P alloys, including the as-plated and annealed deposits, on the aluminum alloy 5052, have been investigated to evaluate the corrosion resistance, thermal stability, electric conductivity, and mechanical characteristics of the coating assemblies. The corrosion behavior of the obtained deposits was examined in a 0.5M H_2SO_4 environment. The crystal structure of the electroless Ni-based alloys was analyzed by X-ray diffraction (XRD), and the surface morphology and microstructure of Ni-P and Ni-Mo-P alloys were examined by field emission scanning electron microscopy (FE-SEM). The experimental results indicate that the corrosion rate of the Ni-Mo-P coating produced at 70°C and pH 7.0 is almost equals that of the Ni-P film in a 0.5M H_2SO_4 environment. It is also found that the Ni-Mo-P ternary alloy reveal excellent thermal stability after annealing at 400°C for 1 hr. The electroless Ni-Mo-P deposits have a potential to apply to precision mould and surface metallization of substrates, based on the excellent performance of the ternary alloys.

Keywords: Ni-Mo-P, electroless, thermal stability, corrosion

1. Introduction

The electroless plating is a non-galvanic type of plating method that deposits the anticipated films on substrates by the autocatalytic reaction of the coated materials without the use of external electrical power. The major advantage of this process is that it provides an even and admirable distribution deposit regardless of workpiece geometry. Another one benefit is the possibility of non-conducting materials metallization by the pre-treatments of substrate sensitization and activation. In addition, the procedure with very low cost (no current, ultra high vacuum or high temperatures are needed) is much simpler and faster. The deposits produced by this method generally show a fine structure with low porosity and have good bonding with the substrate. The electroless deposition is used to prevent the occurrence of corrosion, to make or restore fine dimension, and to improve the surface finish of a particular component. Moreover, this approach is also employed in precision fabrication of the micrometer, sub micrometer, or even nanometer scale components [1-5].

The electroless nickel-based alloys, such as Ni-P and Ni-B, are now widely used as deposited materials for the applications of electronics, mechanical, or chemical production due to the superior characteristics [6-9]. The deposits, in some cases, are expected to produce on precision molds employed at elevated temperature. The deposits used in such conditions may endure the wear and thermal shock. The tribological properties or wear resistance of the Ni-based deposits can be improved by co-depositing technique [10]. Moreover, alloying with refractory metals, such as the molybdenum, is a well known method to improve the thermal stability and enhance the thermal transition temperatures of crystallization transformation of many metals and alloys due to very low mobility of refractory elements in their alloys [11-13]. The physical and chemical characteristics of Ni-P and Ni-Mo-P alloys in the as-deposited and after annealing states, therefore, are worth investigated for the potential of widely applications of these alloys.

2. Experiments

Figure 1.1 shows the electroless deposition procedures of Ni-P and Ni-Mo-P deposits, and the experiments are described in two sections below.

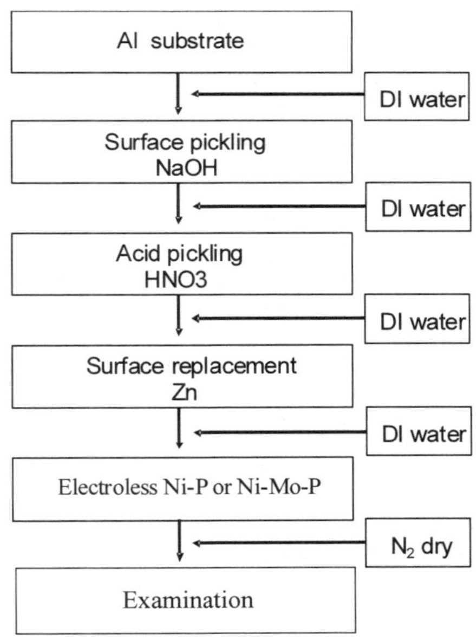

Figure 1.1. The electroless deposition procedures of Ni-P and Ni-Mo-P deposits

2.1 Materials and pre-treatments

The substrate material prepared in this study for the electroless deposition procedures was the aluminum alloy

5052. Prior to the electroless deposition process the specimens of the aluminum alloy were pre-treated following two steps. Step one - The specimens were immersed in a NaOH base solution (pH=12) for 30 min., and cleaned with deionized water. Then, the specimens were immersed in a nitric acid solution (30% HNO_3) for 30 min., and cleaned with deionized water. The specimens were then dipped in the Zn electrolyte for one min. Step two - The specimens underwent the step one were immersed in a nitric acid solution (30% HNO_3) for 20 min., and cleaned with deionized water. Then, the specimens were dipped in the Zn electrolyte for 2 min. After two steps of the Zn replacement procedure, the specimens were immersed into the Ni-P or Ni-Mo-P electroless solution to produce the deposits on the substrates. Table 1.1 lists the plating bath and deposition conditions for preparing Ni-P and Ni-Mo-P deposits.

2.2 Examination

The corrosion behaviors of deposits were investigated by potentiodynamic polarization in a 0.5 M H_2SO_4 environment. The crystal structure of the electroless Ni-based alloys was analyzed by X-ray diffraction (using monochromatic Cu-Kα radiation with a characteristic wavelength of 1.542Å). Field emission scanning electron microscopy (FE-SEM) was employed to examine the surface morphology and microstructure of the coatings. Four-point probe was employed to measure the electrical resistivity, and Vickers hardness tests were performed to measure the micro hardness of Ni-P or Ni-Mo-P deposits produced under different operating conditions.

Table 1.1. The plating bath and deposition conditions for preparing Ni-P and Ni-Mo-P deposits

Chemical compounds	Conditions	
	Ni-P	Ni-Mo-P
$NiSO_4 \cdot 6H_20$	0.114	0.114
$Na_2H_2PO_2 \cdot H_20$	0.283	0.283
$Na_2MoO_4 \cdot 2H_20$	-	4.13×10^{-2}
Main complex salt (ml l^{-1})	40	40
Auxiliary complex salt (ml l^{-1})	0.133	0.133
Additive (ppm)	10	10
KIO_3 (ppm)	20	20
pH	5	7
Temperature (°C)	83 °C	70 °C

3. Results and discussion

To investigate the effects of the addition of Mo into the Ni-P deposit on the structure and properties of the Ni-based alloys, the Ni-P and Ni-Mo-P are produced, and the corrosion resistance, the electric conductivity, the thermal stability, and the hardness of deposits in as-deposited and after annealing (at 400°C for 1 hour) states are examined.

3.1 Corrosion behaviors of as-depoisited Ni-P and Ni-Mo-P coatings

Potentiodynamic polarization was used to investigate the corrosion behaviors of the Ni-P deposit, prepared in an acidic (pH 5) electrolyte, and the Ni-Mo-P deposits, produced in a neutral (pH 7) electrolyte with 4.13×10^{-2} M sodium molybdate, in a 0.5M H_2SO_4 environment. The anodic polarization curves of tested deposits, shown in Figure 1.2, reveal that the corrosion resistance of the Ni-Mo-P is little worse than that of the Ni-P coating. However, the corrosion rates of the as-plated Ni-P and Ni-Mo-P deposits, obtained from the Figure 1.2 and listed in table 2, is almost equivalent.

It is well known that the crystalline state and structure of coatings may influence their corrosion behavior. The FE-SEM topographies of the Ni-Mo-P prepared in a neutral solution and the Ni-P produced in an acidic electrolyte are shown in Figure 1.3. The surface of Ni-P deposit exhibits a dense and fine structure with low porosity and little defects. The surface of the Ni-Mo-P film shows a more coarse structure than that of the Ni-P film. The corrosion media would attack the substrate through grain boundaries of the as-deposited Ni-Mo-P coating. However, the corrosion behavior of deposits can be remarkably improved according as the stuff effect with the passivating element Mo blocked or filled at the grain boundaries [14,15]. Therefore, the corrosion resistance of the Ni-Mo-P deposit is little worse than that of the Ni-P coating.

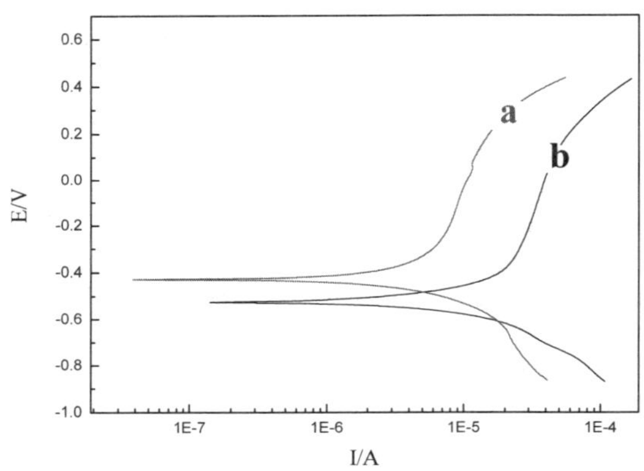

Figure 1.2. Anodic polarization curves of as –deposited Ni-P and Ni-Mo-P coatings in a 0.5M H_2SO_4 solution: (a) the Ni-P alloy, and (b) the Ni-Mo-P alloy

3.2 Corrosion behaviors of annealed Ni-P and Ni-Mo-P deposits

The corrosion behaviors of the annealed Ni-P and Ni-Mo-P deposits were also tested in a 0.5M H_2SO_4 environment using the potentiodynamic polarization. Figure 4 shows the anodic polarization curves of the annealed Ni-P and Ni-Mo-P coatings, and the corrosion rates of the annealed deposits, obtained from the Figure 1.4 are listed in Table 1.2. The corrosion resistance of the annealed Ni-P coating is worse than that of the annealed Ni-Mo-P deposit, although the corrosion resistance of the Ni-P coating is better than that of the Ni-Mo-P deposit before annealing treatment.

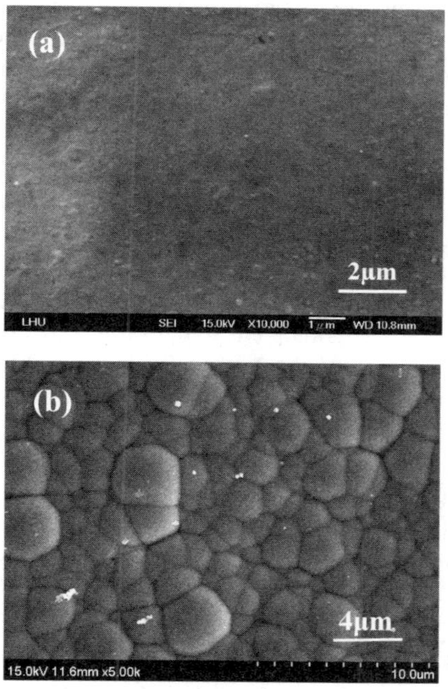

Figure 1.3. FE-SEM topographies of (a) the Ni-P alloy produced at 83°C and pH 5 and (b) the Ni-Mo-P alloy prepared at 70°C and pH 7

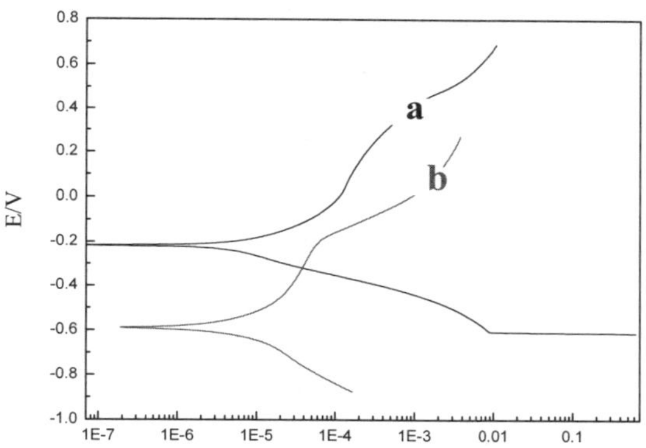

Figure 1.4. Anodic polarization curves of the annealed Ni-P and Ni-Mo-P deposits in a 0.5M H_2SO_4 solution: (a) the Ni-P alloy, and (b) the Ni-Mo-P alloy

Comparing the corrosion rate of the deposits in annealed state with that in as-deposited state, the change of the corrosion rate of the Ni-Mo-P coating is little, but the variation of the corrosion rate of the Ni-P deposit is large. It indicates that the corrosion behavior of the Ni-Mo-P alloy is almost unchanged after annealing at 400°C for 1 hour. However, the structure or the constituent of the Ni-P alloy may undergo an evident alteration during the annealing process. The chemical and physical properties of this deposit, therefore, would be changed with the transformation after annealing.

3.3 The thermal stability and physical characteristics of Ni-Mo-P and Ni-P deposits

The XRD spectra of the Ni-P and the Ni-Mo-P deposits in as-deposited or annealed state are shown in Figure 1.5. In the case of the as-deposited Ni-P alloy, shown in Figure 1.5(a), there is only an extremely broadening peak in the diffraction pattern, revealing that the as-deposited Ni-P alloy prepared in an acidic electroless solution has a near amorphous structure [14]. However, the XRD result demonstrates that the change of structure and constituent resulting from the heat treatment are observed in the annealed Ni-P deposit, and the alloy is composed of Ni, Ni_5P_2, and Ni_3P phases. For another case, the XRD patterns of both the as-deposited and the annealed Ni-Mo-P coatings, shown in Figure 1.5(b), exhibit a crystallized structure with the Ni(111) and Ni(200) diffraction peaks.

The diffraction peaks of the Ni-Mo-P coating in as-deposited state are wider than that in annealed state, indicating that the as-plated Ni-Mo-P coating has a nano-grain structure. Moreover, the annealed Ni-Mo-P coating has recrystallized and grown by coalescence with neighboring nano-grain after the heat treatment, and then the diffraction peaks of this coating in annealed state are sharper than that in as-deposited state. However, there are no apparent change in the constituent and the crystalline state of the Ni-Mo-P alloy after annealing. Therefore, the thermal stability of the Ni-Mo-P alloy is better than that of the Ni-P alloy.

The excellent thermal stability of the Ni-Mo-P alloy can be further confirmed with the hardness test. The measured hardness values, listed in Table 1.2, of the as-plated Ni-P deposit, prepared in an acidic (pH 5) electrolyte, and the Ni-Mo-P deposit, produced in a neutral (pH 7) solution

Table 1.2. The estimated corrosion rates, thermal stability, hardness, and electrical conductivity of the Ni-Mo-P and Ni-P deposits

Condition	Ni-Mo-P	Ni-P
Corrosion Rate (as-deposited)	0.038 mm/ year	0.034 mm/ year
Corrosion Rate (annealing)	0.054 mm/ year	0.96 mm/ year
Thermal Stability	Good	Poor
Hardness Value (as-deposited)	466 Hv	542 Hv
Hardness Value (annealing)	476 Hv	919 Hv
Conductivity	17,488 S / cm	20,746 S / cm

with 4.13×10^{-2} M sodium molybdate, are 542 and 466 HV, respectively. In addition, the hardness of the annealed Ni-P and Ni-Mo-P deposits, also listed in Table 1.2, are 919 and 476 HV, respectively. It indicates that the hardness variation of the Ni-Mo-P deposit is little but that of the Ni-P alloy is large after annealing. The great increase in hardness of the Ni-P alloy results from the formation of intermetallics, the Ni_3P and Ni_5P_2 phases, in the deposit [16]. The intermetallic precipitates have an ordered structure, therefore, the hardness of Ni-P deposit enhances substantially after annealing. However, the brittleness of the Ni-P deposit may increase with increasing the amounts of intermetallic precipitates. Moreover, the dimensions of the Ni-P deposit may change after phase transformation occurring. On the contrary, the Ni-Mo-P deposit reveals good thermal stability after annealing due to the Mo with very low mobility restraining the phase transformation of deposits. Therefore, this Ni-Mo-P alloy coating is more suitable for the application of precision molds than the Ni-P deposit.

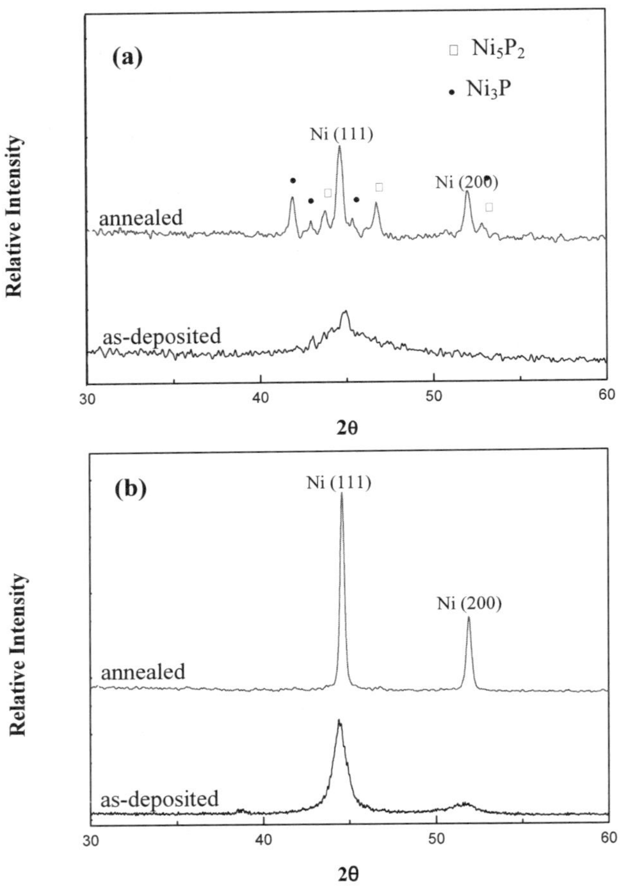

Figure 1.5. XRD spectra of (a) the Ni-P and (b) the Ni-Mo-P alloys in the as-deposited or the annealed states

The electric property of the as-plated Ni-P and Ni-Mo-P deposits are also examined in this study. The values of conductivity, listed in Table 1.2, are in a narrow range of 17.5 kS/cm to 20.7 kS/cm, revealing that the influence of the addition of sodium molybdate in the operating electrolyte on the conductivity of Ni-Mo-P deposits is unapparent.

4. Conclusion

The results obtained from this experiment show that the corrosion rates and the electrical conductivity of the as-plated Ni-P and Ni-Mo-P deposits are almost equivalent, but the corrosion resistance of the annealed Ni-Mo-P is better than that of the annealed Ni-P coating.

The thermal stability of the Ni-Mo-P deposit produced in the neutral electrolyte with 4.13×10^{-2} M sodium molybdate is better than that of the Ni-P deposit prepared in the acidic solution, because the addition of Mo in the deposits restrains the phase transformation of the Ni-Mo-P alloy coating at enhanced temperature. Moreover, the hardness variation of the Ni-Mo-P deposit is little but that of the Ni-P alloy is large after annealing. The great increase in hardness of the Ni-P alloy results from the formation of intermetallics, the Ni_3P and Ni_5P_2 phases, in the deposit. For the consideration of the thermal stability, the Ni-Mo-P alloy coating is more suitable for the application of precision molds than the Ni-P deposit

References

[1] Schultze, J.W. and Bressel, A. Electroche. Acta, 2001 Vol.47, No 5, P3-21.

[2] Homma, T., Komatsu, I., Ttamaki, A., Nakai, H. And Osaka, T. Electroch. Acta, 2001, Vol. 47, No 1, P47-53.

[3] Furukawa, S. and Mehregany, M. Sens. & Actua. A, 1996, Vol. 56, P261.

[4] Takano, N., Hosoda, N., Yamada, T. and Osaka, T., Journal of Electrochemical Society, 1999, Vol. 146, P1407.

[5] Kobayashi, T., Ishibashi, J., Mononobe, S., Ohtsu, M. and Honma, H., Journal of Electrochemical Society, 2000, Vol. 147, No 3, P1046-1049.

[6] Mallory, G.O. and Hajdu, J.B.: *Electroless Plating: Fundamentals and Applications* (AESF Publications, Orlando 1991).

[7] Reidel, W.: *Electroless Nickel Plating* (Redwood Press, 1991).

[8] Tomlinson, W.J. and Carroll, J., J. Mater. Sci., 1990, Vol. 25, P4972.

[9] Luo, S.F., Ma, P.X. and Li, P.X., J. Vac. Sci. Technol. A, 1986, Vol. 4, P2862.

[10] Straffelini, G., Colombo, D. and Molinari, A., Wear, 1999, Vol. 236, No1-2, P179-188.

[11] Wu, F. B., Tien, S.K., Duh, J.G. and Chen, W. Y., Surf. & Coat. Techn., 2004, Vol. 177-178, P312-316.

[12] Tien, S.K. and Duh, J.G.: Thin Solid Films, 2004, Vol. 469-470, P268.

[13] Zhanq, Y., Ma, J., Fanq, Y.K. and Duan, J.G., Trans. Nonfer. Met. Soc. Chi. 2004, Vol. 14, P147.

[14] Osaka, T., Arai, K., Masubuchi, N., Tamazaki, T. and Namikawa, T., Jpn. J. Appl. Phys., 1989, Vol. 28, P866-871.

[15] Wu, Y., Wan, C. C. and Wang, Y. Y., Journal of Electronic Materials, 2005, Vol. 34, No 5, P541-550.

[16] Chen, W.Y. and Duh, J.G., Surf. & Coat. Tech., 2004, Vol. 177-178, P222.

The Mechanism of Amorphous Silicon Thin Films Using XeF Excimer Laser Crystallization

C. C. Kuo[1,3], W.C. Yeh[2], C.P. Hsiao[1] and J. Y. Jeng[1,*]

[1] Department of Mechanical Engineering, National Taiwan University of Science and Technology, Taipei, Taiwan
[2]Department of Electronic Engineering, National Taiwan University of Science and Technology, Taipei, Taiwan
[3]Department of Mechanical Engineering, Mingchi University of Technology, Taipei Hsien, Taiwan
[*]jeng@mail.ntust.edu.tw

Abstract. Melting and recrystallization phenomena of plasma-enhanced chemical vapor deposition (PECVD) of amorphous silicon (a-Si) thin films for both frontside and backside excimer laser crystallization (ELC) have been investigated by in-situ real-time optical reflectivity and transmissivity (TRORT) measurements with nanosecond time resolution. The longest melt-phase duration of 90-nm-thick a-Si thin films for both frontside and backside ELC is 350 ns. The ablation excimer laser fluence of a-Si thin films for frontside ELC is higher than that of backside ELC because SiO_2 films show anti-reflectivity for excimer laser beam. TRORT measurements reveal that the solidification direction of molten Si for both frontside and backside ELC start from the interface of the sample and the recrystallization mechanism for both frontside and backside ELC are the same.

Keywords: excimer laser crystallization, real-time optical reflectivity and transmissivity measurements, amorphous silicon thin films

1. Introduction

Excimer laser crystallization (ELC) of amorphous silicon (a-Si) is an important technique for fabricating high-quality polycrystalline silicon (poly-Si) films for thin film transistor (TFTs) applications such as active-matrix liquid-crystal displays [1]. The interest in ELC is mainly due to its cost-effectiveness, since it allows for the use of inexpensive glass substrates. The solid phases of Si are semiconducting but the liquid phase is metallic [2], so that the values of many physical properties such as optical, electrical and thermal constants are very different for the liquid and solid phase at similar temperatures and pressures. Such abrupt discontinuities in physical properties provide unusual and potent opportunities for in-situ real-time optical measurements of the phase transformation of Si thin films. Accordingly, time-resolved optical reflection and transmission (TRORT) measurements have widely been employed to monitor the laser-induced melting and rapid solidification processes in semiconductors. However, the recrystallization mechanisms involved in the process are not understood well although investigations of ELC of a-Si thin films have been widely performed for the past several years, especially for backside ELC.

In this work, a non-invasive in-situ TRORT [3-4] monitoring system combining two continues-wave He-Ne probe lasers, a digital oscilloscope and three photodiodes is developed to investigate the structural transformation dynamics of PECVD a-Si thin films during ELC. The melt-mediated transformation scenarios of the Si thin films were determined in terms of various excimer laser fluences. After ELC, an extensive microstructural analysis of the resulting poly-Si thin films was performed by Field Emission Scanning Electron (FE-SEM) and high resolution transmission electron microscopy (HR-TEM). According to TRORT, FE-SEM and HR-TEM observations, the recrystallization mechanism of PECVD a-Si thin films is proposed to interpret the formation of grain microstructure and provides a basis for fabrication of single-grain TFTs.

2. Experimental details

Silane-based a-Si thin films of 90-nm-thick were deposited onto a glass substrate covered with a 300-nm-thick SiO_2 layer by PECVD. Figure 1.1 shows the schematic illustration of experimental setup for monitoring structural transformation dynamics of a-Si thin films during ELC using two probe lasers. The a-Si thin films were irradiated by a XeF excimer laser operating at a wavelength of 351 nm with a short pulse of 25 ns in full-width-half-maximum. The repetition rate was kept at 1 Hz to avoid cumulative heating of the samples. During ELC, in-situ real-time TRORT measurements were performed using two continuous wave He-Ne probe lasers operating at a wavelength of 633 nm. A pinhole with a diameter of 0.3 mm was mounted in front of the He-Ne probe laser to increase the resolution of TRORT measurements. He-Ne laser had a 45° of incidence from the normal of the sample's surface and focused on the center of the laser-irradiated spot. The reflected beam from the sample was focused on the fast Si photodiode. The interference filter was mounted in front of the photodiode, allowing only red light from the He-Ne probe laser to contribute to the detected signal. The photodiode is placed in an excellent position where the maximum signal is obtained during ELC. A beam splitter was employed to reflect 10% of the excimer laser beam to a triggering photodiode for triggering the record of reflection spectra using a fast digital storage oscilloscope. The response time of the entire TRORT measurement is 1 nanosecond. The melt-mediated transformation scenarios, melt-phase duration of molten Si, surface melting threshold, complete melting excimer laser fluence and Si film ablation threshold were investigated by the TRORT measurements during ELC. Following ELC, an extensive microstructural analysis of the resulting poly-Si thin films were performed using FE-SEM operated and HR-TEM. Before FE-SEM observation, the crystallized silicon films were Secco-etched for 3 seconds in order to highlight the grain boundaries and intra-grain defects.

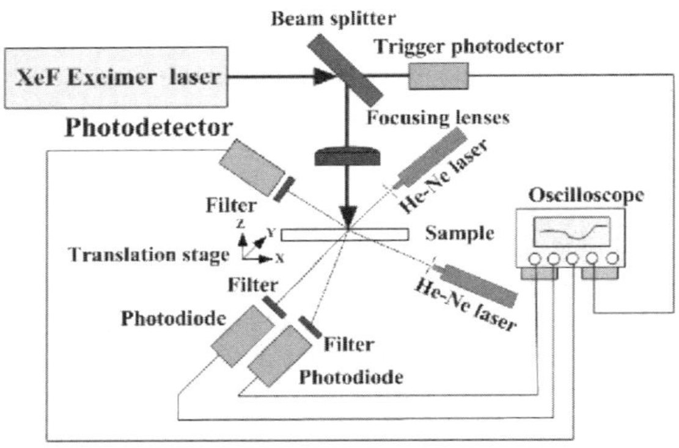

Figure 1.1. Schematic illustration of experimental setup for monitoring structural transformation dynamics of a-Si thin films during ELC using two probe lasers

3. Results and Discussion

Figure 1.2 shows cross-sectional bright field TEM images of excimer laser crystallized Si thin films at various energy densities. Following ELC at laser energy density range from 90 mJ/cm^2 to 700 mJ/cm^2, the experimental results can be divided into five distinct regimes according to different laser energy densities: heating of Si films, partial melting, near-complete melting, complete melting, and ablation of Si films. At a laser energy density of 60 mJ/cm^2, the reflection and transmission signals exhibited an abrupt drop, followed by a relaxation to the initial value indicating that the a-Si thin films did not melt but were merely undergoing heating and cooling during ELC. The changes in the TRORT signals were significantly governed by the increase in the extinction coefficient and the temperature-dependent optical constants of Si thin films [5]. This is also confirmed by the corresponding SEM images because no poly-Si grains were observed in the sample. As the excimer laser fluence is increased above the surface melting threshold of a-Si thin films, two distinct phase transformation sequences take place during ELC. One is the melting of a-Si thin films following the absorption of laser energy and the other is the nucleation and subsequent growth of poly-Si grains as the sample cools. The formation of nuclei is a crucial step for the phase transition.

Figure 1.3 shows melt-phase duration as a function of laser fluence for both frontside and backside ELC. The melt-phase duration increases with increasing excimer laser fluence for both frontside and backside ELC. The longest melt-phase duration of 90 nm thick a-Si thin films for both frontside and backside ELC is 350 ns. This result reveals that the thickness of a-Si thin films is the key factor affecting the melt-phase duration during ELC rather than the approaches of ELC.

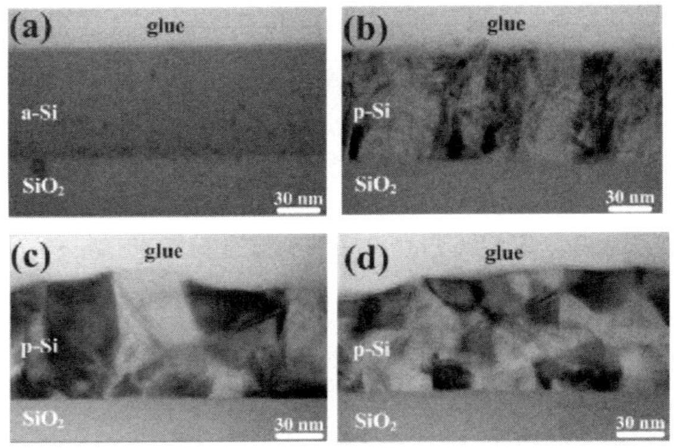

Figure 1.2. Cross-sectional bright field TEM images of excimer laser crystallized Si thin films at various energy densities. (a) E=60 mJ/cm^2. The Si thin films remain amorphous structure. (b) E=100 mJ/cm^2. The Si thin films have completely converted into poly-Si due to explosive crystallization. (c) E=350 mJ/cm^2. Large-grained poly-Si is fabricated due to SLG mechanism. (d) E=500 mJ/cm^2. Fine-grianed poly-Si is fabricated due to homogeneous random nucleation by deep cooling

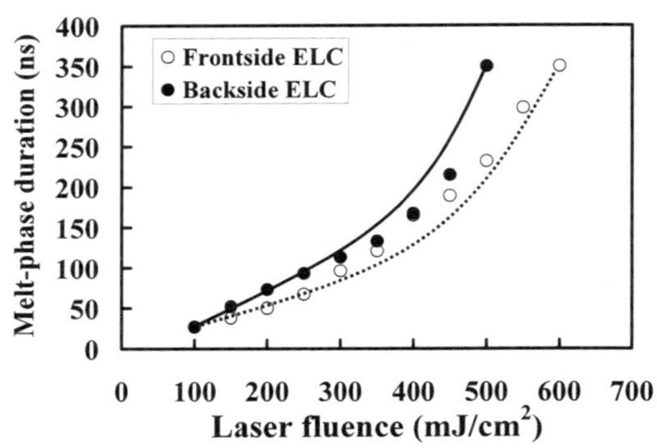

Figure 1.3. Melt-phase duration as a function of laser fluence for both frontside and backside ELC

Figure 1.4 shows that the ablation excimer laser fluence (Eab) and complete melting laser fluence (Ec) of a-Si thin films for frontside ELC is higher than that of backside ELC, while the surface melting threshold of a-Si thin film for frontside and backside ELC are the same. These effects are mainly attributed to the anti-reflectivity characteristic of SiO_2 films for excimer laser beam [6].For a-Si thin films of the same thickness, the lower excimer laser fluences for complete melting and ablation of Si thin films are needed by backside ELC when compared with frontside ELC. Consequently, the backside ELC approach is a good candidate for laser crystallization in terms of energy-saving of excimer laser.

Figure 1.5 shows the frontside reflectivity (FSR), backside reflectivity (BSR) and frontside transmissivity (FST) of the sample during ELC as a function of time at complete melting regime for both frontside and backside ELC. The most striking feature in this figure is the faster resolidification velocity of the surface denoted by L1 than that of the interface denoted by L2 of the silicon film, revealing that the solidification direction of molten Si for both frontside and backside ELC can be deduced from the interface toward the surface. The occurrence of recalescence after solidification is clearly observed at the interface of the sample, illustrated at the second maximum of BSR signal of frontside ELC and FSR signal of backside ELC, respectively. The massive release of solidification heat causes the reheating and partial remelting of a-Si thin films after its complete solidification. This result is in good agreement with the observation of amorphous germanium under picosecond laser irradiation [7].

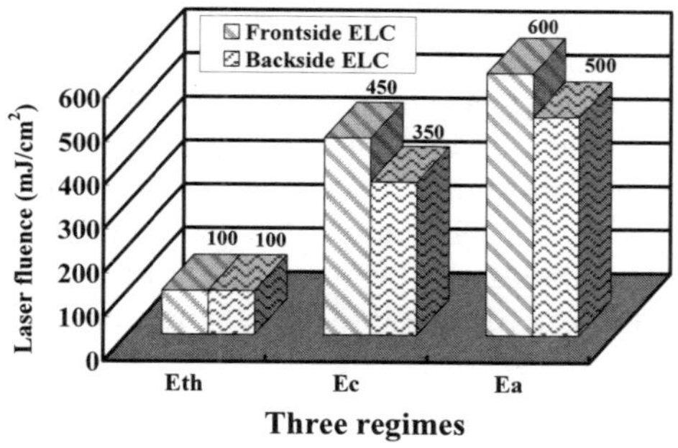

Figure 1.4. Surface melting threshold, complete melting excimer laser fluence and ablation excimer laser fluence of a-Si thin films for both frontside and backside ELC

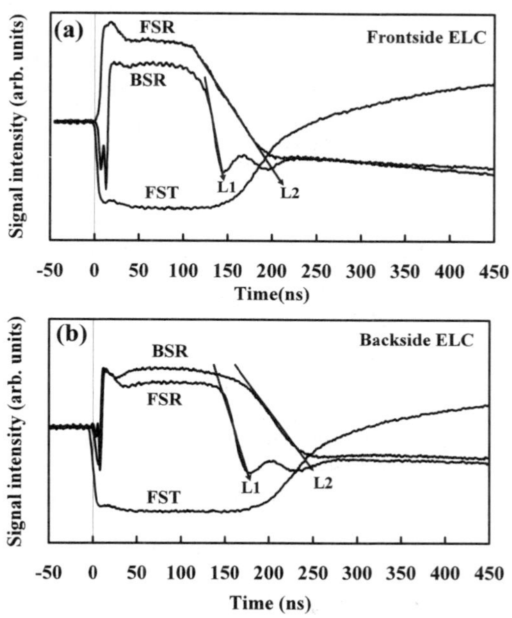

Figure 1.5. FSR, FST and BSR as a function of time at complete melting regime for both (a) frontside ELC and (b) backside ELC

Figure 1.6 shows the SEM images of the Secco-etched poly-Si films in different regimes. The amorphous regions of the silicon thin films have been etched away, leaving only poly-Si on glass substrate. In the partial melting regime, there is an increase in grain size with increase in excimer laser fluence. The grain size, however, reaches the maximum because of super lateral growth (SLG) nucleation in the near-complete melting regime. It is remarkable that the maximum size of the average grain reaches 1 μm in the SLG regime because the lateral growth results from explosive crystallization [8] combined with liquid phase growth. Indeed, this scenario was significantly supported by the observation of Aichmayr et al. [9]. In the complete melting regime, fine-grained poly-Si is observed due to homogeneous random nucleation. When the laser fluence exceeds the ablation excimer laser fluence, some a-Si thin films will evaporate, resulting in poor morphology of the sample. As expected, the measurements show that the dependence of grain size of PECVD a-Si thin films on the crystallization excimer energy densities exhibits characteristics similar to reports on LPCVD a-Si thin films [10], with two distinct regimes (the low energy density and high energy density regimes) clearly discernable. In the low energy density regime, grain sizes slowly increase because the melt depth increases with increasing excimer laser energy density. On the other hand, in the high energy density regime, grain size is not affected by variations in excimer laser fluence since the recrystallization mechanism is changed [11].

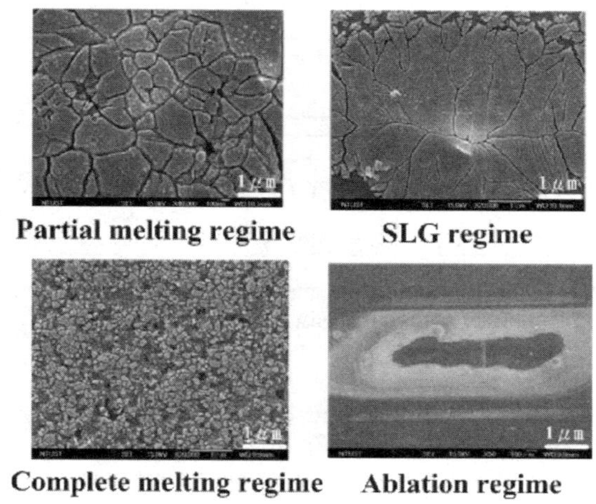

Figure 1.6. FE-SEM images of the Secco-etched poly-Si films in different regimes

According to both TRORT investigations and FE-SEM observations, the recrystallization mechanisms for both frontside and backside ELC upon ELC are the same as summarized in Figure 1.7, revealing the formation of grain microstructure for three distinct regimes. The recrystallization mechanisms can be divided into three regimes, depending on different excimer laser fluences. Owing to a XeF excimer laser pulse inherent with Gaussian beam profile, the molten Si has a steep lateral temperature gradient; the higher temperature within the center molten zone results in a correspondingly longer time to reach the deep supercooling regime where spontaneous nucleation occurs. The excimer laser beams were focused through a signal cylindrical plano-convex lens, resulting in larger lateral temperature gradient in the direction of the two axes of the irradiated rectangular spots. Thus, the crystal growth propagated from outside to the center of the irradiated spots.

The complete melting regime describes the situation that is encountered when the incident excimer laser fluence is sufficiently high to complete melting of the Si thin films. Owing to the amorphous structure of SiO_2, epitaxial regrowth from the surface of the glass substrate is not possible. In the complete melting regime, the recrystallization mechanism is significantly governed by homogeneous nucleation in a deeply supercooled melt, leading to small (smaller than 100 nm), globular grains randomly distributed throughout the Si thin films [12]. The partial melting regime describes the situation that is encountered when the incident excimer laser fluence is sufficient to induce melting of a-Si thin films, but is low enough so that a continuous layer of solid Si remains at the maximum extent of primary melting. In the partial melting regime, the maximum temperature remains close to the melting point of a-Si tin films and the grain size gradually increases from a few tens up to several hundreds of nanometers, depending on different excimer laser fluences.

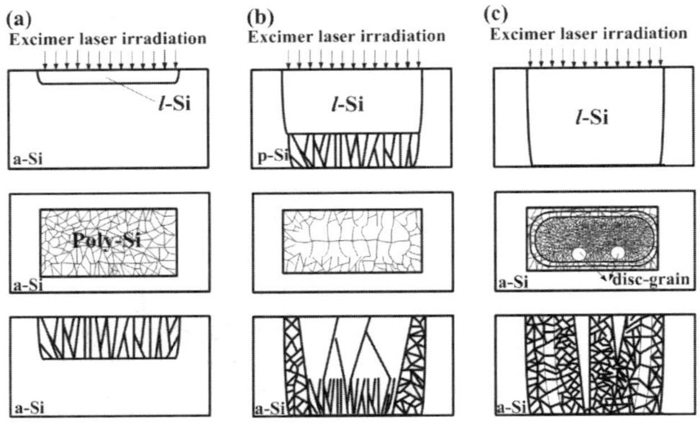

Figure 1.7. Recrystallization mechanisms of Si thin films for three distinct regimes upon ELC. (a) partial melting, (b) SLG and (c) complete melting

Between the partial and complete melting regime, a SLG regime was identified to occur within a narrow process window at the transition between the two distinct regimes. In the near-complete melting regime, the large-grained poly-Si can be obtained. Im et al. [13] have explained the SLG phenomenon, which is a regrowth process commencing from the residual un-melted Si islands in a liquid matrix. The fewer the Si islands that remain, the larger their separation will be. Consequently, the longer they can grow laterally. The maximum attainable grain size is determined by the onset of copious nucleation in the continuously cooling liquid ahead of the crystallization interface. An SLG region can still be observed in the complete melting regime, but the region is very narrow because the excimer beam is a Gaussian profile and the molten Si cools very rapidly [14]. Among the three distinct regimes, the SLG regime is essential for fabrication of single-grain TFTs by backside ELC with a special sample structure named photosensitive capping layer [15]. The range of excimer laser fluence for achieving large-grained poly-Si is defined as the process window. The process window of total excimer energy density is not sufficiently wide in comparison with the pulse-to-pulse fluctuation of a modern commercial excimer laser machine. In general, the energy density of SLG has a very narrow process window, typically about 1.5% [16] and 2.5 % [17]. The process window has been experimentally observed to reach 5 % by TRORT measurements in our previous investigation [18]. These results demonstrate that this recrystallization mechanism of a-Si thin films is sufficient for fabricating single-grain TFTs by backside ELC, which are essential for the device-to-device uniformity of high-performance TFTs.

4. Conclusion

Recrystallization mechanisms of PECVD a-Si thin films upon excimer laser crystallization have been experimentally investigated to interpret the formation of the grain microstructure for both frontside and backside ELC by a comprehensive SEM, cross-sectional TEM and TRORT analysis. The recrystallization mechanisms for both frontside and backside ELC are the same and the recrystallization mechanism of PECVD a-Si thin films sensitively depends on the irradiated excimer laser fluence, which controls the thickness of the molten Si layer and the melt-phase duration. Backside ELC approach is a good candidate for laser crystallization in terms of energy-saving of excimer laser. This result provides a basis for fabrication of single-grain TFTs under appropriate conditions by backside ELC, which are essential for the device-to-device uniformity of high-performance TFTs.

References

[1] Y. F. Chong, K. L. Pey, A. T. S. Wee, M. O. Thompson, C. H. Tung, and A. See, Appl. Phys. Lett. 81(2002)3786.
[2] G. E. Jellison, Jr. and D.H. Lowndes, Appl. Phys. Lett. 47 (1985) 718.
[3] C. C. Kuo, W. C. Yeh, C. B. Chen, and J. Y. Jeng, Thin Solid Films 515 (2006) 1651.
[4] C. C. Kuo, W. C. Yeh, C. B. Chen, and J. Y. Jeng, Mat. Sci. Forum 505 (2006) 337.
[5] G. E. Jellison, Jr. and D.H. Lowndes, J. Appl. Phys. 60 (1986) 841.
[6] H.J. Kim and J. S. Im, Appl. Phys. Lett. 68 (1996) 1513.
[7] J. Siegel, J. Solis, and C. N. Afonso, Appl. Phys. Lett. 75 (1999) 1071.
[8] M. O. Thompson, G. J. Galvin, and J. W. Mayer, Phys. Rev. Lett. 52 (1984) 2360.
[9] G. Aichmayr, D. Toet, M. Mulato, P.V. Santos, A. Spangenberg, and R. B. Bergmann, J. Non-Cryst. Solids 227(1998) 921.
[10] J. S. Im, H. J. Kim, and M. O. Thompson, Appl. Phys. Lett. 63 (1993) 1969.
[11] J. P. Leonard and J. S. Im, Appl. Phys. Lett. 78 (2001) 3454.
[12] S. R. Stiffler, M. O. Thompson, and P. S. Peercy, Phys. Rev. B 43 (1991) 9851.
[13] J. S. Im and H. J. Kim, Appl. Phys. Lett. 64 (1994) 2303.
[14] M. Hatano, S. Moon, M. Lee, K. Suzuki, and C. P. Grigoropoulos, J. Non-Cryst. Solids 266 (2000) 654.
[15] W. C. Yeh and M. Matsumura, Jpn. J. Appl. Phys. 41 (2002)1909.
[16] R. Ishihara, A. Burtsev, and P.F.A. Alkemade, Jpn. J. Appl. Phys. 39 (2000) 3872.
[17] D. Pribat, P. Legagneux, C. Collet, F. Plais, and O. Huet, C. Reita, Proceedings of the Active-Matrix Liquid-Crystal Displays(AM-LCD), Tokyo, Japan ,July 9-10, 1999, p.9.
[18] C. C. Kuo, W. C. Yeh, C. B. Chen, and J. Y. Jeng, Proceedings of 2005 IEEE/ASME International Conference on Advanced Manufacture, Taipei, Taiwan R. O. C., November 28-December 2, 2005, p.337.

Surface Properties of Polypropylene Treated Using Atmospheric Pressure Plasma Jet

C.H. Su[1], T.H. Chen[2], S.H. Yang[3], C.H. Liu[4], S. Lin[5], J.T. Teng[6] and H. Chen[7]

[1]Industrial Technology Research Institute, Hsinchu, 310, Taiwan, R.O.C., such@itri.org.tw
[2]Chung Yuan Christian University, Chungli, 320, Taiwan, R.O.C., g9502305@cycu.edu.tw
[3]National Central University, Chungli, 320, Taiwan, R.O.C., 943404009@cc.ncu.edu.tw
[4]Industrial Technology Research Institute, Hsinchu, 310, Taiwan, R.O.C., stanliu@itri.org.tw
[5]Industrial Technology Research Institute, Hsinchu, 310, Taiwan, R.O.C., shermannLin@itri.org.tw
[6]Chung Yuan Christian University, Chungli, 320, Taiwan, R.O.C., jyhtong@cycu.edu.tw
[7]National Central University, Chungli, 320, Taiwan, R.O.C.,huichen@cc.ncu.edu.tw

Abstract. Polypropylene (PP) films have been widely used in many industrial areas, such as for protective overcoats and food packaging. However, PP film's hydrophobic surface properties induce poor wettability and adhesion; these properties have restrained the application of these films. Many surface modification techniques like wet-chemical treatment, UV irradiation, and plasma (including the atmospheric-pressure plasma, APP) treatment have been applied to films to enhance their hydrophilic properties. Among these technologies, APP treatment has attracted much attention due to its dry process, low vacuum equipment cost, and high productivity.

In this study, the influence of process parameters would be the reactive gas ratio of plasma. To enhance its surface characteristics, treatment of PP film's surface by APP was investigated. The XPS, OES, contact angle analyzer, SEM and AFM were used to examine the effect of process variables on film surface characteristics. It was found that Ar plasma mixing with oxygen has a better lasting aging effect. Moreover, the roughness of films is slightly changed after treatment. Through XPS analysis, we observed that the O/C ratio of PP decreases with an increased exposure time in air. Finally, the relationship among the aging time, surface energy, and roughness of the film was also investigated.

Keywords: atmospheric-pressure plasma, surface modification, surface properties, hydrophobic.

1. Introduction

Among all polymer materials, polypropylene (PP) film is a very versatile polymer material to be used in many industrial fields [1], such as the protective overcoats and food packaging. However, a major problem is that PP film has poor wettability, printability, and adhesion owing to its hydrophobic surface characteristics [2]. This will result in products with poor adhesion of paints, coatings, or adhesive bonding. Therefore, up to the present time, numerous studies have investigated the extent to which the bonding properties can be improved by treating the surface of the substrate in order to alter the surface chemistry [3].

To enhance both wettable and adhesive properties of PP surfaces, a complex and critical process is used in traditional industry. It consists of three steps: (1) acid solvent cleaning to remove dust or organics from surfaces, (2) flaming to increase surface energy, and (3) primer application to reinforce adhesion of the coat on the substrate surfaces. However, this technique generates pollutions due to chemical solvent usage. Therefore, there are some environmental limitations hindering the continuous use of these processes in the labs and industrial facilities.

On the other hand, many studies have been conducted to overcome these problems by modifying the surface properties through ultraviolet irradiation, electron bombardment, or plasma treatment. These techniques increase the surface free energy and wettability of a film surface by incorporating polar functional groups onto the film surface. One of those promising studies is the plasma treatment [4]. However, most of plasma treatment system for cleaning worked at low pressure in the range of several Torr or less. This indicates that all these low-pressure systems require high-vacuum equipments whose costs grow with the degrees of technological processes.

APP provides several advantages compared to the low-pressure plasmas. For the operation of APP, no vacuum equipment is required, thus they represent an efficient and economic alternative to activate polymeric surfaces [5] and may easily be integrated into in-line processes.

In this study, we report the investigations of surface modification phenomena of various reactive gas ratios by APP jet. The aims of our study are to obtain the improvement in the free energy at the surface of a PP film. Initially, Ar gas was used to optimize the processing conditions of the APP jet. Secondly, the surface characteristics such as free energy, composition, morphology, and roughness at the surface of PP films were examined as a function of oxygen or nitrogen content in the gas mixture (Ar/O_2 or Ar/N_2). Finally, the treatment of the PP film surface by pure Ar plasma was compared with those resulted from the uses of mixed gas (Ar/O_2 and Ar/N_2) plasmas in order to determine which is more appropriate for the surface modification of a PP film as far as the surface wettability is concerned.

2. Experimental Processes

An APP jet system and process conditions were shown in Fig. 1.1. Primary input line was fed by the 16-20Hz AC power and the reactive gas with flow rate kept at 30 slm was entered from the upside of the plasma jet as the ionization gas. PP film with 3 x 3 cm^2 as the substrate was put on a movable-stage with the stage moving at 100 mm/s. The system was first discharged with Ar gas maintained for 2 min to stabilize before introducing the mixing gas to the substrate. The mixing gas (N_2/O_2) flow rate varied form 0.5%-1.5% using a mass flow controller. The distance of nozzle-to-sample, applied power and deposition cycle times were set at 5 mm, 555 W, and 3, respectively.

Experimental conditions

Ar APP	Ar=25 slm
O_2/Ar APP	O_2=0.5~1.5%;Ar=25 slm
N_2/Ar APP	N_2=0.5~1.5%;Ar=25 slm

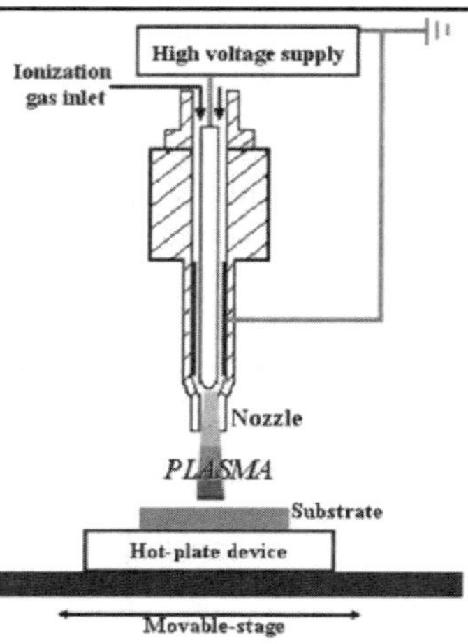

Figure 1. 1. Schematic diagram of the APP treatment system.

The surface composition of PP films was investigated using X-ray photoelectron spectroscope (XPS). Optical emission spectroscopy (OES) was used to investigate the influence of plasma species on the surface modification of PP films. The surface morphology and roughness of the film were observed using atomic force microscope (AFM). Contact angle analyzer was used to achieve the aging effect on PP film after plasma treatment.

3. Results and Discussion

The contact angle aging time of each sample has been measured after plasma treatment. It is known that the decay of contact angle was substantially influenced by the values of plasma parameters considered; i.e., the gas flow rates. However, the contact angle decreases slightly with an increase of the gas flow rate. To identify the plasma variables that provide the best contact angle decay, it was demonstrated by the measurements of contact angles resulting from the use of three reactive gas ratios, as shown in Fig. 1.2. The analysis of variance has shown that the variation of contact angles is relatively large during the first 10 hr after the treatment, while the trend becomes relatively smooth afterward.

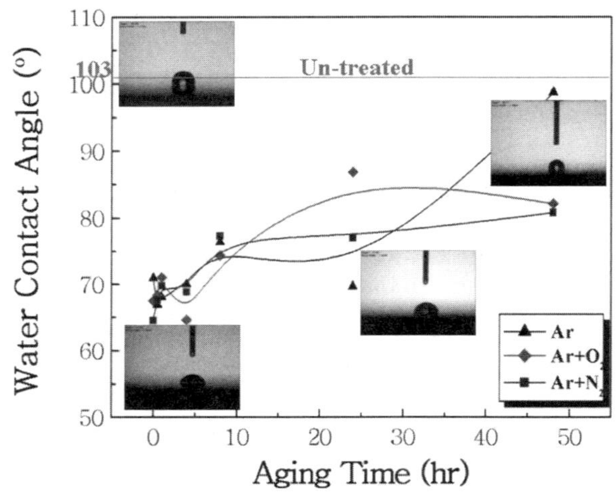

Figure 1.2. Decay of contact angle v.s. time for different reaction gases.

Among all conditions that minimize the angle of contact as a result of the longer aging time, the treatment condition of Ar/N_2 is the most economic one. Comparing all the results with pure Ar as the reactive gas, the original PP film contact angle of 103° is reduced by 42°, as shown in Fig. 1.2. However, its lasting effect of a low contact angle is not as good as that treated by the mixture of Ar/N_2. When small percentage of N_2 is added to Ar, the decay of contact angle remains within 3% of the contact angle value (at 10 hr) for the time duration after 10 hr and up to 48 hr (which is the entire duration for measurement). On the other hand, small percentage of oxygen added as a reactive gas will behavior similarly but with a relatively large variation in its behavior after 10 hr, as shown in Fig. 1.2.

From the viewpoint of the physical changes, the surface morphology and roughness of the film were observed using AFM, as shown in Fig. 1.3. It is very difficult to draw any correlation between the change of RMS roughness and the change of surface free energy with respect to the gas ratio. This is because the change of RMS roughness observed through AFM in the region of 30 x 30 μm remains essentially the same before and after the plasma treatment. It is considered that the untreated PP film's RMS roughness is already a large value, while the structure affecting the surface free energy of PP film is analyzed under macromolecular level. Therefore, we may not conclude that the roughness in a surface is due to the etching by the plasma treatment.

From the viewpoint of the chemical changes, XPS was used to examine the chemical composition on surface after the AP plasma treatment. Fig. 1.4 shows the chemical composition of PP, expressed in terms of the percentage of oxygen/carbon/nitrogen concentration. The obvious change is observed which indicates that an oxygen content of 10.49% in the un-treated PP is increased to a higher oxygen content of 18.94% for Ar/O_2 plasma-treated PP. Namely, after plasma treatment, the percent ratio of O/C increased from 0.12 to 0.24. This, from Fig. 1.4, we can confirm that the polar functional groups containing oxygen are introduced on the PP film surface.

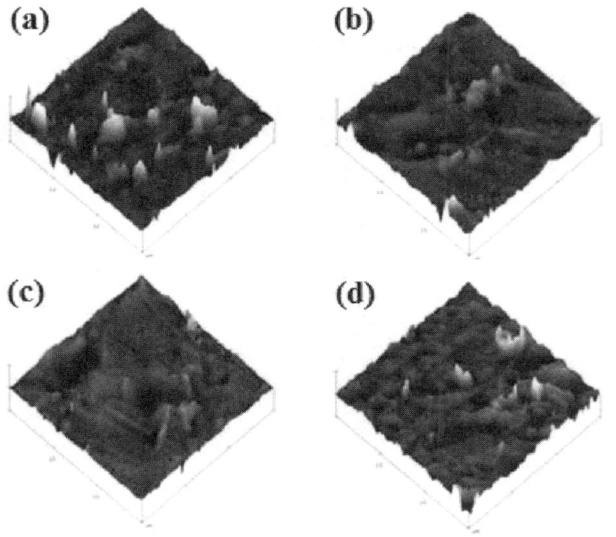

Figure 1.3. The AFM surface morphologies of the PP films treated by APP jet of various gas ratios. (a) un-treated, (b) Ar, (c) Ar/O_2, (d) Ar/N_2

As to the aging effect, Ar/O_2 plasma treated PP substrate after 48 hr was also examined by XPS. As shown in Fig. 1.4, the oxygen contents decrease slightly from 18.94% to 17.6% as the percent ratio of O/C decreases from 0.24 to 0.22 after 48 hr.

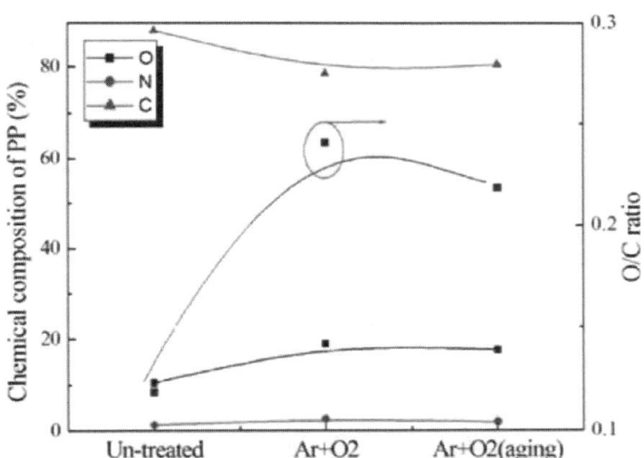

	Un-treated	Ar+O2	Ar+O2(aging)
O	10.49	18.94	17.6
N	1.32	2.59	1.94
C	88.19	78.47	80.46
O/C	0.12	0.24	0.22

Figure 1.4. Chemical composition of PP films treated by APP jet.

As shown in Fig. 1.5, the OES spectra show the emission intensity as a function of the wavelength for plasmas using pure gases of O_2, N_2, and Ar, respectively. It was found that the OH species were detected for all plasma treatments. The results can be used to draw the conclusion that the pure-Ar APP also has hydrophilic treatment effect, in which OH radicals are the primary active species. As to the Ar/O_2 plasma, XPS has presented that the polar functional groups containing oxygen are introduced on the PP film surface. On the Ar/N_2 plasma, the nitrogen functional groups of the polymer surfaces were generated by N_2 or NH_3 plasma treatment.

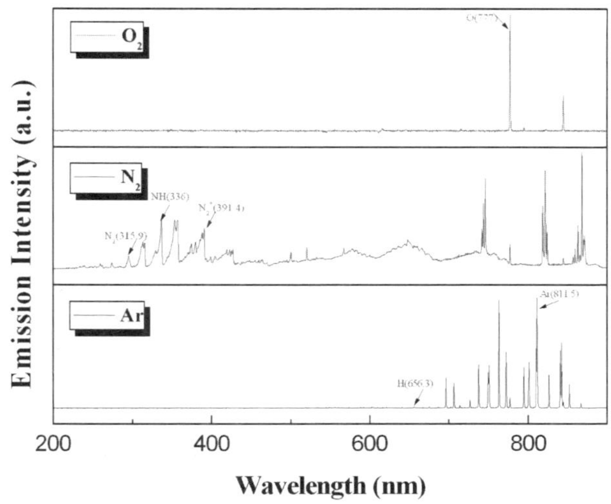

Figure 1.5. OES spectra of O_2, N_2, and Ar via APP jet system.

As shown in Fig. 1.6, the OES spectra observed in Fig. 1.5 reveal here that the treatment by pure N_2 plasma leads to N^*. The reaction on the left side of the figure (an overview) indicates polymer excited states (expressed by P^* in the figure), while the reaction on the right side of the figure indicates the detailed mechanisms of the formation of those components, such as, N_2^+, N_2^*, and O_2 in the air. The reaction mechanisms are (1) N_2^* combines directly with excited states of polymer, (2) O_2 in the air combines with P^*, and (3) N_2^+ recombines with an electron to form an N_2^*, followed by combination of N_2^* with P^*. Resulting from these reactions, the surface of polymer generates many hydrophilic functional group, leading to the formation of more favorable contact angle at the surface relative to that with the plasma using pure Ar gas.

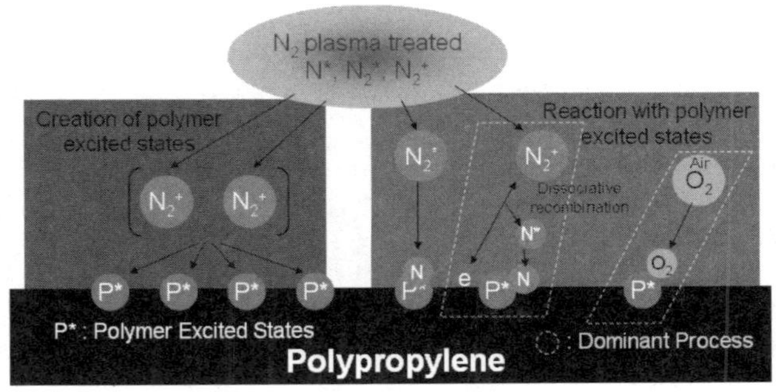

Figure 1.6. Schematic diagram of plasma treated PP[6].

4. Conclusion

It was found that APP can efficiently improve the PP surface wettability. This wettable surface property was observed to be independent of the surface roughness, as measured by the AFM. Moreover, Ar plasma mixing with oxygen has a better lasting aging effect by observing the O/C ratio of PP. Finally, OES spectra reveal that polar functional group induced hydrophilic surface. The relationship among the aging time, surface energy and roughness of the film was invesgated in this study.

Acknowledgements

The authors would like to thank the National Science Conucil of Taiwan, Republic of China for their finacial support.

References

[1] D.L. Cho, K.H. Shin, et al. 2001, J. Adhes. Sci. Technol. Vol. 15 P653-664.

[2] M. Zeniewicz, 2001 J. Adhes. Sci. Technol. Vol. 15 P1769-1785.

[3] Long JW. Painting of polypropylene. Proceedings of Polypropylene in Automotive Applications, 1992. p. 1– 3.

[4] M. Koide, T. Horiuchi, et al. 1998, Thin Solid Films. Vol.316 P65-67.

[5] P. Chaivan, N. Pasaja, et al. 2005, Surf. Coat. Technol. Vol. 193 P356–360.

[6] Y.-H. Choi, J.-H Kim, et al. 2005, Surface & Coatings Technology. Vol.193 P319–324

3

Computer-Aided Engineering

3–1

Development of STEP-NC Compliant Machine Tool Data Model

W. Z. Yang[1], X. Xu[2] and S. Q. Xie[3]

Department of Mechanical Engineering, The University of Auckland, Private Bag 92019, Auckland 1142, New Zealand
[1] wyan041@ec.auckland.ac.nz
[2] x.xu@auckland.ac.nz
[3] s.xie@auckland.ac.nz

Abstract. To implement STEP-NC based manufacturing, a STEP-NC compliant machine tool data model (STEP-NCMtDm) is developed by using EXPRESS modelling lanugage. It models sufficient machine tool data to 'transfer' STEP-NC machining task-level data to machine tool dependent method-level data. The objective and two main technological aspects of developing STEP-NCMtDM are discussed. The model architecture is analyzed. Machine tool data in STEP-NCMtDm are categorized into two groups: "static data" and "dynamic data", to support further system development. A Web-enabled STEP-NCMtDm (a corresponding XML schema) is developed to support Web-based application. STEP-NCMtDm has been implemented to model a milling machine tool and represent its data in a STEP Part 21 file and an XML file to support STEP-NC based manufacturing.

Keywords: STEP-NC, Machine Tools, Data Model, XML Schema.

1. Introduction

STEP-NC based manufacturing is a newly developed digital manufacturing approach to realize an I^4 (informative, intelligent, integrated and interoperated) manufacturing scenario. STEP-NC has several significant advantages when implemented in the whole CNC manufacturing processes: 1) providing a complete and structured data model to represent various product and technical data at different stages of product development processes, 2) supporting standardized data exchange and sharing within integrated manufacturing systems [1]; 3) bi-directional transfer of high level manufacturing information between CAM and CNC systems, which is impossible with G/M code; and 4) using object-oriented concept "workingstep" to encapsulate machining feature information together with machining operation parameters, which makes manufacturing data reusable.

The most important feature of STEP-NC concept is that of 'machine tool independence'. STEP-NC data model mainly portrays data at the Task-level or the "What-to-do" data. Although it is possible to define data at the Method-level or the "How-to-do" data, such as machine tool trajectory, the main aim of STEP-NC is to allow these decisions to be made by a STEP-NC enabled controller. STEP-NC programs may be written once and used on different machine tools provided that the machine tool has the required process capabilities [2]. Therefore, the implementation of STEP-NC is effectively a process of adapting its data model for different CNC systems. That is, the central issue to implement STEP-NC is therefore the 'transition' from the Task-level data to the Method-level data, or from the "What-to-do" data to the "How-to-do" data.

To do this, it is necessary to have in place a complete and readily accessible database of manufacturing resources, describing data such as product data, cutting tool data. Although some machining process data such as machining parameters, machining strategy, etc. have been defined in STEP-NC Part 10, 11 and 12 [5, 6, 7], these data objects are not adequate for modelling machine tool data in support of STEP-NC enabled machining, especially in a dynamical shop floor environment. Thus, to implement STEP-NC based manufacturing, and to accomplish process planning and scheduling tasks, it is essential to build up a STEP-NC compliant machine tool data model that is able to capture all the required machine tool data for supporting STEP-NC enabled machining. In this paper, a STEP-NC compliant Machine tool Data model (STEP-NCMtDm) is described. When combined with the cutting tool data models defined in STEP-NC, STEP-NCMtDm can form a comprehensive Manufacturing Resource Data Model (MRDM) and ultimately provide sufficient information for STEP-NC based manufacturing.

2. Literature Review

There are some efforts in the area of modelling machine tool data. The project 'Modelling of Manufacturing Resources Information' was carried out at the National Institute of Standard and Technology (NIST) in the USA [8]. It specifies two aspects of manufacturing resources: 1) machine tool (milling machines and vertical/horizontal turning machines); and 2) tool assembly (cutting tools, inserts, tool holders, etc.) This model has been considered as a basis to develop the proposed ASME B5.59-2 standard for representing machine tools. The NIST Manufacturing Resources Model and ASME B5.59-2 are developed to principally satisfy the requirements of manufacturing resource venders but not ideal to support process planning. To overcome this, Kulvatunyou et al developed a UML-based manufacturing resource model based on the NIST model [9]. The new model interacts with the Resource Specific Process Planning (RSPP) module to perform process planning tasks for collaborative manufacturing, but it still lacks the capability of integrating CNC machining to form a close loop manufacturing chain.

Ming et al. introduced a PDES/STEP based production resource model, including data for machine, cutting tool, fixture, gauge tool and fixture [10]. The proposed model considered the external resources in support of the collaborative manufacturing environment. With the other two PDES/STEP-based models (part information model and process planning model), the entire information model can support CAPP applications. Lopec-Ortegsa and Ramirez proposed a STEP-based EXPRESS model to present flexible manufacturing resources (i.e. machining processing resources, handling resources, transportation resources and storage resources) in the shop-floor [11]. This data model was implemented in a STEP-based manufacturing information system.

Several standards have been developed for machine tool data modelling. ASME B5 committee has been working on two draft standards to represent machine tool performance data (ASME B5.59-1) and properties' specification (ASME B5.59-2) respectively. These standards mainly aim to support machine tool vendors. Within ISO, STEP Part 240 [12] stipulates a data model for process planning and eight application objects of Units of Functionality (UoF) '*manufacturing_machine_tool_resources*' that are of direct relevance to machine tools [13]. In addition, ISO TC39/SC2 is developing standards for enhancing interchangeability between machine tool components and testing machine tools (i.e. test method, test code, test component, etc.).

None of the above models is fit for supporting STEP-NC based manufacturing. Some models (e.g. those developed by NIST, ASME and ISO) are for machine tool vendors and users but not for manufacturing applications such as process planning, shop floor scheduling, etc. Others (e.g. Ming's and Lopec-Ortegas's models, and STEP Part 240) are not completely compatible with the STEP-NC standards. The proposed STEP-NC machine model is to support STEP-NC based manufacturing.

3. Objective

The purpose of developing STEP-NCMtDm is of two-folds. First, it is to meet the data requirements for implementing STEP-NC data model in process planning and scheduling. More specifically, it helps the user determine the "How-to-do" information for CNC machining based on the "What-to-do" information contained in a STEP-NC program. STEP-NC can be considered as a process plan model. It is not supportive of other activities, such as cost estimation, machine tool selection, etc. Thus, the second goal of STEP-NCMtDm is to support these functionalities of process planning in a distributed manufacturing environment.

4. Technological aspects of STEP-NCMtDm

4.1 Machine tool static data and dynamic data

Machine tool data are classified into two types: static and dynamic. If the data stay unchanged during the lifecycle of a machine tool, they are classified as 'static machine tool data', e.g. machine loading capacity and machine tool dimensions. These data are mainly machine tool's specifications, a gross indication of the machining capability of a machine tool. On the other hand, if the data change during different stages of machine tool usage, or take on different values for different applications, they are classified as 'dynamic machine tool data'. Dynamic machine tool data are the key information for realizing flexible process planning and manufacturing. There may be two causes to the changes of a machine tool's dynamic data. The first is to do with the machine tool wear, which leads to an accuracy drift or changes of mechanical and electrical properties. The second is to do with the changes or re-configuration of a machine tool setup.

The static machine tool data, once modelled, can stay unchanged. For the two types of dynamic machine tool data, different data processing and updating procedures can be followed. The first type of dynamic data need to be updated on a more regular basis due to the factors such as machine tool wear. The second type of dynamic data is updated on an irregular basis.

4.2 Modelling language and Web-based data model

The EXPRESS modelling language [14] is used to model STEP-NCMtDm. It defines an 'entity-attributes' type data representation method and can fully support modelling constraints and functions. STEP standard has defined several implementation methods (through Java, C++, etc.). STEP-NC data models are also developed by using EXPRESS. Thus, use of EXPRESS can ensure the compliance of STEP-NCMtDm with STEP-NC, and STEP-NCMtDm can 'service' the STEP-NC data models to form a comprehensive process planning data model in support of STEP-NC based manufacturing.

STEP Part 21 [15] files are neutral files and can be utilized to transfer data between different CAD/CAPP/CAM systems. However, it is not suited for Web-based data publications and transformations. To overcome this problem, a Web-enable data representation method is required. In STEP Part 28 (second edition) [16], a mapping mechanism between EXPRESS and XML schema is defined. Using this mechanism, the STEP-NCMtDm in EXPRESS format can be converted to an XML schema. The STEP Part 21 file with machine tool data can also be presented in the XML file format for Web-based data exchange and sharing.

5. STEP-NCMtDm

5.1 Architecture of STEP-NCMtDm

STEP-NCMtDm is constructed to model the following five sets of data (Figure 1.1): 1) Machine tool general data; 2) Machine tool component data; 3) Machining cost data; 4) Machine tool performance data; and 5) Machining capability data.

Machine tool general data set represents the general information about a machine tool. The root entity is the abstract ENTITY *machine_tool*. The general information modelled about a machine tool includes its id, its location, and the links to other machine tool technique data. Three sub-entities are modelled to represent milling, turning, and mill-turn machine tools respectively.

Machine tool component data set represents the information about each individual component of a machine tool. It is further divided into machine tool axis data (ENTITY *axis*), spindle data (ENTITY *spindle*), workpiece handling device data (ENTITY *workpiece_handling_device*), cutting tool handling device data (ENTITY *cutting_tool_handling_device*) and auxiliary device data (ENTITY *aux_device*).

Machine tool capability data set contains the information which can be used to determine machining capability of a machine tool. In process planning, these data are the key factors to determine whether a machine tool is suitable for a particular machining job. The machining capability, represented by ENTITY *machining_capability* includes machine tool's operation type capability, table loading capability, machinable workpiece size, machining accuracy capability, axis positioning accuracy capability, feeding capability and maximum cutting depth data.

Machine tool performance data set is used along with some other data, in selecting a proper machine tool for machining. Currently, only machine tool's accuracy data is modelled, which involves machine tool resolution (ENTITY *resolution*) and axis repeatability (ENTITY *repeatablility_accuracy*).

Machine tool cost data set is modelled by ENTITY *machine_tool_cost_information* for estimating machining cost in process planning. It keeps the data such as the average cost per setup (attribute setup_cost) and machining hourly rate (attribute machining_cost_per_hour). The process planning system can use these data to calculate the corresponding manufacturing cost.

5.2 Static data and dynamic data

As discussed in section 1.3.1, machine tool data can be divided into two categories: static and dynamic. Different data acquisition and updating methods are implemented for different categories. The dynamic machine tool data can be further divided into two groups according to their updating frequency, regular or irregular. Classification of static and dynamic data is essential for later model implementation. Tables 1-3 show the three types of data for ENTITIES *machine_tool, travelling_axis* and *machine_tool_capability* respectively.

In general, data classification is based on the following guidelines:

- The machine tool's general information and the data related to machine tool or its components' capability are considered as static data.
- The machine tool data related to the machine tool's performance, such as accuracy data, belongs to the first type of machine tool dynamic data. Frequency of updating this type of data depends on the usage of the machine tool.
- The machine tool data related to cutting tool and workpiece handling devices are of the second type

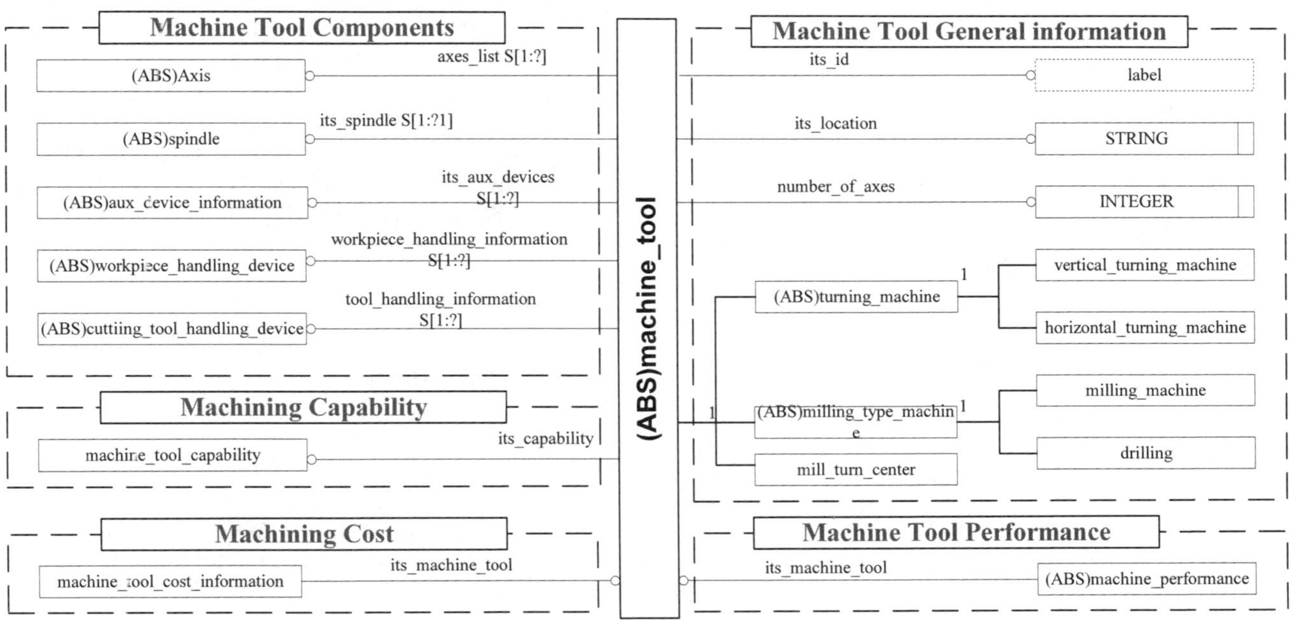

Figure 1.1. Architecture of STEP-NCMtDm

of machine tool dynamic data that needs to be updated on a regular basis. This is due to the fact that different cutting tools handling devices or workpiece handling devices may be utilized for machining a different part. The data are often situational subject to the changes at shop floor.

Table 1.1 Static data and dynamic data in ENTITY *machine_tool*

Static data	Dynamic data	
	Type 1 (regularly updated)	Type 2 (irregularly updated)
its_id	its_capability	tool_handling_information
its_location		
number_of_axis		workpiece_handling_information
axes_list		
its_spindle		its_capability
its_aux_devices		

Table 1.2 Static data and dynamic data in ENTITY *travelling_axis*

Static data	Dynamic data	
	Type 1 (regularly updated)	Type 2 (irregularly updated)
its_name	its_accuracy_information	
its_machine_tool		
its_traveling_range		
its_feeding_range		
rapid_movement_speed		

Table 1.3 Static data and dynamic data in Entity *machine_tool_capability*

Static data	Dynamic data	
	Type 1 (regular updated)	Type 2 (irregular updated)
its_machine_tool	machining_accuracy_information	machinable_workpiece_information
operation_type_information		max_cutting_depth
table_loading_capacity	positioning_accuracy_information	feeding_capability

5.3 Measurement unit

The types of units supported by STEP-NCMtDm are SI units as well as derived or conversion based units defined in ISO 10303 Part 41(ISO 10303-41:1994(E)). If no units are given then, the default values are assumed as defined in Clause 4.2.1 of ISO 14649 Part 10:

- length_measure: millimetre [mm]
- rot_speed_measure: revolutions per second [r/s]
- plane_angle_measure: degrees [°]
- pressure_measure: Pascal [pa]
- speed_measure: meters per second [m/s]
- time_measure: seconds [s]

The default unit for mass_measure (referenced from ISO 10303-41 [17]) is defined as: mass_measure: kilogram [kg]. The power measurement and its measurement unit are defined in ENTITY *power_measure_with_unit* in STEP-NCMtDm. ENTITIES *measure_with_unit* and *si_unit*, and TYPE parameter_value, referenced from STEP Part 41, are utilized to construct ENTITY *power_measure_with_unit*.

6. Web-based STEP-NCMtDm

The EXPRESS-oriented STEP-NCMtDm is not web-enabled. To adapt the distributed manufacturing environment, an XML-oriented STEP-NCMtDm is required to model machine tool data. STEP Part 28 (2nd edition) defines a standardized method to convert EXPRESS data model into a set of XML schema, which can be used to represent machine tool data in a XML document. SETP Part 28 introduces two types of XML schema binding – default XML schema binding (clause 7) and configured XML schema binding (clause 8). This paper utilizes the mapping mechanism corresponding to the default XML schema binding to map STEP-NCMtDm into the corresponding XML schema. These are the standardized mapping mechanisms defined for the EXPRESS elements, such as EXPRESS data type (clause 7.2), defined data type (clause 7.3), entity data type (clause 7.4), entity attributes except DERIVEd attributes and INVERSE attributes (clause 7.6), supertype/subtype relations (clause 7.5), etc.

Figure 1.2 shows a portion of the XML schema (i.e. ENTITY machine_tool) converted from the STEP-NCMtDm EXPRESS data model, and the relationship between *machine_tool* and its subtype ENTITIES *milling_type_machine*, *turning_machine*, and *turn_mill_center*. The header part of the XML schema defines the name space of the STEP-NCMtDm EXPRESS schema and STEP-NC standards. The data part of schema is used to define the main body of the EXPRESS schema of STEP-NCMtDm.

7. Implementation

STEP-NCMtDm has been implemented to model an EMCO Concept Milling 105 milling machine tool [18]. Two types of data file are generated: a STEP Part 21 file (Figure 1.3) and an XML document (Figure 1.4). These files can support adequate machine tool data to meet the data requirements of STEP-NC Part 11 (milling application).

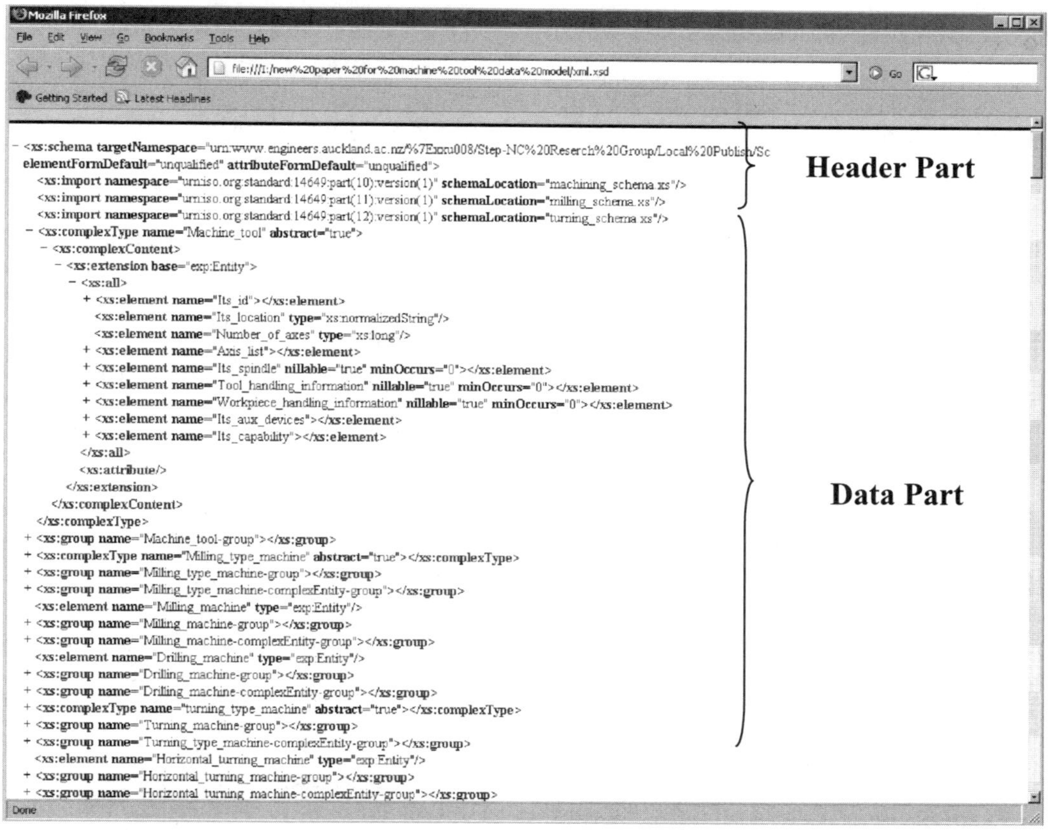

Figure 1.2. XML schema representation of STEP-NCMtDm


```
ISO-10303-21;
HEADER;
FILE_DESCRIPTION (('machine tool database for Emco concept Mill 105','1');
FILE_NAME ('EMCOMILL105.STP', '2006-12-08', (('ALBERT YANG',$), ('XUN WILLIAM
XU', $), ('DEPARTMENT OF MECHANICAL ENGINEERING, UOA'), $, $, 'IMS', $),
FILE_SCHEMA (('machine_tool_information', 'machining_schema', 'milling_schema',
'turning_schema'));
END_HEADER;
DATA;
#1= MILLING_MACHINE(LABEL('CONCPET-MILL-105'), 'MANUFACTURING-LAB',
'3', (#2, #3, #4), $, #15, #12, (#11, #14), $, #7);
#2= TRAVELING_AXIS (AXIS_NAME(x), #1, LENGTH_MEASURE(200), #5,
SPEED_MEASURE(5), #6);
#3= TRAVELING_AXIS (AXIS_NAME(y), #1, LENGTH_MEASURE(150), #5,
SPEED_MEASURE(5), #6);
#4= TRAVELING_AXIS (AXIS_NAME(z), #1, LENGTH_MEASURE(250), #5,
SPEED_MEASURE(5), #6);
#5= CUTTING_FEED_RANGE (SPEED_MEASURE(0), SPEED_MEASURE(5));
#6= RESOLUTION (LABEL(STEP-RESOLUTION), #1, $, '1.5','um');
#7= MILLING_TYPE_MACHINING_SPINDLE (LABEL(SPINDLE1), #8, #4, #9, $, #10, $,
#11, .T., LENGTH_MEASURE(55), '4 2',)
#8= ROT_SPEED_RANGE (ROT_SPEED_MEASURE(150), ROT_SPEED_MEASURE
(5000));
#9= POWER_MEASURE_WITH_UNIT (PARAMETER_VALUE(1.1), #19);
#10= LENGTH_MEASURE_RANGE (LENGTH_MEASURE(95), LENGTH_MEASURE
(245))
#11= COOLANT_INFORMATION (.T., (COOLANT_TYPE(mist), COOLANT_TYPE(spay),
COOLANT_TYPE(flood)), PRESSURE_MEASURE(0.5), .T.);
#12= TABLE (LABEL(TABLE1), MASS_MEASURE(10), (#2, #3), $, $, $, #13);
#13= HOLE_FIXTURE (LABEL(HOLE-FIX-1));
#14= CHIP_REMOVE_DEVICE (.F.);
#15= TURRET (LABEL(10-STATIONS-TOOL-CHANGER), '10', (#16, #17, #18), #18,
LENGTH_MEASURE(50), LENGTH_MEASURE(55), WEIGHT_MEASURE(0.7), .T., 9, 7);
```


Figure 1.3. Potion of the Part 21 File for Concept Mill 105

Figure 1.4. Potion of the XML file for Concept Mill 105

For example, for the Entity *milling_technology* in STEP-NC Part 11, which defines the technological parameters of milling operations, the STEP-NCMtDm based Part 21 file/XML file represents data for the following attributes of this entity:

1) Attributes cutting_speed and spindle: Related to an instance of rotation_speed_range defined in Line #8 (Fig. 1.3). The actual spindle rotation

speed must be within this range. The linear cutting speed range can also be set.

2) Attribute feedrate_per_tooth: Related to Line #5, which defines an instance of feeding_range for the cutter. When multiplying the tooth number, a feeding range is calculated to constrain the feedrate value of a cutting tool.

3) Attribute synchronize_spindle_with_feed: Defined as the ninth parameter in Line #7, which defines whether cutting speed and feed of the tool are synchronized.

The corresponding STEP-NCMtDm based XML file supports the same machine tool data for Web-based STEP-NC manufacturing applications.

8. Conclusions

STEP-NC based manufacturing is about utilizing the STEP and STEP-NC data models to support an informative, intelligent, integrated and interoperable CNC manufacturing scenario. The key for STEP-NC based manufacturing is to build machine tool data model to realize the 'transition' of STEP-NC tasks level data to machine tool dependent method level data. STEP-NCMtDm models various machine tool data to meet the requirements of STEP-NC compliant process planning and manufacturing activities. It has also been converted into the XML schema, making it Web-enabled.

Machine tool data in STEP-NCMtDm are classified as either 'Static data' or 'Dynamic data'. This classification reflects the nature of different machine tool data, and their different acquisition and updating methods to make efficient and effective machine tool data management. The machine tool data can be exchanged or shared by using either STEP Part 21 files or XML documents.

The proposed data model can be easily extracted to support other types of machine tools and CNC machining applications. Further work includes re-instating the constraint information in STEP-NCMtDm that was lost when using STEP bindings to map its EXPRESS data model into the XML schema. Also under development is a platform on Internet that utilizes STEP-NCMtDm for facilitating on-line bidding process for manufacturing tasks.

References

[1] X. Xu and Q. He, 2004, Striving for a total integration of CAD, CAPP, CAM and CNC, Robotics and Computer-Integrated Manufacturing 20 (2):101-109.

[2] X. Xu, H.Q. Wang., J. Mao, S.T. Newman, T.R. Kramer, F.M. Proctor, and J.L. Michaloski, 2005, STEP-compliant NC research: The search for intelligent CAD/CAPP/CAM/CNC integration, International Journal of Production Research 43(17):3703-3743.

[3] ISO/FDIS 14649-111.2:2004(E), Industrial automation systems and integration -- Physical device control -- Data model for computerized numerical controllers -- Part 111: Tools for milling machines.

[4] ISO 14649-121:2005(E), Industrial automation systems and integration -- Physical device control -- Data model for computerized numerical controllers -- Part 121: Tools for turning machines.

[5] ISO 14649-10:2004(E), Industrial automation systems and integration – Physical device control -- Data model for computerized numerical controllers -- Part 10: General process data, 2nd edition.

[6] ISO 14649-11:2003(E), Industrial automation systems and integration -- Physical device control -- Data model for computerized numerical controllers -- Part 11: Process data for milling, 1st edition.

[7] ISO 14649-12:2005(E), Industrial automation systems and integration -- Physical device control -- Data model for computerized numerical controllers -- Part 12: Process data for turning, 1st edition.

[8] Jurrens, K., Fowler, James, and M. B. Algeo, Modeling of manufacturing resources information, requirements specification, NISTIR 5707, National Institute of Standards and Technology, Gaithersburg, MD, 1995.

[9] Kulvatunyou, B., Wysk, R. A., Cho, H., and Jones, A, 2004, Integration framework of process planning based on resource independent operation summary to support collaborative manufacturing, International Journal of Computer Integrated Manufacturing 17 (5):377-393.

[10] Ming, X. G., Mak, K. L., and Yan, J. Q., 1998, PDES/STEP-based information model for computer-aided process planning, Robotics and Computer-Integrated Manufacturing 14 (5-6):347-361.

[11] Lopez-Ortega, O., and Ramirez, M., 2005, A STEP-based manufacturing information system to share flexible manufacturing resources data, Journal of Intelligent Manufacturing 16 (3):287-301.

[12] ISO 10303-240:2005(E), Industrial automation systems and integration -- Product data representation and exchange -- Part 240: Application protocol: Process plans for machined products.

[13] Xu, X. W., 2004, Machine Tool Data in the STEP-NC Environment – Machine Tool Data in Current Standards (Report No. 1), Intelligent System Division, Manufacturing Engineering Laboratory, National Institute of Standards and Technology, Gaitherburg, MD 20899.

[14] ISO 10303-11:2004(E), Industrial automation systems and integration -- Product data representation and exchange -- Part 11: Description methods: The EXPRESS language reference manual, 2nd edition.

[15] ISO 10303-21:1994(E), Industrial automation systems and integration -- Product data representation and exchange – Part 21: Clear text encoding of the exchange structure.

[16] ISO TC184/SC4/WG11 N258, ISO/DIS 10303-28:2006(E), Industrial automation systems and integration -- Product data representation and exchange -- Part 28: Implementation methods: XML representations of EXPRESS schemas and data.

[17] ISO 10303-41:1994(E), Industrial automation systems and integration -- Product data representation and exchange – Part 41: Integrated generic resources: Fundamentals of product description and support, 1st edition.

[18] http://www.emco.at/pcmill_105.php?changelang=en#.

Geometric Modelling and NC Machining of Scroll Profiles Using Circular Biarcs Fitting

J.N. Lee[1] and R.S. Lee[2]

[1]Cheng Shiu University, Taiwan, 833, jengnan@csu.edu.tw
[2]National Cheng Kung University, Taiwan, 701, mersl@mail.ncku.edu.tw

Abstract. Scroll configuration plays a very important role in the design of a scroll compressor. In order to comprehend the concepts of design and improve the flexibility of machining, the objectives of this paper are the regeneration of scroll profile, the study of the curve fitting and circular interpolation. Since the sample geometry is unknown, reverse engineering is used to obtain a sample scroll configuration. Then, by analyzing the configuration, the design paramters are retrieved. The geometry of the scroll is derived using conjugate surface theory. A deviation analysis between the theoretical curve and the actual curve is implemented to modify the design parameters. In order to generate a continuous smooth toolpath, a circular biarcs fitting method is adopted to approximate the toolpath of the scroll profile. The developed biarcs fitting method eliminates the problem in conventional straight-line approximation toolpath which leaves ridges on the finished surface of the workpiece. The toolpath generated by the proposed method is verified through the solid cutting simulation and the trial-cut on a three-axis machine.

Keywords. scroll type compressor, reverse engineering, involute curve, conjugate theory, biarcs fitting.

1. Introduction

The scroll compressor has the characteristics of reliability, high efficiency, simplicity, low level of noise/vibration when compared with other types of compressors. It has been increasingly popular in refrigeration and air conditioning equipments. The scroll configuration is complex with involute curve and high precision is required (Figure 1.1). The design and manufacture of scroll configuration is complicated. The concept of a scroll machine was established by Cruex [1] at the beginning of the 20th century. However, was not implemented until the 1970's when scroll configuration machining technique became possible with the advent of NC machining technology. Since then, research and development of scroll compressor has increased substantially.

Lee and Wu [2] proposed a general design process to profile a scroll compressor. The design process is developed based on geometric and kinematic considerations. Li et al. [3] investigated the effects of involute curves of circles, squares, and line segments on the performance of the scroll compressor, such as geometrical parameters, leakage line length, and various gas forces acting on the orbiting scroll. Jiang et al. [4] presented the integration approach of computer-aided design (CAD) and computer-aided manufacturing (CAM) for the design and manufacture of refrigeration scroll compressors. The approach is based on concurrent engineering and is implemented in Pro/ENGINEER software and Visual C++ programming.

The geometry of scroll wrap is generated with a cylindrical end mill using a special machine and software. The kinematic relations of the cutting tool with respect to the workpiece are based on the property of the involute curve. This cutting method is called generating method [5]. By using this method, the worn tool cannot be reused after dressing; thus, the flexibility of choosing a cutting tool is limited. With the increasing use of computers in design and manufacturing, procedures for generating the scroll profile can be obtained by general purpose CAD/CAM software. However, commercial CAD/CAM software generates the numerical control code by linear interpolation and single circular interpolation. Schönherr [6] investigated the biarc-curve fitting technique used in NC machining. The objective was to minimize the difference of curvatures between two adjoining circular arcs. The biarcs fitting method eliminates the problem in conventional straight-line approximation toolpath where it leaves ridges on the finished surface of the workpiece.

To understand the concepts of design and obtain the flexibility of the machining process, the objectives of this paper are the regeneration of the scroll profile and a study of the curve fitting process. Firstly, reverse engineering is used to obtain a sample scroll configuration and this configuration is analyzed to obtain design parameters. Then, the geometry of the scroll is derived using conjugate surface theory. In order to generate a continuous smooth toolpath, a circular biarcs-fitting method is adopted to approximate the toolpath of the scroll profile. Figure 1.2 shows the flowchart of cutter location file generated for a scroll profile.

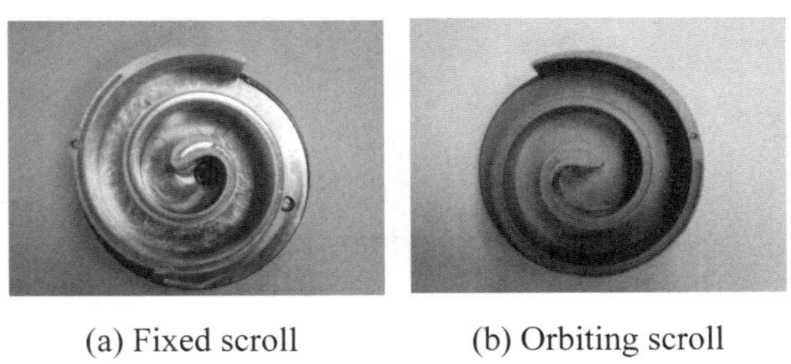

(a) Fixed scroll (b) Orbiting scroll

Figure 1.1. Illustrations of scroll profiles

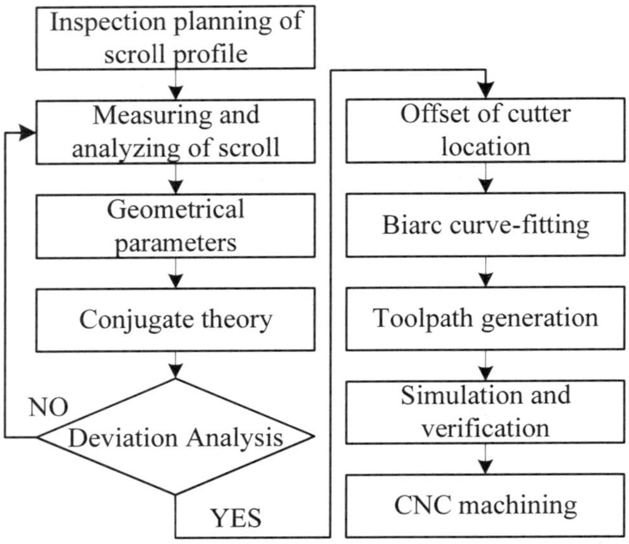

Figure 1.2. Flowchart of CL file generation for scroll profile

2. Geometry of a Scroll Wrap

2.1 Involute Curve

As shown in Figure 1.3, an involute curve is generated by the origin of $(OXYZ)_a$ coordinate system which is defined on the base circle. The radius of base circle is r_b and the involute angle is ϕ. Coordinate systems $(OXYZ)_1$ and $(OXYZ)_a$ are the reference and movement coordinate system respectively. The Y_a axis of $(OXYZ)_a$ coordinate system rolls over the base circle. The involute curve is traced out by the origin of $(OXYZ)_a$ coordinate system. By applying the homogeneous coordinate transformation matrix, the relative position of the coordinate system $(OXYZ)_{a'}$ with respect to $(OXYZ)_1$ can be derived. The equation of involute curve can be obtained as follows:

$$^1R^{(c)} = {}^1M_{a'}R_{a'} \tag{1.1}$$

where the designation $^1M_{a'}$ indicates that the coordinate transformation is performed from $(OXYZ)_{a'}$ to $(OXYZ)_1$.

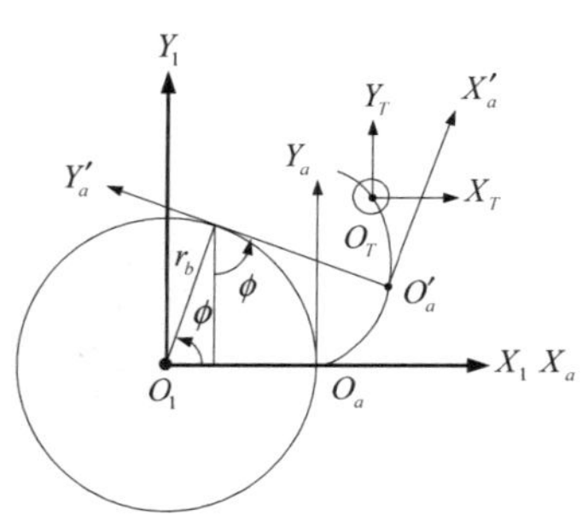

Figure 1.3. The coordinate system of an involute curve

2.2 Profile of Scroll Wrap

As shown in Figure 1.4, the scroll rotates a cycle with a uniform angular velocity, while the tool translates a wrap pitch. The scroll wrap with involute curve is generated in accordance with the kinematics relations of the tool with respect to the workpiece. The relationship between the scroll profile and the tool is established by the conjugate theory. The conjugate theory describes the constraints at instantaneous contact points as follows [7]:

1. The contact points must coincide with respective conjugate surfaces.
2. The conjugate surfaces have common normal at contact point.
3. The component of relative velocity at contact points is equal to zero along the common normal.

The equation of the scroll profile can be derived according to the differential geometry, homogeneous coordinate transformation and conjugate theory. Referring to Figure 1.3, surface equation ^1R and normal vector ^1n of scroll profile in the $(OXYZ)_1$ coordinate system can be expressed as:

$$^1R = {}^1M_T \cdot {}^TR \tag{1.2}$$

$$^1n = {}^1M_T \cdot {}^Tn \tag{1.3}$$

The superscript T in ^TR and ^Tn indicate the tool and unit normal vector in the $(OXYZ)_T$ coordinate system.

The meshing equation is derived based on the third constraint of conjugate theory. The relationship between relative velocity and normal vector can be expressed as:

$$^1n^{\mathrm{T}} \cdot \frac{d\,{}^1R}{dt} = 0 \tag{1.4}$$

The equation of meshing is the necessary condition for the existence of the envelope to the family of curves.

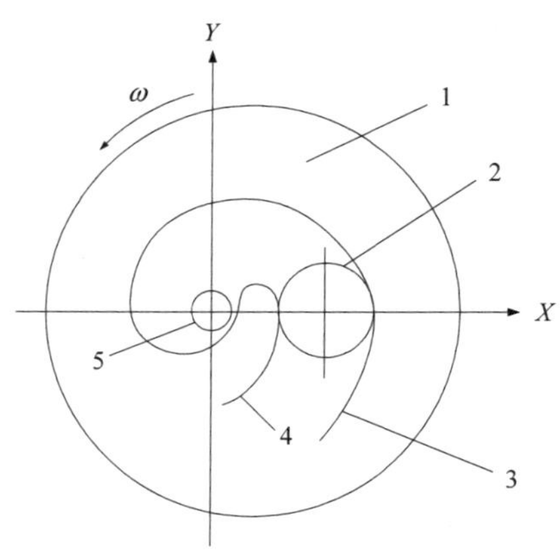

1. Scroll wrap, 2. Tool, 3.inner profile, 4. Outer profile, 5. Base circle

Figure 1.4. Generation of a Scroll profile

2.3 Modification of the Scroll Profile

Since the geometry of the sample profile is different from the derived involute curve and a favourable scroll profile should be of high efficiency, light weight and easy to manufacture, the scroll wrap geometry needs to be modified. Referring to Figure 1.5, the procedure for circular arc modification is described as follows.

1. Determine the involute angle (ϕ) for point A.
2. Derive the coordinates of points A, B, D and E.
3. Establish the line equations of $\overline{AD}$ and $\overline{BE}$.
4. Derive the coordinates of arc centres (O_1 and O_2).
5. Calculate the distance L_r between O_1 and O_2 .

Figure 1.5 shows the situation of $L_r = R_1 + R_2$ for the circular arc modification. It is tangential at the intersection point of the two arcs, so the profile is smooth and continuous. The geometric shape is a so-called Perfect Meshing Profile (PMP).

In the case that $L_r < R_1 + R_2$, as shown in Figure 1.6, the intersection point of the inner and outer curves is kept continuous only to the zeroth order. It is not easy to establish the toolpath during the machining process.

In the case that $L_r > R_1 + R_2$, as shown in Figure 1.6, the gap of adjoining arcs is bridged by a linear segment. The modified curve is kept continuous to the first order and is a practical scroll profile.

Through the analysis of the obtained surface data, the geometry of the sample profile is obviously modified by the case $L_r > R_1 + R_2$. There is no inflexion point in the modified scroll wrap.

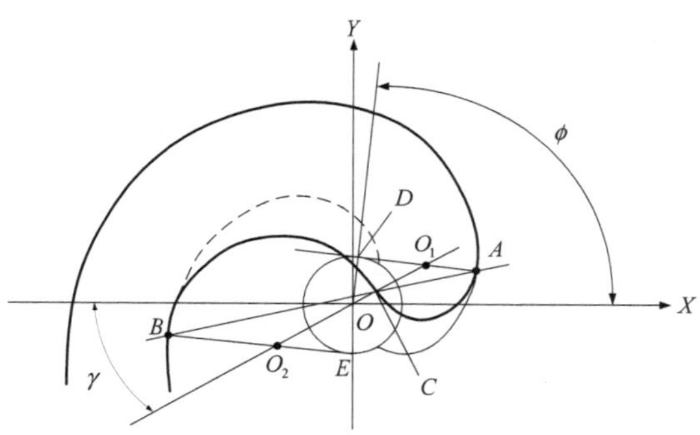

Figure 1.5. Circular arc modification for involute curve ($L_r = R_1 + R_2$)

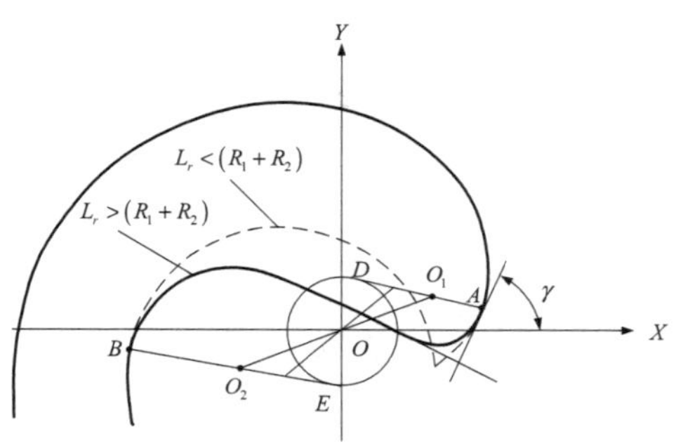

Figure 1.6. Other modifications for the involute curve ($L_r < R_1 + R_2$ and $L_r > R_1 + R_2$)

3. Biarcs Curve Fitting

The control modulus in numerically controlled machines usually provides only straight line and circular arc interpolations. Linear segments are often used because of their simplicity in implementation and computational efficiency. The resultant surface finish is poor because of the lack of tangency in piecewise linear curves. The simple circular arc fitting without the consideration of tangential continuity at junction points results in great changes of the tangent direction. The biarcs curve fitting ensures the first order continuity between adjoining segments.

In biarcs curve fitting, two mutually tangential circular arcs should fit between any two data points. Six conditions are required to form these two arcs. Among them, five conditions are known. Each arc must pass one data point with a prescribed unit tangent vector in data point. The two arcs must have a common tangent at a junction point. To obtain unique biarcs curve, a valid sixth condition which guarantees a smooth curve fit is necessary. In this paper, the condition that minimizes the difference in radii of the two arcs is chosen.

In the case of no-inflection point, as shown in Figure 1.7, the coordinates of junction point and centre points of biarcs are presented as follows:

$$\begin{bmatrix} x_m \\ y_m \end{bmatrix} = \begin{bmatrix} x_{c1} + R_1 \sin(\phi_1 + \beta_1) \\ y_{c1} - R_1 \cos(\phi_1 + \beta_1) \end{bmatrix} \quad (1.5)$$

$$\begin{bmatrix} x_{c1} \\ y_{c1} \end{bmatrix} = \begin{bmatrix} x_1 - R_1 \sin\phi_1 \\ y_1 + R_1 \cos\phi_1 \end{bmatrix} \quad (1.6)$$

$$\begin{bmatrix} x_{c2} \\ y_{c2} \end{bmatrix} = \begin{bmatrix} x_2 - R_2 \sin\phi_2 \\ y_2 + R_2 \cos\phi_2 \end{bmatrix} \quad (1.7)$$

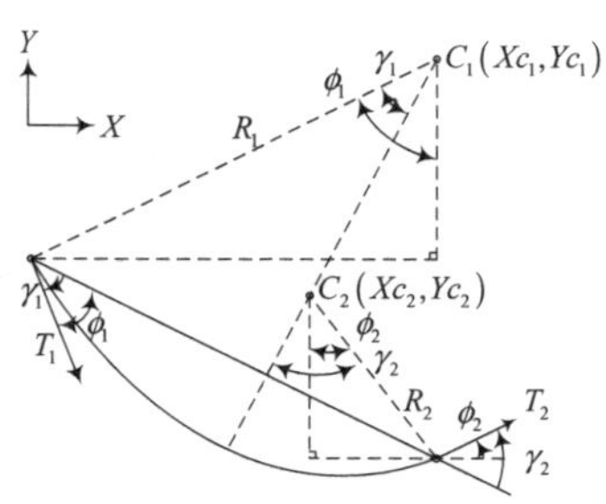

Figure 1.7. Difference in radii of the two arcs

$(\beta_1 = \gamma_1, \beta_2 = \gamma_2)$

4. Results and Discussion

4.1 Analysis for Geometric Parameters

The design parameters are retrieved and confirmed by the analysis of the obtained surface data. The geometry of the scroll is derived from conjugate surface theory. To fit the sample profile, the conjugate surface of thescroll has to be modified. A deviation analysis between the theoretical curve and the actual curve is implemented to modify the design parameters. Figure 1.8 shows the deviation between the theoretical curve and the actual curve. The deviation in the segment of circular arc modification is smaller than $0.01mm$. The result is reasonable due to the existance of the tolerance.

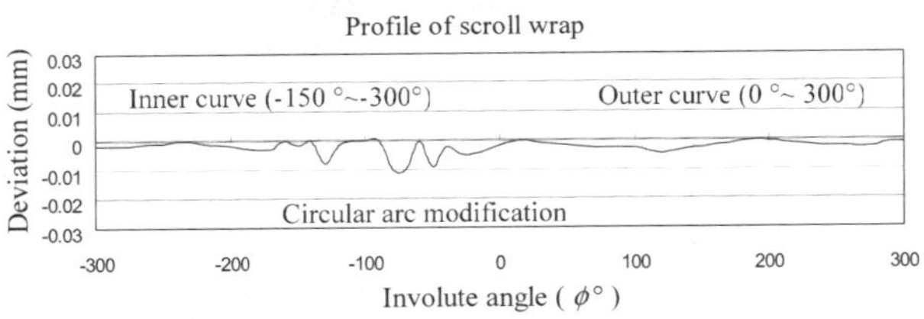

Figure 1.8. Deviation analysis of scroll profiles

4.2 Toolpath Simulation and Trial-Cut

To avoid collision between all machine tool components, the generated toolpath was verified before actual machining by carrying out a solid cutting simulation. Figure 1.9 presents the scroll profile machined by the proposed method with the cylindrical end mill simulated by VERICUT® software. The CNC machine tool and controller were modelled to perform realistic 3D simulation. After the simulation and verification, the generated toolpath was converted into a NC program by the postprocessor. A three-axis machine tool was used to perform the actual machining. Figure 1.10 shows the machined product with the scroll profile.

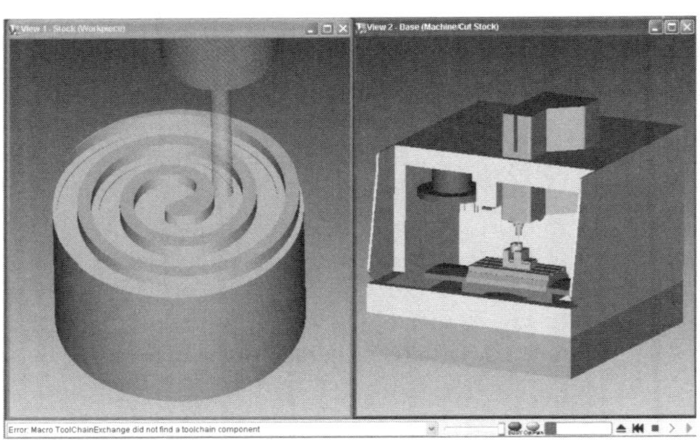

Figure 1.9. Simulation of the scroll profile cutting using the proposed method

Figure 1.10. Machining of scroll profile

5. Conclusion

In order to comprehend the concepts of design and improve the flexibility of the machining, this paper regenerates the scroll profile based on the involute curve composed of circular arcs. The design parameters are retrieved and confirmed by reverse engineering the part. The geometry of the scroll is derived in accordance with the conjugate surface theory. Through a deviation analysis between the theoretical and actual curves, the design parameters are revised and verified. In order to generate a continuous smooth toolpath, a circular biarcs fitting method is adopted to approximate the toolpath of the scroll profile. The cutting simulation results show that the machining process works well and interference does not occur. The proposed methodology can be used to automate the programming of tool paths for machining of a scroll profile.

Acknowledgement

The authors would like to thank the support of the National Science Council, Taiwan, R.O.C. under grant NSC 94-2212-E-230-011.

References

[1] Cruex, L., 1905, Rotary Engine, US Patent 801 182.
[2] Lee, Y. R. and Wu, W. F., 1995, On the Profile Design of a Scroll Compressor, International Journal of Refrigeration, Vol.18, No.5, P308-317.
[3] Li, L.S., Shu, P.C. and Yu, Y.Z., 1997, The Effect of Scroll Wraps on The Performance of Scroll Compressors, International Journal of Refrigeration, Vol.20, No.5, P326-331.
[4] Jiang, Z., Harrison, D. K. and Cheng, K., 2003, Computer-aided Design and Manufacturing of Scroll Compressors, Journal of Materials Processing Technology, Vol.138, P145-151.
[5] Lee, R. S. and Lee, J. N., 2002, A New Method of Tool Orientation Determination by Enveloping Element for 5-axis Machining of Spatial Cam, International Journal of Production Research, Vol.40, No.10, P2379-2398.
[6] Schönherr, J., 1993, Smooth Biarc Curves, Computer-aided Design, Vol.25, No.6, P365-370.
[7] Litvin, F.L., 1994, Gear Geometry and Applied Theory, PTR Prentice Hall, Englewood Cliffs, New Jersey.

3–3

Establishing Parametric Design Management to Facilitate Activity-based Costing

C.L. Liu[1] and R. Holland[2]

[1]Department of Mechanical Engineering, National Taiwan University, Taipei, clliu@ntu.edu.tw
[2]Department of Design and System, Brunel University, Uxbridge, Ray.Holland@brunel.ac.uk

Abstract. Parametric design is usually involved in technical analysis with an emphasis on mathematics. In management, even though some of the activities can be quantified, others may be appeared to be fuzzy concepts. This work attempts to establish a means to facilitate cost estimation by using parameters and equations to link design, manufacture, and management activities as a whole. All items under consideration must therefore be quantified so that they can be treated as parameters for further manipulation. The inputs of the system are primary parameters, usually related to design activity, and the derived parameters are produced from equations which describe the relationship among these parameters or empirical data. The design, manufacture, and management parameters are thus transformed to a unified measure, the cost. These primary parameters and empirical data in equations can be revised and adjusted dynamically so that an activity-based costing is obtained in the form of a spreadsheet. The parametric design management system containing all activities as a whole is thus supporting comparison, decision-making and planning. From this base a total design management approach is enabled.

Keywords: parametric design, design management, activity-based costing

1. Introduction

A preliminary study of the International Engineering and Technology Accreditation was held by the Taiwanese National Science Council in 2003. The goal of this project is to improve the quality of education, raise the competitiveness of the graduates, get consensus of the local society and encourage international cooperation. One of the sub-projects is to send questionnaires to the graduates and collect their opinions and views to what the courses and activities they experienced during their undergraduate and graduate studies. The department of Mechanical Engineering at National Taiwan University collected 62 questionnaires returned from the alumni and some results are interesting. The results show that the basic theories and application capabilities are important and the department provides quite good quality and quantity of related courses for undergraduate and graduate programmes. As for practical engineering, multi-field integration and management, more than 50% alumni express rather higher expectations than the department had provided. Besides, for engineering courses they prefer more knowledge of management to be integrated in their education ethos. Such capabilities as planning, management, integration, communication, leadership, and problem definition were frequently quoted in their questionnaire opinions. The graduates with mechanical engineering backgrounds cover almost every industry and key business sector and their views reflect trends in the mechanical engineer's future and are worthy of respect.

Generally speaking, design is a kind of integration about information and management integrates resources. They are quite similar in terms of integration. Engineering students are inherently good at design. The hypothesis may be that a design engineer is easily trained to be a manager due to their capability of integration. This work suggests a means to link design, manufacture, and management with parameters and pave the way for mechanical engineering students to management.

2. Costs and Prices

Every product on the shelf has a price. The price assigned to a product may be based on many phases such as the market conditions, supply and demand, financial situation, season, location, quantity, production means, etc. However, some of them can be attributed to the cost determined rather than the profit determined. Profit may be come from raising the price if high price is accepted by the customer. Otherwise profit must be induced by lowering the cost. The price of most products is determined by the market. Therefore, profit is the cost subtracted by the price. However, price and cost fluctuate every day. If only a price is presented to the managerial board without the support of the information of how the cost is derived, they may not be able to judge whether the price is reasonable or not and how much the profit may be. Moreover, whenever new predictions or adjustments are required a new spread sheet is expected as soon as possible. Thus a prompt response to cost and price estimation system is needed. This generates two key questions. The first question is who should provide the estimation system? The answer seems to be the designers because they design the product and know every detail and reasons of change of the product. The second question would be who is supposed to use the cost and price estimation system. It is no doubt the manager. Therefore, the designer and the manager should work together to address the cost and price estimation problem and the design management.

3. Design and Management

Before asking designers what they do, it is first necessary to ask "what is design?" There may be no absolute definitions of design that will please everyone. However, there is no doubt that design is everywhere. Generally, design is viewed as a kind of technical work such to develop a product or service designed by trained designers. There are many other active participants including project managers, company directors, market managers; even environment politicians will influence the birth of the product [1]. The design work is no longer solely technical work but managerial work too.

The next question is "what is management?" Management may be regarded as a social process. Again there is no universally agreed definition but broad acceptance that such functions as to forecast and plan, to organize, to command, to coordinate and to control with a significant modern emphasis on motivation and empowerment [2]. Comparing management with design, there emerges something in common. They are both a kind of process, one is social and the other is technological. They always need to plan ahead and to forecast that which has not yet happened. To a manager, a lot of conflicts interweave among different departments. As for a designer, many trade-offs aggregate in fields of technology. So both managers and designers are responsible for solving problems and coordinating and reconciling arguments.

From another point of view, as design activity is everywhere, some researchers have suggested that people who are not working behind the draught table or in the front line in the design studio are participating in design activities but may not be aware of his/her contribution to design. These indirect but associate participants are practicing silent design [3]. The activity of silent design thus pervades to all departments of the company. The design department, however named, thus becomes one of the links of the managerial chain. In other words, the organizational position of the design department is under the managerial structure and the design is deemed to exist for the purpose of management.

In the nineteenth century, a number of new production technologies were developed and used in the industry such as time study, standardization and mass production, etc. In the twentieth century, more technologies have boomed, such as organization, production control, queuing and inventory theories, human factors, material handling, reliability, operational research, statistical analysis, automation, decision theory, simulation, total system design, etc. How to direct these technologies and steer people working in these technological departments to create wealth for the company is a challenge to the management level [4]. The firm's strategy made by the top managers is one thing but to communicate smoothly with engineers is another thing. The gap needs to be bridged with care and new thinking tools. This work seeks to focus their communication within defined parameters so as to avoid divergence.

Designer or manager both work with products. The problem is how they work together? Many products are similar in function and even similar in performance. But their shape, look and image can make a dramatic difference. That is seen as a primary function of marketing. However, a product is usually created by the designer, and the market investigator must pass the crucial information in the form of a brief to the designer and audit the result very carefully. That is a part of design management. However, designer is usually invited to design a product only in the middle stage of a new product development process when the whole specifications have been finalized. The designer becomes a follower of instructions and feels no responsibility or finds it difficult to engage with the commercial success or failure of the new product development. Some researchers suggest that the designer should get involved as early in the process as possible. If the designer may be brought in at the beginning, the stage of idea generation, and go through the concept development, marketing strategy and business analysis, then the new product is definitely sure to fit the market expectations [5].

Good design costs money and design faults cost money too. Spending money on design is a commitment to spend more money later [6]. A lot of expenditures are decided in the design stage. That is one main reason why design is important from the managerial point of view. The profit earned or lost is largely affected by product design. Also, design integrates many things such as operation, maintenance, replacement, modification, marketing, selling, distribution, etc. These all relate to cost and that is why design needs to be managed. Based on the consideration of safety, the strength and deformation are analyzed by mechanical engineers. Proper selection of material and product life expectation in terms of reliability and ecology will be the main theme of engineering design which determines part of the direct cost [7].

Many types of data processing software may be used as the calculation tool for parametric design management. One of the easiest to source and use to do parametric design management may be Excel. It has been used for the mechanical design and draft of a transmission system where Excel and AutoCAD are integrated for the project [8].

Costs of manufacturing operations include mainly direct labor, material, and overhead [9]. Each cost happens because an activity is being executed. Once the activities of the manufacturing process have been accomplished the cost of the product can to be counted. Several activity-based costing methods have been presented and referring to many aspects, such as how to choose the most costly items and to investigate the possibilities of getting cost information from the product design, purchasing and manufacturing processes in a manufacturing company [10]; to allocate a structure designed specifically for projects in a generic model [11]; to place accurate and reliable economic values to a material handling manufacturing facility so as to provided a quantified investment justification approach that can benefit management [12]; to help cost reduction learning or comparison in mechanical workshops [13].

A paradigm for fundamental parameters and rules presents parametric design management and how parameter values can be changed to revise costs in a spreadsheet calculation.

4. Design with Parameters

A round steel shaft of a machine is designed as shown in Fig. 1.1. A sheave may be fit in the middle and applies a load F on the shaft of length l. It is a pure mechanical problem and what we want to know is the safety of the design. No management matter is involved in this early stage.

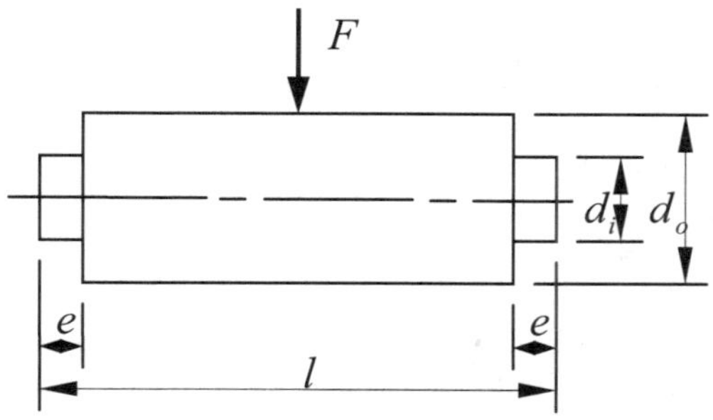

Figure 1.1 A simply supported shaft loaded in the middle

This problem is solved as follow. The single force F is loaded in the middle as a simply supported beam. The two shouldered ends (d_i by e) of the shaft are ground to fit bearings so as to support the load F equally from the two ends. The strength of material will give the relationships [14] among dimensional parameters as

$$\sigma_{\max} = \frac{8F(l-e)}{\pi d_o^{3}} \tag{1}$$

where σ_{max} is the maximum working stress in the middle of the shaft. As F, l, e and σ_{max} are given the diameter of the shaft d_o can be computed as

$$d_o = \left[\frac{8F(l-e)}{\pi\sigma_{\max}}\right]^{1/3}. \tag{2}$$

Where the units of dimensions are in mm and the loading is in kgf. The dimension d_i can be decided by the empirical shoulder factor $k_{shoulder}$ times the shaft outer diameter d_o, that is

$$d_i = k_{shoulder} \cdot d_o. \tag{3}$$

The volume of the shaft v_{shaft}, one of the useful parameters, is thus derived as

$$v_{shaft} = \frac{\pi[(d_o)^2(l-2e)+(d_i)^2\cdot 2e]}{4} \tag{4}$$

The weight of the shaft w_{shaft} is the product of the density of the shaft material ρ_{steel} and the volume v_{shaft}. That is

$$w_{shaft} = \rho_{steel} \cdot v_{shaft}. \tag{5}$$

The above can be regarded as design with parameters. The geometric parameters such as dimensions of the shaft and quantified mechanical properties are included and linked by equations.

The parameters used in this work can be classified in two kinds: primary and derived. The primary ones are like independent variables such as F, l, e, σ_{max}, and ρ_{steel}. As the derived ones are functions of the independent and some other derived parameters such as d_0, d_i, v_{shaft}, and w_{shaft}, the change of parameters can only be done on the primary ones and the derived ones follow the change. The factor denoted by k, such as $k_{shoulder}$, is also regarded as a parameter. This kind of parameter is either an empirical value or a value obtained from an empirical equation including primary and/or derived parameters.

5. Process Planning with Parameters

Once the whole dimensions and properties of the shaft have been derived in the design stage, the manufacturing process follows. The shaft is supposed to fit in a pulley hole at the middle and two bearings at both ends, grinding must be the last operation that fulfills tight tolerances. Heat treatment to increase the strength of the shaft is required before grinding. For this bar shaped machine part, rough and fine turning operations are first applied to the raw material. An allowance is introduced to enlarge the shaft diameter d_0 so as to assure enough material to be cut to size in case the bar is not so straight. A parameter for diameter allowance k_{da} is introduced so that the diameter of raw material d_{raw} would be

$$d_{raw} = k_{da} \cdot d_o. \tag{6}$$

Another parameter k_{la} is also introduced for length allowance. Thus the raw material l_{raw} is

$$l_{raw} = k_{la} \cdot l. \tag{7}$$

The direct costs of a product usually include material and operation. In this shaft case, material cost is charged by weight and operation cost is the product of operation time and rate of operation, including labor. The weight of the raw material w_{raw} can be calculated by the product of the density ρ_{steel} and the volume of the raw material v_{raw}. The volume and the weight of the raw material are

$$v_{raw} = \frac{\pi(d_{raw})^2 l_{raw}}{4}, \tag{8}$$

$$w_{raw} = \rho_{steel} \cdot v_{raw}. \tag{9}$$

The operations of this shaft consist of turning, heat treatment and grinding. Turning cost usually can be calculated on chip removal basis. The chip removal volume from turning v_{chip} can be calculated roughly by subtracting the product volume v_{shaft} from the raw material volume v_{raw}

$$v_{chip} = v_{raw} - v_{prod}. \tag{10}$$

The weight of the removed chip w_{chip} is thus

$$w_{chip} = \rho_{steel} \cdot v_{chip}. \tag{11}$$

As in this shaft case, heat treatment cost is sometimes charged by the weight and the weight for heat treatment is nearly the same as the product weight w_{shaft}

The last operation is grinding and the rate is usually based on the time consumption which is proportional to the area of the grinding surface times the number of grinding cycles. Since dimensions of the shaft for grinding are almost the same as the finished product, and the grinding surface area $A_{g\text{-}surface}$ is almost the same as

$$A_{g-surface} = \pi(d_o(l-2e)+d_i(2e)). \tag{12}$$

The total grinding process will take number of grinding cycles k_{cycle} and the total grinding area A_{grind} thus can be calculated as

$$A_{grind} = k_{cycle} \cdot A_{g-surface} \cdot \qquad (13)$$

6. Management with Parameters and Costs

From a management point of view, decision making is based on comparison, and comparison must be done on the same unit, such as cost in this case.

The cost for either material or operation is the product of the unit cost and the quantity. Thus the total cost of the shaft C_{shaft} including material C_{matl} and operations C_{turn}, C_{ht}, and C_{grind} is

$$C_{shaft} = C_{matl} + C_{turn} + C_{ht} + C_{grind} \qquad (14)$$

and the expansion of Eq. (14) is

$$C_{shaft} = w_{raw} \cdot u_{steel} + v_{chip} \cdot u_{turn} + w_{prod} \cdot u_{ht} + A_{grind} \cdot u_{grind} \qquad (15)$$

where u_{steel}, u_{turn}, u_{ht} and u_{grind} are the parameters representing the unit costs of material and operations respectively.

In producing the product, the indirect costs, so called overhead, are counted such as storage, packing, etc. The cost estimation can be calculated through parameters as well. A total management including design, manufacture, costing thus can be processed.

The above mentioned costs are first calculated using these primary parameters and adjusted by factors collected from the past experiences. As for the parameters such as marketing prosperity, customer's preference, timing of the sale, those fluctuate dramatically day by day. The marketing manager can also add them in the form of parameters and related with other parameters as a form of equation. Thus, the sale price can be calculated through a spreadsheet dynamically.

7. Conclusions

Product cost is the sum of costs of activities as specified. Parametric design is probably one of the most effective ways to assist management through quantified results. Parameter definition and development of their relationships constructs a clear scope and boundary to the parametric design management system. Some parameters are accompanied with factors that adjust the results dynamically. Every primary parameter is assigned with a value and derives the cost which reflects every activity in spreadsheets.

Product development is a trade-off kind of work. Design, manufacture and management change in every moment and every stage that make costs fluctuate. The manager needs a quick response to market. Parametric design management integrates design, production, and all functions concerned in a whole quantified system. The manager thus can review and adjust each item and check the result, facilitating decision making and better cost-benefit control.

Acknowledgements

The authors would like to thank the National Science Council of Taiwan, ROC to support this research under the contract number NSC-95-2221-E-002-434.

References

[1] R. Birmingham, G. Cleland, R. Doiver and D. Maffin, 1997, Understanding Engineering Design: Context, Theory, and Practice, Prentice Hall

[2] C. A. Cole, 1993, Management: Theory and Practice, 4th edition, DP Publications

[3] P. Gorb and A. Dumas, 1987, "Silent design," Design Studies, Vol. 8, No. 3

[4] T.M. Khalil, 2000, Management of Technology: the Key to Competitiveness and Wealth Creation, McGraw-Hill

[5] M. Bruce and R. Cooper, 1997, Marketing and Design Management, International Thomson Business Press

[6] B.W. Jones, 1982, Inflation in Engineering Economic Analysis, John Wiley & Sons

[7] D.A. Marca and C.L. McGowan, 1993, IDEF0/SADT, Business Process and Enterprise Modeling, Eclectic Solutions Corporation, San Diego

[8] M.P. Groover, 2001, Automation, Production Systems, and Computer-Integrated Manufacturing, Prentice-Hall, p. 51

[9] K. Tornberg, M. Jamsen, J. Paranko, 2002, "Activity-based costing and process modeling for cost-conscious product design: A case study in a manufacturing company," Journal of Production Economics, Vol. 79, No. 1, p. 75-82

[10] T. Raz and D. Elnathan, 1999, "Activity-based costing for projects," International Journal of Project Management, Vol.17, No. 1, pp. 61-67

[11] G. Loan and W. G. Sullivan, 1999, "Use of activity-based costing and economic value analysis for the justification of capital investments in automated material handling systems," International Journal of Production Research, Vol.37, No. 9, pp. 2109-2134

[12] M. C. Andrade, Pessanha R.C. Filho, A. M. Espozel, L. O. A. Maia and R. Y. Qassim, 1999, "Activity-based costing for production learning," International Journal of Production Economics, Vol.62, No. 3, pp. 175-180

[13] K. LaScola Needy, H. Nachtmann, N. Roztocki, R.C. Warner, and B. Bidanda, 2003, "Implementing activity-based costing systems in small manufacturing firms: A field study," Engineering Management Journal, Vol.15, No. 1, pp. 3-10

[14] Dubbel, Handbook of Mechanical Engineering, 1994, edited by W. Beitz and K.-H Kuttner, English edition edited by B. J. Davies, translation by M. J. Shields, Springer-Verlag, pp. B8-B15

An Enhanced Graph and Artificial Intelligence-based Approach for Assembly Sequence Planning

Y. Y. Hsu[1], W. C. Chen[2], P. H. Tai[3] and M. D. Tsai[4]

[1]Chung Hua University, Taiwan, 300, yyhsu@chu.edu.tw
[2]Chung Hua University, Taiwan, 300, wenchin@chu.edu.tw
[3]Chung Hua University, Taiwan, 300, d09403015@chu.edu.tw
[4]Chung Hua University, Taiwan, 300, m09408007@chu.edu.tw

Abstract. Research in assembly planning can be categorised into graph-based, knowledge-based and artificial intelligence approaches. The main drawbacks of the above approaches are as follows: the first is time-consuming; in the second approach it is difficult to find the optimal solution; and the third approach requires high computing efficiency. To tackle these problems, this study develops a novel integrated approach with some graph-based heuristic working rules and neural networks-based practices to assist the assembly engineers in generating and predicting a best and most effective assembly sequence. In the first stage, the Above Graph and transforming rules are used to create a correct explosion graph of the assembly models. In the second stage, a three-level relational model, with geometric constraints and assembly precedence diagrams (APDs), is generated to create a complete relational model graph, an incidence matrix and a globally optimal assembly sequence. In the third stage, Knowledge Fusion (KF) programming language and back-propagation neural network (BPNN) engine are employed to validate the available assembly sequences. Two real-world examples are utilised to evaluate the feasibility of the proposed model in terms of differences in assembly sequences. The results show that the proposed model can promptly generate feasible assembly sequences, facilitates assembly sequence optimisation and allows the designers to recognise contact relationships and assembly constraints of three-dimensional components in a virtual environment.

Keywords: assembly sequence planning, Above Graph, assembly precedence diagrams, back-propagation neural network

1. Introduction

In general, assembly involves the integration of components and parts to create a product or system through computer-aided design and manufacturing (CAD/CAM) systems. Assembly planning is a crucial design step for generating a feasible assembly sequence. Traditional assembly planning is manual and based on the experience and knowledge of industrial engineers. However, manual analysis does not allow the feasibility of assembly sequences to be easily verified. Good Assembly Sequence Planning (ASP) has been recognised as a practical way of reducing operational difficulties, the number of tools and the working time [1].

De Fazio and Whitney (1987) adopted the concept of Bourjault (1984) to generate a complete set of assembly sequences. They generated sequences in two stages – creating the precedence relations between liaisons or logical combinations of liaisons in a product and verifying the liaison sequence [2]. Homen de Mello and Sanderson (1991) proposed the directed AND/OR graphs to create feasible assembly sequences [3]. In addition, Kroll (1994) used graph-based procedures with conventional representations to reduce the number of sorting operations required. He then extended his previous approach from uniaxial assemblies to triaxial assemblies and presented a set of rules for resolving conflicts between multiple parents and multiple offspring [4]. However, in practice most assembly companies use semi-automatic systems to generate an assembly plan and employ 2D cross-sectional views to represent their heuristic models [5].

The other approach to assembly planning is "assembly by disassembling," i.e., an assembly sequence results from systematically disassembling the final product and reversing the disassembling sequence [6]. This approach usually employs contact-based features to represent the precedence relationships of the product. A designer can successively assign the assembly relations to form the assembly plan based on the precedence diagram. However, contact-based precedence diagrams cannot effectively express the complexity of the assigned assembly relations. An effective assembly plan must include other graphs, such as explosion graph, relational model graph, incidence matrix, assembly precedence diagram (APD), etc. In reality, few experts or engineers know exactly how to derive a correct explosion graph, draw a complete relational model graph or incidence matrix among the components, or determine a complete APD to generate an optimal assembly sequence [7].

Knowledge-Based Engineering (KBE) systems conduct powerful CAD/CAM applications involved in the engineering knowledge (i.e., Knowledge Fusion) in recent decades, which are becoming a critical part of business strategies [8]. In addition, numerous researchers have employed Artificial Intelligence (AI) tree searches or graph search methodologies to generate assembly sequences [9,10]. Unfortunately, search space increases explosively when the number of components in a design grows. To relieve this combinational complexity, heuristic rules and Genetic Algorithms (GAs) have been used in the searching process [11]. Other studies have used Hopfield and back-propagation neural network (BPNN) as means to generate optimum or sub-optimum assembly sequences [12,13].

This study proposes a three-stage integrated approach with some heuristic working rules to assist the planner to obtain an optimal assembly plan. In the first stage, the Above Graph and transforming rules are used to create a correct explosion graph of the assembly model; the two graphs can be used to represent the correct geometric constraints among the assembly parts. In the second stage, a three-level relational model is developed to generate a complete relational model graph (RMG) and the incidence matrix. The relational model graph can be advanced and transformed into an Assembly Precedence Diagram (APD), which is used to describe the assembly precedence relations of parts. Based on these graphs, the designer can easily find feasible sequences and evaluate the difficulty of assembly. In the third stage, a case study of a toy car is utilised to evaluate the feasibility of the proposed model in terms of the differences in underlying assembly characteristics and to generate a near-optimal assembly sequence according to the defined performance criteria.

2. The working concepts and procedures

The working concepts and procedures of the proposed approach are shown in Figure 1.1. Initially, detailed data was input from a 2D engineering drawing and related assembly information into a CAD assembly package. Then, the correct explosion graph was developed using the transforming rules. Finally, the relational models were generalised to represent the assembly precedence relations, and an evaluating mechanism was then employed to find a feasible solution. The planning process was recursive until the defined criteria was satisfied. The main outputs of the graph-based assembly planning were the complete relational model graph and the APD. In addition, Figure 1.2 presents the knowledge-based engineering (KBE) system and a robust BPNN engine dedicated to promptly creating the optimal assembly sequence.

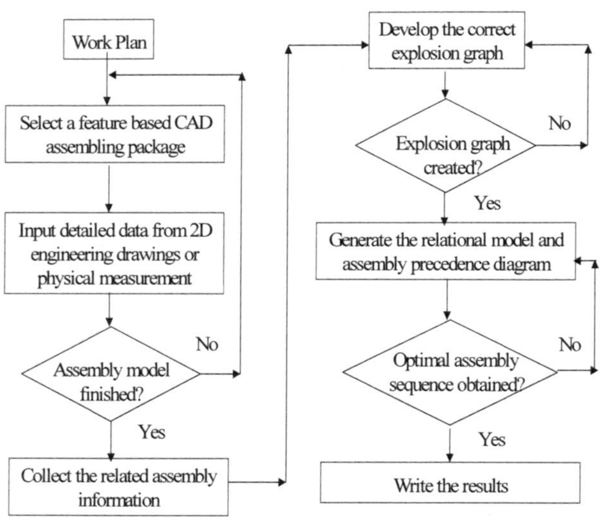

Figure 1.1. Working concepts and procedures.

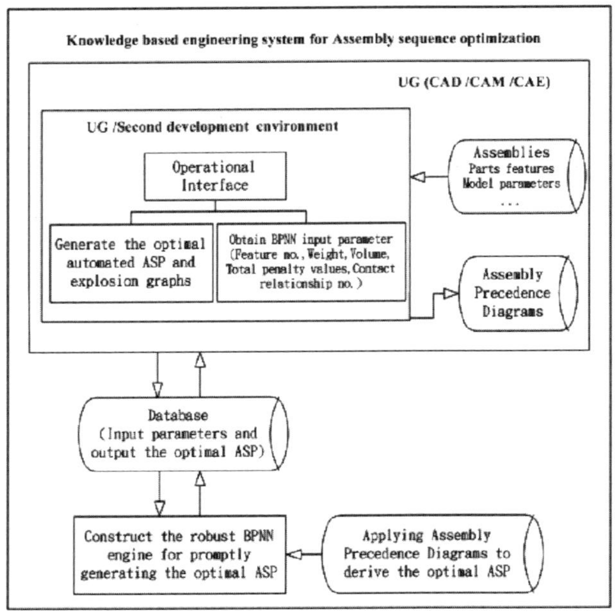

Figure 1.2. KBE model for assembly sequence optimisation

3. A Case Study of Assembly Planning

In this section, a toy car is used as an example to demonstrate the generation procedures of assembly planning.

3.1 Creating exploded view, RMG and APD

The exploded view can be directly created from the Above Graph, which possesses the contact relationships of a spatial structure. Figure 1.3 shows the parts list, assembly codes and the exploded view. The validity of each exploded view can be confirmed by contact relationships of the spatial structure and Above Graphs. Applying a correct exploded view allows us to derive the exact assembly plans. For brevity, the detailed planning steps are omitted. Figure 1.4 shows the complete relational model graph (RMG) and APD for the proposed case study.

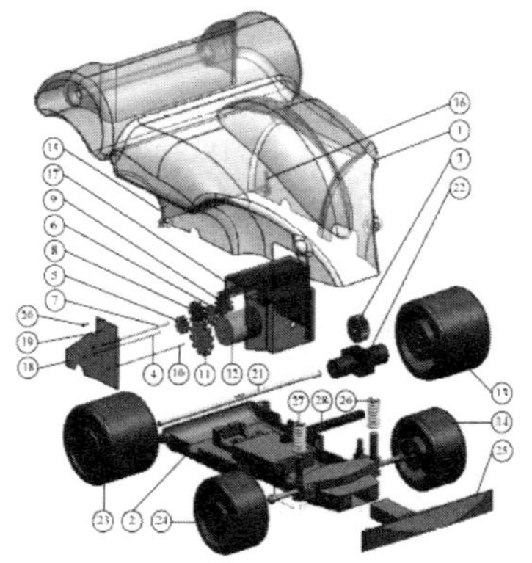

No	Parts' name
1	MB(MainBody)
2	CP(ChassisPan)
3	DG(DriveGear)
4	GS1_1(GearSet1_1)
5	GS1_2(GearSet1_2)
6	GS1_3(GearSet1_3)
7	GS2_1(GearSet2_1)
8	GS2_2(GearSet2_2)
9	GS2_3(GearSet2_3)
10	GS3_1(GearSet3_1)
11	GS3_2(GearSet3_2)
12	PO(Power)
13	LBW(LeftBackWheel)
14	LFW(LeftFrontWheel)
15	BS1(BaseScrew1)
16	BS2(BaseScrew2)
17	PP1(PowerPack1)
18	PP2(PowerPack2)
19	PPS1(PowerPackScrew1)
20	PPS2(PowerPackScrew2)
21	RA(RearAxis)
22	RD(RearDiff)
23	RBW(RightBackWheel)
24	RFW(RightFrontWheel)
25	SL(Spoiler)
26	SP1(Spring1)
27	SP2(Spring2)
28	SR(SteeringRack)

Figure 1.3. The parts list and exploded view of a toy car

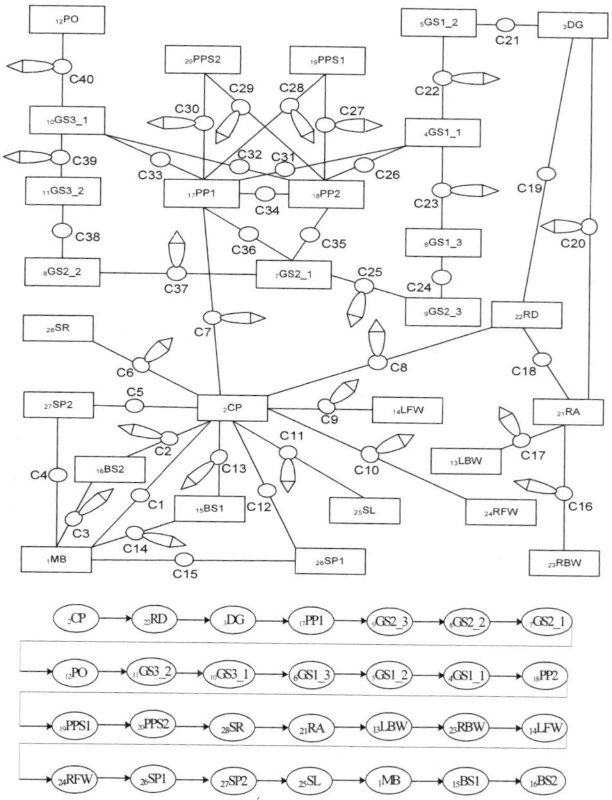

Figure 1.4. The complete RMG and APD of a toy car

3.2 Assembly sequence generation using back-propagation neural network

Step 1. Creating the data structure of neural networks

In this study, a toy car is used as a training sample and a motorbike is employed as a verifying sample. Figure 1.5 shows parts list, assembly codes and exploded view of a motorbike. The characteristics of each assembly part include the number of assembly incidence (AI), total penalty value (TPV), feature number (FN), weight and volume. These characteristics are commonly regarded as the larger the better for the assembly sequence priority. The optimal sequence results with information on five characteristics of a toy car and a motorbike are shown in Tables 1.1 and 1.2, respectively.

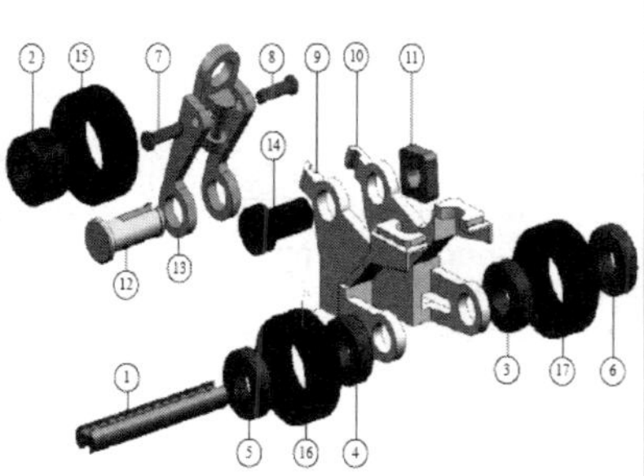

No	Part name
1	MA(MotorbikeAxle)
2	MB1(MotorbikeBearing1)
3	MB2_1(MotorbikeBearing2_1)
4	MB2_2(MotorbikeBearing2_2)
5	MB3_1(MotorbikeBearing3_1)
6	MB3_2(MotorbikeBearing3_2)
7	MH1(MotorbikeHandlE1)
8	MH2(MotorbikeHandlE2)
9	MMB1(MotorbikeMainBody1)
10	MMB2(MotorbikeMainBody2)
11	MN(MotorbikeNut)
12	MPN(MotorbikePin)
13	MPE(MotorbikePlate)
14	MS(MotorbikeScrew)
15	MW1(MotorbikeWheel1)
16	MW2(MotorbikeWheel2)
17	MW3(MotorbikeWheel3)

Figure 1.5. The parts list and exploded view of a motorbike

Table 1.1. The optimal sequence of a toy car

Optimal Assembly Sequence	Parts	AI	TPV	FN	Weight	Volume
1	$_2$CP	19	47	9	981.88	125415.99
2	$_{22}$RD	4	8	10	31.422	11246.393
3	$_3$DG	5	8	27	4.8336	3452.5738
4	$_{17}$PP	10	29	11	83.641	29935.982
5	$_9$GS2_3	3	5	22	1.9571	1397.9177
6	$_8$GS2_2	3	5	22	1.124	802.85347
7	$_7$GS2_1	6	16	1	3.0744	392.69908
8	$_{12}$PO	2	3	2	56.343	20165.608
9	$_{11}$GS3_2	3	5	26	2.2803	1628.7664
10	$_{10}$GS3_1	6	16	1	3.0744	392.69908
11	$_6$GS1_3	3	5	22	1.0797	771.23028
12	$_5$GS1_2	3	5	22	0.8731	623.61418
13	$_4$GS1_1	6	16	1	3.0744	392.69908
14	$_{18}$PP2	8	22	11	17.663	6321.7644
15	$_{19}$PPS1	4	6	3	0.1266	14.993677
16	$_{20}$PPS2	4	5	3	0.1166	14.893677
17	$_{28}$SR	7	4	4	27.576	3522.2622
18	$_{21}$RA	7	13	3	29.789	3804.9837
19	$_{13}$LBW	2	5	7	308.9	219936.4
20	$_{23}$RBW	2	3	7	307.9	219926.4
21	$_{14}$LFW	2	3	9	176.9	119224.45
22	$_{24}$RFW	2	3	9	166.9	119214.45
23	$_{26}$SP1	2	6	3	9.9946	1286.4786
24	$_{27}$SP2	2	6	3	9.9936	1276.4786
25	$_{25}$SL	2	3	2	234.01	83756.141
26	$_1$MB	7	17	28	932.5	333750.12
27	$_{15}$BS1	4	10	3	2.3821	303.98645
28	$_{16}$BS2	4	10	3	2.3721	302.98645

Table 1.2. The optimal sequence of a motorbike

Optimal Assembly Sequence	Parts	AI	TPV	FN	Weight	Volume
1	$_9$MMB1	5	13	20	7.35	7697.0419
2	$_{13}$MPE	9	19	4	5.19	5176.4575
3	$_{10}$MMB2	5	15	20	6.78	7696.0419
4	$_{14}$MS	3	7	2	1.53	2297.2896
5	$_{11}$MN	3	7	3	0.78	856.50444
6	$_{17}$MW3	1	9	4	8.13	7296.1419
7	$_3$MB2_1	4	23	3	1.2	1931.2918
8	$_6$MB3_2	3	16	5	1.41	1892.1789
9	$_1$MA	8	52	2	3.32	2907.5579
10	$_{16}$MW2	2	9	4	8	7296.1419
11	$_4$MB2_2	4	18	3	1.18	1931.2918
12	$_5$MB3_1	3	5	5	1.4	1892.1789
13	$_2$MB1	4	11	4	2.49	3841.3824
14	$_{15}$MW1	2	4	4	7.99	7295.1419
15	$_{12}$MPN	3	12	5	1.28	1619.5462
16	$_7$MH1	2	3	3	0.16	231.60587
17	$_8$MH2	2	3	3	0.17	230.60587

Step 2. Experimental results

The toy car and the motorbike can be decomposed into 28 parts and 17 parts, respectively. Each part of the toy car and the motorbike has five characteristics parameters: value of assembly incidence (AI), total penalty value (TPV), feature number (FN), weight and volume, which are used as network input parameters, and one expected network output adopts the ranking number of the optimal assembly sequence.

When applying BPNN (5-20-1) to the presented model, the relationship between RMSE and iteration epochs for the toy car training case can be obtained and is shown in Figure 1.6. In addition, Figure 1.7 shows a comparison between the actual sequential ranking values and the predicted sequential ranking values for the motorbike testing case.

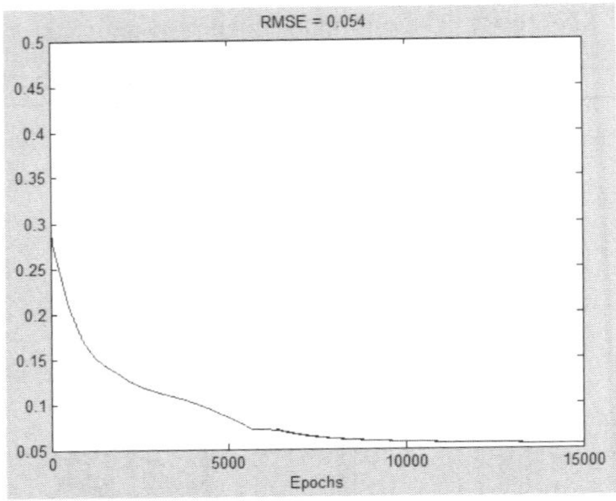

Figure 1.6. RMSE versus iteration epochs for training BPNN

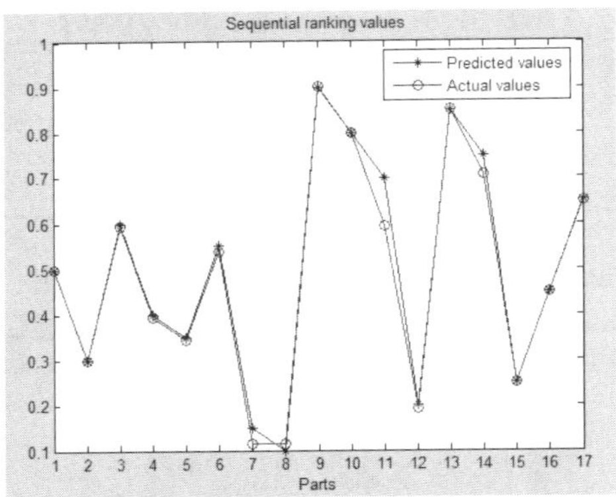

Figure 1.7. Predicted and actual values for testing case

4. Conclusion

Theoretically, an assembly plan can be optimised based on the factors of shortest assembly time and assembly sequence optimisation. However, these are uncertain factors prior to the determination of the optimised assembly scheme and the completion of the jig and fixture.

The proposed model adopts a three-stage integrated assembly planning approach to express the complexity of assembly relations and to evaluate the feasibility of respective assembly sequences in the design phase. The experimental results for the case study verify the feasibility of the proposed approach, which facilitates the DFA in potential applications of 3D component models to assist manual or automatic assembly in a virtual environment, and allows the designer to recognise the relative position, geometric constraints and relationships of the 3D components using the following graph-oriented methods: the Above Graph, APD and relational model graph. The planner can further generate a correct explosion graph and construct an incidence matrix for validating the assembly relations through applying the Above Graph and relation models. In addition, this three-stage integrated approach can effectively promote the quality of the generated assembly plan and facilitate assembly sequence optimisation via the robust BPNN engine.

Acknowledgements

Financial support from the National Science Council, Taiwan, ROC, under contract NSC95-2622-E-216-014-CC3 and Chung Hua University, under contract CHU-95-M-016.

References

[1] Lai, H.Y., & Huang, C.T. 2004, A Systematic Approach for Automatic Assembly Sequence Plan Generation, International Journal of Advanced Manufacturing Technology, Vol.24, P752-763.

[2] Bourjault, A. 1984, Contribution á Une Approche Méthodologique de L'assemblage Automatisé: Elaboration Automatique des Séquences Opératoires, Ph.D. Thesis, Faculté des Sciences et des Techniques de l'Université de Franche-Comté, France.

[3] Homen de Mello, L.S., & Sanderson, A.C. 1991, Representations of Mechanical Assembly Sequences. IEEE Transactions on Robotics and Automation, Vol.7, No.2, P211-227.

[4] Kroll, E. 1994, Intelligent Assembly Planning on Triaxial Products. Concurrent Engineering: Research and Applications, Vol.1, No.2, P311-319.

[5] Lin, A.C., & Chang, T.C. 1993, An Integrated Approach to Automated Assembly Planning for Three-dimensional Mechanical Products. International Journal of Production Research, Vol.31, No.5, P1201-1227.

[6] Lee, S. 1989, Disassembly Planning by Subassembly Extraction, Proceedings of the third ORSA/TIMS Conference on flexible manufacturing systems, P383–388.

[7] Chen, R.S., Lu, K.Y., & Tai, P.H. 2004, Optimizing Assembly Planning Through a Three-stage Integrated Approach. International Journal of Production Economics, Vol.88, P243-256.

[8] Kulon, J., Broomhead, P. & Mynors, D.J. 2006, Applying Knowledge-based Engineering to Traditional Manufacturing Design. International Journal of Advanced manufacturing Technology, Vol.30, P945-951.

[9] Homen de Mello, L.S., & Sanderson, A.C. 1991, A Correct and Complete Algorithm for The Generation of Mechanical Assembly Sequence, IEEE Transactions on Robotics and Automation, Vol.7, No.2, P228-240.

[10] Chen, R.S., Lu, K.Y., & Tai, P.H. 2004, Optimization of Assembly Plan Through a Three-stage Integrated Approach, International Journal of Computer Applications in Technology, Vol.19, No.1, P28-38.

[11] Marian, R.M., Luong, L.H.S., & Abhary, K. 2003, Assembly Sequence Planning and Optimisation Using Genetic Algorithms. Applied Soft Computing, Vol.2, No.3, P223-253.

[12] Chen, C.L.P. 1990, Neural Computation for Planning and/or Precedence-constraint Robot Assembly Sequences, Proceedings of the International Conference on Neural Networks, P127-142.

[13] Hong, D.S., & Cho, H.S. 1995, A Neural Network Based Computational Scheme for Generating Optimized Robotic Assembly Sequences, Engineering Application Artificial Intelligence, Vol.8, No.2, P129-145.

3–5

The Development of an Agent-based Distributed Surface Machining System

Y.C. Kao[1], M.S. Chen[2], J.S. Lia[3]

[1]National Kaohsiung University of Applied Sciences, Taiwan,yckao@cc.kuas.edu.tw
[2]National Kaohsiung University of Applied Sciences, Taiwan, mau@cc.kuas.edu.tw
[3]National Kaohsiung University of Applied Sciences, Taiwan, ck291107@yahoo.com.tw

Abstract. This paper focuses on the development of an agent-based intelligent distributed surface machining system structure through traditional manufacturing engineering activity analysis and systematic software engineering method. The advantages of a distributed system structure include ease of manipulation, high expandability, flexibility and the achievement towards the goal of networked manufacture.

The proposed system has presented the feasibility in applying software agent-based technology in constructing a distributed freeform surface machining environment. Several agents have been developed in the proposed system: (1) preliminary evaluation agent, (2) sophisticated evaluation agent, and (3) post processor agent. Preliminary evaluation agent is in evaluation on the surface machining based on surface curvatures. Sophisticated evaluation agent could be accessed if the surface could not be machined by three axis milling machine based on the calculation results from preliminary evaluation, this agent can suggest either four or five axis machine that could be applied for the surface milling operation based on visibility cone. Post processor agent takes care of generation of the NC codes for milling operation. This agent can also convert five-axis NC codes among different five-axis milling machines when the machine's mechanism could be defined. These agents communicate via pre-defined performatives underlying knowledge query and manipulation language (KQML) for the surface machining capability. The developed system has then been successfully implemented in forming a distributed shoe surface machining system. This means agent technology is very promising in enhancing the communication of a distributed system to be able to communicate with higher level knowledge rather than only in remote function call through platform independent networking architecture.

Keywords: CORBA, Agent, CAD/CAM/CAPP, Distributed Intelligent Manufacturing

1. Introduction

Owing to the advanced development of networking technology and continuous enhancement of computer and communication, innovative applications can result from the synergy of these technologies and the traditional mechanical design and manufacturing processes. These applications could greatly increase the global competitive advantages of the manufacturing industry. To name only a few, collaborative CAD/CAM/CAE [1,2] and experiment, networked CNC monitoring and control, remote machining, assembly, diagnosis, etc., have been paid great attention in industrialized regions such as the USA, Japan, and Europe [3]. That is to say, the integration technology of traditional CAD/CAPP/CAM has been matured enough, been promoted and applied in the manufacturing industry in enhancing productivity and international competitive advantages [4]. This has resulted from advanced broadband networking, intelligent computing, collaborative work and distributed manufacturing trend [5]. The synergy with domain knowledge base, inference engine and parametric design have been the supporting technologies [6].

Furthermore, the application of distributed object-oriented technology in manufacturing technology is getting matured and paid more and more attention. For example, Lo [7] has applied intelligent agent technology in automating plastic injection molding, Sher [8] has also studied the networking interface on Internet manufacturing. Cheng [9] has also adopted object-oriented analysis and design technology and used C++ in developing a CORBA-based distributed computer integrated manufacturing system for a semi-conductor factory. Aguirre [3] utilized OOA/OOD/OOP, UML, MMS and CORBA in developing a distributed networked manufacture controlling system in C++; Young [10] utilized UML in building virtual machine abstract and decomposing manufacturing systems into small unit so as to enhance lean production control and shorten the time required in changing processes change; Satake and Hayash [11] proposed that distributed object could be used in constructing general purpose innovative and adaptive software platform in linking networked manufacturing systems. Gressier-Soudan [12] worked on using CORBA and JAVA in achieving the control of distributed manufacturing control systems; Lei [13] used agent based methodology in integrating CORBA based distributed environment in developing an intelligent distributed controlled machining center.

With computing intelligence being incorporated, software agents originated from 1990s have been paid great attention in continuing enhancement extending distributed intelligent manufacturing technologies of CAD/CAPP/CAM [14]. For example, Feng et al. [15] applied intelligent agents in building intelligent agents on process planning, maximum material removal rate (MMRR) model, tool life model, and NC machining interface.

Intelligent agent technology has been applied in this research and a prototype agent-based integration framework to enable the use of machining process knowledge with higher-level manufacturing application to optimize process performance has been demonstrated. However, an integrated distributed machining system based on platform transparent and intelligent agent has not been explored yet.

2. System Architecture in Distributed Manufacturing Application

2.1 CORBA Based Distributed System Structure

This paper describes the developed remote collaborative CAD/CAM/CAPP manufacturing system underlying CORBA and Agent based architecture in achieving an agent-based distributed surface machining system. There are eight services in the developed three-tired system and the configuration is shown in Figure 1.1. The services object management mechanism showed in Figure 1.2.

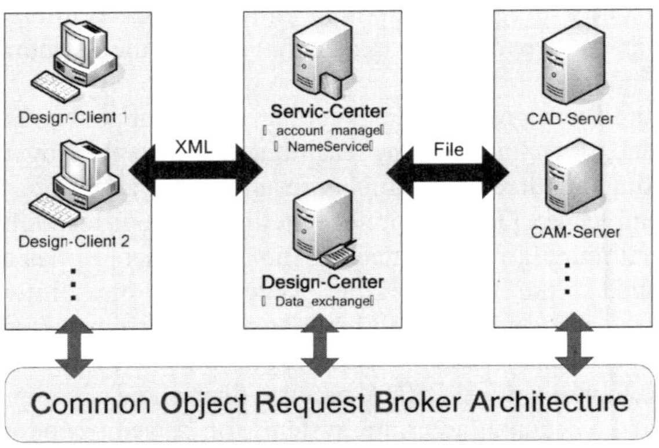

Figure 1.1. Three-tiered CORBA-based distributed remote collaborative manufacturing system

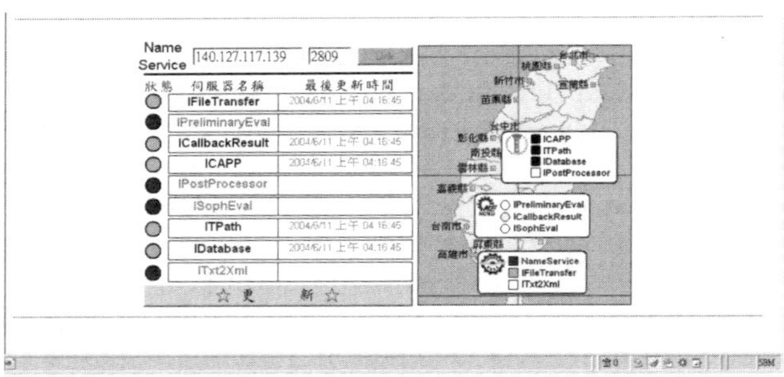

Figure 1.2. The object management mechanism

2.2 Operation Mechanism of the Developed Distributed System

To realize the developed system, the distributed system operation mechanism allocates the engineering activities in "surface machining" into three geographically dispersed locations in Taiwan: the National Cheng-Kung University (NCKU) in Tainan, the Da-Yeh University in Chang-Hwa (DYU), and the National Kaohsiung University of Applied Sciences (KUAS) in Kaohsiung. Each location consists of various functional modules that were linked through Taiwan Academic Network (TANET) underlying CORBA architecture. A user can upload a freeform surface file (CAD geometry) through web-based interface, and triggers CAD/CAPP/CAM processes. Machining procedures and cutter location files could be generated. The systematic flow includes tasks of collaborative viewing and modification, preliminary evaluation, process planning, toolpath planning, detailed evaluation, postprocessing, cutting process emulation, etc. The related tasks are explained in Table 1.1.

Table 1.1. Tasks of the Developed System

System Tasks	Function
Task 1: Collaborative viewing and modification	A client can upload CAD file (in IGES format) to the CAD Parser server so that the ICAD Parser module can interpret the file. The interpreted information will be sent back to the client for collaborative display and revision.
Task 2: Preliminary evaluation	The interpreted CAD file based on IGES information can be uploaded to the Preliminary evaluation server for the suggestion of clamping angle and machine tool configuration (in a text file "evalreport.txt") for triggering "Processing planning" mechanism.
Task 3: Processing planning	Once the text file "evalreport.txt" is received the manufacturing process planning will be conducted and CAPP work sheet will be generated. The work sheet will be sent back to the client and "tool path planning" will be actuated.
Task 4: Toolpath Planning	This task proceeds tool path planning according to the cutting tool in the work sheet, generated in Step 3, and the CAD file. Sophisticated evaluation task will be called in evaluating feasible tool information. The CL file will then be generated and the postprocessor task will be triggered.
Task 5: Sophisticated Evaluation	This task accepts the call from task 4 "tool path planning", evaluates the CAD file and tool information, and generates feasible cutting angle ranges (tile and lead angle).
Task 6: Post processor	The CL file generated by Task 4 will be post-processed in this task. NC codes will be generated by this task and sent back to the client.

3. Agent-Based Remote Collaborative Surface Machining System

3.1 Agent-based system structure

The system structure of the agent-based distributed surface machining system is shown in Figure 1.3. There are nine agents including "User Agent", "Preliminary Evaluation Agent", "Sophisticated Evaluation Agent", "Virtual CAD Agent", "Virtual CAM Agent", "CAPP Agent", "Tool Path Agent", "Post Processor Agent)", and "KM Agent". The "User Agent" serves as the client's communication interface with the service Agents through KQML (Knowledge Query Manipulation Language).

The preliminary evaluation agent and sophisticated evaluation agent were allocated in NCKU; the process planning agent, tool path planning agent, and post processing agent were allocated in DYU; KM agent was allocated in KUAS serving as naming service, file transfer, file format transformation and process knowledge base management. Agent-based mechanism was adopted in implementing remote collaborative distributed surface machining structure. The information flow between agents is shown in Figure 1.4.

3.2 Agent Communicaton Language KQML

The Knowledge Query and Manipulation Language (KQML) is a language and an associated protocol to support the high level communication among intelligent agents.

This language can be thought of as consisting of three layers: a communication layer; a message layer; and a content layer. The format of a KQML message is shown in Table 1.2.

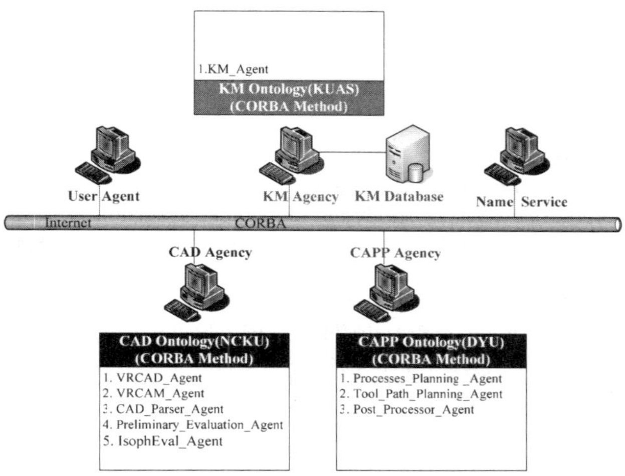

Figure 1.3. Structure of the agent-based distributed surface machining system

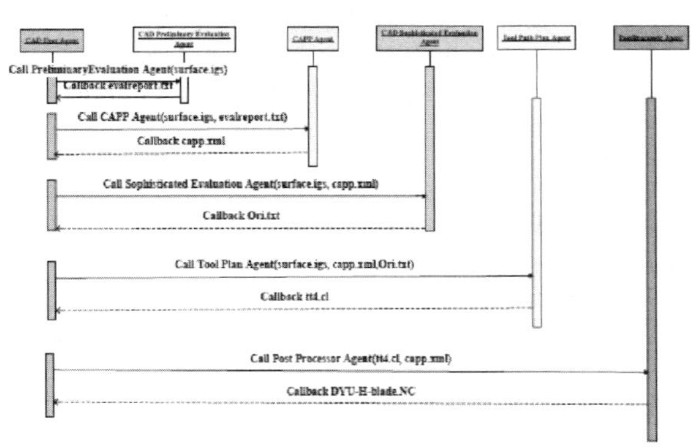

Figure 1.4. Information flow between agents in the developed system

The message in the example starts with the word "ask-one", which is the action (performative) intended for the message. The remainder of the message contains keywords needed for the message and communication layers. Keywords used in KQML message are defined as follows:

Table 1.2. Format of a KQML message

```
(ask-one
        :sender        User_agent
        :receiver      Virtual CAM agent
        :content       ?machining Time
        :ontology      Simulation ontology
        :language      KQML
        :reply-with    "machtime_Request"
)
```

- sender: agent sending the message.
- receiver: agent receiving the message.
- content: context specific information describing the specifies of this message.
- reply-with: identifier to be used by a message reply to this message.
- in-reply-to: identifier of the message that triggered this message submission.
- language: language for interpreting the information in the content field of this message.
- ontology: identifies the ontology to interpret the information in the content field of this message.
- to: final recipient; used when a message is sent using intermediary agents.
- form: original sender; used when a message is sent using intermediary agents.

3.3 Implementation of Agent-Based System

To realize the developed system, the agent-based distributed system is implemented and explained in the following steps:

1. Create CAD/CAPP/CAM agency and related agents, ontology as shown in Figure 1.5.
2. (IDL) underlying CORBA architecture. The compiler "Idltojava" can then be used to convert the IDL file (contains remote methods and attributes) into Java classes and linked with the Agentbuilder. If the Ontology is created by other languages such as C++ and Visual Basic, the remote corba methods and attributes can be output in DLL (Dynamic Link Library) and linked through Java JNI (Java Native Interface).
3. Define protocols of the Agency. The protocols include "Role", "State", and "State transition" settings, etc.
4. Import the protocols into the Agency.
5. Import the Project Accessory Class (PAC) into the agency so as to share the remote methods and attributes.
6. Define rules for the Agent's inference. The KQML performatives can be used to define the behaviors of the inference.
7. Execute the behavioural activities of the agents. Firstly, every agent needs to register their services to the "Naming Service" underlying CORBA architecture and then be initiated. User's agent will

then be started and the inference rules will be accessed and actuated sequentially.

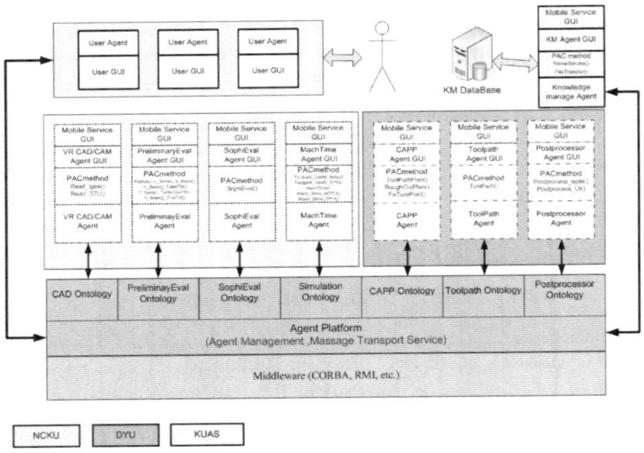

Figure 1.5. System ontology and agents of agency

4. Conclusion

This research has successfully implemented an Agent-based remote collaborative distributed surface machining structure underlying CORBA architecture. The developed system has also been realized and implemented; reliable connection has been tested by linking three geographically dispersed locations: KUAS, NCKU, and DYU. Domain knowledge can be incorporated in the developed system through rule-based agent architecture. This means artificial intelligent can be implemented in the developed distributed system. Major contributions of the developed system include; (1) software agent based remote collaborative distributed surface machining structure has been established, and (2) prototypical system has been constructed and feasibility has been successfully evaluated. Unified Modeling Languages (UML) can be used for further integration towards reusable agent-based distributed system on freeform surface machining.

Acknowledgements

The authors appreciate the grand supported by the National Science Council through the project NSC 95-2212-E-151-049. This project is a sub-project of an integrated project, therefore, thanks also extend to the integrated project leader Prof. Rong-Shean Lee, and the other sub-project leader Chen-Hua She.

References

[1] Kao, Y.C.; Lin, Grier C.I., "CAD/CAM collaboration and remote machining", Computer Integrated Manufacturing Systems Volume: 9, Issue: 3, July, 1996, pp. 149-160.

[2] Y.C. Kao and Grier C.I. Lin, "Development of a Collaborative CAD/CAM System", Journal of Robotics and Computer-Integrated Manufacturing, Volume 14, Issue 1, pp. 55-68, 1998

[3] Omar Aguirre; Richard Weston; Francisco Martin; Jose Luis Ajuria, "MCSCARCH: An architecture for the dev elopment of manufacturing control systems", Int. J. Production Economics, 62 (1999) pp. 45-59.

[4] Xu, X.W.; He, Q., "Striving for a total integration of CAD, CAPP, CAM and CNC", Robotics and Computer-Integrated Manufacturing Volume: 20, Issue: 2, April, 2004, pp. 101-109.

[5] Aguirre, Omar; Weston, Richard; Martín, Francisco; Luis Ajuria, Jose, "MCSARCH: An architecture for the development of manufacturing control systems", International Journal of Production Economics Volume: 62, Issue: 1-2, May 20, 1999, pp. 45-59.

[6] George Rzevski, "A Framework for Designing Intelligent Manufacturing Systems", Computer in Industry, Volume:34, 1997, pp211-219.

[7] Goh, W.T.; Zhang, Z., "An intelligent and adaptive modelling and configuration approach to manufacturing systems control", Journal of Materials Processing Technology Volume: 139, Issue: 1-3, August 20, 2003, pp. 103-109.

[8] Chong-Ping Sher, Shang-Liang Chen, "Intelligent Internet Machining Parameter Creation and Machining Parameters Generation for 3D-Free Surface", Master Thesis, National Cheng Kung University, 2001

[9] Cheng, Fan-Tien; Eric Shen; Jun-Yan Deng, "Development of a Distributed Object-Oriented System Framework for the Computer-Integrated Manufacturing Execution System", International Conference on Robotics & Automation, May 1998, pp. 2116-2119.

[10] K.W. Young, R. Piggin and P. Rachitrangsan, "An Object-Oriented Approach to an Agile Manufacturing Control System Design", The International Journal of Advanced Manufacturing Technology, 17, pp.850-859, 2001

[11] T. Satake and A. Hayashi, "Development of Software Platform for Constructing Generative and Adaptive Manufacturing System", Third International Conference on Industrial Tools, Rogaška Slatina, Maribor, Slovenian, 22-26 April 2001, pp. 325-328

[12] Robert K.L. Gay, P.K Seet., and B.S. Lee, "Network-based Concurrent Design Environment for Distributed-Based CAD", Computing & Control Engineering Journal, December 1993, pp. 253-260

[13] Ming Lei; Xiaohong Yang; Mitchell M. Tseng; Shuzi Yang, "Design an intelligent machine Center - Strategy and practice", Mechatronics, 8 (1998), pp. 271-285

[14] J. Zhang , F.T.S. Chan , P. Li, "Agent- and CORBA-Based Application Integration Platform for an Agile Manufacturing Environment", The International Journal of Advanced Manufacturing Technology Volume 21, Number 6, March, 2003, pp.460-468.

[15] Feng, Shaw C.; Stouffer, Keith A.; Jurrens, Kevin K., "Manufacturing planning and predictive process model integration using software agents", Advanced Engineering Informatics Volume: 19, Issue: 2, April, 2005, pp. 135-142.

4

Design of Optical Components

4–1

The Improvement of Plastic Aspheric Lens for the Application of Compact Camera Module

S.C. Yen[1], Y.C. Chien[2] and K.M. Chang[3]

[1]Hon-Hai Precision Ind. Co., Ltd., Taipei 236, Taiwan, Shih-Chieh.Yen@foxconn.com
[2]Hon-Hai Precision Ind. Co., Ltd., Taipei 236, Taiwan, Yang-Chang.Chien@foxconn.com
[3]Hon-Hai Precision Ind. Co., Ltd., Taipei 236, Taiwan, Keng-Ming.Chang@foxconn.com

Abstract. The requirement for plastic injection of ultra-high precision optical components is critical. Therefore, the precision of optical components has to meet the optical specification. We compensate the core inserts of aspheric lenses and the refine conditions of optical lens injection to improve the forming precision of optical lenses. The key issues of manufacturing plastic optical lenss are thermal shrinkage of the lenses and the decenter of optical axes for the lens module. In the process of plastic lens injection, the shrinkage of a lens is due to temperature variation of the mold and the lens.

Because of this, the shrinkage will cause an asymmetric surface profile. The performance of optical lens modules will be affected by the decenter of the optical axes. According to the research, the performance of aspheric lens injection is based on the mold design. In order to meet the optical design requirements, we study diamond turning tool radius compensation, the coordination origin calibration of diamond tool, and the aspheric parameter modification to improve the precision of aspheric core inserts. At the same time, we decrease the decenter of the mold by modifying the mold's structure. The plastic aspheric lens had been improved and raised the performance of optical lens modules.

Keywords: injection molding, aspherical, decenter

1. Introduction

The lens of compact camera modules for cellular phones requires a high precision in the dimension and the surface curvature. For miniaturization, the aspheric lens can be manufactured by plastic injection, which improves the optical performance and lowers the production costs. The optical performance of injection molded plastic lenses is related to, first, the form precision and, second, the machining of the core insert [1,2].

Therefore, this study focused on, first, the thermal shrinkage of plastic lenses and, second, compensating the form error of core insert in order to improve the form precision of aspheric plastic lens.

2. Ultra Precision Lens Injection Mold and Aspheric Lens Measuement and Compensation

The aspheric lens is frequently applied for minimized and low cost lens modules. The quality and yield rate of the aspheric lens are influenced by the form precision of the aspheric lens, the optical axes decenter of the aspheric lens and its center thickness.

2.1 The form precision of the aspheric lens

To minimize the aberration, the aspheric design is applied to the lens design of the compact camera module. The aspheric function of a lens is:

$$Z = \frac{cr^2}{1+\sqrt{1-(1+k)c^2r^2}} + a_1r^2 + a_2r^4 + a_3r^6 + a_4r^8 + +a_5r^{10} + a_6r^{12} \ldots\ldots \qquad (1)$$

where, Z is the sag of the aspheric from corner to center, k is the conic constant, c is the surface curvature, r is the distance from the center of the aspheric surface, and a_n is the aspheric parameter of the aspheric function.

In the progress of the lens injection molding, the melted optical plastic material fills the mold cavity with high velocity and under high pressure. After the material temperature is below the glass transition temperature, the injection lens is taken out from the mold. In the lens injection molding cycle, the temperature gradient changes rapidly. Therefore, the optical material exhibits serious shrinkage, because of which the form precision of the aspheric lens is affected significantly. Furthermore, the ray path cannot follow the design ray path.

The form error of the lens normally should be smaller than 2 μm, for high-end or compact lens modules it should be smaller than 1 μm. In general, we use the contact-type probe to measure the form-error of the aspheric surface. For CCMs (compact camera modules), the diamond tip radius of the contact-type probe is 1 to 10 μm by polishing. And the roundness of the probe tip is smaller than 10 nm.

2.2 The optical axes decenter of the aspheric lens

When the optical axes decenter or tilt, the path of a ray would be changed. The relative position between the two aspheric surfaces of the lens would also change.

Assuming the aspheric lens Z_2 is decentered from a nearby lens Z_1 by the distance Δr, the form equation of Z_1 is the same as given in equation (1.1), while Z_2 will be calculated according to equation (2).

$$Z_2 = \frac{c(r+\Delta r)^2}{1+\sqrt{1-(1+k)c^2(r+\Delta r)^2}} + a_1(r+\Delta r)^2 + a_2(r+\Delta r)^4 + a_3(r+\Delta r)^6 + a_4(r+\Delta r)^8 + a_5(r+\Delta r)^{10} + a_6(r+\Delta r)^{12} \ldots\ldots \quad (2)$$

Therefore, the ray tracing through the aspheric lens would obviously be influenced. Because there are optical axes decentering, the practical performance of the lens module is reduced and would deviate from the original optical design of the lens module.

2.3 The parameters of lens injection molding process and the shrinkage compensation of lens injection

During the process of lens injection molding, the temperature changes rapidly. In the temperature gradient field, the optical plastic would have serious and asymmetric shrinkage. Thus, the form precision of the aspheric surface would be insufficient.

Using the traditional design of an injection mold, the shrinkage can be estimated by the shrinkage rate of the plastic material injected. In practice, this kind of shrinkage can be compensated and the dimension error can be reduced.

The concept of shrinkage compensation can be partially applied to the optical aspheric lens injection molding. The shrinkage compensation is only calculated for linear dimensions. But the thermal shrinkage compensation of the aspheric lens cannot be simply calculated using common CAD/CAE packages, because the aspheric surface is a 3D profile. Therefore, the aspheric surface shrinkage of aspheric lenses must be calculated in three dimensions.

In the optical design of aspheric lenses, there are high order terms in the aspheric function. Thus, we use the aspheric function to calculate the thermal shrinkage of the optical plastic.

We define the parameters that would be influenced by temperature in the aspheric function, such as ***c*** and ***r***, and assume the shrinkage rate to be α. Thus, assuming the linear dimension is L, the shrinkage rate α would make the linear dimension ***L*** becoming ***L(1+α)*** at high temperatures. Then, the parameters ***Z***, ***c***, ***r*** would become ***c(1+α)***, ***r(1+α)***. Consequently, the original aspheric function would be changed to:

$$Z_{Mold} = \frac{[c(1+\alpha)]r^2}{1+\sqrt{1-(1+k)[c(1+\alpha)]^2[r(1+\alpha)]^2}} + a_1[r(1+\alpha)]^2 + a_2[r(1+\alpha)]^4 + a_3[r(1+\alpha)]^6 + a_4[r(1+\alpha)]^8 + a_5[r(1+\alpha)]^{10} + a_6[r(1+\alpha)]^{12} \ldots\ldots \quad (3)$$

For different kinds of material, the compensation methods for thermal shrinkage are different.

In some optical design cases, if the thermal shrinkage of the optical plastic is small or the form of the aspheric surface is approximated as a typical spherical surface, we can use only the surface curvature ***c*** to compensate for the form error of the aspheric surface.

In conclusion, after using the thermal shrinkage analysis method and the compensation of the aspheric surface, we can obviously reduce the form error of the aspheric surface. Before compensating for the thermal shrinkage, the form error is between 5 to 20 μm. Compensating for the thermal shrinkage, the form error of the aspheric surface can be reduced to between 0.5 and 2 μm by properly compensating the aspheric core insert of the injection mold.

3. Experiments and Results

According to the error analysis discussed above, the aspheric form precision is affected by the tool radius error and the tool origin error in the X-axis. If these two sources of error are combined together, the effect of the sum of both errors on the aspheric form precision would be as illustrated in Figures 1.1 and 1.2:

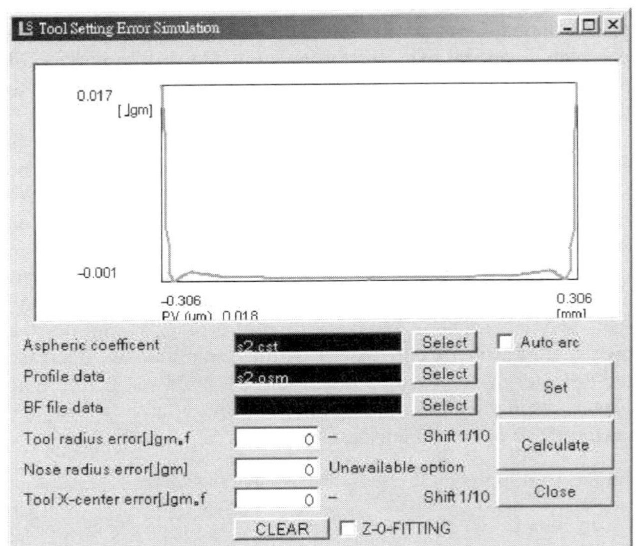

Figure 1.1 Diamond tool radius and tool original setting in X-axis without error

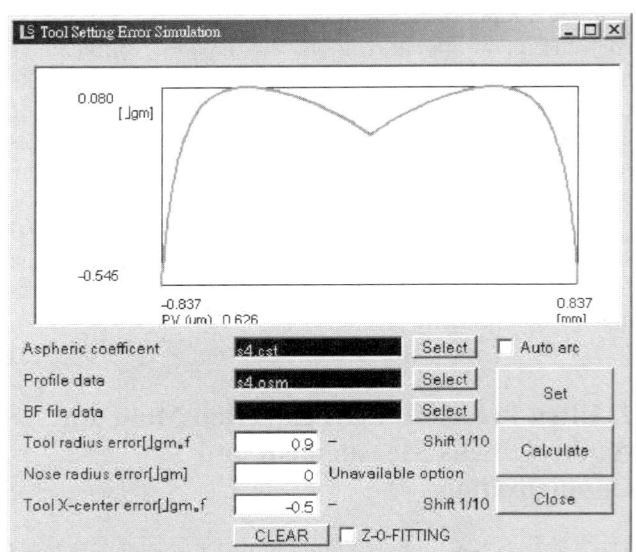

Figure 1.2 Diamond turning tool radius and tool original setting in X-axis with error

Aspheric form precision could meet the specification of optical design by error compensation and modification. Therefore, we could also find the same form precision on the ejected lens. With such a high precision lens, we can promote our optical-mechanical product yield remarkably. Figures 1.3 to 1.6 are the measurement results of the aspheric profiles of two core inserts. High-precision molding (form precision < 0.5 μm, surface roughness < 10 nm) could be achieved by proper error compensation and modification to meet the optical specifications. Therefore, using this extra-high precision core insert, we could eject the plastic lens with high transparency and high stability. Otherwise, we might need to modify the ejecting parameters like temperature, ejecting velocity and filling pressure, etc. Only when we optimize all these parameters, high-precision optical lenses can be produced.

Form Error (P-V=0.2202um)

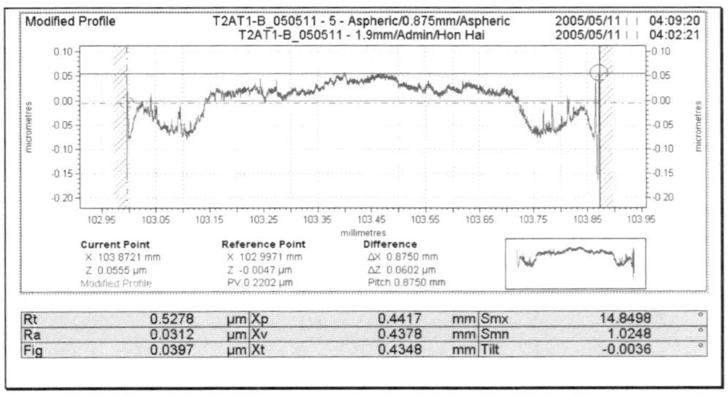

Figure 1.3 Form error of core insert A

Surface Roughness (Ra=6.9nm)

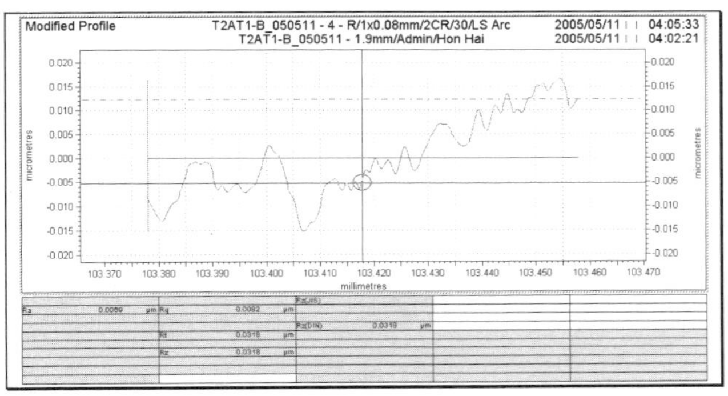

Figure 1.4 Roughness of core insert A

Form (P-V=0.2972um)

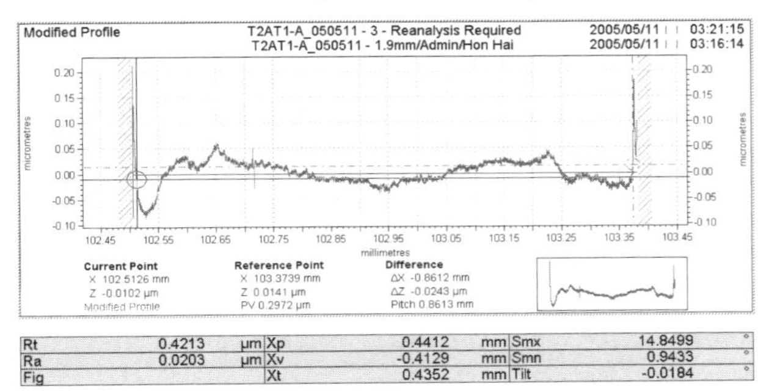

Figure 1.5 Form error of core insert A

Surface Roughness (Ra=8.8nm)

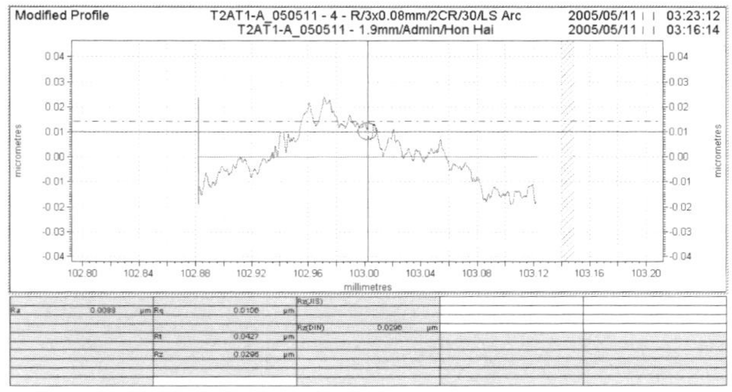

Figure 1.6 Roughness of core insert A

4. Conclusion

This study focused on the improvement of one of the most crucial components of a compact camera module, the lens. Due to the requirements of miniaturization and cost reduction of compact camera modules, there are many considerations with respect to the optical design which have to be taken into account.

Therefore, the aspheric plastic lens was introduced to improve the optical performance. Aspheric plastic lens can decrease astigmatism and module size comparing to the traditional spherical lens, but there are more severe requirements on injection molding, such as form precision and decentering.

The factors decreasing the optical performance are as follows:

- Machining error;
- Tool position error in Y axis;
- Tool position error in X axis;
- Tool radius error;
- Roundness of tool tip;
- Form error of lens injection;
- Precision of mold clamp;
- Precision of lens curvature;
- Decenter of injection lens;
- Air space between lenses;
- Error of the shrinkage during the injection molding process;
- Lens module assembly and test;
- Dimension tolerance of parts;
- Optical module examination of the lens module; and
- Reliability test of lens module

During compact camera module development, optical design, mechanical design, mold design, plastic lens injection, glass lens grinding, lens module assembly, and optical examination should be carefully reviewed and controlled. Thus, the quality and yield of product can be improved.

References

[1] A. Krulikowski, Tolerance Stacks A Self-Study Course, Effective Training, 1992.

[2] E. K Thorburn, Concept and Misconceptions in the Design and Fabrication of Optical Assemblies,Optomechanical Systems Design, Proceedings of the SPIE Vol. 250, PP 2-7, 1980.

Design of O-ring Driven Liquid-Filled Lens

D. Shaw[1], C. W. Lin[2]

[1]National Tsing Hua University, R.O.C. (Taiwan), dishaw@pme.nthu.edu.tw
[2]National Tsing Hua University, R.O.C. (Taiwan), chihwei@webmail.pme.nthu.edu.tw

Abstract. The purpose of this study is to design and analyze an o-ring driven liquid-filled lens (ODLL). This lens is composed of a liquid chamber, a membrane and an o-ring device. The o-ring is on the top of the membrane and is used to change the curvature of the membrane. When the actuators move the ring up or down evenly, the curvature of the membrane is changed and the focus can be varied according to the magnitude of movement. The direction of the light can also varied by tilting the o-ring. The mathematical model of the membrane is developed to find the deformation of the membrane under different o-ring movement. Effective focal length and direction of the light can be obtained by substituting the aspherical equation of the shape of the membrane deformation into ZEMAX software. Finally, a possible precision design is demonstrated and discussed.

Keywords: o-ring driven liquid-filled lens, membrane development, mathematical model

1. Introduction

Many optical devices are required to alterate the direction of the light and the focal length. Customarily, glass or plastic is the most common material used to make the optical lens, due to the rigidity of the material the optical features of the lens is fixed. To change the focal length and direction of the light, an optical system is usually designed for this purpose. In the system, several actuators are used to control the lens position, the volume of the system is large and need to be miniaturized. A lens which can change the focal length is a good idea to acomplish this requirement. Several different types of lenses, which can adjust the focal length, have been developed before. One of those lenses is the liquid-filled lens. Usually, liquid-filled lenses are composed of a membrane covered chamber and penetrated by aqueducts. After filling the chamber with liquid, the membrane curvature can be controlled by pressure and the focal length can be changed.

A number of studies has been conducted on liquid-filled lenses. In 1993, Sugiura [1] analyzed the gravity effect and aberrations of a fabricated liquid-filled lens. In 1996, Rawicz [2] developed and experimentally confirmed the mathematical model of the liquid-filled flexible lens based on theoretical evaluation of absolute values of the tension in the film. In 2003, Chronis [3] indicated a Polydimethylsiloxane (PDMS)-based tunable liquid-filled microlens array fabricated by soft lithographic techniques. The focal length of lenses can be dynamically adjusted by pneumatically controlling the pressure within the microfluid network. In 2003, Krupenkin [4] demonstrated a focal length and lateral position tunable liquid microlens by controlling the electrode voltage to adjust the focal length and lateral position of a transparent conductive droplet placed on a dielectric substrate. Zhang [5] introduced a fluidic adaptive PDMS lenses with wide range focal length adjusted and controlled by syringe pump. In 2004, Zhang [6] demonstrated plano-convex and biconvex fluidic adaptive PDMS lenses. Jeong [7] reported a tunable micro doublet PDMS lens, which consisted of a tunable-filled lens and a solid negative lens. Chen [8] designed a new PDMS microlens with variable focusing ability, which consisted of a 3D convex diaphragm, chamber, and microfluidic channel. In 2005, Shaw and Lin [9] used ANSYS Fluid Structure Interaction (FSI) module for the transient analysis of single and two entrances of liquid-filled lens, and Shaw [10] discussed the nonlinear effect of the shape of a membrane on optical parameters. Kuqano [11] exhibited the application of liquid pressure varifocus lens in YAG laser processing. In 2007, Shaw and Sun [12] analysized the optical properties of liquid-filled lenses with different membrane shapes.

The traditional liquid-filled lens needs a pump or syringe-like devices to inject liquid into the chamber to increase the pressure inside the chamber to expand the shape of the lens's membrane to a convex shape. However, chamber and pump are connected by conduits, hence leakage occures after repeatedly operations. In addition to the leakage problem, the use of syringes also confines the system' operating speed. To solve these problems, an o-ring driven liquid-filled lens is designed and evaluated in this study.

2. Simulation Model Construction

The o-ring driven liquid-filled lens is ameliorated from the original design; a thin ring is designed on the top of a membrane. By moving or tilting the ring in contact with the membrane, the membrane deforms symmetrically or asymmetrically. The change of the membrane's curvature can vary the focal length and can tilt the optical axis as shown in Figure 1.

The mathematical model of a membrane can be used to estimate the curvature under different loading conditions, and the shape of the membrane is substituted into ZEMAX to calculate the focal length and appraise the image performance. Here, the model of an o-ring driven liquid-filled lens is considered as the superposition of the concentrically loaded and the uniformly loaded equations of a circular plate with clamped edges. In Figure 1.2, a is the radius of the membrane, b is the radius of the o- ring, P is the loading on the ring, and q is the hydraulic pressure.

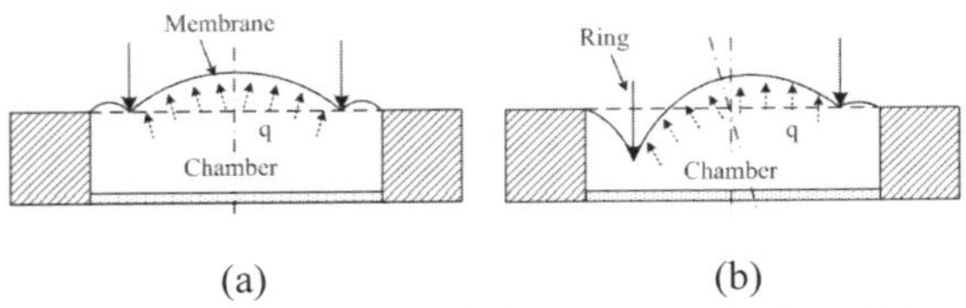

Figure 1.1. (a) symmetrical; (b) asymmetrical conditions

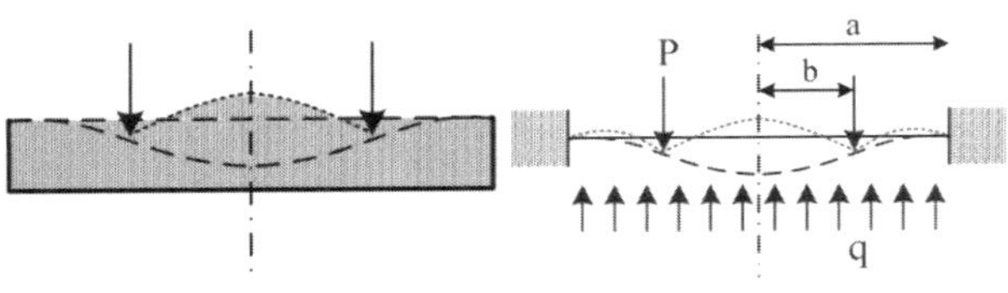

Figure 1.2. The scheme of o-ring driven liquid-filled lens

The deflection w_1 for the inner portion of a circular plate concentrically loaded with clamped edge [13] can be expressed as:

$$w_1 = \frac{P}{8\pi D}\left(\left(b^2 + r^2\right)\log\frac{b}{a} + \frac{\left(a^2 + r^2\right)\left(a^2 - b^2\right)}{2a^2} \right) \tag{1}$$

The deflection w_2 for the circular plate uniformly loaded with clamped edge [13] can be expressed as:

$$w_2 = \frac{q\left(a^2 - r^2\right)^2}{64D} \tag{2}$$

$$D = \frac{E \times t^3}{12 \times \left(1 - v^2\right)}$$

where w_1 and w_2 are the deformation in the distance r with respect to the membrane's center, E is the modulus of elasticity, v is Poisson's ratio, t is the membrane thickness, and D is its flexural rigidity. Superposition the equations of (1) and (2), the inner portion deflection w of the o-ring driven liquid-filled lens is:

$$w = \frac{P}{8\pi D}\left(\left(b^2 + r^2\right)\log\frac{b}{a} + \frac{\left(a^2 + r^2\right)\left(a^2 - b^2\right)}{2a^2} \right) - \frac{q\left(a^2 - r^2\right)^2}{64D}, 0 < r < b \tag{3}$$

According to the measurement results, the hydraulic pressure q is proportional to the loading P on the ring. The membrane material in this study is Polydimethylsiloxane (PDMS), which has good physical properties, including excessive hydrophobicity, flexibility, thermal and UV stability. In this study, the PDMS membrane is using SYLGARD 184 silicone elastomer, base & curing agent of Dow Coming Corp., and mixed with a 10:1 weight ratio. After curing at 75°C for two hours, the Young's modulus of the PDMS is 1.5MPa, Poisson's ratio is 0.49, and refractive index is 1.43. The injecting transparent liquid is water, and the refractive index is 1.33. Here, parameters, a is 10 mm, b is 5 mm, t is 0.5 mm (Figure 1.3), and the material properties are substituted into the equation, the deformation profile of o-ring driven liquid-filled lens under different loads can be plotted as shown in Figure 1.4.

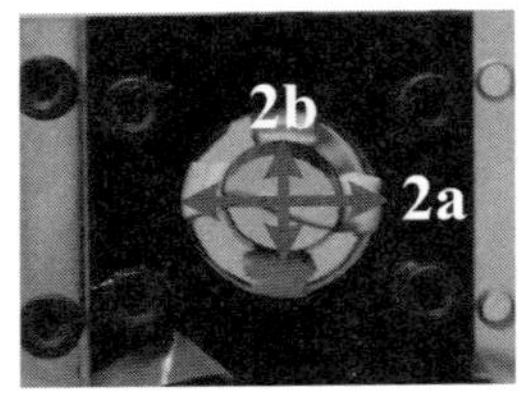

Figure 1.3. An o-ring driven liquid-filled lens prototype

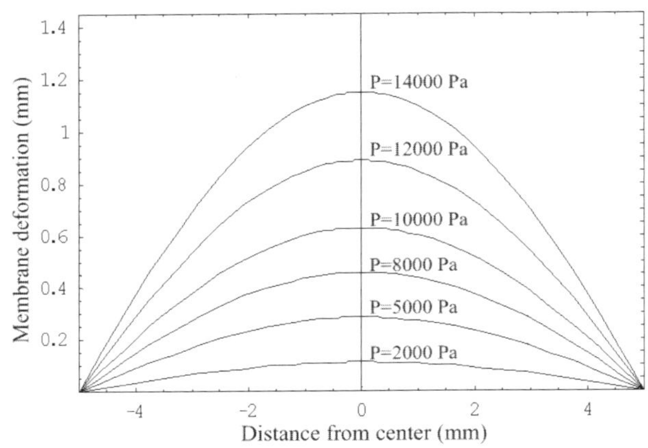

Figure 1.4. The membrane deformation of o-ring driven liquid-filled lens

The even aspherical equation can be fitted from the membrane profile, and substituted into ZEMAX optical software to calculate the focal length and image performance. In Figure 1.5, the effective focal lengths under different ring loadings are shown. According to the results, the payloads putting on the ring and effective focal length are inversely proportional.

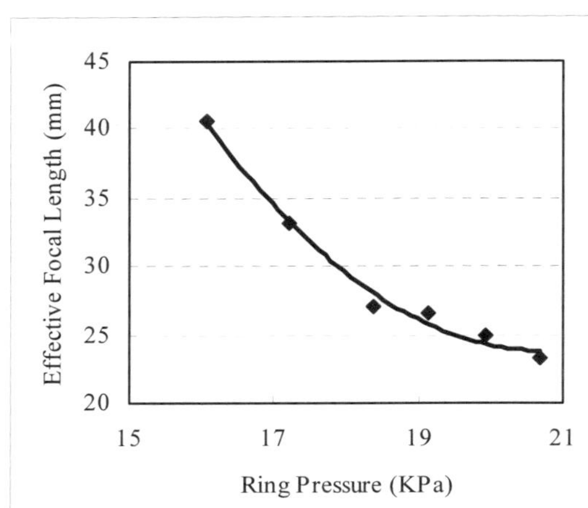

Figure 1.5. The focal length of o-ring driven liquid- filled lens under different loadings

3. Driven Mechanism Design

An o-ring driven liquid-filled lens is indicated and analyzed before. However, for many applications in precision optical fields, smaller size lenses and more accurate actuators are needed. Consequently, a possibility precision design is demonstrated in this study. The configuration of the device is shown in Figure 1.6, where two actuators, PI (P-280.30), are fixed on a frame to control the tips of O-ring driving frames. These two actuators can operate synchronously or asynchronously to push on the o-ring contacted with the membrane. Therefore, symmetrical or asymmetrical conditions can be achieved to change the focal length and light direction of the lens.

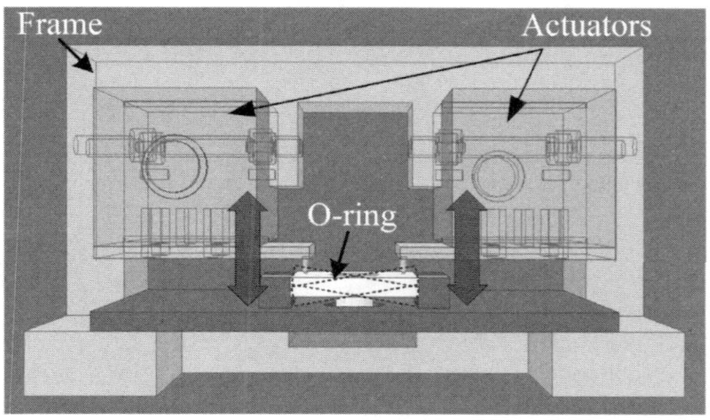

Figure 1.6. The scheme of an o-ring driven liquid-filled lens

The maximum travel length of the actuator, PI (P-280.30), is 100 μm. If a laser beam is passing through the o-ring driven liquid-filled lens, this maximum pushing-on depth can be used to estimate the scanning range ±s. The optical geometric of the tilt o-ring is shown in Figure 1.7, where the distance from lens is denoted as L, the ring diameter is varied from 2 mm to 5 mm. Hence, the scanning distance can be calculated, and the results are listed in Table 1.1.

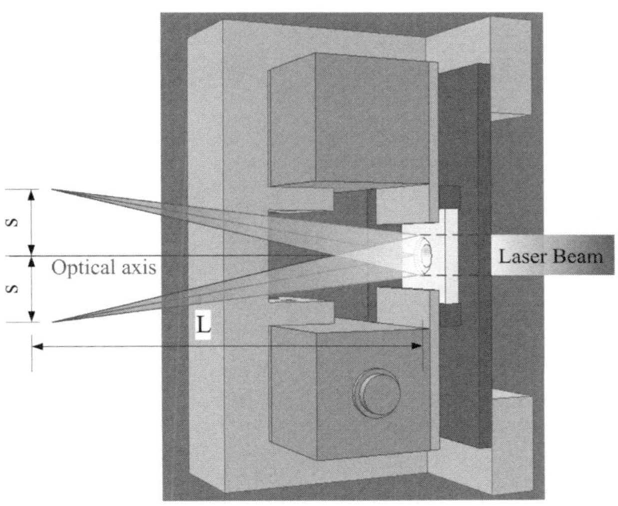

Figure 1.7. The geometrics of the tilt o-ring driven liquid-filled lens

Table 1.1. Scaning range of o-ring driven liquid-filled lens under different tilt angles and lens diameters

Depth (mm)	Lens Diameter (mm)	Tilt Angle (degree)	(s) Scanning range (mm)		
			(L) Distance from lens (mm)		
			10	50	100
0.1	5	1.15	0.2	1	2
0.1	4	1.435	0.25	1.25	2.5
0.1	3	1.91	0.33	1.67	3.33
0.1	2	2.86	0.5	2.5	5

The movement of the ring is controlled by actuators. According to different applications, the type of the actuators can be piezoelectricity, pneumatic, electric and hand-controlled. If a larger design is used in one-time use cameras (OTUC) to change the focal length, a hand-controlled device is needed for low cost consideration.

4. Conclusions

An o-ring driven liquid-filled lens (ODLL) is proposed in this study. In order to overcome the drawback of traditional liquid-filled lenses, a thin ring is designed on top of a membrane with a sealed chamber. The design of the o-ring allows the membrane to have two degrees of freedom, i.e. translation and rotation. The o-ring can make the shape of deformation approach to a paraboloid surface and cause better image performance. The design using aqueducts not only has problems with leakage but also has fast operation speed. Changing the tile angle of the ring, the direction of light can be controlled.

The deformation equation of the o-ring driven liquid-filled lens is also derived. The effective focal length can be computed by submitting the aspherical equation into the ZEMAX optical software. According the results, the shortest effective focal length of 10 mm diameter ring is 23 mm at 20.7 KPa ring pressure. The beam diameter is very small. The coma aberration is neglected.

Finally, a possible precision design is developed and the scanning range of 2 mm diameter lens under a maximum tilt angle of 2.86° ranges from ±0.5 mm to ±5 mm at a distance of 10 mm to 100 mm from the lens. One application of this design could be in laser controlled systems, such as laser scanning systems.

References

[1] N. Sugiura and S. Morita, "Variable-focus liquid-filled optical lens," *Applied Optics,* vol. 32, pp. 4181-4186, 1993.

[2] A. H. Rawicz and I. Mikhailenko, "Modeling a variable-focus liquid-filled optical lens," *Applied Optics,* vol. 35, pp. 1587-1589, 1996.

[3] N. Chronis, G. L. Liu, K.-H. Jeong, and L. P. Lee, "Tunable liquid-filled microlens array integrated with microfluidic network," *Optics Express,* vol. 11, pp. 2370-2378, 2003.

[4] T. Krupenkin, S. Yang, and P. Mach, "Tunable liquid microlens," *Applied Physics Letters,* vol. 82, p. 316, 2003.

[5] D.-Y. Zhang, V. Lien, Y. Berdichevsky, J. Choi, and Y.-H. Lo, "Fluidic adaptive lens with high focal length tunability," *Applied Physics Letters,* vol. 82, p. 3171, 2003.

[6] D. Y. Zhang, N. Justis, V. Lien, Y. Berdichevsky, and Y.-H. Lo, "High-performance fluidic adaptive lenses," *Applied Optics,* vol. 43, 2004.

[7] K. H. Jeong, G. L. Liu, N. Chronis, and L. P. Lee, "Tunable microdoublet lens array," *Optics Express,* vol. 12, 31 May 2004.

[8] J. Chen, W. Wang, J. Fang, and K. Varahramyan, "Variable-focusing microlens with microfluidic chip," *J. of Micromechanics and Microengineering,* vol. 14, pp. 675-680, 2004.

[9] D. Shaw and C. W. Lin, "The Transient Analysis of Fluid-Membrane Interaction of Multi-Fluid-Entrances Liquid Lens," in *The 5th IWAOIM* Beijing, 2005.

[10] D. Shaw and T. E. Sun, "Effect of nonlinear behavior of elastic membrane of liquid filled optical lenses," in *The 5th IWAOIM* Beijing, 2005.

[11] R. Kuwano, T. Tokunaga, Y. Otani, and N. Umeda, "Liquid Pressure Varifocus Lens," *Optical Review,* vol. 12, pp. 405-408, 2005.

[12] D. Shaw and T. E. Sun, "Optical properties of variable-focus liquid-filled optical lenses with different membrane shapes," *Optical Engineering,* vol. 46(2), 2007.

[13] S. Timoshenko and S. Woinowsky-Kerieger, *Theory of Plates and Shells*: McGraw-Hill, 1959.

Surface Modification for Advanced Light Guide Plate for TFTLCD Display

Ga-Lane Chen[1], and Tai-Cherng Yu[2]

[1]Hon-Hai Precision Ind. Co., Ltd., Taipei 236, Taiwan, galane.chen@foxconn.com
[2]Hon-Hai Precision Ind. Co., Ltd., Taipei 236, Taiwan, tai-cherng-yu@foxconn.com

Abstract. Techniques for performing uniform and high brightness is very important for TFTLCD display. The backlight module with light guide plate (LGP) is the key to achieve uniform and high brightness. Surface modification on light guide plate (LGP) can enhance the brightness and uniformity of TFTLCD display. A novel design, which integrates the merits of diffraction optics and forming technology, is proposed in this research for the higher brightness of LGP. A light guide plate with diffraction grating has been designed. The light emitting surface has a plurality of diffraction grating units. Each diffraction grating unit has a strong diffraction portion and a weak diffraction portion. They are orthogonal each other. Area ratios of the strong diffractive portions in the diffraction grating units proportionally increase with increasing distance away from the light incident surface. Two LGP designs with diffraction optics have been demonstrated. One is for large size TFTLCD displays with CCFL as the light source. Another one is small size TFTLCD displays with LEDs as the light source. Both designs with diffraction optics on LGP can increase the overall efficiency of utilization of light with high brightness and uniformity.

Keywords: Light Guide Plate (LGP), TFTLCD, injection molding, diffraction grating, light brightness.

1. Introduction

A typical liquid crystal display requires a backlight module in order to be able to provide uniform illumination. Figure 1.1 shows the configuration of an TFTLCD display with a light guide plate (LGP) and backlight module. The performance of the backlight module greatly depends on a light guide plate (LGP). The modification of emitting surface and reflective surface can vary the brightness and uniformity. In this paper, we discuss three pattern designs by surface modification for LGP.

The first one is conventional technique with dots pattern design. Conventional techniques with dot patterns have relatively low brightness. The mold insert of light guide plate (LGP) is made by chemical etching process. The diameter and depth of dots were usually used as the process parameters to change the brightness and light uniformity. However, chemical etching has difficulty in controlling the profile and depth of the dots. The brightness of LGP by dot pattern design is therefore relatively low.

The second design for LGP is V-cut pattern. V-cut pattern can be mechanically created with parallel grooves on the surface of mold insert. However, any surface defects or vibration on tool cutting can cause the failure of V-cut patterns.

The third design is the diffraction pattern. Our novel design, which integrates the merits of diffraction optics and forming technology, is proposed in this research [1] for the higher brightness of LGP. The light guide plate used in this design is based on utilizing the phenomenon of diffraction based on wave optics of light [2]. A light guide plate with diffraction grating has been designed. It can provide much higher and more uniform brightness over an entire illuminated surface than the brightness level achievable by the prior art [3-5] and can assure longer battery life through reduced power consumption for the light source.

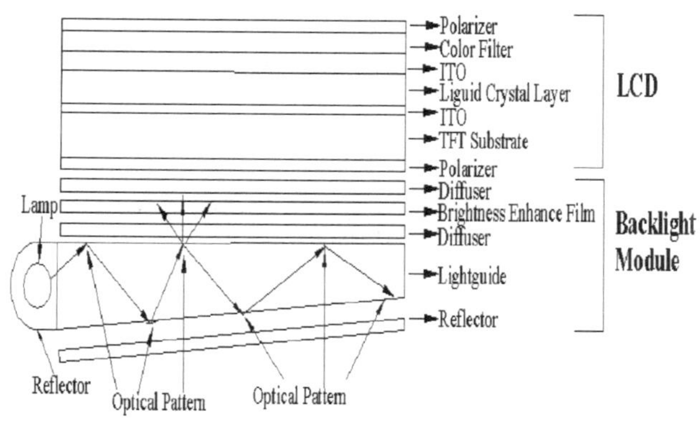

Fig. 1.1 Configuration of TFTLCD Display with LGP and Backlight Module

2. Pattern Design and Process for Fabricating Light Guide Plate (LGP)

Light guide plate (LGP) with patterns of dots and V-cut can be fabricated in three manufacturing steps. In the first step, the optical design needs to be designed. In the second step, the mould and core insert have to be machined by following the optical design. Then, the mold and core insert are installed on the injection molding system. Finally, the transparent light guide plate (LGP) can be made with the materials such as polycarbonate, or PMMA.

Figure 1.2 depicts the pattern design for light guide plate for TFTLCD. Figure 1.2 (a) shows the dots (also called as micro dots) pattern for LGP. Figure 1.2 (b) shows the V-cut pattern (also named as prisms pattern), and Fig. 1.2 (c) shows the diffraction pattern.

Light guide plates with the configurations of dots (micro dots) and V-cut (prisms) have been developed. However, the uniformity provided by dots is relatively low compared with LGP having gratings. This is because the grating LGP can be more precisely configured to correspond to the wavelenth band of visible light beams, thereby accurately controlling the uniformity of transmission of the light beams.

The light emitting surface has a plurality of diffraction grating units. Each diffraction grating unit has a strong diffraction portion and a weak diffraction portion. They are orthogonal each other. Area ratios of the strong diffractive portions in the diffraction grating units proportionally increase with increasing distance away from the light incident surface. V-cut pattern can be mechanically created with parallel grooves on the surface of mold insert. Any surface defects or vibration on tool cutting can cause the failure of V-cut patterns.

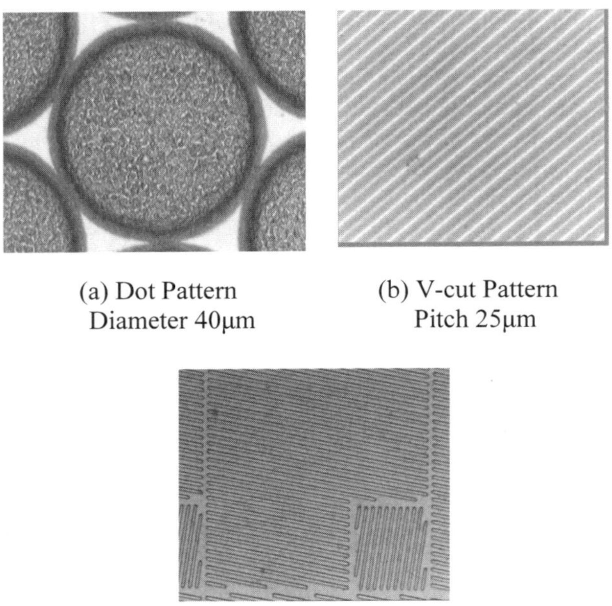

(a) Dot Pattern
Diameter 40μm

(b) V-cut Pattern
Pitch 25μm

(c) Diffraction Pattern
Pitch 1.5μm

Fig. 1.2 Three Pattern Design for Light Guide Plate for TFTLCD

3. Design of Diffraction LGP for Large and Small Sizes TFTLCD Displays

3.1 Diffraction LGP for Large Size Display

Our diffraction LGP comprises a light incidence surface for receiving light, a light-emitting surface for emitting light, and a bottom surface. The diffraction pattern has a plurality of diffraction grating units. Each diffraction grating units comprises a strong diffractive portion 1S and a weak diffractive portion 1W. Grating directions of the strong diffractive portion 1S is substantially perpendicular to a direction of light beams received by the diffraction grating units. Grating directions of the weak diffractive portion 1W is substantially parallel to a direction of light beams received by the diffraction grating units. Therefore, the grating directions of the strong and weak diffractive portions are orthogonal to each other as shown in Figure 1.3 with CCFL (cold cathode fluorescent lamp) as the light source for the large size TFTLCD display. Each diffraction grating unit comprises a grating part and a non-grating part.

The ratio of the area of the strong diffractive portion to an area of the weak diffractive portion in each unit becomes progressively greater with increasing distance away from the light incidence surface as shown in Fig. 1.3. The units 1 in three regions I, II, III sequenced along a direction away from the light incidence surface have different configurations and different area ratios. The diffraction grating units further from the light incident surface have larger strong diffractive portions so that they have stronger diffracting capability.

Since the intensity of light beams decreases with increasing distance of propagation, the quantity of light beams received by the each units decreases with increasing distance away from the light incidence surface. The configuration of the diffraction grating units compensates the location of each unit with a corresponding diffraction capability. This design assures the uniformity of LGP for large size TFTLCD display.

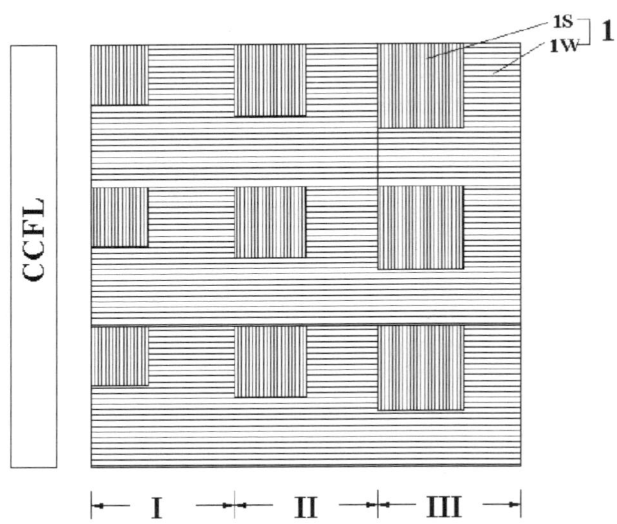

Fig. 1.3. Diffraction Grating Design for for Light Guide Plate for Large Size TFTLCD Display with CCFL as the light source

3.2 Diffraction LGP for Small Size Display

In Figure 1.4, we have designed diffraction LGP for small size TFTLCD display by using single LED as the light sources. Each diffraction grating unit comprises a grating part and a non-grating part. This transparent LGP has a wedge cross-section. The main reason for this is that the light beams emitted by LED are quite different from the light beams emitted by CCFL.

Propagation of the light beams emitted by LED is approximately over a range of angles covering three directions as shown on the Fig. 1.4. The light-emitting surface is divided into three regions I, II, III sequenced along a direction away from the light incidence surface. The units 2 at the upper, middle, and lower positions have three different grating directions respectively, each being perpendicular to a main direction of incoming light beams. In each unit 2, the grating direction of the weak diffractive portion 2W is perpendicular to that of the strong diffractive portion 2S. Thus light beams incident on the weak diffractive portion 2W in oblique directions are diffracted. The ratio of the area of the strong diffractive portion 2S to an area of the weak diffractive portion 2W in each unit becomes progressively greater with increasing distance away from the light incidence surface as shown in Fig. 1.4.

Since the diffraction grating units further from the light incident surface have larger strong diffractive portions, they have a stronger diffracting capability. Since the intensity of light beams decreases with increasing distance of propagation, the quantity of light beams received by each unit decreases with increasing distance away from the light incidence surface. The configuration of the diffraction grating units compensates the location of each unit with a corresponding diffraction capability. This design assures the uniformity of LGP for small size TFTLCD display with LED as the light source.

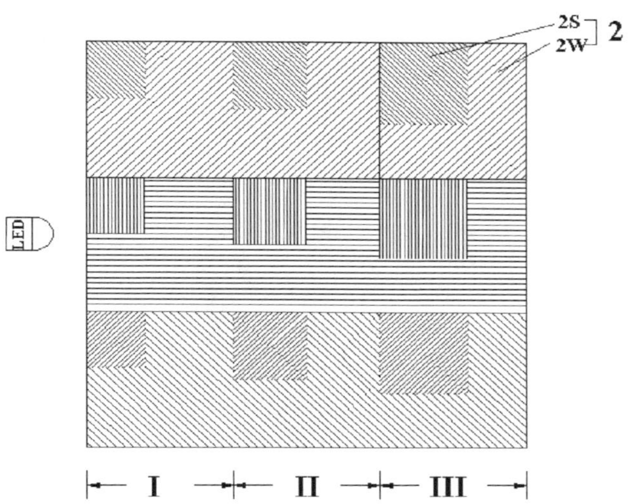

Fig. 1.4. Diffraction Grating Design for Light Guide Plate for Small Size TFTLCD with LED as the light source

4. Forming Experiments and Process Technologies

A diffraction light guide plate was designed which utilizes the diffraction phenomenon of wave optics of light. This novel design providse much higher and more uniform brightness over the entire illuminated surface than the level of brightness that is achievable by the prior dsign principles. It assures longer battery life through reduced power consumption with the same light source.

The traditional way to make LGP is using "scattering optics design with injection molding". In this paper, we use diffraction optics and special forming process to fabricate LGP. New forming technologies with hot embossing and nano-imprinting have been used. Also we use "stamper" technology with micro-electronics process to replace "traditional mold insert by machining". In this study, six different process approaches have been studied for LGP:

(I) Design with Diffraction Optics → Mask → Silicon Photolithography → Ni Stamper → Hot Embossing
(II) Design with Diffraction Optics → Mask → Silicon Photolithography → Nano-imprinting
(III) Design with Diffraction Optics → Mask → Silicon Photolithography → Si Stamper → Hot Embossing
(IV) Design with Diffraction Optics → Mask → Silicon Photolithography → Si Stamper→Nano-imprinting
(V) Design with Diffraction Optics → Mask → Nickel Stamper → Hot Embossing
(VI) Design with Diffraction Optics → Mask → Nickel Stamper → Nano-imprinting

Based on our study, Processes (I) and (II) are good for most LGP designs. Since Ni has a good fracture toughness, it can be used under higher embossing pressure. Ni stamper has a long lifetime. Process (III) and (IV) are good for the optical design with high "aspect ratios" or optical pattern with significant depth. However, we have to pay attention to the loading force during embossing, since silicon is relatively brittle. To have low manufacturing cost, Process (V) and (VI) are the better solutions.

By using LIGA process, we can grow thick Ni layer after masking process. Ni stamper can be grown with thickness from 0.5 mm to 2 mm without pinholes and surface defects by electrode-less NiP plating process.

5. Application of the Proposed Process Technology

There are many applications on our process technology:

1. Our process technology can be used for the products of "diffraction light guide plate" for backlight module of TFTLCD display.
2. Our forming and process technologies can be used in the products of microelectronics, optics, optical storage, optical display, and optical communication.
3. Hot embossing and nanoimprinting technologies can be used on the components for 3C industries. It includes
4. MEMS (micro-electromechanical-systems), MOEMS (micro-optoelectromechanical systems), and many of the next generation products.

6. Conclusions

Light guide plate is the critical component for TFTLCD display. Surface modification on LGP is the key to achieve uniform and high brightness for TFTLCD. Our novel design, which integrates the advantages of diffraction optics, and advanced forming technologies have increased the overall performance of light guide plate.

Morereover, our study shows the potential for fabrication of optical-mechanical components for applications in the output and input devices for 3C industries. The forming and process technologies have potential to be applied on MEMS, MOEMS, and many of the next generation products.

References

[1] Ga-Lane Chen, Charles Leu, and Tai-Cherng Yu, United States Patents US 7,085,056 B2, Aug. 1, 2006

[2] Leo Levi, Applied Optics, Wiley Series in Pure and Applied Optics, 1983

[3] S. Ochiai, United States Patents US 5, 703,667, Dec. 30, 1997

[4] T. Ishikawa, United States Patents, US 5,799,338, July 14, 1998

[5] J. Lee, United States Patents, US 6,164,790, Dec. 26, 2000

4–4

Optical Design for Three Mega Pixel Mobile Phone Camera

Ga-Lane Chen[1], Chun-Yu Lee[2], Adams Yuan and Patrick Chang

[1]Hon-Hai Precision Ind. Co., Ltd., Taipei 236, Taiwan, Galane.chen@foxconn.com
[2]Hon-Hai Precision Ind. Co., Ltd., Taipei 236, Taiwan, Chun-Yu.Lee@foxconn.com

Abstract. We have designed a phone lens module of compact size which is less than 5 mm thick and has an excellent image performance. For cost and production considerations, plastic lenses instead of glass/plastic hybrid or all glass design are chosen. The sensor pixel size is 1.75 um, resolution limit are 280 lp/mm for Nyquist frequency and 140 lp/mm for Black/White and full color images respectively. The resolution performance (MTF, modulation transfer function) of this design can reach 50% or higher at half of Nyquist frequency within 90% image field. The RMS wavefront error is also less than 0.2 at all field. TV distortion is only 0.2%.

Keywords: Spherical Aberration, Distortion, Modulation Transfer Function

1. Introduction

Recent advancement in plastic molding technology has boosted active use of aspherical lens system for highter order abberation correction. A typical and successful example is the lens module for mobile phone, which has become the main stream on the market for its compactness, light weight, low cost and excellent image quality

The design of such lens system is a very challenging work. The factors of image quality and all other constraints must be taken into consideration simultaneously during optical design. The definition of image quantities include the parameters of spectral range, image format and image sensor pixel size. Constraints include lens size, plastic lens molding limitation, lens surface measurement, specifications of image sensor, lens operation environment and others.

To reduce the manufacturing cost, all plastic aspherical lenses have been designed.

2. System Specifications and Constraints

In the present case, the specifications and constraints are listed as follows:

1. Sensor Basic Information:
1.1 Pixel size: 1.75 um square
1.2 Active pixels: 2064*1552
1.3 Chief rate angle: 23 degree

2. Lens Request:
2.1 Total track length (TTL : 4.90 mm
2.2 Flange back length (FBL): > 1.2mm
2.3 F#: 2.8
2.4 Max image circle: > ψ 4.9 mm
2.5 Field of View: 60 ° (For sensor diagonal)
2.6 TV-Distortion: < 1%
2.7 Relative illumination: > 55%
2.8 Lens elements: 4 elements

3. Optical Design Procedure

3.1 Fundamental Configuration

The fundamental configuration is modified by Cooke triplet structure, separating the third element into a positive power and a negative power lens as shown in Figure 1.1.

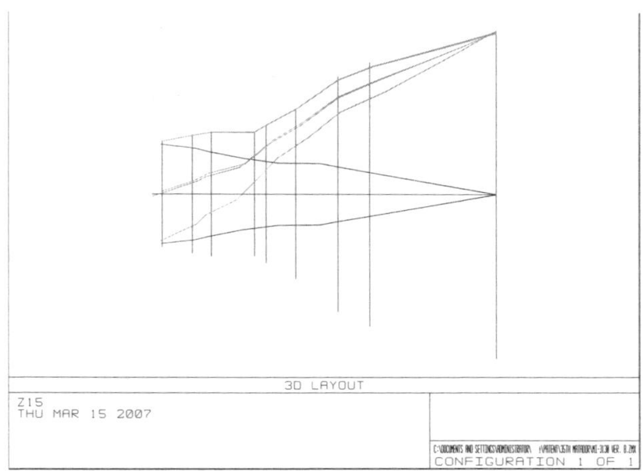

Figure 1.1 Cooke triplet lens separate the third element into a positive power lens and a negative power lens.

Due to the large field angle, the longitudinal and lateral colors are difficult to correct. Chromatic aberration of the chief ray can be well corrected by the proper choice of optical plastic material. Fig 1.2 and Fig 1.3 showed the longitudinal aberration and lateral aberration diagram respectively. Our design has reduced the longitudinal and lateral aberration (particularly secondary) by choosing proper dispersion material for negative power lens.

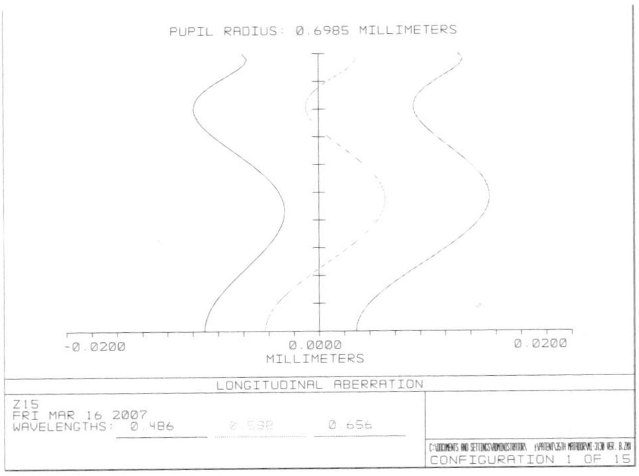

Figure 1.2 Longitudinal aberration diagram

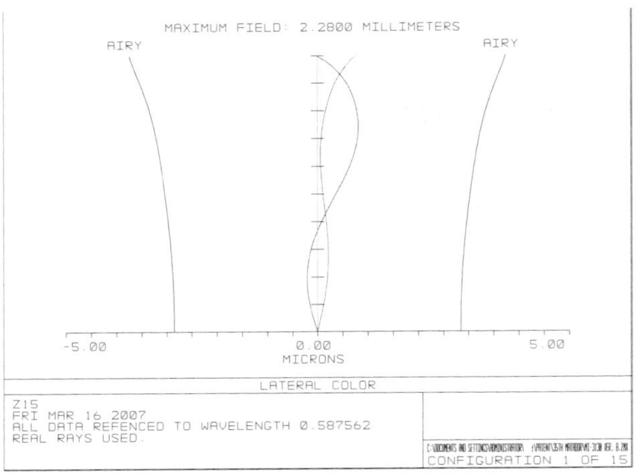

Figure 1.3 Lateral aberration diagram

Figure 1.4 depicts the transverse aberration (TA) and OPD plots for this case. In order to reach minimum spot size, and minimum wave front error, the aspheric coefficients are added to correct and perfect the optimization.

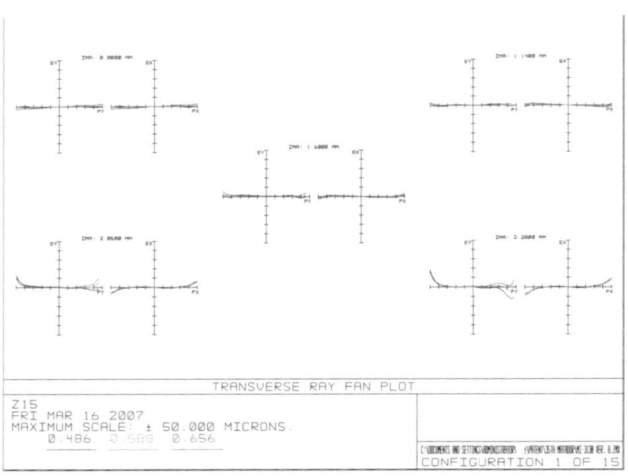

Figure 1.4 Ray fan diagram

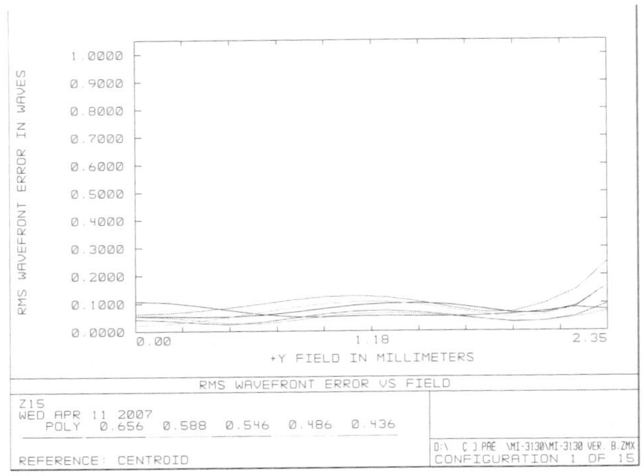

Figure 1.5 RMS vs. Field

Figure 1.5 shows the RMS wavefront error for all field for this case. The effective image height for the sensor is 2.2 mm and we can find that the RMS wavefront error for all field and all wavelength is less than 0.2, which will guarantee a balanced and uniform image quality.

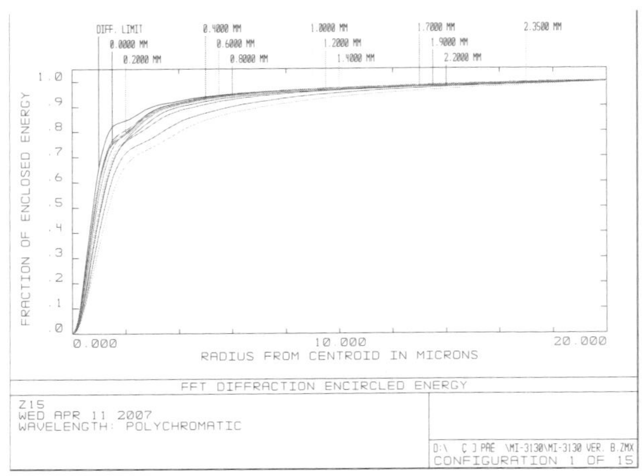

Fig. 1.6 Diffraction Encircled Energy

Figure 1.6 shows diffraction encircled energy for this case. The spot size and the energy concentration was also well optimized. For all field, almost 80% energy is concentrated within the area of two pixel size.

Figure 1.7 shows astigmatic field curves and distortion for this case. In order to minimize distortion and field curvature, the higher order aspheric coefficients were added in the rear lenses to smooth the field curvature. For camera and projector lens design, TV distortion is very important. In this case, TV distortion of phone camera lens is less than 0.5%.

Figure 1.8 shows Seidel coefficient for this case. The aberrations were well optimized and balanced.

Fig 1.9 is color filter arrangement for image sensor. It is also called "Bayer Pattern". For this case, pixel size is 2.0 μm and the resolution limits of the sensor are 280 lp/mm (Nyquist Frequency) and 140 lp/mm for B/W and full color image respectively. We often use half and one-third of Nyquist Frequency for resolution inspection to understand the lens performance.

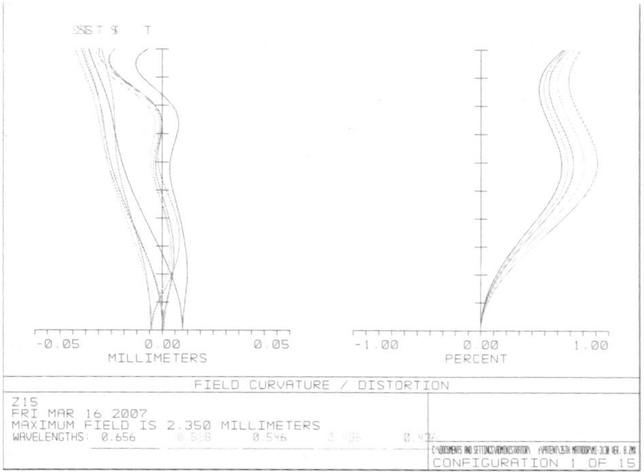

Figure 1.7 Astigmatic field curvature and distortion

Surf	SPHA S1	COMA S2	ASTI S3	FCUR S4	DIST S5	CLA (CL)	CTR (CT)
STO	0.000000	0.000000	0.000000	0.000000	0.000000	0.000000	0.000000
2	-0.016452	0.013360	0.019677	0.031952	0.082981	-0.002454	-0.003941
3	-0.082479	-0.061854	0.020282	0.014247	-0.065045	-0.002479	0.003177
4	0.116996	0.066654	-0.008180	0.009373	0.093733	0.003023	-0.009859
5	-0.032966	-0.025941	-0.012552	-0.038037	-0.207455	0.001899	0.012839
6	0.026076	0.077945	0.150098	-0.038992	0.326440	0.001661	-0.000648
7	-0.004697	-0.053646	-0.199255	0.049680	-0.578518	-0.002115	-0.002185
8	-0.024759	-0.079693	-0.281999	0.031120	-0.707234	-0.000086	-0.004326
9	0.012445	0.058130	0.305095	-0.044622	1.015526	0.000087	0.004452
10	0.000000	0.000000	0.000000	0.000000	0.000000	0.000000	0.000000
11	0.000000	0.000000	0.000000	0.000000	0.000000	0.000000	0.000000
12	-0.000193	0.000683	-0.002419	0.000000	0.008569	0.000096	-0.000339
13	0.000041	-0.000144	0.000510	0.000000	-0.001807	-0.000020	0.000072
IMA	0.000000	0.000000	0.000000	0.000000	0.000000	0.000000	0.000000
TOT	-0.005988	-0.004508	-0.008744	0.014722	-0.032811	-0.000389	-0.000758

Figure 1.8 Seidel Coefficients

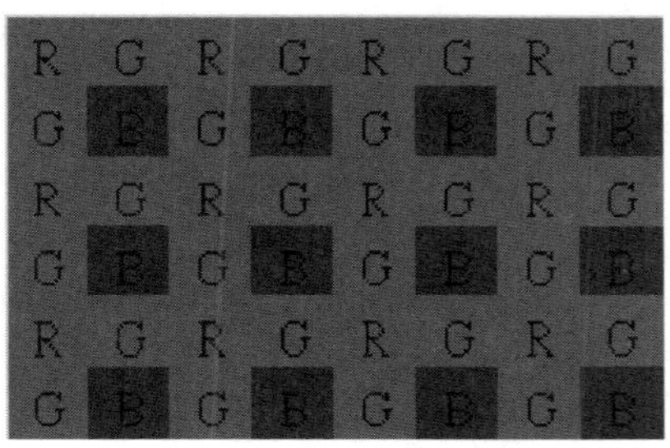

Figure 1.9 Bayer pattern arrangement

The balance condition is empirically determined and optimized as far as MTF is concerned. Figure 1.10 shows that the MTF curves with different image high by optimizing spot size, and wavefront. Low frequency response is another important design target. Spatial frequency response is very important for optical performance.

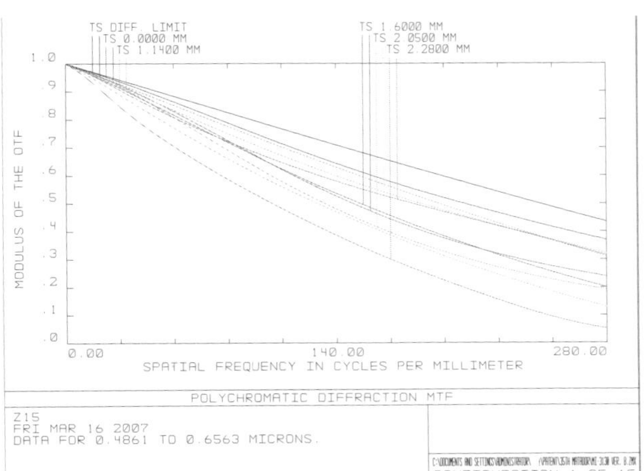

Figure 1.10 MTF curves for different image height

Fig 1.11 is the modulation transfer function (MTF) with 140 line pairs per mm with through focus. The more concentrative curve means the larger depth of focus.

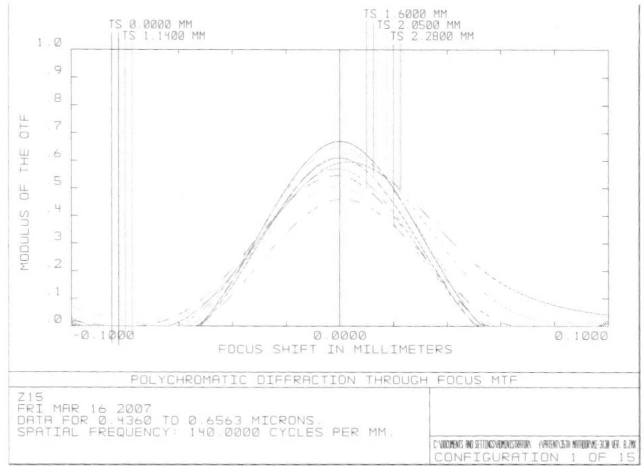

Figure 1.11 MTF 140lp/mm through focus

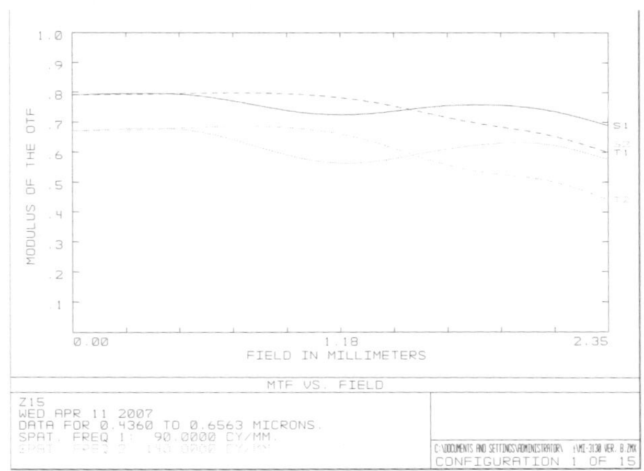

Figure 1.12 MTF vs. Field at 90 lp/mm and 140 lp/mm

Fig 1.12 is the modulation transfer function (MTF) vs. field at half (140 lp/mm) and one-third (90 lp/mm) of Nyquist Frequency. The MTF performance at half frequency is larger than 50%, which is almost within the effective area of the sensor.

4. Conclusion

We have designed a phone lens with compact size which is smaller than 5 mm and shows excellent image performance. For the concerns of cost and production, we have successfully used all plastic lenses instead of glass/plastic hybrid or all glass design.

Based on our optical design we have fabricated molds and core inserts for lens with ultra precision machining. With ultra precision core insert at roughness average (Ra) 2-5 nm. We have manufactured lens module with high yield and good optical performance.

References

[1] Lee; *Chun-Yu*, Tsai; Ming-Chiang, Chiang; Tsung-Wei, Patents US 7,079,330 B2, Jul. 18, 2006
[2] Dr. Michael E. Harrigan, "Effective Use of Aspheres in Lens Design", Proc. of SPIE Vol 0554, 1985 International Lens Design Conference, ed. Taylor, Moore (Jun 1985)
[3] Ian Marshall, "Moulded plastic lens design", Proc. of SPIE Vol. 2774, Design and Engineering of Optical Systems, ed. J J Braat (Aug 1996)
[4] A. Offner, "MTF Maximization at Low Spatial Frequencies", Applied Optics, Vol8, 1969
[5] P. J. Reardon and R. A. Chipman, "Maximum Power of Refractive Lenses: a Fundamental Limit," Submitted to Optics Letters, June 1990
[6] J.W. Goodman, Introduction to Fourier Ontics, McGraw-Hill, San Francisco, 1968.
[7] W. J. Smith, Modern Lens Design, McGraw-Hill, New York, 1992.
[8] Gregory Hallock Smith, Lens Design, California, 1998
[9] Robert E. Hopkins, Optics Design, New York, 1987
[10] 8. R. Kingslake, "Applied Optical and Optical Engineering", chp3, Academic Press, Vol III, 1965
[11] Geary, J. M., "Response of Long Focal Length Optical System to Thermal Shock", Optical Engineering Vol.19, No.2 , pp.233-239, March/April 1980
[12] Jamieson, T. H.," Thermal Effects in Optical Systems", Optical Engineering Vol.20, No.2, 1981.
[13] Marcano, A. and Escalona, R., "Measurement of Stationary Thermal Nonlinear Refraction Light Wavefront Distortions by Image Processing", Rev. Science Instrument Vol.64, No.8, 1997.
[14] Paul, R. Y., " Optomechanical Design", Optical Engineering Vol.20, No.2, July 1992.
[15] Stapp, J. L., "Thermal Distortion Test Facility" , Optical Engineering Vol.34, No.2, 1995.
[16] M. Heckele, "Hot embossing – The molding technique for plastic microstructures," Microsystem Technologies, 1998.
[17] Rong-Seng Chang, "Optical design and assembly by intelligent robot", Chung Shan Institute of Science and Technology
[18] T. Brendel, US patent, US2731884, Triplet-type photographic objective Lens, 1956.
[19] E. Tronnier, US patent, US3176582, Three lens photographic objective, 1965.
[20] R. Kingslake, US patent, US3418039, Triplet covering a wide field, 1968.
[21] J. M. Geary, Introduction to lens design with practical ZEMAX Examples, Willmann-Bell, 2002.
[22] H. H. Hopkins, Wave theory of aberrations, Oxford University Press, 1950.
[23] W. T. Welford, Aberrations of optical systems, Boston: Adam Hilger, 1986.
[24] R. E. Fisher, Optical system design, New York: McGraw-Hill, 2001.
[25] E. Hecht, Optics, 4th Edition, San Francisco: Addison Wesley, 2002.

5

Electro-physical and chemical Processes

Micro-Electric Discharge Machining in Standard Setup

Satyajit Chatterjee[1] and Santanu Das[2]

[1]Indian Institute of Technology, Kharagpur - 721302, satyajit_ch@mech.iitkgp.ernet.in
[2]Kalyani Government Engineering College, 741235, sdas_me@rediffmail.com

Abstract. Micro-Electric Discharge Machining is an alternative method to fabricate microstructures due to its distinct advantages. However, tools for electric discharge machining (EDM) of lower diameters face problems related to manufacturing due to its low strength, difficulty in maintaining alignment and positioning accuracy. The cost of the required setup and accessories to undergo a Micro-EDMing process is low to make this technique popular in industries.

In this paper, first, a brief overview of the micro electric discharge machining is presented. Next, the parametric dependence of die-sinking type micro-EDMing process using the typical tool holder attachment made is discussed. It is found that micro-EDMing can be performed in a standard EDM set-up satisfactorily to show its industrial applicability.

Keywords: Micro electric discharge machining, die-sinking type,

1. Introduction

In recent years, Electric Discharge Machining (EDM) is being explored to develop micro-features as it needs small processing force and can achieve high-aspect-ratio maintaining high level of precision and accuracy. Micro-EDM is often preferred over other micro-manufacturing operations due to its low set up cost [1]. Burr formation is negligible and additional finishing is, in general, not required.

However, the process experiences some problems [2,3] related to manufacturing and handling of miniaturized electrodes and parts, and adopting suitable flushing technique, which should be addressed.

A number of investigations aim at optimizing material removal rate (MRR), tool wear rate (TWR) and surface quality achieved of the EDM and micro-EDM processes [4-6]. The accuracy and repeatability of positioning of the machine employed is a major source of error. A surface detection error introducing variation in the spark gap is often found. Other errors include temperature instability error [4], random error due to flushing condition, material inhomogeneity [5], etc.

Electrode wear reduces the cross section as well as the form of the micro-electrode. Electrode wear becomes a major issue when employing electrodes with micro-features in die-sinking as the combination of micro-features and macro-features on one electrode will introduce different wear ratios [7,8]. When eroding down to a fixed depth, the real depth of the hole will be significantly smaller. Bleys et al. [1] proposed a method to achieve a specific depth in this case to compensate for wear of the electrode by constant electrode feeding in the Z-axis.

For die-sinking type micro-EDMing, Piezoelectric/ electrostrictive actuators can be used in micro-feeding and ultra-precision positioning to eliminate feeding error, because they have features of fine resolution, rapid response, high generative force and ease of miniaturisation [3].

In this paper, some designs of tool holders are first proposed to perform micromachining operations in a standard die-sinking setup for EDMing. Details of experimental investigations using the designed tools, and discussion on the observations made are presented next.

2. Experimental Investigation

2.1 Experimental Setup

In the present work, micro-hole machining has been done in a conventional die-sinking type EDM machine (Electronica make, model: T-3822/RR-15) with electrolytic copper electrodes. The small diameter electrodes were produced from thicker wires through wire drawing operation.

Designing of tool holders made are shown in the Figure 1.1 (a, b, c). At one end of it, there is a cylindrical shank portion, which will be held at the gripper of the machine tool. On the other side of it, thin electrode wires are brazed at the tool holder face precisely. A counter-prism attachment fitted to an X-Y table is used to hold the wire straight, and then brazing is done into the pre-drilled tool head. The thicker electrode is held by a threaded engagement with the extended portion of the shank. The alignment of the electrode face is made true by using a filler gauge set and the perpendicularity of the thin wire electrodes are also checked.

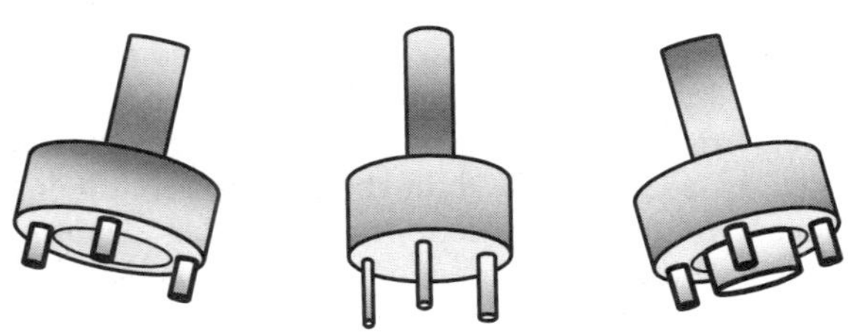

(a) (b) (c)

Figure 1.1. Typical tool holding attachments made

2.2 Experimental Conditions

Experiments have been performed at the following settings of the machining process parameters for analyzing the machining characteristics of the micro-Electric Discharge Machining process. The controlled parameters are given in Table 1.1.

Electrode diameters are varied three times with 0.92, 0.46 and 0.34 mm diameters for making holes in 1.33 mm low carbon steel plates. Experiments have been performed with pressure flushing using two nozzles from both the sides of the workpiece maintaining a nozzle gauge pressure of 0.2 kgf/sq.cm. Some experiments have also been done without any flushing to compare the effects of flushing during the EDMing operation. Experiments are repeated three times at each condition to justify the repeatability of the experimental observation.

EDMed holes have been put under a tool maker's microscope to observe shape and nature of the holes and to measure major and minor diameters of each hole both at entry and exit side. Throughout the experimental procedure, electrode diameters were very closely monitored to determine the changes in diameter. This was done with the help of a digital vernier caliper (Mitutoyo, Japan).

Table 1.1. Controlled parameters

Sl. No.	Parameters	Range of Control
1	Set Current	1, 2, 4 Ampere
2	Pulse-on Time	200, 500, 1000 microsec
3	Flushing Pressure	0(OFF), 0.2 kg / sq.cm

3. Experimental Results and Discussion

From the observations made, to find out the characteristic effects of the process, a number of graphical representations have been made.

Variations in taper angle, tool wear and kerf with a change in EDM process variables (set current and pulse-on time) are observed as shown in Figures 1.2 to 1.7. Kerf is the difference between the radius of the EDMed hole and the wire-type tool radius, whereas taper angle is measured on the EDMed hole from one end to the other without considering the overcuts at the entry and exit sides. Each of the data points are averaged over the three numbers of data obtained from repeated experiments.

From the microscopic observation of the EDMed holes, it was noticed that there are irregularities in circularity of the micro holes. Current density and excessive tool wear may cause problems in the process of machining by deforming circularity and cylindricity. In this work, flushing is done by supplying dielectric fluid through a nozzle and through the preset up-down movement of the tool electrode. Some slag deposition may occur at the hole internal outline and this may cause modification in the hole circularity. The irregularity in the slag thickness may have occurred because of strong micro-explosions, which causes slag desegregation from the outline. Kaminski and Capuano [5] also noticed this fact and proposed that, because of strong micro-explosions occurring during machining, the vibration amplitude of the electrode increases, causing the lack of micro-hole circularity.

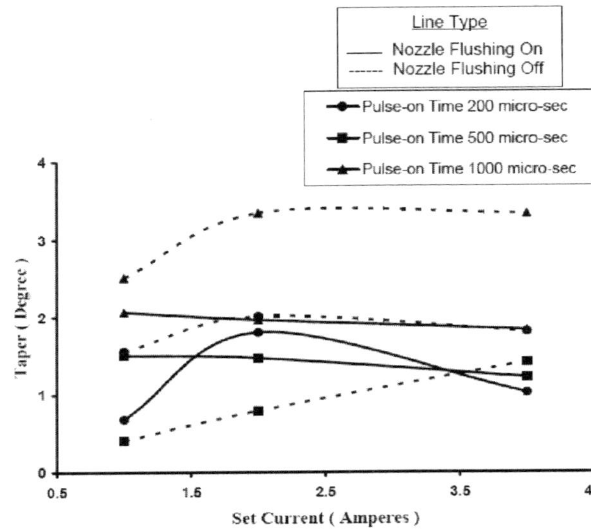

Figure 1.2. Effect of set current, pulse-on time and nozzle flushing condition on taper along the depth of the hole (Electrode diameter: ∅0.92 mm)

The taper along the depth of the hole has been measured for different machining conditions with a constant value of voltage and varying current and pulse-on time. The influence of the current on taper (measured in degrees) with varying pulse-on time and flushing condition (on/off) are shown in Figure 1.2. Also, the influence of the pulse-on time on taper with varying flushing condition and constant set current (1 Amp) are shown in Figure 1.4. Here, two different sets of electrodes with different diameters are used.

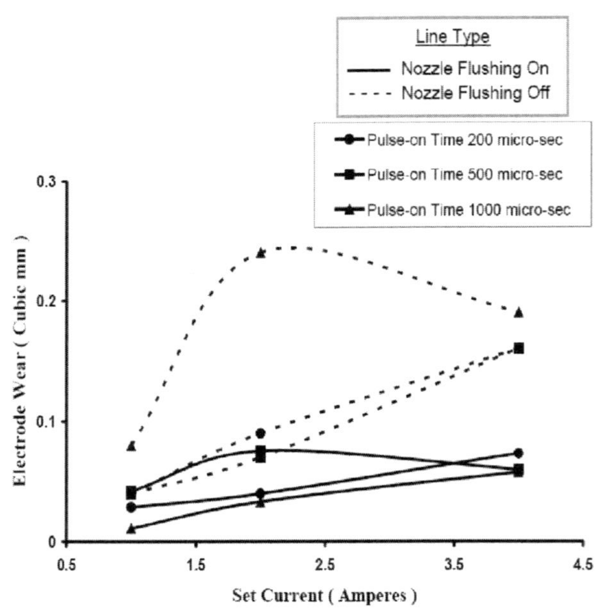

Figure 1.3. Effect of set current, pulse-on time and nozzle flushing condition on electrode wear (Electrode diameter: ∅0.92 mm)

The kerf along with the depth of the hole has been measured for different machining conditions with a constant value of voltage and varying current and pulse-on time. The influence of the current on kerf (measured in mm) with varying pulse-on time and flushing conditions are shown in Figure 1.6. Also, the effect of the pulse-on time on kerf with varying flushing conditions and constant set current of 1 Amp is shown in Figure 1.7. Here, two different sets of electrodes with 0.46 mm and 0.34 mm diameters are used.

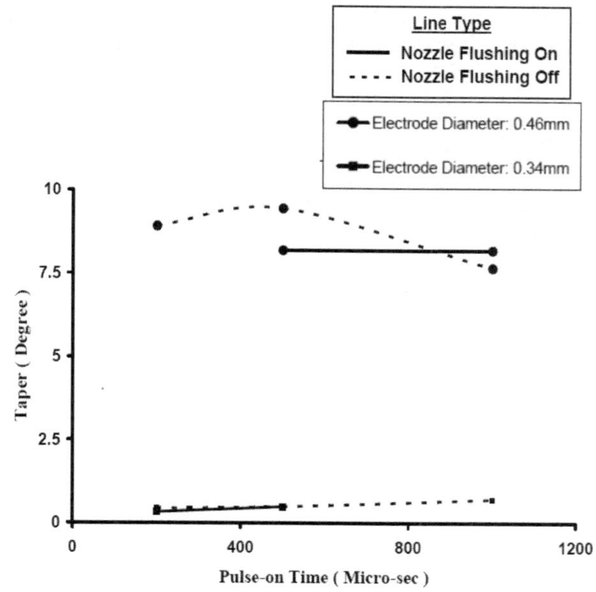

Figure 1.4. Effect of pulse-on time and nozzle flushing condition on taper along the depth of the hole for set current of 1 Amp

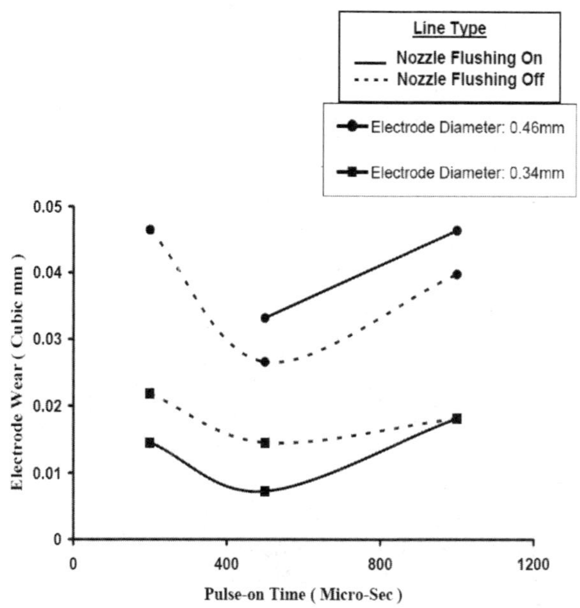

Figure 1.5. Effect of pulse-on time and nozzle flushing condition on electrode wear for set current on 1 Amp

The influence of the current on the electrode wear (measured in cubic mm) with varying pulse-on time and flushing condition (on/off) is shown in Figure 1.3. Also, influences of pulse-on time on electrode wear with varying flushing conditions and constant set current (1 Amp) are shown in Figure 1.5. In this case, two different sets of electrodes with different diameters are used.

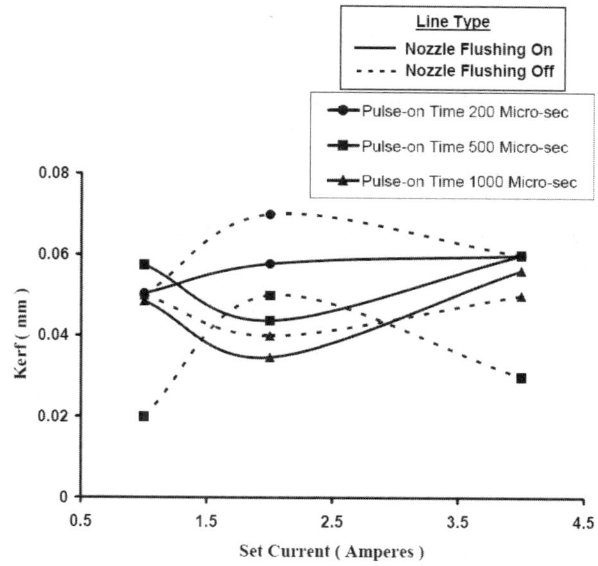

Figure 1.6. Effect of current, pulse-on time and nozzle flushing condition on kerf (Electrode diameter: Ø0.92 mm)

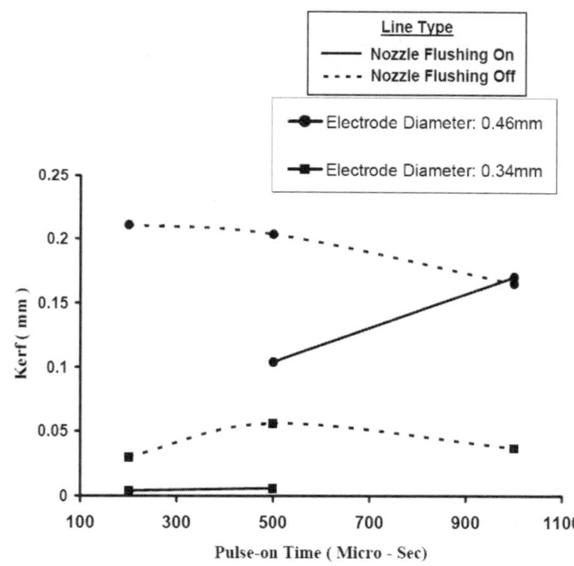

Figure 1.7. Effect of pulse-on time and nozzle flushing condition on kerf for the set current of 1 Amp

3.1 Discussion on Taper

The general tendency of taper (measured in degrees) generated when flushing through the nozzle is on has been seen (according to Figure 1.2 to be decreasing with the increase of set current (measured in Amperes). When flushing through nozzle is kept off, the tendency of taper has been seen to be increasing on the whole with the set current. It can be inferred that, there may be some effects of nozzle flushing on taper in this set of experiments. In Figure 1.2, it is observed that corresponding to pulse-on time of 200 microseconds, first, taper angle increases with the increase in set current from 1 Amp to 2 Amp, and with further increase in set current to 4 Amp, taper angle decreases considerably. The flow from the nozzle might have had problems to reach deep inside the EDMed hole, and improper flushing might have resulted from this.

Almost constant values of taper with respect to pulse-on time and nozzle flushing at constant set current of 1 Amp are seen in Figure 1.4 with two different sets of electrodes of different diameters (∅0.46 and ∅0.34). One hole was not formed properly in both cases (corresponding to electrode diameter ∅0.46 and pulse-on time 200 microseconds as well as corresponding to electrode diameter ∅0.34 and pulse-on time 1000 microseconds). This may be due to an alignment problem, which disabled the electrode to penetrate through the workpiece in the desired fashion. This may also be due to a lack of rigidity of the electrode that got bowed against the pressure of flushing coming out of the nozzle. Observations during the experiment suggested that these things took place.

3.2 Discussion on Electrode Wear

Figure. 1.4 shows that tool wear (measured in cubic mm) increases with the increase of set current irrespective of flushing conditions. Tool wear at the higher set current and lower pulse-on time was maximum. The figures on the whole prove the same thing. Kaminski and Capuano [5] also observed this to happen. Distortions from this are seen in the plots, and they could have been caused due to any type of instability occurred during machining, e.g. in case of the plot corresponding to pulse-on time of 1000 microsecond and without flushing (Figure 1.3), it shows a large deviation from the trend of the other two graphs. During the experiment with set current 2 Amp (without flushing), tendencies of arcing observed supports the inference made above.

Trends of electrode wear with respect to pulse-on time and nozzle flushing (constant set current (1 Amp)) are shown in Figure. 1.5 with two different sets of electrodes with different diameters (∅0.46 and ∅0.34). As one hole was not formed in case of electrode diameter ∅0.46, plots have been made from two available data points.

3.3 Discussion on Kerf

The general tendency of kerf (measured in mm) generated irrespective of flushing conditions (on/off) has been seen expectedly (Figures 1.6 and 1.7) to increase slightly with the increase in set current (measured in Amperes) with lower values of pulse-on time (measured in microsecond (μs)) in most cases. Difficulty in reaching the narrow hole by the dielectric fluid may have resulted in the deviation observed. In fact, the gap voltage controls the gap between the two electrodes in EDMing process, and this ultimately controls the spread of plasma, which is responsible for material removal by creating pock marks on the workpiece [5,6]. In addition, higher set current and lower pulse-on time results in high material removal rate (MRR) and this may cause higher kerf when the side wall of the wire electrode is also engaged in EDMing the workpiece. In these sets of experiments, it was attempted to set the gap voltage to a constant value.

Differences in overcut values, in entry and exit side, are observed to be different in every case. This might have been caused by the occasional change in gap voltage during the run. Tendency of instability in the EDMing operation may be the reason of this.

The tendency of kerf with respect to pulse-on time and nozzle flushing (constant set current) are shown in Figure 1.7 with two different sets of electrodes with different diameters (∅0.46 and ∅0.34). As one hole has not been formed in each case, so in every case, graph has been drawn from the available two results of each.

4. Conclusion

From the experiments carried out on micro-EDMing using a specially designed tool holders, following conclusions may be drawn;

- It has been found to be effective for machining micro-holes indicating the opportunity for its applications in manufacturing industry.
- To guarantee circularity of the hole, standardisation of the process with suitable cleaning method is needed.
- Formation of taper can be reduced with the selection of appropriate flushing technique.
- Rigidity of the electrode is to be enough for accurate hole making.

References

[1] Bleys, P., Kruth, J. P., et al. 2002, Real time tool wear compensation in milling EDM. Annals of the CIRP, Vol.51, No.1, P157-160

[2] Pham, D. T., Dimov, S.S., et al. 2004, Micro- EDM: Recent developments & research issues. Journal of Materials Processing Technology, Vol.149, P50-57

[3] Chatterjee, S., Das, S. 2005, Characterisation of micro-electric discharge machining. 20th Indian Engineering Congress, P68, December, Kolkata

[4] Higuchi, T., Furutani, K., et al. 1991, Development of pocket-size-discharge machine. Annals of the CIRP, Vol.40, No.1, P203-206

[5] Kaminski, P. C., Capuano, M. N., 2003, Micro hole machining by conventional penetration electrical discharge machine. International Journal of Machine Tools and Manufacture, Vol.43, No.11, P1143-1149

[6] Ho, K. H., Newman, S. T., 2003, State of the art electrical discharge machining EDM. International Journal of Machine Tools and Manufacture, Vol.43 P1287–1300

[7] Miller, S.F., Shih, A.J., et al. 2004, Investigation of the spark cycle on material removal rate in wire-electric discharge machining of advanced materials. International Journal of Machine Tools and Manufacture, Vol.44, P391-400

[8] Yan, B. H., Huang, F. Y., et al. 1999, Micro hole machining of carbide by electric discharge machining. Journal of Materials Processing Technology. Vol.87, P139-145

Development of *in situ* Monitoring and Control of Micro-EDM Process

S.H. Yeo[1], W. Kurnia[2], P.C. Tan[3], M. Mushan[4]

[1]Nanyang Technological University, Singapore 639798, mshyeo@ntu.edu.sg
[2-4]Nanyang Technological University, Singapore 639798

Abstract. Research work has shown much interest in the tool wear monitoring and the system control of micro-EDM process. Offline measurement, touch sensor, and laser sensor were used to obtain the tool length during machining, hence the wear. However, the time taken and the inaccuracy introduced by the offline method and electrode movement for the usage of sensors have become the major setback. A reliable process monitoring of the tool wear in real time is highly crucial since the requirements of the accuracy of the machined features are more demanding in the micro scale. This paper proposed a development of online monitoring of the tool wear in micro-EDM process based on the crater prediction using a developed theoretical model. The TWR approximation value is then used to determine the compensation required in the Z-Axis direction to improve the accuracy in blind hole fabrication. The verification are performed experimentally by comparing the machining results with and without compensation. The presented results show that this method is able to improve the error within 1% to 7.6% of the hole depth setting.

Keywords: Micro-EDM, Tool Wear Ratio, Tool Wear Compensation

1. Introduction

Micro-EDM is the process where the material is removed by the thermal energy created between the tool and workpiece. Since the material is also removed from the electrode, the tool will eventually wear out after machining. This will introduce machining inaccuracies and minute imperfections are magnified in the micro and sub micro scale. The electrode size and shape will change progressively during machining, resulting in a shorter hole depth than the depth setting and a tapered hole. Hence, it is necessary to monitoring strategy to compensate any tool wear.

The measurements of tool wear were done by moving the electrode from the workpiece to the reference point where a touch or laser range sensor is used to measure the tool length (i.e. end wear) [1, 2]. However, the time taken and the inaccuracy introduced during the electrode movement have become the major setback for this method. Dauw [3] and Bleys et al. [4] used a mathematical model of the sparking efficiency to estimate the tool wear online based on the monitoring of the sparking condition. Unfortunately, the development in the micro-EDM process is still limited due to the accuracy involved.

In order to compensate for the tool wear effect on the accuracy, Bleys et al. [4, 5] used the tool wear approximation results based on the pulse analysis for the anticipated down-movement compensation. Yu et al. [6] used a very small layer thickness for the layer-by-layer machining path to ensure that the wear only occurred at the face of the electrode and the electrode shape is restored after each layer. A theoretical model to determine the tool path is reported [7]. Pham et al. [8] proposed the use of multiple electrode strategy using a number of different milling paths by the redressed electrode after each path.

In this paper, an online monitoring of the tool wear in micro-EDM process is developed based on the crater geometry calculation using a developed theoretical model. The model is used to calculate the temperature distribution of the electrode and workpiece using the real time measurement data of voltage and current from the discharge gap. The craters are numerically calculated and used to define the tool wear ratio (TWR). The TWR results are then used as the input compensation of the tool wear in the Z-axis direction. To verify the tool wear compensation method, experimental work is carried out with and without the compensation. A comparative analysis is then performed to study the results.

2. Numerical Methodology

For the crater calculation, a two dimensional heat transfer model for a material subjected to a disk heat source on its surface is used. The energy is considered to be uniform over the heat flux radius. The heat flux radius for the anode and cathode for the discharge energy range of 5 μJ to 150 μJ are determined as follows [9]:

$$R_{anode} = 0.0284.t_d^{\ 0.9115}\ [\mu m] \tag{1.1}$$

$$R_{cathode} = 0.0425.t_d^{\ 0.0895}\ [\mu m] \tag{1.2}$$

Where, R_{anode} and $R_{cathode}$ are the heat flux radius at anode and cathode respectively, and t_d is the pulse-on-time in nanoseconds.

By introducing dimensionless variables $u = \frac{r}{R}$, $w = \frac{z}{R}$, $\tau = \frac{\alpha.t_d}{R^2}$, and $\theta(u,w,\tau) = \frac{K_t.T(r,z,t)}{q.R}$; the temperature distribution solution in the vertical axis is given as follows [10]:

$$\theta(0,w,\tau)=2\sqrt{\frac{\tau}{\pi}}\exp\left(-\frac{w^2}{4\tau}\right)\left[1-\exp\left(-\frac{1}{4\tau}\right)\right] + \sqrt{w^2+1}\,erfc\left(\frac{\sqrt{w^2+1}}{2\sqrt{\tau}}\right)-w\,erfc\left(\frac{w}{2\sqrt{\tau}}\right) \quad (1.3)$$

Where r is the radial distance, z is the vertical distance, R is the heat flux radius of the anode or cathode, α is the average thermal diffusivity of the material [m^2/s], t_d is the pulse-on-time [s], K_t is the average thermal conductivity [W/m.K], *erfc* is the complementary error function, and the heat flux q is given by:

$$q=\frac{C.V_a.I_a}{\pi.R^2} \quad (1.4)$$

Here, V_a is the average voltage during pulse-on-time [V], I_a is the average current during pulse-on-time [A], and C is the combination of the fraction of energy transferred and fraction of molten area for the anode or cathode. The C value for the anode and cathode are found to be C_a=0.39 and C_c=0.14 respectively [9].

The temperature distribution solution in the radial axis is given by:

$$\theta(u,0,\tau)=I_1-I_2 \quad (1.5)$$

The first integral of equation (1.5) is a steady state term while the second integral is the transient term. An exact solution of the first integral is given by [11]:

$$I_1=\begin{cases}\frac{2}{\pi}E(u) & ;\ 0<u<1\\ \frac{2}{\pi} & ;\ u=1\\ \frac{2u}{\pi}\left[E(u^{-1})-(1-u^{-2})K(u^{-1})\right] & ;\ u>1\end{cases} \quad (1.6)$$

Here, K and E are complete elliptic integrals of the first and second kind respectively.

Following the approach taken by Beck [12], the solution for the transient term is given as:

$$I_2=\frac{1}{2\sqrt{\pi\tau}}\sum_{i=0}^{\infty}\left\{\frac{(-1)^i i!}{(2i+1)(4\tau)^i}\sum_{j=1}^{i+1}(i-j+2)\left[\frac{u^{j-1}}{(j-1)!(i-j+2)}\right]^2\right\} \quad (1.7)$$

Combining the calculation results of (1.6) and (1.7) into (1.5), the dimensionless temperature distribution along the radial axis of the material can be calculated.

The boundary where the material will be expelled from the superheated and molten area is defined as the crater's edge temperature, which is given as:

$$\theta_{crater}=\frac{K_t(T_m-T_0)}{q.R} \quad (1.8)$$

Here, T_0 is the ambient temperature [K].

The vertical and radial temperature distributions are calculated by using equation (1.3) and (1.5), which are plotted together with the crater's edge temperature (θ_{crater}) to obtain the values of u and w. These values are then used to approximate the diameter (d), depth (h) and volume (V_{crater}) of the micro crater where the crater volume is given as:

$$V_{crater}=\frac{\pi d^2 h}{8}+\frac{\pi h^3}{6} \quad (1.9)$$

3. Experimental Setup

The experimental work was conducted using a commercially available micro-EDM system, MG-E72W, as the platform. AISI 4140 alloy steel was used as the workpiece with the dimension of 10 mm cube. As for the electrode, tungsten rod with the diameter of 300 μm was used. During machining, the voltage and current between the electrode and the workpiece were measured using the Tektronix P5205 voltage probe and Tektronix CT-1 current probe. The data waveform was captured for the anode and cathode crater volume calculation. Figure 1.1 illustrates the basic experimental setup used in the experimental work. The properties of the workpiece and electrode used in the experiments are provided in Table 1.1.

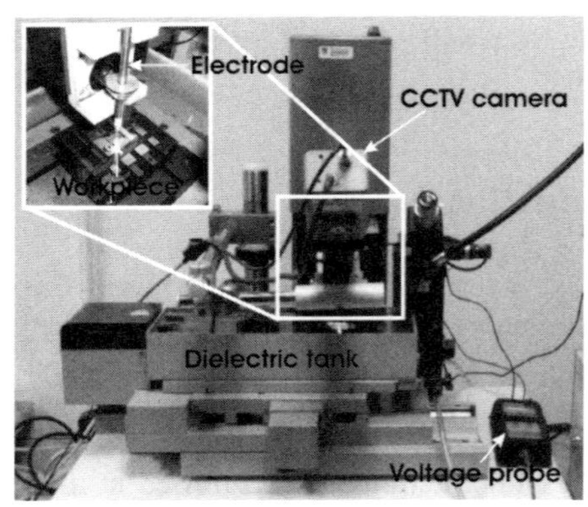

Figure 1.1. Basic experimental setup

Table 1.1. The physical properties of AISI 4140 and tungsten

Properties	AISI 4140	Tungsten
Average density	7600 kg.m^3	19300 kg.m^3
Average thermal conductivity	38.95 W/m.K	130.2 W/m.K
Average specific heat	473 J/kg.K	141.3 J/kg.K
Melting temperature	1688 K	3387 K

The experiments were conducted to verify the analytical approximation of the tool wear monitoring method as compared to the experimental TWR. After machining, the tool electrode and workpiece hole profiles were measured to obtain the volume removed from both materials for the experimental TWR calculation. Five replications were used for each parameter setting to ensure the reliability of the data. Table 1.2 shows the process parameters used in the experiments.

Table 1.2. Experimental parameters

Workpiece	AISI 4140
Electrode	Tungsten
Dielectric fluid	Synthetic oil (Daphne Cut HL-25)
Voltage input	110 Volts
Capacitor	3300 pF
Polarity (workpiece)	Positive
Depth setting	250/500/750/1000 µm

The analytical TWR results were then used along with the workpiece volume to be removed for each depth setting to determine the necessary wear compensation in the Z-Axis direction. The wear was compensated by correcting the initial depth setting (Z_i) with the tool end wear (Z_{ew}), resulting in the programmed depth setting $Z_p = Z_i + Z_{ew}$. The experiments were conducted using similar experimental setting to study the performance of the compensation method.

4. Results and Discussion

4.1 Verification of the Tool Wear In-situ Monitoring

In this section, the analytical TWR approximation using the crater geometry calculation results is compared with the experimental TWR. Based on the crater calculation results, the analytical TWR is determined according to the following:

$$\text{Analytical TWR} = C_{TWR}\frac{V_{el}}{V_{wp}} \qquad (1.10)$$

Where V_{el} is the crater volume removed from the electrode [µm^3], V_{wp} is the crater volume removed from the workpiece [µm^3], and C_{TWR} is the correction factor. The correction factor is used to take into account the effect of using tungsten as the electrode since cobalt, which has low melting temperature, is used as the binder. Therefore, the use of the melting temperature value during the electrode crater calculation might have to be compensated by the existence of cobalt, increasing the portion which will be removed from the electrode. Based on the experimental results, the C_{TWR} is found to be 1.782.

The occurrence of tool wear on the electrode after the experiments is illustrated in Figure 1.2. The total volume removed from the electrode is calculated based on the shape difference between the initial and final tool. The ratio of this value to the volume of material removed from the workpiece is defined as the experimental TWR.

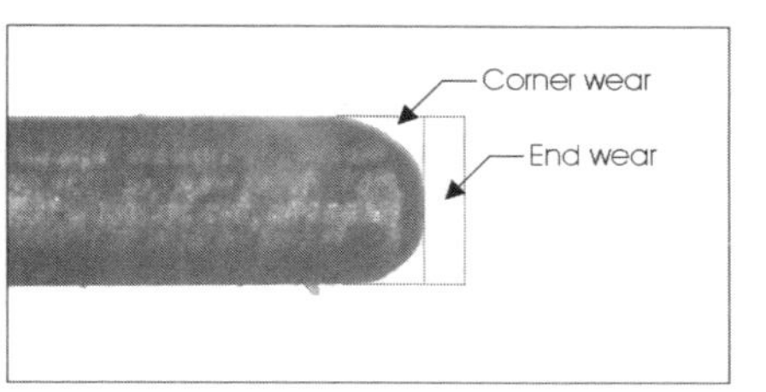

Figure 1.2. Electrode tool wear. Depth setting: 1000 µm

The comparison between the experimental and analytical TWR is depicted in Figure 1.3. Although the plot seems to vary with the depth setting, the variations from 250 µm to 1000 µm depth setting are small and can be considered to be fairly constant. As compared to the experimental results, the analytical results are lower by 15% and 8% at 250 and 750 µm depth setting, respectively. However, the analytical results overestimate the experimental values by 20% and 24% at 500 and 1000 µm depth setting, respectively. Based on the presented results, the proposed model for the micro-EDM in-situ tool wear monitoring is able to provide a close approximation to the experimental results.

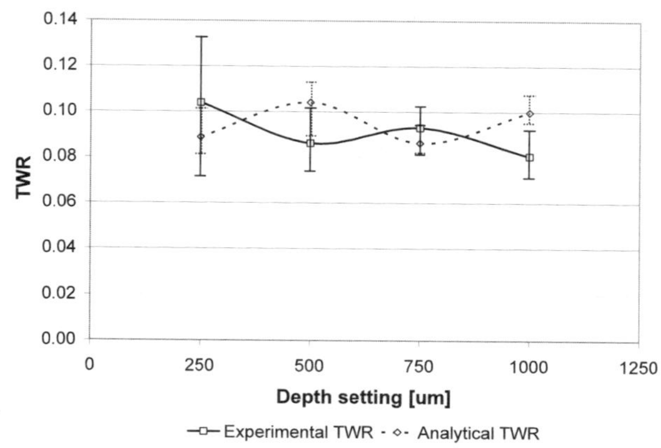

Figure 1.3. Comparative analysis between the experimental and analytical TWR

4.2 Tool Wear Compensation

Based on the analytical TWR calculation results, the Z-axis compensation is applied to the micro-EDM process. Due to the complexity of the tool wear shape on the tool wear, the compensation is calculated by assuming that the wear only occurs at the end of the electrode and the corner wear is neglected. The typical machining results of this method are illustrates in Figure 1.4.

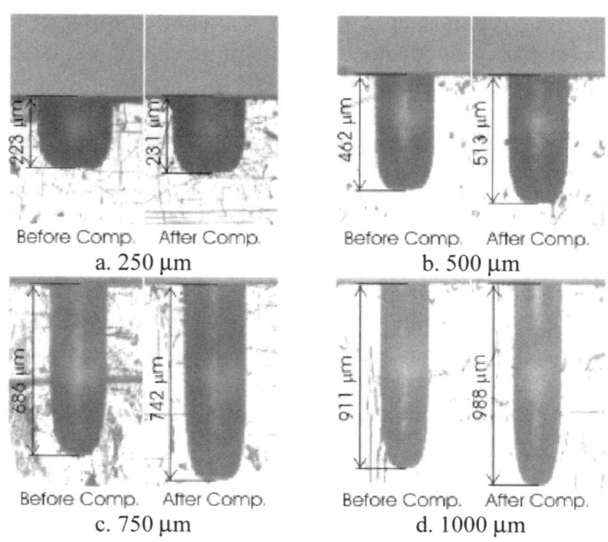

Figure 1.4. Compensation results for various depth setting

Based on the experimental results, the compensation method is able to provide a better accuracy in terms of the hole depth as compared to the original process (see Figure 1.5). The original process results in 7.5% to 10.6% error in terms of the final hole depth. By using the Z-Axis compensation, the errors are improved to be within 1% to 7.6%.

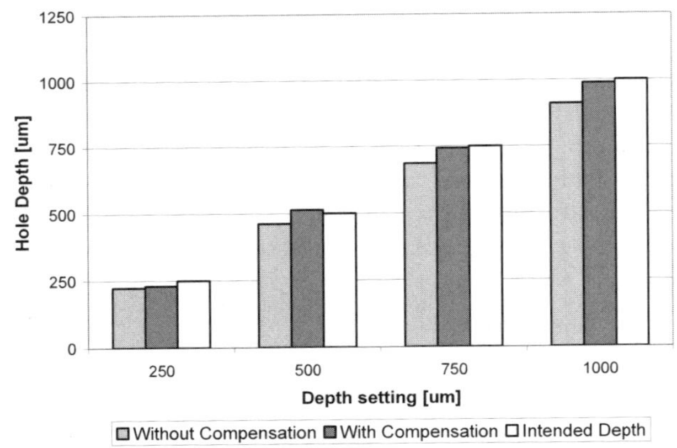

Figure 1.5. Comparative analysis between the original process and the compensated machining results

Although the compensation method are able to improve the error by more than 28% relative to the original process, the machining results still differ by approximately 7.5 μm to 19 μm as compared to the intended hole depth. The plausible reason for this is due to the effect of side over-cut. During the calculation of Z-axis compensation, the hole diameter is assumed to be similar with the electrode diameter. However, based on the experimental results, the hole diameter is larger by approximately 6.3% (i.e. 19 μm) due to over-cut. By taking this factor into account, the compensation method will be able to give smaller error. The future work will emphasize on the improvement of this method by taking into account the effect of corner wear and side over cut.

5. Conclusion

An online monitoring of the tool wear in the micro-EDM process has been developed based on the crater prediction using a developed theoretical model. The craters from the electrode and workpiece are numerically calculated and used to define the tool wear ratio (TWR). The TWR results are then used in the calculation of the necessary tool wear compensation in the Z-axis direction. The experimental work is conducted by varying the hole depth setting to verify the tool wear monitoring and the Z-axis compensation method.

The comparison between the experimental and analytical TWR shows that the proposed in-situ tool wear monitoring method is able to provide a close approximation to the experimental results. The application of the compensation method using the analytical TWR results presents a better accuracy in terms of the hole depth as compared to the original process. The original process results in 7.5% to 10.6% errors in terms of the final hole depth. By using the Z-axis compensation, the errors are improved to be within 1% to 7.6%.

References

[1] Kaneko, T. and M. Tsuchiya, Three-dimensional numerically controlled contouring by electric discharge machining with compensation for the deformation of cylindrical tool electrodes. Precision Engineering, 1988. 10(3): p. 157-163.

[2] Mizugaki, Y. Contouring electrical discharge machining with on-machine measuring and dressing of a cylindrical graphite electrode. in Proceedings of the 1996 IEEE 22nd International Conference on Industrial Electronics, Control, and Instrumentation, IECON. Part 3 (of 3), Aug 5-10 1996. 1996. Taipei, Taiwan: IEEE, Los Alamitos, CA, USA.

[3] Dauw, D. On the derivation and application of a real-time tool wear sensor in EDM. 1986. Jerusalem, Isr: Verlag Technische Rundschau, Bern, Switz.

[4] Bleys, P., J.-P. Kruth, B. Lauwers, A. Zryd, R. Delpretti and C. Tricarico, Real-time tool wear compensation in milling EDM. CIRP Annals - Manufacturing Technology, 2002. 51(1): p. 157-160.

[5] Bleys, P., J.-P. Kruth and B. Lauwers, Sensing and compensation of tool wear in milling EDM. Journal of Materials Processing Technology, 2004. 149(1-3): p. 139-146.

[6] Yu, Z.Y., T. Masuzawa and M. Fujino, Micro-EDM for three-dimensional cavities - development of uniform wear method. CIRP Annals - Manufacturing Technology, 1998. 47(1): p. 169-172.

[7] Narasimhan, J., Z. Yu and K.P. Rajurkar, Tool wear compensation and path generation in micro and macro EDM. Papers Presented at NAMRC 32, Jun 1-4 2004, 2004. 32: p. 151-158.

[8] Pham, D.T., S.S. Dimov, S. Bigot, A. Ivanov and K. Popov, Micro-EDM - Recent developments and research issues. Journal of Materials Processing Technology, 2004. 149(1-3): p. 50-57.

[9] Yeo, S.H., W. Kurnia and P.C. Tan, Electro thermal modelling of anode and cathode in micro-EDM. Journal of Physics D: Applied Physics, 2007 [Accepted for publication].

[10] Carslaw, H.S. and J.C. Jaeger, Conduction of heat in solids. 2nd ed. 1986: Clarendon Press.

[11] Thomas, P.H., Some conduction problems in the heating of small areas on large solids. Quarterly Journal of Mechanics and Applied Mathematics, 1957. 10: p. 482-493.

[12] Beck, J.V., Large time solutions for temperatures in a semi-infinite body with a disk heat source. International Journal of Heat and Mass Transfer, 1981. 24(1): p. 155-64.

Improvement of Roundness of Cylindrical Parts Using Hybrid Electrochemical Smoothing and Roller Burnishing Process

T.A. El-Taweel[1], and S.J. Ebeid[2]

[1] Menoufia University, Faculty of Engineering, Shebin El-Kom, Egypt, tahaeltaweel@yahoo.com
[2] Ain Shams University, Faculty of Engineering, Cairo, Egypt, drsami@menanet.net

Abstract. A novel finishing process, which integrates the merits of electrochemical smoothing (ECS) and roller burnishing (RB) is proposed in this research for improving the roundness error of cylindrical parts. This process can be used after various turning operations. Through simple equipment attachments, electrochemical smoothing-roller burnishing (ECS-RB) can follow the machining process on the same turning machine. To explore the optimum combinations of the ECS-RB process parameters in an efficient and quantitative manner, the experiments were designed on the basis of the response surface methodology (RSM) technique. The effect of ECS-RB parameters namely, burnishing force, applied voltage, inter-electrode gap and workpiece rotational speed on the roundness error was studied. Experimental results indicate that the combined process effectively improves the roundness error. Therefore, the combination of ECS and RB is a feasible process by which it is possible to minimize geometrical errors.

Keywords: Electrochemical smoothing, Roller burnishing, Hybrid machining process, Roundness error, Response Surface Methodology

1. Introduction

The surface of engineering parts is a significant design specification that is known to have considerable influence on properties such as wear resistance and fatigue strength. Surface roughness and roundness play an important role on the required tolerance and fit especially during part assembly. Processes by which surfaces are finished differ in their capabilities concerning finishing action, mechanical and thermal damage, and residual stresses.

Recent hybridization techniques are applied to electrochemical machining (ECM) to improve its performance; the most adequate of these are a combination with EDM, USM and recently the latest is laser-ECM [1-3]. The reasons for developing hybrid machining/finishing processes are to make use of the combined or mutually enhanced advantages, and to avoid or reduce some adverse effects that the constituent processes produce when they are individually applied. In mechanical hybridization the mechanical interaction with the workpiece material is able to change the conditions of the anodic dissolution process. The main factor to be considered is the mechanical de-passivation of the surface, through which thin layers of oxides and other compounds are removed from the anode, thus causing the surface dissolution and the smoothing process more intensively. The mechanical-chemical effect, that is present in micro-cutting processes, consists of changes in chemical potential at places where dislocations occur and results in plastic deformations, which are caused by the mechanical action. For example electrochemical grinding (ECG) where the material removal phenomenon consists of two different processes, electrochemical and mechanical actions has been studied [4].

The ECM of aluminium alloys usually reduces their fatigue strength. This reduction may be explained by the absence of a strain-hardened layer, electrolytic erosion and hydrogen pick-up [5]. The fatigue properties can be restored if surface strengthening follows ECM. Electrochemical turning and roller burnishing processes have been combined together to produce a high-quality surface finish for metals. Ebeid and El-Taweel [6] investigated the feasibility of improving surface quality via a novel combined process of electrochemical turning (ECT) with roller burnishing (RB). This combined process effectively improved the surface finish, imparting compressive residual stress to a surface, and eliminates the pitting caused by ECT. This, in turn, improved the fatigue structural integrity of components.

In the present work, a novel hyprid finishing process, which integrates the merits of ECS and RB is proposed for improving the roundness error of Al-Zn-Mg alloy cylindrical parts. To explore the optimum combinations of the ECS-RB process parameters in an efficient and quantitative manner, the experiments were designed on the basis of the response surface methodology (RSM) technique. The effect of ECS-RB parameters namely, burnishing force (F), applied voltage (V), inter-electrode gap (S) and workpiece rotational speed (N) on the roundness error (RO) of cylinrical parts was studied.

2. Experimental work

2.1 Electrochemical turning-roller burnishing system

The schematic diagram of a hybrid electrochemical smoothing-roller burnishing (ECS-RB) system is shown in Fig. 1.1. The effective tool as shown in Fig. 1.2 is attached symmetrically to two non-conductive ZrO_2 rollers (18 mm outside diameter, 22 mm width). The effective tool has been correctly positioned and fixed on the center lathe tool post. Adjusting the ZrO_2 rollers controls the inter-electrode gap between the electrode surface and the workpiece surface (S). Therefore, the combined process has both the ECS ability by arc cathode surface and the RB to generate a plastic deformation layer owing to the smooth ZrO_2 rollers that burnish the ECS surface directly.

Table 1.1. ECS-RB machining conditions

Working condition	Description
Workpiece material	Al-Zn-Mg alloy
Tool material	Stainless steel
Applied voltage (V)	8–24 volt
Electrolyte pressure (P)	0.3 MPa
Rotational speed (N)	300–1100 rpm
Inter-electrode gap (S)	0.1-0.5 mm
Burnishing force (F*b*)	100 - 500 N
Electrolyte	20 % $NaNO_3$
Electrolyte flow rate	5.2 lit/min
Machining time (t)	5 min

The burnishing force was calibrated and controlled through a specially designed half-bridge strain gauge dynamometer. Al-Zn-Mg aluminum alloy was selected as the workpiece material in these experiments. Generally, this alloy is widely used in the aerospace industry. The experimental conditions of ECS-RB machining conditions are summarized in Table 1.1. The geometrical accuracy of components was tested by measuring the roundness value using the "MITUTOYO, RA-100, RE-1 Round Test" measuring instrument. For obtaining the roundness values three measurements were performed in each cut, and peak-to-valley values (maximum deviation) were used as an overall measure of the roundness error.

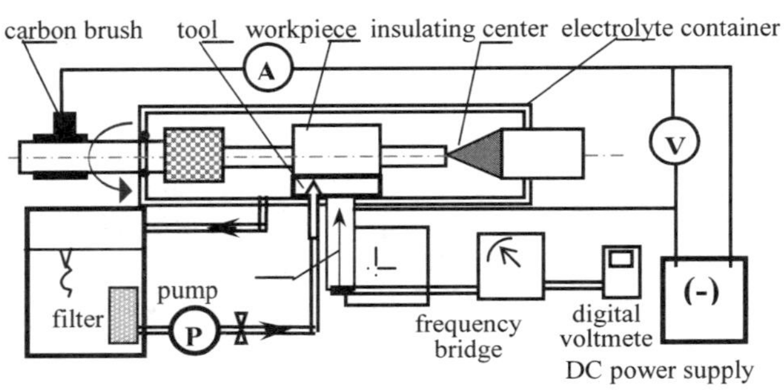

Fig. 1.1. Schematic of hybrid ECS-RB system

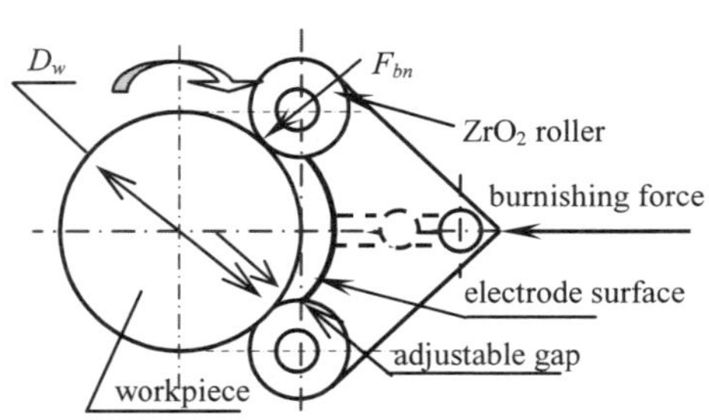

Fig. 1.2. Tool-workpiece

2.2 Design of experiments

With a proper designed experiment such as central composite design, it is possible to determine, with a much-reduced number of experiments, the effect of changing any one variable with the same accuracy as if only one factor has been varied at a time, and the interaction effects between the factors [4,8]. The present investigation studied the effects of burnishing force (F), applied voltage (V), inter-electrode gap (S) and rotational speed (N) on the roundness error (RO) during the ECS-RB. The experiments were designed on the basis of the response surface methodology (RSM) technique. A 2^k factorial with central composite-second-order ratable design was used (in this case k = 4). This consists of $n_c = 2^k = 16$ corner points at +1 level, $n_a = 2k = 8$ axial points at $\gamma = +2$, and a centre point at zero level repeated 5 times to estimate the pure error [9].

3. Mathematical modeling

The general second order polynomial response surface mathematical model, which is given below, is considered to determine the parametric influences on the various response criteria.

$$Y_u = b_o + \sum_{i=1}^{k} b_i X_{iu} + \sum_{i=1}^{k} b_{ii} X_{iu}^2 + \sum_{j>i}^{k} b_{ij} X_{iu} X_{ju} + e_u \quad (1)$$

where, Y_u is the corresponding response, X_{iu} is the coded values of the ith process parameters for the uth experiment, n is the numbers of machining parameters, b_i, b_{ii} and b_{ij} are second order regression coefficients and e_u is the experimental error of the uth observation.

Equation 1 has been evaluated and the mathematical model based on the RSM has been developed by utilizing test results of response obtained through the entire set of experiments by using the computer software MINITAB-13. The response surface methodology based analysis has been done to establish the mathematical model through the development of mathematical relationship between roundness and the important process parameters, burnishing force (X_1), applied voltage (X_2), inter-electrode gap (X_3) and rotational speed (X_4). Based on the roundness test results obtained from the planned experiments, the values of the different constants of the Eqn. 1 are obtained. The mathematical relationship for correlating the roundness and the considered working parameters has been established as follows:

$$RO = 3.01 - 0.156 X_1 + 0.744 X_2 - 0.458 X_3 - 0.432 X_4 + 0.187 X_1^2 + 0.129 X_2^2 - 0.135 X_3^2 - 0.114 X_1X_2 + 0.36 X_1X_3 - 0.088 X_1X_4 - 0.08 X_2X_3 + 0.055 X_2X_4 + 0.063 X_3X_4 \quad (2)$$

This mathematical model has been obtained to reflect the independent, quadratic and interactive effects of the various process parameters on the RO in ECS-RB process. It can be noted from the above final equation that some coefficients were omitted. These coefficients are non-significant according to Students t-test. The analysis of variance (ANOVA) for the final model indicates that the adequacy of this model was established. The calculated values of F-ratio for the lack of fit are found to be lesser than the standard value of the F-ratio for RO. The second order regression model is adequate at 99.3% confidence level to represent the relationship between RO and process parameters of the model, which correlates the various process parameters with RO, can adequately be represented through the RSM for ECS-RB operation.

4. Experimental results and discussion

4.1 Effect of working parameters on roundness

The effect of burnishing force on roundness error at various applied voltages, inter-electrode gaps, and workpiece rotational speeds can be assessed from Figs 1.3 to 1.5. It is expected that the ECS-RB process causes a change in the roundness of finished parts. It can generally be seen from these figures that an increase in burnishing force leads to a decrease in roundness to a certain limit (300 N) at all levels of applied voltage, inter-electrode gap, and workpiece rotational speed. This is due to, firstly, the deforming action of the ZrO_2 rollers, which increases with increasing burnishing force, secondly, the increase of the ZrO_2 rollers pressure on the workpiece surface, compressing the most asperities, and, thirdly, the increasing metal flow that causes the filling of more voids and/or valleys that exited in the subsurface layer due to electrochemical action. Beyond the limit of 300 N, the increase in burnishing force causes an increase in roundness. The repeating plastic deformation action on the metallic surface, as the burnishing rollers passes along the workpiece surface, will lead to an increase in work-hardening of the already affected layers of the deformed surface. Accordingly, when the applied burnishing force exceeds a certain limit, and because of the high amount of work-hardening that is developed in the affected metallic surface layers, flaking will occur on the surface. This leads to a deterioration of the workpiece surface, which will in turn cause an increase in the roundness error.

Applied voltage is found to be the most significant parameter influencing the RO. It is clear that as the applied voltage increases the roundness error increases (Fig. 1.3). This is because the effect of applied voltage is related to the current density and the inter-electrode gap [10]. Any increase in voltage increases the current density in the gap between electrode surface and workpiece surface, which in turn enhances the anodic dissolution of the work material. This results in a higher material removal rate and produces a larger roundness error. The deterioration in the roundness error is attributed partly to the stray current effect. A combination of high burnishing force with high-applied voltage leads to a substantial increase in roundness deviation.

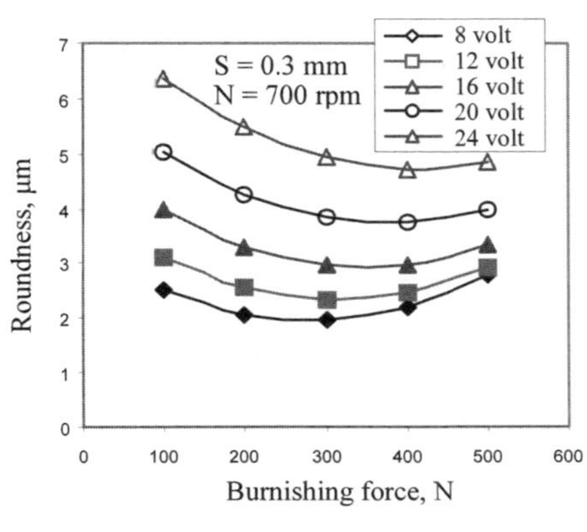

Fig. 1.3. Effect of burnishing force on roundness at different applied voltages

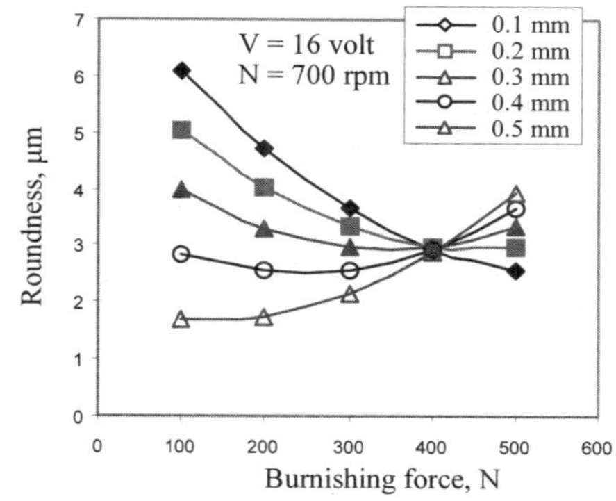

Fig. 1.4. Effect of burnishing force on roundness at different inter-electrode gaps

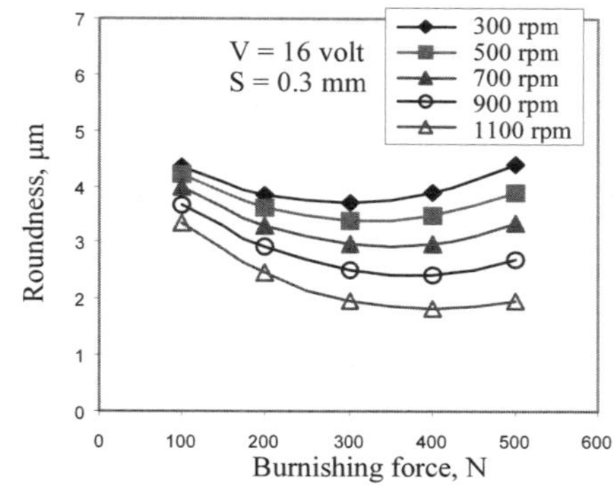

Fig. 1.5. Effect of burnishing force on roundness at different rotational speeds

There is an interaction between burnishing force and inter-electrode gap, as shown in Fig. 1.4. It can be realized that the combination between a low burnishing force and a high inter-electrode gap value results in a considerable reduction in machined roundness error. This is because repeating the ECS-RB process on the same workpiece at low burnishing force leads to an increase in the surface structural homogeneity, which results in an improvement in the roundness error. When carrying out the ECS-RB process at a low value of the inter-electrode gap, where the current density and hence stray current attack increases, an increase in the burnishing force leads to an increase in the roundness error.

From Fig. 1.5 it is also clear that as the rotational speed increases the roundness error decreases. In this combined process, the ZrO_2 rollers produce a force, which compresses the projection peaks into the indentation valleys, thus forming a smooth finished surface. The nature of the ECS process provides a second explanation. When the rotational speed is high under certain inter-electrode gap, a smaller thickness is removed per revolution. When the process stops, a step with a hight of ΔR will remain at that position where the electrode was at the particular moment. This step extends over the entire width of the electrode.

Figure 1.6 shows the effect of the inter-electrode gap on the roundness error at various rotational speeds. Increasing the inter-electrode gap value within the permissible range (matching with dissolution rate of the work material) results in a lower roundness error. Increasing the inter-electrode gap value from 0.1 mm to 0.5 mm improved the roundness error by about 1.79 times.

The reason for this is follows: A further increase in the inter-electrode gap decreases the current density in the gap zone and, therefore, results in a minimum stray current effect with smoother current density, which reduces the roundness error. Figure 1.7 displays a roundness profile comparison of the finished surfaces by ECS and hybrid ECT-RB processes. This figure demonstrates that the surface roundness of the ECS cross-section profile improved substantially by combining ECS and RB, due to the fact that the ZrO_2 roller inserted into the tool electrode contacted the machined surface directly.

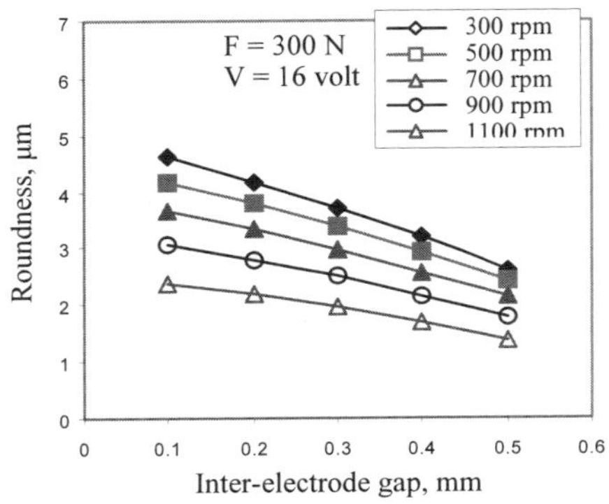

Fig. 1.6. Effect of inter-electrode gap on roundness at different rotational speeds

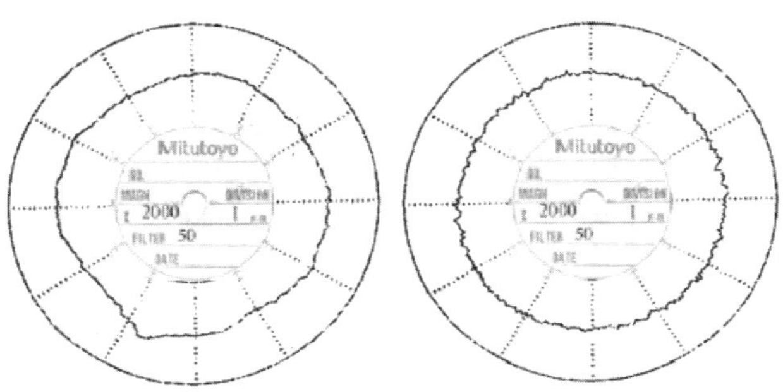

(a) ECS, RO = 6.14 µm (b) ECS-RB, RO = 1.17 µm

Fig. 1.7. Roundness profile comparison of the finished surfaces by ECS and ECS-RB processes (run No. 22)

4.2. Optimization and confirmation experiment

An optimization technique of RSM based on the developed second order response surface equation can be obtained. The Steepest Ascent Optimization technique is applied for moving sequentially along the path of steepest accent, that is, in the direction along the optimum values of the responses. Usually, the path of Steepest Ascent is the line through the center of region of interest and normal to the fitted response surface [4]. This combination has been evaluated with the help of MINITAB. In this study, after determining the optimum conditions and predicting the response under these conditions, a new experiment was designed and conducted with the optimum levels of the process parameters, to predict and verify the improvement of the RO. The predicted roundness error (0.46 µm) was compared with the observed roundness error (0.52 µm) and a good agreement with about 11 % deviation was obtained. Based on the experimental confirmation the improvement of the RO error from the initial process parameters to the optimal process parameter combination is about 65%. Hence, the geometrical error is improved significantly by using the optimal condition. The experimental results confirm the effectiveness of response surface model for optimal finishing parameters using the ECS-RB process.

5. Conclusions

The proposed combined electrochemical turning-roller burnishing process has been found to offer substantial gain in the geometrical errors of machined components. Imparting roller burnishing with a burnishing force of up to 300 N substantially improves the roundness error compared to electrochemical smoothing. Applied voltage has a significant effect on the roundness error. Higher applied voltages yield higher roundness errors. An increase in inter-electrode gap yields better roundness error. Experimental results have indicated that the interaction between burnishing force an inter-electrode gap have a significant effect on the roundness error. The geometrical error is improved by 65 % using the optimal conditions compared to the initial process parameters.

References

[1] J. Kozak, and K.P. Rajurkar, 2000, Hybrid machining process evaluation and development. Keynote Paper, Proc. 2nd Int. Conf. on Machining and Measurements of Sculptured Surfaces, Poland, P501-536.

[2] J. Kozak and K. E. Oczoś. 2001, Selected problems of abrasive hybrid machining, J Mater Process Technol, Volume 109(3), P360-366.

[3] P.T. Pajak, A.M. De Silva, et al. 2006, Precision and efficiency of laser assisted jet electrochemical machining, Precision Engineering, 30 (3), P288-298.

[4] T.A. EL-Taweel, 2007, Modelling and analysis of hybrid electrochemical turning-magnetic abrasive finishing of 6061 Al/Al_2O_3 composite, Int J Adv Manuf Technol, Accepted for publication.

[5] H. Luo, J. Liu, et al. 2005, Investigation of the burnishing process with PCD tool on non-ferrous metals, Int J Adv Manuf Technol, 25, P454-459.

[6] S. J. Ebeid and T. A. El-Taweel. 2005, Surface improvement through hybridization of electrochemical turning and roller burnishing based on the Taguchi technique, Proc. IMechE Part B: J. Engineering Manufacture, Vol. 219, P423-430.

[7] E. Rumyantsev, and A. Davydov, 1989, Electrochemical machining of metals. Mir Publication, Moscow.

[8] A. Hofsted, and J. Brekel. 1970, Some remarks on electrochemical turning. CIRP, XVIII, P93-106.

[9] D. C. Montgomery, 2001, Design and analysis of experiments, Wiley, New York.

[10] J.A. McGeough. 1988, Advanced Methods of Machining, Chapman and Hall.

Workpiece Surface Hardness as an Indicator of Process Regime in Peripheral Electrochemical Grinding

J. Atkinson

The University of Manchester, School of Mechanical, Aerospace and Civil Engineering, Manchester, M60 1QD, UK
john.atkinson@manchester.ac.uk

Abstract. Peripheral electrochemical grinding can operate under four main regimes depending on the process settings. The operating regime is determined by the relative contributions of electrochemical dissolution and mechanical abrasion, which can be controlled by the level of applied voltage. Current, overcut and surface roughness have all been shown to be indicators of the transition voltages between operating regimes. The present paper presents experimental evidence to show how surface hardness changes after machining AISI 01 tool and die steel in the different regimes. A marked change in surface hardness is seen to occur between the totally abrasive regime and the totally electrochemical regime.

Keywords: ECM, Electrochemical Grinding.

1. Introduction

Electrochemical grinding (ECG) is a hybrid process that combines mechanical removal by means of abrasive action with electrochemical dissolution. Its origin [1] lay in the need to grind hard materials like tungsten carbide without excessive loss of industrial diamond from the wheel. The addition of the electrochemical action alleviated the loss of diamonds at a time when that abrasive was at a premium. Other technological advantages apply as follows:

- The process imparts a scouring action to the surface of the workpiece (anode), tending to remove films that may otherwise inhibit dissolution.
- The process can produce compressive residual stresses in the workpiece surface and sub-surface when mechanical abrasive action is present.
- Rapid roughing of very hard metals is possible as with normal ECM.
- The abrasive grits serve to control accuracy during finishing passes.
- The non-conductive abrasive grits ensure against a short circuit condition.

In order that the advantages of ECG are properly realised in the finishing mode, it is important that the process is carefully controlled. This is most readily achieved by controlling the applied voltage [2]. Figure 1.1 shows, exaggeratedly, how the inter-electrode gap varies around the wheel. As the applied voltage is increased, the grit/workpiece contact region reduces. The leading radial inter-electrode gap is denoted by Y_e in Figure 1.1. It is also shown that an overcut may exist, when the grits at the bottom dead centre of the wheel are no longer in contact with the finished workpiece surface. Four process regimes can be identified in ECG. These correspond to different ranges of applied voltage that must be identified for appropriate control. Regime I corresponds to total mechanical removal when no electrolytic dissolution is possible. This occurs in the range $0 < V < \eta$. Where V is the applied voltage and η is the sum of the electrode potentials and over-potentials

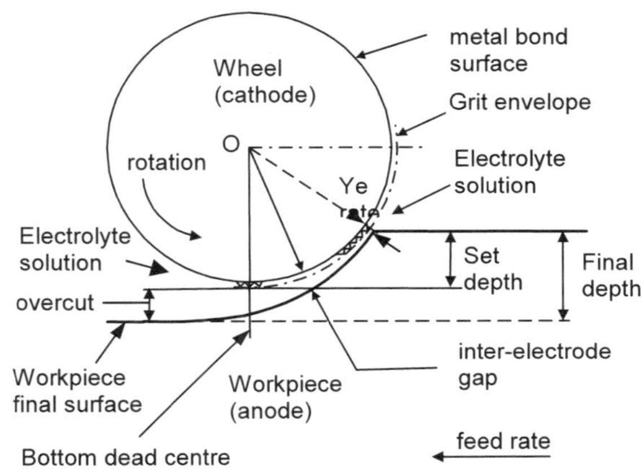

Figure 1.1. Wheel/workpiece configuration in ECG

In regime II, electrolytic action assists the mechanical abrasive action, but is insufficient to develop an overcut. Thus, regime II relates to the voltage range $\eta < V < \Omega$. Ω represents a critical value of applied voltage such that at $V=\Omega$, the abrasive grits at the bottom dead centre of the wheel are only just in contact with the finished surface of the workpiece. The voltage range, $\Omega < V < \zeta$ corresponds to a regime in which mechanical abrasion occurs towards the leading portion of the tool/workpiece inter-electrode gap, yet there is only electrolytic action taking place directly under the centre of the wheel. This defines regime III. ζ is a critical value of applied voltage at which the wheel is only just in contact with the workpiece. Regime IV relates to total electrolytic action with no mechanical assistance. Here, $V > \zeta$ and the condition $Ye > g$ applies, where g is the protrusion height of the grits above the conductive bond surface of the wheel. Figure 1.2 shows conceptually how the three main transition voltages can be found from current, overcut and surface roughness data. Thus current variation defines the I/II transition. Overcut distribution defines the II/III transition, whilst the surface roughness characteristics define the III/IV transition.

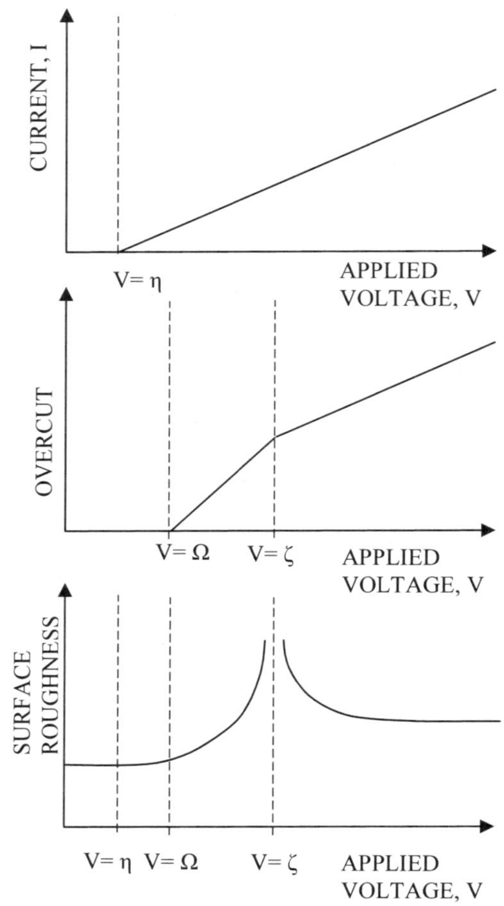

Figure 1.2. Conceptualised graphs showing transition voltages in peripheral ECG

2. ECG equipment and procedures

A view of the ECG equipment is shown in Figure 1.3. A voltage-controlled DC supply was available from a transformer with auto-variac voltage control and a Westinghouse rectifier set rated at 1000 Amp when set for 32 Volt maximum.

The electrochemical grinding machine consisted of a converted Cincinnati tool and cutter grinder having a slip ring arrangement to the rear of the main spindle for the negative potential connection. The whole machine frame was made cathodic to inhibit corrosion. The positive lead was taken to the work table.

The work table was fitted with a Hydrocheck device that allowed the table feedrate to be set and maintained constant. Electrolyte supply was from a settling tank equipped with thermostatic control so that the electrolyte solution temperature was held within +/-1°C. The electrolyte was pumped to the working zone by means of a Monopump, via a Cuno-cartridge filter system. The machining area was enclosed and there was fan extraction of gases.

The workpieces were oil-hardening tool and die steel, AISI 01, supplied as ground flat stock 12.7 mm wide and 0.794 mm thick, with a tolerance on thickness of +/- 0.0064 mm. Before performing the tests, the material was heated slowly up to 650°C and then allowed to cool to room temperature within the furnace. The 229 mm long strips were then stretched axially by 5% on a 50 mm central gauge length to ensure that the specimens were fully stress relieved. A Lindley extensometer with a least-division of 0.0013 mm was used to control the extension. Following stretching, each specimen was cut to a length of 89 mm disposed symmetrically about the gauge length. In order to establish typical mechanical properties for the material, four tensile tests were continued to destruction. The results indicated that a mean yield stress of 607 MN/m^2 and a modulus of elasticity of 2.06 x 10^5 MN/m^2.

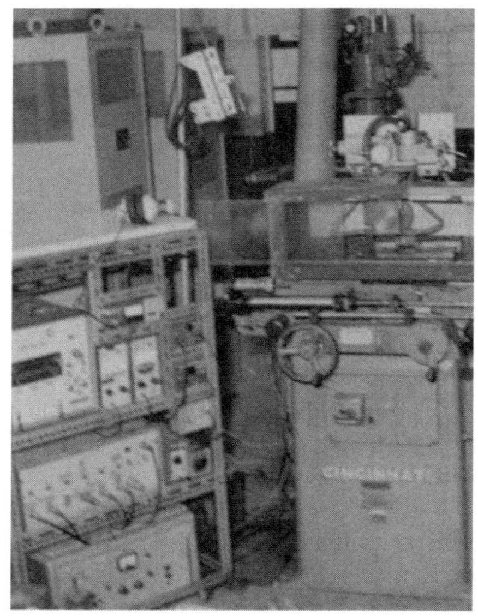

Figure 1.3. ECG equipment

The grinding wheel was of plain rectangular form steel blank having a diameter of 152 mm and a 12.7 mm face width. The abrasive consisted of diamond grits (100/120 mesh BS sieve grading) electro-deposited around the periphery of the wheel and with a 1.6 mm side-face wrap. Grit concentration was 0.00116 $carat/mm^2$. Nominal wheel speed was 2150 rpm revolving in the up-cutting mode. The wheel was cleaned for 45 seconds prior to each test with a UNI WA 120 LV aluminium oxide abrasive stick. Grit protrusion above the wheel bond was typically 0.064 mm. The set depth of cut was 0.076 mm and table feed rate was 0.533 mm/s.

The electrolyte was a 10% by weight aqueous solution of sodium nitrite delivered through a perspex nozzle, closely-fitted around the wheel (see [3] for details), at an average flow rate of 0.107 l/s. Its conductivity was 0.12 $\Omega^{-1}cm^{-1}$ at a tank operating temperature of 21.5^{0}C. The pH value was 8.9.

3. Hardness survey

Hardness measurements were performed directly on the upper surfaces of the AISI 01 specimens with a Vickers diamond pyramid hardness testing machine. For each specimen, hardness readings were recorded prior to machining and subsequent to machining. The main purpose of the tests with AISI 01 material was the determination of residual stress distribution. Therefore, surface preparations, such as polishing, were avoided so as not to risk altering the stresses in the specimens. For the same reason, handling, with the possibility of accidentally distorting the specimens, was kept to a minimum by limiting the number of hardness tests.

The Reichert Company [4] distinguishes between three indentation load levels in hardness testing, these are as follows.

- Micro hardness at loads below 0.2 kgf
- Low-load hardness at loads between 0.2 kgf and 3 kgf
- Macro hardness at loads from 3 kgf upwards.

Micro hardness tests generally relate to the hardness of single crystals or grains and can measure the hardness at accurately located sites. However, such local hardness variations were not of immediate importance to the present investigation.

Low-load hardness tests measure hardness over a larger surface area but results can be dependent on load. The low-load range is usually adopted for hardness testing of thin layers and films. Further, the low-load technique can be influenced by the prevailing surface topography. The present investigator carried out some tests (outside the theme of the present work) and detected a possible correlation between CLA value and apparent low-load hardness (1 kgf indentation load).

Blain [5] has simulated the effect of residual stresses on surface hardness. This researcher considered that the degree of hardness of a body should not be taken as its resistance to penetration, as in conventional tests, but rather its resistance to impingement, ie. the maximum pressure which can be locally applied to a specimen without causing plastic deformation. Using a polished ball indenter, Blain measured the maximum pressure which did not cause local yielding of a specimen surface stressed in tension. The measurement technique was also performed on the reverse side of the same specimen where compressive surface stresses were present. A substantial difference in the two pressure values was recorded indicating that tensile stresses reduce the maximum pressure whereas compressive stresses cause an increase. The experiment was repeated using a conventional Rockwell hardness tester, measuring the indentation hardness in the normal way. In this case, no hardness difference was detected. Nelsson [6] refers to the work of other investigators and confirms that "The size and signs of residual stresses in a surface govern its resistance to plastic flow". One may conclude that, although a macro hardness survey may not reveal the apparent hardness resulting from residual stresses at the outer skin of a specimen, a hardness value more representative of the sums of surface and sub-surface work hardening effects, arising from previous processes, will be recorded. Further, when incidental hardness tests do not permit surface preparation, macro-hardness impressions are less likely to be influenced by surface topography since a larger indentation area results.

For the present work, a standard Vickers pyramid indentation load of 10 kgf was used which generally complies with the preceding considerations. Care was taken to ensure that the specimens were supported in such a way that they were not distorted during the indentation test.

4. Discussion

Figure 1.4 shows a plot of the mean surface hardness following ECG versus applied voltage. Each data point is the mean of five indentations. Although there is some scattering of results, certain features are discernible. These are a uniform, slightly falling, hardness of about 288 Vickers pyramid number from 0 volt to 3 Volt, followed by a rapid fall in hardness which becomes less rapid until, between 5 Volt and 6 Volt, the curve levels off at a Vickers pyramid number of 217. Above 6 volt the scatter is large, although there is a possibility that the hardness value would be constant. The curve described is shown as a broken line in Figure 1.4 since the distribution of the points are subject to some degree of scatter. However, the line has been drawn with reference to the plot of Figure 1.5 which is described in the following text.

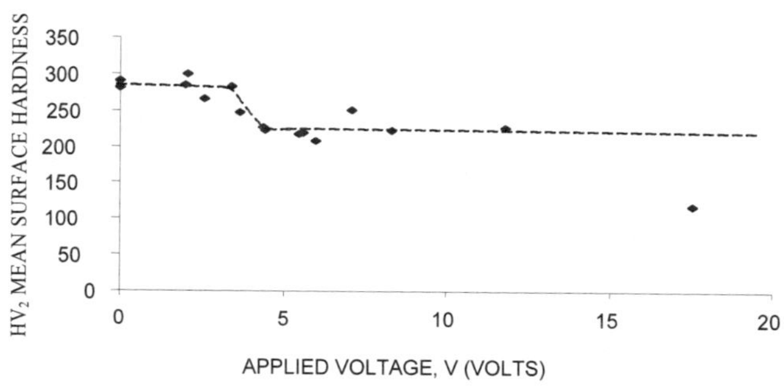

Figure 1.4. Mean surface hardness after ECG, vs. applied voltage (Vickers Pyramid, 10 kgf load)

A more accurate description of the effect of the ECG process is obtained by considering the change in the mean surface hardness, ie. mean surface hardness after ECG, HV_2 – mean surface hardness before ECG, HV_1. These differences are plotted against applied voltage in Figure 1.5 and the following characteristics are evident when considered in terms of increasing voltage.

a) 0 Volt to 3.5 Volt: gradual fall in hardness change (increase in hardness, ie. $HV_1 < HV_2$)
b) 3.75 Volt to 5 or 6 Volt: sudden fall in value of hardness change followed by a steadily reducing trend until hardness change becomes constant (hardness increase followed by apparent softening).
c) Above 6 Volt: apparently constant hardness change (apparent softening effect, ie. $HV_1 > HV_2$).

It is possible to suggest hardness features that may correspond to transition phenomena. The range 0 to 3.5 Volt (Figure 1.5) shows no sudden variations in hardness change (apart from a single result at 2.6 Volt which is likely to be a scatter effect) and so the regime I to regime II transition cannot be very precisely located in this range. The striking drop in hardness difference in the range 3.75 to 5/6 Volt could be compared with the changing mechanical contribution defined as regime III. Therefore, with the aid of the hardness results, the regime II to regime III transition might be put at an applied voltage of 3.75 Volt and that of the regime III to regime IV transition range, at 5 to 6 Volt.

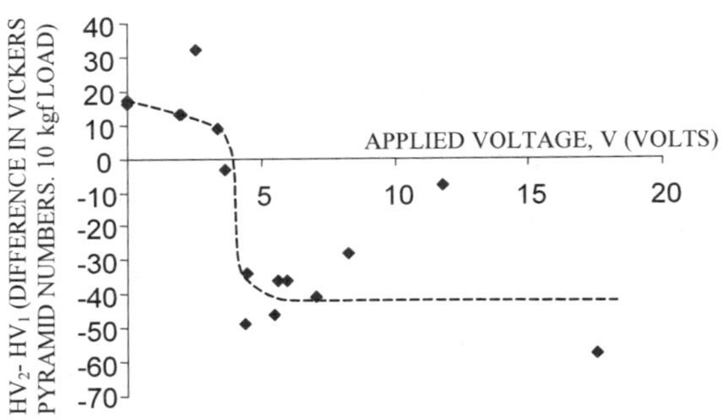

Figure 1.5. Change in hardness resulting from ECG vs. applied voltage

It is clear that the interplay between mechanical and electrochemical proportions of removal is probably responsible for the observed changes in hardness. The specific manner in which this operates is not immediately obvious from the present results. However, there are two possible phenomena which may apply. One of these is work hardening due to plastic deformation of the finished surface and sub-surface by the cutting mechanism between grit and workpiece and by the normal (vertical) pressure between the wheel and the work when electrochemical contribution is zero or very small. In combination with the mechanical effects, electrochemical action will influence the work hardness by varying amounts. Kahles and Field [7] point out that ECM produces a soft layer in a majority of aerospace alloys. It is thus possible that electrochemical contribution, in the ECG process, may operate to reduce the work-hardening effect produced by the contribution of mechanical working. The Kahles and Field observation may point to another mechanism causing softening due to increased electrolytic action at the increasing values of applied voltage. The present results appear to support this softening premise. Whether or not this is the correct explanation for the observations, the data must still be viewed in terms of the reduced contribution of mechanical action which results when electrochemical contribution is increased. Thus hardness data can be used, to a greater or lesser extent, as an indicator of transition voltages between operating regimes.

On grinding under total mechanical conditions (regime I), maximum work hardening occurs and this remains fairly constant (Figure 1.4) in the applied voltage range, $0 < V < \eta$. Such a constant hardness increase is not in evidence in Figure 1.5 and possibly suggests that the value of η is smaller than the lowest, positive applied voltage investigated, ie 2 Volt. This would place the regime I to regime II transition voltage within the range, 0 to 2 Volt. As electrochemical assistance increases in regime II, the work-hardening effect of the mechanical cutting action is gradually reduced until an overcut is initiated. At this point (approximately 3.75 Volt) mechanical abrasion no longer has direct contribution to work-hardening the final surface. However, plastic distortion can still be achieved in the final surface from the influence of mechanical action in upper regions of the machining zone. Thus, as mechanical abrasion retreats from the bottom of the wheel to the position of the leading gap, there is a corresponding fall in surface hardness. This hardness reduction is most rapid immediately after the transition from regime II to regime III. Finally, at the end of regime III, removal becomes totally electrochemical and a more constant hardness change pertains. Scatter of the three data points above V = 8 Volt does not permit verification of this last feature and obviously more points would be required. Here, the plot has been continued as a horizontal line (Figure 1.5).

Since overcut is dependent on voltage, it was conceivable that the distribution of Figure 1.4 had been revealed by naturally machining to increasing depths under the control of an increasing voltage. To test this, one stress-relieved specimen was subjected to an incremental layer removal process as an anode in an electrolytic cell. This is described fully in [8]. At the commencement of, and after, each incremental layer removal, hardness tests were performed on the exposed surface. A least two indentations were measured at each new level. The results are plotted in Figure 1.6 as hardness number versus depth below the original surface. Apart from some minor scatter in the data, there is virtually no change in the hardness to the final depth investigated, i.e. 0.114 mm. In terms of overcut and set depth in the ECG process, this depth would correspond to an applied voltage of 6 Volt (see [2]). Thus the range of layer removal is adequate for comparison with the most significant features of Figure 1.4 and shows that the distribution therein is mainly the result of the ECG process. The scatter in the absolute hardness data is partly due to variations in the initial workpiece hardness values.

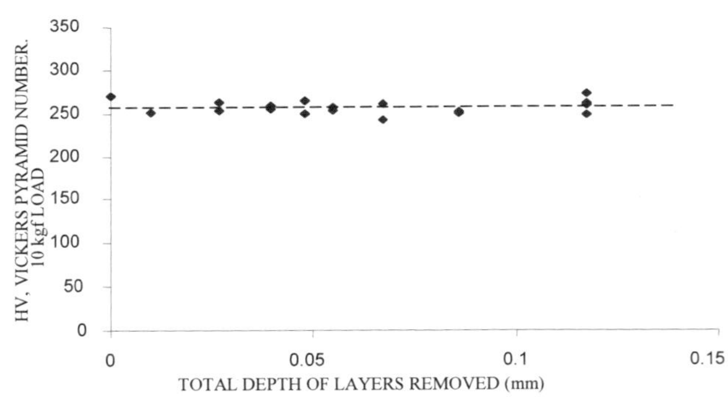

Figure 1.6. Hardness vs. depth below surface for specimen in un-machined, stress-relieved state

The limited number of results, discussed in the foregoing text, suggest that surface hardness data may give an indication of the contributions of the prevailing removal mechanisms, but is especially indicative of the regime II to regime III and regime III to regime IV transitions.

Verification of the preceding deductions can be obtained from previous published work ([2] and [9]) relating to similar operating conditions. Here, current data reveal a I/II transition of approximately zero volt. Overcut data yield a II/III transition of between 3 to 3.75 Volt, whilst surface roughness values identify a III/IV transition at 5.7 Volt. Thus, these results add confirmation to the transitions identified from the hardness data.

As a further confirmative test, a specimen subjected to ECG at 3.41 Volt was tested for hardness at different depths below the finished surface using the layer removal technique. Figure 1.7 shows the results. The data are almost representative of the transition point at which overcut is just zero and mechanical action is just available at the finished surface. It can be seen that the hardness is greatest at the machined surface and that it decreases to a depth of 0.038 mm, whereupon the hardness is relatively constant. In comparison with Figure 1.6, it is clear that a definite hardening had occurred and that this was initiated at the machined surface.

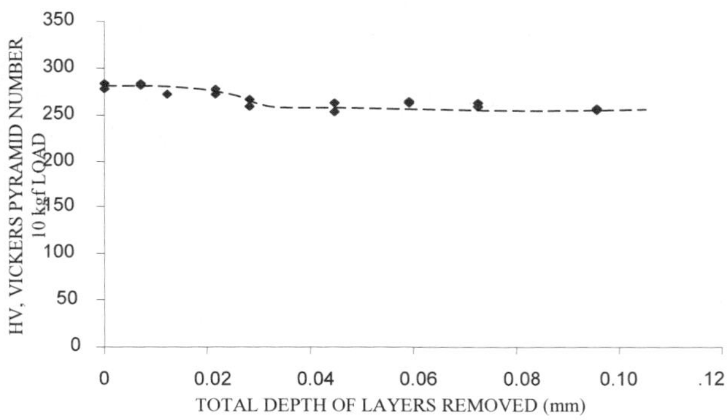

Figure 1.7. Hardness vs. total depth removed following ECG at 3.41 Volt

Surface appearance, as revealed in micro-photographs, affords a qualitative appreciation of the numerical values recorded in the surface roughness survey. Figure 1.8 presents micro-photographs (at x32 magnification) of the machined surfaces. The regime I to regime II transition is indeterminate from these photographs. However, the regime II to regime III transition is seen as occurring between 3.65 Volt and 4.4 Volt. This is because the fine lines at 3.65 Volt are indicative of abrasive grit contact with the workpiece, whilst at 4.4 Volt, rougher striations appear which are indicative of a channelled flow pattern resulting from the presence of grits determining the electrolyte flow in the leading part of the inter-electrode gap.

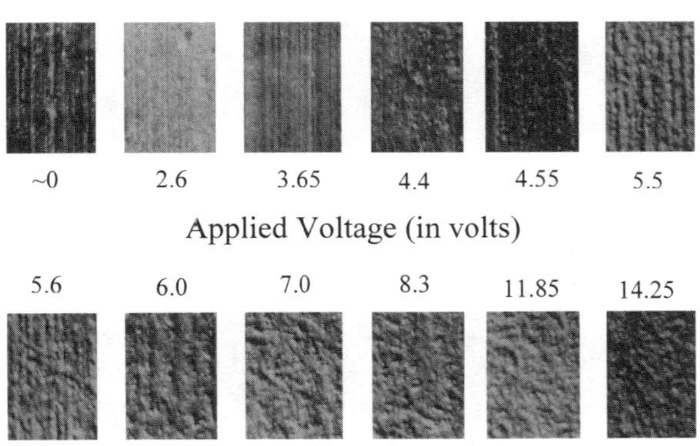

Figure 1.8. Surface appearances at x32 magnification for various applied voltages

These striations disappear somewhere between 5.6 Volt and 7.0 Volt, whereupon the photographs show no further directionality of surface characteristics as voltage further increases. This indicates that only ECM is in operation, and hence this voltage range corresponds to the regime III to regime IV transition.

Table 1.1 gives a comparison between transition voltages obtained from different sources. Careful examination of these data shows general correspondence, but also allow the transitions to be narrowed down to: I/II = just greater than zero Volt, II/III = approximately 3.75 Volt, III/IV = 5.7 to 6.0 Volt.

Table 1.1. Comparison of transition voltages obtained from different sources

Source	Regime transition Voltages (Volts)		
	I/II	II/III	II/IV
Current	~0		
Overcut		3 to 3.75	
Surface roughness			5.7
Surface appearance		3.65 to 4.4	5.6 to 7
Hardness	0 to 3.5	> 3.75	6

5. Conclusions

- Whenever mechanical action is present in the ECG process, surface work-hardening can occur, but the amount of this is controlled by the proportion of electrochemical contribution which reduces the mechanical action and by the strain hardening characteristics of the work material.
- It is a possibility that electrochemical action may operate to reduce the work – hardening effect of the work material.
- When removal is by total mechanical abrasion, maximum work-hardening occurs and is independent of applied voltage.
- As applied voltage is increased such that overcut remains zero but removal occurs with electrochemical assistance, the effect of work-hardening gradually reduces since the proportion of mechanical removal is gradually reduced.
- When an overcut is initiated, there is a rapid fall in hardness and this becomes less steep as voltage increases. This is because the finished workpiece surface is no longer cut directly by the grits. Some hardening still occurs, however, at the finished surface due to the influence of mechanical action in the upper regions of the machining zone.
- After complete grit/workpiece separation, when totally electrochemical removal is in operation, hardness appears to be constant at a minimum value since mechanical work-hardening is no longer possible.
- Surface appearance, as revealed by micro-photographs, assists in understanding the hardness data.
- When mechanical action contributes to machining the final surface, fine pitch, longitudinal grooves, created by the grits, are in evidence and this helps in identifying the applied voltage at overcut initiation.

- Coarser pitch and rougher, longitudinal grooves are in evidence when the electrolyte flow channel is separated into discrete streams as mechanical action retreats around the wheel with increasing voltage.

References

[1] Storey OW, (1953) Electrolytic grinding or machining of metals. Journal of the Electrochemical Society 100: 125C-126C.

[2] Atkinson J, Noble CF, (1977) Repeatability and residual stresses resulting from peripheral electrochemical grinding. Proceedings of the International Symposium for Electrical Machining.

[3] Geddam A, Noble CF, (1972) Peripheral electrochemical grinding with a formed wheel. Proceedings of the 13th Machine Tool Design and Research Conference, University of Birmingham.

[4] Anon, Micro Hardness. (Booklet). C. Reidhert, Optische Werke AG, Vienna, 17 Hernalser Hauptstrasse 219, Austria.

[5] Blain PA, (1957) Influence of residual stress on hardness. Metal Progress 71: 99-100.

[6] Nelsson LE, (1962) Wear of machine tool slideways. Advances in Machine Tool Design and Research, Proceedings of the 3rd International Machine Tool Design and Research Conference, University of Birmingham 443-460.

[7] Kahles JF, Field M, (1971) Surface integrity guidelines for machining. Society of Manufacturing Engineers Technical Paper IQ71-240.

[8] Atkinson J, Noble CF, (1978) Residual stresses in workpieces after peripheral electrochemical grinding. Proceedings of the 19th MATADOR conference, UMIST, Manchester 525-532. pub. Macmillan Education Ltd.

[9] Atkinson J, Noble CF, (1981). The surface finish resulting from peripheral electrochemical grinding. Proceedings of the 22nd International Machine Tool Design and Research conference, UMIST, Manchester 371-377 pub. Macmillan Education Ltd.

Design of Compound Processes of Electrochemical Finishing and Rolling-Levelling following Turning

P.S. Pa

National Taipei University of Education, Taiwan, (R.O.C.), myhow@seed.net.tw

Abstract. A new finishing method that uses an effective electrode and a nonconductive roller to execute the design of compound processes of electrochemical finishing and rolling-levelling following turning is investigated. The submitted processes can be used for various turning operations. Through simple equipment attachment, electrochemical finishing and rolling-levelling can follow to execute the finishing process on the same machine. Among the factors affecting electrochemical finishing, the performance of rolling-levelling combined with electrochemical finishing is primarily discussed. In the experiment, the electrode is used with continuous and pulsed direct current. The experimental parameters are feed rate of electrode and roller, rotational speed of the workpiece, die material, electrical current rating, pulsed period, roller geometry, and electrode geometry. An adequate workpiece rotational speed produces better finishing. The design change of the electrode from a semicircle to a wedge-form with small end radius causes the electrolytic products and heat bringing away more rapidly and performs the best finishing. The finishing effect of pulsed direct current is slightly better than using continuous direct current. However, the use of pulsed current raises the machining time and cost. An effective and low-cost finish process through the design electrode and using the rolling-levelling assistance in the electrochemical finishing after the turning process make the surface of the workpiece smooth and bright is presented.

Keywords: Compound Processes, Electrochemical Finishing, Rolling-Levelling, Design electrode, Pulsed Period, Turning

1. Introduction

Burnishing is a cold working process which produces a fine surface finish by the planetary rotation of hardened rollers over a bored or turned metal surface. Since all machined surfaces consist of a series of peaks and valleys of irregular height and spacing, the plastic deformation created by roller burnishing is a displacement of the material in the peaks which cold flows under pressure into the valleys. The result is a mirror-like finish with a tough, work hardened, wear and corrosion resistant surface. Lapping and Honing is eliminated [1]. Simple ball-burnishing and roller-burnishing tools were used for the experimental work, these tools being quite similar in their design principles. The performance of the tools, together with the effects of the burnishing force and the number of burnishing tool passes on the surface roughness and surface hardness of commercially available aluminum and brass, were studied [2]. Burnishing is a super-finishing, it is a plastic deformation process, which is becoming more popular as a finishing process. The lubricant, force, speed, and feed have significant effects on surface roughness and surface hardness. A set of rollers is used to roll on the component surface with adequate pressure. As a result all the pre-machined peaks get compressed into valleys thus giving a mirror-like surface finish. A pre-machined surface roughness of 0.63-0.75 μm (by turning) can be finished up to 0.11 μm (by burnishing) and improved micro hardness is obtained [3].

The application of electro-polishing is suitable for difficult-to-machine materials, such as ceramic and cermet. Those as die components, plastic or press dies, wire-drawing dies, and optical units, this technique can be applied to electrolytic components (silicon chips, VLSI/ULSI chips). Electrochemical machining (ECM) was presented by Faraday in 18th century. He made use of the elastic energy and chemical to remove materials [4]. However, the potential for internal hole polishing is yet to explore. The electrochemical technique is based on the electrochemical reaction between an electrode and a workpiece. The main difficulty lies in the design of tool electrode considering the complicated process of metal removal. In addition, hole machining through electrochemical process can improve the precision of hole with the appropriate control of machining conditions or the electrode geometry [5]. Datta [6] showed that the gap width between electrode and workpiece directly influences the current condition and the discharge dregs of the electrolyte. The experimental result of Mileham et al. [7] proves that the quality of the machined surface will be influenced by the main factors including current density, flow rate of electrolyte as well as the gap width in the electrochemical machining. Shen used $NaNO_3$ as the electrolyte to proceed the electro-polishing on die surface. The result showed that the surface roughness of workpieces decreases with increased current density, flow rate and concentration of electrolyte [8]. Schuster et al. showed that the machining resolution is down to a few micrometers by applying ultra short pulses of nanosecond duration, and micro structures can be machined by ECM [9]. Varies types of electrode were also developed for electro-polishing, including the disc-form and arrow-head [10,11]. In ECM, good surface quality of the workpiece was obtained through the arrangement of the experimental conditions [12]. However, the major difficulty of the electrochemical finishing is the cost and the compensation design of finish-tool. Thus, the design of finish-tool plays the major role [13]. The application of rolling-leveling assistance to electrochemical finishing and the design of the finish-tool are crucial for the finish process of turning. In the current study, the most effective finish process and the advantage of low cost equipment with the compound processes of electrochemical finishing and rolling-leveling following turning is investigated.

2. Experimental Setup and Processes Parameters

The development of an effective design of compound processes of electrochemical finishing and rolling-levelling following turning is illustrated in Figure 1.1. It is based on a scientific method of design process with an analysis includes design considerations, design development, detailed design, and experimental analysis to construct the final design result [14].

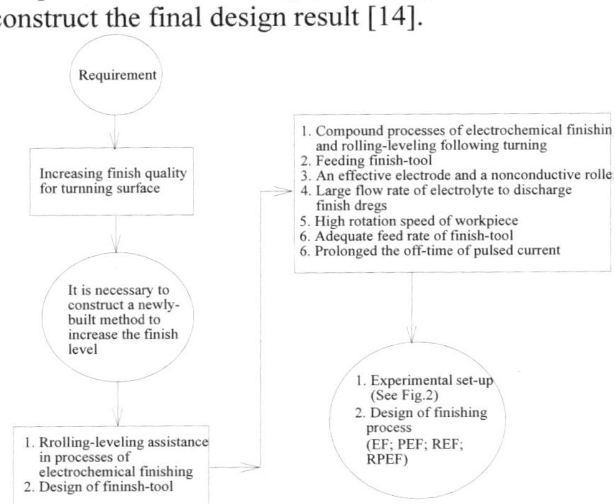

Figure 1.1. Design Process with a Basis of Analysis [10]

The system schematics and configuration of finish-tool and workpiece of the experiment setup are shown in Figures 1.2 and 1.3 respectively.

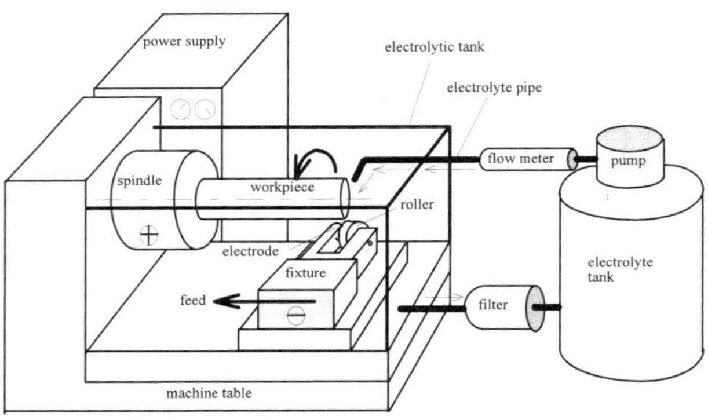

Figure 1.2. Experimental Set-Up

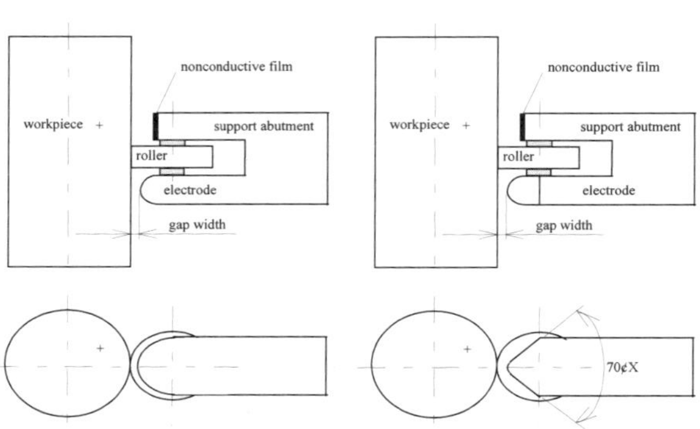

Figure 1.3. Confizuration of finish-tool and workpiece

The material of the workpiece uses ASTM D2. The chemical composition is shown in Table 1.1.

Table 1.1. Chemical composition of workpieces (Wt %)

(Wt %)	Fe	C	Si	Mn	P	S
ASTM D2	88.65	1.40	0.40	0.30	0.02	0.03
	Cr	Mo	Al	V	Cu	Ni
	8.20	0.80	/	0.20	/	/

The dimension of the workpiece is 30.02 mm in diameter and 150mm in length after fine turning. The amount of the reduction of the diameter after compound processes of electrochemical finishing and rolling-leveling is 0.02mm. The initial average surface roughness of the workpiece afer precise turning is around 3.5μm. The electrolyte is $NaNO_3$ of 15%wt. The temperature of the machining is maintained at 40±5□. The gap width between electrode and workpiece is 1.0, 2.0, 3.0, 4.0 and 5.0mm. The flow rate of electrolyte is 4, 6, 8, 10, 12, and 14 L/min. The rotational speed of workpiece is 600, 800, 1000, and 1200rpm. The current rating is 15, 20, 25, 30, 35, and 40A. The feed rate of the finish-tool ranges from 20 to 120 mm min^{-1}. The pulsed period (on/off time) is 100ms/100ms, 100ms/200ms, 100ms/300ms, 100ms/400 ms and 100ms/500ms. The experiment relates to the design finish tool imposed with electrochemical finishing (EF), pulsed electrochemical finishing (PEF), compound processes of electrochemical finishing and rolling-levelling (REF), and compound processes of pulsed electrochemical finishing and rolling-levelling (RPEF). All workpieces after different finish processes is measured by the surface roughness measurement (Hommel T500, the accuracy is within ±5% after standard correction).

3. Results and Discussions

Figure 1.4 shows that the larger the flow rate is, more rapid by the electrolytic depositions and heat can be brought away, the surface roughness of workpiece is improved. As a result, the use of large electrolytic flow rate is advantageous.

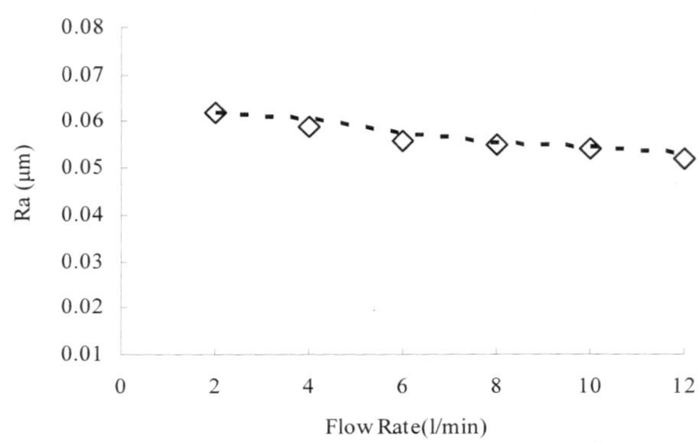

Figure 1.4. Compound processes of electrochemical finishing and rolling-leveling at different flow rate of electrolyte (ASTM D2, $NaNO_3$, 15%wt, 40A, 20mm/min, Continuous DC)

Figure 1.5 shows that an adequate gap width between the electrode and workpiece produces a better finish.

Small gap width takes less time for the same amount of material removal, but the discharge of electrolytic depositions from the gap is difficult, so that the finish effect is reduced. Large gap width takes a longer time since the effect of electrochemical finishing is limited. As far as the stable operation of electrochemical finishing and grinding is concerned, the gap width of 2mm is more effective in the current experiment and is suggested for the next-stage test.

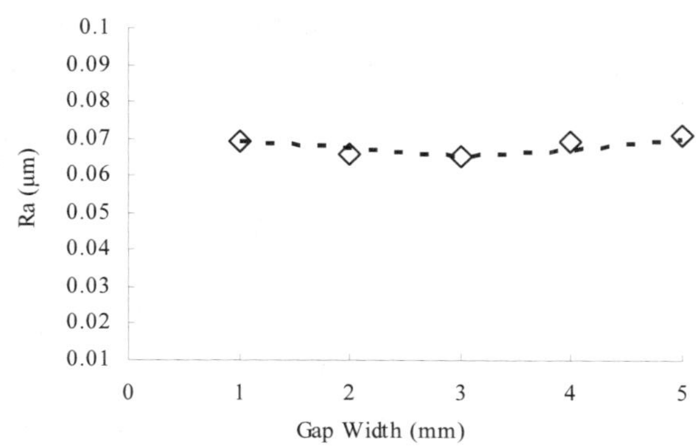

Figure 1.5. Compound processes of electrochemical finishing and rolling-leveling at different gap width between electrode and workpiece (ASTM D2, $NaNO_3$, 15%wt, 10 Lmin, 40A, 120mm/min, Continuous DC)

The rotation effect of the workpiece is shown in Figure 1.6. The range between 1000 and 1200rpm is suggested. Over 1000rpm, the rotation effect contributes much to effective flushing, while high-speed rotation produces a better finish effect to the electrochemical finishing and the rolling-levelling.

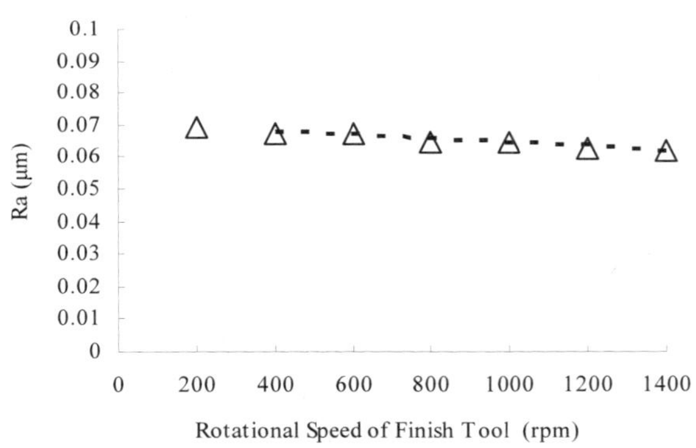

Figure 1.6. Compound processes of electrochemical finishing and rolling-leveling at different rotational speed of finish tool (ASTM D2, $NaNO_3$, 15%wt, 10 Lmin, 40A, 120mm/min, Continuous DC)

Figure 1.7 shows that a good finish is achieved through an adequate combination of current rating and feed rate of the finish-tool for the compound processes of electrochemical finishing and rolling-levelling (REF). At a constant current rating, the finish-tool has an optimal feed for the best surface finish. Fast feed reduces the power delivered to a unit area of the workpiece surface, and slow feed increases it. The former could not supply sufficient electrochemical power, while the latter could cause the problem of dregs discharge. In order to reach the same removal amount of 0.02mm for ASTM D2 from compound processes of electrochemical finishing and rolling-levelling, the following combination of parameter values is suggested: 15A and 20mm/min, 20A and 40mm/min, 25A and 60mm/min, 30A and 80mm/min, 35A and 100mm/min, and 40A and 120mm/min for the finish process. Figure 1.8 shows the effects of the pulsed direct current. Longer off-time is slightly more advantageous, because the discharge of electrochemical finishing dregs during the off-time is more complete. But the machining time is longer and the cost is raised.

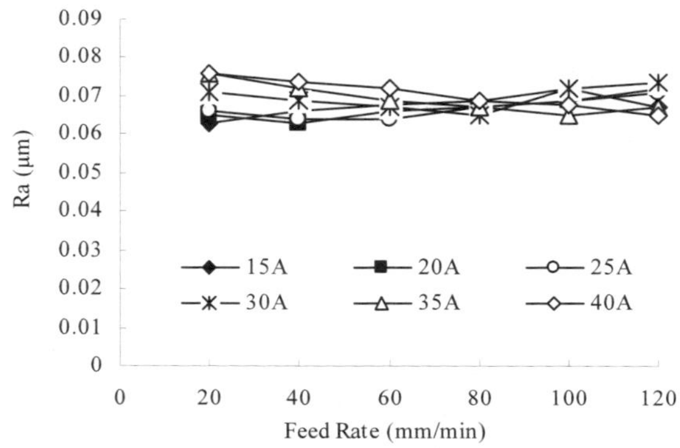

Figure 1.7. Compound processes of electrochemical finishing and rolling-leveling at feed rate of finish tool through different current rating (ASTM D2, $NaNO_3$, 15%wt, 10 Lmin, 40A, 120mm/min, Continuous DC)

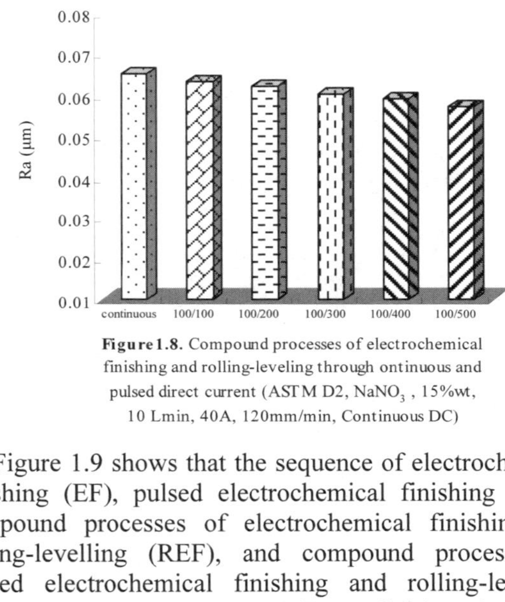

Figure 1.8. Compound processes of electrochemical finishing and rolling-leveling through ontinuous and pulsed direct current (ASTM D2, $NaNO_3$, 15%wt, 10 Lmin, 40A, 120mm/min, Continuous DC)

Figure 1.9 shows that the sequence of electrochemical finishing (EF), pulsed electrochemical finishing (PEF), compound processes of electrochemical finishing and rolling-levelling (REF), and compound processes of pulsed electrochemical finishing and rolling-levelling (RPEF) performance among different finish-tools (Types A and B). Type B gives the best surface finish. The design change of the electrode from a semicircle (type A) to a wedge-form with small end radius (type B) provides more open space for dregs discharge and causes the electrolytic products and heat bringing away more rapidly and benefits to the whole finish effect. The results show that the sequence of the finish effect of EF, PEF, REF, and RPEF performance among different finish-tools remains the same. The surface roughness is obviously further reduced than without the assistance of rolling-levelling. Electrochemical finishing (EF) using rolling-levelling is obviously superior to the use of pulsed direct current.

Figure 1.10 shows the average contribution of compound processes of pulsed electrochemical finishing and rolling-levelling (RPEF) to surface finish improvement through the use of pulsed current (11%), rolling-levelling (75%), and electrode geometry (12%). In summary, the design of the compound processes through rolling-levelling produces the most influential parameter in this study. The compound processes of electrochemical finishing and rolling-levelling (REF) without pulsed current is also recommended, while the use of pulsed current is of limited advantage, particularly when the increased finish time and cost of pulsed-power supply equipment is considered.

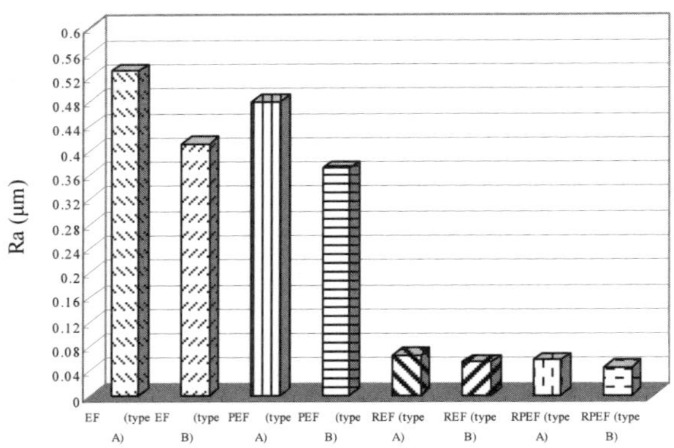

Figure 1.9. Effects of evaluation of the finishing effect of four processes (ASTM D2, $NaNO_3$, 25%wt, 12 L/min, 40A, 120mm/min, 100ms /500ms)

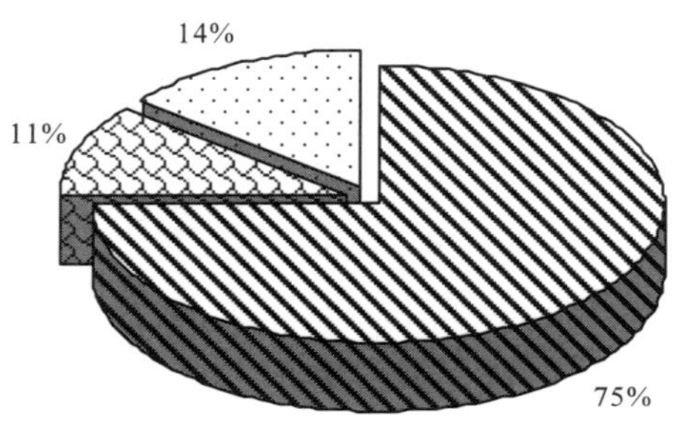

Figure 1.10. The contribution pie of surface finish improvement of design finish-tool (ASTM D2, Type B, $NaNO_3$, 25%wt, 12 L/min, 40A, 120mm/min, 100ms /500ms)

4. Conclusion

The compound processes of electrochemical finishing and rolling-levelling following turning by a design finish-tool that require a shorter time to make the workpiece surface smooth and bright. The effective design finish-tool with an effective electrode and a nonconductive roller has an optimal value for higher current density and provides larger discharge space, which produces a smoother surface. The use of large electrolytic flow rate is advantageous to the finish effect. High rotational speed of a workpiece produces a better finish. The finish effect is better with longer off-time because discharge of finish dregs becomes easier. The design change of the electrode from a semicircle to a wedge-form with small end radius causes the electrolytic products and heat bringing away more rapidly and performs the best finishing. The rolling-levelling following electrochemical finishing can obviously improve the finish effect, and take the advantage of the low cost equipment as well in the turning process.

Acknowledgement

The current study is supported by National Science Council, contract 95-2411-H-152-014- and 95-2622-E-152-001-CC3.

References

[1] Hassan, A.M. and Maqableh, A.M. 2000, The effect of initial burnishing parameters on non-ferrous components, Journal of Materials Processing Technology, Vol.102, P115-121.

[2] Hassan, A.M. 1997, The effects of ball- and roller-burnishing on the surface roughness and hardness of some non-ferrous metals, Journal of Materials Processing Technology, Vol.72, P385-391.

[3] Shirsatu, M., and Ahujab, B. 2004, Parametric analysis of combined turning and ball burnishing process, Indian journal of engineering & materials sciences, Vol.11, No.5, P391-396.

[4] McGeough, J.A. 1974, Principles of Electrochemical Machining, P1-10.

[5] Jain, V. K., and Nanda, V.N. 1986, Analysis of Taper Produced in Size Zone during ECD, Precision Engineering, Vol. 8, P27-33.

[6] Datta, M., and Landolt, D. 1981, Electrochemical Machining Under Pulsed Current Conditions, Elector. Acta, Vol. 26, P899-907.

[7] Mileham, A.R., Harrey, S.J., and Stout, K.J. 1986, The Characterization of Electrochemically Machined Surfaces, Wear, Vol.109, P207-214.

[8] Shen, W.M. 1995, The Study of Polishing of Electric Discharge-Machined Mold With ECM, M.Sc. Thesis, National Yunlin Institute of Technology, Taiwan.

[9] Cagnon, L., Kirchner, V., Kock, M., Schuster, R., Ertl, G., Gmelin, W.T., and Kuck, H. 2003, Electrochemical micro machining of stainless steel by ultra short voltage pulses, Z. Phys. Chem. 217, P299-313.

[10] Hocheng, H., and Pa, P.S. 2003, Electropolishing of Cylindrical Workpiece of Tool Materials Using Disc-Form Electrodes, Journal Materials Processing Technology, Vol.142, P203-212.

[11] Hocheng, H., and Pa, P.S. 2004, The Arrow Head Electrode in Electropolishing of External Cylindrical Surface, International Journal of Materials and Product Technology, Vol.20, P312-326.

[12] Kim, B.H., Ryu, S.H., Choi, D.K., and Chu, C.N. 2005, Micro electrochemical milling, Journal of Micromechanics and Microengineering, Vol.15, P124-129.

[13] Pa, P.S. 2006, Design of Completely Inserted and Feeding Electrode for Female Screw in Electrochemical Finishing, Key Engineering Materials, Vols. 329, P231-236.

[14] Rao, S.S. 1996, Engineering Optimization. John Wiley and Sons.

6

Forming

6–1

Investigation of Micro Sheet Metal Forming of Ball Retainer Using Finite Element Analysis

R.S. Lee[1], C.H. Chen[2], Y.F. Sung[3]

[1] Department of Mechanical Engineering, National Cheng Kung University, Tainan, Taiwan, 701, R.O.C., mersl@mail.ncku.edu.tw
[2] Department of Mechanical Engineering, National Cheng Kung University, Tainan, Taiwan, 701, R.O.C., viatorpicket@gmail.com
[3] Metal Industries Research & Development Centre, Kaohsiung, Taiwan, 811, R.O.C., yvonne@mail.mirdc.org.tw

Abstract. In this paper, the thickness/grain size ratio was proposed as an indicator for studying the size effect of micro deep drawing of a ball retainer. Experiments with AISI 304 stainless steel with 0.06mm thickness were performed by heat treatment and tensile test. Flow stresses at various thickness/grain size ratios were obtained. The commercial finite element software LS-DYNA was used to simulate the deep drawing process of the ball retainer. The simulation results agree well with experimental results. The effective stress on the lower fringe of ball hole decreases as the thickness/grain size ratio decrease. Effective strain and thickness shows less sensitivity with the thickness/grain size ratio according to the comparison. It determined that the thickness/grain size ratio would be an important indicator for process design as it corresponds to a heat treatment condition.

Keywords: Micro sheet metal forming, ball retainer, Finite Element Analysis

1. Introduction

Deep drawing process has been well developed at the macro scale, However, the development of micro forming for sheet thickness less then 0.1mm has gained much attention only recently. The effect of material and process parameters on mechanical properties in micro sheet metal forming has not been entirely clarified. When the blank miniaturizes to micro scale, the grain size effect can not be ignored because the defects included in the grain boundary become significant and the micro sheet with only a small number of grains on the thickness direction tends to fail in a brittle mode. Micro deep drawing characteristics of micro sheet steels of below 0.2 mm thickness was studied by Saoteme et al. [1] and the relation of punch diameter and drawability without large amount of blank holder pressure was obtained. The influence of grain size on thin metal sheet with thickness ranging from 0.17 to 2 mm was investigated by Raulea et al. [2] and the yield strength and maximum load decrease with a decreasing number of grains over the thickness in tension test and bending test. However, the tensile strength was obtained in reduction of the sheet thickness at a constant grain size and there was no further study about thicknesses below 0.1 mm. Experimental investigation of micro deep drawing was carried out by Vollertsen et al. [3] and the limit drawing ratio was determined and the relation between friction coefficient and the applied pressure in micro deep drawing was found.

Various parameters affect the metal flow and the quality of the product. Furthermore, the coupling of these parameters makes the problem more complicated. The product accuracy in two-stage cylindrical cup deep drawings with 0.3mm thickness blank were investigated by Manabe et al. [4] and the cup wall geometry, thickness distribution and surface roughness were investigated using both experiment and finite element analysis. The tool roughness was studied but the effect of grain size and thickness to deep drawing process was not investigated.

Fig.1.1 shows the components of ball bearing used on penny-motor. The ball retainer marked with the circle, having the centre hole diameter 1.61 mm, outer diameter 2.234 mm, height 0.813 mm and the thickness 0.06 mm. Deep drawing of ball retainer should have enough precision to meet the requirements for assembly of ball bearing. When scaling down the sheet metal forming process to such dimension, using only thickness or grain size to explain the variation of flow stress is not sufficient. In this paper, the thickness/grain size ratio was proposed as an indicator to study the size effect of the micro deep drawing process.

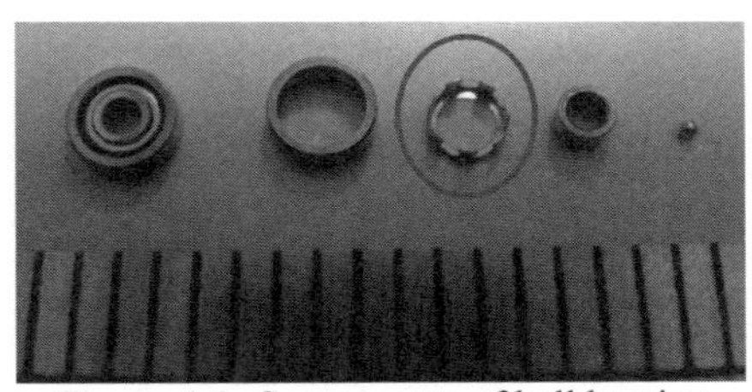

Figure 1.1. Components of ball bearing

2. Experimental Process

2.1 Heat Treatment

The grain size of AISI 304 stainless steel was changed by heat treatment. The heat treatment process consists of heating the material to a temperature of about 1000°C and holding it for a period of time. After this, the material is quickly cooled. The problem of heating the material to such a high temperature is surface oxidation. This problem must be resolved since the blank has only a thickness of 0.06 mm and the oxidation layer would have great influence on the mechanical property. Vacuum heat treatment was used for preventing surface oxidation. The heat treatments were carried out in a vacuum furnace with a vacuum of 10^{-3} Torr at temperatures of 900°C, 950°C, 1000°C, and 1025°C for 3 min duration. Temperature and duration would influence the grain size.

2.2 Tensile Test

Tensile tests were carried out to obtain the flow stress and the ultimate strain of AISI 304 stainless steel by SHIMADZU AG-IS universal testing instrument. The dimensions of the tensile test specimen based on ASTM E8 standard are given in Fig 1.2. Tensile velocity was set at three different values of 10, 100 and 1000 mm/min, respectively for as-fabricated blank and 100 and 1000 mm/min for heat treated blank.

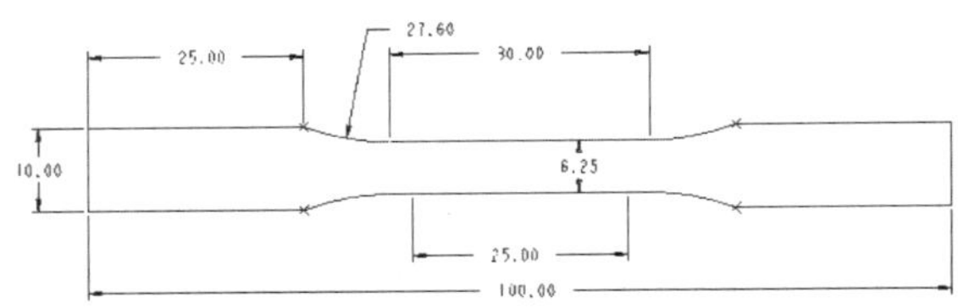

Figure 1.2. Dimensions of the tensile test specimen

2.3 Experiment of Ball Retainer

The ball retainer in Fig 1.1 has the required centre hole with diameter 1.61mm, and ball hole with diameter 0.5mm. There are three passes to deform the ball retainer, including fine blanking, deep drawing and sizing. The roundness of the centre hole and ball holes after deep drawing process are the critical requirements of the retainer. The experiment was performed in Metal Industries Research & Development Centre and the dimensions of the deep drawing blank are shown in Fig-1.3.

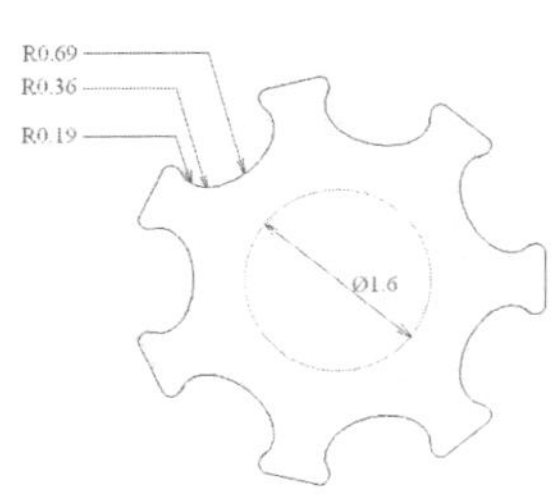

Figure 1.3. Dimensions of the deep drawing specimen

3. Finite Element Analysis

The commercial finite element software LS-DYNA was applied to the ball retainer deep drawing process modelling. Fig 1.4 is a schematic view of the finite element mesh describing the punch, the die, and the blank profile. The blank has seven blades around the 1.6mm centre hole, as shown in Fig 1.3. By moving the punch, the blank is pushed down to the bottom die. The dimensions of the punch, the blank and the die are given in Table 1.2.

Barlat-Lian's plasticity material model is suitable for modeling sheets with anisotropic material under plane stress conditions. The material model allows the use of the Lankford parameters for definition of anisotropy. The influence of the element type in the sheet metal forming finite element simulation on calculation precision and computation efficiency is important. Belytschko-Tsay (BT) element is the mostly used shell element in sheet metal forming simulation. The element size must be considered since the dimensions of the blank are quite small in micro deep drawing. The maximum element size of the blank is 0.03 mm and the total mesh number is 5128. The numbers of rigid tooling elements are 1261 for the punch and 3217 for the lower die. The Coulomb friction coefficient in the micro deep drawing simulation is 0.125. Flow stress data applied in the simulation was obtained in the tensile test for different thickness/grain size ratios. The detail settings of simulation parameters are shown in Table 1.3.

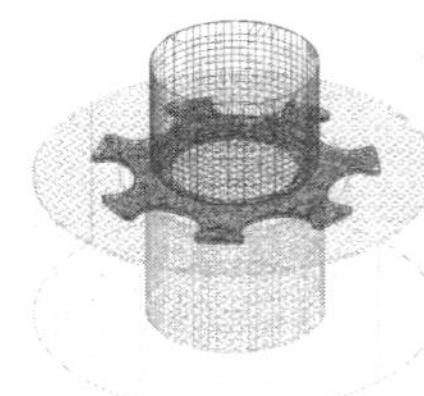

Figure 1.4. FEM model of deep drawing test

Table 1.2. Chemical composition of stainless steel blank

Punch Eedge Radius (mm)	0.14
Punch Diameter (mm)	2.1
Blank Thickness (mm)	0.06
Die Radius (mm)	0.3
Die Mold Diameter (mm)	2.235

Table 1.3. Parameters settings for simulation

Punch Velocity(mm/min)	100	1000
Element Type	Belytschko-Tsay Shell Element	
Mesh Size(mm)	Tooling - max:0.1, min:0.05	
	Blank - max:0.03, min:0.01	
Friction Coefficient(μ)	0.125	

4. Results and discussion

4.1 Thickness/Grain Size Ratios

Electrical Corrosion was applied to obtain microstructure since AISI 304 has high corrosion resistance. The composition of the etchant is 30% nitric acid and etching condition is 6V DC for 10 seconds.

Grain size measurement is based on Heyn's linear intercept from ASTM E112 standard. A line was drawn to intercept the grain on the picture of microstructure; the average grain size was equal to the length of the line divided by the number of grains that the line intercepted. Thickness/grain size ratios are calculated by grain size divided by thickness. Table 1.4 shows the grain size and thickness/grain size ratio at different heat treatment conditions.

Table 1.4. Grain size and thickness/grain size ratio at different heat treatment conditions

Heat treatment condition	Grain size (μm)	thickness/grain size ratio
1025°C	46.17	1.30
1000°C	28.27	2.12
950°C	19.28	3.11
900°C	13.05	4.59

4.2 Tensile Test

The fracture on the as-fabricated blank of the tensile test result displays about 45°, but the necking area is small. The phenomenon occurred because the strain hardening accumulated and the ductility decreased as the strip was rolled. The flow stresses at different velocities are shown in Fig 1.6. Tensile velocity influences the flow stress. The ductility is better when the tensile velocity is higher since the rise in temperature caused by a high deformation rate does not have enough time to dissipate across the thin blank and then soften the material. Table1.5 shows the results of the tensile test. The tensile test with a tensile velocity of 1000 mm/min has the smallest maximum load and largest maximum displacement.

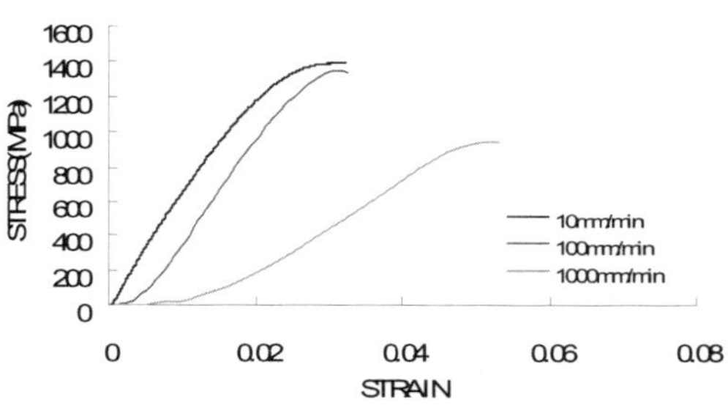

Figure 1.6. Flow stress of as-fabricated blank at different tensile velocities

Table 1.5. Tensile test results

Tensile velocity(mm/min)	10	100	1000
Tensile time(sec)	5.4	0.56	0.18
Maximum load(kN)	0.536	0.538	0.363
Maximum displacement	0.86	0.857	1.366
Elongation (%)	1.034	1.034	1.055

Fig 1.7 shows the flow stress at different tensile velocities (heat treatment 1025°C, 3 min duration). The flow stress of high tensile velocity has higher yield stress but has almost the same curve after yielding. This explains the blank with heat treatment is less sensitive to the deformation rate. Heat treatment can remove residual stresses and soften the material, so that the ultimate stress without heat treatment is much larger than that with heat treatment. As-fabricated sheet material has severe stain hardening after rolling, as shown in Figs. 1.8 and 1.9. Multilayer grains in the thickness direction would display the defects of the structure at several grains and grain boundaries and restrain dislocation growth and make the flow stress higher. However, the effective stress would be smaller if there are only one or two layers in thickness direction. Strain hardening not only reduces formability but also increases the load of the press. Also, the quality and precision may not be good enough to meet the requirement of micro fabrication and the life of dies would be shortened by wear.

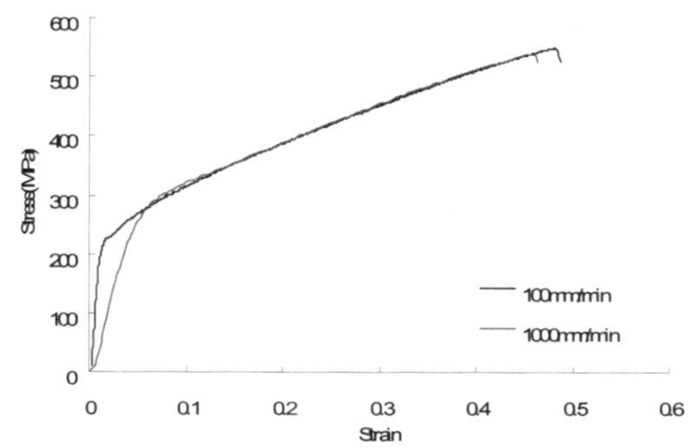

Figure 1.7. Flow stress curves at different tensile velocities

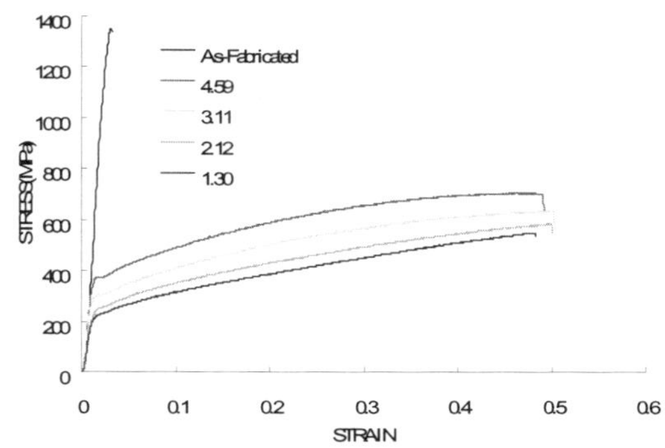

Figure 1.8. Flow stress curves at different thickness/grain size ratios (tensile velocity: 100 mm/min)

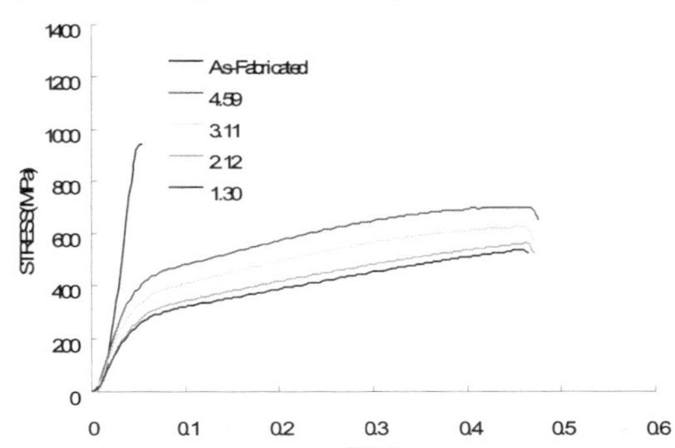

Figure 1.9. Flow stress curves at different thickness/grain size ratios (tensile velocity: 1000 mm/min)

4.3 FEM results

The designed diameter of the centre hole is 1.61 mm, and the diameter of the centre hole is 1.674 mm after deformation and exhibit irregular heptagon. The centre hole increased and lost roundness because of metal flow and anisotropy, as shown in Fig. 1.10(a). The diameter of the ball hole is 0.523 mm as a desired diameter is 0.50 mm and the total height of ball retainer is 0.858 mm.

The simulation result of the as-fabricated blank is shown in Fig 1.10(b). The average diameter of the centre hole is 1.686 mm and it displays an irregular heptagon, which is consistent with the experimental result. The average diameter of the ball hole is 0.518 mm and the roundness is good. The height of the retainer is 0.856 mm, which conforms to the result of the experiment.

Applying finite element analysis to micro sheet deep drawing appears to be feasible since there is a good level of agreement between the contour obtained by the finite element analysis and the experiment.

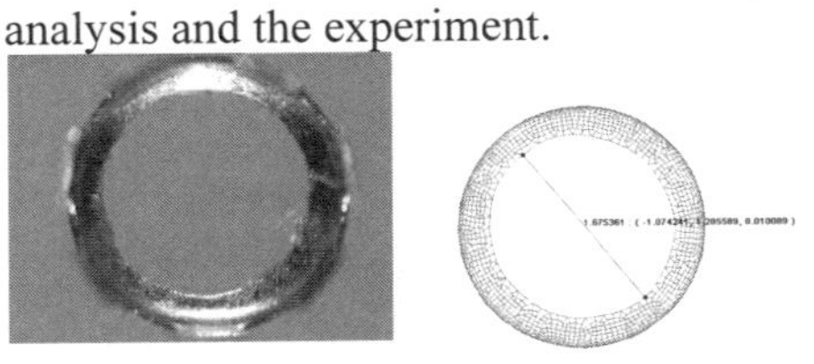

(a) (b)

Figure 1.10. Centre hole shape after deep drawing: (a) experimental result and (b) simulation

Different heat treatments can form different grain sizes and different flow stress curves can also be obtained with tensile test. The finite element software LS-DYNA does not have the function to define the grain size. Therefore, different flow stress curves were used to represent different grain sizes. Figure 1.11 shows the diameters of the centre hole and ball hole of the heat treated part (1000°C, 3 min). Except for better formability, anisotropy would be removed as well by heat treatment. The measurements are close to the desired dimensions mentioned before, and the roundness of the centre hole after deformation is better than in the as-fabricated blank. The comparison between different thickness/grain size ratios in Table 1.6 reveals that the diameters of the centre hole and ball holes are consistent with the designed diameters when the thickness/grain size ratio is 3.11.

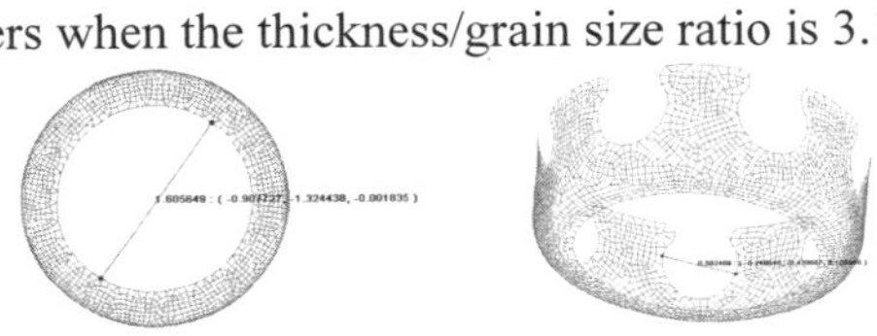

(a) (b)

Figure 1.11. Diameters of (a) centre hole and (b) ball hole of 1000°C, 3 min heat treated part

Table 1.6. Product dimensions at thickness/grain size ratios

Thickness/grain size ratio	Centre hole diameter(mm)	Ball hole diameter(mm)
1.30	1.604	0.508
2.12	1.605	0.504
3.11	1.612	0.503
4.59	1.617	0.503

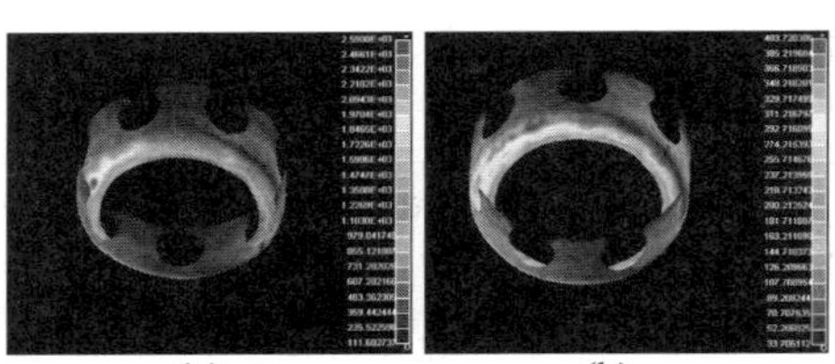

(a) (b)

Figure 1.12. Effective stress distribution of (a) as-fabricated part and (b) 1000°C, 3 min heat treated part

The effective stress distribution of as-fabricated part and the heat treated part are shown in Fig 1.12. It can be seen that although the lower fringe of the ball hole has the highest effective stress, the value of the as-fabricated part is much higher. Besides, the heat treated part shows a more uniform distribution of the effective stress than the as-fabricated part. Table 1.7 shows the simulation results of various heat treatment conditions. Effective stress on the lower fringe of the ball hole decreases as the thickness/grain size ratio decreases and the effective strain on lower fringe of the ball hole exhibits only a slight change as the thickness/grain size ratio changes. Although the flow stress shows less sensitivity to the deformation rate, the local stress would be influenced by the tensile velocity.

Table 1.7. Simulation result at various heat treatment conditions

Thickness/ grain size ratio	Tensile velocity (mm/min)	Effective stress on lower fringe of ball hole (MPa)	Effective strain on lower fringe of ball hole
4.59	100	586.1	0.310
	1000	555.5	0.333
3.11	100	475.2	0.336
	1000	462.7	0.321
2.12	100	421.1	0.318
	1000	403.7	0.319
1.30	100	398.3	0.316
	1000	400.7	0.316

5. Conclusions

The flow stress of as-fabricated material has significant difference for various tensile velocities as a result of material softening by temporary adiabatic. The simulation results have good agreement with the experimental results. From the simulations at different heat treatment conditions, it was found that effective stress on the lower fringe of the ball hole decreases as the thickness/grain size ratio decreases. However, the effective strain on the lower fringe of the ball hole exhibits only a slight change as the thickness/grain size ratio changes. The thickness/grain size ratio has been demonstrated to be an important indicator for process design. A further study of micro sheet metal forming for various thickness values and the influence of thickness/grain size ratio to flow stress and micro sheet metal is under way.

References

[1] Saotome Y, Yasuda K, Kaga H, (2001) Microdeep Drawability of Very Thin Sheet Steels. Journal of Materials Processing Technology 113:641-647

[2] Raulea LV, Govaert LE, Baaijens FPT, (2001) Size Effects in Processing of Thin Metal Sheets. Journal of Materials Processing Technology 115:44-48

[3] Vollertsen F, Hu Z, Niehoff HS, Theiler C, (2004) State of the Art in Micro Forming and Investigations Into Micro Deep Drawing. Journal of Materials Processing Technology 151:70-79

[4] Manabe K, Shimizu T, Koyama H, (2007) Evaluation of Milli-Scale Cylindrical Cup in Two-Stage Deep Drawing Process. Journal of Materials Processing Technology, doi:10.1016/j.jmatprotec. 2006.11.164.

Design of Titanium Alloy Superplastic Blow-forming in Ellip-cylindrical Die Using Taguchi Method

Dyi-Cheng Chen*, Jyun-You Lin, Ming-Wei Jheng, Jhih-Ming Chen

Department of Industrial Education and Technology, National Changhua University of Education, Changhua 500, Taiwan, *ROC*
*dyi3510@ms46.hinet.net

Abstract. This study employs the commercial DEFORM™ 3D finite element analysis software to simulate superplastic blow-forming of titanium alloy (Ti-6Al-4V) sheet in an ellip-cylindrical die. The titanium sheet is assumed to be a rigid-plastic material with homogeneous and isotropic properties. The model of the ellip-cylindrical die and titanium sheet is constructed using three-dimensional solid elements. Simulations are performed to investigate the effects of shear friction, sheet temperature, die height and length of the die's short axis on the thickness, effective stress, effective strain and critical damage characteristics of the formed sheet. The Taguchi method is then applied to optimize the processing parameters for the superplastic blow-forming process such that the thickness of the formed titanium sheet is optimal average distribution. The simulation results confirm the effectiveness of the Taguchi method in identifying the optimal blow-forming processing parameters for the current Ti-6Al-4V titanium alloy sheet.

Keywords: superplastic blow-forming, ellip-cylindrical die, Ti-6Al-4V titanium alloy sheet, Taguchi method

1. Introduction

Certain polycrystalline materials, such as 8090 Al-Li alloy, Ti-6Al-4V alloy and so forth, exhibit superplasticity behaviour under particular strain rate and temperature conditions, i.e. they undergo extensive tensile plastic deformation prior to failure. The characteristics of superplasticity, namely a low flow stress accompanied by a large elongation, are exploited in a variety of 'superplastic forming' (SPF) processes, including forging, extrusion, blow-forming and so on [1,2]. Cornfield and Johnson [3] employed theoretical and experimental techniques to analyze the superplastic deformation of thin circular diaphragms subjected to one-sided hydrostatic pressure. Holt [4] showed that the bulging of a superplastic alloy sheet as the result of uniaxial tension or compression could be predicted analytically by the expression $\sigma = K\dot{\varepsilon}^m$, where σ is the flow stress, $\dot{\varepsilon}$ is the strain rate, and K and m are material parameters. Hwang *et al.* [5] used the commercial DEFORM™ finite element (FE) software package to calculate the evolution of the pressurization profile and sheet thickness distribution during blow-forming of 8090 Al-Li sheet. Senthil Kumar *et al.* [6] investigated the superplastic deformation behavior of AA7475 aluminum alloy during blow forming in a circular die using a simple theoretical model and numerical simulations based on ABAQUS FE code. Hwang *et al.* [7] applied a mathematical model based upon the finite-difference method to examine the plastic deformation behaviour of 8090 Al-Li sheets during blow-forming in a conical closed die.

Hwang *et al.* [8] employed a finite-difference method to analyze the non-uniform thinning effect in the free-bulged region of sheet metal during blow-forming in a circular closed die. Bonet *et al.* [9] presented an algorithm for calculating the pressure cycle in FE simulations of superplastic forming. Kim *et al.* [10] developed a combined FE analysis and optimization technique to predict the initial blank thickness distribution required to obtain a uniform final thickness in superplastic blow forming. Yarlagadda *et al.* [11] developed FE models to optimize superplastic forming conditions and to predict product forming time. Applying these models, the authors presented numerical results for deformed shapes, stress-strain distributions and thickness evolution of blow-formed Al-Ti alloy sheets. Huang *et al.* [12] employed ABAQUS FE simulations to determine blow-forming processing conditions required to optimize sheet thickness distribution of the final product and then validated the predicted results experimentally. Lee and Huh [13] applied FE code to analyze the superplastic-forming/diffusion-bonding (SPF/DB) process for four-sheet sandwich components.

The current study employs 3D rigid-plastic DEFORM™ FE code to analyze the effects of the processing conditions and the die geometry on the thickness distribution, effective stress, effective strain and critical damage value of Ti-6Al-4V alloy sheets blow-formed in an ellip-cylindrical die. The Taguchi method is then applied to establish the processing conditions which optimize the thickness average distribution of the formed alloy sheet.

2. Taguchi Method

The Taguchi method is one of the most well known and widely applied of all robust design methodologies. The principal benefit of the Taguchi approach is that it provides a detailed understanding of the individual and combined effects of various design parameters based upon a minimum number of experimental (or simulation) trials. The basic objectives of the Taguchi method are to establish the optimal combination of design (or processing) parameters and to reduce variations in the product (or process) quality by rendering the parameter design robust to the effects of noise.

The Taguchi method employs generic signal-to-noise (S/N) ratios to quantify the quality of the product or process design. Depending on the particular characteristics of the design problem, different S/N ratios may be applicable, including "lower is better" (LB), "nominal is best" (NB) or "higher is better" (HB).

In optimizing the superplastic blow-forming process, the aim is to establish the processing and die geometry parameters which maximum value for minimum thickness position of the formed titanium sheet. Accordingly, the HB S/N ratio is applicable, i.e.

$$S/N = -10\log[\frac{1}{n}\sum_{i=1}^{n}\frac{1}{y_i^2}], \qquad (1)$$

where n is the number of experimental or simulation repetitions performed under the same design parameter conditions, y_i is the simulation or experimental result, and i indicates the number of design parameters arranged in the Taguchi orthogonal array (OA).

Having carried out the trials prescribed in the OA, the resulting S/N data are analyzed using the analysis of means (ANOM) statistical method. The ANOM results are used to produce S/N response tables and S/N response graphs from which the optimal design parameters can then be determined.

The mean S/N ratio of any factor (e.g. Factor A) at level m is given by

$$M_{Am} = \frac{1}{n_A}\sum_{m=1}^{L_A}(S/N)_{Am}, \qquad (2)$$

where n_A is the number of appearances of Factor A in the OA table and $(S/N)_{Am}$ is the S/N ratio of Factor A at level m.

In the Taguchi method, a design parameter (i.e. factor) is considered to be significant if its influence on the design solution is large compared to the experimental error, as estimated using the analysis of variance (ANOVA) statistical technique. If a design parameter is found to have a high level of significance, it implies that this particular parameter plays a key role in determining the optimal solution of the design problem. The ANOVA analysis is performed using the following formula:

$$SS_{total} = \sum_{i=1}^{n}\sum_{j=1}^{r}n_{ij}^2 - nr\eta_m^2, \qquad (3)$$

$$SS_{factor} = \frac{nr}{L}\sum_{k=1}^{L}(\eta_k - \eta_m)^2, \qquad (4)$$

$$DOF = L - 1, \qquad (5)$$

$$V_{factor} = \frac{SS_{factor}}{DOF}, \qquad (6)$$

$$F_{factor} = \frac{V_{factor}}{V_{error}}, \qquad (7)$$

where SS_{total} is the total sum of squares, SS_{factor} is the factorial sum of squares, n is the number of experiments, r is the number of specimens taken, DOF is the number of degrees of freedom, V_{factor} is the variance of the factor, F_{factor} is the F ratio of the factor, η_m is the mean S/N ratio of the factor at level m, and η is the mean of the factor.

3. Simulation Process

In this study, FE simulations were performed to investigate the respective effects of various superplastic blow-forming processing conditions, i.e. shear friction factor, sheet temperature, depth of die, and length of the die's short axis on thickness distribution of the formed titanium sheet, effective stress, effective strain and critical damage value. The temperature-dependent stress-strain relationship for current Ti-6Al-4V titanium alloy is summarized in Table 1.1. The simulations are based on the following assumptions: (1) the die is a rigid body; (2) the titanium alloy sheet is a rigid-plastic material; and (3) the friction factor between the titanium sheet and the ellip-cylindrical die is constant. The invariable die geometry parameters and the strain rate applied during the blow-forming process are presented in Table 1.2.

Table 1.1. Stress-strain relationship for Ti-6Al-4V alloy; Flow Stress = f (Temperature, Strain Rate 0.1/s, Strain) (MPa)

Strain \ Temperature	700°C	850°C	1000°C
0	310.12	153.99	31.21
0.025	318.46	160.26	31.97
0.05	326.79	168.59	33.69
0.1	330.92	180.86	37.48
0.3	324.66	210.21	40.99
0.5	320.52	213.59	41.62

The current simulations were performed using the commercial implicit FE DEFORM™ 3D code. The FE element code is based on a flow formulation approach using an updated Lagrangian procedure. The nonlinear equations describing the plastic deformation of the titanium sheet are solved using a direct iteration method and the Newton-Raphson method. In the solution procedure, the direct iteration method is used to generate initial estimates of the deformation and the Newton-Raphson method is then employed to obtain a rapid convergence to the final result. The convergence criteria are specified as a velocity error norm of $\|\Delta\mathbf{v}\|/\|\mathbf{v}\| \le 0.005$ and a force error norm of $\|\Delta F\|/\|F\| \le 0.05$, where $\|\mathbf{v}\|$ is $(\mathbf{v}^T\mathbf{v})^{1/2}$.

Figure 1.1 presents the meshed models of sheet and die before the blow-forming process, during forming and after forming, respectively. Note that due to its inherent symmetry, only one-quarter of the sheet/die model is shown. In the simulations, it is assumed that the sheet is firmly attached to the periphery of the die, and thus no metal flows from the periphery regions of the sheet into the die cavity during the forming process. The simulation conditions in Figure 1 are as follows: t (thickness of sheet) = 2mm, a (long axis of the die) = 40mm, b (short axis of the die) = 20mm, R (die entry radius) = 5mm, r (die root radius) = 2mm, h (die height) = 9mm, m (friction factor) = 0.02, T (temperature of Ti-6Al-4V alloy sheet) = 850°C, and strain rate = $0.1s^{-1}$.

Figure 1.2 presents product appearance, effective strain distribution and effective stress distribution of the titanium alloy sheet following superplastic flow-forming in an ellip-cylindrical die under the conditions specified above. Figure 1.2(a) shows that the product is sound for the ellip-cylindrical die. Furthermore, Figure 1.2(b) reveals that the maximum effective strain is induced at the near of the die root radius. Finally, Figure 1.2(c) shows that the maximum effective stress occurs at the complete under face of the ellip-cylindrical product.

Table 1.2. Constant parameters in simulations of superplastic blow-forming in ellip-cylindrical die

Materials	Thicknes of sheet (t)	Long axis (a)	Die entry radius (R)	Die root Radius (r)	Strain rate
Ti-6Al-4V	2mm	40mm	5mm	2mm	0.1(1/s)

t=2mm, a=40mm, b=20mm, R=5mm, r=2mm, h=9mm, m=0.02, T=850°C, Strain rate=$0.1s^{-1}$

4. Taguchi Design

In general, the basic objective of Taguchi design methodology is to optimize a product or process design such that it becomes insensitive to unavoidable sources of external noise. In the current Taguchi design process, the principal robustness characteristic is the minimum thickness, t_{min}, under titanium alloy superplastic blow-forming.

4.1 Factor selection

In minimizing the thickness of the blow-formed sheet, the factors of interest include shear friction factor, length of the die's short axis, temperature of the titanium alloy sheet, and height of the die. As shown in Table 1.3, these four factors were each assigned three levels, and hence the simulation trials were arranged in an $L_9(3^4)$ Taguchi OA.

4.2 Analysis of means

Table 1.4 presents the S/N ratio data for the nine simulation trials arranged in the OA, while Tables 1.5 and 1.6 present the factor response data for the minimum thickness (t_{min}) and the S/N ratio, respectively. The factor response data presented in Table 1.6 are plotted in graphical form in Figure 1.3. In the present Taguchi design, the optimal processing conditions are indicated by the highest S/N ratios. Therefore, from Figure 1.3, it can be seen that the optimal processing conditions for the superplastic blow-forming of Ti-6Al-4V titanium alloy sheet are as follows: Factor A3: friction factor $m = 0.02$; Factor B1: short axis length $b = 40$ mm; Factor C2: sheet temperature $T = 850$°C; and Factor D1: die height $h = 9$ mm.

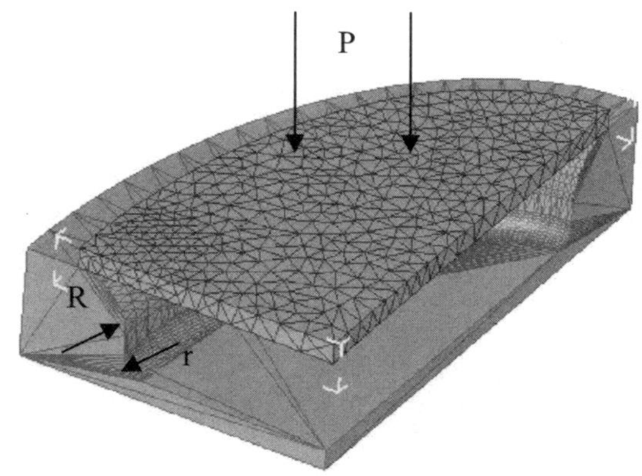

(a) before forming (step 1)

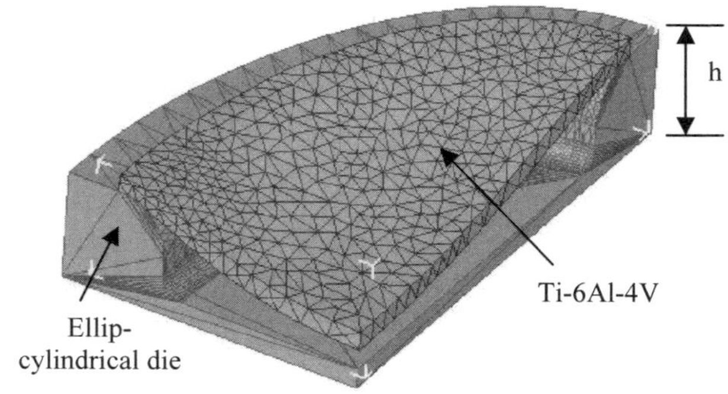

(b) during forming (step 8)

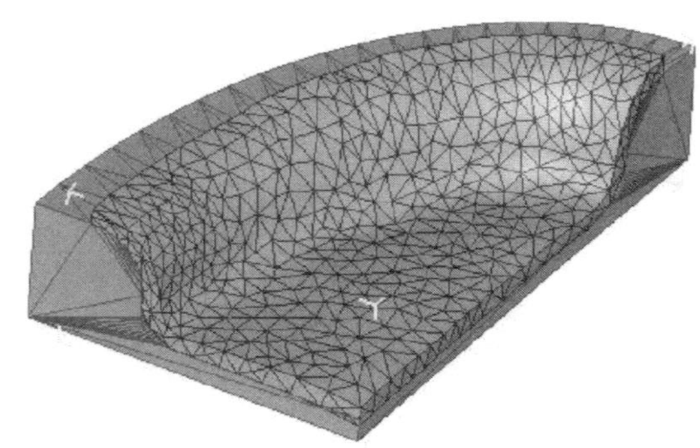

(c) after forming (step 100)

Figure 1.1. Finite element model of titanium sheet and ellip-cylindrical die

4.3 Analysis of variance

The S/N data obtained from the nine simulation trials were investigated using the ANOVA statistical analysis method based on Eqs. (3) to (7). The corresponding results are presented in Table 1.7. The high confidence (at least 99%) and variance of Factors A, B and C indicate that the friction factor, the short axis length, and the sheet temperature exert a significant effect in minimizing the thickness of the formed sheet. However, the die height has relatively less influence.

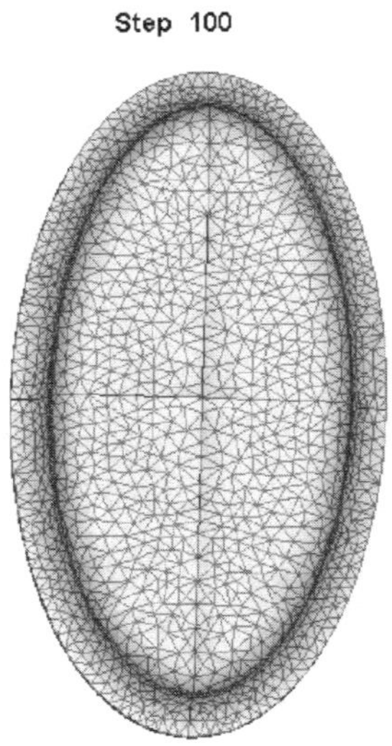

(a) Appearance of product

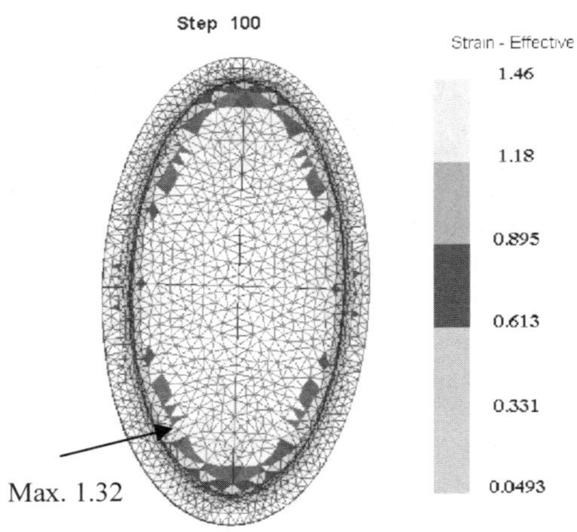

(b) Effective strain

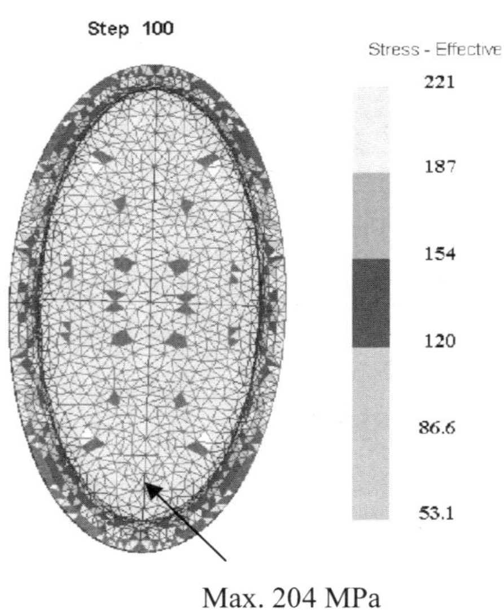

(c) Effective stress

Figure 1.2. Appearance, effective strain and effective stress in blow-forming titanium alloy sheet

4.4 Confirmation experiment

In the Taguchi methodology, a confirmation experiment is conducted using the optimal design parameters identified from the ANOVA analysis to confirm that the identified parameter values do indeed improve the quality of the design. Accordingly, the superplastic blow-forming of the current titanium alloy was simulated using the following processing conditions: A3, B1, C2 and D1. The results indicated that the corresponding minimum thickness and S/N ratio were 1.55 mm and 3.81 (dB), respectively. Comparing these results with those presented in Table 4, it is apparent that the optimal processing conditions yield an improvement in the original results, because the value, 1.55 mm, was optimal average thickness. Hence, the effectiveness of the Taguchi design methodology in optimizing average thickness distribution the blow-forming process is confirmed.

Table 1.3. Design factors and levels for titanium alloy superplastic blow-forming in ellip-cylindrical die

Factors	Description	Level 1	Level 2	Level 3
A	m (friction factor)	0.5	0.2	0.02
B	b (short axis) (mm)	40	30	20
C	T (temperature) (^{o}C)	700	850	1000
D	h (height of die) (mm)	9	10	11

Table 1.4. S/N ratio data for Taguchi trials

Experiment number	A	B	C	D	t_{min} (mm)	S/N ratio (dB)
1	1	1	1	1	1.05	0.42
2	1	2	2	2	1.04	0.34
3	1	3	3	3	0.80	-1.94
4	2	1	2	3	1.43	3.11
5	2	2	3	1	1.30	2.28
6	2	3	1	2	0.83	-1.62
7	3	1	3	2	1.46	3.29
8	3	2	1	3	1.27	2.08
9	3	3	2	1	1.15	1.21
Mean of the total sum					1.15	1.02

Table 1.5. Factor response table for minimum thickness t_{min}

Control factor	A	B	C	D
Level 1	0.96	1.31	1.05	1.17
Level 2	1.19	1.20	1.21	1.11
Level 3	1.29	0.93	1.19	1.16
Effects	0.33	0.38	0.16	0.06
Rank	2	1	3	4

Table 1.6. Factor response table for S/N ratio

Control factor	A	B	C	D
Level 1	-0.39	2.27	0.29	1.30
Level 2	1.26	1.57	1.55	0.67
Level 3	2.19	-0.78	1.21	1.08
Effects	2.58	3.05	1.26	0.63
Rank	2	1	3	4

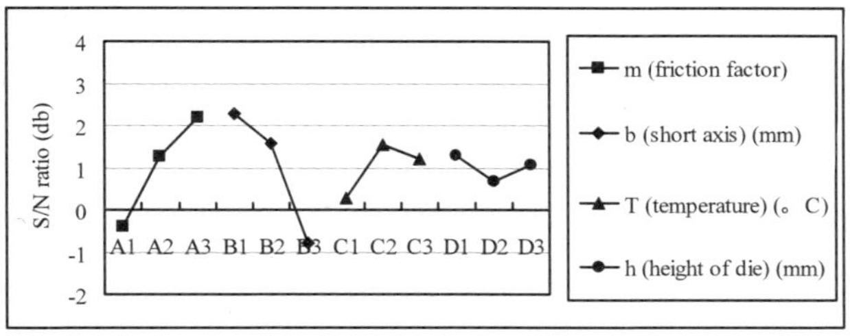

Figure 1.3. S/N response graph for superplastic blow-forming of titanium alloy in ellip-cylindrical die

5. Conclusions

This study has used commercial DEFORM™ 3D FE code to investigate the superplastic blow-forming of Ti-6Al-4V titanium alloy in an ellip-cylindrical die. The processing parameters have been optimized using the Taguchi design method. In general, the results have shown that: (1) the friction factor, the short axis of the die and the temperature of the Ti-6Al-4V alloy sheet all have a significant effect on the value of the minimum thickness, t_{min}, of the formed sheet; (2) the height of the die has relatively less influence on the value of the minimum thickness, t_{min}; and (3) the maximum effective strain is induced near die entry radius and die root radius of the formed sheet, while maximum effective stress is induced at the complete forming face.

Overall, the simulation results confirm the effectiveness of the Taguchi robust design methodology in deriving the optimal processing parameters for superplastic blow-forming of titanium alloy sheet.

Table 1.7. Analysis of variance (ANOVA) results for Taguchi trials

Factors	Description	SS	DOF	Variance	F	Confidence	Significant?*
A	m (friction factor)	0.344	2	0.172	68.8	100%	Yes
B	b (short axis) (mm)	0.459	2	0.230	92.0	100%	Yes
C	T (temperature) (°C)	0.0912	2	0.046	18.4	99.9%	Yes
D	h (height of die) (mm)	0.0126	2	0.006	2.4	85.39%	No
Error		0.0225	9	0.0025	S=0.05		
Total		0.9293	17	*Note: At least 99% confidence			

References

[1] Miller, Q.S., White, J., in Superplasticity in Aerospace, Heikkenen, H.C., McNelley (Eds.), T.R., AIME, PA, Warrendate, 1988, USA, P211.

[2] Al-Naib, T.Y.M., Duncan, J.L., 1970, Superplastic Metal Forming, Int. J. Mech. Sci., Vol.12, P463-477.

[3] Cornfield, G.C., Johnson, R.H., 1970, The Forming of Superplastic Sheet Metal, Int. J. Mech. Sci., Vol.12, P479-490.

[4] Holt, D.L., 1970, An Analysis of the Bulging of a Superplastic Sheet by Lateral Pressure, Int. J. Mech. Sci., Vol.12, P491-497.

[5] Hwang, Y.M., Lay, H.S., Huang, J.C., 2002, Study on Superplastic Blow-forming of 8090 Al-Li Sheets in an Ellip-cylindrical Closed-die, International Journal of Machine Tools & Manufacture, Vol.42, P1363-1372.

[6] Senthil Kumar, V.S., Viswanathan, D., Natarajan, S., 2006, Theoretical Prediction and FEM Analysis of Superplastic Forming of AA7475 Aluminum Alloy in a Hemispherical Die, Journal of Materials Processing Technology, Vol.173, P247-251.

[7] Hwang, Y.M., Yang, J.S., Chen, T.R., Huang, J.C., Wu, W.U., 1998, Analysis of Superplastic Blow-forming in a Conical Closed Die, Int. J. Mech. Sci. Vol.40, No.9, P867-885.

[8] [Hwang, Y.M., Yang, J.S., Chen, T.R., Huang, J.C., 1997, Analysis of Superplastic Sheet-metal Forming in a Circular Closed-die Considering Non-uniform Thinning, Journal of Materials Processing Technology, Vol.65, P215-227.

[9] Bonet, J., Wood, R.D., Collins, R., 1994, Pressure-control Algorithms for the Numerical Simulation of Superplastic Forming, Int. J. Mech. Sci., Vol.36, No.4, P297-309.

[10] Kim, Y.H., Lee, J.M., Hong, S.S., 2001, Optimal Design of Superplastic Forming Processes, Journal of Materials Processing Technology, Vol.112, P166-173.

[11] Yarlagadda, P.K.D.V., Gudimetla, P., Adam, C., 2002, Finite Element Analysis of High Strain Rate Superplastic Forming (SPF) of Al-Ti Alloys, Journal of Materials Processing Technology, Vol.130-131, P469-476.

[12] Huang, A., Lowe, A., Cardew-Hall, M.J., 2001, Experimental Validation of Sheet Thickness Optimisation for Superplastic Forming of Engineering Structures, Journal of Materials Processing Technology, Vol.112, P136-143.

[13] Lee, K.S., Huh, H., 1999, Simulation of Superplastic Forming/diffusion Bonding with Finite-element Analysis using the Convective Coordinate System, Journal of Materials Processing Technology, Vol.89-90. P92-98.

6–3

The Cold Orbital Forging Die and Process Design of a Hollow-ring Gear Part

J.J. Sheu[1] and C.H. Yu[2]

[1]National Kaohsiung University of Applied Sciences, Kaohsiung, Taiwan, 807, jjsheu@cc.kuas.edu.tw
[2]National Kaohsiung University of Applied Sciences, Kaohsiung, Taiwan, 807, xian88999@yahoo.com.tw

Abstract. In this paper, the cold orbital forging die is designed to make a hollow ring gear part. The forging process with suitable preform design is proposed to obtain the sound cold orbital forging gear part. Different preform designs have been studied to find out the deformation characteristic of the orbital forging process. The dimensions of height and outer diameter of a preform design are considered as the main design parameters. The 3D FEM simulation has been adopted to evaluate the process and die design, and predict the forging loads. The Cockroft-Latham damage criterion has been adopted to evaluate the quality of the product. The upsetting tests of workpiece SCM415 have been carried out to obtain the flow stress model. A PXW-200 orbital forging press has been adopted to do the experimental tests. The top die is rotating in the orbital motion way, and the bottom die only has translational motion. The circular locus is chosen to guarantee the gear shape can be forged precisely. The conventional closed die forging results have been compared with the orbital forging. The forging load of conventional forging is about three times of the orbital forging. The orbital locus shows a better ability of die filling at the finishing than the conventional forging. The simulation results are in good agreement with the experimental tests. The von Mises stress, effective strain, and damage value of the workpiece have been studied to develop a better forging die and process design.

Keywords: orbital forging, ring gear, FEM, preform design.

1. Introduction

In the orbital forging process, a conical upper die is rotating about an inclined axis with respect to the vertical axis of the press. The contact area between the upper die and the billet is smaller than the conventional forging process, as a result, the forging load is reduced about 15-25% [1]. The main deformation mechanism of the orbital forging is very complex due to the periodically local loading and continuous shifting of contact area.

Many researchers had studied and explored the parameters of the orbital forging process using trial-and-error experiments [2-5]. The experiments carried out included the backward extrusion of a cylindrical cup, and upsetting of a flanged rod. The ring workpiece had been studied using the photo-plastic experiments. Numerical simulations had also been adopted to predict the deformation behaviour of the orbital forging process, such as, the finite element method and upper-bound method [6-13]. The products under investigation including the cylindrical billet, ring workpiece, spoke forging, electrical components, and pint parts. The main products in the previous studies are focused on the solid geometries, except the simple ring component. There are some defects that occurr in the orbital forging process, such as center thinning and the mushroom effect [14-16]. It takes efforts of design to prevent these defects.

In this study, a ring gear forging has been studied via both FEM simulation and experimental tests. A mandrel has been designed at the tip of the punch to prevent the inner material flow as mentioned in the divided flow method[17]. Conventional forging is also examined to demonstrate the advantage of orbital forging process. Experimental tests have been carried out and compared with the numerical results.

2. The Part and the Designed Forging Die

The ring gear part of material SCM-415 is shown in Figure 1.1. The cold orbital forging process with lubrication was chosen for the finish forming and full annealing was carried out after preforming. The preform design is necessary for such a complex part geometry. The gear shape is crucial and die filling problems should occur if the die and process has not been properly designed. The die material adopted is SKD-11 and the stress ring material is SKD-61. The designed die set is shown in Figure 1.2 which has been installed on a PXW-200 orbital forging press for the experimental tests. The FEM program DEFORM 3D is used to simulate the forging process.

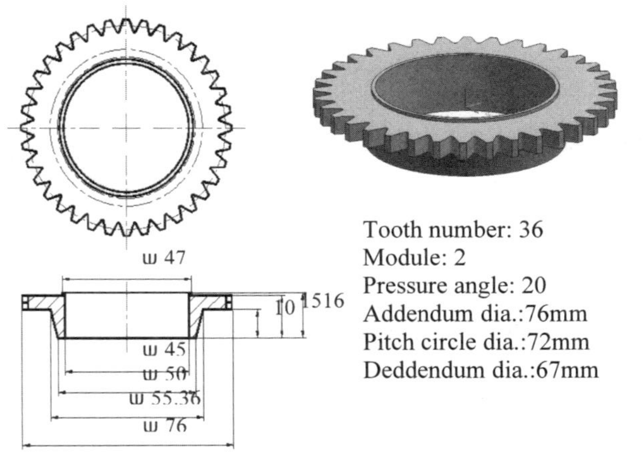

Figure 1.1. Geometry and dimensions of ring gear part

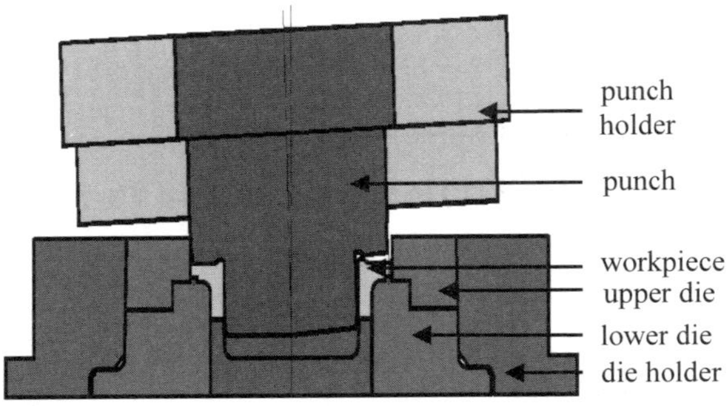

Figure 1.2. Die set of the ring gear orbital forging

3. The Preform and Forging Process Design

Conventional forging and orbital forging are compared to find a better solution for the ring gear part prolduction. The complete forging process is shown in Figure 1.3. The workpiece has been fully annealed after the preforming stage to get rid of the accumulated strains and stresses. The material properties, especially the flow stress, of the annealed SCM-415 should be determined via the experimental tests. There are three types of preform design to decrease the forging loading and avoid the forming defects . The preforms are designed in different heights and diameters to obtain better material flow control. These preforms A, B, and C are shown in Figure 1.4 and designated as H9.5xD69.7, H14xD69.7, and H14xD74, repectively. The 'H' parameter represents the height of the lower ring, and the 'D' parameter indicates the outmost diameter of the preforms.

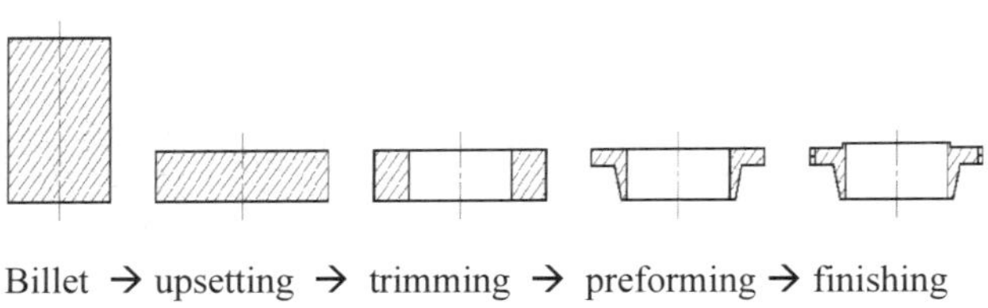

Figure 1.3 The complete forging process of the ring gear part

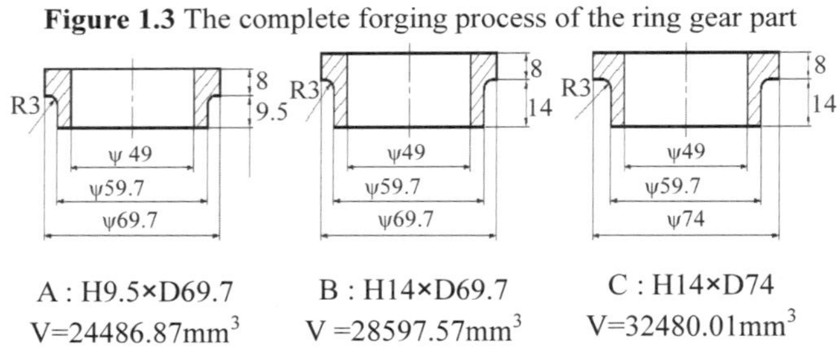

Figure 1.4. The preform designs of the ring gear forging process

4. FEM simulation and Experiments Setup

In this paper, the FEM code DEFORM 3D is adopted for the forging process simulation. The FEM model of the tools and workpiece is shown in Figure 1.5. The forming load and the stress/strain distribution of the workpiece can be predicted. The Cockroft-Latham damage criterion [18] is applied to estimate defect tendency of the forged product. The critical damage factor, C, is defined by equation 1.1

$$\int_0^{\bar{\varepsilon}_f}\left(\frac{\sigma^*}{\bar{\sigma}}\right)d\bar{\varepsilon}=C \qquad (1.1)$$

where

σ^* is the maximum principal tensile stress, $\bar{\sigma}$ is the maximum principal tensile stress, $\bar{\varepsilon}_f$ is the fracture strain. The forging conditions and simulation models are shown in Table 1. The workpiece has been modelled as a plastic part, the other parts of the die set are assumed to be rigid during the forging process. There are four types of motion locus, circle, straight line, mult-leaf, and spiral, can be chosen in the PXW-200 press (Figure 1.6). The circular type of locus has been set for convenience of experiment control. The press is hydraulically driven with the maximum capacity of 200 tons. The feed per revolution of the die can be controlled from 0.2-2 mm. The rocking angle can be adjusted from 0 to 2 degrees. The rotation speed of the punch is adjustable from 30-300 rpm with dial. The forging conditions are translated to the suitable parameters of FEM simulation. The assembled tool set is shown in Figure 1.7 and installed on the PXW-200.

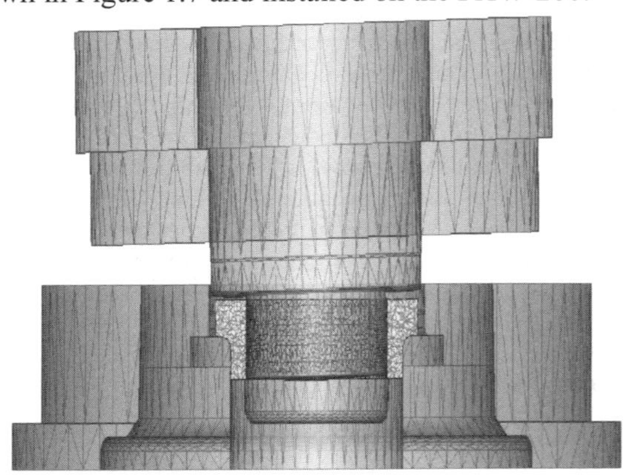

Figure 1.5. The FEM model of the ring gear forging

Table.1. The conditions of forging experiemtns and models of simulation

Orbital forging conditions	Motion type	Circular locus
	rocking angle	2°
	Feed per rev.	0.2 mm/rev
	Revolution speed	100 rpm
Models and parameters of the simulation	Workpiece	Element = 74701 Material = SCM415 Object type = plastic
	Punch	Angular speed = 10.472 (rad/s)
	Die	Axial speed = 0.333 (mm/s)
	Friction models and coefficients	punch = 0.1 (coulomb) die and ejector = 0.1 (constant shear friction)

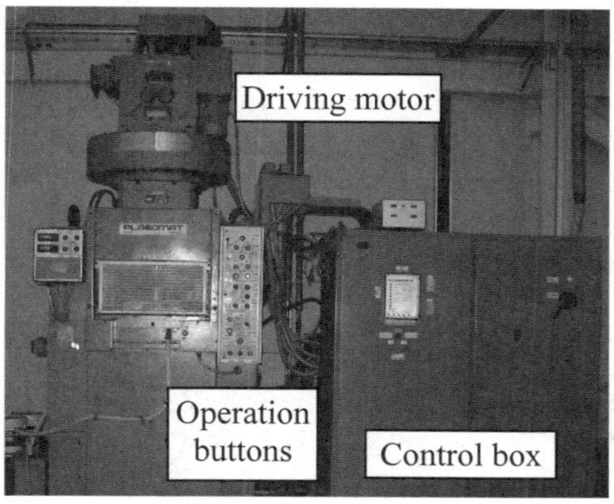

Figure 1.6. The PXW-200 orbital forging press with capacity of 200 tons

Figure 1.7. The assembled die set of the orital forging test

In order to obtain the material model of the SCM-415, an universal material testing machine has been employed for the cylindrical compression testing. The dimensions of the specimen for the compression testing are 10 mm in diameter and 20 mm in height. The force-displacement curve has been recalculated to obtain the true strain-stress curve. The flow stress model and the testing data are shown in Figure 1.8.

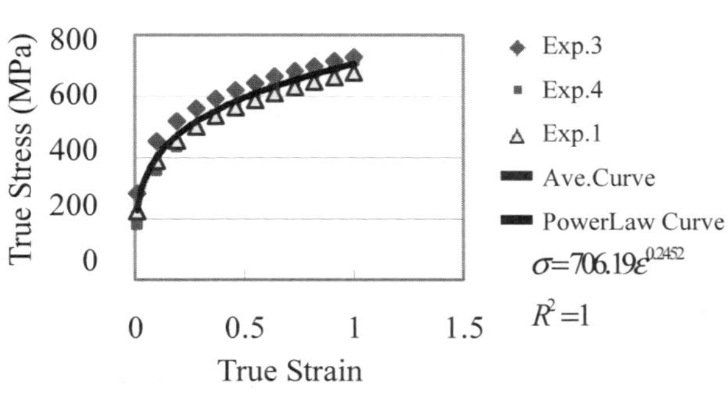

Figure 1.8. The flow stress model of the workpiece and the upsetting test data

5. Results and Discussions

The different types of forging and preforms are compared to find out suitable die and process design for the ring gear production.

5.1 Comparison of the Conventional and Orbital forging processes

The forging loads of the conventional and orbital forging processes using A-type preform are shown in Figure 1.9. The load of the conventional forging process increases dramatically at the final die filling stage and up to the level of 600 tons. While the orbital forging process shows a lower increase of foring load with respect to the die stroke and only requires 65 tons. The lower forming load will give a longer tool life and save the production cost.

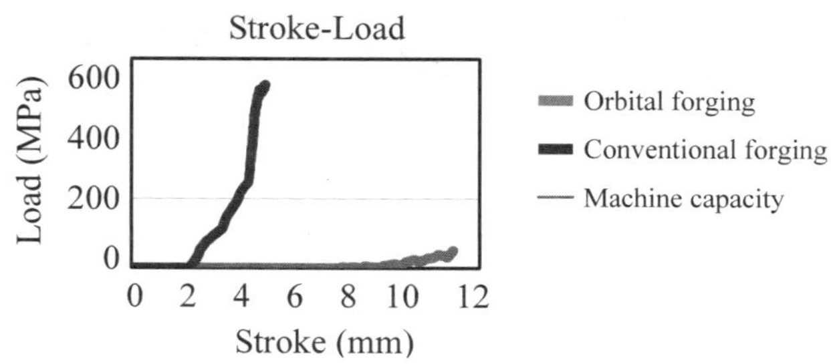

Figure 1.9. The forging loads of the orbital forging and conventional forging

5.2 Forging Load Comparison of the Different Preform Designs

The forging loads of the orbital forging process using the A, B, and C preforms are shown in Figure 1.10. The maximum load of the C-type preform with larger outmost diameter and bigger height of lower ring is higher than the other designs. The larger preform design seems difficult for the material flow and results in a higher forging load. The smallest preform is the best design not only for the reason of saving material but also for the sake of lowering tool loads.

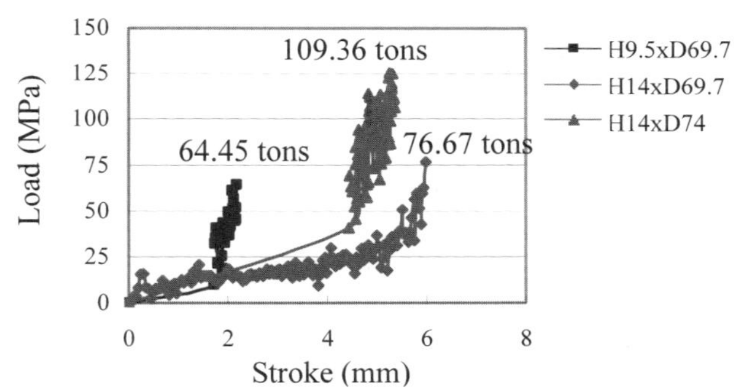

Figure 1.10. The forging loads of the orbital forging process using different prefom designs

5.3 Deformation Comparison of the Experimental and the Simulation Results

The preform design A (H=9.5, D=69.7) is adopted for the forging experiments. Figure 1.11 shows the deformed workpiece at different die strokes of the experimental results and the FEM prediction. The tendency of die filling predicted is similar to the experimental results. It demonstrates the simulation result of the deformation process is feasible.

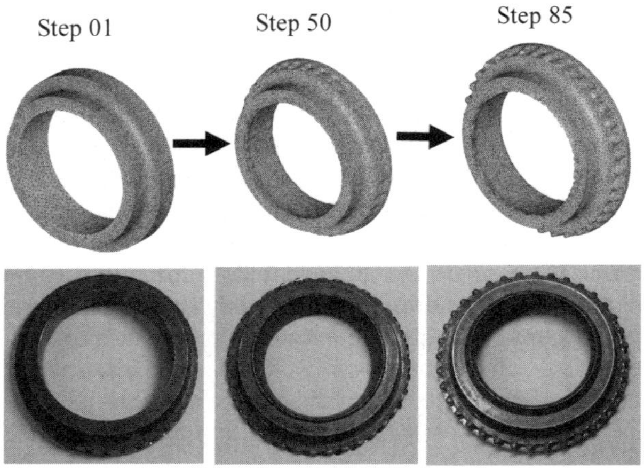

Figure 1.11. The FEM predicted workpiece deformation and the experimental results

5.4 Strain and Damage Value of Different Preform Designs

The effective strain and damage value distributions of the different preform designs are given in Figure 1.12 and 1.13, respectively. The preform design B has the higher maximum effective strain and damage value than the other two designs. The preform design A has the lowest strain and damage.

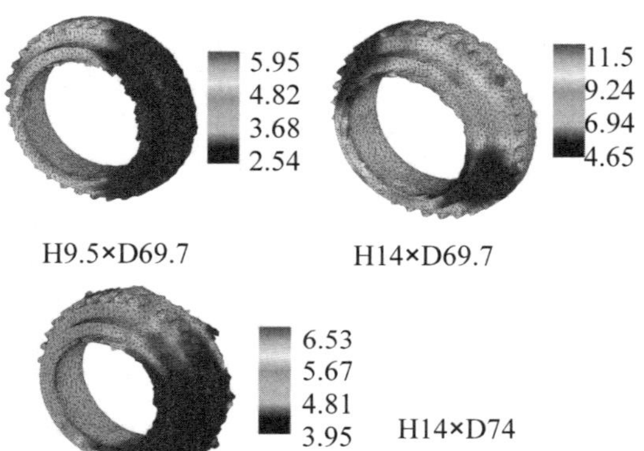

Figure 1.12. The effective strain distribution of the different preform designs

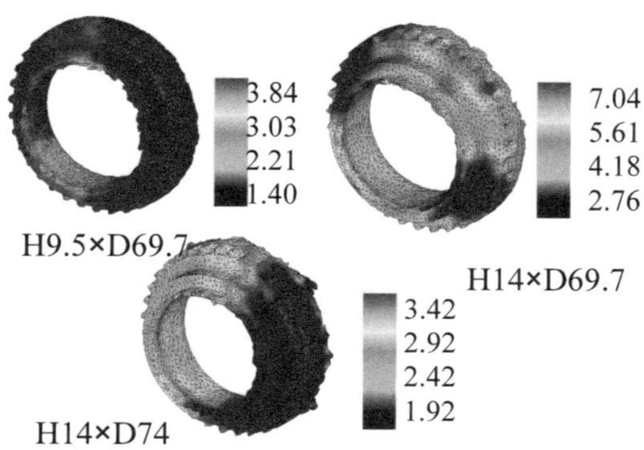

Figure 1.13 The damage value distribution of the different preform designs

6. Conclusion

The orbital forging process of a ring gear part has been simulated successfully by using FEM code which demonstrates the similar deformation of workpiece. The forging load of the conventional forging process is 9 times of the orbital forging process. The preform design with a smaller volume is a better design for both material flow and material saving. The decrease of the outmost diameter of preform results in a higher maximum effective strain and damage value. The maximum effecitve strain and damage value are increased with the increase of the height of the lower ring of the preform.

Acknowledgements

The authors would like to thank the financial support of the NSC of Taiwan. The man powers and experimental facility supports of the NEW KAILUNG GEAR & MACHINERY CO., LTD are also thanked. The granted project number is NSC-94-2622-E151-019-CC3.

References

[1] Liu, G., Zhang, L.B., et al. 2004, J. of Materials Proc. Tech. Vol. 150, P56-61.
[2] Canta, T., Frunza, D., et al. 1998, J. of Materials Proc. Tech. Vol. 80-81, P195-198.
[3] Rusz, S. and Dyja H. 2004, J. of Materials Proc. Tech. Vol. 158, P604-608.
[4] Wang, G., Jing, G., et al. 2005, J. of Materials Proc. Tech. Vol. 169, P108-114.
[5] E. Kalinowska-Ozgowicz, et al. 1997, J. of Materials Proc. Tech. Vol. 64, P215-222.
[6] J. Oudin, et al, 1985, Int. J. Mech. Sci. Vol. 27, No. 11/12, P761-780.
[7] Wang, G. et al. 1997, J. of Materials Proc. Tech. Vol. 71, P293-304.
[8] S. Choi, et al., 1997, J. of Materials Proc. Tech. Vol. 67, P78-82.
[9] Wang G., et al., 1999, J. of Materials Proc. Tech. Vol. 95, P112-115.
[10] Shijian Y. et al., 1999, J. of Materials Proc. Tech. Vol. 86, P252-256.
[11] Li, G. et al., 2001, J. of Materials Proc. Tech. Vol. 113, P40-45.
[12] Wang, G. et al., 2002, Finite Elements in Analysis and Design Vol. 38, P1151-1164.
[13] Hung-Kuk Oh and Seogou Choi, 1997, J. of Materials Proc. Tech. Vol. 66, P101-106.
[14] Hung-Kuk Oh and Seogou Choi, 1997, J. of Materials Proc. Tech. Vol. 68, P23-26.
[15] Zhou, D.C. et al., 1992, J. of Materials Proc. Tech. Vol. 33, P471-479.
[16] Liu, G. et al., 2004, J. of Materials Proc. Tech. Vol. 151, P178-182.
[17] Choi, J.C. and Choi, Y., 1999, Int. J. of Machine Tools and Manufacture. Vol. 39, P1575-1588.
[18] Fang, Gang. Et al., 2002, J. of Materials Proc. Tech. Vol. 122, P 249-254.

Semisolid Extrusion of Wrought Magnesium Alloy AZ61 and Its Mechanical Properties

S. Sugiyama[1], J. L. Kuo[2], S. H. Hsiang[3] and J. Yanagimoto[4]

[1]The University of Tokyo, Tokyo, Japan, 153-8505, sugi@iis.u-tokyo.ac.jp
[2]National Taiwan University of Science and Technology, Taipei, Taiwan
[3] National Taiwan University of Science and Technology, Taipei, Taiwan, shhsiang@mail.ntust.edu.tw
[4]The University of Tokyo, Tokyo, Japan, 153-8505, yan@iis.u-tokyo.ac.jp

Abstract. The basic characteristics of extrusion tests on semisolid wrought magnesium alloy AZ61 are investigated. The selected temperature range is from hot extrusion at 300 °C up to the semisolid temperature. Other process parameters investigated are extrusion ratio, which was varied from 6.3 to 39, and extrusion speed which was varied from 3 mm/s to 150 mm/s. Results are expressed in terms of forming load and process limit by comparing hot extrusion with semisolid extrusion. Moreover, the microstructure and hardness of the extruded products are investigated and correlated with the mechanical properties.

Keywords: Wrought magnesium alloy, AZ61, Semisolid, Extrusion

1. Introduction

Magnesium alloys are widely used in several industrial fields, and currently, the demand for such alloys is increasing significantly. Examples of use include highly popular electronic gadgets such as cellular phones, digital cameras and personal portable computers, the cases of which are made of magnesium alloys formed through a semisolid die cast or by thixomolding process. Recently, wrought magnesium alloy has been used to satisfy the demand for high-tensile-strength and high-elongation materials and because of the shorter manufacturing process, which does not require deburring during press forming [1]. Even though there are many reports dealing with the semisolid processing of magnesium alloys, most of them discuss cast magnesium alloys [2]; a very limited number of reports on the extrusion of wrought magnesium are available [3-6].

In this study, semisolid extrusion using wrought magnesium alloy AZ61 is carried out to investigate the effect of extrusion temperature, extrusion (ram) speed and extrusion ratio on the extruding and product characteristics.

2. Experimental Condition and Methods

Commercially available wrought magnesium alloy AZ61, prepared by hot extrusion, is used in the experiment. AZ61 is turned to have a 10 mm diameter and 10 mm height for use as an extrusion billet. The chemical composition is shown in Table 1.1. It has already been demonstrated that the start and end points of solidification are respectively 575°C and 410°C from the cooling curve of the alloy.

A compression tester with a high-temperature pressure-quenching function is used for the extrusion. The die, container and punch are set in the compression tester, and the billet, including its container, is heated in an induction heating furnace in nitrogen atmosphere (Fig. 1.1). The extrusion is carried out under a temperature range from hot extrusion at 300°C up to the semisolid temperature of 550°C, an extrusion speed range from 3 mm/s to 150 mm/s and an extrusion ratio from 6.3 to 39 by a direct extrusion process. The process does not use a lubricant. The extruded products are left to cool naturally in air.

Table 1.1. Chemical composition of AZ61

Al	Zn	Mn	Fe	Ni	Cu	Si
6.400	0.740	0.3500	0.0010	0.0012	0.0029	0.0150
Pb	Ca	Sn	Zr	Other Total		Mg
0.001	0.001	<0.001	-----	0.30		Bal

The microstructure and Vickers hardness are then tested for the extruded products. As an etching solution for microstructure observation, a mixture of 50 ml ethanol, 5 g picric acid, 5 ml acetic acid and 10 ml water is used.

Table 1.2. General chart of AZ61 extrusion

Billet	
Material	AZ61
Size /mm	∅10-10
Semisolid Temp. range / °C	410-575
Extrusion	
Container Size / mm	∅10-20
Die size / mm	∅1.6, 2, 4 – B.L.5
Extrusion Temp. / °C	300, 400, 460, 500, 520, 550
Extrusion ratio	6.3, 25, 39
Extrusion Speed / mm^{-1}	3, 30, 150
Atmosphere	Nitrogen

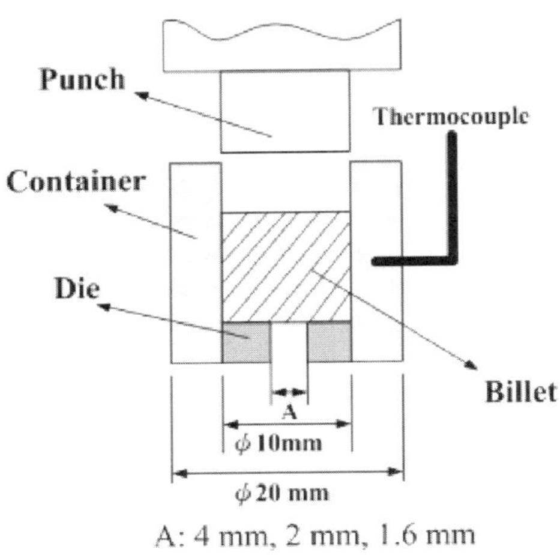

Figure 1.1. Diagram of the extrusion assembly

3. Experimental Results and Discussion

The microstructure of the wrought alloy AZ61 is shown in Fig. 1.2. (a) The grain size is 20-30 μm, and the shape is slightly angular. In contrast, the microstructure in the semisolid state at 525°C, although quenching was done at the temperature, has a grain size of 50-80 μm and a spherical shape.

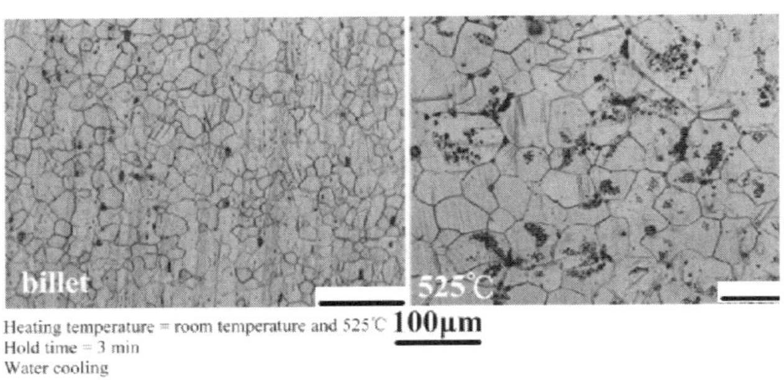

Figure 1.2. Microstructure of the raw material and semisolid state

The relationship between stroke and extrusion pressure is shown in Fig. 1.3. The extrusion ratio in this case is 6.3, and the extrusion speed is 3 mm/s. The decrease in extrusion pressure that accompanies the extrusion is slight for semisolid extrusion, and the extrusion pressure of the semisolid extrusion is about half that of hot extrusion.

Figure 1.4 shows example of the extruded products. The surface quality of the product extruded between 300°C and 500°C is good. However, there are cracks in the top of the product extruded at 520°C, because of the excess liquid component. Figure 1.5 shows the effect of the extrusion speed on the surface quality of the product. For the extrusion at over 30 mm/s, transverse cracks are generated in the product surface. This is because the extruded material is not sufficiently cooled in the die. The microstructures of hot- and semisolid-extrusion products are shown in Fig. 1.6.

The cross section and longitudinal section of the central portion along the length of the product are observed. The microstructure becomes coarser with increasing extrusion temperature. Grain flow is observed in the longitudinal section of the product extruded at 300°C. There is no significant change in the internal structure of the semisolid-extruded products on cross section or the longitudinal section.

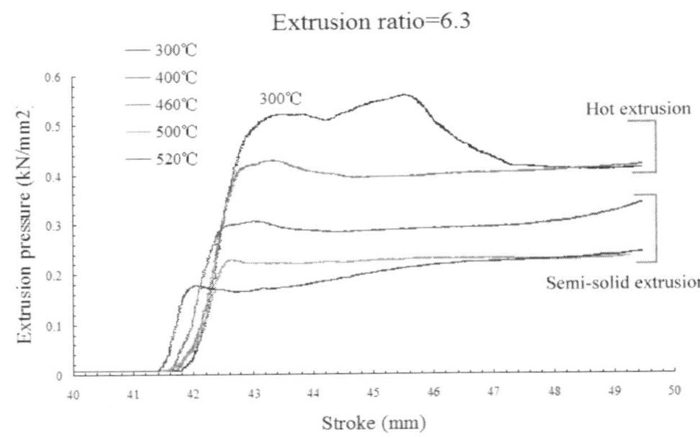

Figure 1.3. Extrusion pressure – stroke curves

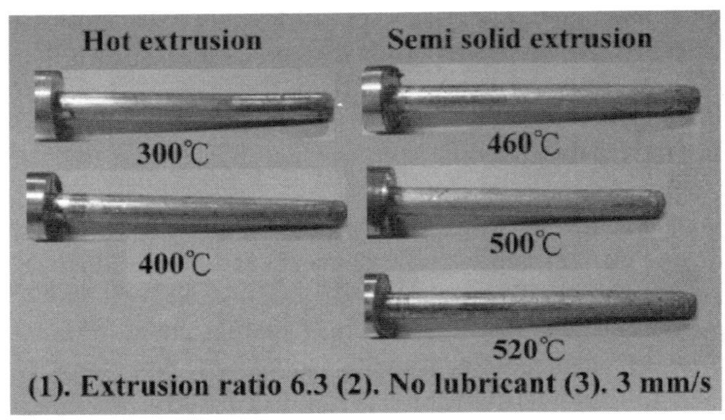

Figure 1.4. Surface property of extruded products

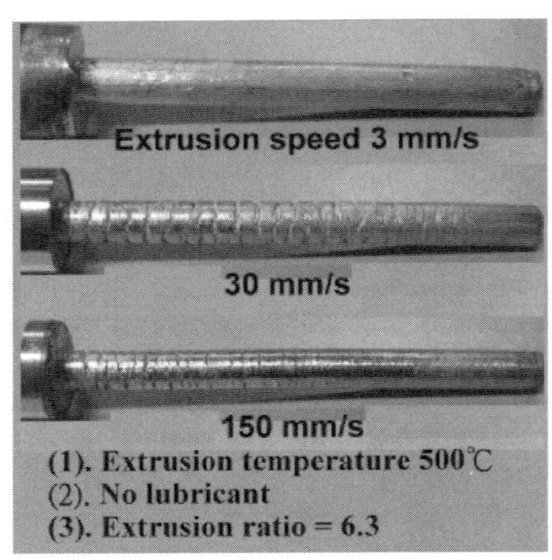

Figure 1.5. Effect of extrusion speed on surface quality

The microstructures at each portion on hot- and semisolid-extrusion products are shown in Fig. 1.7. The numbers from 1 to 4 correspond to the positions in the figure. During hot extrusion, the microstructure becomes smaller with the progress of the extrusion; however, during semisolid extrusion, there is little change in microstructure.

Figure 1.8 shows the measurement result of Vickers hardness of the product. The hardness was measured in radial direction at 1 mm intervals on the cross section of the products. Heat treatment was not given to the products. The following findings are obtained from the measurements.

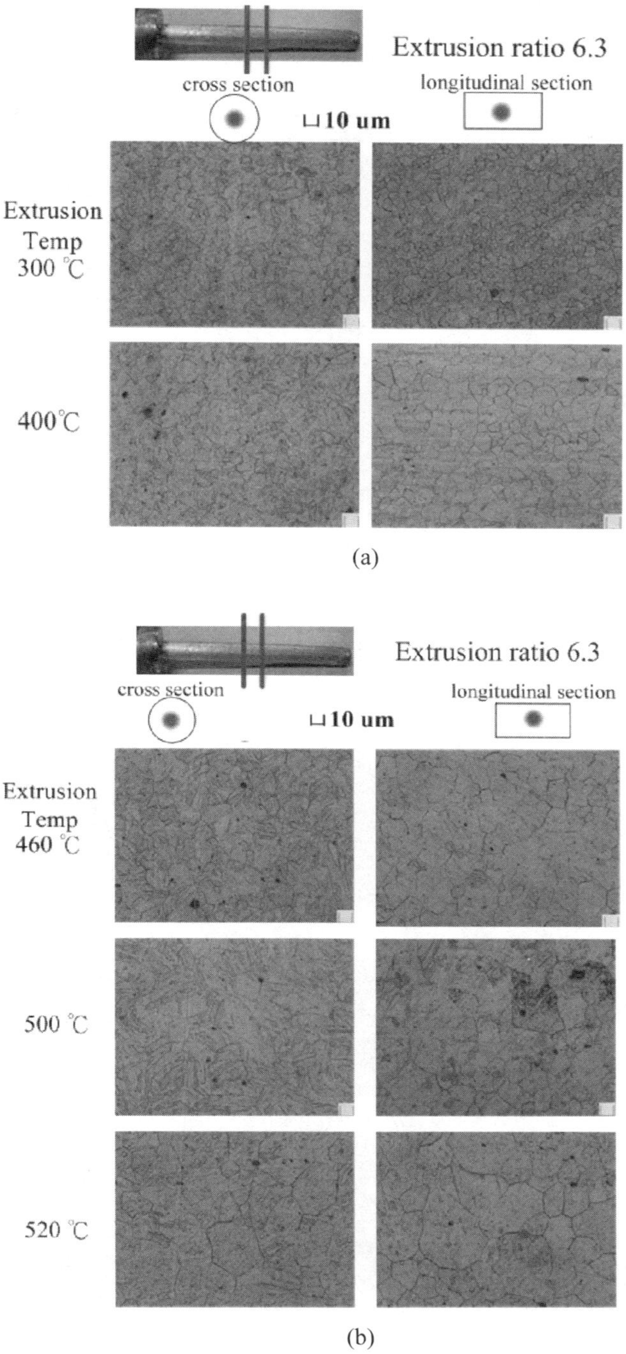

Figure 1.6. Microstructures of extruded products: (a) using hot extrusion, (b) using semi solid extrusion

1. The hardness of the hot-extrusion product is greater than that of the semisolid-extrusion product by about 10%, except for the result for 520°C.
2. The central part of the cross section of the semisolid-extrusion product is slightly softer than the peripheral part.

As a possible reason why the hardness result for 520°C deviates from the trend of hardness, at this temperature, a precipitate appears to form in the grain or grain boundary; the details of this will be the subject of a future study.

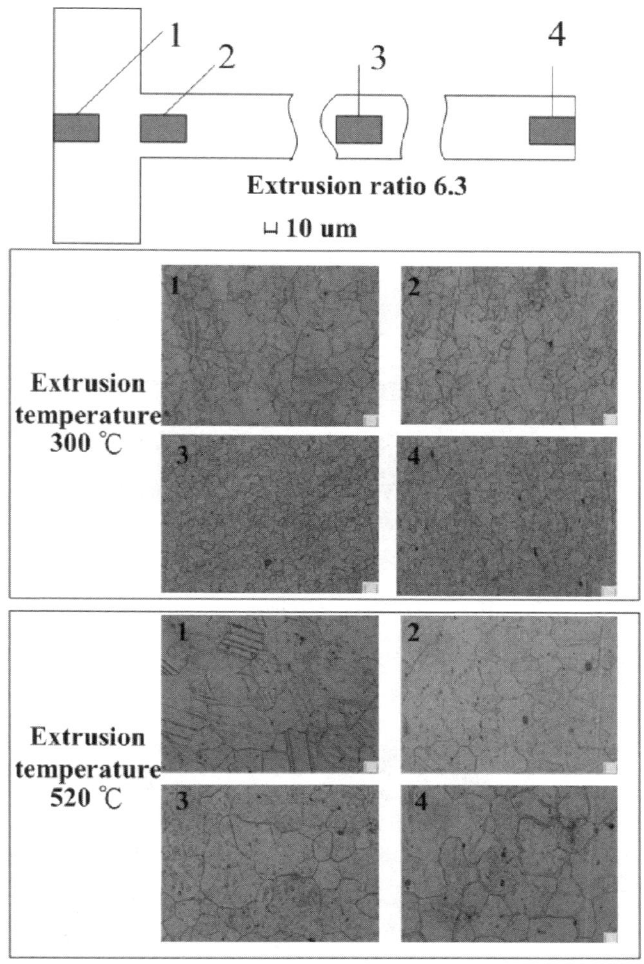

Figure 1.7. Microstructures at each position in extrusion direction

4. Conclusion

In this paper, we focused the semisolid extrusion of wrought magnesium alloy AZ61. In this study, extrusion force, the microstructures of the products and hardness distribution of the products were investigated. The results are summarized as follows:

- High-quality products were formed by semisolid extrusion.
- The semisolid-extrusion force was roughly half that of the hot-extrusion force.
- In the semisolid extrusion, the control of the surface quality of the product becomes difficult with increased extrusion speed.

- The size of the microstructure of the semisolid-extrusion product was 10 times that of the hot extrusion product.
- The hardness of the semisolid-extrusion product at 500°C was about 10% less than that of hot-extrusion product.

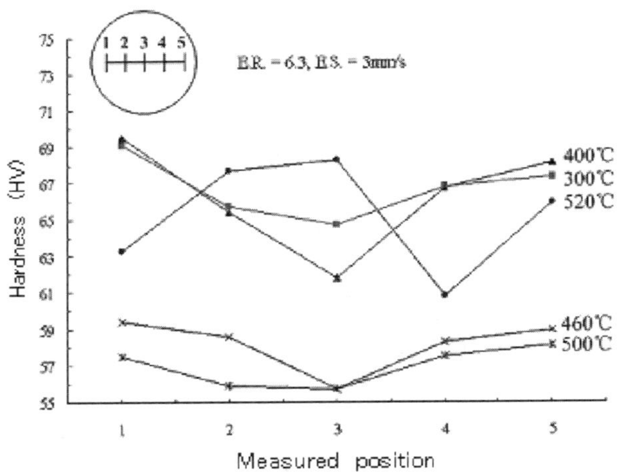

Figure 1.8. Vikers hardness of products

References

[1] S. H. Hsiang, J. L. Kuo, 2003, J. Mater. Process Techno. 140, pp. 6-12.

[2] Y. Doi, Y. Uetani, T. Yamazaki, K. Himi, 2002, The 7th Int. Conf. on Semi-Solid Proc., pp.771-776.

[3] J. Aguilar, M. Fehlbier, G. Rettemeier, P. R. Sahm, W. Michaeli 2002, The 7th Int. Conf. on Semi-Solid Proc., pp.605-610.

[4] M. Fehlbier, J. Aguilar, E. Schaberger-Zimmermann, A. Buhrig-Polaczek, P.R. Sahm, 2002, The 7th Int. Conf. on Semi-Solid Proc., pp.521-526.

[5] Z. Ji, S. Sugiyama, J. Yanagimoto, 2006, Solid State Phenomena Vols. 116-117, pp.159-162.

[6] J. L. Kuo, S. Sugiyama, S. H. Hsiang, J. Yanagimoto, 2006, Int. J. Adv. Manuf. Technol. 29, pp.670-677.

Improving of the Productivity and the Quality of a Manufacturing Robotized Cell for MIG/MAG Welding

A. Barcellona[1], M. Bruccoleri[1], C. D'Onofrio[1], D. Palmeri[1] and R. Riccobono[1]

[1]University of Palermo, Viale delle Scienze 90128, Palermo, Italy, barce@dtpm.unipa.it, manbru@dtpm.unipa.it, cdonofrio@dtpm.unipa.it, palmeri@dtpm.unipa.it, rr@dtpm.unipa.it

Abstract. During the last years, globalization and the necessity to reduce production costs have pushed industry to employ more and more manufacturing robotized cells. This trend has been assimilated by arc welding too. The use of robotic systems allows to increase the performance of productive systems by means of reduction in lead-time, reproducibility of processes and increase in quality. However, process automation needs pre-scheduling of manufacturing cell movements and also a setup of different welding parameters, such as power provided by the welding generator, arc length and robot speed, all of which allows to get a joint without defects. An online scheduling of robotic systems brings certainly a delay in production times, whereas offline scheduling allows to maintain the desired production without any production stop. An interface that allows to easily translate the coordinate points established by a simulation software (such as Robcad) in some complete programs that can be sent to the controller of the manufacturing cell, has been developed. Besides, in order to improve the joint's quality, an analysis of the joint quality has been carried out by macrographic analysis. The employment of image analysis techniques on the obtained macrographs allowed to translate the morphology of each joint in terms of a matrix. After different morphologies have been distinguished in numerical terms and correlated to the joint's quality, these data have been introduced as input to the interface. The output, constituted by the program for the controller of the cell and the best welding parameters, under such conditions, has allowed to increase both the cell's productivity and the quality of the joints. Therefore, during offline scheduling, the initial phase of loop control of the joint morphology allows to evaluate the best welding parameters both in terms of productivity and joint quality. The same interface allows also to plan the best parameters for different welded joints with varying thickness and typology, which makes the system more eflexible and maintains a high level of productivity.

The whole system has been implemented on a robotized cell GMAW Gas Metal Arc Welding Comau, that foresees a welding robot with six degrees of freedom and a position system with other six freedom degrees. Different T fillet joints have been realized at using of weathering steel sheets 5 and 6 mm thick.

Keywords: MIG/MAG welding, robotised cell, productivity, process parameters

1. Introduction

The need to interact with the environment characterized by high competition induces industrial environments towards the introduction of flexible robotic manufacturing cells. This also applies to Gas Metal Arc Welding (GMAW) production environments. The gas metal arc welding is increasingly employed for fabrication in many industries. The process is versatile, since it can be applied to position welding, it can be easily automated and can easily be integrated into robotized production centres. The advantages of this process have motivated many researchers to study the GMAW process in detail.

This manufacturing cell typology consists of one or more welding robots and systems for positioning the elements in a cell. The proposed approach is based on offline programming, simulation and automatic generation of the control program, both in terms of robot positioning and selection of welding parameter. The main goal of this paper is to present an offline approach for programming a robotic manufacturing cell in such a way that the control software program is automatically generated after having simulated the given sequence of movements offline in a user-friendly environment. Moreover, the offline programming foresees the selection of the best set of welding parameters for the given joint geometry [1-4].

For such purposes, an actual test-bed has been considered as a case study.

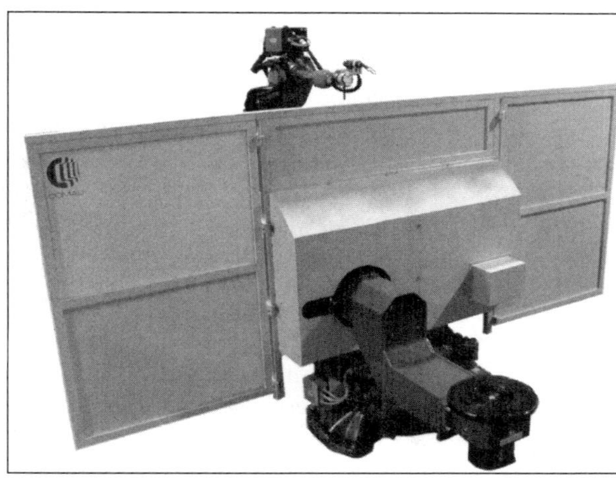

Figure 1.1. Robot Smart Six 6-1.4 and Smart PTDOrb 250 COMAU

The robotic manufacturing system which has been considered for practical application of off-line programming and automatic robot program generation is a COMAU robotized welding cell composed by a Robot Smart Six 6-1.4 ARC and a orbital double lathe positioner Smart PTDOrb 250 COMAU (see Fig.1.1). The robot is anthropomorphous with six axis and a load at wrist of 6 kg, while the positioner presents a handling capacity of 250 kg and five axis. This robotic cell is used for welding a high variety of workpieces in small batches.

The specific welding operation requires to maintain the orientation of the torch with respect to the weld position and, in large groove welding operations, this also requires multiple layers to fill. For these reasons, the online teaching approach is not a suitable approach for small batches production.

On the other hand, an offline programming approach can be used to drastically reduce inefficiencies; the robotized cell can continue processing workpieces while the programmer realizes the new sequence of movements and operations that will be subsequently loaded into the robot's controller. The program, which is written offline, should be tested before launching the actual production run and simulation tools have to be used to support the programmer in the control program development phase. As an example of such kind of tools, eM-Workplace® by Tecnomatix Inc. is a software platform to simulate robotic environment. In this paper it is shown how this simulation capability can be employed to support the control and synchronization activities of a welding robotic cell.

A set of invariant geometrical attributes that characterize the fusion zone shape were selected. Thereafter, a classification of this zone depending on the welding current was made. For each welding condition, that in this study means different sheets thickness, an optimum parameter combination in order to obtain good quality welds was selected.

2. GMAW process for weathering steels

In GMAW welding, the electric arc is generated in a protective gas atmosphere, between a fusible wire and the sheets to be welded. Gases can be either inert or active. The heat of the arc fuses the base material and the wire, which forms the weld material. The wire is fed out constantly from an automatic feeder [5-7].

Weathering steel (or Cor-Ten Steel) is a very common material that has been used for many years for steel construction in the field of transportation. The chemical composition of weathering steel is able to promote, after exposure in the atmosphere, the formation of a stable and protective oxide layer reducing remarkably the corrosion rate. For this material the best way to produce a weld with corrosion resistance equal to or better than the base metal is to use an "over matching" filler metal. In this study the experimental tests were performed on sheets with differents thickness of weathering steel using as filler material an over-matching wire FILCORD 48.

The yield strength of base material is equal to 350 MPa whereas for the filler material is equal to 420 MPa. The welding performances were evaluated by means of Vickers micro-hardness tests. The best condition was detected in the maximum of the over-matching effect. The base material was characterized by means of microstructual analysis afther proper grinding, polishing and etching procedures (Fig.1.2).

In this way the different welding zones have been highlighted. This was the starting point for the subsequent image analysis process in order to select an indicator sensitive to the joint quality at varying of the welding current.

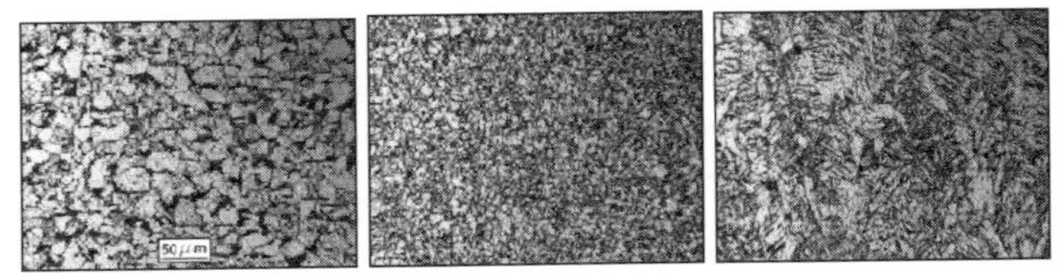

Figure 1.2. Microstructure of base material, heat affected zone and fusion zone

3. Interface for control position

Two approaches exist for developing proper control software for robotized welding cells: online teaching or offline programming. Online teaching is accomplished by acquiring workpiece positions and robot movements and creating the robot sequence of movements for a given production order afterward. In general the operator uses a Programming Terminal that allows manually commanding movements of the robot, programming it, performing and modifying movements step by step, and it is also useful for robotic systems monitoring.

Although online teaching is a very simple approach for welding robot programming, it results useful when it is possible to reuse or to modify a program that are created for welding workpieces that are very similar in dimension and size, i.e. in those situations where only few modifications in positioning the object are required.

If a completely new object has to be processed by the robotized cell, a new program has to be written and this activity, being very time-consuming, causes a critical loss in production efficiency since, for obvious reasons, the robot cannot be used for production purposes during the programming activity. In a robotic welding cell, new arc-welding operations need to be programmed.

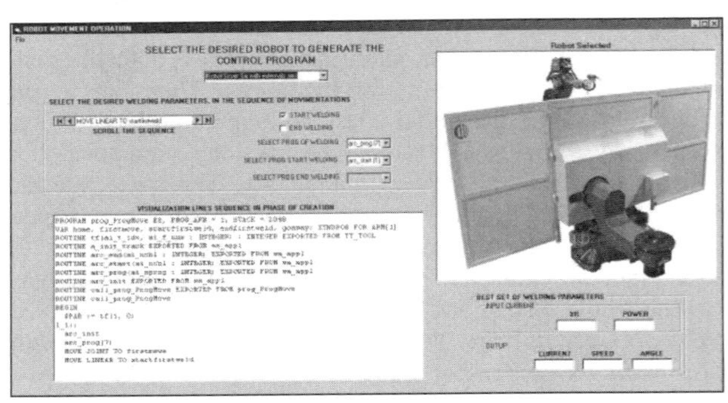

Fig. 1.3. Interface to select how to convert robot program and the welding current value

The conversion of the program, which has been generated by eM_Workplace, into the proprietary language suitable for being recognized by the robot controller has to be done by the user-friendly interface, which has been developed (see Fig. 1.3). This system allows the user to create a modality to convert the program of a "default" controller in a specific robot controller program. At the beginning, the user has to select the name associated to the robot, with relative number of axes and number of external axes. Then the user has to write the header of the program that the interface must generate at the beginning of every program. Once the process of insertion of each element is finished, the interface processes and translates the program in a way which is interpreted by the Comau robot controller. In the same window it is possible to select the best welding parameters for each welding process by means of the introduction of the invariant geometrical attribute value.

4. Interface for control of welding parameters

Prediction of fusion in welding is essential for obtaining adequate joint qualities and increasing productivity. Intelligent automated welding processes require adaptive control technologies that can predict weld fusion in a simple and easy manner [8-10].

The control of the welding properties, quality and productivity in the GMAW process is not an easy task because of the complexity of relationships between the process parameters and the fusion zone geometry. In fact it is difficult to etablish a mathematical model that can predict the optimum welding condition for a given joint geometry. Furthermore, the geometry of the fusion zone plays an important role in determining the mechanical characteristics of the weld. The purpose of this study is to present a practical methodology for the prediction of fusion zone shape in welding. In order to realize a production-friendly predictive model, the present study adopts a method that obtains a prediction of the best welding parameters by employing a semi-empirical procedure that, starting from systematic experiments, is able to select the welding parameters from the evaluation of the fusion zone morphology. The experimental welding process employed in this study is T-joint fillet with gas metal arc welding (GMAW), with the following conditions: welding power range equal to 20% to 100%. The used values were 20 – 35 – 37 – 40 – 45 – 50 – 70 – 80 – 90 and 100%. In this first step of the study, the other process parameters, i.e. torch speed and feeding wire, were maintained at the constant level equal to the optimum level previously determined. Therefore, only the influence of the welding current was taken into account. With such aim, different T-fillet joints were produced using different current values. In order to select an invariant geometrical attribute that characterizes the fusion zone's shape, for each considered power value, five different cross sections of the joint were made. The macrographic analysis of each specimen, after the etching procedure, allows to evaluate the fusion zone geometry, as shown in Fig. 1.4. The difference in the contrast level between the heat affected zone and the fusion zone has been used in the image analysis step in order to evaluate the extension of the fusion area as mean values for five specimens.

The macrographic images at the same magnification were converted into an RGB image with only three-colour level, red, green and yellow. In fact, considering mathematical operations, the whole fusion zone was split in three different regions, as shown in Fig. 1.5. Two regions are constituted by the fusion zone in each sheet, whereas the third one is the fusion zone outside the sheets. For each welding parameter the mean extension of each area has been evaluated. In such way it has been observed the connexion between the area value and the best welding results highlighted by means of the evaluations of the over matching effect. Therefore, a set of micro-hardness tests were performed in order to evaluate the mean micro hardness for each region, as shown in Fig. 1.6 [11, 12].

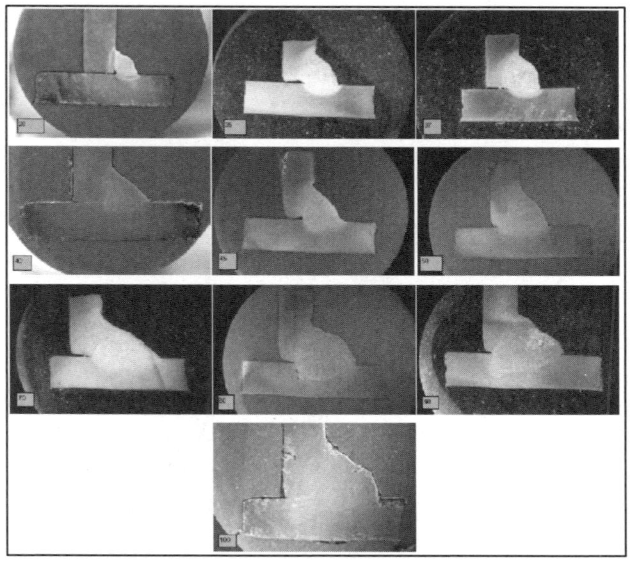

Figure 1.4. Cross section weldig morphology using different welding power

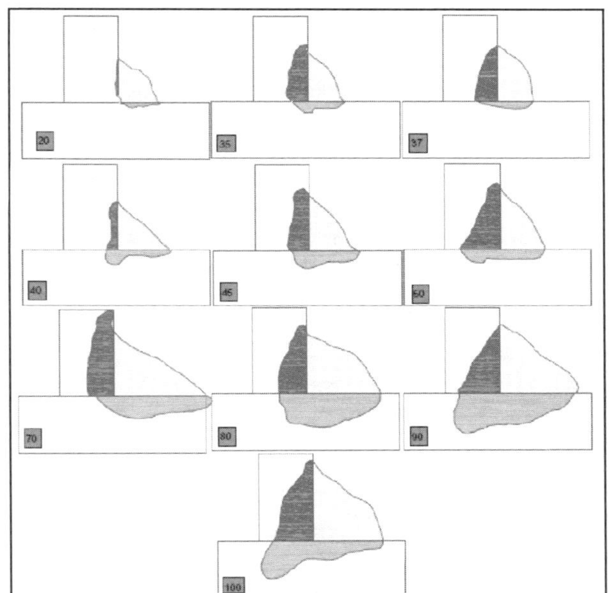

Figure 1.5. Cross section fusion zone morphology at varying of the power level

5. Results

The evaluation of the area value for each region of the fusion zone has highlighted the occurrence of a minimum size of the red area in the best welding condition, determined by the highest micro-harness value of the fusion zone (see Figs. 1.7 and 1.8).

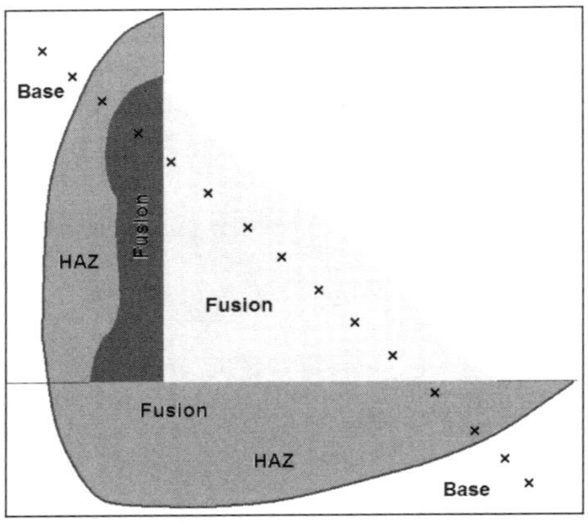

Figure 1.6. Micro-hardness evaluation of different zones

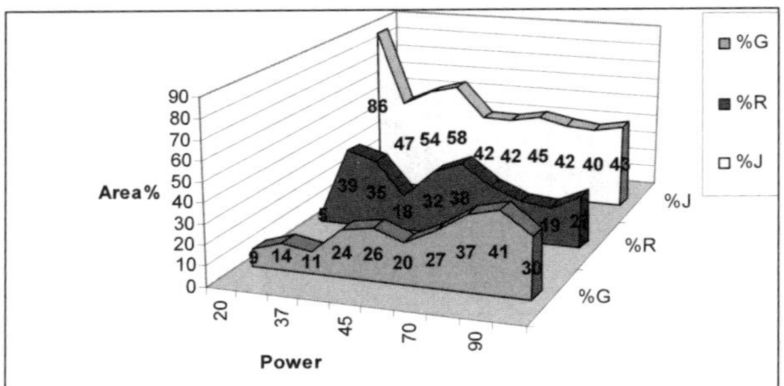

	20	35	37	40	45	50	70	80	90	100
%G	9	14	11	24	26	20	27	37	41	30
%R	5	39	35	18	32	38	27	21	19	27
%J	86	47	54	58	42	42	45	42	40	43

Figure 1.7. Evaluation of the area of three regions at varying of the power level

The observation that the influence of the sheet thickness is relevant on the HAZ shape, but determines only an offset of the fusion zone shape, allows to make a subroutine that can predict the proper best current value only by means of the evaluation of one red area for a chosen current value.

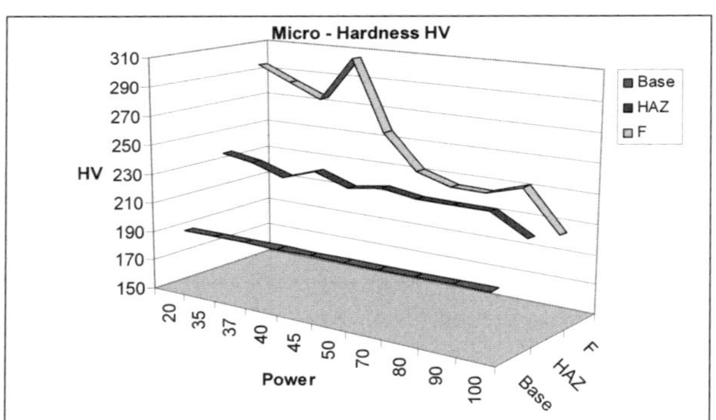

	20	35	37	40	45	50	70	80	90	100
Base	189	189	189	189	189	189	189	189	189	189
HAZ	237	233	226	233	226	230	226	226	226	213
F	295	285	275	306	257	233	226	226	233	206

Fig. 1.8. Mean micro-hardness mesurements at varying of the power level

In fact, the trend of the red area value provides the best current level at the variation of the sheet thickness by shifting the profile.

6. Conclusions

In this paper, an interface that allows to translate easily the coordinate points established with a simulation software in some complete programs that can be sent to the controller of the manufacturing cell, has been presented. This interface is also able to select the best welding current for different thickness by means of the offline evaluation of an invariant geometrical attribute value for only one joint.

References

[1] Kim C.S., Hwang H.S., Han J.M. and Han H.Y, (2002), Module-Base Off-Line Programming for Welding Robots of Panel Blocks. Proceedings of the ASC2002, Singapore.

[2] Kim C.S., Hong K.S., Han H.Y.S., Kim S.H. and Know S.C., (2004), PC-Based Off-line Programming using VRML for Welding Robots in Shipbuilding. Proceedings of IEEE Conference on Robotic, Automation and Mechatronics, Singapore.

[3] Konukseven E.I. and Abidi A., (2004) Development of Man Machine Interface Software For an Industrial Robot. in Proc. VECIMS 2004, Boston MA USA.

[4] Cui M., Dong Z. and Tian Y., (2004) Simulation and Execution of Event-Driven Robot Teaching in a Virtual Environment. Proceedeings of 5thWorld Congress on Intelligent Control and Automation, China.

[5] Berger U., Lepratti R. and May M., 2005, "An Approach for the Automatic Generation of Robot Paths from CAD-Data", Proceedings of ETFA, Italy, pp 291-297.

[6] Chan S.F. and Kwan R., (2003), Post-processing methodologies for off-line robot programming within computer integrated manufacture. Journal of Materials Processing Technology 139, pp 8-14.

[7] D'Onofrio C. and Bruccoleri M., (2006), A user-friendly control system to easy reconfigure a manufacturing cell. Proceedings of the IEEE ICIT Conference, Mumbai, INDIA, pp. 1642-1647.

[8] Kim D., Kang M., Rhee S., (2005), Determination of optimal welding conditions with a controlled random search procedure. Welding Journal 125 – 130-s.

[9] Okui N., Ketron D., Bordelon F., Hirata Y., Clark G., (2007), A Methodology for Prediction of Fusion Zone Shape. Welding Journal 35 – 43-s.

[10] Kim I. S., Son J. S., Park C. E.,. Kim I. J and Kim H. H., (2005), An investigation into an intelligent system for predicting bead geometry in GMA welding process. Journal of Materials Processing Technology,Vol. 159, Issue 1, Pages 113-118.

[11] Kim I.S., Jeong Y.J., Son I.J., Kim I.J., Kim J.Y., Kim I.K. and Prasad K.D.V. Yaragada, (2003), Sensitivity analysis for process parameters influencing weld quality in robotic GMA welding process. Journal of Materials Processing Technology, Vol. 140, Issues 1-3, Pages 676-681.

[12] Kim I.S., Chang-Eun P., Hoon C. Y., Young-Jae J., Kim I. K., Kim J. Y., Son J. S., (2001), A study on development of a new algorithm for predicting the process variables in GMA welding processes. JSME international journal. Series C, vol.44, n° 2, pp. 561-566.

Surface Finish of the Injection Molding Tool Steel PX4 Using Ball Burnishing, Nickel-Plating, and Ball Polishing Processes

F.J. Shiou and C.M. Cheng

National Taiwan University of Science and Technology, Taiwan, 106, shiou@mail.ntust.edu.tw, M9003912@mail.ntust.edu.tw

Abstract. The objective of this study is to improve the surface roughness of PX4 plastic injection mold steel using ball burnishing, nickel-plating, and ball polishing surface finish processes. The test specimens were first fine milled and then burnished by applying the optimal ball burnishing process parameters. The surface roughness value, Ra, of the specimens can be improved from about 1.0 µm on average to 0.086 µm on average, after ball burnishing. In order to protect the metal from oxidization, the burnished specimens were plated with nickel, to a thickness of about 10 µm. The surface roughness of the plated specimens was further improved by ball polishing. The optimal ball polishing parameters were determined after conducting Taguchi's L18 matrix experiments, analysis of variation (ANOVA), and the full factorial experiments. The optimal flat surface polishing parameters for the plastic injection mold steel PX4 were as follows: rotation speed of 6,000 rpm using Al_2O_3 with grid no. 8000 (3 µm in diameter), a feed rate of 20 mm/min, a stepover of 100 µm, and a depth of penetration of 120 µm (polishing force of 1.05 N). Using these parameters for ball polishing, the surface roughness value, Ra, of the burnished and plated specimens was improved from about 0.074 µm on average to 0.02 µm on average.

Keywords: ball burnishing, nickel-plating, ball polishing, Taguchi's method, analysis of variation (ANOVA)

1. Introduction

Polymers have increasingly replaced metallic components in industrial applications due to their specific characteristics, such as resistance to corrosion, resistance to chemicals, low density, ease of manufacture etc. Injection molding is one of the important forming processes for plastic products. The surface finish quality of the plastic injection mold is an essential requirement since it affects directly the appearance of the plastic product. Finishing processes such as grinding, polishing and lapping are commonly used to improve the surface finish of the plastic injection mold.

Tool steel grade AISI P20 is popular for mold construction for large plastic injection products in the field of automobile and domestic appliances since it is relatively inexpensive. The material used in this study was PX4 tool steel (equivalent to AISI P20, PDS5 tool steel) developed by Daido Co. in Japan, which was appropriate for long life use of thermoplastics mold [1]. The hardness of this material is about HRC 30-33.

One specific advantage of this material is that after machining, the mold can be directly used for further finishing processes without heat treatment due to its special pre-treatment. One of the drawbacks of the PX4 and PDS5 tool steels is ease of corrision after executing the polishing process.

Recently, there have been some studies carried out on the surface finish of mold tool steel, and in particular to identify the optimal parameters of ball burnishing. This process results in a plastic deformation on the workpiece surface by using a tungsten carbide ball or a roller, to improve the surface roughness, surface hardness, and fatigue resistance [2-5]. The optimal plane surface burnishing parameters for the plastic injection mold steel PDS5 were found to be a combination of grease lubricant, tungsten carbide ball, burnishing speed of 200 mm/min, burnishing force of 300 N, and a feed of 40 µm [6]. The improvement of the surface roughness through burnishing generally ranged between 40% and 90% [2-6].

Mounted polishing tools have been widely used in conventional mold and die finishing industries. The geometric model of mounted polishing tools for automated surface finishing processes is described in [7]. A finishing process model of spherical polishing tools for automated surface finishing systems has been developed in [8]. Polishing speed, depth of penetration, feed rate, wheel properties such as abrasive material and abrasive grain size, are the dominant parameters for the spherical polishing process, as shown in Fig. 1.1.

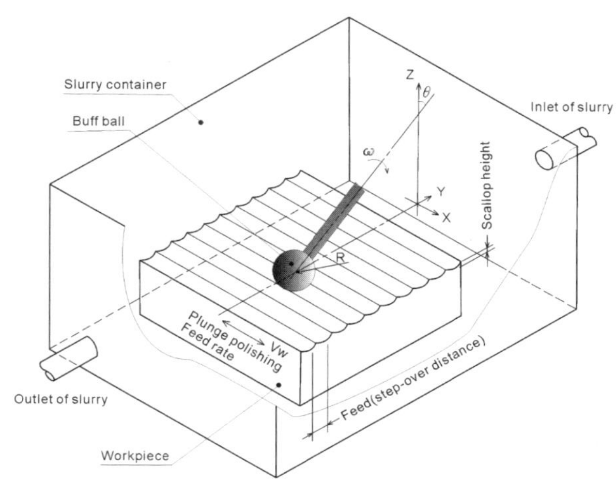

Figure 1.1. Schematic diagram of the spherical polishing process

To overcome oxidization of the PX4 tool steel after having executed the polishing process, the burnished test specimens were plated with nickel to a thickness of about 10 µm. The optimal spherical polishing parameters for the nickel-plated injection mold steel PX4 have not been investigated based on the literature survey. The aim of this study was to determine the optimal spherical polishing parameters of the burnished and nickel-plated mold steel on a machining centre.

2. Planning of matrix experiments

2.1 Configuration of the Taguchi's orthogonal array

The effects of several parameters on a process can be determined efficiently by conducting matrix experiments using Taguchi's orthogonal array [9]. As mentioned, the main spherical polishing parameters having significant effects on surface roughness, are the speed of the polishing ball, abrasive material, abrasive grain size, feed, stepover distance and the depth of penetration etc. Consequently, the speed of the polishing ball, abrasive material with grid no. 6000 for the polishing ball of wool with diameter 10 mm, feed rate, stepover distance and the depth of penetration, were selected as five experimental factors (parameters) and designated as factors A to E (Table 1.1) for spherical polishing in this research.

Table 1.1. Factors and their levels for the matrix experiments

Factor \ Level	1	2	3
A. Revolution (rpm)	2000	10000	18000
B. Abrasives (4~5μm)	SiC	Al_2O_3	Diamond
C. Feed (mm/min)	20	60	100
D. Stepover (μm)	20	60	100
E. Penetration (μm)	60	120	240

Three levels (settings) for each factor were configured to cover the range of interest and identified by the digits 1, 2, and 3. Three types of abrasive materials, namely silicon carbide (SiC), aluminium oxide (Al_2O_3), and diamond, were selected and investigated. Three numerical values of each factor were determined based on the pre-study results. The L_{18} orthogonal array was selected to conduct the matrix experiment for five 3-level factors of spherical polishing process.

2.2 Definition of the data analysis

Engineering design problems can be divided into the-smaller-the-better type, the-nominal-the-best type, the-larger-the-better type, the signed-target type, etc. [9]. The signal-to-noise (S/N) ratio is used as the objective function for optimizing a product or process design. The surface roughness value of the ground surface via adequate combination of polishing parameters should be smaller than that of the original surface. Consequently, the spherical polishing process is an example of the-smaller-the-better type problem. The S/N ratio, η, is defined by the following equation [9],

$$\eta = -10\log_{10}\left[\frac{1}{n}\sum_{i=1}^{n} y_i^2\right] \quad (1.1)$$

y_i: observations of the quality characteristic under different noise conditions

n : number of experiment

The optimization strategy of "the-smaller-the-better" problem is to maximize η defined by equation (1.1). Levels that maximize η will be selected for the factors that have a significant effect on η. The optimal conditions for spherical polishing can then be determined.

3. Experiments and Results

3.1 Setup for experiments

Fig. 1.2 shows the setup for the spherical polishing experiments. The test specimens with flat surfaces were designed and manufactured so that they could be mounted on a dynamometer to measure the machining force. The surface to be burnished or polished was divided into three zones so that the average burnished or polished surface roughness value could be calculated from that of the three zones. The PX4 specimens were roughly machined and then mounted on the dynamometer to carry out the fine milling and ball burnishing processes on the 3-axis machining centre made by Yang-Iron Co. type MV-3A, equipped with the NC-controller of FUNUC Co. type 0M [10].

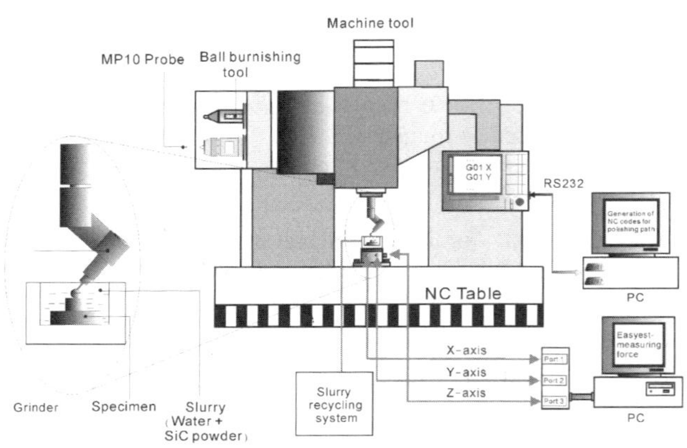

Figure 1.2. Setup for the experiments to determine the optimal spherical polishing parameters

The surface roughness values, Ra, of the fine-milled specimens were about 1.0 μm. The test specimens were further burnished using the optimal ball burnishing process parameters, namely a combination of tungsten carbide ball, burnishing speed of 200 mm/min, burnishing force of 300N, feed of 40 μm, and grease as the lubricant. The surface roughness value, Ra, of the specimen could be improved from about 1.0 μm on average to 0.086 μm on average after the ball burnishing process. The surface roughness of test specimens was measured using the Hommelwerke T4000 surface roughness measuring equipment made in Germany. In order to protect the metal from oxidization, the burnished specimens were taken off and plated with nickel, to a thickness of about 10 μm. The surface roughness value, Ra, of the specimens could be improved from about 0.086 μm on average to 0.074 μm on average, after the plating process. An MP10 touch-trigger probe made by Renishaw Co. was also integrated with the machining centre tool magazine to measure and determine the coordinated origins of the specimens to be machined. The NC codes needed for both ball burnishing and spherical polishing paths were simulated and generated by PowerMILL CAM software. These codes can be transmitted to the CNC controller of the machining centre via RS232 serial interface.

3.2 Results of the matrix experiments

Table 1.2 summarizes the measured polished mean surface roughness value, Ra, and the calculated S/N ratio of each L_{18} orthogonal array based on equation (1.1) after the 18 matrix experiments had been executed. The surface roughness value, Ra, of the specimens could be improved from about 0.074 μm on average to 0.030 μm on average after the polishing process. The average S/N ratio for each level of the five factors can be obtained by taking the numerical values listed in Table 1.2. The average S/N ratio for each level of the five factors is shown graphically in Fig. 1.3. They are the separate effects of each factor, and commonly called main effects.

Table 1.2. Polished surface roughness values and S/N ratios of PX4 specimens of matrix experiments

Exp. No.	Factor/Level					S/N ratio	Ave. Ra
1	A1	B1	C1	D1	E1	29.536	0.0322
2	A1	B2	C2	D2	E2	33.876	0.0189
3	A1	B3	C3	D3	E3	27.987	0.0389
4	A2	B1	C2	D3	E2	31.698	0.0256
5	A2	B2	C3	D1	E3	35.229	0.0171
6	A2	B3	C1	D2	E1	33.442	0.0200
7	A3	B1	C3	D2	E1	28.721	0.0356
8	A3	B2	C1	D3	E2	33.585	0.0210
9	A3	B3	C2	D1	E3	27.299	0.0400
10	A1	B1	C2	D2	E3	28.721	0.0356
11	A1	B2	C3	D3	E1	29.805	0.0311
12	A1	B3	C1	D1	E2	27.969	0.0611
13	A2	B1	C1	D3	E3	32.149	0.0244
14	A2	B2	C2	D1	E1	30.526	0.0289
15	A2	B3	C3	D2	E2	31.431	0.0267
16	A3	B1	C3	D1	E2	28.895	0.0356
17	A3	B2	C1	D2	E3	32.467	0.0233
18	A3	B3	C2	D3	E1	33.883	0.0200

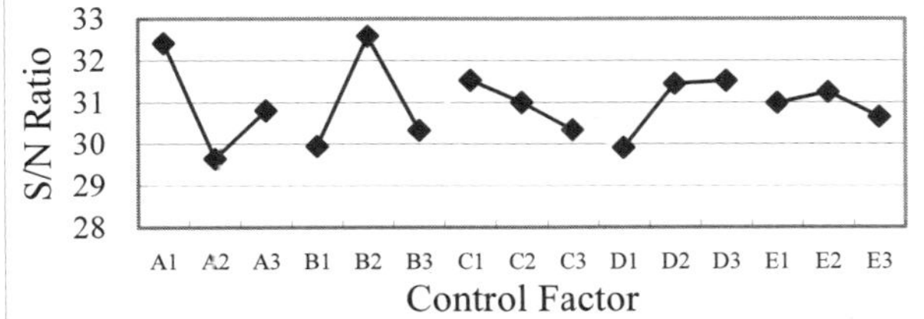

Figure 1.3. Plots of control factor effects

3.3 Combination of the optimal level for each factor

The goal in the spherical polishing process is to minimize the surface roughness values of burnished and nickel-plated specimens by determining the optimal level of each factor. Since –log is a monotonic decreasing function, it implies that we should maximize the S/N ratio. Consequently, we can determine the optimal level for each factor as being the level that has the highest value of η. According to Fig. 1.3, the combination of the optimal level for each factor is A1B2C1D3E2. Therefore, based on matrix experiments, the optimal revolution of the elector grinder was 2,000 rpm, the abrasive material was aluminum oxide (Al_2O_3) with grid no. 6000, the optimal feed was 20 mm/min., the optimal stepover distance was 100 μm, and the depth of penetration was 120 μm (polishing force of 1.05 N). Revolution of the electric grinder and the abrasive material were dominant factors based on the analysis of variance (ANOVA) and F-test. Full factorial experiments of these two factors, each with two appropriate levels, were further carried out to determine the optimal polishing parameters. As a result, the optimal revolution of the electric grinder was 6,000 rpm, and the abrasive material was aluminium oxide (Al_2O_3) with the grid no. 8000 (3 μm in diameter). The surface roughness value of Ra = 20 nanometers (nm) on average could be obtained using the determined optimal ball polishing parameters. A surface roughness value of R_a = 0.96 nm (R_{max} = 14.3 nm) measured using an atomic force microscope within the area of 5x5 um^2, was obtainable using the ball polishing process with the optimal polishing parameters (Fig. 1.4).

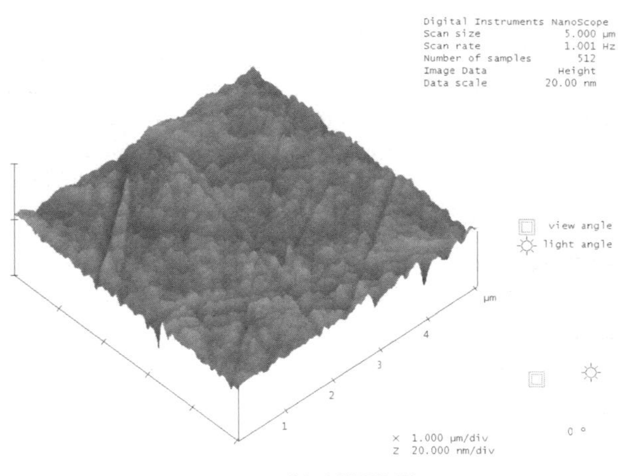

Figure 1.4. Surface roughness (R_a = 0.96 nm, R_{max} = 14.3 nm) of the polished specimen measured using an atomic force microscope within the area of 5X5 um^2

Applying the optimal flat surface polishing parameters to a fine milled, burnished, nickel-plated mold cavity of an aspheric lens, the surface roughness value, Ra, on the 3D tested part can be improved from about 1.2 μm on average to 30 nm on average measured with a KEYENCE VK-8550 color laser 3D stereomicroscope (Fig. 1.5).

Figure 1.5. Fine-milled, burnished, nickel-plated and polished surface of a mold cavity and a mold insert for an aspheric lens

4. Conclusion

Surface finish of PX4 mold steel using ball burnishing, nickel-plating, and spherical polishing processes, was investigated in this work. Based on the results of Taguchi's L18 matrix experiments, the optimal flat surface spherical polishing parameters for the plastic injection mold steel PX4 were as follows; revolution of 6,000 rpm, the abrasive aluminum oxide (Al_2O_3) with grid no. 8000 (3 µm in diameter), a feed rate of 20 mm/min., a stepover distance of 100 µm, and a depth of penetration of 120 µm (polishing force of 1.05 N). The surface roughness value, Ra, of the nickel-plated specimens was improved from about 0.074 µm on average to 20 nm on average using the optimal spherical polishing parameters. The surface roughness improvement of the tested specimens on burnished surfaces was about (1.0-0.086)/1.0 = 91.4%, that on the nickel-plated surface was about (0.086-0.074)/0.086 = 14%, and that on the polished surface was about (0.074-0.020)/0.074 = 73%. Applying the developed surface finish processes and their individual optimal parameters sequentially to a fine milled surface mold cavity of an asperic lens, the final surface roughness value of R_a = 30 nm on average was possible.

Acknowledgments

The authors are grateful to the National Science Council of the Republic of China for supporting this research under grant NSC 94-2212-E-011-012.

References

[1] http://www.daido.co.jp/english/products/tool/plasticmold.html
[2] Loh NH, Tam SC, (1988) Effect of ball burnishing parameters on surface finish-A literature survey and discussion. Precision. Engineering. 10(4):215-220.
[3] Klocke F., Liermann J., (1996) Roller burnishing of hard turned surfaces. Int. J. Mach. Tool M. 38(5):419-423.
[4] López de Lacalle LN, Lamikiz A, Muñoa J, Sánchez JA, (2005) Quality improvement of ball-end milled sculptured surfaces by ball burnishing. Int. J of Mach. Tools M. 45:1659-1668.
[5] Luca L, Neagu-Ventzel S, Marinescu I, (2005) Effects of working parameters on surface finish in ball-burnishing of hardened steels. Precision Engineering. 29:253-256.
[6] Shiou FJ, Chen CH, (2003) Determination of optimal ball burnishing parameters for plastic injection moulding steel. Int. J. Adv. Manuf. Tech. 3:177-185.
[7] Chen CCA, Yan WS, (2000) Geometric model of mounted grinding tools for automated surface finishing processes. Proceedings of the 6th International Conference on Automation Technology, Taipei, May, P43-47.
[8] Chen CCA, Duffie NA, Liu WC, (1997) A finishing model of spherical grinding tools for automated surface finishing systems. Int. J. Manuf. Sci. & Prod. 1(1):17-26.
[9] Phadke MS, (1989) Quality engineering using robust design. Prentice-Hall, Englewood Cliffs, New Jersey.
[10] Yang Iron Works, (1996) Technical handbook of MV-3A vertical machining centre, Taiwan.

7

Fuel Cell Technology

The Performance Analysis of a Portable Charger System Using Direct Methanol Fuel Cells

Y. D. Kuan[1], S. R. Lee[2], S. M. Lee[3], M. F. Sung[4], H. S. Chiou[5]

[1]Department of Mechanical Engineering, Technology and Science Institute of Northern Taiwan, Taiwan, 112, ydkuan@tsint.edu.tw
[2] Department of Mechanical Engineering, Technology and Science Institute of Northern Taiwan, Taiwan, 112, srlee@tsint.edu.tw
[3] Department of Aerospace Engineering, Tamkang University, Taipei 251, lee@cfd.aero.tku.edu.tw
[4] Department of Mechanical and Electro-Mechanical Engineering, Tamkang University, Taiwan 251, song221@gmail.com
[5] Department of Electrical Engineering, Technology and Science Institute of Northern Taiwan, Taiwan, 112, hschiou@tsint.edu.tw

Abstract. This paper discusses the design of a portable charger which uses a printed circuit based (PCB) Direct Methanol Fuel Cells (DMFCs) as a power source. The charger consists of three DMFC modules, one pump, one fuel tank and a DC-DC voltage converter. The power output was controlled under 5 volts via parallel connection USB ports. The DMFC charger served simultaneously as the power source for a cell phone charger, the a cycling pump for operating the DMFC, one LED light, and one MP3 player with a pair of speakers. To evaluate the reliability and efficiency of the DMFC power source, a data acquisition (DAQ) system connected to a real-time monitor assessed the current output from the DMFC system running under different fuel flow rates with different power levels. The results provide important information for the control algorithm to obtain fuel flow rate control in the DMFC system with increased reliability and efficiency in the present portable charger system with PCB-based DMFCs.

Keywords: DMFC, PCB, Power Source, Charger

1.1 Introduction

In recent years, studies on substitute energy sources are rapidly increasing. The fuel cell is one of the attractive power sources for the new generation technology. Fuel cells are electrochemical devices that convert chemical energy directly into electrical energy [1]. Among the different types of fuel cells, the direct methanol fuel cell (DMFC) has the most potential for lower power applications, because it offers high power and energy density, low emissions, operation at or near ambient conditions, fast and convenient refueling and a potential as a renewable fuel source [2,3]. Although the DMFC has the major advantage of fuel cell simplicity, which makes it more acceptable for portable electronics, there are two major considerations for commercialization: the methanol crossover through the membrane and low activity anode of methanol oxidation catalysts limit the DMFC performance and applications due to high cost [3, 4]. Many DMFC portable application prototypes have been demonstrated: powering cellular phones by Energy Related Devices (RED) in 1999, a camcorder powered by DMFC by Smart Fuel Cell (SFC) in 2001 [5], Notebook PC powered by DMFC by NEC [6] and Toshiba [7]. Intel also claimed the need for fuel cells in the near future to support today's battery [8].

The printed circuit board is a novel technology for a fuel cell stack, first presented by O'Hayre et al. [9] for portable fuel cell arrays (PCBFC). The main advantages of a PCBFC include increasing design flexibility, low cost, higher power density, easier integration into devices because of its' flat design, and improved packing form factors. These advantages give the PCBFC mass production potential. In the PCBFC structure, the anode and cathode board consists of a thin cooper layer laminated onto a substrate, acting as a current collector. Each copper plate must be coated with a material that prevents corrosion and with hole-arrays for reactants and products to move out to the MEA [9, 10].

In this paper, a small PCB-based DMFC pack was designed and demonstrated as a portable power supply providing power to multiple devices. The fuel flow rate is discussed and the individual power consumption of each device is investigated.

1.2 Experimental Setup

The study investigates the power performance of a direct methanol fuel cell and designs an integrated system to supply the required power dissipations to 3C products such as a mobile phone, MP3 player with a pair of speakers, LED light, mini-pump and LED light, as shown in Figure 1.1. The 5 V Mini USB for the charge rating power in the mobile phone is selected. Three sets of 4-cell DMFC modules (12 cells total) shown in Figure 1.2 and a DC-DC transformer are controlled to reach the 5 V output requirement. The basic transformer diagram shown in Figure 1.3 comprises two capacitors and a 7805 transistor. It transforms the fuel cell output voltage to reach 5V. A 5 V pump was selected to pump the methanol fluid. For a good stable and anticorrosive 5V pump, the pump flow rate was kept very low. The flow rate performance of the methanol fluid for 2 watts of DMFC is also investigated in this paper. Figure 1.4 shows the power transmission flowchart for the present system. The power is generated by DMFC pack, the output voltage is controlled by a DC-DC convertor to keep around 5 volts, and the current is sent to an USB port, and distributed to the devices.

Two methods for controlling voltage and current are generally applied in lithium, nickel-hydrogen batteries and fuel cells. The battery or fuel cell performance is measured by the voltage or current at different loads. The experimental platform is built with a programming current loading instrument, CHROMA, and a 488.2 GPIB control card supplied by National Instrument Comp. (NI). The program is developed with the assistance of Labview to control and record measurements. The SCXI instrument from NI Comp. is used to carry on automatic and multi-point voltage and current measurements for the three sets of fuel cell modules.

The experimental measurement requires three steps to process. The first step studies the methanol flow rate to the performance of a single fuel cell module (four cells). The flow rate is set to a range between 2 cc/min. and 20 cc/min. The CHROMA instrument supplies the voltage in the range of 0.8 to 2.4 V. The fuel cell currents are recorded for the different voltages, respectively. The next step is to analyze the performance of different flow rates and load voltages from 2.4 to 7.2 V in the fuel cell module. The output power in the same fuel cell is measured to analyze the fuel cell module voltage. The power dissipations of the mobile phone, MP3 player, methanol pump and LED light are simultaneously measured by the integrated system. The fuel cell output power loss is also discussed with the difference in voltage and current measured by the load instrument and fuel cell output power.

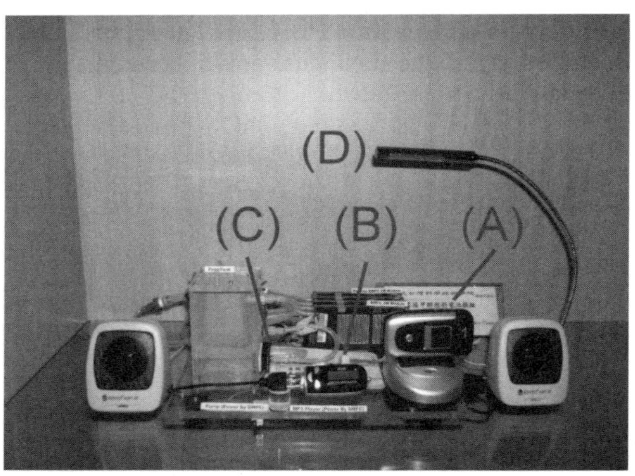

(A) cell phone charger, (B) MP3 player with speakers, (C) mini-pump, (D) LED light

Figure 1.1. The small DMFC power supply and peripheral system

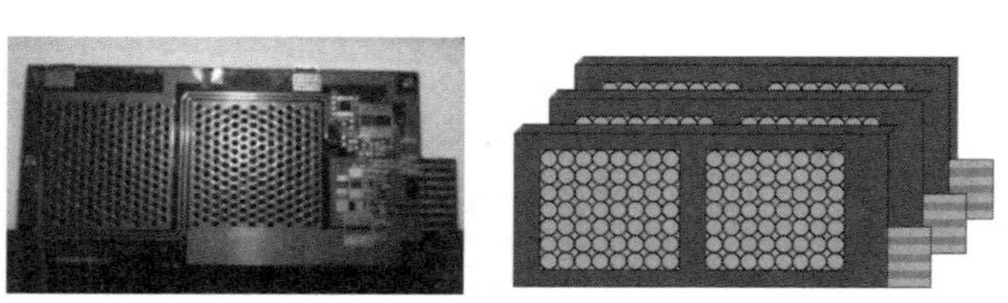

Figure 1.2. The DMFC modules adopted in the study

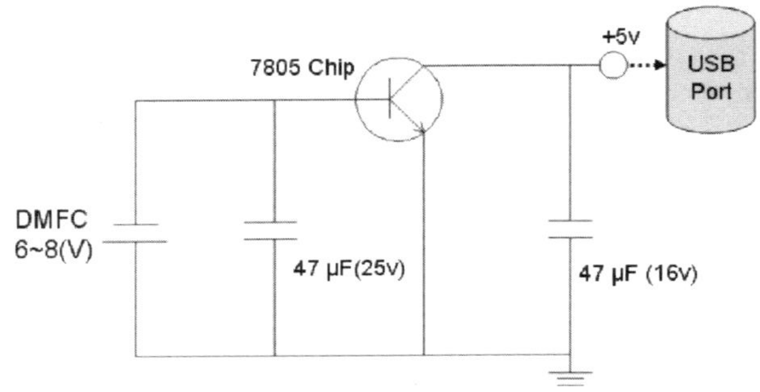

Figure 1.3. Illustration of a DC-DC converter

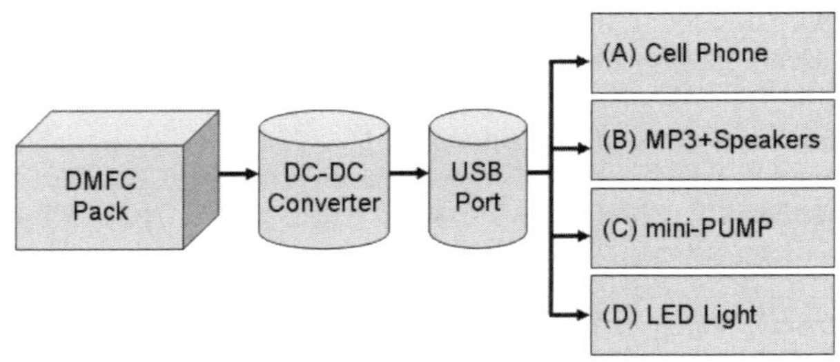

Figure 1.4. Flowechart of power transmission

1.3 Experimental Results

The DMFC modules were evaluate under different fuel flowrates. Figure 1.5 shows one of the DMFC performance curves. The results show that the lower fuel flow rate, the lower the cell performance and the best performance occurred at a fuel flowrate of 2 cc/min. The output current density was about 0-145 mA/cm^2 and the corresponding output power density is 0-115 mW/cm^2. Figure 1.6 is the average cell performance of a DMFC pack composed of three DMFC moudles in series and the fuel flowrate were controlled between 2-20 cc/min. The results show that smaller fuel flow rates produce higher DMFC performance. The average out current density is 0-140 mA/ cm^2 and the corresponding output power density is 0-11-5 mW/cm^2. On the basis of these data, a mini-circulation pump was selected to transmit the fuel with about 2 cc/min flowrate.

As described above, the power distributed from the small DMFC power supply to the devices was through a standard USB port. The USB voltage specification is about 4.5-5 Volts. The liquid fuel was simultaneously supplied to the three DMFC modules in parallel arrangement via the single mini-pump through the fuel tank, as shown in Figure 1.7.

Figure 1.8 shows the power output of DMFC pack and power consumption of each device during operation. The DMFC power generation was about 680-720 mW and the device power consumption range of each device was between 135-165 mW. Table 1.1 shows the average power output of the DMFC pack and the power consumption of these devices. The results show that the average DMFC pack power output was 603.6 mW, the power consumption of the cell phone charger, mini-pump, MP3 player + a pair of speakers, and the LED light were 141 mW, 152.1 mW, 142.2 mW, 160.8 mW, respectively.

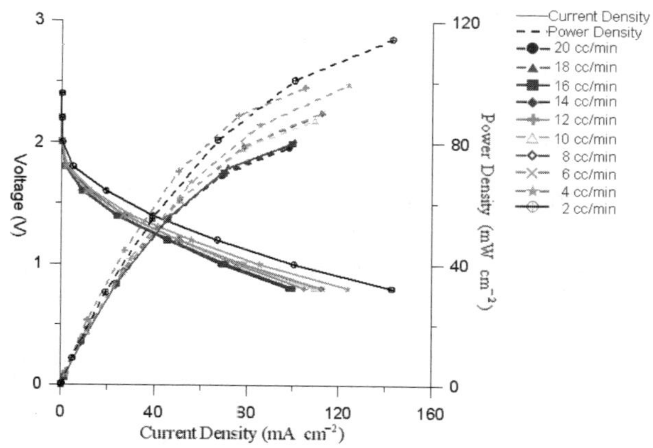

Figure 1.5. Cell performance of a single DMFC module

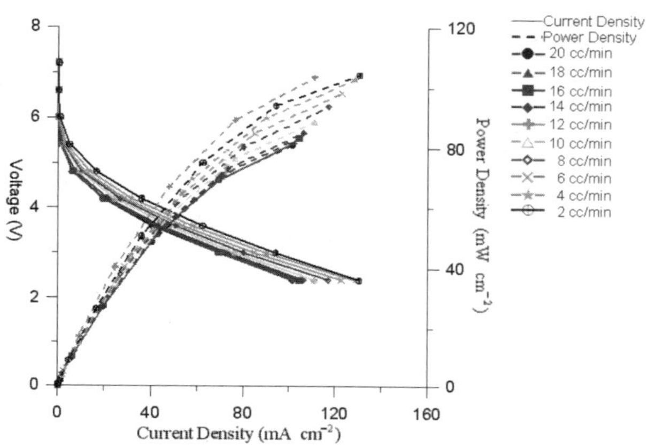

Figure 1.6. Average cell performance of DMFC pack composed by 3 DMFC modules in series

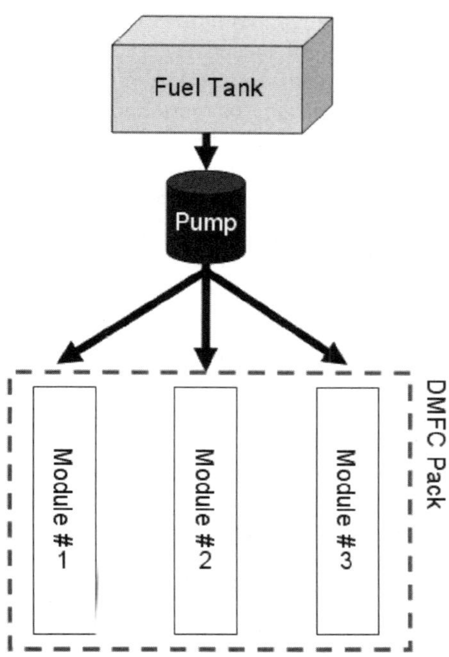

Figure 1.7. Illustration of fuel supplement

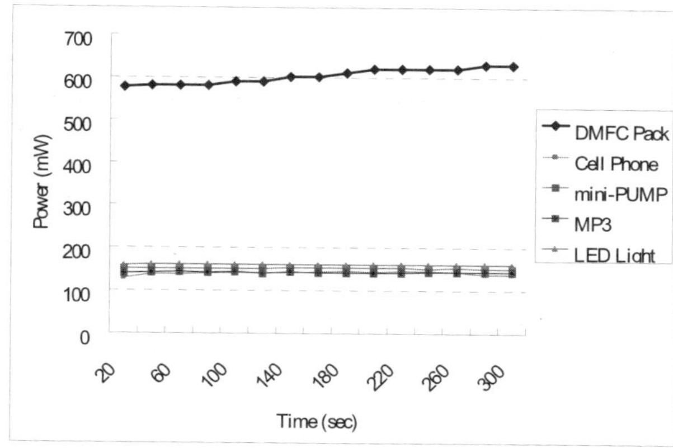

Figure 1.8. Power output of DMFC pack and power consumption of each device

Table 1.1. Average power output and consumption of DMFC pack and devices

Items	Average Power Output/consuming (mW)
DMFC Pack	**603.6**
Cell Phone	**141.0**
mini-PUMP	**152.1**
MP3+Speakers	**142.2**
LED Light	**160.8**

1.4 Conclusion

A small DMFC power system was succesfully developed that could provide power simultaneously to several devices via a DC-DC convector and standard USB port. The present power system was composed of a pcb-based DMFC pack that contains three DMFC modules. Each module has four 3.5 mm x 3.5 mm MEAs. The single DMFC module test shows a suitable low fuel flow rate could provide good cell performance.

A simular finding was obtained in the DMFC pack performace test, connecting three DMFC modules in series. The power distribution shows the average power output of the DMFC pack is 603.6 mW and average power consumptions of the devices are in the range of 141 mW-160.8 mW.

Acknowledgement

The authors sincerely acknowledge finicial support provided by the National Science Concil under grant no. NSC 95-2221-E-149-013-.

References

[1] FUEL CELL Handbook, 5th Edition, EG&G Services, Parson, Inc., Science Applications International Corporation, 2000.

[2] Birgersson et al. 2003, Reduced Two-Dimensional One Phase Model for Analysis of the Anode of a DMFC, J. of the Electrochemical Society, Vol. 150, No. 10, , PA1368-A1376.

[3] J. P. Meyers, and J. Newman, 2002, Simulation of the Direct Methanol Fuel Cell II- Modeling and Data Analysis of Transport and Kinetic Phenomena, J. of the Electrochemical Society, Vol. 149, No. 6, PA718-A728.

[4] J. Nordlund and G. Lindbergh, 2002, A Model for the Porous Direct Methanol Fuel Cells Anode, J. of the Electrochemical Society, Vol. 149, No. 9, P A1107-A1113.

[5] G. Hoogers, 2003, Portable Applications, Chapter 9 at Fuel Cell Technology Handbook, CRC Press.

[6] Foley D. 2003, NEC Unveil Notebook PC with Built-in Fuel Cell, NEC Press Contacts in Japan.

[7] http://www.toshiba.co.jp/about/press/2003_03/pr_j0501.htm/ 2003.

[8] C. Xie, 2003, Direct Methanol Fuel Cell (DMFC) System for Portable Communication Applications: Prospects and Challenges, Proceedings of 2003 Taipei Power Forum, Taipei, Taiwan, December 1-3, 2003, PF7-1 –F7-8.

[9] R. O'Hayre et al., 2003, Development of Portable Fuel Cell Arrays with Printed-Circuit Technology, J. of Power Sources, Vol. 124, P459-472.

[10] A. Schmitz et al., 2003, Planar Self-Breathing Fuel Cells, J. of Power Sources, Vol. 118, P162-171.

A Novel Integration Approach for Combining the Porous Silicon on a Micro Fuel Cell

C.Y. Lee*, S.J. Lee, C.W. Chuang, R.D. Huang

Department of Mechanical Engineering, Fuel Cell Research Center, Yuan Ze University, Taoyuan, Taiwan, R.O.C.,
*cylee@saturn.yzu.edu.tw

Abstract. This investigation employs porous silicon as the gas diffusion layer (GDL) in a micro fuel cell by micro-electro-mechanical-systems (MEMS) fabrication technology. Pt is then deposited on the surface of the porous silicon, to enhance the fuel cell conductivity. The porous silicon with different pore sizes of 10 μm, 30 μm, 50 μm to test performance for GDL, and the fuel cell performances are discussed. The GDL was replaced by porous silicon and used in a proton exchange membrane fuel cell (PEMFC). Dry etching is applied to a 500 μm-thick layer of silicon to yield fuel channels with a depth of 450 μm and a width of 200 μm. Accordingly, the GDLs of the fuel cell are fabricated using macro-porous silicon technology. Porous silicon was fabricated by deep reactive ion etching (DRIE). The experimental shows the results with different pore size, fuel flow rate and operation condition. The best performance was 9.37 mW/cm^2 with dry gas flow rates of H_2/O_2 at 30/30 ml/min and 10 μm hole.

Keywords: Porous silicon, GDL, MEMS.

1. Introduction

In recent years, the size of fuel cells has been minimized. They have been adopted in portable electronic products, such as cellular phones and PDAs. Silicon-based substrate is very compatible with MEMS technology. In fuel cells, porous silicon has been used as the gas diffusion layer to replace traditional carbon cloth or carbon paper [1,2]. It is also used to form proton exchange membranes [3].

Electrochemical etching in hydrofluoric acid has been studied since 1956. In 1990, Lehmann characterized porous silicon in detail. In 1996, Lehmann examined the formation of an array porous silicon structure whose etching depends on electrolyte concentration, electrolyte temperature, silicon doping density and current density. Such structures are classified into three regimes by the average dimensions of the porous silicon. The microporous regime has mean dimensions of under 2 nm; the mesoporous has dimensions 2 nm to 50 nm, and the macroporous has dimensions of over 50 nm. Kleimann formed a macroporous array which was 42 μm wide and 200μm deep. He showed that porous silicon etching technology could be applied to make a structure with a high aspect ratio structure at lower cost than associated with deep reactive ion etching (DRIE).

2. Design and fabrication

Wet etching technology was utilized to form the fuel channels in a micro fuel cell and then electrochemical etching was employed to make the gas diffusion layer from porous silicon. Therefore, in this study, the pore size and depth were controlled. After the porous silicon was formed, the Pt metal was deposited on the pore surface as the catalyst of the fuel cell and the conductivity of the silicon was increased. Part of the Pt metal layer formed as the thermal sensor.

2.1. Flow field design

Hoogers [4] shows the performance of the serpentine flow field on a fuel cell which was demonstrated to be better than the other flow field configurations (meshed and interdigitated) in some cases, because the fuel (gas or liquid) was pushed strongly to flow all around the active area of the fuel cell. Accordingly, the serpentine flow field was utilized in the design herein, as presented in Figure 1.1

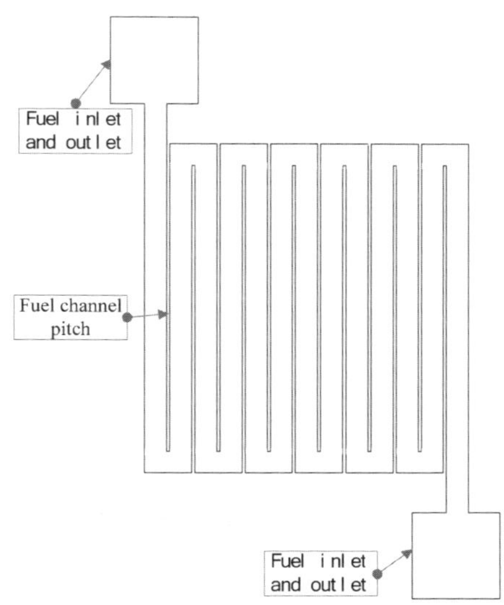

Figure 1.1. Form of flow channel

An N-type thickness of 525±25 μm, and a (100)-oriented double-side polished wafer was chosen. After the low pressure chemical vapor deposition (LPCVD) oxidation of Si_3N_4 on the silicon wafer (5000 Å thick), one side of the silicon was photolithographically processed. Reactive ion etching (RIE) was then applied to transfer the pattern in Figure 1.1, following the wet KOH etching process. This process is used to etch a silicon layer with a thickness of 450 μm. The remaining thickness of the silicon forms the gas diffusion layer, with a thickness of 50~70 μm, and a width of 200 μm. Figure 1.2(a)-(c) show the details of the process. After this process, the porous silicon is fabricated.

2.2. Fabrication of porous silicon

Porous silicon with an anodic bias was employed photolithography and DRIE etched onto it to form the porous silicon on the side. The applied anodic bias and the doping density strongly affect the pore size of the porous silicon. As stated above, N-type silicon with a resistivity of 1-20 Ω-cm was employed.The approaches of other investigations [5,6] were used to design square pores of side 10 μm, and form fuel channels with vertical walls. The etching time and current density were the important parameters. In the proposed design, square pores of size 10 μm are fabricated by coherent porous silicon technology [5,6]. Coherent porous silicon technology can easily control the porosity and the pore diameter of the porous silicon could be easily controlled. The porous silicon was formed first using lithography to transfer the pattern, and then using RIE technology. Figure 1.2 present the porous silicon fabrication process flow. The flow field process was conducted as in Figure 1.2(a)-(c), the other side of the wafer was patterned lithographically, 10 μm square at a pitch of 15 μm covered the defined area, forming a gas diffusion layer 200 μm (Figure 1.2(d)), then was transferred in the Si_3N_4. KOH wet etching on (100)-oriented silicon is anisotropic etching, and transforms 10 μm squares into inverted pyramids. Si_3N_4 was removed from the fuel field side of the wafer by RIE, after the fuel channel was formed. The side of the wafer was connected via an Al metal ring to the anode side of the power supply. To make sure the porous silicon through to flow field, we adopted the DRIE to reach the purpose have through-holes. Figure 1.2(e) showed porous silicon through to flow filed. A diameter of 10 μm was patterned by lithography technology on top of each flow channel. The pattern (200 μm × 13140 μm) was transferred by Si_3N_4.

2.3. Integrating catalyst and thermal sensor into gas diffusion layer

After the porous silicon was produced, the wafer was metallized on the porous silicon side with a layer of Ti/Pt (20 nm/70 nm). The Pt acted as the current collector and thermal sensor. Physical vapor deposition (PVD) was utilized to deposit Pt metal and wet etching was adopted to form the conductive layer and thermal sensors. Photoresist was employed as the etching mask, ensuring that the Pt metal remained on the surface of the porous silicon. The etching mask was used in lithography process by an exposing process. Figure 1.2(e)-(h) display the process flow in detail. The micro thermal sensor was fabricated, as presented in Figure 1.3.

The experiment used temperatures of 20~46°C. The resistance varied at 1.649~1.694 kΩ. Experimental results demonstrate that the temperature is almost linearly related to resistance, as shown in Figure 1.4.

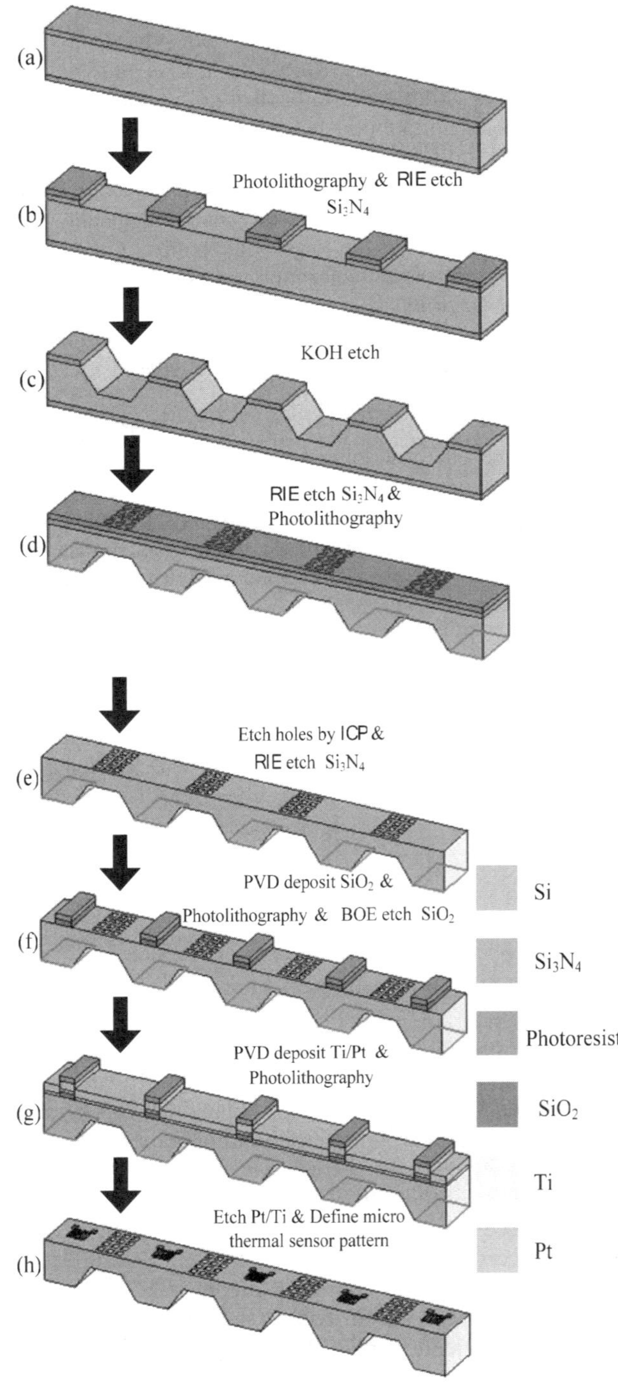

Figure 1.2. Fabrication flowchart

3. Results and discussion

In this investigation, three sizes of pores in silicon, 10μm, 30 μm, and 50 μm were used. The GDL of the fuel cell was replaced with three different porous silicon. The three different porous silicon porous silicon was fabrication by DRIE. When a fuel channel formed, the porous silicon, a through-hole, and they also have conductive layer. The Pt metal layer also formed the thin film thermal sensor on the ridge of the porous silicon side, as presented in Figure 1.5. Fuel cell performance with a thin film thermal sensor on the anode electrode is compared with a fuel cell without a thin film sensor tested with dry H_2/O_2 at room temperature. The following results are measured with an Ion power membrane electrode assembly (MEA) with gas flow rates of H_2/O_2 at 30/30 ml/min and overall reaction area of 1.82 cm^2.

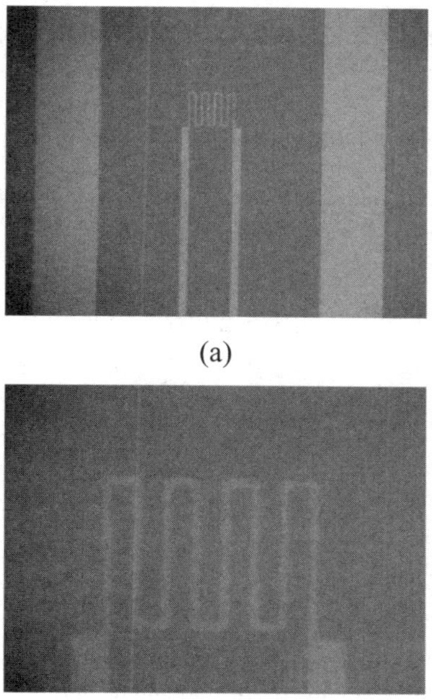

Figure 1.3. Micro thermal sensor pattern and microscopic photograph: (a) the overall vision, (b) enlarge the pattern

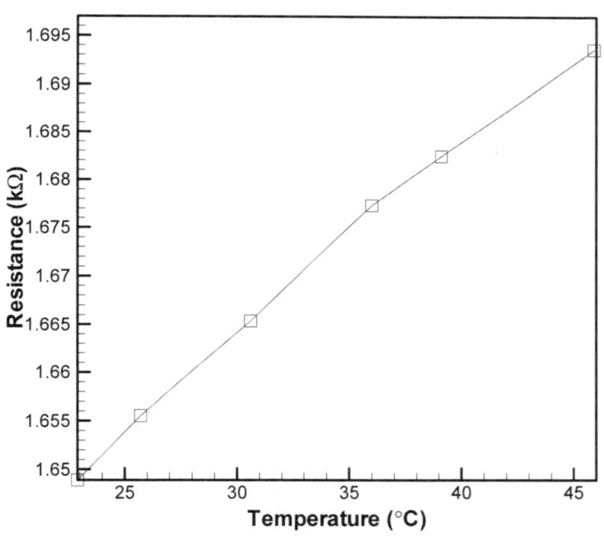

Figure 1.4. The resistance vs. temperature curve

Maximum power density of approximately 2.15 mW/cm^2 at 302 mV was obtained for a fuel cell with a thermal sensor on the anode electrode, as shown in Figure 1.6. The performance of the electrode without a thermal sensor was 9.37 mW/cm^2 at 502 mV, as shown in Figure 1.7. The performance of the electrode with a micro thermal sensor is not as high as that without a micro thermal sensor. Production of the micro thermal sensor likely decreased the conductive area. The fuel cell temperature during operation was 26°C gaining by thermocouples that touched the electrode backside. In situ monitoring of temperature during the fuel cell operation indicated that the resistance measured by a resistance meter was 1.6645 kΩ, and the temperature was 30.3°C, as shown in Figure 1.4.

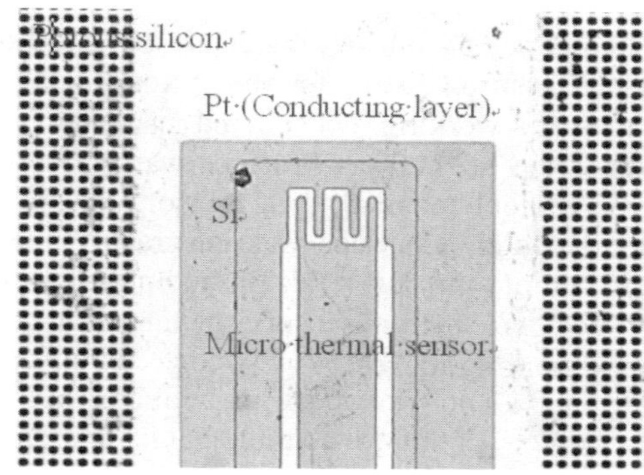

Figure 1.5. Optical microscopic photograph of porous silicon combined with micro thermal sensor

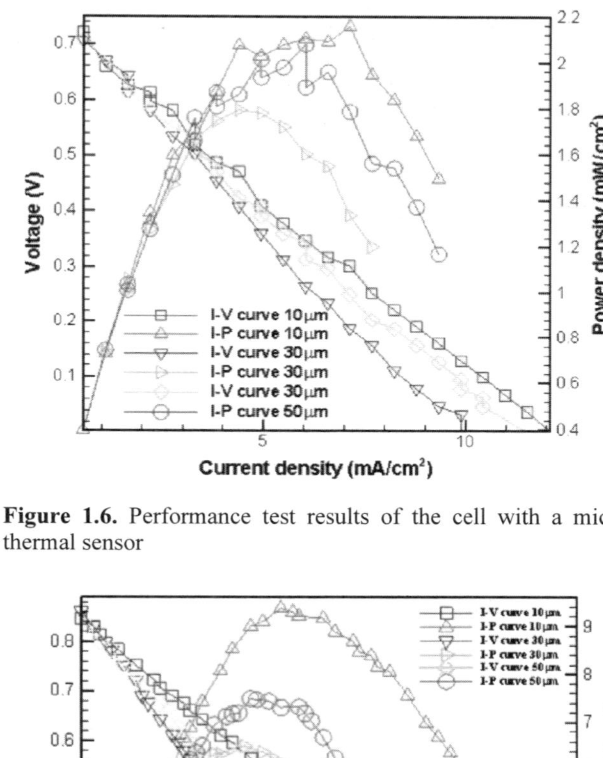

Figure 1.6. Performance test results of the cell with a micro thermal sensor

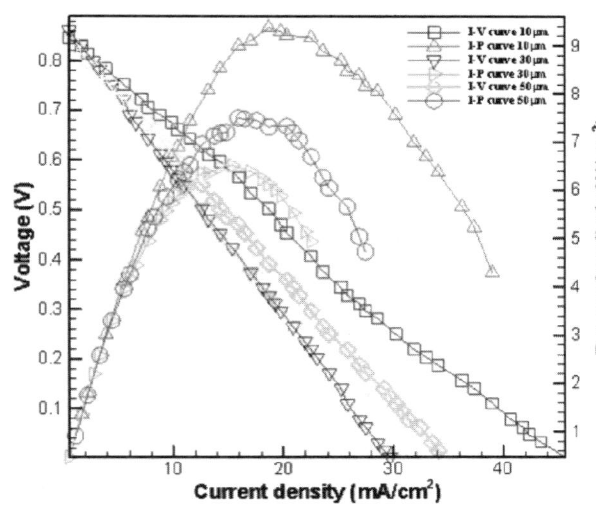

Figure 1.7. Performance test results of the cell without a micro thermal sensor

4. Conclusions

Porous silicon was fabricated on the defined area of the flow filed. Hence, one silicon wafer was verified to combine the flow field with the gas diffusion layer. The porous silicon was fabricated with pores with diameters of 10 μm, 30 μm, and 50 μm. As the catalyst was deposited, the thermal sensor was formed. Additionally, the performance of a working fuel cell and inside temperature were measured. Fuel cell performance was 9.37 mW/cm^2 at 502 mV without micro thermal sensors on the anode electrode; the anode electrode was integrated with 10μm porous silicon into the fuel field. Based on the calibration curve for the micro thermal sensors, the temperature of the MEA in the working fuel cell was 30.3°C.

Acknowledgement

This work was accomplished with much needed support and the authors would like to thank the Ministry of Education of R. O. C. and YZU Fuel Cell Center through the grant No. 0950026846 and National Science Council of R. O. C. through the grant NSC 95-2221-E-155-043 for their financial support of this research project.

The authors also like to thank Professors Shih Hung Chan, Ay Su, Fangbor Weng, Guo Bin Jung of the Department of Mechanical Engineering, Yuan Ze University for their valuable advice and assistance in experiment. In addition, we would like to thank the YZU Fuel Cell Center and NTU NMES Research Center for providing access to their research facilities.

References

[1] Meyers J. P., Maynard H. L.: 2002, Design Considerations for Miniaturized PEM Fuel Cells, Journal of Power Sources, Vol.109, P76-88.

[2] D'Arrigo G., Spinella C., et al.: 2003, Fabrication of Miniaturized Si-based Electrocatalytic Membranes," Materials Science and Engineering C, Vol.23, P13-18.

[3] Pichonat T., Manuel B. G.: 2005, Development of Porous Silicon-based Miniature Fuel Cells, Journal of Micromechanics and Microengineering , Vol.15, P s179-s184.

[4] H. Gregor, 2003, Fuel Cell Technology Handbook, CRC Press, New York.

[5] Lehmann V., Gösele U.: 1991, Porous Silicon Formation: A Quantum Wire Effect, Appl. Phys. Lett., Vol.58, P856-858.

[6] Lehmann V.: 1993, The Physics of Macropore Formation in Low Doped N-Type Silicon, J. Electrochem. Soc., Vol.140, P2826-2843

8

Laser Processing

8–1

High-performance Laser Technological Processing Systems for Scientific and Industrial Applications

Y.V. Chugui[1], V.P. Kiryanov[2] and A.G. Verkhogliad[3]

Technological Design Institute of Scientific Instrument Engineering, Siberian Branch of the Russian Academy of Sciences (TDI SIE SB RAS), 41, Russkaya str., Novosibirsk, 630058, Russia
[1] chugui@tdisie.nsc.ru
[2] kiryanov@tdisie.nsc.ru
[3] verhog@tdisie.nsc.ru

Abstract. The novel results of the TDI SIE's activity in the field of Laser technologies for high performance manufacturing processing of 3D and 2D objects are presented. We have developed a multifunctional Laser technological system LSP-2000 for precision processing 3D arbitrary shaped large scale components for cutting, welding and ablation within (3×3×0.6 m^3 working field with position error less than 10 µm), as well as Laser Image Generator CLWS-300 for synthesis of 2D structure with arbitrary topology and micron accuracy (working field up to 300 mm). Peculiarities and perfomances of these systems are presented.

Keywords: Laser Image Generator, technical Laser processing, cutting, welding, ablation

1. Introduction

Laser technologies find wider and wider applications [1, 2] in processing components along arbitrary contours, including formation of products of arbitrary geometry without change of tools, as well as synthesis of very high precision microstructures. In order to resolve different actual problems we have developed two Laser technological systems: one of them is a multifunctional technological complex LSP-2000 for component processing, the other one is a Laser Image Generator CLWS-300 for the synthesis of 2-dimensional structures.

As for Laser processing of large scale arbitrarily shaped components it takes a beam positioning error below 10 µm. Such problems arise, e.g., in creating large-scale antennas (of several meters in diameter) with a given directional pattern when it is required to perform micro-profiling of previously formed coal-plastic surfaces, coated with a thin metal film. This technology demands precision equipment to ablate metal a film according to the calculated configuration. For millimeter band antennas the formed topology should differ from the calculated one by no more than several tens of microns over the entire antenna area. Our Laser system LSP-2000 satisfies all strict requirements for precision processing of arbitrary shaped large-scale components. The system carries out the following operations: cutting of metal components with a thickness of up to 6 mm, welding of metal components with a thickness of up to 2 mm, and micro-profiling of surfaces by means of ablation. The size of the components can be up to 3×3×0.6 m^3 and the deviation of a single synthesized element position from the given one is below 10 µm at any point of the surface.

The Laser Image Generator CLWS-300, developed together with the Institute of Automation and Electrometry (SB RAS), operates in the polar coordinate system. It provides the synthesis of 2D structures with arbitrary topology and micron resolution (working area up to 300 mm diameter). This generator has the best precision performances, first of all, for axisymmetric structures forming. The CLWS-300 has fast rotation of photosensitive layer and slow displacement of recording head in radial direction. Laser beam forms the latent image of component topology due to the corresponding control of the focused irradiation power (with 10^4 discrete light levels under the irradiation interruption frequency up to 5 MHz). The positioning system on radial direction has a resolution of 0.6 nm and provides a position uncertainty of the recording head of about 10^{-7} over a measuring distance of 200 mm.

The main applications for the Laser Image Generator CLWS-300 are presented. Using this generator ensures angle measuring topology synthesis with angular position of elements less than 1″.

The peculiarities, structure, performance and the experimental results of the multifunctional Laser technological system LSP-2000 and the Laser Image Generator CLWS-300 are presented.

2. Laser Technological System LSP-2000

The purpose of the multifunctional Laser system LSP-2000 is processing and treatment of 3D articles: cutting, welding and surface micro-profiling with ablation process. This system is equipped with two Nd-YAG Laser. The robotics for the Laser head positioning and CNC control interface are provided for processing and treatment of parts with an arbitrary topology. The system spatial working space is about 3×3×0.6 m^3. Inside this space all types of Laser processing operations can be performed with a contour displacement accuracy of about ±10 µm for any point in the trajectory. The picture of the system is presented in Figure 1.1. The LSP-2000 was developed as a universal Laser processing system with a unique combination of certain parameters/technical characteristics [3]. These characteristics are listed below.

- **Multifunctionality.** It means the ability to perform a range of operations. Each operation using customised Laser processing heads requires an individual positioning accuracy, different positioning speeds and movement patterns depending on the operation. This requirement was fulfilled using two types of Lasers. The first Laser (MLTI-500, produced by Tulamashzavod) for cutting and welding has the following specifications: the pulse repetition frequency is 300 Hz, average power output is 500 W, and the Laser wave length is 1.064 μm. Its purpose is cutting and welding of metals with a thickness of less then 6 mm. The second Laser, used for the ablation of thin metal films on dielectric surfaces, has a pulse repetition frequency of 300 Hz, high pulses power (> 10^7 W), a Laser wavelength of 0.532 μm, and an average power output of 10 W.

Figure 1.1. Laser system LSP-2000

- **Large processing working space.** Our system has a working space of 3×3×0.6 m^3. The absolute positioning accuracy in this working space is better than 20 μm. A special XYZ movement stage was developed in order to move the tool (Laser processing heads, etc.) to any point in the working space with the required accuracy. The stage movements are controlled by a special CNC system. The positioning of each stage uses feedback-control and is based on linear incremental optical sensors, which provide the required stage position information (position error 1 μm).
- **The ability to process components with arbitrary surface shape.** For this purpose the processing head can operate with 5 degrees of freedom. The processing head can be moved by the XYZ carriage for bringing the head to the required processing region. Also, two more possible movements are added in the polar coordinate system. The head can rotate azimuthally in the processing region, thus it allows us to orient the head perpendicularly to any surface element of the arbitrary 3D object, which is being processed.
- **Long term parameter stability.** One of the main tasks of the current system is micro profiling of large-size components. The expected time machining a single component is approximately 30 hours. The system is designed for operating 24 hours a day. At the same time it is required to keep high reproducibility and high processing accuracy during the entire service life of the current system without any additional tuning. The long term stability of this device was achieved by using linear motors providing an air gap between sliding parts. This air gap, supplied with compressed air, allows for a frictionless motion of mechanical parts according to the pneumatic bearing principle.

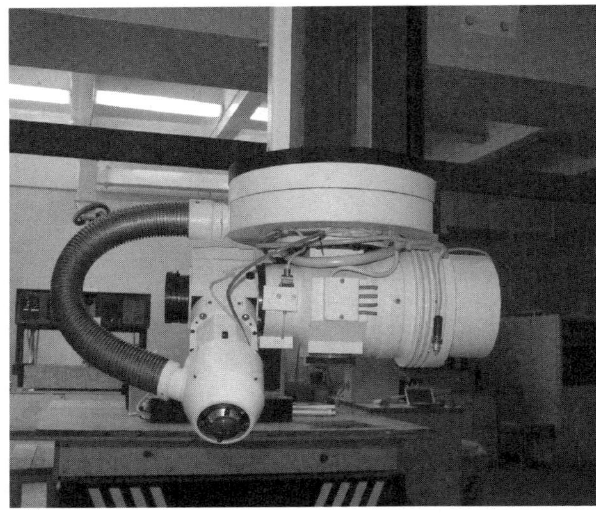

Figure 1.2. Working Laser processing head

The LSP-2000 system's specifications are: Coordinate table travel range and maximum size of the processing parts is 3×3×0.6 m^3 Geometrical shapes of processing parts are arbitrary. Uncertainty of system positioning in start-stop regime is ≤2 μm. Uncertainty of system positioning along arbitrary movement loop is ≤10 μm. Maximum head movement speed is 10 m/min. The material for ablation is Al, NI-Cr alloy. The weight of the optical head is ≤6 kg. Both Ti and stainless steel can be Laser cut and welded. The maximum cutting thickness for Ti is 6 mm.

At the moment, the LSP-2000 is under industrial operation in China.

3. Laser Image Generator CLWS-300

At the moment, standard photolithographic equipment and routine practice are used under the works on developing the new elements of integral or diffraction optics, micromechanics as well as optical writing [4]. This equipment and these methods allow us to create rather simple structures with binary profiles on small surfaces (about the size of a microcircuit crystal), but they cannot provide the specific diffraction optic requirements. The peculiarities of these requirements are as follows: the minimal size of structures can reach up to 0.5 microns under the overall dimensions of the synthesized elements equal to 300 mm, the topology element errors must not exceed 1/4 - 1/20 of the minimal size of the structure, the profile height of the working slope should vary smoothly from zero up to tens of microns while the profile height of non-working slope must be terminated as sharp as possible. These requirements make themselves evident when one must synthesize wide aperture kinoform elements, including on curvilinear surfaces. The majority of these requirements in scanning type devices can be realized rather simply. The direct beam writing scanning methods are the most universal and widely used in science, including microelectronics, optics, and radio engineering. These methods are based on scanning of the surface covered by the registering layer using a focused Laser beam modulated in amplitude according to requirements of synthesized topology. Thus, the writing beam travels on the given trajectories. These trajectories are direct lines (for rectangular coordinate system), or circles and spirals (for polar coordinate system).

Instead of the traditional rectangular coordinate system, which is widely used for micro image writing in serial devices (produced in the microelectronic industry), we have used the polar (cylindrical) coordinate system. It allows us to achieve high scanning speeds and, as a consequence, short writing cycles, absence of joints, an opportunity for spheres writing, technical decisions simplicity as well as functional universality.

Together with the Institute of Automation and Electrometry (SB RAS), we have developed a commercial version of the Laser Image Generators CLWS-300 operating in the polar coordinate system (Fig. 1.3). Since 1995 we have been producing the CLWS-300 for Russia and abroad [4].

Figure 1.3. The Laser Image Generator CLWS-300

Writing of images in these generators is carried out in circular scanning mode. The specific feature for this mode is the presence of two motions, namely: continuous quick rotation of photosensitive layer and slow displacement in radial direction of the writing head. In the latest CLWS-300 models two writing technologies have been used: writing of latent images in chromium films and classical writing technology on photoresists. The advanced software of these devices allows us to synthesize different structures, including ones with arbitrary topology in polar and cartesian coordinate systems. Laser Image Generators that that are capable of synthesizing arbitrary structures on curvilinear optical axisymmetric surfaces are incorporated into the model sequence [6].

The latent image in chromium films and in photoresists is formed by the Laser beam due to the power control of focused radiation (interruption in time and stabilization on a given level). In this case it ensures a frequency interruption up to 5 MHz and forming up to 10^4 discrete power levels. These characteristics were obtained due to corresponding configuration of optical writing channels. The structural scheme of the CLWS-300 is shown in Figure 4. The Laser Images Generator consists of four main units, namely: optical writing channel, linear displacement unit, spindle unit and control electronic unit. In its turn the optical writing channel includes a powerful Argon Laser system for transportation and modulation of Laser radiation, writing autofocusing head, that focuses the Laser beam on the surface of photosensitive layer. The Argon Laser of INNOVA-90 or INNOVA-305 of Coherent Firm (USA) is mostly used as writing Laser in CLWS-300. Two acoustic-optical modulators, turning mirrors and radiation collimators are incorporated into the Laser radiation transportation and modulation system.

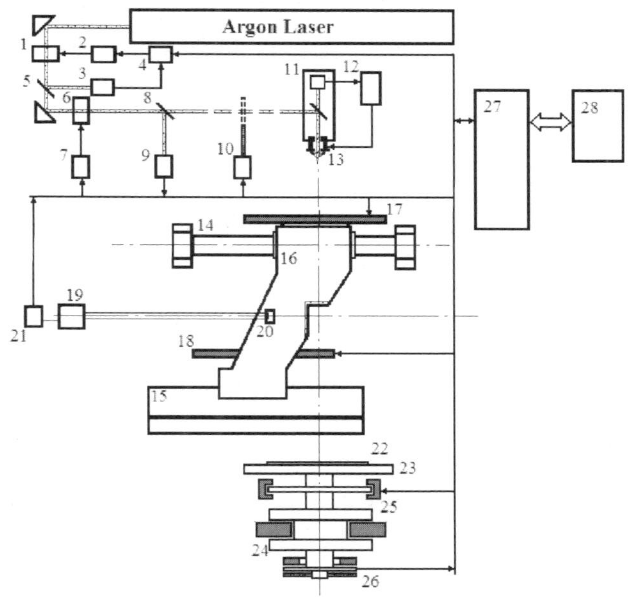

Figure 1.4. Structural scheme of CLWS-300. 1 - "Slow" AOM; 2, 7 - digital regulators; 3, 8 - photodetectors; 4 - comparator; 5, 8 - beam splitters, 6 - "fast"AOM, 10 - shutter; 11 - record head; 12 - AF digital regulator; 13 - executive element; 14 - round guide; 15 - flat guide; 16 - portal; 17,18 - linear engines; 19 - interferometer; 20 - reflector; 21 - Laser-interferometric gauge; 22 - substrate with a photosensitive layer; 23 - chuck; 24 - rotor; 25 - angular position gauge; 26 - engine; 27 - control system; 28 - manager computer

The presence of two consequently switched acoustic-optic modulators allows us to realize the forming of 10^4 discrete power levels. Such a wide dynamic range of adjustment of power radiation directed to the surface of photosensitive layer is required due to the corresponding change of linear velocity movement of layer on different writing radii (from 10 m/s on the periphery of working field up to 20-30 μm/s in the center of rotation). Thus, the modulator (1) is part of the output radiation intensity stabilization system at the level given by the control computer. For this purpose a part of the modulator output signal is transformed into an electric signal using a special photodetector (3), which one compares to the fixed voltage proportional to a necessary level of power radiation supplied to a photosensitive layer surface. The comparison result is sent to the digital regulator (2). This regulator controls the level of the electric signal submitted on an electric input of the piezoelectric converter of the acousto-optic modulator (1). Thus, the modulator is covered a deep negative feedback and forms a stabilization system of power radiation at a given level.

The modulator (6) is used for the fast interruption of Laser radiation stabilized at the required level using modulator (1). For this reason the modulator (6) is not covered by constant negative feedback. The stability of the modulator (6) coefficient transmission is inspected (once every 10 min) with the help of the photoreceiver (9) but in mode of writing interruption. For this purpose the shutter (10) is used. However, the modulator is also incorporated into the control process on power radiation level supplied to the surface of photosensitive layer (22). In cases when it is required to fulfill the writing of continuous profiles in photo-resistant films on small writing radii, one can realize the control with wide-pulse modulation of the supplied power radiation.

The writing head (11) incorporated into the optical channel works with flat carriers as well as with curvilinear ones. Figure 1.4 presents the writing head scheme for structures recording on flat carriers. It includes the gauge of a defocusing signal and an executive element (13). The digital regulator (12), located on the carriage of the linear displacements (16), accepts the signals from the gauge of a defocusing signal and operates by the work of an executive element (13).

The writing head (11) is mounted on the carriage of linear displacements (16) providing a travel range up to 200 mm that includes 150 mm for a working range of displacements and 50 mm for the positioning on the physical reference point used at control operations. The mobile portal of the carriage moves on a round guide (14) and a flat guide (15) using linear motors (17) and (18) located symmetrically to the center of mass of the carriage's (16) mobile portal. Positioning the carriage portal is controlled by a Laser interferometer (19); one reflector of it is located on a portal. Resolution of the Laser-interferometric displacement gauge (1) is 0.6 nm. As a whole the control system of the writing head displacement provides a head position uncertainty of about 10^{-7}.

Substrate with a photosensitive layer (chromium or photoresist) is placed in a chuck (23) located at the top end face of an aerostatic spindle (24). The rotor of the spindle is set in rotation by motor (26). The rotor spindle rotation control system includes also the gauge of an angular position (25) ensuring at the speed of 10 rev/c the formation of not less than $4 \cdot 10^6$ angular discrete values for one rotation.

All control contours by executive mechanisms of the Laser Images Generator are fulfilled on the basis of digital regulators included in a control system (27). The managing computer (28) is interacting directly with this system.

There are four main application fields for the Laser Image Generator of CLWS-300 model: the synthesis of diffraction optical elements, the forming of high precision angle measuring structures (raster, scales, limbs, encoder disks, grids, etc.), the formatting of optical memory master-disks and the synthesis of micromechanical elements.

The wave front repeatability is up to $\lambda / 20$ when the diffraction optical elements are synthesized. The angular position error is less than 1″, which is ensured when the structure's topology synthesis is fulfilled by the Laser Image Generator CLWS-300. The Laser Image Generator CLWS-300 provides the optical memory master-disk formatting of all existing standards. The characteristic line size is reached up to 1 μm when the micromechanical element topology is fulfilled.

4. Conclusion

We have developed and produced two Laser technological systems, LSP-2000 and CLWS-300, for high performance manufacturing processing of 3D and 2D objects.

The Laser technological system LSP-2000 satisfies all strict requirements of precision processing arbitrary shaped large-scale components. The system carries out the following operations: cutting of metal components with a thickness of up to 6 mm, welding of metal components with a thickness of up to 2 mm, and microprofiling of surface by means of ablation. The size of the components can be up to $3 \times 3 \times 0.6$ m^3 and the deviation of a single synthesized element position from the given one is less than 10 μm at any point on the surface. A component with the maximal size can be processed in no more than 24 hours. The LSP-2000 was developed as a universal Laser processing system with a unique combination of certain parameters / technical characteristics. At the moment, LSP-2000 is used in industrial environment in China.

We have developed and created a commercial version of the Laser Image Generator CLWS-300, which operates in the polar coordinate system. It provides synthesis of 2D structures with arbitrary topology and micron accuracy (working field up to 300 mm): the angular position error is less than 1″ and the characteristics line size is equal to 1 μm. Laser Image Generators of this type have been delivered to Germany, Italy, China and Russia and are employed successfully in industrial environment here. In Germany, Italy and China they are used only for synthesis of elements diffraction optics while in Russia they are used successfully for producing high precision angular measuring structures.

References

[1] Introduction to idustrial Laser material processing, 1998, Rofin Sinar Laser, Hamburg, P.93.

[2] E.B. Kley. 1997, Lithographic Techniques in Diffractive Optics, Diffractive Optics'97, EOS Topical Meeting, Digests Series, Vol.12, P.72-73.

[3] V.S. Bazin, A.G. Verkhogliad, et al. 2005, Multifunctional precision Laser technological system for processing large-scale articles with an arbitrary topology, Optoelectronics, Instrumentation and Data Processing, Vol.41, No6, P.89-94.

[4] V.P. Kiryanov. 1997, Laser Setup for Flat Optical Components Fabrication with Submicron Resolution. Proc. SPIE, Vol.3091, P.66-70.

[5] V.M. Gurenko, L.B. Kastorsky, A.V. Kiryanov, V.P. Kiryanov, S.A. Kokarev, V.M. Vedernikov, A.G. Verkhoglad. 2002, Laser writing systems CLWS-300/C-M for microstructure synthesis on axisymmetric 3D surfaces, Proc. SPIE, Vol.4900, P.320-325.

Evaluation of Minimum Distance for Split Rectangular Beam for Glass Cutting

S. Nisar[1], S. Safdar[2], M A Sheikh[3] and L. Li[4]

[1]Univeristy of Manchester, UK, 62370, salman.nisar@postgrad.manchester.ac.uk
[2]Univeristy of Manchester, UK, 63828, s. safdar@postgrad,manchester.ac.uk
[3]Univeristy of Manchester, UK, 63802, mohammad.a.sheikh@manchester.ac.uk
[4]Univeristy of Manchester, UK, 63816, lin.li@manchester.ac.uk

Abstract. Laser cleaving of brittle materials using control fracture technique has great advantage over other processes. In this technique, thermal stresses are used to induce the crack and the material is separated along the cutting path by extending the crack. It is believed that split beam shapes can generate bi-axial stress state which may help in successful parting of the glass sheets. Precise thermal and stress distribution can be induced in this way and can significantly optimise the process. In this study, the minimum possible distance for split rectangular beam has been evaluated through numerical modelling. The temperature distributions are obtained by using finite element software ABAQUS.

Keywords: Control Fracture Technique (CFT), Laser Beam Shapes, Bi-axial stress, Finite-Element-Method.

1. Introduction

With the advancement of the glass industry in recent times, glass has become an important engineering material in architecture, medical, automotive, flat panel display and electronics industry. In glass fabrication, cutting is the fundamental process unlike other processes which includes coating, polishing-grinding, peeling etc. Desired shape and size of the glass can only be achieved through accurate and precise cutting technique.

The conventional method of glass cutting with the help of a metal or diamond point tool has been used for decades. In this method, the glass surface is first marked and scribed using diamond point tool and then mechanical force is applied to break the glass along the scribing path. Irregular cut path, poor surface finish and large material deformation is usually observed when the above method is used for glass cutting. It has also been reported that the strength of the glass reduces to 60% on average by the use of mechanical cutting [1].

Different techniques like laser scribe and break, control fracture technique etc exists for glass cutting from which some are being used in industries and some are still in their developmental stages. Control fracture technique has great potential in laser glass cutting because it uses less laser power and high cutting speed as compared to other laser cutting methods. It was pioneered by Lumley [2] in 1968. In this method, the cutting is done below glass transition temperature T_g which results in smooth edges and requires no further cleaning or grinding.

In control fracture technique, the crack initiates due to heat of the laser beam. The stress near the laser spot is compressive due to high temperature. After the passage of the laser beam, compressive stresses relax and induce local residual tensile stress. If these stresses exceed the fracture toughness value of the glass the crack initiates.

Tsai and Liou [3] investigated the cutting technique of controlled fracture and mentioned that it can be divided into three stages. In the first stage, fracture is initiated due to tensile stresses. The stable crack growth was observed in the second stage and, in the third stage, the crack extended unstably due to the complete domination of tensile stresses. It has been reported that the velocity of crack tip and laser beam spot remains constant.

Tsai and Lin [4] noted that the specimen width plays a key role in control fracture technique. The crack opening displacement in narrow specimen is faster as compared to wider substrates. Therefore, the cutting speed is high while cutting narrow specimens.

Dekker et al. [5] proposed a similar method of controlled fracture for cutting glass. A small scratch was formed at the start of the cut which helped in initiating the crack and led to the separation of the glass. Georgi [6] proposed a novel approach in which two concave mirrors were used on either side of the laser beam and glass substrate in order to obtain a higher rate of absorption.

Controlled fracture technique has been improved by introducing water-air mixture along the cutting path to generate tensile stresses and to maintain the temperature below the glass transition temperature [1, 7]. However, coolant adds extra cost and complexity to the process.

The laser cutting technique using controlled fracture mainly depends on laser wavelength and the spot size. As glass has quite low absorption, a suitable laser (wavelength) must be employed for this technique. The spot size is normally kept large at a defocused distance because of low transition temperature of the glass as to minimise melting and evaporation. Glass thickness of 3 mm has been successfully cut through by using control fracture technique [8].

In control fracture technique, it is difficult to control the crack propagation inside the glass because of its fracture characteristics and adhesions. The major hurdle for this process is the generation of precise thermal and stress fields. In this regard different beam geometries have been used to cut the glass. Donut mode, i.e. TEM_{01}, and beam shapes like U and V were used to cut the glass [9, 10]. The idea of using different beam shapes for laser glass cutting is to generate stress distribution which can allow high cutting speeds and ensure a stable crack growth.

Shakeel [11] has investigated the effect of a donut laser beam shape on tube bending by numerical modelling and highlighted that a bi-axial stress state occurs with a donut beam shape. It was also highlighted that the effect of the beam geometry will be more significant in materials with low thermal diffusivity, e.g. ceramics, glass, etc.

Two beam configurations, i.e. solid and split rectangular beams, have been investigated in this study. A solid rectangular beam produces a conventional temperature distribution whereas the temperature distribution in the case of a split rectangular beam will produce a bi-axial stress state. The easiest way to produce a split rectangular beam is to put a mask in the path of a solid rectangular beam. This would cause the solid rectangular beam to split up into two rectangular beams. It is therefore imperative to calculate the mask width which would be sufficient to create a bi-axial stress state in the cut path. In the present work, the minimum distance for split rectangular beams has been evaluated where bi-axial tensile stresses may exist for the successful cutting of the glass.

2. Finite Element Analysis

A 3D tranisent thermal analysis was carried out on a soda lime glass with a commerical finite element package ABAQUS. A speciment of 200 mm length, 100 mm width and 6 mm thickness has been selected for the simualtion. For simplicity, the thermal-physical properties of the material are assumed to be temperature independent which were provided by the Pilkington as shown in Table 1.1

The parameters used for the simulation with a moving surface heat source are summarised in Table 1.2. The temprature field T(x,y,z,t) is calculated using the transient solution procedure. The general three-dimensional heat diffusion equation can be expressed as follows:

$$k\left(\frac{\partial^2 T}{\partial^2 x}+\frac{\partial^2 T}{\partial^2 y}+\frac{\partial^2 T}{\partial^2 z}\right)+Q=\rho c\frac{\partial T}{\partial t} \tag{1}$$

where k is the thermal conductivity, ρ is the density, c is the specific heat and Q is the laser power per unit area. The initial temperature of the substrate was kept at 25°C. The effect of natural convection coefficient has been neglected [3].

Several models were developed and an iterative procedure was employed to predict the working width of the mask which may generate a bi-axial stress state. Only three models have been discussed in this study. In the first stage, a defocused solid rectangular beam with a spot size of 6×5 mm has been modelled. In the second stage, 0.06 mm split distance was used in the simulation, as this was the minimum mask width available for beam splitting. The area of the spot size for splitted rectangular beams was kept constant while the distance between them was varied. After several attempts, it was found that a mask width of 1 mm can produce a low temperature zone in between the two split beams, which may generate a bi-axial stress state.

Table 1.1. Thermal-physical properties of soda lime glass provided by the manufacturer

Glass type	Density (Kg m^{-3})	Thermal conductivity (W m^{-1} K^{-1})	Specific heat (J kg^{-1} K^{-1})
Optifloat	2500	1.03	770

3. Results and Discussion

A conceptual representation of solid beam geometry is shown in Figure 1.1 [12]. It can be seen this figure that tensile stresses occur behind the laser beam where as a compressive stress state exists in the region of high temperature. If a split rectangular beam is used, the high temperature zone of the both beams will result in a relatively low temperature zone at the reference point. As there will be compressive stresses around the high temperature zone of the split rectangular beam, the reference point will therefore experience bi-axial stress state (tensile pull) as shown in Figure 1.2.

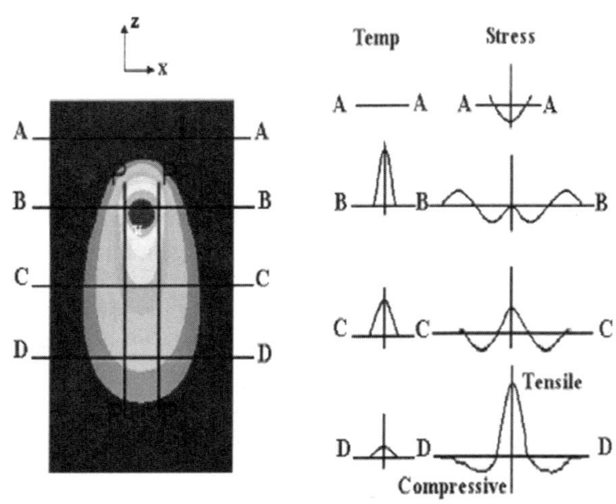

Figure 1.1. Conceptual behaviour of solid beam shape [12]

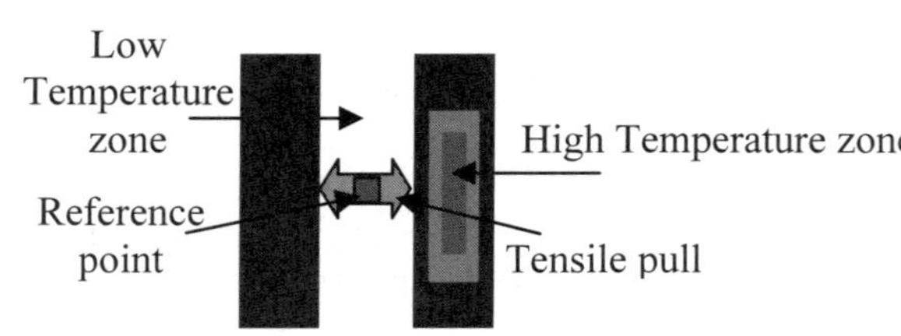

Figure 1.2. Transient bi-axial stress state for a split rectangular beam (derived from [11])

The criteria for the suitable width of the mask is considered to be a distance where the maximum temperatures zones are safely apart and, therefore, create a low temperature zone in between the split beams. The high temperature zone of the split beams results in a bi-axial tensile pull which will result in the successful separation of the glass.

Figure 1.3 shows the temperature profile of the soda-lime silicate glass following the scanning of the solid rectangular laser beam. The highest temperature occurred on the surface of the substrate was around 440°C.

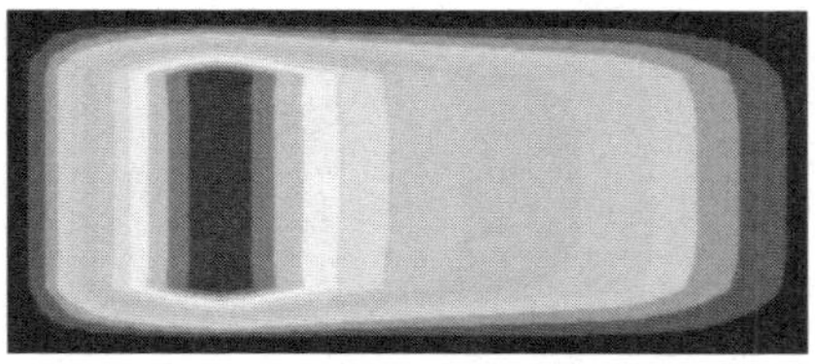

Figure 1.3. Isotherms on the top surface of a solid rectangular beam

The second model was simulated with a split distance of 0.06 mm between the two rectangular split beams by keeping the area constant. It can be seen in Figure 1.4(a) that the high temperature zones have diffused into the middle, as indicated by the centre dashed-line. This temperature distribution is unlikely to produce a bi-axial stress state.

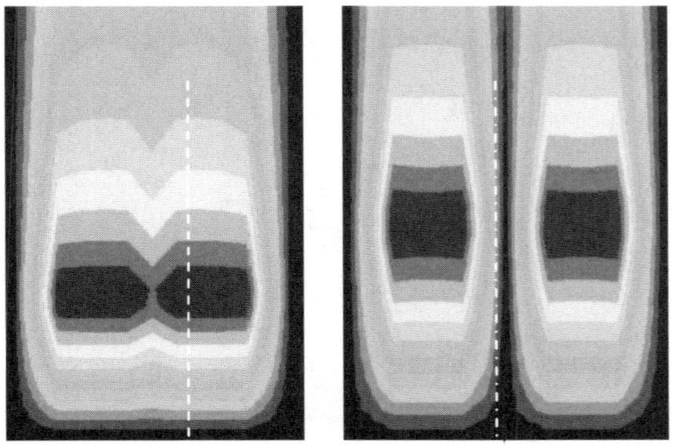

(a) (b)

Figure 1.4. Isotherms on the top surface of a solid rectangular beam: (a) rectangular beam with distance 0.06 mm and (b) 1 mm

Table 1.2. Parameters used for the simulation

Laser power (absorbed)	58 W
Dimension	200×100×6 mm
Scanning speed	50 mm/sec
Laser beam size	6 X 5 mm
Step-time	0.0125 s

Figure 1.4(b) shows the split beam configuration with a split distance of 1 mm. It can be seen that high temperature zones for the split rectangular beams have not defused. The centre dashed-line, as can be seen from Figure 1.4b, is therefore at a relatively low temperature, because of which it is likely that a bi-axial tensile stress will occur.

4. Conclusion

In this paper, a commercial FE package ABAQUS is used to predict the working width of the mask. A three-dimensional model of the soda-lime glass was employed to evaluate the suitiable width of the mask, which may generate a bi-axial stress state. It was found that a mask width of 1 mm is suitable to generate a bi-axial tensile stress state as no temperature diffusion between the two split beams was observed.

The current work is just an initial study, a comprehensive modelling and experimental work is planned for the future. These will include:

1. The experiments with same parameters will be conducted with and without splitting rectangular beam and the results will be verified with FEM.
2. Thermal distribution inside a glass along with the stress analysis will be carried out using numerical modelling technique to further investigate the behaviour of splitted beam shapes in glass cutting.
3. Investigation into thickness of glass above 6 mm will also be conducted.

References

[1] A.B. Zhimalov, V.F. Solinov, V.S. Kondratenko and T.V. Kaplina, 2006, Laser Cutting of Float Glass during Production, Science for Glass Production, Vol. 63, No.10, P3-5.

[2] R Lumley, 1968, Controlled Separation of Brittle Materials using a Laser, American Ceramic Society Bulletin, Vol.47, No.4, P385.

[3] C.H. Tsai, C.S. Liou, 2003, Fracture Mechanism of Laser Cutting with Controlled Fracture, ASME Journal of Manufacturing Science and Engineering, Vol.125, P519-528.

[4] C.H. Tsai, B.C. Lin, 2004, Laser Cutting of Silica Glasses by using the Controlled Fracture Technique, Proceedings of 23rd International Congress on Applications of Lasers and Electro-Optics, P34-41.

[5] J.N. Dekker, M.H. Zonneveld, 1989, Thermal Severing. The Cutting of Brittle Materials by Thermally Induced Fracture, Advances in Fracture Research. Proceedings of the 7th International Conference on Fracture, Vol.4, P2825-2834.

[6] S. Georgi, 2003, Separating and Structuring of Brittle Material by the Use of Laser Radiation, Proceedings of SPIE--the International Society for Optical Engineering, Vol.5063, No.1, P426-431.

[7] N. Paterson, 2004, On the Numerical Modelling of Laser Shearing of Glass Sheets used to Optimize Production Methods, Proceedings of the Institution of Mechanical Engineers. Part C, Mechanical engineering science, Vol. 218, No.1, P1-12.

[8] Black, M.A. Finucane, 1996, CO2 laser cutting of stained glass, The International Journal Advanced Manufacturing Technology, Vol.12, No.1, P47-59.

[9] Krause, Loch, 2002, Mathematical Simulation in Glass Technology, Springer-Verlag Berlin Heidelberg, ISBN 3-540-43204-3.

[10] C. Hermanns, 2000, Laser Cutting of Glass, Proceedings of SPIE--the International Society for Optical Engineering Vol.4102, P219-226.

[11] S. Safdar , L. Li, M.A. Sheikh , Z. Liu, 2005, The Effect of Beam Geometry on Stress Distribution in Laser Bending of Tubes, Proceedings of 24th International Congress on Applications of Lasers and Electro-Optics, ICALEO 2005, P54-63.

[12] S. Safdar, M.A. Sheikh, L. Li, Z. Liu, The Effect of Non-Conventional Laser Beam Geometries on Stress Distribution and Distortions in Laser Bending of Tubes, Journal of Manufacturing Science and Technology, ASME. (In-Press)

Nonlinear Penetrating Behavior in High Energy Beam Drilling

J.E. Ho[1] and H.T. Young [2]

[1]National I-Lan University, I-Lan, Taiwan, R.O.C;26047 email: jeho@niu.edu.tw
[2]National Taiwan University,Taipei, Taiwan, R.O.C:10617 email:hyoung@ntu.edu.tw

Abstract. The geometrical keyhole scale produced in high energy beam drilling is approximately several hundred micrometers in size. Detecting its physical transport in such a narrow cavity is extremely difficult, let alone understanding the associated energy distribution in the workpiece. To describe nonlinear behavior in the penetration process, this study develops a 2D quasi-steady velocity model based on the enthaply theory. In the application of separation method, the penetration velocity can be expressed as an exponential function of liquid-thickness and thermal properties. Setting up non-uniform grids in the calculating domain enables successful prediction of nonlinear penetration behaviors such as efficiency, mass flow rate, liquid film, and energy transport versus incident energy density. Even the continous formation of the drilling cavity can also be reasonably simulated. Compared with the experimental data of Allmen, the present model shows a good agreement for copper drilling, where the relative errors between the calculated and experimental data are no more than 20%.

Keywords: high energy bean drilling, 2D quasistatic velocity model, drilling cavity

1. Introduction

Nonlinear penetration behavior is an important phenomena in high energy beam drillings, including the Laser beam (L.B) and electron beam (E.B) methods. However, describing the rapid reactions between the vapor atoms at the cavity base seems so complex that a thorough under-standing of energy transport inside the workpiece is still lacking. Two important effects in L.B drilling have been widely discussed; one is the integrated whole depth in terms of energy absorbed, and the other is the extracted material volume per input power. The former is not as practical because it depends on specific energy distribution parameters. The latter, defined as penetrating efficiency, can be applied to any desired beam energy in this study. Previous studies have discussed simplified models of penetrating velocity and energy density, such as the pure melting model [2] and single evaporating theory [3]. Both proposed mechanisms, which exhibit a linear relation between the penetration velocity and energy density, seemed to only describe drilling behavior in lower energy densities. Allmen [1] made direct measurements of penetrating efficiency, and those experimental results showed that a non-linear relationship was observed during the energy intensity range between 5 MW/cm^2 and 20MW/cm^2. Chiou and Wei [4] developed an axial symmetrical quasi-steady model to calculate the fluid flow of the liquid layer by considering the surface tension as the driving force; a surprising result showed that the calculated evaporation rate was only 1/1000 of the melting mass. Ho and Yong [5] proposed a 1D model to describe the nonlinear behavior in a high energy beam. Their analytical solution was expressed as an exponential function of the thermal properties. This approach presented an excellent agreement in higher input energy density; but for lower energy density, the predicted value was overestimated. Ho [6] employed the Monte Carlo method to trace the reflection of an L.B in the cavity and successfully estimated the heat absorption of a drilling hole under a high energy beam.

Most of these previously released papers concentrated on the linear drilling behavior and the discussion of nonlinear variation in penetration characteristics is often neglected. This paper focuses on nonlinear changes in the drilling process. The approach adopted neglects the convective term by estimating a small Pelect number in the liquid film and assuming the absence of friction drag at the drilling front as the boundary layer does not grow enough to influence the laminar flow.

2. Analysis

A thin layer of molten metal flows along the wall near the base of a vapor-filled cavity as illustrated in Fig. 1.1. An rz coordinate system is assumed to move into the solid at a steady penetration velocity, which is relevant for penetrating times greater than 0.1 μs, as shown by the measurements of von Allmen.

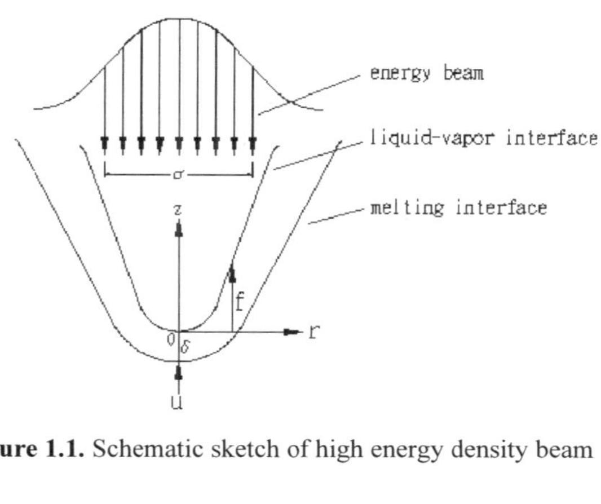

Figure 1.1. Schematic sketch of high energy density beam

2.1 Governing Equations and Boundary Conditions

The formulation of the enthalpy equation with an axial symmetrical, quasi-steady state in both molten and solid zones can be expressed as:

$$\rho u \frac{\partial h}{\partial z} = \frac{1}{r}\frac{\partial}{\partial r}\left(k_i r \frac{\partial h}{\partial r}\right) + \frac{\partial}{\partial z}\left(k_i \frac{\partial h}{\partial z}\right) \quad (1)$$

$$h = \begin{cases} c_s T & T < T_m - \varepsilon \\ c_s T + \frac{h_{sl}}{2\varepsilon}(T - T_m + \varepsilon), & T_m - \varepsilon \le T \le T_m + \varepsilon \\ c_l T + h_{sl}, & T_m + \varepsilon < T \end{cases} \quad (2)$$

$$-k_l \frac{\partial T}{\partial z} = h_c (T - T_\infty) \quad at \quad z = 3\sigma \qquad \frac{\partial T}{\partial r} = 0 \quad at \quad r = 0$$

$$k_s \frac{\partial T}{\partial z} = h_c (T - T_\infty) \quad at \quad z = -3\sigma \qquad T = T_\infty \quad at \quad r = \infty \quad (3)$$

where σ is the distribution radius that defines the region in which 40% of the incident energy is deposited ,T_∞ is the ambient temperature,k_i represents the thermal conductivity in both phases, and the enthalpy function h smoothes the discontinuous enthalpy at the solid-liquid interface. Penetration velocity u can be determined at r = 0 while neglecting radial conduction at the cavity base by scaling analysis (Ho [5]).

$$u = \left(\frac{2\alpha_l}{r} l_n\left(\frac{1+cl}{cl}\right)\right)^{1/2}\left(\frac{2\Delta p}{\rho}\right)^{1/4} \quad (4)$$

where $cl = \frac{(K_s T_m - \rho \alpha_s h_{sl})}{K_c (T_0 - T_m)}\left(\frac{\alpha_l}{\alpha_s}\right)$

$$\Delta p = Ae^{-\frac{B}{T_0}} - Ae^{-\frac{B}{T}}$$

Here, T_o is the bottom temperature of the cavity, T_m is the melting temperature, and α_l and α_s indicate the thermal diffusivity in liquid and solid phase, respectively.

An additional boundary condition, effective surface pressure is balanced with the surface tension given by Knight [7], is necessary to determine the surface temperature while the vapor-liquid interface is specified.

$$\beta p_b \exp\left[\frac{h_{lg}}{R_g}\left(\frac{1}{T_b} - \frac{1}{T}\right)\right] = \gamma\left(\frac{1}{R_1} + \frac{1}{R_2}\right) \quad (5)$$

where β=0.55 was also calculated by Knight, considering the thermodynamics non-equilibrium at the evaporating surface. The surface tension γ at the bottom of the cavity is assumed to be a linear function of temperature along the free surface. R_1 and R_2 give the curvature radius along the vapor-liquid interface.

2.2 Mass Conservation

As far as the mass flow conservation in liquid film $\dot{m}_l + \dot{m}_v = \rho u \pi r^2$ is concerned, the melting rate $\rho u \pi r^2$ at the cavity base will be shared by the evaporation rate $\dot{m}_V$ and the flow rate $\dot{m}_l$ carried away due to the pressure difference at the bottom of the cavity. $\dot{m}_l$ can be approximately expressed as $2\pi r \delta v$ while $v = \left(\frac{2\Delta p}{\rho}\right)^{1/2}$ was estimated by Bernoulli's equation with the absence of friction drag set at r = 0.

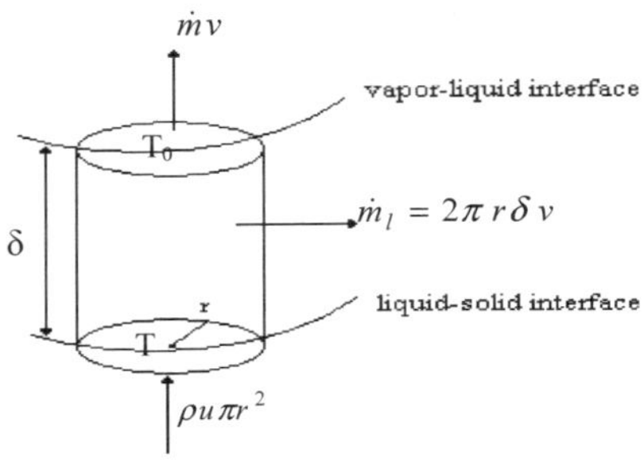

Figure 1.2. Flow distribution

The evaporation rate $\dot{m}_V$ developed by Langmuir [8] leads to

$$\dot{m}_v = \frac{\alpha p_b}{\sqrt{(2\pi R_g T)}} \exp\left[\frac{h_{lg}}{R_g}\left(\frac{1}{T_b} - \frac{1}{T}\right)\right] \quad (6)$$

where h_{lg} is the latent heat of evaporation, R_g is the specific gas constant, and P_b is the equilibrium vapor pressure at the boiling point. Knight's accommodation coefficient α=0.816 was used to account for back-scattering of evaporating ions.

2.3 Energy Conservation at Vapor-liquid interface

The beam energy is balanced with the heat conduction and evaporation energy at the vapor-liquid interface, yielding

$$q \exp\left(-\frac{r^2}{2\sigma 2}\right) = -k\left(\frac{\partial T}{\partial r}\frac{df}{dr} - \frac{\partial T}{\partial z}\right) + \dot{m}_v h_{lg} \quad (7)$$

Where q is the energy density at r = 0 and f indicates the location of free surface.

2.4 Penetration Efficiency

The penetration efficiency η is defined as the extracted material volume per unit input power at r = 0.

$$\eta = \frac{u}{q} \quad (8)$$

where $q = k_l \frac{\partial T}{\partial Z} + \dot{m}_v h_{lg}$

3. Numerical Procedure

The discrete form of equations (1) and (2) with boundary conditions (3)-(5) was obtained using the central finite differences. A grid of 40×30 nodal points ensured the independence of the solution on the grid. The nodal points in both the r- and z- directions were not uniformly distributed, and had a greater concentration of points near the vapor-liquid interface. For comparison with the experimental results, a grid of 60×45 nodal points was used. Key steps for solving this problem are as follows.

(i) Specify the shape of the cavity first, and then calculate the temperature distribution along the vapor-liquid interface from equation (5).
(ii) Estimate the penetration velocity from equation (4).
(iii) Solve the enthalpy equations (1) (2) with boundary conditions (3)-(5) using the successive relaxation method with a relaxation factor of 1.25 until the temperature converges to a relative error limit of 0.5%.
(iv) Estimate a new cavity shape from equation (7). Compare the newest shape and last shape of the cavity; if the relative error is more than 5%, update the latest geometry and repeat steps (i) ~ (iv).
(v) Determine the evaporation rate m_v, input energy density q, and penetration efficiency η from equations (6) (8) respectively.
(vi) Give another input power q, repeat steps (i) ~(vi).

4. Results and Discussion

The workpiece was chosen to be copper, and all the figures provide dimensional coordinates for comparison with Allmen's experimental data [1].

Figure 1.3 illustrates the penetration velocity versus energy density. At the lower energy density (< $7*10^{10}$ W/m^2), the penetration velocity increases slightly from 2.3 m/s~ 4 m/s, and exhibits a nonlinear increase from 4 m/s~ 15 m/s during the critical energy density ($8*10^{10}$ W/m^2 ~ $10*10^{10}$ W/m^2). That is because the radial conduction energy ratio drastically decreases from 25% to 3%, as indicated by Fig 1.5, which provides extra energy in the axial direction and liquid film to strengthen the penetrating potential. As for the higher energy density (>$10*10^{10}$ W/m^2), the penetration velocity approaches a stable value (about 15 m/s) which corresponds to the constant ratio of energy dissipated in the radial conduction and melting during the same energy density interval in Fig. 1.5. Comparing with Allmen's experiments, the results are in close agreement with each other and the maximum relative error is about 20% at an energy density of $7*10^{10}$ W/m^2.

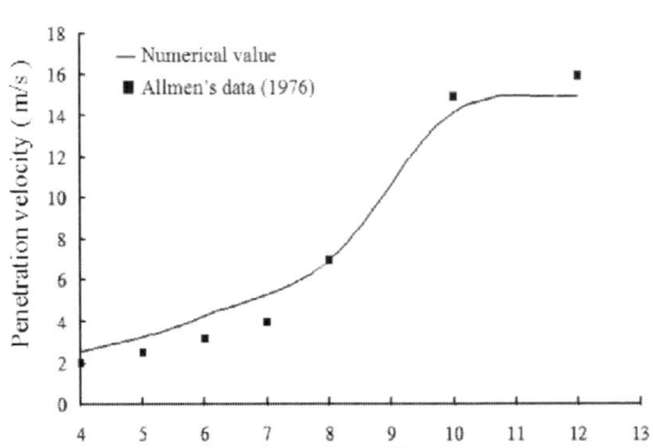

Figure 1.3. Penetration velocity vs. energy density in copper drilling

Concerning the variation in penetration efficiency with different energy densities, Fig. 1.4 shows that a poor drilling efficiency of about $6*10^{-11}$ m^3/J abruptly climbs up to $1.5*10^{-10}$ m^3/J as the energy density varies from $4*10^{10}$ W/m^2 to $10*10^{10}$ W/m^2 where the penetration velocity in Fig. 1.3 quickly increases from 2 m/s ~15 m/s with a faster growth than the energy density (<$10*10^{10}$ W/m^2) dose. The opposite result ,i.e., a slow decrease of efficiency from $1.5*10^{-10}$ m^3/J to $1.3*10^{-10}$ m^3/J, occurs when the energy density is more than $10*10^{10}$ w/m^2, while the stable value 15 m/s in drilling velocity grows slowly than the increment of input energy density. Both tendencies, in Figs. 1.3 and 1.4, exhibit a similar distribution in penetration velocity, efficiency, and input power density. Moreover; the maximum drilling efficiency $1.5*10^{-10}$ m^3/J occurring at $10*10^{10}$ W/m^2 in Fig. 1.4 is produced when the drilling velocity reaches the stable value in Fig. 1.3. Comparing the numerical and experimental results shows that the agreement is good in most energy regions, especially in higher energy densities (>$9*10^{10}$ W/m^2) where the maximum relative error of 15% emerges at an input energy density of $10*10^{10}$ W/m^2.

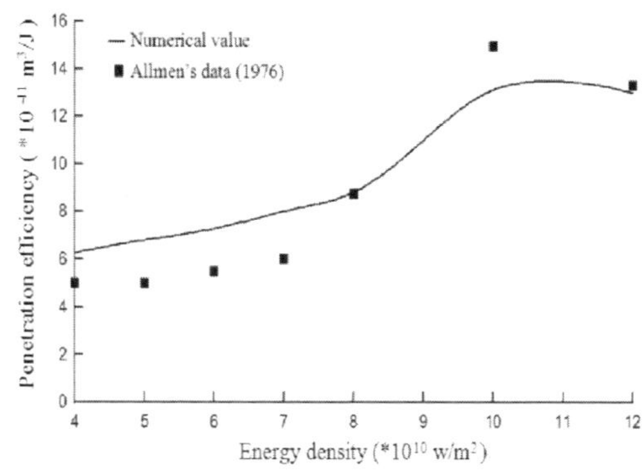

Figure 1.4. Penetration efficiency vs. energy density in copper drilling

The incident energy can be dissipated in evaporation, melting, convective term, axial (radial) conduction and the increase in internal energy of workpiece.

Figure 1.5 shows that the energy ratio absorbed in evaporation, convective term are no more than 1% and can be neglected without causing significant error, which agrees with the previous assumption. As for the melting energy ratio, a slight increase from 10% ~ 22% prevails throughout the working period ($4*10^{10}$ W/m^2 ~$12*10^{10}$ W/m^2) and remains constant to the end. Moreover; the radial conduction ratio quickly decays from 20% ~ 3% during the same energy density interval above. This implies that the less energy absorbed in melting with more heat conducted to the radial direction will weaken the penetration potential and widen the cavity opening (<$7*10^{10}$ W/m^2); otherwise, a narrow cavity with a higher penetration efficiency will be predicted during the energy density $8*10^{10}$ W/m^2~$10*10^{10}$ W/m^2, where the radial conduction ratio has been reduced to 3% as in Fig. 1.5

In view of the sensible heat in liquid film, it almost takes most of input energy to raise the liquid from melting temperature to the cavity base temperature and the ratio increases from 32% to the peak value 58% at the critical energy density $8*10^{10}$ W/m^2, where the extra increase of incident energy in the liquid film is absorbed to strength drilling potential and causes a significant change in penetrating behavior as described in Figs. 1.3 and 1.4. As to the energy transferred and used to preheat solid material "axial conduction" is concerned, its ratio can be estimated by unity minus the total of dissipated ratio in Fig. 1.5 for individual input power density, and the value nearly keeps 30% of incident power during the entire working region.

Another worthy mentioned is the ratio of evaporating mass to melting flow , it can be calculated by dividing melting, evaporating latent heat to corresponding latent ratio in Fig. 1.5 ,and the result tells that evaporating mass is just only 0.6% ~ 0.3% of melting flow throughout the working period. From this, continuous formation of the drilling cavity can be proven to be determined by the action of melting and flow carried to radial direction due to the pressure gradient instead of the evaporating phenomena.

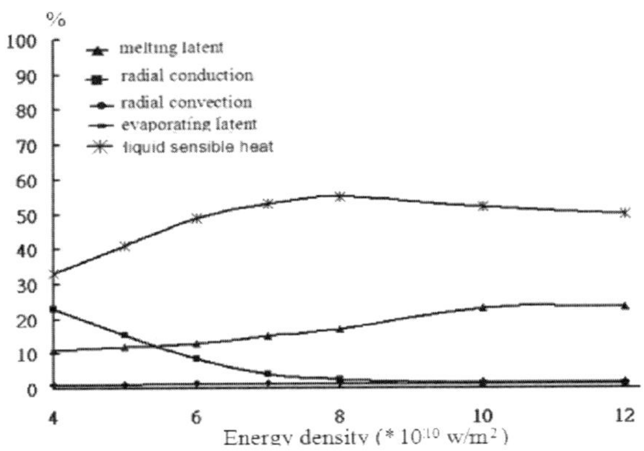

Figure 1.5. Various energy dissipation ratios vs. energy density in copper drilling

In Fig. 1.6, the liquid thickness slightly decreases from $28*10^{-6}$ m to $22*10^{-6}$ m during lower energy densities ($4*10^{10}$ W/m^2 ~ $7*10^{10}$ W/m^2), and then drastically drops to $4*10^{-6}$ m around the critical energy density of $8*10^{10}$ W/m^2. This sharp decay indicates that the extra more heat absorbed due to the higher temperature gradient in the liquid film will cause a remarkable change in drilling behavior, which is also agreement with the discussion in Figs. 1.3 and.1.4.

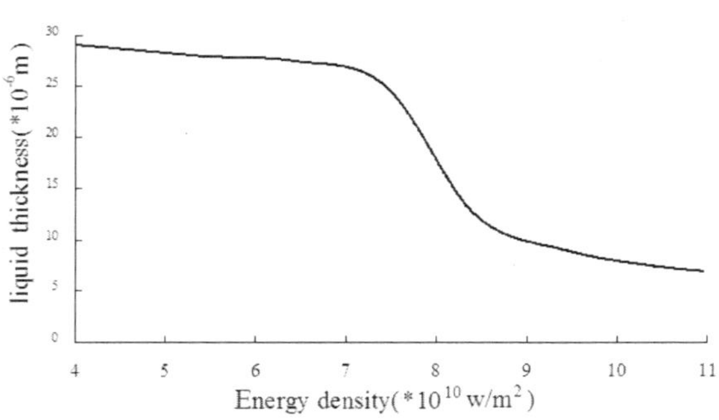

Figure 1.6. Variation in liquid thickness of energy density in copper drilling

5. Conclusions

The discussions made above lead to the following important conclusions:

(1) Since the evaporation ratio accounts only 3/1000 in this study, the cavity formation is apparently caused by the action of melting, and the flow carried to radial direction instead of evaporating phenomena.

(2) Larger portion of radial conduction and less liquid sensible heat weaken the penetration velocity and, hence reduce drilling efficiency. Otherwise, overall penetration behavior improves as the efficiency rises.

(3) Due to the small ratio of convective energy and latent heat in evaporation, the assumption made by neglecting these terms in this study without significant errors is reasonable.

(4) During the critical energy density range, the sharp decrease in liquid thickness conducts more heat absorbed in liquid film, resulting in an abrupt increase in penetration velocity. This increase helps penetration efficiency reach its maximum value.

(5) Detailed analysis of various energy transfer ratios in this study will be useful to understand the nonlinear behavior in the penetrating process.

Acknowledgements

Support for this work by National Science Council of the Republic of China under grant no. No. NSC 95-2221-E-002-233- is gratefully acknowledged.

References

[1] M. von Allmen, Laser drilling velocity in metals, J. Appl. Phys. 47, 5460-5463 (1976).

[2] J. F. Ready, Effects due to absorption of laser radiation, J. Appl. Phys. 36,462-468 (1965).

[3] N. A. Ol'Shanskii, Movement of molten metal during electron beam welding, Svar. Proiz. 21, 12-14(1974).

[4] P. S. Wei and L. R. Chiou, Molten metal flow around the base of a cavity during a high-energy beam penetrating process, J. Heat Transfer110, 918-923(1988).

[5] J.E.Ho and H.T.Young,The Analysis on Penetrating Efficiency in high-energy beam Drilling, Asia Pacific Conference on Optics Manufacture (2007)

[6] Je-Ee Ho and Ching-Yen Ho, Computing the Absorption of a Drilling or Welding Hole for Laser by Using Monte Carlo Method, Journal of the Chinese Society of Mechanical Engineers, Vol.27, No.1, pp.61-67 (2006).

[7] C. J. Knight, Transient vaporization from a surface into vacuum, AIAAJ. 20, 950-954 (1982).

[8] I. Langmuir, The vapor pressure of metallic tungsten, Phys. Rev. 2, 329-342 (1913).

[9] P. S. Wei, and C. Y. Ho, Beam focusing characteristics effect on energy reflection and absorption in a drilling or welding cavity of paraboloid of revolution, International Journal of Heat and Mass Transfer 41, 3299-3308 (1998).

Analysis of Enhanced Thermal Conductivity Approach for Predicting Melt Pool Geometry and Temperature Distribution for Laser Melting Processes

S. Safdar[1], F. Qureshi[2], L. Li[1], M.A. Sheikh[1]

[1] LPRC, School of MACE, University of Manchester, Sackville Building, Sackville Street, United Kingdom, M60 1QD.
[2] College of E&ME, National University of Sciences and Technology (NUST), Peshawar Road, Rawalpindi, Pakistan.

Abstract. Modelling of laser manufacturing processes involving laser melting is an extremely important activity, for controlling and optimising these processes. However, modelling of these processes is not straight-forward due to Marangoni and buoyancy convection within the melt pool. Detailed CFD models are required to accurately predict the melt pool geometry and temperature distribution. CFD modelling of melting processes requires much greater expertise and also requires longer computational time. To simplify and speed up the modelling process, many researchers have used the enhanced thermal conductivity approach to account for melt pool convection. Instead of solving the intricate Navier-Stokes equations, only the energy equation is solved by enhancing the thermal conductivity beyond the melting temperature, to predict the melt pool geometry and temperature distribution. However, researchers have used the values of enhanced thermal conductivity from the literature without any validation. Moreover it is also not established whether the enhanced thermal conductivity approach is able to predict accurately the melt pool geometry and temperature distribution. This paper presents an analysis of the enhanced thermal conductivity approach for laser melting of mild steel. The finite volume method has been used to simulate the transient effects of a moving beam for laser melting of mild steel (EN-43A). Experimental melt pool geometry has been compared with the CFD model and the enhanced thermal conductivity model.

Keyword: Melt pool geometry, temperature distribution, laser manufacturing processes, CFD models

1. Introduction

Many laser manufacturing processes involve laser melting of material, e.g. selective laser melting, welding, brazing, soldering, glazing, surface alloying, cladding, etc. Modelling of these processes is an extremely important activity for controlling these processes by predicting the melt pool geometry, thermal gradients, cooling rate, microstructure, and stress distribution, etc. However, modelling of processes involving laser melting is not straight-forward due to Marangoni and buoyancy convection within the melt pool.

It is well established that heat transfer in the melt pool is influenced more by the fluid flow than the heat conduction [1]. Fluid flow within the melt pool is dominated by thermocapillary or Marangoni forces and hence conduction-based models will not be able to accurately predict the laser melting process particularly in metals [1]. Thus any modelling attempt to investigate the effects of beam geometries on laser melting of metals should take into account melt pool convection due to thermocapillary and buoyancy forces. Detailed CFD models are required to accurately predict the melt pool geometry and temperature distribution.

CFD modelling of melting processes requires much greater expertise and also requires longer computational time. Moreover, CFD models are also prone to errors due to uncertainties in the values of surface tension co-efficient and viscosity of material in the fluid state [2]. Model agreement can be improved by adjusting the values of surface tension coefficient and viscosity, but to reach the correct figure a very long computational time is required. To simplify and speed up the modelling process, many researchers have used the enhanced thermal conductivity approach to account for melt pool convection. Instead of solving the intricate Navier-Stokes equations, only the energy equation is solved by enhancing the thermal conductivity beyond the melting temperature, to predict the melt pool geometry and temperature distribution. This approach has been used in the past for modelling laser cladding, laser welding [3-5], etc. However, researchers have used the values of enhanced thermal conductivity from the literature. These values range from 2 times the original values to 420 times [3-5]. Due to this huge bandwidth it is difficult to pick an accurate value from the literature. Moreover, even if careful selection is made by looking at the materials and processing conditions, it is not known whether it would give the desired matching between the experimental and the predicted melt pool profile and temperature distribution. There is very little available literature where a comparative study has been made to see the accuracy of the enhanced thermal conductivity approach. Thus there is a need to study the enhanced thermal conductivity approach and see its effectiveness.

This paper presents an analysis of the enhanced thermal conductivity approach for modelling laser melting processes. A finite volume model has been constructed using a commercial software, FLUENT, to simulate the transient effects of a moving beam for laser melting of mild steel (EN-43A). Melt pool convection, induced due to the Marangoni and buoyancy forces, has been taken into account in the CFD model, whereas the enhanced thermal conductivity approach has been used for the conduction model. Experimental melt pool geometry has been compared with the CFD model and the enhanced thermal conductivity model to assess the effectiveness of the enhanced thermal conductivity approach.

2. Finite Volume Model

A 3D finite-volume model, based on the Reynolds Averaged Navier Stokes Equations and the Enthalpy-Porosity technique is set-up in FLUENT, for modelling of the laser melting process. The laser beam was input as a surface heat flux. Area sector approach [6] was used to simulate a moving laser beam instead of a moving wall approach. Necessary subroutines were added to handle the spatially varying heat flux to simulate a moving laser beam. For the normal heat transfer and enhanced thermal conductivity approach, same model was used but only the energy equation was solved. In the enhanced thermal conductivity model, thermal condcutivity values were enhanced by three and four times their original values. The enhanced thermal conductivity k' is defined by Eq 1.

$$k' = \alpha k \tag{1}$$

where k is the normal thermal conductivity value at the corresponding temeprature and α is the enhancement factor which is defined as

$$\alpha = \begin{cases} 1 & \text{if } T < T_{liquidus} \; \& \; T_{solidus} \\ \\ \textit{multiplying factor} & \text{if } T > T_{liquidus} \end{cases}$$

where $T_{liquidus}$ is the temperature at which the solid formation begins and $T_{solidus}$ is the temperature at which full solidification occurs.

Energy conservation equation for the complete single domain in enthalpy-porosity technique is written in terms of enthalpy and temperature (Eq 2) [7]:

$$\frac{\partial}{\partial t}(\rho H) + \nabla.(\rho \bar{\upsilon} H) = \nabla.(k \nabla T) + \bar{S} \tag{2}$$

where H is enthalpy, ρ is density, $\bar{\upsilon}$ is the fluid velocity and $\bar{S}$ is the surface heat source. For enahnced thermal conductivity model k is replaced by k' in Eq 2.

Enthalpy of material is computed as the sum of sensible enthalpy (h) and latent heat (ΔH) (Eq 3).

$$H = h + \Delta H \tag{3}$$

where

$$h = h_{ref} + \int_{T_{ref}}^{T} C_p dT$$

h_{ref} = Reference enthalpy

T_{ref} = Reference temperature

C_p = Specific heat at constant pressure

Latent heat content (ΔH) can be written in terms of latent heat of the material L (Eq 4) [7]:

$$\Delta H = \beta L \tag{4}$$

where β is the liquid fraction, which is defined as

$$\beta = \begin{cases} 0 & \text{if } T < T_{solidus} \\ \dfrac{T - T_{solidus}}{T_{liquidus} - T_{solidus}} & \text{if } T_{solidus} < T < T_{liquidus} \\ 1 & \text{if } T > T_{liquidus} \end{cases}$$

Thermocapillary or Marangoni convection is treated as shear stress applied at wall. Shear stress applied at the wall is given by Eq 5 [7]:

$$\tau = \frac{d\sigma}{dT} \nabla_s T \tag{5}$$

where τ is the shear stress, $d\sigma/dT$ is the surface tension gradient with respect to temperature and $\nabla_s T$ is the surface gradient. Shear stress given by Eq 8 is then applied to momentum equations.

The governing equations are finally solved simultaneously using a segregated solver and employing SIMPLE algorithm to obtain temperature and velocity fields. Because of the complexity of the model simplifying assumptions have been made which are:

1. Liquid metal flow within the melt pool is Newtonian, incompressible and laminar.
2. As the power density is low and there is no external gas pressure, the free surface is assumed planar. Similar laser processes involving melting, have been modeled in the past with the assumption of planar free surface with reasonable accuracy.
3. Radiation effects are neglected

A circular laser beam with a diameter of 3.34 mm was used for the model. The power distribution of fiber optic coupled high power diode laser was close to top-hat; therefore a uniform power distribution was used in the model. The power density was set at *5.5 kW/cm*^{2} and the processing of a single surface track length of *20 mm* was modelled. The scanning speed of laser was set at *5 mm/s.*

The material used was bright drawn mild steel EN-43A (0.45-0.55 wt % C, 0.05-0.5 wt % Si, 0.70-1.00 wt% Mn, 0.060 wt %S, 0.060 wt % P), as supplied (average hardness of 220 HV). The size of the work piece was *50x50x10 (mm*^{3}*)*. The ambient temperature was set at *20℃*. Temperature dependent material properties were used for the model [8, 9].

3. Experimental Set-up

The laser source used was a *1.5 kW* fiber coupled high power diode laser (LDL 160-1500) operating in the range of *800 - 980 nm,* with a top hat (approximately uniform) power distribution and a circular laser spot.

The sample was sand blasted to improve the absorptivity of the laser beam. Ocean Optics SD 2000 fibre optic spectrometer was used to measure the reflectivity of the investigated material (EN-43A) at the abovementioned laser wavelengths. It was found to be between *41-43 %* at room temperature.

A CNC table was used to move the material under the laser beam, and the table velocity was set at *5 mm/sec.* The input power was adjusted according to the material absorptivity and fiber coupling in order to maintain the power density at 5.5 kW/cm^2. A single colour Impac Pyrometer, requiring input of material emissivity, was used for temperature measurement in this study. The upper limit of pyrometer was *2000 °C* (2273 *°K)* while the lower limit was *250 °C (523 °K)*. The temperature sensitivity at the lower limit was poor. Emissivity of material increases significantly close to melting temperature due to changing surface conditions and oxidation. Emissivity of molten mild steel of 0.50 (taken from literature [10]) was therefore supplied as input to the pyrometer for accurate measurement of surface temperature. The temperatures recorded by the pyrometer below the melting temperature were therefore inaccurate and were not considered; only maximum surface temperature was recorded. Fig 1.1 shows the schematic arrangement of experimental set-up

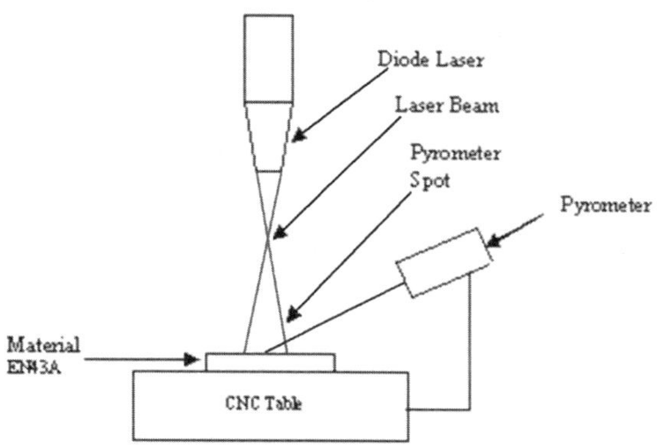

Figure 1.1. Schematic representation of the experimental set-up

4. Results and Discussion

To analyse the effects of the enhanced thermal conductivity approach for predicting the melt pool geometry and the temperature distribution, a comprehensive post-processing analysis was carried out on all the four models (i.e. CFD, heat transfer and two enhanced thermal conductivity models). The global co-ordinate system for the model was in the middle and on the top of the work-piece. Three orthogonal paths were created namely 'Along' (path along the scanning direction), 'Cross-sec' and 'Depth' for the purpose of analysis. A liquidus line has been superimposed on all the graphs to quickly see the temeprature distribution above the melting temeprature. Some of the results have also been compared with the experimental results.

Fig. 1.2 shows the temeprature distribution on the 'Along' path. Melt pool length can be estimated by looking at the temperature distribution above the liquidus line. It can be seen that the melt pool length predicted by all the four models is approximately similar. The maximum temperature predicted by the normal heat transfer model *(Circle HTx1)* is *2615 K,* whereas the maximum temperature predicted by the CFD model *(Circle conv)* is *1946 K*, which is much closer to the experimentally measured temperature of *1992 K*. The enhanced thermal conductivity models predict a temperature in between the CFD and the heat transfer model. The maximum temperature predicted by the four times model is *2069 K*, which is less than the maximum temperature predicted by the three times model (*i.e. 2127 K*), mainly due to the increased heat diffusion above the liquidus temperature. Though the maximum temeprature predicted by the enhanced thermal conductivity models is less than the normal heat transfer model, but the temeparature profile remains similar to the normal heat transfer model.

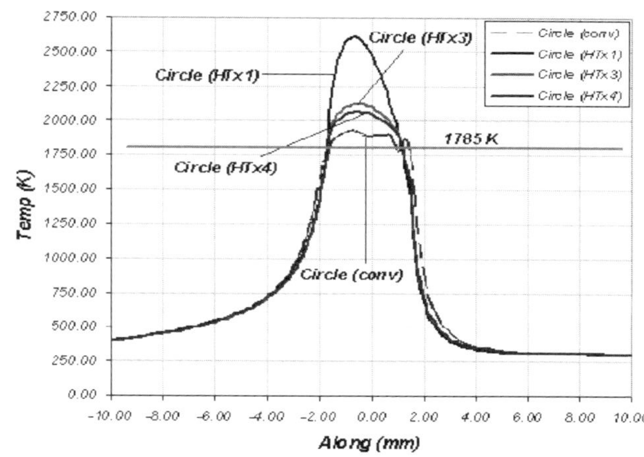

Figure 1.2. Temp vs distance on the 'Along' path

Fig. 1.3 shows the temperature distribution on the 'Cross-sec' path. Melt pool width can be estimated by looking at the temperature distribution above the liquidus line. It can be seen that the melt pool width predicted by all the models is approximately similar.

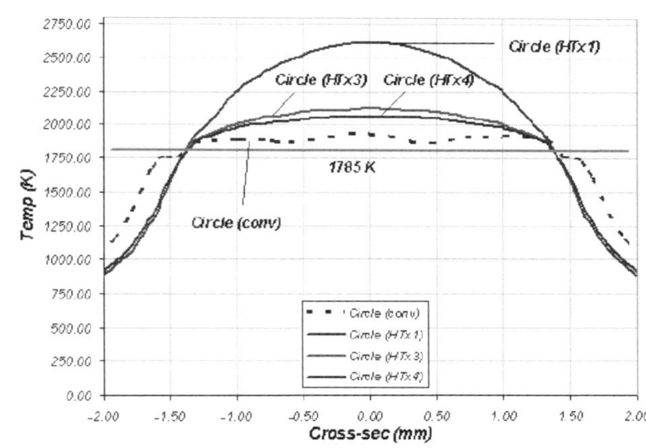

Figure 1.3. Temp vs distance on the 'Cross-sec' path

By looking at figures 1.2 and 1.3, it can be suggested that if the enhanced thermal conductivity value is further increased, the maximum surface temperature will approach that predicted by the experiments and the CFD model.

Fig. 1.4 shows the temperature distribution on the 'Depth' path. Melt pool depth can be estimated by looking at the temperature distribution above the liquidus line. Unlike the 'Along' and the 'Cross-sec' paths, the melt pool depth predicted by the CFD model and the heat transfer models are significantly different. The melt pool depth predicted by the heat transfer and the enhanced thermal conductivity models is approximately five times higher as predicted by the fluid flow model (Table 1.1). This shows that increasing the thermal conductivity does not change the melt pool depth. This is mainly due to the energy balance. The melt pool volume predicted by the heat transfer models (i.e. the normal heat transfer model and the enhanced thermal conductivity models) remains the same. The only effect of increased thermal conductivity is that it clips the maximum surface temperature due to the increased heat diffusion. Thus, it can be said that the enhanced thermal conductivity model is not able to accurately predict the melt pool depth.

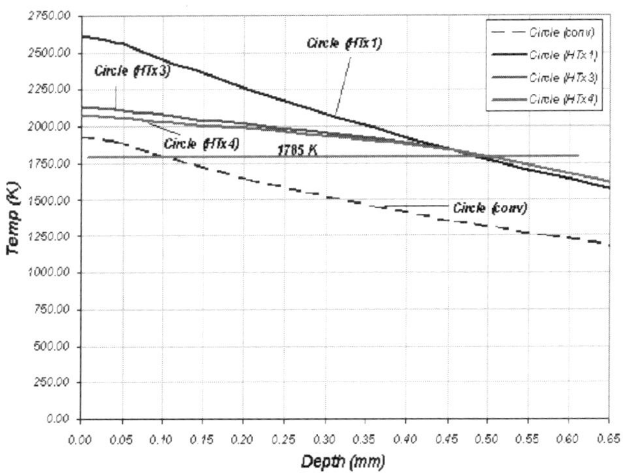

Figure 1.4. Temp vs distance on the 'Depth' path

Fig. 1.5 shows the comparison between the experimental melt pool profile and the cross-sectional melt pool profiles predicted by the CFD, heat transfer, and enhanced thermal conductivity models. It can be seen that the experimental pool profile is flatter and is in better agreement with the CFD model. The melt pool profile predicted by the normal heat transfer model and the enhanced thermal conductivity models is having a meniscus profile. Accurate prediction of melt pool profile is very important as the solidification microstructure would exist within the melt pool. Moreover, the orientation of the solidification microstructure is dictated by the melt pool profile. Thus if the predicted melt pool profile is not accurate the predicted orientation of the solidification microstructure will also be inaccurate.

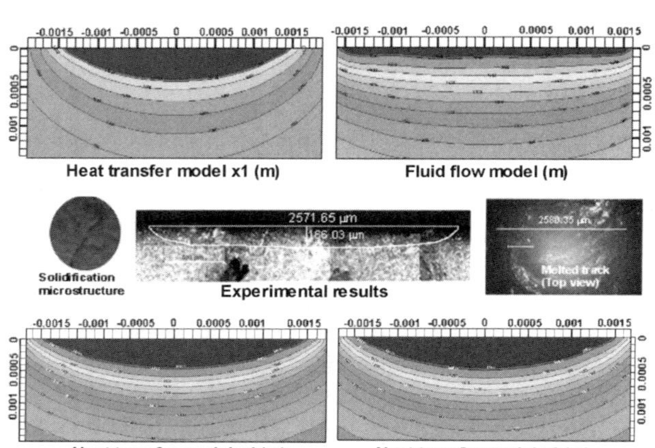

Figure 1.5. Schematic representation of the experimental setup

Table 1.1. Comparison of experimental and modelling results

Model	Max Temp (K)	Width (mm)	Depth (mm)
Experiment Results	*1992*	*2.57*	*0.166*
CFD Model	*1946*	*3.08*	*0.132*
Heat Transfer Model	*2615*	*2.84*	*0.48*
Enhanced thermal Conductivity model × 3	*2127*	*2.84*	*0.47*
Enhanced thermal Conductivity model × 4	*2069*	*2.84*	*0.47*

5. Conclusions

This paper investigated the enhanced thermal conductivity approach to artificially simulate the fluid flow within the melt pool. A CFD model, normal heat transfer model and two enhanced thermal conductivity models were developed using the finite volume method.

It was found that the enhanced thermal conductivity approach gives reasonable results on the work-piece surface. However, it is not able to accurately predict the melt pool depth and melt pool profile, which are of significant importance for any process involving laser melting.

References

[1] Mills, K. C., Keene, B. J., Brooks, R. F., and Shirali, A., 1998, "Marangoni Effects in Welding," in *The Royal Society*, pp. 911-925.

[2] Eustathopoulos, N., Dervet, B., and Ricci, E., 1998, "Temperature Coefficient of Surface Tension for Pure Liquid Metals," *Journal of Crystal Growth*, vol. 191, pp. 268-274.

[3] Toyserkani, E., Khajepour, A., and Corbin, S., 2004, "3-D Finite Element Modeling of Laser Cladding by Powder Injection: Effects of Pulse Shaping on the Process," *Optics and Laser in Engineering*, vol. 41, pp. 849-867.

[4] Zhang, W., Kim, C.-H., and DebRoy, T., 2004, "Heat and Fluid Flow in Complex Joints During Gas Metal Arc Welding-Part Ii: Application to Fillet Welding of Mild Steel," *Journal of Applied Physics*, vol. 95, pp. 5220-5229.

[5] Kim, C.-H., Zhang, W., and DebRoy, T., 2003, "Modeling of Temperature and Solidified Surface Profile During Gas-Metal Arc Fillet Welding," *Journal of Applied Physics*, vol. 94, pp. 2667-2679.

[6] Safdar, S., Li, L., Sheikh, M. A., and Schmidt, M. J., 2006, "A Thermal History Analysis of Surface Heating of Mild Steel with Different Laser Beam Geometries," *Journal of Mechanical Engineering Science, IMechE Part C*, vol. (in press).

[7] Fluent, "Users Guide," vol. Vol 1,2 &3, pp. 2005.

[8] Beranger, G., Henry, G., and Sanz, G., *The Book of Steel*, Intercept Limited, 1996.

[9] Kumar, A., Zhang, W., and DebRoy, T., 2005, "Improving Reliability of Modelling Heat and Fluid Flow in Complex Gas Metal Arc Fillet Welds-Part I: An Engineering Physics Model," *Journal of Physics D:Applied Physics*, vol. 38, pp. 119-126.

[10] "Emissivity of Common Materials," 2005,http://www.omega.com/literature/transactions/volume1/emissivitya.html

9

Manufacturing

Towards Collaborative Manufacturing in Small and Medium Manufacturing Enterprises

H.W. Lin[1], S. Nagalingam[2] and G. Lin[3]

[1]CAMR, University of South Australia, Mawson Lakes, South Australia, 5068, Australia, hao.lin@postgrads.unisa.edu.au
[2]CAMR, University of South Australia, Mawson Lakes, South Australia, 5068, Australia, sev.nagalingam@unisa.edu.au
[3]CAMR, University of South Australia, Mawson Lakes, South Australia, 5068, Australia, grier.lin@unisa.edu.au

Abstract. This paper aims to provide information, knowledge, and a new decision-making approach, which will adequately prepare Small and Medium Manufacturing Enterprises (SMMEs) toward participation in Collaborative Manufacturing (CM). Through CM, SMMEs can leverage competitiveness by establishing a unique niche in the process of delivering a total manufacturing solution to the global market. Hence, this paper provides a realistic and practical approch to managers of SMMEs, who are venturing into CM.

Keywords: SMMEs, Collaborative Manufacturing, total manufacturing solution

1. Introduction

Collaborative Manufacturing (CM) is a new business paradigm that requires manufacturing enterprises to establish close relationships and exploit each other's core competencies in order to improve the overall business performances [1]. This paradigm is especially suitable for Small and Medium Manufacturing Enterprises (SMMEs), as it is a promising gateway for them to compete against large enterprises. With scarce resources focused on maintaining their core competencies, SMMEs lack the ability to deliver total solutions, either a product or service, in an optimal manner [2]. Consequently, they have a desire to strategically collaborate with other business and manufacturing partners who are strategic to maximise performance over their entire product life cycles [3]. CM is however complex. The management of business inter-relationships, allocation of resources, and the coordination of manufacturing activities throughout networks of manufacturers are some of the reoccurring problems that could overwhelm decision-makers of SMMEs [2]. Furthermore, it has been identified that next generation of interoperable systems are required to improve the effectiveness of data, information, and knowledge management [4]. These systems must also be featured with function-rich tools to accurately support collaborative decision-making activities [5]. These issues hinder many SMMEs in using the CM approach. The purpose of this paper is to serve as a guide for SMME managers in preparing their enterprises for the participation of CM. First, the paper identifies the benefits of CM and determines what positive strategic outcomes can be achieved from this new manufacturing paradigm. Second, the critical issues that SMMEs may encounter while participating in CM are summarised. These issues include the working culture of the enterprise (lack of openness), sharing of sensitive information, resource re-allocation, necessity for specific computer network architecture, necessity for specific software tools, necessity for competent information technology experts, and others. Based on the authors' previous work and current research, suitable solutions are suggested. As part of a solution, the authors propose a collaborative decision analysis framework to support the collaborative decision-making activities.

2. Collaborative manufacturing (Benefits and strategic outcomes)

In CM, total manufacturing solutions are achieved through collaboration of multiple enterprises. Each enterprise provides a specific manufacturing phase within an entire value chain, and every enterprise actively promotes its specialities for the optimal accomplishment. The outmost important strategic objective for a CM participant is to continuously refine its niche in the global manufacturing value chain and consequently gain access to as many CM Networks (CMN) as possible. Since business partners are dedicated to addressing some selective supportive processes, SMMEs can better focus their resources on improving the efficiency and effectiveness of their core competencies in order to establish global competitiveness. Furthermore, the ability to deliver customer demands is no longer restricted by the internal capacity and capability of a single SMME, but rather an exhaustive combination of all core competencies contributed by every participant of the collaborative network. The CM effort has the flexibility to orchestrate unique processes to address specific manufacturing requirements in a more responsive manner. Overall, a participant of CM can expect to achieve the following positive outcomes:

- Increase global market share: Collaboration with regional manufacturers, sales agents, and marketing experts enable SMMEs to readily gain access to unfamiliar markets and extends their network to attract more customers.
- Improve customer service: CM processes have the flexibility to be customised to deliver results that best fulfil customers' demand. For example, a collaborative production process can be re-routed

to decrease the overall delivery cost or to increase product quality. Furthermore, regional participants can assist in understanding corresponding markets, in order to tailor unique products and services.

- Lower production cost: As selective supporting processes are outsourced to more competent collaborative partners, the costs of maintaining those processes are eliminated and the overall production cost is reduced. Furthermore, CM allows SMMEs gain access to the benefits of cheap labours that are readily available in some parts of the world.
- Improve product differentiation: Product distinction can be developed as participants of the collaborative design process promote their core competencies toward the final design of the product. Research and development lead-time is reduced as participants' contributions quickly aggregate a vast knowledge base to ensure compatibility between multi-disciplinary requirements, and that overall product specifications are fulfilled.
- Decrease manufacturing lead-time: The extended capacity that result from CM ensures manufacturing lead-time is reduced.
- Increase manufacturing throughput: CM limits SMMEs to repetitively and efficiently perform their core competencies, which consequently increases the overall throughput.

3. Issues with collaborative manufacturing

The key issue with CM is to determine the best approach for orchestrating the complex inter-relationship between participating SMMEs, toward the achievement of collaborative objectives. The orchestration process must establish relationships between a overwhelming number of manufacturing variables, to explore the possible alternatives, and to settle on the alternative that offer most desirable returns to all participants. The realisation of this process posses many challenges, which are discussed in the following sub-sections.

3.1 Organisational culture

There are critical issues in facilitating CM in SMMEs that value the conventional organisational cultures. Conventionally, SMMEs are self-oriented and prefer to work in a 'black box'. Most of their business activities with associates can be categorised as simple sell-buy relationship. There is no synergy analysis between SMMEs and their business partners to exploit potentials in case of collaboration. Thus, manufacturing enterprises aim to invest in all necessarily business processes, which enable the delivery of total solutions to their customers in the most profitable manner possible. SMMEs view these know-how's as valuable intellectual properties and assets that generate revenue for their business, and they are reluctant to share these know-how's with business partners in order to protect their privilege in the associated manufacturing processes. Abandoning the uncompetitive processes of some SMMEs maybe viewed a loss of security and flexibility, as the processes cannot be orchestrated as freely and dynamically as if they were all under the control of one enterprise to address customer demands. Furthermore, as fulfilment of collaborative objectives take priority over local objectives at enterprise level, it is difficult to justify the appropriateness in profit and risk sharing. Finally, as employees have developed habits in their pattern of work, it is difficult to motivate them to make changes in their roles in preparation for collaborative manufacturing.

3.2 Data, information, and knowledge sharing

The appropriate sharing of data, information, and knowledge is critical supporting the operation of CM, however the process has proven a sensitive and challenging task. Before deciding what information is to be shared, the source SMME must consider what to share and to what extent, what are the chances the information and knoweldge will be mis-used at the receiving end, and how the enterprise can recover from subsequent miss-use. After deciding what information to share, the source SMME must select presentation formats and include cognitive interpretation on the data to ensure the transferring process is done as coherently and accurately as possible. After this is done to support decision-making and other management processes, the quality of associated data, information, and knowledge must be evaluated to maintain their values for future purpose. Finally, SMME must include a plan to professionally manage and archive the large volume of data, information and knowledge required to support CM.

3.3 Legacy systems

The use of legacy systems is extensive throughout SMMEs, as software tools are either purchased or developed in-house to address specific needs by functional units. SMMEs lack the systematic approach to ensure that all their systems are interoperable, and that work processes can flow interchangeably through different systems. The custom software tools developed in-house are often not fully tested for 'bugs'. This makes them unreliable and could potentially cripple business processes. Furthermore, the authors have observed during some on-site visits, that SMME employees lack the qualification to use software packages supplied by leading vendors, such as database systems, finance systems, and project management systems. Missuse of these systems could lead to archiving of 'junk' data, that cannot be used to properly support management activities. Finally a key challenge lies in the integration of the legacy systems to closely support the CM processes.

4. Necessities for collaborative manufacturing

After analysing the potential issues present in CM for SMMEs in Section 3 of this chapter, this section identifies and lists the necessities for SMMEs in preparation for CM.

4.1 Organisational culture

The following organisational-culture issues must be taken into consideration to support the participation in CM:

- Identify the true core competencies of the enterprise, and outsource supporting processes to more competent business partners whenever possible.
- Acquire multi business partners for every outsourced process to ensure reliability, flexibility, and create space for competition.
- Establish agreement with business partners on the terms for collaboration, which include the sharing of risks, commitments to deliver, and compensate when failed commitments. Legal bonding in terms of contract is strongly advisable at first, but the relationship must shift to trust as quickly as possible.
- Fulfil collaborative objectives before local objectives at functional unit or enterprise levels.
- Share data, information, and knowledge just enough to support collaborative processes.
- Respect the terms given by source enterprises on the usage of data, information, and knowledge.
- Restructure resources to optimise collaborative processes.
- Dedicate personnel to take on the role of CM management, and information technology support to maintain the complex interoperable system.
- Perform group management and decision making processes virtually on web-based applications.
- Use on-line applications to analyse and make decisions at real-time.

4.2 Interoperable framework

An interoperable framework is required to integrate the legacy systems existed in all the participants of the CM network. To be qualified to enter the interoperable system arena, legacy system must be free of 'bugs' that would affect the normal function of the collaborative process. Furthermore, the corresponding users of legacy systems must have adequate skills to confidently operate the systems to support the operation of their manufacturing activities. In our previous study [4], several interoperable frameworks were analysed to determine their appropriateness toward supporting CM. It has been identified that Web Services architecture is the most suitable option. This conclusion is supported by Shen and et. al.'s work [6]. For the selection of interoperable framework, it is recommended that the following points are condisered:

- Business connectivity: To ensure affordable connections for all CM participants, and enable rapid implementation and ease-of-use capabilities.
- Business benefits: To facilitate coherence of knowledge management across all CM participants, and seek to achieve convergence through voluntary compliance in order to minimise the impact on the participants while maximising the benefits.
- Financial considerations: To provide a technology roadmap for choosing and implementing reliable, scaleable and secure systems and ensuring that the information technology investment justifies the rate of technological changes and business requirement.
- E-Government standards: To comply with existing government standards wherever they are available and appropriate.
- Forward outlook: To grow in parallel with the technological developments and establish a process to grow the framework.
- Usage limitations: To provide a high level or minimum basis for interoperability within the CM participants and should not impose standards, but rather collects and reflects standards in use.
- Product support: To provide flexibility in the selection of software solution vendors, preferably being vendor and product neutral.
- Technical support or staff experience: To ensure minimum staff training is required to support and operate the system, and that professional support is readily available.

4.3 Collaborative decision analysis framework

In order to support the decision-making processes in CM, our previous work [7] has proposed a collaborative decision analysis framework (depicted in Figure 1.1). The steps shown in this figure is explained in detail as follows:

- Step 1: Decision-makers retrieve important operational and performance information, identify environmental and manufacturing parameters, and assign a desired target for each parameter. A rule is then imposed on each parameter where a certain degree of deviation from its target would trigger a decision-making process. It is important that the entire set of parameters monitored could aggregate to convey the manufacturing environment of the CMN, and that any potential problem and or opportunity could be detected from the observation of these parameters.
- Step 2: This step generates a formal statement for the decision-problem to be solved. The definition would include gathering all associated data and information required for analysing the decision, the objectives of the CMN, the uniqueness of the decision, the knowledge domains the decision belongs to, and who is responsible for executing this decision.
- Step 3: Based on the decision-problem statement, the most expected scenario is defined, and a set of decision alternatives is selected for consideration. This information is then forwarded to all the entities that would be affected by this decision.

- Step 4: The Internatove Meta Goal Progamming (IMGP)-based decision analysis framework assumes that every entity has a set of analysis tools and/or models to analyse its local decision variables and parameters using the information received from Step 3, and then propose a set of local objectives.
- Steps 5 to 9: These steps constitute an interactive process and a mathematical algorithm involved in the setup an overall decision-problem model, analysis of the model, and the evaluation of such model. In our previous work, a meta-goal programming solver tool was developed on the Web Services plateform to solve decision models. This tool is driven by the LindoAPI [8] optimisation engine. Detailed description of this process, and the result of a hypothetical analysis can be referenced from our previous work [7].
- Step 10: The model solution is verified with the real problem and/or opportunity, and if found successful, the decision is confirmed and forward to all entities to allow the execution of the evaluated decision.
- Step 11: As the decision is being implemented, its performance is constantly monitored, so that any negative deviation from the desired target can be detected and addressed at the earliest possible time.

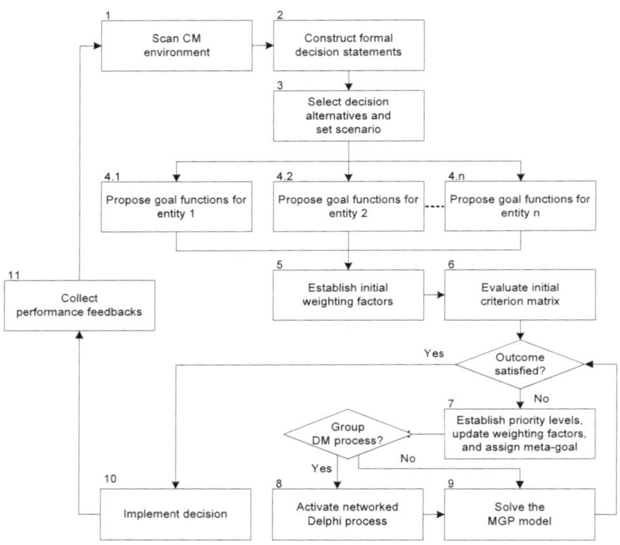

Fig 1.1. Collaborative decision analysis framework

5. Conclusion

This paper sumarises the benefits, issues, and neccessities of participating CM for SMMEs. Furthermore, a framework is proposed to address the problem of recursive and complex collaborative decision-making processes. For SMMEs that are planning to participate in CM, this paper can be used as a brief guide on what can be expected, and what should be done in order to particiapte in a CMN. For SMMEs that have already participated in a CMN, this paper can be viewed as a benchmark to identify what else can be done to further exploit the benefits of being in CMN. In future work, the decision analysis framework will be used to model and solve other types of decision problems commonly found in CM practices. It is expected that subsequent results will further verify that this proposed framework is suitable for supporting a wide spectrum of decision problems, and thus is a valuable tool for SMMEs, especially toward the participation in a CMN. Ultimately, our study will lead to a complete suite of decision support tools being developed to support CM and virtual manufacturing.

Acknowledgement

Authors acknowledge the funding provided by Australian Research council for this linkage research project (project number LP0347965)., and all the supports provided by industry partner, Dynek Pty Ltd, South Australia, Australia.

References

[1] Danilovic, M., Winroth, M., (2005). A tentative framework for analyzing integration in collaborative manufacturing network settings: a case study. Journal of Engineering and Technology Management, 22 (1-2), 141-158.

[2] Bilbao, N., del Pozo, D., Lopez, D.J.M., Etxaniz, I. (2004) of Conference. The collaborative Manfuacturing Approach. in IEEE International Conference on Industrial Informatics. IEEE.

[3] Johansen, K., Comstock, M., Winroth, M., (2005). Coordination in collaborative manufacturing mega-networks: A case study. Journal of Engineering and Technology Management, 22 (3), 226-244.

[4] Chiu, M., Lin, H.W., Nagalingam, S.V., Lin, G.C.I., (2006). Inter-operability framework towards virtual integration of SMEs in the manufacturing industry. International Journal of Manufacturing Technology and Management, 9 (3/4), 328-349.

[5] Lin, H.W., Nagalingam, S., Chiu, M., Lin, G. (2004) of Conference. Development of a Web-based Three-tier Decision Support System for Virtual Enterprises. in The 14th International Conference on Flexible Automation and Intelligent Manufacturing. Toronto, Canada: National Research Council of Canada.

[6] Shen, W., Li, Y., Hao, Q., Wang, S., Ghenniwa, H. (2005) of Conference. Implementing Collaborative Manufacturing with Intelligent Web Services. in The Fifth International Conference on Computer and Infomration Technology. IEEE.

[7] Lin, H.W., Nagalingam, S.V., Lin, G. (2007) of Conference. Manufacturing Decision-support using Interactive Meta-Goal Programming. in International MultiConference of Engineers and Computer Scientists. Hong Kong: International Association of Engineers.

[8] Lindo Systems Inc. Website, 2006. LindoAPI http://www.lindo.com/, assessed 28th April 2006

Application of Virtual CIM in Small and Medium Manufacturing Enterprises

N. Zhou[1], S. Nagaligam[2], G. Lin[3]

[1]University of South Australia, Australia, 5095, Ning.Zhou@postgrads.unisa.edu.au
[2]University of South Australia, Australia, 5095, sev.nagalingamisa.edu.au
[3]University of South Australia, Australia, 5095, Grier.Lin@unisa.edu.au

Abstract. Worldwide cooperation among manufacturing companies is increasingly gaining importance to face emerging challenges in manufacturing. The traditional Computer Integrated Manufacturing (CIM) systems cannot satisfy the needs of global market as they are deployed only within an enterprise. Therefore, a more flexible and comprehensive integrating methodology is required to overcome distance barriers, facility sharing problems and communication obstacles. These issues lead to the concept of Virtual CIM (VCIM).Applying VCIM concepts to today's enterprises means establishing real-time connections in an effective manner among components of VCIM systems, many of which can have inconsistent data formats between different enterprises,. With the development in Information and Communication Technology (ICT) tools, distributed departments of an organisation or even different companies can be brought together to cooperate with each other. In this paper, how VCIM technologies can help collaborating manufacturing companies, especially the Small and Medium Manufacturing Enterprises (SMMEs) is analysed. This research shows that VCIM can be implemented as a distributed object system over pervasive internet without needing huge investment in ICT infrastructure. It also demonstrates that the WWW can be used to share various data and applications to control facilities in VCIM systems.

Keywords: Virtual CIM, VCIM, Small and Medium Manufacturing Enterprises, Web services, Agent based systems

1. Introduction

Manufacturing enterprises face challenges from worldwide competitors and struggle to extend their business worldwide in today's economic climate. Therefore, these enterprises tend to use fully integrated manufacturing systems so that they can have the capabilities to rapidly respond to constantly changing customer requirements and produce high quality products in shortest possible time with lowest possible cost. However, Small and Medium Manufacturing Enterprises (SMMEs) are having difficulties in achieving this flexibility and competing with large companies, since SMMEs often do not have enough resources. By integrating manufacturing resources of many partner enterprises (which may be located at different regions), these enterprises can form a globally integrated SMME network and achieve a competitive edge [1].

Virtual Computer Integrated Manufacturing (VCIM), which is a network of interconnected global Computer Integrated Manufacturing (CIM) systems, extends the traditional concept of CIM from a local and centralised system to worldwide co-operation [2]. Today, companies are expanding their business boundaries locally and internationally, merging or co-operating with others across geographical regions. As a result, departments of a company or several companies that are involved in a same project may be located far away from each other or even on different continents. However, it is almost impossible to deploy a proper and dedicated Information and Communcaition Technology (ICT) infrastructure to support such a co-operative environment. It is very difficult to efficiently utilise all available facilities and share common data in a real-time environment. One feasible solution to the distributed and cooperative situation is to set up a common database to manage availability of facilities and data that can be used in a common project. Users can obtain access by querying the database with their requirements. However, two drawbacks exist in this approach. Firstly, if the required ICT infrastructure (compatible systems with a network connection) is not available between the corresponding departments it may be not be possible to deploy a centrally located database and access it through the network. Secondly, even if the central database can be developed and applied, the management and manipulation of the shared facilities may be neither dynamic nor automatic, which means the database cannot dynamically report whether a specific facility is working, inactive or waiting for requests nor automatically start the facility, and the system cannot be used in a real time environment. In these types of systems, users need first to find out whether a facility can complete a certain task. Then, they need to apply other mechanisms to determine the availability and initiate operations remotely or on the local system. These drawbacks will impact the efficiency and effectiveness of using expensive ICT tools and negatively impact on co-operation.

2. SMME issues

SMMEs play a significant role in pioneering new technologies, markets, and creation of knowledge based industries, all of which are important for future growth and jobs of many countries. In addition, they are often characterised by niche specialist markets in which they have expertise. They succeed by providing high levels of responsiveness and personalised service. In many instances, they can offer lower priced products compared to other large enterprise, because SMMEs have less overheads and their labour and management force are the same. However, when SMMEs grow and develop a typical functional structure, they need to develop efficiencies within their total processes to remain competitive. Many growing SMMEs see their profit margin drop as these inefficiencies mount.

Global competition and today's open markets are driving the enterprises to introduce high quality products and services economically and efficiently. A VCIM system is a strategic move that requires manufacturing enterprises to establish close relationships in order to exploit each others' core competencies for the betterment of the SMME network.

Participation in a VCIM is especially challenging for SMMEs. Since the VCIM activities involve complex operations and these participants of a VCIM system are distributed worldwide, the SMMEs must overcome global boundaries in terms of distances, time, regulatory constrains, as well as cultural and political differences, for reaching mutually-beneficial agreements on how to optimise the customer order fulfilment process. In addition, some of these partnerships are dynamic and becoming virtual representing the transient status of the collaboration [3]. The issues that affect SMMEs are:

1) How can a customer agent find and select a suitable facilitator agent?
2) How to interlink local resource planning issues with VCIM resource planning issues?
3) How to automatically decompose each new order request?
4) How to select suitable resources with a multiple criteria approach to include cost, time, quality, friendship, credit, and delivery reliability?
5) How to allow multiple orders simultaneously?
6) How to allow customers to modify a product schedule with their changing demands?

3. Current developments

Intelligent agents have been developed by many researchers as an important framework for interconnecting distributed manufacturing systems. However, earlier approachs are neither reliable nor efficient to address the SMMEs' issues. To optimise resource sharing and to provide a dynamic integration, an agent based VCIM architecture has been developed.

3.1 Agent based system, Agent based architecture

To implement an agent based VCIM system, three categories of agents have been identified [4]. These agents are Facilitator agents, Customer agents, and Resource agents. Facilitator agents are designed to act as coordinators to route the information flow across the VCIM agent community. Customer agents are designed to provide interfaces for customers to participate in the VCIM system. Finally, Resource agents are designed as an agent interfaces to encapsulate distributed manufacturing functional entities and connect them with the agent community. The agent functionalities and responsibilities are described in our earlier research outputs [1, 3-5]. The summarised activities of these agents are as follows:

Facilitator agent: Users send request messages to this agent to participate in VCIM and get a reply. This agent decomposes the request message from users as a set of subtasks and sends subtasks requests to related Resource agents. Then, it receives proposal messages for subtasks from Resource agents, combines the proposals as production schedules, and seeks an optimised schedule. Resource agents register their information such as functionalities, name, and address in a database of this agent. To have these capabilities, knowledge on agent communication and strategies for manufacturing resource planning and scheduling are stored in this agent.

Customer agent: Users input request messages consisting of information such as product name, required due date/time, delivery place and others by using the agent's Graphical User Interface (GUI), get replies for the Facilitator agent and displays the results to users.

Resource agent: This agent is responsible for further decomposing tasks and for local resource planning and optimisation. To have these capabilities, knowledge for agent communication and functional records such as the time and cost for this agent to complete a subtask are identified and stored in the database of this agent. Each Resource agent maintains an independent working schedule and renews this schedule when a new task is confirmed.

By connecting a Facilitator agent, a Customer agent and many Resource agents through the Internet, these agents can form a basic multi-agent VCIM system. Nevertheless, a real VCIM system includes many Facilitator agents, Customer agents and Resource agents while the functionalities of Resource agents may include design, manufacture, delivery, material supply and others.

In our VCIM concept, Resource agents can include a manufacturing organisation or a unit that has an authority to plan internal resources to meet a customer order. With this definition, a Resource agent can be any of the following or a combination of: an assembly workshop, a manufacturing system, a group of delivery facilities, a material provider, a design department, a branch company or a small enterprise. The Resource agent usually has a group of functional entities. In addition, when manufacturing resources at lower levels are encapsulated as sub multi-agent systems, a Resource agent hierarchy for VCIM system can be formed.

Generally, each manufacturing organisation has its own advantages, partner relationships, resource selection criteria, and other elements related to manufacturing planning. Our agent-based VCIM architecture uses these realistic issues related to the advantages of an organisation, as a particular knowledge for partner resource selection, and dictates this knowledge through Facilitator agents. We consider a VCIM system is formed when an organisation has built a Customer agent to provide a user's interface and a Facilitator agent to coordinate the Resource agents shared by multiple Facilitator agents (that may have different knowledge on resource planning depending on the local issues and situation), to invoke VCIM services. By means of these approaches, when many organisations join together as a network of organisations to work together and achieve a competitive edge by building Facilitator agents and Customer agents for a VCIM system and enable participation of their Resource agents, an agent based architecture is formed.

With this proposed architecture, the VCIM system can potentially achieve more reliability, efficiency and needed flexibility. In this architecture, all agents (except sub-agents) are connected through the Internet and new Facilitator agents, Customer agents, and Resource agents are permitted to connect with the existing structure when and where necessary. This feature enables a VCIM system to have the necessary flexibility in extending information processing capability, customers' convenience, and manufacturing capability of the partner organisations. In addition, each Facilitator agent has the capability to communicate and negotiate with all other Customer agents and Resource agents. Therefore, when an error occurs or information is blocked in one Facilitator agent, a Customer agent can select and use other Facilitator agents to invoke the VCIM services. This alternative selection of the Facilitator agent to process a task allows the information flow in a more reliable manner than a centrally controlled system that can have a possible communication bottleneck.

3.2 Web services – a review

Today, Web Services (WS) are receiving much attention in meeting the Integration requirements for the networks. Although, many can confuse a "Web Service" with accessing a Webpage, the W3C consortium [6] clearly defines a WS as: a software system identified by a Uniform Resource Identifier (URI) [7], whose public interfaces and bindings are defined and described using XML that can be accessed over any TCP/IP network using the SOAP standard for integration, the WSDL standard for self-description, and the UDDI standard for registry and discovery within a public or private directory [6]. Some of the key terminologies used in WS are [8]:

- XML: It is a widely recogonised abbrevation for Extensible Markup Language and a text format similar to HTML used to store data. Each XML file can be a database and readable by a human.
- SOAP: It is an abbrevation for Simple Object Access Protocol. SOAP provides a message format for Web servers to transfer XML.
- UDDI: UDDI stands for Universal Description, Discovery and Integration. Primarily it is a directory for locating XML entities and integrating them into the Web Services Interoperability standard to faciliatate WS infrastructure.

In simple terms, a WS can be considered as a universal server/client architecture [9] that allows different systems to communicate with each other by open standard protocols. WS minimize the complexity in software developments associated with server/client applications by eliminating the code dependencies between client and server. WS is not a Web server but a service model. It uses the same HTTP protocol because:

- A text format is human readable and easy to maintain.
- HTTP is friendly to firewalls. For security issues, most of network administrators are afraid to open ports other than an HTTP enabled ones.
- The same Web server can be used to host a WS. The only difference is transferring a SOAP message instead of HTML.
- It is esay to output data. Converting from XML to HTML is simple and allows real time data access from Web page for all people.

Currently, we are in the process of migrating our previously developed agent based VCIM system to WS. The reasons for this approach are :

- WS is a flexible approach to removing the firwall restrictions in communications between VCIM agents residing in distributed locations of partner enterprises.
- The code-independency in WS ensures the interoperability of different SMMEs' manufacturing systems in spite of which software/hardware vendor they choose. Since WS is achieved by open standards of XML and SOAP and can substitute the earlier agent message communication language.
- The low cost apporch of utilising WS would enable SMMEs to participate in a VCIM system without resorting to huge investments in ICT or the need for upgrading their existing systems.
- WS is a server/client model for communication that means the existing agent model will not be affected except for the communication layer among SMMEs. A WS add-on is only needed as there is no need to replace the whole VCIM architecture.

Therefore, WS implementations in VCIM will provide a low cost, reliable Internet communication model for globally distributed SMMEs from different ICT network domains to participate dynamically and collaboratively.

3.3 Interlinking web services and agent based systems

There are two alternative approachs for implementing a WS with an agent based system: 1) utilsing a WS wrapper for current software agents, 2) rewriting the agent activation codes so that agents will have WS ability. However, having a new "Add-on" to integrate into the existing multi-agent VCIM system is our current approach because WS needs a Web server to run, it is a simple migration. WS is not optimised for every agent to carry on a web server. To implement this proposed model (Figure 1.1), the following steps are needed.

1. Develop a proxy agent. The current VCIM model consists of Customer, Facilitator and Resource agents: In order to intergrate these agent communications into WS, we propose to introduce a new proxy agent. This new agent will interpret all agent communication messages, reformat them into XML, and transfer them to the Internet. This agent should have the capability to convert XML into agent messages and transfer it to other agents. The proxy agent will be bi-directional between agent message and SOAP message.
2. Establish a local directory. According to W3C standards, a UDDI directory is recommended to store all the details of each agent. Those agents must register on the UDDI directory when they start.

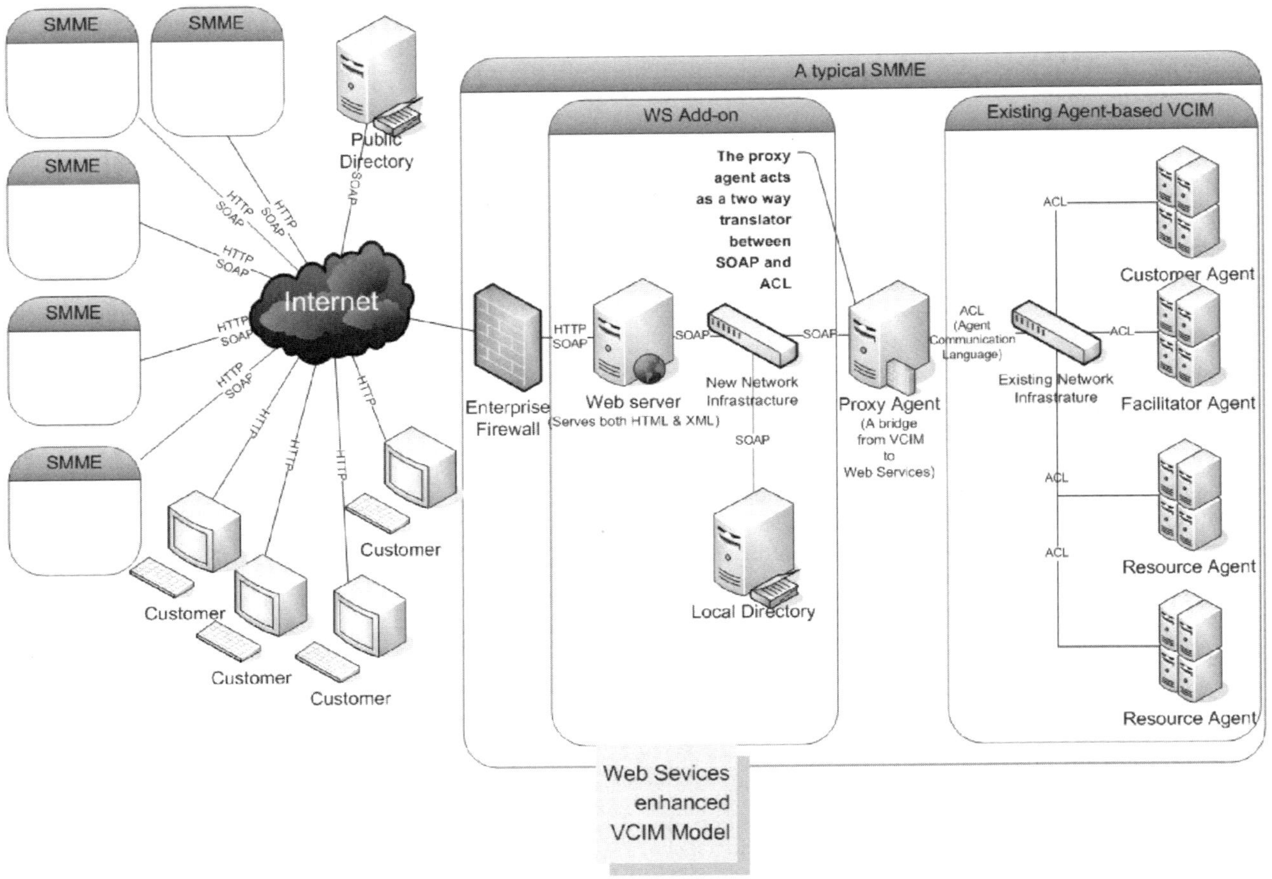

Figure 1.1 Web Sevices enhanced VCIM Model

Because old agents cannot perform SOAP, the proxy agent will aid them to complete the task.

3. Implement a web server. Behind the enterprise firewall, a Web server needs to be established to perform tasks such as facilitating customers order entry, relaying SOAP message from the Proxy agent and enabling agent communication over the Internet.
4. Establish a global directory. For agents to find other agents from different SMMEs, a global UDDI directory is needed to share the information. Each SMME needs to submit a copy of its local directory and later dynamically update the list.

4. Conclusions

VCIM is a way forward to the SMMEs in many countries, where globalisation has impacted dramatically on manufacturing industry. VCIM provides a facility to share resources of partner enterprises that are geographically distributed and provides a competitive edge for SMMEs to be on an equal footing with large organisations that have abundant resources to meet the challenges imposed by the globalisation. This article provides a practical and simple approach for the SMMEs to embrace the VCIM concept and fully implement it or mutual benefit of the network of SMMEs.

References

[1] Wang, D., Nagalingam, S. V., and Lin, G. C. I., Development of an agent-based Virtual CIM architecture for small and medium manufacturers. Robotics and Computer Integrated Manufacturing, 2007. 23(1): p. 1-16.

[2] Nagalingam, S. V. and Lin, G. C. I., Latest developments in CIM. Robotics and Computer-Integrated Manufacturing, 1999. 15(6): p. 423.

[3] Wang, D., Nagalingam, S. V., and Lin, G. C. I., Development of a parallel processing multi-agent architecture for a virtual CIM system. International Journal of Production Research, 2004. 42(17): p. 3765.

[4] Wang, D., Nagalingam, S. V., and Lin, G. C. I. Implementation approaches for a multi-agent Virtual CIM System. in 9th International Conference on Manufacturing Excellence (ICME – 2003). 2003. Melbourne, Australia.

[5] Wang, D., Nagalingam, S. V., and Lin, G. C. I., A Novel Multi-Agent Architecture for Virtual CIM System,. International Journal of Agile Manufacturing System, 2005. 8(8): p. 69-82.

[6] Booth, D., et al. Web Service Architecture. 11 February 2004 [cited 2005 20 February 2007]; Available from: http://www.w3.org/TR/ws-arch/.

[7] Berners-Lee, T. F., R.; Irvine, U.C.; Masinter, L. (1998) RFC 2396 - Uniform Resource Identifiers (URI): Generic Syntax. Network Working Group Volume,

[8] Marwan Sabbouh SJ, D. A., Paul Silvey, Paul Denning, Workshop on Web services, in World Wide Web Consortium. 2001.

[9] Group, A. D., Web Services 2002: Market Milestone Report, in A Delphi Group White Paper. 2002.

Effectiveness Measurement of Facilities Layout

D. Raman, S. V. Nagalingam, B. W. Gurd and G. C. I. Lin

University of South Australia, Australia, 5095, Dhamodharan.Raman@postgrads.unisa.edu.au
University of South Australia, Australia, 5095, Sev.Nagalingam@unisa.edu.au,
University of South Australia, Australia, 5095, Bruce.Gurd@unisa.edu.au
University of South Australia, Australia, 5095, Grier.Lin@unisa.edu.au

Abstract. Measuring effectiveness in an existing facilities layout is a prerequisite in order to initiate any action that will improve layout effectiveness. Available methods on measurement of layout effectiveness help measure it in respect to only material handling cost and select an effective layout from a set of available alternatives. Nevertheless, factors such as empty travel of material handling equipment, layout flexibility and area utilisation contribute significantly towards the layout effectiveness. Thus, it is necessary to have a measurement model to determine the facilities layout's effectiveness by considering all significant factors. Hence, a measurement model considering a set of three layout effectiveness factors namely Facilities Layout Flexibility (FLF), Productive Area Utilisation (PAU) and Closeness Gap (CG) is developed in our research. The proposed model will enable the decision-maker of a manufacturing enterprise to analyse a layout in three different aspects; based on which they can make decision towards productivity improvement. Due to the space limiation, this paper only discusses about the measurement of the CG. The CG is developed in respect to the objective of bringing closer the highly interactive facilities/departments. The CG presented in this paper extends other related works by incorporating numerous aspects of layout that include: empty travel of material handling equipment, information flow, personnel flow and equipment flow.

Keywords: Facility Layout, Effectiveness, Facilities Layout Flexibility, Productive Area Utilisation, Closeness Gap

1. Introduction

Facilities layout is one of the key areas, which significantly contribute towards manufacturing productivity in terms of cost and time, in a manufacturing system. Tompkins [1] stated that 10 - 30% of the material handling cost (MHC) can be reduced by having an effective facilities layout. Benjaafar [2] argued that a layout has a direct impact on the operational performance, as measured by manufacturing leadtime, throughput rate, and work-in-process (WIP). A facilities layout has a life cycle which consists of phases such as design, implementation, growth, maturity and obsolescence [3]. In each of this phase, many decisions have to be taken which include: evaluation and selection of an effective layout, production planning, scheduling and the decisions related to the modification in an existing design due to the changes in the market demand [4-6]. It is estimated that 8% of the gross national product of the United States of America has been spent on new facilities in each year and it excludes the expenses on the modification of existing facilities [1]. To support the above decision making process suitable measurement models are required, since performance measurement provides inputs to a decision making problem.

As a result, we developed a measurement model that extends other related works through incorporating three key performance factors: Facilities Layout Flexibility (FLF) [7], Productive Area Utilisation (PAU) [8] and Closeness Gap (CG) . These factors are developed in view of the objectives of a layout that include [9]: 1. Minimising MHC, 2. Improving flexibility for arrangement and operation, 3. Utilising the available area most effectively, and 4. Minimising overall production time. The CG considers minimising the MHC and the total production time by bringing closer the highly interactive facilities/departments (Objectives 1 and 4). The FLF considers improving flexibility for arrangement and operation (2nd objective). Whereas, improving the effectiveness of the utilised area is considered in developing the PAU (3rd objective).

2. Literature Review

Based on the characteristics of production demand, the facilities layout problem can be categorised into two types: (i) single period layout – when the production demand is almost constant, (ii) flexible/robust layout – when there is a variation in production demand. Most of the available methods help analyse a single period layout problem by considering MHC in an objective function. In this school of thought, Yaman et al. [10] argued that an effective layout is the layout with minimum travelling distance between processes and they developed a heuristically based sorting method with an objective of minimising MHC. Mak et al. [11] developed a method based on genetic algorithms to determine the effective layout by minimising MHC within a manufacturing plant. In a nested facilities layout problem, Mckendall-Jr et al. [12] arranged the irregular shaped departments within a shop floor to minimise the MHC. Abdinnour-Helm and Hadley [13] utilised Tabu search based heuristics for multi-floor layout problem to reduce the inter/intra-floor MHC. On the other hand, Raoot and Rakshit [14], Dweiri [15], Deb and Bhattacharyya [16] considered qualitative and quantitative factors in Fuzzy logic based approaches.

In all these approaches, in search of an effective layout, researchers tried to locate the departments and facilities which have a large amount of interaction as close to each other as possible. By bringing the highly interactive departments and facilities closer, a significant reduction in total movement within shop floor can be achieved. Consequently, the reduced movement minimises corresponding production time and MHC. However, there are three general shortcomings in those methods. The shortcomings are:

1. The MHC does not include the empty travel of MHE, despite the fact that the empty travel has a significant contribution towards minimising the MHC, WIP and lead time,
2. The total flow between the departments as well as the flow between the facilities within a department has not been considered together. However, these two kinds of flows have to be integrated to assess the MHC at an enterprise level, and
3. For a given layout, only an absolute value for MHC can be obtained and this absolute value gives no indication in assessing the possibilities for further improvement in the existing layout.

Therefore, to highlight the importance of placing the highly interactive departments/facilities closer to each other and to address the identified shortcomings, we developed the factor CG.

3. Closeness Gap

The CG measures the extent to which the existing layout varies from best possible theoretical layout in terms of a closeness value. The closeness value between any two facilities is a product of step value and interaction value between the corresponding two facilities. The step value is defined as relative position between any two facilities and it is measured in relation to the minimum number of passes a MHE has to have between the two facilities in the layout through the normal material flow path. Whereas the interaction values are measured in terms of the material flow, information flow, personnel flow and equipment flow. Sum of the closeness values of each pair of facilities and departments in the existing layout is defined as the current closeness value. The current closeness value is then compared with the values of best layout in order to measure the CG. Whereas the best layout is the one which brings closer the highly interactive facilities and department that are available in the existing layout (described in section 1.3.3). In our method, the CG is measured in two levels: (1) intra-functional level which considers the flow among the facilities within a department, for each department at the enterprise and (2) inter-functional level in which the flow among departments in an enterprise is considered (Figure 1). Hence, the measurement of CG at enterprise level (CG_e) should include the intra-functional level and the inter-functional level. The CG_e is formulated as a geometric mean of closeness gap for intra-functional flow $(CG_{Intra})_d$ and inter-functional flow (CG_{Inter}) as given in Equation 1.

$$CG_e = \left[\left(\prod_{d=1}^{D}(CG_{Intra})_d\right) * CG_{Inter}\right]^{1/D+1} \quad (1)$$

3.1 Intra - functional flow

The closeness gap due to intra-function flow in the department 'd' (CGIntra)d measures how extensively the closeness value for the existing layout (CCVd) differs

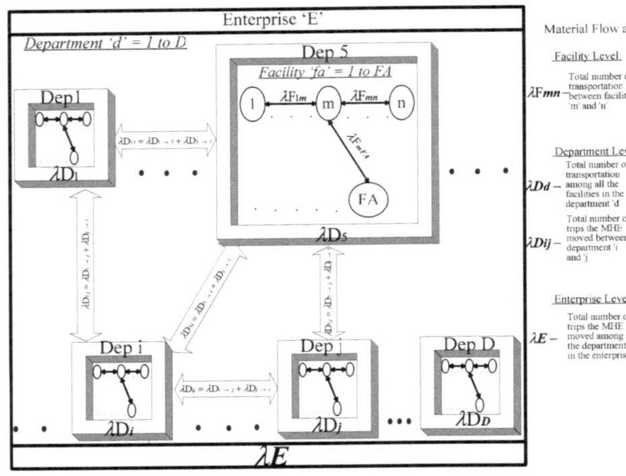

Figure 1.1. Intra and Inter-functional flow in an enterprise

from the closeness value for the best layout (BCV_d) and it is calculated using Equation 2.

$$(CG_{Intra})_d = {(CCV_d - BCV_d)}/{CCV_d} \quad (2)$$

$$CCV_d = \sum_{m=1}^{FA-1}\sum_{n=m+1}^{FA}(SV_{mn}) * (INV_{mn}) \quad (3)$$

With;

FA Total number of facilities in the department under evaluation,
INV_{mn} Interaction value between facilities 'm' and 'n', and
SV_{mn} Step value between facilities 'm' and 'n'.

The interaction value is determined by various factors which include the quantity of material, information, personnel, and equipment flow. However, quantity of material flow is the only factor that could significantly impact on intra-functional flow and all other factors have insignificant contribution towards the interaction value in an intra-functional flow situation. Other researchers [17, 18] have considered annual quantity of material flow while calculating the interaction value. Nevertheless, their approach may not represent an actual value as bigger size materials may need more frequent transportation though the total quantity of material flow is low. In contrast, smaller size materials may need less frequent transportation, despite the total quantity being high. Hence, instead of directly using the quantity of material flow, total number of transportation required between two facilities 'λF_{mn}' should be used to determine the interaction value 'INV_{mn}'.

$$\lambda F_p = \left(V_p / TBSF_p\right) \quad (4)$$

$$\lambda F_{mn} = \sum_{p=1}^{P}(\lambda F_p) * (p_{mn}) \quad (5)$$

$$\lambda D_d = \sum_{m=1}^{FA-1}\sum_{n=m+1}^{FA}\lambda F_{mn} \quad (6)$$

$$INVmn = \left(\lambda F_{mn} / \lambda D_d\right) \quad (7)$$

Where;

$TBSF_p$ - Average transportation batch size of part 'p', and
V_p - Annual production demand for part 'p'.
p_{mn} = 1, if part 'p' flows between facilities 'm' and 'n', = 0, otherwise

3.2 Inter-functional flow

Similar to the intra-functional flow, the CG for inter-functional flow (CG_{Inter}) is calculated using Equation 8.

$$CG_{Inter} = \left(CCV_E - BCV_E\right) \Big/ CCV_E \tag{8}$$

$$CCV_E = \sum_{i=1}^{D-1} \sum_{j=i+1}^{D} (SV_{ij}) * (INV_{ij}) \tag{9}$$

In contrast to intra-functional flow, all the factors that are discussed in Section 1.2.1 have an impact in an inter-functional flow situation. Further, the indirect flow due to empty transportation is considered in our approach, because in an inter-functional situation the MHE may have a significant number of empty travels. Minimisation of such empty travel has been found to be very important in reducing cost, WIP and leadtime [2]. Empty travel can be reduced by bringing closer the departments that are frequently visited by MHE, independent of actual direct material flow between them. Hence, the factors that determine the interaction value for inter-functional flow are: Material handling equipment travel (Loaded travel and Empty travel), Information flow, Personnel flow, and Other equipment flow (other than an MHE). The interaction values due to each of the above factors are calculated individually ($INVEF_{ij}$, $INVIF_{ij}$, $INVME_{ij}$, $INVPF_{ij}$) and then they are integrated in order to get an overall value of INV_{ij} through weighted multi-criteria decision making method as given in Equation 10.

$$INV_{ij} = \left(INVME_{ij} * WME\right) + \left(INVIF_{ij} * WIF\right) + \left(INVPF_{ij} * WPF\right) + \left(INVEF_{ij} * WEF\right) \tag{10}$$

Material handling equipment travel [$INVME_{ij}$]. In our extended method, interaction value due to MHE is calculated by considering both the loaded travel $(INV_{LT})_{ij}$ and the empty travel $(INV_{ET})_{ij}$ as shown in Equation 11.

$$INVME_{ij} = \left[(INV_{LT})_{ij} + (INV_{ET})_{ij}\right] \Big/ \left[\sum_{i=1}^{D-1} \sum_{j=i+1}^{D} (INV_{LT})_{ij} + (INV_{ET})_{ij}\right] \tag{11}$$

The interaction value due to the loaded travel between two departments 'i' and 'j' $(INV_{LT})_{ij}$ can be calculated from total number of trips 'λD_{ij}' the MHE moved between corresponding departments as shown below.

$$\lambda D_{j \to i} = \sum_{P=1}^{N} (\lambda D_P) * (p_{j \to i}) \tag{12}$$

$$\lambda D_{i \to j} = \sum_{P=1}^{N} (\lambda D_P) * (p_{i \to j}) \tag{13}$$

$$\lambda D_{ij} = (\lambda D_{i \to j}) + (\lambda D_{j \to i}) \tag{14}$$

$$\lambda E = \sum_{i=1}^{D-1} \sum_{j=i+1}^{D} \lambda D_{ij} \tag{15}$$

$$(INV_{LT})_{ij} = \left(\lambda D_{ij} \big/ \lambda E\right) \tag{16}$$

With;

λD_p Total number of trips the MHE moved for 'p'
λE Total number of trips the MHE moved among departments
N Total number of parts produced in the enterprise,
$p_{i \to j}$ = 1, if part 'p' flows from the departments 'i' to 'j' = 0, otherwise
$p_{j \to i}$ = 1, if part 'p' flows from departments 'j to 'i' = 0, otherwise

Whereas, the interaction value due to empty travel between departments 'i' and 'j' $(INV_{ET})_{ij}$ is calculated in terms of empty travel between the two departments and the total empty travel in the enterprise.

$$(INV_{ET})_{ij} = \left((EmptyTravel)_{ij} \Big/ \sum_{i=1}^{D-1} \sum_{j=i+1}^{D} (EmptyTravel)_{ij} \right) \tag{17}$$

In a discrete material handling system, materials are transported among different departments using a general purpose MHE such as a forklift. In such material handling system, the MHE travels empty from an available department (department of its previous delivery) to a requesting department (from where it has to pick-up materials for next delivery). Therefore, the empty travel from 'i' to 'j' is determined by two factors: total number of instances the MHE is available at department 'i' (PA_i), and total number of instances the MHE is requested by department 'j' (PR_j). Values for the two factors can be obtained by calculating the total number of trips the MHE traveled from available and requesting departments, respectively. Similarly, the empty travel from department 'j' to 'i' can also be calculated. Addition of empty travels from 'i' to 'j' and 'j' to 'i' gives the total number of empty travel of MHE between the departments 'i' and 'j'. Therefore,

$$(EmptyTravel)_{ij} = (PA_i * PR_j) + (PA_j * PR_i) \tag{18}$$

With;

$$PA_i = \sum_{d=1}^{D} \left(\lambda D_{d \to i} \big/ \lambda E\right) \quad \text{and} \quad PR_j = \sum_{d=1}^{D} \left(\lambda D_{j \to d} \big/ \lambda E\right) \tag{19}$$

$$PA_j = \sum_{d=1}^{D} \left(\lambda D_{d \to j} \big/ \lambda E\right) \quad \text{and} \quad PR_i = \sum_{d=1}^{D} \left(\lambda D_{i \to d} \big/ \lambda E\right) \tag{20}$$

Information, personnel, and equipment flow. The interaction value for information, personnel and other equipment flow can be evaluated using either qualitative method or quantitative method. However, selection of the evaluation method is subject to availability of quantitative data for these three factors. If the data for the quantity of information (QIF_{ij}), personnel (QPF_{ij}) and equipment flow (QEF_{ij}) are available then their corresponding interaction values can be estimated as given below.

$$INVIF_{ij} = QIF_{ij} \Big/ \sum_{i=1}^{D-1} \sum_{j=i+1}^{D} QIF_{ij} \tag{21}$$

$$INVPF_{ij} = QPF_{ij} \Big/ \sum_{i=1}^{D-1} \sum_{j=i+1}^{D} QPF_{ij} \tag{22}$$

$$\text{INVEF}_{ij} = QPF_{ij} \Big/ \sum_{i=1}^{D-1} \sum_{j=i+1}^{D} QPF_{ij} \quad (23)$$

3.3 Best closeness value

From the discussion given in Sections 1.3.1 and 1.3.2, it is evident that to calculate both BCV_d (for Equation 2) and BCV_E (for Equation 8), we need to find a best arrangement for the facilities and departments in the existing layout. There exist various algorithms which could generate a best layout [1]. However, these algorithms are time consuming and often generate solution far below the best solution, and they are not suitable to find the best layout, especially when the layout effectiveness is measured. An appropriate alternative to directly calculate both the BCV_d and BCV_E is by computing a lower bound of CCV_d and CCV_E, respectively (where CCV_d and CCV_E are a function of step values and interaction values as given in Equations 3 and 9). Mathematically, the lower bound, which is the lowest inner product of two vectors, is obtained by sorting the vectors in opposite direction. Therefore, if the interaction and step values for intra-functional flow (INV_{mn} and SV_{mn}) are sorted in descending and ascending order respectively and then multiplied, the LB of the CCV_d is obtained. This LB of the CCV_d is used as the BCV_d. Similarly, the INV_{ij} and SV_{ij} values are sorted in descending and ascending order respectively and then multiplied in order to get BCV_E. In reality, obtaining BCV_d and BCV_E in a layout may be impossible due to practical limitations. However, the BCV_d and BCV_E obtained using the LB gives us the best theoretical reference value to measure the closeness gap in an existing layout.

4. Conclusion

Measuring the performance level of a facilities layout is essential to support various decisions making processes. However, it is apparent that the existing methods are appropriate mostly for the design phase and have considered only the MHC to determine the effectiveness of a layout. Hence, in order to measure the effectiveness of an existing facility layout and to overcome the identified drawbacks, we developed three factors such as Facilities Layout Flexibility (FLF), Productive Area Utilisation (PAU) and Closeness Gap (CG). The proposed CG extends other related works by encompassing several critical features including the empty travel of material handling equipment, equipment flow, information flow and personnel flow.

The proposed model provides an index value for the performance of a layout. However, it might not be sufficient to utilise directly in certain decision making process like planning. Hence, a potential expansion of the current work would be to convert the performance index into cost functions, which is a self-justification factor in any decision making process.

References

[1] Tompkins, J.A., et al., (2003) *Facilities Planning*. Third ed., New Jersey: John Wiley & Sons.

[2] Benjaafar, S., (2002) *Modeling and Analysis of Congestion in the design of Facility Layouts.* Management Science, 48(5): p. 679-704.

[3] Leung, Y.-T. and Suri, R., (1990) *Performance Evaluation of Discrete Manufacturing System.* IEEE Control System Magazine, 10(4): p. 77-86.

[4] Bitran, G.R. and Morabito, R., (1996) *Open Queueing Networks: Optimisation and Performance Evaluation Models for Discrete Manufacturing Systems.* Production and Operations Management, 5(2): p. 163-193.

[5] Govil, M.K. and Fu, M.C., (1999) *Queueing theory in manufacturing: A survey.* Journal of Manufacturing Systems, 18(3): p. 214-40.

[6] Muthaiah, K.M.N. and Huang, S.H., (2006) *A review of literature on manufacturing systems productivity measurement and improvement.* International Journal of Industrial and Systems Engineering, 1(4): p. 461-84.

[7] Raman, D., Nagalingam, S., and Chiu, M. (2005) *A Fuzzy Rule Based System to Measure Facility Layout Flexibility.* in *18th International Conference on Production Research.* Fisciano (SA), Italy: University of Salerno.

[8] Raman, D. and Nagalingam, S.V. (2006) *Productive area utilisation - Towards measuring the effectiveness of facilities layout.* in *4th International Conference on Manufacturing Research.* Liverpool John Moores University, UK.

[9] Francis, R.L., McGinis, L.F., and White, J.A., (1992) *Facility layout and location - An analytical approach.* Englewood Cliffs, NJ: Prentice Hall.

[10] Yaman, R., Gethin, D.T., and Clarke, M.J.,(1993) *An effective sorting method for facility layout construction.* International Journal of Production Research, 31(2): p. 413-427.

[11] Mak, K.L., Wong, Y.S., and Chan, F.T.S., (1998)*A genetic algorithm for facility layout problems.* Computer Integrated Manufacturing Systems, 11(1-2): p. 113 -127.

[12] Mckendall-Jr, A.R., Noble, J.S., and Klein, C.M., (1999) *Facility layout of irregular-shaped departments using a nested approach.* International Journal of Production Research,. 37(13): p. 2895-1914.

[13] Abdinnour-Helm, S. and Hadley, S.W., (2000) *Tabu search based heuristics for multi-floor facility layout.* International Journal of Production Research,. 38(2): p. 365-383.

[14] Raoot, A.D. and Rakshit, A., (1991) *A fuzzy approach to facilities lay-out planning.* International Journal of Production Research,. 29(4): p. 835-857.

[15] Dweiri, F., (1999) *Fuzzy development of crisp activity relationship charts for facilities layout.* Computer & Industrial Engineering, 36: p. 1-16.

[16] Deb, S.K. and Bhattacharyya, B., (2005) *Fuzzy decision support system for manufacturing facilities layout planning.* Decision Support Systems, 40(2): p. 305-314.

[17] Adel El-Baz, M., (2004) *A genetic algorithm for facility layout problems of different manufacturing environments.* Computers & Industrial Engineering, 47: p. 233-246.

[18] Hu, M.H. and Wang, M.-J., (2004) *Using genetic algorithms on facilities layout problems.* International Journal of Advanced Manufacturing Technology, 23: p. 301 - 310.

High-level Business Processes for Agile B2B Collaboration

K. Mous[1], K.L. Ko[2], S.G. Lee[3], P.S. Tan[4] and E.W. Lee[5]

[1]Nanyang Technological University, MAE, 50 Nanyang Avenue, Singapore 639798, karel@ntu.edu.sg
[2]Nanyang Technological University, MAE, 50 Nanyang Avenue, Singapore 639798, L060044@ntu.edu.sg
[3]Nanyang Technological University, MAE, 50 Nanyang Avenue, Singapore 639798, msglee@ntu.edu.sg
[4]Singapore Institute of Manufacturing Technology, 71 Nanyang Drive, Singapore 638075, pstan@simtech.a-star.edu.sg
[5]Singapore Institute of Manufacturing Technology, 71 Nanyang Drive, Singapore 638075, ewlee@simtech.a-star.edu.sg

Abstract Increasingly, more businesses are collaborating to leverage on the competencies of potential partners in furtherance of common goals. Although best sourcing is the mantra of the day, many businesses are finding it more efficacious to form business alliances supported by IT-enabled business processes.

In today's dynamic, fluid business environment, short term collaborations are frequent. The problem with short term collaborations is that common (public) business processes have to be revised every time a change is encountered. This is cumbersome as it is not good for business processes to change every time a change takes place. It is better to define public business processes at a higher level of abstraction so they can be altered with minimal impact on the company's private processes.

This paper discusses business process concepts and how a public business process for B2B collaboration can be formulated, into which the private business processes of collaborators can simply be "plugged". An industrial example of a semiconductor wafer fabrication plant is discussed.

The authors believe that this approach lays the foundation for automating the search for and selection of potential business collaborators in truly agile B2B collaboration.

Keywords: Business Process, B2B Collaboration, Public Process, Private Process.

1. Introduction

Everywhere in the world, people work together in organizations, ranging from one-person shops to big multi-national enterprises.

Within every organization, people perform a sequence of activities / tasks to accomplish a certain goal. This logical sequence of timely activities is called a *process*. When the set goal is driven by the objectives of a company, the process is often called a *business process*. Since all activities within an organization are part of some business process, business processes are really the backbone of every organization.

These days, many organizations engage in Business to Business (B2B) collaboration to capitalize on each others' best practices to accomplish common goals. Some B2B collaborations only last for a short time, because the business environment is rapidly changing. To facilitate collaboration between organizations, their business processes are often integrated for seamless information exchange.

Many business process modeling techniques have been proposed [1] and many standards developed to facilitate B2B collaborations. Some of these standards include ebXML, BizTalk, and RosettaNet.

With the advent of the Internet, web services have been gaining ground, because they can link different systems together, especially when supported by BPEL engines [2].

There is no standard approach to integrating business processes at the present moment. As collaborations are becoming more and more short-term, business processes have to be rapidly re-configured to accommodate new business partners. This is very difficult and costly to do because if the business coupling is rigid.

Integration for short term collaboration is even more difficult due to the many different standards used. Since it is wishful thinking to hope that one day one standard will prevail, other solutions will have to be found to facilitate business process integration for short term collaborations.

In this paper, we discuss how high level business processes address these issues and how collaborators may communicate with one another.

Section 1.2 discusses business process concepts while section 1.3 describes how a public business process for B2B collaboration can be defined.

2. Organizing Business Processes

Let us now examine the two generic types of business process and what this means for a B2B collaboration.

2.1 Public and Private Business Processes

The business processes which are internal to an organization are often called "private processes" and are invisible to the outside world. The outside world sees this collection of private processes as one public process. A public process, on the other hand, is visible to all parties outside the organization.

Figure 1.1. shows the symbolic relationship between a public and a private process.

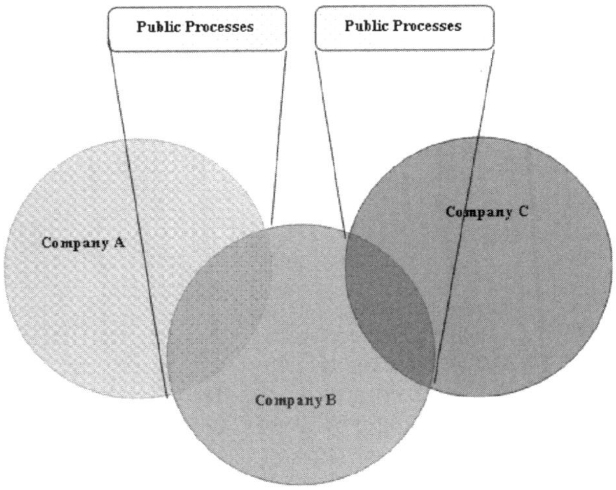

Figure 1.1. Public and Private Business Processes

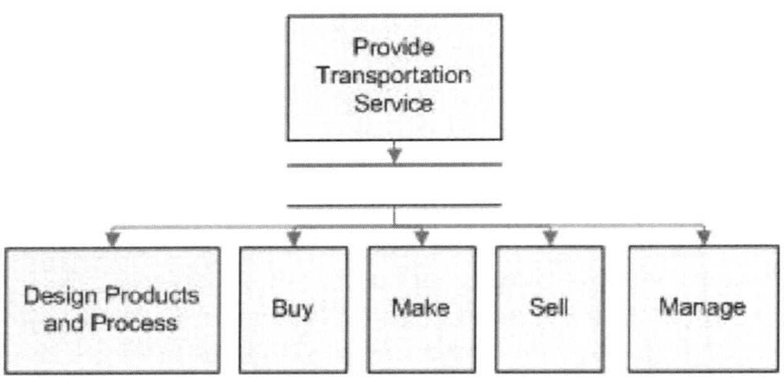

Figure 1.2. The Business Process of UPS

An example of private and public business processes is shown in figure 1.2. for the case of UPS [3]. The business process described at the highest level - "Provide transportation service" - is also the overall business goal of UPS. This is UPS' public process since it can be seen by all stakeholder. This overall biz goal is decomposed into five other (sub) processes; "Design product and process", "Buy", "Make", "Sell" and "Manage a business", which are not visible to parties outside UPS. These sub processes are therefore "Private Processes".

2.2. Public and Private Processes in B2B collaborations

When two companies collaborate with each other, their business processes are not merged into one new process but only linked together. Every company retains its private business processes. Sometimes, the business processes of its collaborators become sub processes of a higher level business process. This higher level business process is a "public" process because parts of the process are implemented by collaborators but the overall process is visible to all collaborators.

To illustrate the concept of public and private processes for B2B collaboration, consider the business processes of Amazon, schematically displayed in figure 1.3. This figure only displays the sub business processes relevant to this discussion.

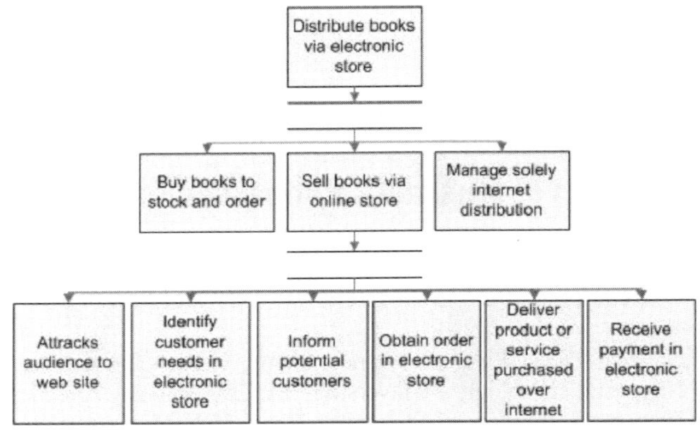

Figure 1.3. The Business Processes of Amazon

The highest level business process describes the ultimate business goal of Amazon which is: "*Distribute books via electronic store*".
This process can be decomposed into three other sub business processes: "*Buy books to stock and order*", "*Sell books via online store*" and "*Manage solely internet distribution*". These, in turn, can be broken down into several sub processes: for example, the process "*Sell books via online store*" consists of six components.

Suppose Amazon would like UPS to handle the delivery of the ordered goods. Towards that end, UPS will implement Amazon's "*Deliver product or service purchased over the internet*" business process. The "*Sell books via electronic store*" process and all its components becomes a public process visible to both UPS and Amazon.

UPS can only see the "*Sell books via online store*" business process. UPS may even be able to see the "*Distribute books via electronic store*" business process as well, but nothing more than that. Amazon in turn has no idea how UPS implements its "*Provide Transportation Service*" process.

3. Formulating a Public Business Process

How is a public business process for B2B collaboration defined? A public business process has three characteristics: the level of abstraction, the process decomposition and the communication between collaborators. We will discuss all these aspects.

3.1. The Appropriate Level of Abstraction

The challenge is to describe the public process in such a way that it is neither too abstract nor too detailed so that collaborators can implement a part of the public process to accomplish the goal of the collaboration.

As can be seen in the UPS and Amazon examples, a business process can be decomposed into several levels of (sub) business processes. As a matter a fact, the number of levels is actually not limited, but the case studies described in the MIT Business Process Handbook have no more than 10 levels [3].

The first level is always the grand business goal of an organization or collaboration. When this grand business goal is decomposed to the second level, there are really only five different but associative sub processes: *Design*, *Buy*, *Make*, *Sell* and *Manage*. Not every organization has all these 5 sub processes. Because second level processes can be implemented in many ways, business processes at the third level are not quite dissimilar.

Except for the first level, a public B2B business processes can be defined on any level, depending on the scope of the collaboration. For instance, Amazon might decide to completely outsource its *Buy* process to a collaborator. In that case the public process is defined at level 2. On the other hand, if Amazon decides to appoint a collaborator like UPS to deliver its goods, then the public process is defined at level 3.

Thus, it is not possible to define criteria by which the level of abstraction of a public process is determined. It all depends on the nature and scope of the collaboration, and sometimes management policy.

3.2. Decomposing a Business Process

It is possible define guidelines for the decomposition of a public business process. These guidelines are as follows:

3.2.1. Maximum Cohesion, Minimum Coupling

The first, and probably the most important, guideline is "Maximum Cohesion, Minimum Coupling". This means that every component of the public process should be stand-alone, i.e. it does not need to call another component at the same level during its standard execution. It is possible though that processes need to invoke each other during execution due to exceptions that might occur at runtime, but this should not be the norm. At runtime however, information can be requested from and exchanged between different collaborators. The components of a business process should only be coupled with each other at the beginning and the end of the execution.

This can be clearly seen in the Amazon example. The third level is decomposed into 6 components, all of which run independently by themselves and don't need to invoke each other during execution. Note also that all components at the third level form stand alone links in a chain of components implementing the "*Sell books via electronic store*" process at the second level.

The components in the above example have the maximum cohesion, since they are all related to each other, leading towards a higher goal. Since none of the components needs to invoke another component during its execution there is also minimum coupling between the individual components of business processes.

3.2.2. Pre conditions

To enable a smooth execution of the overall public process, the pre conditions for each component have to be defined. These pre conditions determine which component is to be executed first, and which information and resources are required for the execution.

As can be seen in the Amazon example, a product can't be delivered before the order is confirmed. Thus "*Deliver product or service purchased over internet*" can't be executed before "*Obtain order in electronic store*". To deliver a product, the actual product needs to be available (resource) along with the information related to the order and the customer.

3.2.3. Post conditions

For each component, the post conditions after execution - information, a material or simply a state - will also have to be defined.

Information can for example be the details of a customer and the order placed. Material can for example be an engine that is assembled from different parts. A state can simply be "*Product is delivered*".

3.2.4. Add Value

Finally the last guideline is that each component must add value to the overall business process. When a component contributes little or nothing to the public process, there is no point in having it.

3.3. Communication

Although companies collaborate with one another, they always remain distinct, independent companies. Since information is exchanged, communications needs have to be considered when designing the business processes.

In this paper, it will suffice to describe the exchange of messages between business processes. We do this *via* UML sequence diagrams like those by Bauer [4] in figure 1.4. for the Amazon example.

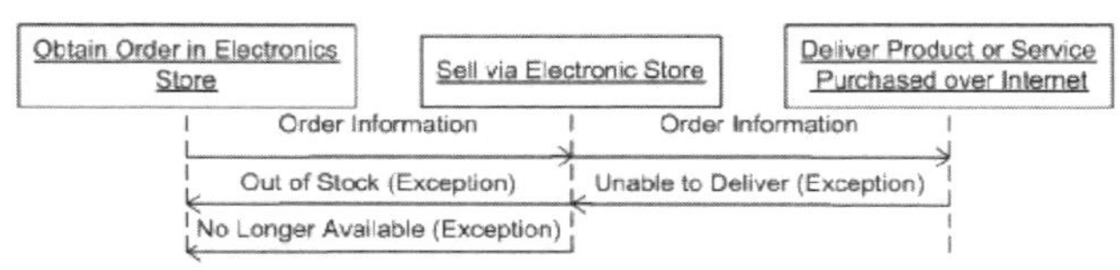

Figure 1.4. Message exchange

The process "*Deliver Product or Service Purchased over Internet*" needs order information from "*Obtain Order in Electronics Store*" to deliver the ordered product.

Notice that information is passed between the two processes through their parent processes, "*Sell via Electronic Store*". The reason for this is twofold. In the first place it would be unmanageable, especially in short term collaborations, if business processes were to connect to each other directly. In the second place, it might be necessary for an action to be first taken at a higher level.

In the Amazon example, a product has to be ordered first, before it can be delivered. Ordering a product however, is not done by any component of the "*Sell via Electronic Store*" process. Buying products is done by the "*Buy books to stock and to order*" business process, which is a process at level two. In this case the "*Sell via Electronic Store*" process should prompt the "*Buy books to stock and to order*" to make the needed purchase.

Communications between processes, as described in this section, is very high level and therefore only useful when designing a public process at the conceptual level. If we are designing a public process defined in BPEL [5], communication has to be described in more detail as defined by Bussler [6].

4. An Industrial Example

The industrial example is from a semiconductor wafer fabrication plant in Singapore. The partial breakdown of its core business process of "*Producing DRAM Wafers*" is displayed in Figure 1.5.

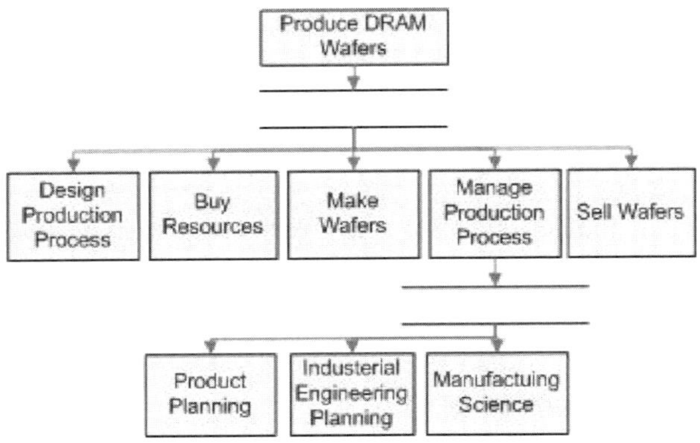

Figure 1.5. The planning Business Process of a semiconductor wafer fabrication plant in Singapore

The business goal of the wafer fabrication plant is to produce DRAM wafers. This is also the highest level business process. The highest level business process can be decomposed into the sub processes "Design", "Buy", "Make", "Sell" and "Manage" as discussed earlier.

The "Manage" business process is largely the responsibility of the Planning Department, and consists of 3 sub processes: "Production Planning", "Industrial Engineering Planning" and "Manufacturing Science".

4.1 Maximum Cohesion, Minimum Coupling

"Production planning" schedules the wafer lots and coordinates the production. "Manufacturing Science" formulates the best scheduling algorithms subject to manufacturing resources and constraints, while "Industrial Engineering Planning" ensures that the plant capacity is fully utilized and that the location of the equipment is consistent with the manufacturing line. As can be seen, the above 3 sub processes are loosely coupled, yet have maximum cohesion.

4.2 Pre Conditions and Post Conditions

The pre-conditions of the 3 sub processes are well defined. An example of a pre-condition for the production planning sub process is that the production target has to be set. Another example of pre conditions for Industrial Engineering sub processes are the plant capacity and the capacities of the new and old equipment.

The post-conditions of the 3 sub processes allow their parent process "*Manage Production Process*" to function as expected. Examples of these post conditions are optimized algorithms, proper floor plans, and efficient production schedules for the different DRAM products.

4.3 Add Value

The "Manufacturing Science" sub process adds value to production scheduling in the "Production Planning" sub process because of its optimized algorithms.

High capacity utilization and good shop floor layout by the "Industrial Engineering Planning" sub process will in turn add value to the other 2 sub processes in the "Planning" business process.

In conclusion, all sub processes are loosely coupled, maximally cohesive, and add value to one another and also to their parent business process.

5. Conclusion

The concept of private (internal) and public (external) business processes is presented. A public business process has three characteristics: the *level* of abstraction, the method of decomposition into sub-processes and the communications between collaborators.

The level of abstraction of a public process is determined by factors such as the nature and scope of the collaboration, and sometimes management policy. The authors are unaware of any prevailing guidelines to define a public process and neither do they think it is possible to do so.

Four guiding principles by which a business process may be decomposed are discussed and illustrated by a case study.

Some preliminary thoughts on the mode of communications / message exchange between business processes are presented.

Acknowledgements

The authors gratefully acknowledge the support of Singapore's Agency for Science, Technology & Research (A*STAR) under research grant # 0521160077.

References

[1] Havey, M. (2005) Essential Business Process Modeling, O'Reilly.
[2] Oh JY, Jung J-Y, Cho NW, Kim H, Kang S-H. Integrated Process Modeling for Dynamic B2B Collaboration Lecture Notes in Computer Science. 2005, Vol 3683, P602-8.
[3] The MIT Process Handbook Project. Massachusetts Institute of Technology 1993 - 2003. http://ccs.mit.edu/ph/
[4] Bauer B, Müller JP, Roser S. A Model-Driven Approach to Designing Cross-Enterprise Business Processes. Lecture Notes Computer Science. 2004, Vol 3292, P544 – 55.
[5] Andrews T, Curbera F, Dholakia H, Goland Y, Klein J, Leymann F, et al. Business Process Execution Language for Web Services 1.1. 2003.
[6] Bussler C. The Application of Workflow Technology in Semantic B2B Integration. Distributed and Parallel Databases: Kluwer Academic Publishers 2002, P163–91.

9–5

Configuring an Optimal Supply Chain of New Products Using Analytical Target Cascading

T. Qu[1], G.Q. Huang[2], S. Bin[3], H.P. Chen[4]

[1,2,3]The University of Hong Kong, Hong Kong, P.R. China,
quting@hku.hk, gqhuang@hku.hk, binsheng@hku.hk
[4]The University of Science and Technology of China, P.R. China, hpchen@ustc.edu.cn

Abstract. This paper presents how to apply Analytical Target Cascading (ATC), a multi-level hierarchical optimization method, in solving Supply Chain Configuration (SCC) problems. Through partitioning the supply chain by stages, the original complex problem of configuring the whole supply chain is converted into several simpler sub-problems of configuring stages. These sub-problems are solved in a concurrent and consistent manner and the ATC coordinating mechanism is responsible for assembling their results into the final optimal supply chain configuration. This kind of decentralized optimization is especially important if autonomous decision rights of the participating enterprises are required in the SCC problem. The general procedure for applying ATC is illustrated with a notebook computer SCC problem and the satisfactory optimization results are obtained.

Keywords: Analytical target cascading, Supply chain configuration, Autonomy, Hierarchical modelling.

1. Introduction

Supply Chain Configuration (SCC) aims to design a unique supply network through choosing one from the alternative options for each processing stage, so that certain system performance indicators can be optimized [1]. The options may refer to alternative raw materials suppliers, alternative manufacturing/assembling processes and alternative delivery modes for procurement stages, manufacturing/assembling stages and delivery stages respectively. The performance indicators often include the total supply chain cost, customer service level, etc.

Some popular optimization methods have been applied for solving SCC problems. For example, Graves and Willems [1] and Huang et al. [2] applied Dynamic Programming (DP) and Genetic Algorithm (GA) respectively to configure an optimal supply chain for a notebook computer. These methods are mostly all-in-one (AIO) methods which require the decisions of the whole supply chain be centrally made in a common decision model. Although acceptable computational effectiveness has been achieved, two intrinsic limitations of AIO methods are hard to overcome. First, they can not deal with SCC problems with large numbers of stages. When the number of decision variables exceeds a critical value, the improvement of optimization algorithms will have little effect on maintaining the optimization efficiency. Second, they can not separate the decision process of each individual stage. If different stages are to be configured by different autonomous enterprises, forcing them to submit the decision details to a public decision model is sometimes impractical.

This paper will apply Analytic Target Cascading (ATC) for solving SCC problems. The overall configuration targets are defined at the top-level stage (most downstream stage) and cascaded down to the stages at lower levels (upstream stages). In this way, all the stages can be isolated from each other and configured independently using their respective decision models. As a result, no matter how many stages are involved in the supply chain, the configuration of each stage can be considered in its own decision model with an independent computing resource. Therefore, the above two limitations of AIO methods could be naturally addressed.

The remainder of this paper is organized as follows. Section 1.2 outlines an overview of ATC and summarizes a five-step procedure of applying ATC. The five steps are illustrated with a notebook computer SCC problem in Sections 1.3 to 1.7 to demonstrate how the ATC process can be applied in the optimal configuration of supply chains. Conclusions are presented in Section 1.8.

2. Overview of Analytical Target Cascading (ATC)

ATC is a multi-level hierarchical optimization method for translating the overall design targets of a system to the specifications of the constituent elements that are decomposed to the multilevel hierarchy (Kim, et al., 2000). The coordinating mechanism offered by ATC is responsible for assembling the local designs of these elements consistently into the final optimal system design. Let us use the hierarchical model in Figure 1.1 to explain a typical target cascading process.

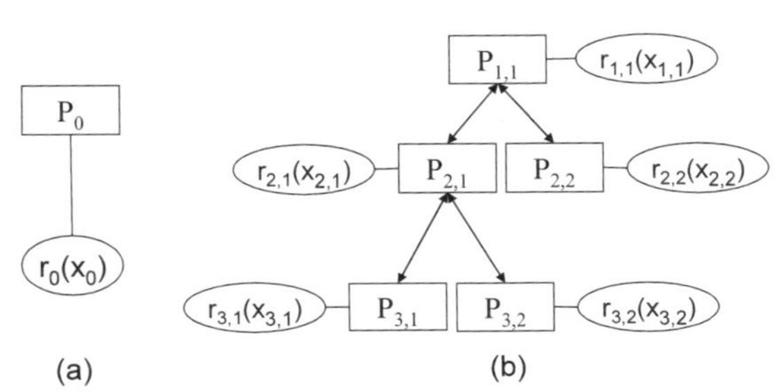

Figure 1.1. Hierarchically partitioned ATC model

Firstly, the original system, P_0, is partitioned into a three-level ATC hierarchical model composed of five ATC elements, $P_{1,1}$, $P_{2,1}$, $P_{2,1}$, $P_{3,1}$, $P_{3,2}$, as can be seen in part (a) and (b) of the figure. Each of these elements has its own analysis model indicated as $r_{i,j}(x_{i,j})$ indicated. The subscript m,n denotes this is the nth element in the mth level. Then, key links between these elements are identified based on the relationships of analysis models to reflect their design dependencies. Through embedding the deviation minimizing terms of the key links into the sub-problem formulation of each element, functional dependencies are established in the whole hierarchy. From the top-level element, sub-problems are solved and design targets for the child elements are sequentially cascaded down. If the cascaded targets are not achievable in an element, adjustment of targets is required to the parent element resulting in an iterative target cascading process. If the element can achieve the targets, target cascading will proceed to lower levels in a coordinated fashion until the bottom level is reached. This process will continue until the whole system converges to a consistent design.

Together with the final results analysis, the above process has been summarized to a five-step procedure of applying ATC: (1) ATC modeling, (2) key links identification, (3) ATC element formulation, (4) ATC solution and (5) results analysis. The following five sections will explain how to solve SCC problems with this procedure.

3. Hierarchical Modeling of Supply Chain

The first step of applying ATC is to partition the supply chain system into a hierarchical model. A supply chain for assembling notebook computers is given in Figure 1.2. It is adapted from the SCC problem solved by GA [3]. The 13 stages are numbered in hierarchical index (i.e. $stage_{i,j}$ represents the jth stage in i^{th} level) and the corresponding products are given below. In this paper, we assume that each stage is undertaken by a different enterprise which enjoys autonomous decision right. Each enterprise should select a suitable option such that the US demand could be filled with minimal cost in zero service time. For simplicity, $stage_{1,1}$ and $stage_{0,1}$ are combined to a whole as $stage_{1,1'}$.

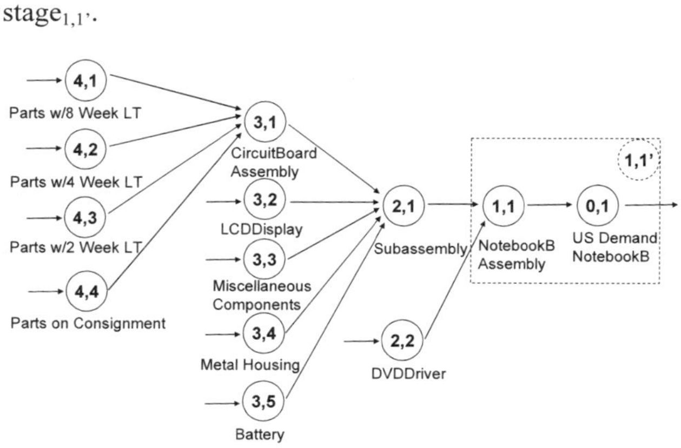

Figure 1.2. A notebook computer supply chain

Due to the hierarchical structuring of stages, this supply chain could be easily converted to an ATC hierarchical model through by stage partitioning. Therefore, 12 elements are generated in the ATC model.

Despite the different decision right allotment, this decentralized SCC problem shares the same mathematical models and parameter settings with Huang et al. Specifically, the market demand for this note book computer follows a normal distribution ($\mu = 125$ and $\sigma = 80$); the service level (α) in each stage is 0.98; the annual holding cost rate (h) is 40%; all the enterprises operate for 360 days per year. The options of each stage are differentiated by their processing times (T), and production costs (c), and the detailed information is given in the Appendix.

4. Key Links Identification among Supply Chain Stages

The second step of applying ATC is to identify the key links between stages, including responses which are common variables shared by parent and child elements as well as linking variables which are common variables shared by sibling elements (Kim, et al., 2000). The identification of responses starts from the top level element, stage1,1'. As the customer requires that the total supply chain cost (TC) and service time (ST) be minimized, $TC_{1,1'}$ and $ST_{1,1'}$ will be identified as two responses. In the analysis models of $TC_{1,1'}$ and $ST_{1,1'}$, three variables related to $stage_{2,1}$ are used, namely $TC_{2,1}$, $CC_{2,1}$ and $ST_{2,1}$. Hence, they are identified as responses of $stage_{2,1}$. $CC_{2,1}$ refers to the cumulative cost of $stage_{2,1}$. This process will continue to the most upstream stages until all responses are identified.

The identification of linking variables is often based on certain practical relationships of the sibling elements. For this SCC problem, each stage hopes that all its upstream stages share the same service time (ST), so as to avoid unnecessary inventory. Thus ST will be identified as linking variables among sibling stages.

5. ATC Formulation for Individual Stages

The third step of applying ATC is to formulate sub-problem for each element. Generally, ATC formulation comprises three parts. The first part is objective function (Minimize), which is composed of deviation terms between current element and both its upper-level and lower-level elements. The second part is analysis models (Where), which are response functions of the design variables, including local variables and key links. The third part is constraints (Subject to). Apart from the constraints of the SCC mathematic models, ATC also appends tolerance constraints for coordinating lower-level key links.

According to their relative position in the supply chain, stages are categorized into three sets: demand stage (top-level element), assembly stages (intermediate-level elements), and procurement stages (bottom-level elements). Since the problem formulations for the stages in the same set are mathematically alike, only one typical stage will be selected from each set for illustration.

(1) ATC Formulations for the Demand Stage

Stage$_{1,1'}$ is the only demand stage in the supply chain. The sub-problem in stage$_{1,1'}$ could be stated as: minimize the deviation between the demand stage and demand targets coming from customers subject to SCC constraints and tolerance constraints that coordinate with the upstream stages. The complete formulation is as follows:

Minimize $w_{1,1'}^{TC}(TC_{1,1'}-t^{TC})^2+w_{1,1'}^{ST}(ST_{1,1'}-t^{ST})^2+\sum_{i=1}^{5}\varepsilon_{(1,1'),i}$

With respect to

$ST_{v1,1}, ST_{1,1}, SO_{1,1}, c_{1,1}, T_{1,1}, ST_{0,1}, SO_{0,1}, c_{0,1}, T_{0,1}, TC_{2,1}, TC_{2,2},$
$CC_{2,1}, CC_{2,2}, \varepsilon_{(1,1'),1}, \varepsilon_{(1,1'),2}, \varepsilon_{(1,1'),3}, \varepsilon_{(1,1'),4}, \varepsilon_{(1,1'),5}$

Where

$$TC_{1,1'}=SCCC_{1,1}/(\mu_{1,1}H_{1,1})+SCCC_{0,1}/(\mu_{0,1}H_{0,1})+TC_{2,1}+TC_{2,2}$$
$$ST_{1,1'}=ST_{0,1}$$
$$SCCC_{0,1}=h_{0,1}[CC_{0,1}(\lambda\mu_{0,1}/2+\alpha_{0,1}\sigma_{0,1}\sqrt{ST_{1,1}+T_{0,1}-ST_{0,1}+\lambda})+(c_{0,1}/2+CC_{1,1})\mu_{0,1}T_{0,1}]+H_{0,1}c_{0,1}\mu_{0,1}$$
$$SCCC_{1,1}=h_{1,1}[CC_{1,1}\alpha_{1,1}\sigma_{1,1}\sqrt{ST_{v1,1}+T_{1,1}-ST_{1,1}}+(c_{1,1}/2+CC_{2,1}+CC_{2,2})\mu_{1,1}T_{1,1}]+H_{1,1}c_{1,1}\mu_{1,1}$$
$$CC_{0,1}=c_{0,1}+CC_{1,1}$$
$$CC_{1,1}=c_{1,1}+CC_{2,1}+CC_{2,2}$$

Subject to

$$\sum_{SO_{0,1}\in S(0,1)}T_{(0,1),SO_{0,1}}y_{(0,1),SO_{0,1}}-T_{0,1}=0$$
$$\sum_{SO_{0,1}\in S(0,1)}c_{(0,1),SO_{0,1}}y_{(0,1),SO_{0,1}}-c_{0,1}=0$$
$$\sum_{SO_{1,1}\in S(1,1)}T_{(1,1),SO_{1,1}}y_{(1,1),SO_{1,1}}-T_{1,1}=0$$
$$\sum_{SO_{1,1}\in S(1,1)}c_{(1,1),SO_{1,1}}y_{(1,1),SO_{1,1}}-c_{1,1}=0$$
$$y_{(0,1),SO_{0,1}}=\begin{cases}1 & \text{If } SO_{0,1} \text{ is selected}\\ 0 & \text{Otherwise}\end{cases},\ \sum_{SO_{0,1}\in S(0,1)}y_{(0,1),SO_{0,1}}=1$$
$$y_{(1,1),SO_{1,1}}=\begin{cases}1 & \text{If } SO_{1,1} \text{ is selected}\\ 0 & \text{Otherwise}\end{cases},\ \sum_{SO_{1,1}\in S(1,1)}y_{(1,1),SO_{1,1}}=1$$
$$ST_{v1,1}+T_{1,1}-ST_{1,1}\geq 0\ ,\ ST_{v1,1},ST_{1,1}\geq 0 \text{ and integer}$$
$$TC_{2,1},TC_{2,2},CC_{2,1},CC_{2,2}\geq 0$$
$$(TC_{2,1}-TC_{2,1}^{3})^2\leq\varepsilon_{(1,1'),1}\ ,\ (TC_{2,2}-TC_{2,2}^{3})^2\leq\varepsilon_{(1,1'),2}$$
$$(CC_{2,1}-CC_{2,1}^{3})^2\leq\varepsilon_{(1,1'),3}\ ,\ (CC_{2,2}-CC_{2,2}^{3})^2\leq\varepsilon_{(1,1'),4}$$
$$(ST_{v1,1}-\max\{ST_{2,1}^{3},ST_{2,2}^{3}\})^2\leq\varepsilon_{(1,1),5}$$
$$0<\varepsilon_{(1,1)',i}\leq g_{(1,1)',i}(I)\quad i=1,2,3,4,5$$

t^{TC} and t^{ST} are the demand targets of customers. They are set to zero to achieve the minimization objectives of total cost and service time. For the two target deviation terms, weighting coefficients, $w_{1,1'}^{TC}$ and $w_{1,1'}^{ST}$, are attached to manifest the customer preference. Five tolerance variables, $\varepsilon_{(1,1),1}$, $\varepsilon_{(1,1),2}$, $\varepsilon_{(1,1),3}$, $\varepsilon_{(1,1),4}$ and $\varepsilon_{(1,1),5}$ are used to keep upstream linking with stage$_{2,1}$ and stage$_{2,2}$.

(2) ATC Formulations for the Assembly Stages

There are two assembly stages in this supply chain, namely stages$_{2,1}$ and stages$_{3,1}$. The sub-problems of these stages could be stated as: minimize the deviations between the assembly stages and targets from their downstream stages subject to SCC constraints and tolerance constraints that coordinate with the upstream stages. Take stage$_{2,1}$ as an illustration, the complete formulation is as follows:

Minimize $w_{2,1}^{TC}(TC_{2,1}-TC_{2,1}^{1})^2+w_{2,1}^{CC}(CC_{2,1}-CC_{2,1}^{1})^2+w_{2,1}^{ST}(ST_{2,1}-ST_{2,1}^{1})^2+\sum_{i=1}^{11}\varepsilon_{(2,1),i}$

With respect to $ST_{v2,1}, ST_{2,1}, SO_{2,1}, c_{2,1}, T_{2,1}, TC_{3,1}, \ldots, TC_{3,5}, CC_{3,1}, \ldots, CC_{3,5}, \varepsilon_{(2,1),1}, \ldots, \varepsilon_{(2,1),11}$

Where

$$TC_{2,1}=SCCC_{2,1}/(\mu_{2,1}H_{2,1})+\sum_{k=1}^{5}TC_{3,k}\ \ ST_{1,1'}=ST_{0,1}$$
$$CC_{2,1}=c_{2,1}+\sum_{k=1}^{5}CC_{3,k}$$
$$ST_{2,1}=ST_{2,1}$$
$$SCCC_{2,1}=h_{2,1}[CC_{2,1}\alpha_{2,1}\sigma_{2,1}\sqrt{ST_{v2,1}+T_{2,1}-ST_{2,1}}+(1/2c_{2,1}+\sum_{k=1}^{5}CC_{3,k})\mu_{2,1}T_{2,1}]+H_{2,1}c_{2,1}\mu_{2,1}$$

Subject to

$$\sum_{SO_{2,1}\in S(2,1)}T_{(2,1),SO_{2,1}}y_{2,1,SO_{2,1}}-T_{2,1}=0$$
$$\sum_{SO_{2,1}\in S(2,1)}c_{(2,1),SO_{2,1}}y_{(2,1),SO_{2,1}}-c_{2,1}=0$$
$$y_{(2,1),SO_{2,1}}=\begin{cases}1 & \text{If } SO_{2,1} \text{ is selected}\\ 0 & \text{Otherwise}\end{cases},\ \sum_{SO_{2,1}\in S(2,1)}y_{(2,1),SO_{2,1}}=1$$
$$ST_{v2,1}+T_{2,1}-ST_{2,1}\geq 0,\ ST_{v2,1},ST_{2,1}\geq 0 \text{ and integer}$$
$$TC_{3,k},CC_{3,k}\geq 0\ ,\ (TC_{3,k}-TC_{3,k}^{4})^2\leq\varepsilon_{(2,1),k}$$
$$(CC_{3,k}-CC_{3,k}^{4})^2\leq\varepsilon_{(2,1),k+5}\ ,\ k=1,2,3,4,5$$
$$(ST_{v2,1}-\max\{ST_{3,1}^{4},ST_{3,2}^{4},ST_{3,3}^{4},ST_{3,4}^{4},ST_{3,5}^{4}\})^2\leq\varepsilon_{(2,1),11}$$
$$0<\varepsilon_{(2,1),i}\leq g_{(2,1),i}(I)\quad i=1,2,\ldots,11$$

This formulation is similar to that of stage$_{1,1'}$. The targets for the responses come from the optimization results of stage$_{1,1'}$, instead of the customer demand.

(3) ATC Formulations for the Procurement Stages

There are nine procurement stages in this supply chain, namely stage$_{2,2}$, stage$_{3,2}$, stage$_{3,3}$, stage$_{3,4}$, stage$_{3,5}$, stage$_{4,1}$, stage$_{4,2}$, stage$_{4,3}$, stage$_{4,4}$. The sub-problem of these stages could be stated as: minimize the deviations between the procurement stages and targets from their downstream stages subject to SCC constraints. Take stage$_{3,2}$ as an illustration, the complete formulation is as follows:

Minimize $w_{3,2}^{TC}(TC_{3,2}-TC_{3,2}^{2})^2+w_{3,2}^{CC}(CC_{3,2}-CC_{3,2}^{2})^2+w_{3,2}^{ST}(ST_{3,2}-ST_{3,2}^{2})^2$

With respect to $ST_{3,2}, SO_{3,2}, c_{3,2}, T_{3,2}$

Where

$$TC_{3,2}=SCCC_{3,2}/(\mu_{3,2}H_{3,2})$$
$$CC_{3,2}=c_{3,2}$$
$$ST_{3,2}=ST_{3,2}$$
$$SCCC_{3,2}=h_{3,2}c_{3,2}(\alpha_{3,2}\sigma_{3,2}\sqrt{T_{3,2}-ST_{3,2}}+\mu_{3,2}T_{3,2})+H_{3,2}c_{3,2}\mu_{3,2}$$

Subject to

$$\sum_{SO_{3,2}\in S(3,2)}T_{(3,2),SO_{3,2}}y_{(3,2),SO_{3,2}}-T_{3,2}=0$$
$$\sum_{SO_{3,2}\in S(3,2)}c_{(3,2),SO_{3,2}}y_{(3,2),SO_{3,2}}-c_{3,2}=0$$
$$y_{(3,2),SO_{3,2}}=\begin{cases}1 & \text{If } SO_{3,2} \text{ is selected}\\ 0 & \text{Otherwise}\end{cases},\ \sum_{SO_{3,2}\in S(3,2)}y_{(3,2),SO_{3,2}}=1$$
$$T_{3,2}-ST_{3,2}\geq 0,\ ST_{3,2}\geq 0 \text{ and integer}$$

$Stage_{3,2}$ has no upstream stage, thus the tolerance variables are omitted from objective function, leaving only target deviation terms exist to coordinate with its downstream stage.

6. ATC Solution

Two kinds of issues should be considered during the ATC solution process, namely the local optimization algorithm and the overall solution sequence. ATC has no restriction on the decision details of each individual element, yet the solution sequence should guarantee a convergent ATC process. In this SCC problem, we adopt genetic algorithm (GA) for solving the sub-problem of each stage and "Lower levels converge first" as solution sequence.

7. Result Analysis

The ATC process takes five iterations to reach a convergence with maximum inter-element tolerance of 0.01%. Table 1.1 lists both the optimization results by ATC and by AIO approach (Huang et al., 2005). As can be seen, ATC achieves the same results as those by AIO approach

Table 1.1. Optimization results of AIO and ATC approach

Stage	AIO Results		ATC Results	
Index	SO	ST	SO	ST
$Stage_{4,1}$	3	10	3	10
$Stage_{4,2}$	2	10	2	10
$Stage_{4,3}$	1	10	1	10
$Stage_{4,4}$	1	0	1	0
$Stage_{3,1}$	1	30	1	30
$Stage_{3,2}$	1	30	1	30
$Stage_{3,3}$	1	30	1	30
$Stage_{3,4}$	1	30	1	30
$Stage_{3,5}$	1	30	1	30
$Stage_{2,1}$	1	35	1	35
$Stage_{2,2}$	1	35	1	35
$Stage_{1,1}$	1	36	1	36
$Stage_{0,1}$	2	0	2	0

8. Conclusion

This paper has introduced how to apply analytical target cascading (ATC) for solving supply chain configuration (SCC) problems. A five-step application procedure is proposed: (1) hierarchical modeling of supply chain; (2) key links identification among stages; (3) sub-problem formulation of individual stages; (4) ATC solution; (5) result analysis.Besides providing comparable optimization effectiveness, ATC has advantage over all-in-one (AIO) approaches in that it has no limitation to the number of stages involved in the SCC process and offers enough decision autonomy for the participating enterprises.

Acknowledgements

This research is partially supported by National Natural Science Foundation of China (#70629002).

References

[1] Graves, S.C., Willems, S.P. 2001. Optimizing the supply chain configuration for new products. Working Paper, Leaders for Management Program and A.P. Sloan School of Management, MIT.

[2] Huang, G.Q., Zhang, X.Y., Liang, L. 2005. Towards Integrated Optimal Configuration of Platform Products, Manufacturing Processes, and Supply Chains. Journal of Operations Management, 23, 267-290.

[3] Kim, H.M., Michelena, N.F., Papalambros, P.Y., Jiang, T. 2003. Target Cascading in Optimal System Design. Journal of Mechanical Design (Transactions of the ASME). 125(3), 474-480. September,2003.

Appendix - Options of each stage

Table 1.2. Options of each stage

Stage	Option	Processing Time	Production Cost
$Stage_{4,1}$	1	40	$130.00
	2	20	$133.25
	3	10	$134.91
	4	0	$136.59
$Stage_{4,2}$	1	20	$200.00
	2	10	$202.50
	3	0	$205.03
$Stage_{4,3}$	1	10	$155.00
	2	0	$156.93
$Stage_{4,4}$	1	0	$200.00
$Stage_{3,1}$	1	20	$120.00
	2	5	$150.00
$Stage_{3,2}$	1	60	$300.00
	2	5	$350.00
$Stage_{3,3}$	1	30	$200.00
$Stage_{3,4}$	1	70	$225.00
	2	30	$240.00
$Stage_{3,5}$	1	60	$40.00
	2	20	$45.00
$Stage_{2,1}$	1	5	$120.00
	2	2	$132.00
$Stage_{2,2}$	1	40	$15.00
	2	5	$16.50
$Stage_{1,1}$	1	1	$30.00
$Stage_{0,1}$	1	5	$12.00
	2	1	$20.00

An Agent-based Inference System for FMS Dispatching

C.L. Kuo[1], C.C. Ku[2] and J.P. Tsai[3]

[1] Industrial Technology Research Institute (ITRI), Taiwan, ChunLiangKuo@itri.org.tw
Department of Industrial Engineer and Management, National Chiao Tung University, Taiwan
[2] Industrial Technology Research Institute (ITRI), Taiwan, ccku@itri.org.tw
[3] Industrial Technology Research Institute (ITRI), Taiwan, JuiPinTsai@itri.org.tw

Abstract. In the era of mass customization and ever-changing market, it is important for enterprises to develop the dynamic and quick-response planning techniques in their flexible manufacturing systems (FMSs). This research paper presents an agent-based inference system, called Intelligent Decision Inference System (IDIS), to support existing FMS for production dispatching. When system integration is required between two separate operating systems, there are many issues to be taken into consideration, including system communication method, data flow design, and application program control. In the past, the most universal method was to integrate databases with two or more systems. However, this method had a main weakness of "data confusion and inconsistent". Therefore, this research combines the agent, JESS, and related IT technologies to systematically integrate the IDIS with FMS. The agents can easy communicate and deliver the messages, data, or information among the different systems and automatically scan the signals of the programmable logic controller (PLC)/sensors periodically. The method applies web service techniques as a standard information exchange mechanism and the agent communication language (ACL) of the Foundation for Intelligent Physical Agents (FIPA) for multi-agents to communicate/coordinate with each other. The agents, implemented in Java Agent Development Framework (JADE), are designed to decide whether a pallet should be moved in the shop floor of the FMS. Finally, one explicit example about how the cooperative multi-agent supporting the IDIS and FMS integration is presented. The unique contribution of this research is in the integration of expert systems and an agent framework to apply in real-time shop floor control such as FMS.

Keywords: Multi-Agent System, Flexible Manufacturing System (FMS), Production Dispatching, JESS

1. Introduction

In the past years, a rule-based inference system structure was developed, called the Intelligent Decision Inference System (IDIS) building on the JAVA Expert System Shell (JESS), has been developed to support collaborative design, maintenance, remote equipment diagnosis and shop flow control systems [1-4]. JESS has also been integrated with a PLC (programmable logic controller), RFID (radio frequency identification) and smart phone in the production process. All these cases obtain feasible results by implementing JESS in the manufacturing system control [5,6].

In this IDIS and FMS integration case, transfer and exchange data and information within two different systems via the database. It was a practicable method let the IDIS to support FMS dispatching, and gained the gratifying results.

However, this platform ran a long time; it was sometimes occurred some defects. The FMS control system was suddenly broken off by the unknown reasons; this main defect maybe was caused by frequently accessing the database.

Besides, when the conveyable data was huge and various, the handshaking flags were complex as the handshaking mechanism. Therefore, it was important to try to find a new IT technology to convey data with a different system. The agent technology can provide the solution to solve these problems. So, the previous integrated structure was modified and a new structure of the IDIS for FMS dispatching was developed to replace the medium database using agent technology. The new integrated structure includes three agents: IDIS Agent, FMS Agent and Control Agent to deal with the conveyance jobs.

These agents work co-ordinately and autonomously in real time scenarios. In order to convey any change in real time, the system uses RFID and PLCs to transform the physical processes into information flow. Receiving data from PLCs and RFID middleware, the event conveys all changes to others agents autonomously. This would realize a real-time collaborative command and control environment. In addition, by the interaction of these intelligent agents, the FMS dispatching decision is made by the IDIS automatically. Then practised, tested and verified this feasibility study of the new structure. The whole integrated structure is faster and more accurate than human decision making and through the medium data transfer for better collaboration effectiveness and quality.

2. Lecture Review

The platform consists of different service agents which coordinate each other in order to support information communictation/transference within IDIS and FMS.

2.1 Intelligent Decision Inferenc System

The inference kernel of IDIS was JESS inference engine. It was a rule-base inference mechanism to solve and support decision-making problems. The rule-based expert system (RBES) consisting of Knowledge Base (Rule

Base), Fact Base (Working Memory), and Inference Mechanism (Rule Engine) [7].

The Knowledge Base contains domain knowledge, useful for problem solving in rule formation. The knowledge is acquired from experts or documented sources. All the rules are constructed in accordance with the 'If…Then' syntactical structure, as well as with natural spoken language. Hence, the Knowledge Base starts with a rule-base, which contains all of the appropriate knowledge encoded into 'If-Then' rules.

The fact base contains a set of facts, which can be entered by users or created automatically by the system. Initially the system may not contain any data, assertions or known information (facts) which will require to be input. It examines all the rule conditions 'If' and determines a subset, a conflict set and the rules whose requirements are satisfied by the working memory. With respect to this conflict set, after the source of the conflict and the appropriate conflict resolution strategy are identified, one or more of the rules are applied. When the rule is applied, any actions specified in the rules 'then' clause are carried out. These actions can modify the working memory, the rule-base itself, or they can do just about anything else the system programmer decides to include.

The inference mechanism (Rule Engine), the interpreter of a rule-based expert system, uses an iteractive match-select-act cycling model. This loop of firing rules and actions continues until one of the two conditions is met. If there are no more rules whose conditions need to be satisfied or if a rule is fired whose action specifies the program should terminate, then the cycle stops. The cycle resumes again when another rule is applied.

The main advantages of RBES [8] are as follows: It is a more natural approach to retrieve knowledge from experts' dictations, writings and experiences; it is modular which permits simple construction; it is also easy to debug and to maintain; and it has a finite syntax and the ability to explain rules. Nowadays, RBES has been widely applied in several domains.

Java Expert System Shell (JESS) is a rule engine and scripting environment written entirely in Sun's Java^{TM} language by Ernest Friedman-Hill [9,10]. It provides the basic elements of an expert system which include a fact-list that provides a template for both knowledge and application, a knowledge base that contains all the rules and an inference engine that controls overall execution of rules. JESS has been developed from similar clip language which is also a simple and powerful scripting language, which can access all Java's APIs and allow for the building of multiple expert system applications in a single platform. JESS uses an enhanced version of the Rete algorithm to process rules. Rete is a very efficient mechanism for solving the difficult many-to-many matching problems. JESS has also become a popular expert tool for the development of intelligent systems. Figure 1.1 shows the architecture of JESS and how all the parts fit together [11], including 1) Pattern Matcher, which decides what rules and when to fire, 2) Agenda, which schedules the order in which activated rules will fire, and 3) Execution Engine, which is responsible for firing rules and executing other code.

Consequently, when executing the inference process, the rule base and the working memory, all the information is loaded into the inference engine and then compared using a pattern matcher. Then these matched rules are activated into a conflict set. Afterwards, the conflict set resolves into agenda (called conflict resolution) and, then, the execution engine fires in the agenda. The engine fires only once on any given set of facts in the working memory. The inference engine cycles recursively until all the rules are satisfied. Using JESS provides the essential capability to integrate easily with other Java programs through its well-defined API (Application interface).

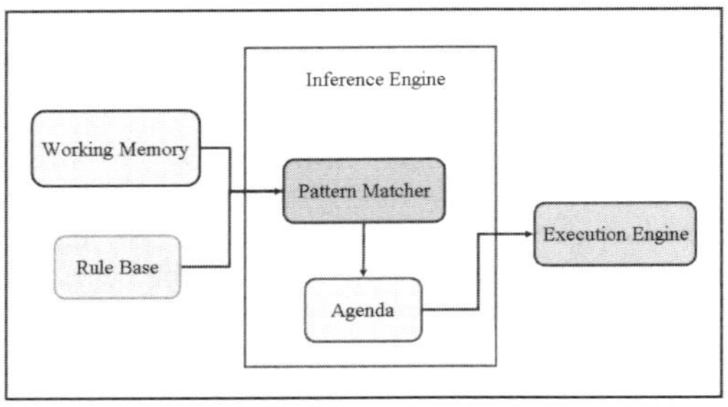

Figure 1.1. The architecture of JESS [11]

2.2 Agent Technology

The 'agent' is a software abstraction, an idea, or a concept, similar to 'OOP' (Object-Oriented Programming) technology such as methods, functions, and objects. The concept of an agent provides a convenient and powerful way to describe a complex software entity that is capable of acting with a certain degree of independence or autonomy in order to accomplish tasks on behalf of some knowledge or representation of a user or another program. But unlike 'OOP', which is defined in terms of methods and attributes, an agent is defined in terms of its behavior. Various scholars have proposed different definitions of agents, these commonly include concepts such as reactive, autonomous, goal-oriented, persistence, communicative, learning, mobile, flexible, character, and so on [12,13].

In this paper, agent technology is the most useful tool to analyze the information conveys and communicates within IDIS and FMS. An agent communication model is used to identify the interaction model among agents within different system. The use case diagram of communication mold between IDIS and FMS is shown in Figure 1.2. There are tree agents in the use case diagram, including IDIS agent, FMS agent and controller agent.

A basic scenario starts from a FMS agent announcing a task requesting FMS status. The controller agent then accepts and scans the FMS status signals. After receiving FMS information, the FMS agent starts and sends the inference request to IDIS agent. Therefore, the IDIS agent receives inference request and FMS status information, starts inference process. Until inference process completes, the IDIS agent sends this messages and inference result to the FMS agent, which receives this messages and informs the Controller agent to trigger the RGV (rail guided vehicle) real moves the pallet from current Buffer to next buffer in real time. Moreover, the FMS agent starts the FMS animation process at the same time.

When an agent received the request from the other agent as soon as did itself tasks. The IDIS agent, FMS agent and Controller agent communicates and conveys details as shown in Figure 1.3.

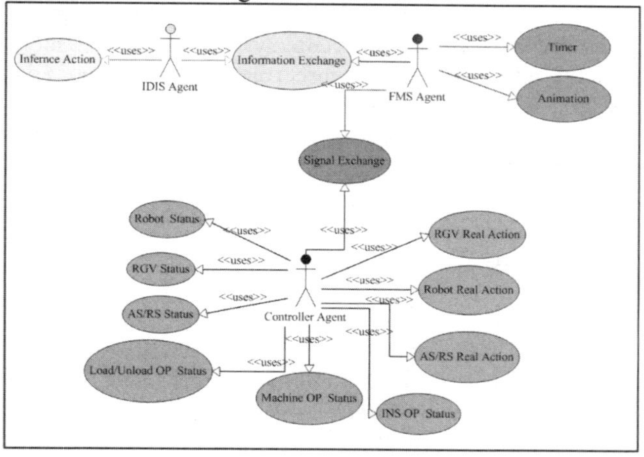

Figure 1.2. Use case diagram of the FMS and IDIS integrated

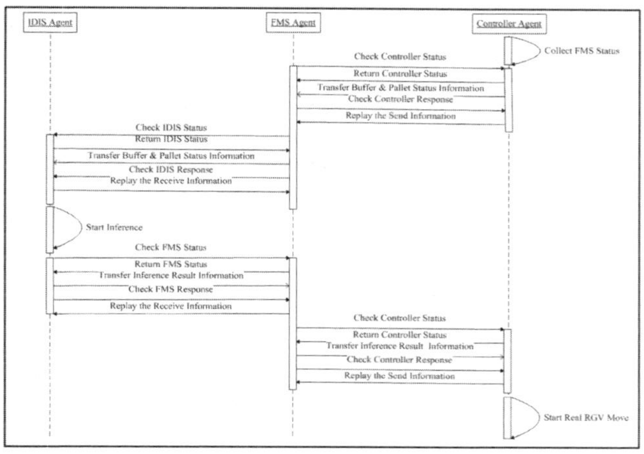

Figure 1.3. Agents' interaction model of the FMS and IDIS

2.3 Flexible Manufacture System (FMS)

In general, a FMS production system consists of computer controlled machines, automatic parts delivery and material handling systems, all of which are controlled by a supervisory computer. The machines, material handling system, and computers are very flexible, which permits the FMS to produce many different products in low volumes.

Flexible and reliable systems are the key factors. With the help of FMS, a manufacturer is able to meet market requirements and achieves high productivity. Furthermore, controlled by computers, the FMS is a combined robot, transport and organized system of complex machines, which assist in the performance of accurate automatic and rapid functions. It uses continuous manufacturing actions and can produce low cost products. Moreover, the FMS system console carries out the workflow and related functions including scheduling jobs for the machine tools, sending instructions to the automated vehicle system for transportation and downloading part-programs.

3. System Implementation

The prototype that combines the IDIS and FMS using the agent technology. It is designed and implemented to substitute for the job-dispatching function of FMS using the inference mechanism in the manufacturing environment. The integrated solution, with its structure and methodology is used in the modeling process which incorporates the knowledge base is as follows.

3.1 System Architecture

The main concept is that the FMS dispatching rules can easy to be added, edited and modified via IDIS' tools and user-friendly interface. Furthermore, information are communicated and conveyed within IDIS and FMS via IDIS agent, FMS agent and Controller agent to support job-dispatching decisions in the manufacturing environment.

The test platform system consists of four Computer Numerical Controlled (CNC) machines, two robots, one inspection station, an automated storage / retrieval system (AS/RS), a rail guided vehicle (RGV) for transporting, and apparatuses for loading/unloading [14]. This FMS test platform as an illustration platform to test and verify the integrated method and structure of the solution (as shown in Figure 1.4).

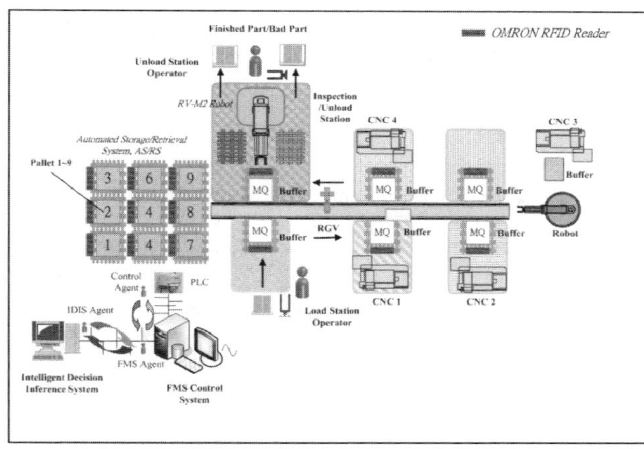

Figure 1.4. Layout of the FMS and IDIS

In this structure includes three agents, these agents are based on the Java Agent Development Environment (JADE) [15]. JADE simplifies the implementation of multi-agent. The system architecture of 'IDIS agent', 'FMS agent' and 'Controller agent' within IDIS and FMS base on JADE are shown on the Figure 1.5. The 'IDIS Agent' handled the input/output of the IDIS inference and communication/transferred information with 'FMS Agent'. The 'FMS Agent' dealt with the FMS status, part, process and communication/transferred information with 'FMS Agent' and 'Control Agent'. The 'Control Agent' coped with the real-time production process information collection and communication/transferred information with 'FMS Agent'.

3.2 Case Study

The whole process of operation in this solution as follows: When the system begins, the 'FMS Agent' informs the 'Controller agent' to scan FMS status signals then sends the information and request the 'IDIS agent' to trigger IDIS inference process. When the IDIS produces inference result, the 'IDIS agent' sends the inference result and informs 'FMS agent' to starts the FMS animation. And that requests the 'Controller agent' to trigger RGV real moves the pallet form current buffer to next buffer in real time. The FMS controller real time display of the RGV movement path via the animation is also provided (see the Figure 1.6) [1]. In this way, the entire operation can be monitored. Figure 1.7 shows the photograph of the operation of the test platform at the FMS laboratory of the NTUT [1].

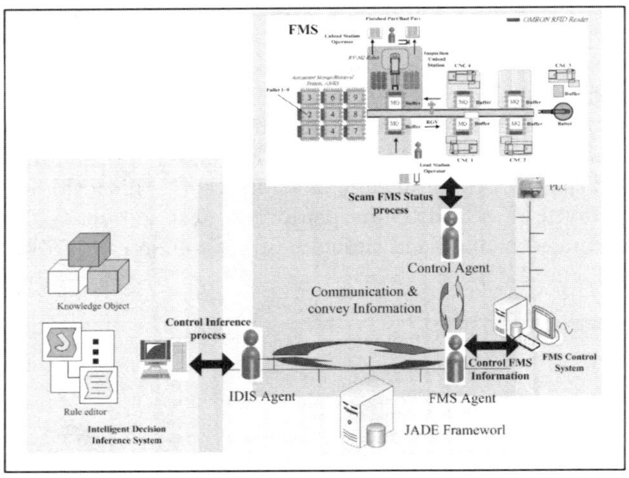

Figure 1.5. The system architecture of agents

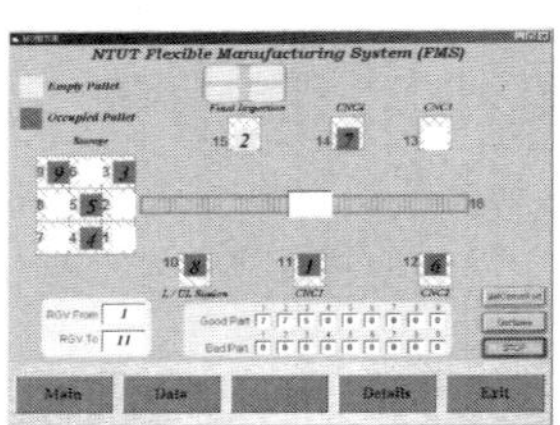

Figure 1.6. An animation of the FMS in progress [1]

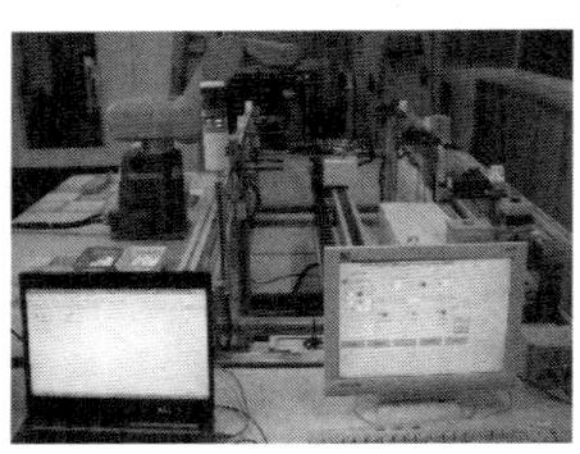

Figure 1.7. The scene of the whole system [1]

4. Conclusion

This paper present an application integration environment to enhance the FMS operation efficiency. IDIS has tested in FMS and performance analysis shows the system has achieved its dispatch objectes. The design and implementation of this project is a successful integration of the following key elements 1) through agent framework of coordination for dispatch planning and execution. 2) Tools: Jess, JESS, JGraph, Java, and agent technologies. 3) Unique methodology that support dynamic engineering knowledge managemenst.

There are three major contributions in the research. 1) this research presents a systematic process to establish the decision model of IDIS to support FMS dispatching. 2) this paper demonstrates a methodology to convey and commnicate information within IDIS and FMS via agents. 3) in the user aspect, the IDIS system provides a more friendly interface to configure rules. In the future, the research will focus on the development of extended applications of agent and IDIS. Especially, the multi-level agent applies in IDIS.

References

[1] Kuo, C.L., Ku, C.C., Tsai, J.P., Hung, Benson C. W., Chang, W.C. and Chen, K.Y., 2006, Development of Rule Base Inference Oriented FMS Dispatcher Using the JESS Platform, 36th International Conference on Computers and Industrial Engineering, June 20-23, Taipei, ROC.

[2] Chiang, A.T.A., Trappey, A.J.C., and Ku, C.C., 2004, Using knowledge-based intelligent reasoning to support dynamic collaborative design, Proceedings, APIEMS, December 12-15, Gold Coast, Queensland, Australia.

[3] Yao, Y.-H., Trappey, A.J.C., Ku, C.C., Lin, Gilbert Y.P., Tsai, J.P., Ho, P.-S., and Hung, Benson, 2005, Develop an intelligent equipment maintenance system using cyber-enabled JESS technology, International Conference on Hands-on Intelligent Mechatronics and Automation, IEEE, July 10-12 2005, Taipei, Taiwan., pp. 927-932.

[4] Chiang, A.T.A., Trappey, A.J.C., and Ku, C.C., 2005, Using knowledge-based intelligent reasoning to support dynamic collaborative design, International Journal of Advanced Manufacturing Technology (Submitted.)

[5] Trappey, A. J. C., Trappey, C., Lin, Gilbert Y.P., Ho, P.-S., Ku, C.C., and Kuo, C.L., 2006, Design an Integrated Intelligent Equipment Maintenance and Production Dispatching System Using the JESS Platform, Proceedings, International Conference on WECAM 2006 , July 11-14, Gold Coast, Australia.

[6] Chen, Kai-Ying, Luo, Peter Y. F. and Tseng, Wei-Rung, 2006, Modular PAC Software Design Based on Color Petri Net, 2006 IIE Annual Conference and Exposition.

[7] Yao YH, Trappey AJC, Ku CC, Lin GYP, Tsai JP, Ho PS & Hung Benson. (2005) Develop an intelligent equipment maintenance system using cyber-enabled JESS technology. Proceedings, International Conference on Hands-on Intelligent Mechatronics and Automation, IEEE, July 10-12 2005, Taipei, Taiwan., 927-932.

[8] Jess. the Rule Engine for the Java Platform. (2006) Retrieved January 5, 2006, from http://www.jessrules.com/jess/docs/index.shtml

[9] Ernest J. Friedman-Hill, Jess, The Rule Engine for the Java Platform, http://herzberg.ca.sandia.gov/jess/, 21 November 2003.

[10] Ernest Friedman-Hill, "JESS IN ACTION—Rule-Based Systems in Java," MANNING, 2003.

[11] Reichgelt H. (1991) Knowledge representation, an AI perspective. Norwood, NJ: Ablex Publishing Corporation.

[12] Stan Franklin and Art Graesser 1996, Is it an Agent, or just a Program? A Taxonomy for Autonomous Agents, Third International Workshop on Agent Theories, Architectures, and Languages, Springer.

[13] Wooldridge, Michael and Nicholas R. Jennings (1995), Agent Theories, Architectures, and Languages: a Survey, in Wooldridge and Jennings Eds., Intelligent Agents, Berlin: Springer, 1-22

[14] Chen, K. Y., 1997, Design of shop floor control systems, Journal of the Mechatronic Industry, 177, 151-160.

[15] JADE. JADE Programming Guide, Documentation, (2006) http://jade.tilab.com/doc/index.html

Investigation of Machines Sequencing Flexibility in Traditional and Flexible Manufacturing Systems (FMS) Using Genetic Algorithms (GA).

S.Noor[1], M.K. Khan[2], I.Hussain[3]

[1]NWFP, University of Engineering & Technology, Pakistan, saharnoor@gmail.com
[2]University of Bradford, United Kingdom, m.k.khan@bradford.ac.uk
[3]NWFP, University of Engineering & Technology, Pakistan, ihussain_pk@yahoo.com

Abstract. Machine Sequencing Flexibility (MSF), a measure of alternate machine sequences in which the operations of a job can be performed, is very important especially in the context of today's competitive manufacturing environment where responsiveness to the market determines the survival of the industries. In a scheduling scenario where ***N*** numbers of jobs are to be scheduled on ***M*** number of machines and each job has to visit each machine, the problem becomes an Open Shop problem if MSF is 100%.
In this research, a scheduling methodology, based on GA, is proposed which integrates planning (machine sequencing) with shop floor scheduling (job sequencing) assuming 100% MSF. The methodology works for both Traditional Manufacturing (TM) and FMS. It takes the machine sequences for ***N*** number of jobs as chromosomes (process plan). The chromosomes of the given number of population size are evaluated for shop floor scheduling using a heuristic to sequence the jobs for the machine loading. A solution space of the chromosomes is searched through the use of GA for the best/optimal chromosome (machines sequence) for minimization of the Makespan. Various benchmark problems have been solved through the proposed scheduling methodology in TM and FMS environments and the results have been compared with other research results. These comparisons indicate that the presented GA-based scheduling methodology has performed extremely well.

Keywords: Genetic Algorithms, Machine Sequencing Flexibility, FMS Scheduling.

1. Introduction

Machine Sequencing Flexibility (MSF) is very important in the context of today's competitive manufacturing environment where responsiveness to the market determines key performance of manufacturing organizsation. The MSF exists both in Traditional as well as Flexible Manufacturing environments.

To understand the concept of MSF, consider Figure 1 which shows both FMS and TraditionalManufacturing (TM) shops. In TM systems, Figure 1(a), a part having four operations and where the precedence relations between the operations of the part exist {O1, O2, O3, O4}, the only machine sequence for the part will be {M1, M2, M3, M4}. If there is no precedence relation between the operations of the part, then there are 24 machines sequences (4!). This will be the case where 100% MSF is assumed. This sequencing flexibility is inherent in the job structure [1].

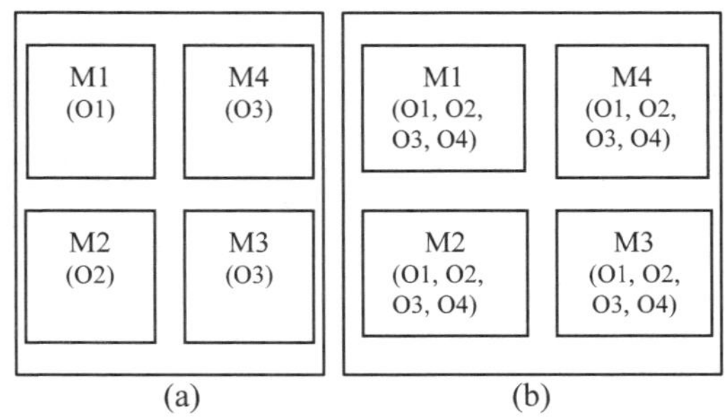

Figure 1.1. Comparison between (a) Traditional Manufacturing and (b) FMS

If there are many operations in a job, it becomes difficult to have 100% MSF because of unavoidable precedence relations between some operations. The sequencing flexibility can be realistically improved without changing the precedence relations of operations if identical flexible machining centers are used in FMS. The use of identical flexible machining centres has been driven by the rapid growth in demand of quantity, quality and variety of products and is considered as one of the key ways to achieve organizational flexibility [2]. FMS can have a set of identical machines, as in Figure 1(b), which are highly automated and can perform any operation on any part in contrast to the traditional job shop manufacturing. The same part as discussed earlier can have 24 machines sequences without relaxing the precedence relations {O1, O2, O3, O4}. For example, if there is a machines sequence {M2, M3, M1,M4},the part will go to M2 for operation O1, then go to M3 for O2, than M1 for O3 and finally M4 for O4. It can also follow a machines sequence {M1, M2, M4, M3} without relaxing the precedence relations between the operations.

From the above discussion, it is clear that MSF exists both in conventional and modern manufacturing systems such as FMS and the problem of operational scheduling becomes as complex as the Open Shop Scheduling Problem (OSSP), where ***N*** numbers of jobs are to be scheduled on ***M*** number of machines and each job has to visit each machine, if MSF is 100% [3,4]. This scheduling scenario is more complex than the Job Shop Scheduling Problem (JSSP) since it has a wider solution space.

Operation scheduling consists of sequencing of machines for a job (process planning) and job sequencing on machines (shop floor scheduling) and these two steps are normally not considered together. The processing planning stage affects the shop floor scheduling performance, especially when there is an inherent possibility of alternate process plans or machines sequencing flexibility. Many researchers have urged the integration of these two stages [5,6].

In the present research, a scheduling methodology, based on a GA approach is proposed which integrates planning (machine sequencing) with shop floor scheduling (job sequencing) assuming 100% MSF, both for traditional manufacturing and as well as FMS. The proposed methodology is applied to a number of benchmark scheduling problems [6] and a comparison of the results is made to ascertain the performance of the proposed methodology.

2. Methodology

The GA-based methodology proposed by [5], as a Hybrid Scheduling Tool (HST), is modified and extended. The problem under consideration is tackled as OSSP which is similar to the JSSP except that there is no precedence relation between the sequences of machines (operations) for jobs. Figure 2 illustrates the three main stages i.e. Stage 1- Creation of Process Plans, Stage 2- Evaluation of Makespan for process plans and Stage 3- the Evolution of of the best Process Plans.

Stage -1

Stage-1 is the planning stage. In this stage with 100% MSF, a process plan or chromosome is represented by a matrix of permutations of number of machines where the row indices indicate the part (job)'s number and the column indices show the operation's number. The integer inside the cell represents the machine number on which the operation of the part will be performed. For example, if there are three jobs, three machines and each part has to visit each machine, one of the possible sets of sequences of machine for the parts (Process Plan) is shown in the Table 1 of Figure 1.2. Such permutation type chromosomes (Process Plans) are randomly initialized as per desired size of population of chromosomes.

Stage -2

In this stage, the chromosome (semi active schedule) is transformed (evaluated) to an active schedule through an algorithm/heuristic where the first operation of Job 1 (J1) is scheduled first, then second and so on for all parts. Each operation under consideration is allocated to the earliest available position on the corresponding machine. This process is continued until all the operations of all jobs are scheduled. The algorithm takes a given population of chromosomes and evaluates their Makespans.

Stage-3

In this stage, the Makespan of a chromosome, obtained at Stage-2, is used as an objective function of the GA evolution process. The Makespan values for a population size are scaled to obtain the fitness values for the selection of parents using a ranking function from the GA Tool Box [9]. The selection of parents from a population pool can be achieved using variety of selection methods. In this case, the Stochastic Universal Sampling (SUS) function is used for the selection of parents to be crossed over and mutated for reproduction of the off- springs [9].

The mutation and cross over processes have been adopted as proposed in [5]. The Cross over of chromosomes of the problem is to interchange the rows of the parent chromosomes at a single random cutting point. For example, in Table 1.3, there are two chromosomes and let us suppose that there is a random cross over cutting point equal to Row 3.

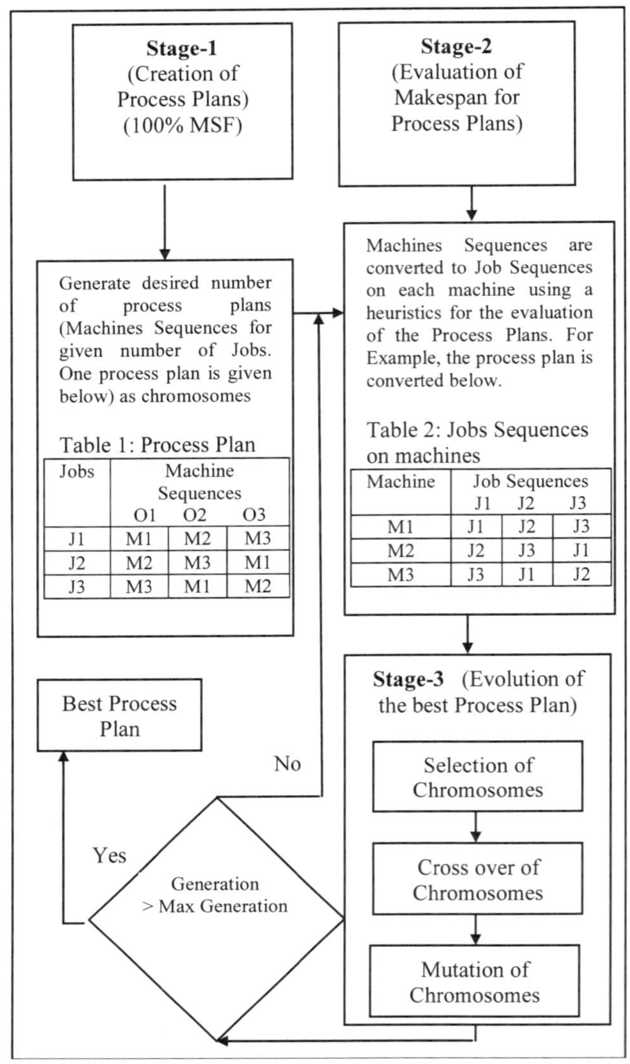

Figure 1.2. Solution Approach to the Scheduling Problem

In Table 1.3, the bold integers in Chromosome 1 are above the random cross over point at Row 3 and the underlined integers are below the cross over point. Similarly, for Chromosome 2, the integers of smaller font size are above the cross over point, and bold italic underlined and larger font size integers are below the cross over point in Chromosome 2. In the cross over of the parent chromosomes, the lower parts of the chromosomes (Rows 4 and 5) will be interchanged as shown in Table 1.4.

In the mutation process, a random cell, but not the last cell of a chromosome, is selected and the integer of the cell is interchanged with the integer of the next cell. For example, in Table 1.5, the bold undelined Integer 1 in Row 1 is selected and interchanged with the Integer 3 in the next cell of the same row. This process is repeated for each row of a chromosome.

Table 1.3. Parent Chromosomes

Parent Chromosome 1				Parent Chromosome 2			
2	**4**	**1**	**3**	1	3	4	2
4	**2**	**1**	**3**	4	3	1	2
2	**1**	**3**	**4**	4	1	3	2
3	2	1	4	***2***	***3***	***4***	***1***
4	1	3	2	***1***	***2***	***4***	***3***

Table 1.4. After Cross Over

Parent Chromosome 1				Parent Chromosome 2			
2	**4**	**1**	**3**	1	3	4	2
4	**2**	**1**	**3**	4	3	1	2
2	**1**	**3**	**4**	4	1	3	2
2	***4***	***4***	***1***	3	2	1	4
1	***2***	***4***	***3***	4	1	3	2

Table 1.5. Chromosome before and after Mutation

Parent Chromosome 1				Parent Chromosome 2			
2	4	**1**	3	2	4	3	**1**
4	**2**	1	3	4	1	**2**	3
2	1	3	4	1	**2**	3	4
2	***3***	*4*	*1*	*2*	*4*	***3***	*1*
1	***2***	*4*	*3*	*1*	*4*	***2***	*3*

3. Results and Discussion

The performance of the methodlogy was compared against a variety of forty two benchmark Job Shop Scheduling (JSS) Problems i.e la01-la35, abz7-abz9 and yn1-yn4 under the GA parameters of Cross over Rate =0.8, Mutation Rate =0.1, Population Size 150 and Max Generation =200 with 100% MSF assumption. The percentage solution gaps for the lower bound measure are calculated for both TM and FMS and compared with the results for the same problems reported by [5] with zero MSF (JSS enviroment). The methodology has not only achieved the best/optimal Makepsan for all the 42 problems easily, but has also shown improvement in terms of the Makespan.

Figure 1.4 shows those 22 JSS benchmark problems which have shown imrovement in the Traditional Manuafacturing enviroment for their Makespan values. For these 22 benchmark problems, the percentage Makespan improvement ranges from approximately 2% to a maximum of 24%. The Makespan of the benchmark problem la16, for example has been improved from 945 to 718 units of time which makes about 24% improvement.

As discussed earlier that 100 % MSF in TM may not be realistic and the portrayed improvements might not be materialized. However, it has been observed that multiple solutions (multiple process plans) of the same Makespan for a problem can be obtained which can be an opportunity for the designer of the operations to choose one of the best solutions or process plans in terms of technological feasibility at the planning stage. There might be a process plan which can be easily implemented with a little effort of changing the sequence of operations for the product, but it will still be hectic to accrue all of the possible benefits of assumption of 100% MSF in TM. This is why the quicker response to the market in TM is hard to achieve, hence, the advantage with the FMS to meet the requirements of modern day competition.

Figure 1.5 shows the potential improvements by the proposed methodlogy in the FMS environment.The 100% MSF assumption in FMS scheduling is more realistic as compared to the TM because of the nature of FMS. The methodology shows improvement in the Makespan values of **all** the forty two benchmark problems in the range of 2.25% to 24 %. The optimal Makespan of the problem lao1, for example which was 666 units of time in the pure JSS (no MSF), has been reduced to 574 value in the FMS environment, which is about 13% improvement.

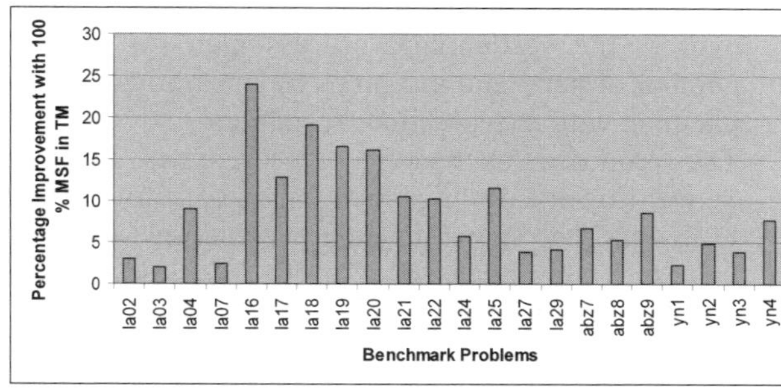

Figure 1.4. Potential improvement by the proposed methodlogy in Traditional Manufacturing with 100% MSF

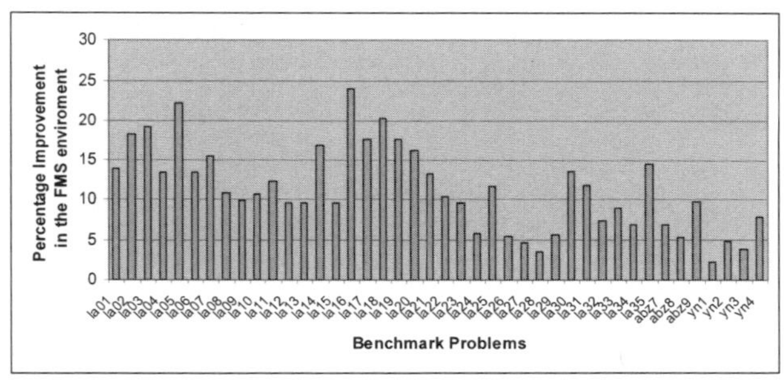

Figure 1.5. Potential improvement by the proposed methodlogy in the FMS environment.

Comparing the solutions of the methodology in TM and FMS environment, Figure 1.6 shows the percentage improvement of the Makespan values in the FMS environment over the TM which ranges between 0.15% to 22.25%.

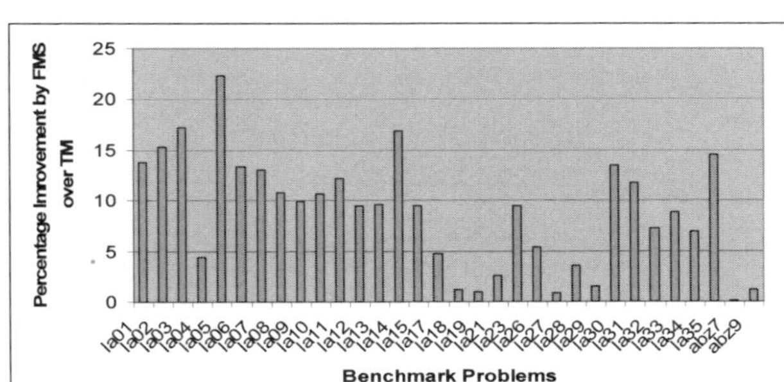

Figure 1.6. Potential improvement by the proposed methodolgy in the FMS environment over TM.

The performance of the proposed methodology in terms of the quality of the solutions is satisfactory but its final solution is expected to be further refined by incorporating some local search techniques as proposed in literature, for example, Tabu Search [10].

The CPU solution time has not been measured in the computer runs of the proposed methodology for the given set of the problems but it can be easily icorporated in the methodology for obtaining the CPU solution time in future work.

4. Conclusion

1. A GA-based methodology has been developed which evolves the best/optimal process plan for a given number of parts and machines by integrating process planning with the shop floor scheduling.
2. The proposed methodology incorporates Machine Sequencing Flexibility,which exists both in TM and FMS and has shown an improvemnet in Makespan values in the range of 2% to 24% for 42 JSS benckmark problems.
3. Incorporating MSF in FMS is more realistic and has shown improvement when compared to TM, in the range of 0.15% to 22%.
4. The performance of the methodology is expected to be improved further if its final solution is refined by local search techniques such as Tabu Search.

References

[1] Rachamadugu R.,Nandkeolyar U., 1993, Scheduling with Sequencing Flexibility, Decision Sciences, Volume 24, Number 2, P315-341.

[2] Chan F.T.S, Chan H. K. and Lau H. C .W., 2002, The state of the Art in Simulation Study on FMS Scheduling: A comprehensive Survey, the International Journal of Advanced Manufacturing Technology, 19, P830-849.

[3] Gonzalez M.J. and Sahni S., 1976, Open Shop Scheduling to minimize finish time, Journal of the Association for Computing Machinery, 23, P655-679.

[4] Bitran G.R., Dada M., Sisan L.O., 1982, Simulation Model for Job Shop Scheduling (working paper). A.P. Sloen School of Management, Massachusetts Institute of Technology.

[5] Noor S, Khan M.K., Hussain I, Ullah Irfan , 2006, Scheduling Tool for the Flexible Manufacturing Systems using Hybrid Genetic Algorithm, 2nd International Conference on Emerging Technologies (IEEE), Pakistan 13-14 November.

[6] Jain A. S. and Meeran S., 1999, Deterministic Job-Shop Scheduling: Past, Present and Future, European Journal of Operational Research 113, P390-434.

[7] Geyik Faruk., Cedimoglu Ismail Hakki, 2004, A Review of the Production Scheduling Approaches Based on Artificial Intelligence and the Integration of Process Planning and Scheduling ",http://www1.gantep.edu.tr/~fgeyik/SwissCadCam99.pdf, August.

[8] Kumar M., Rajotia S., 2003, Integration of Scheduling with Computer Aided Process Planning", Journal of Materials Processing Technology, 138, P297-300.

[9] Chaperfield A., Flemming P., Pohlhein H., Fonseca C., 2001, Genetic Algorithm MATLAB Tool Box –User's Guide" Version 1.2, Department of Automatic Control and Systems Engineering, University of Sheffield.

[10] Morshed M.Sarwar, 2006, A Hybrid Model for Job Shop Scheduling, PhD Thesis, University of Birmingham, UK.

10

Metal Cutting

Effect of Tool Coatings on the Performance of Small-diameter Drills in Drilling Deep Holes with Higher Cutting Speed

R. Heinemann[1] and S. Hinduja[2]

School of Mechanical, Aerospace and Civil Engineering, The University of Manchester, Sackville Street, Manchester, M60 1QD, UK
[1]robert.heinemann@manchester.ac.uk, [2]sri.hinduja@manchester.ac.uk

Abstract. Machining with higher cutting speeds is one of the most welcome advances in metal cutting today, since it is, among other benefits, associated with a significant reduction in process time. However, certain prerequisites are inevitable in order to make a particular process using certain cutting tools eligible for realising higher cutting speeds, because the rise in cutting speed is usually associated with increased cutting temperatures, very often resulting in a substantial decrease in tool life. This paper investigates the extent to which tool coatings can enable small diameter Co-HSS twist drills to achieve a satisfactory performance in deep-hole drilling of carbon steel using minimum quantity lubrication. It is shown that tool coatings that combine a high hot-hardness with a smooth surface are very important in order to ensure a satisfactory tool life of the Co-HSS twist drills employed with higher cutting speeds.

Keywords: Small diameter twist drills, deep holes, high-speed cutting, tool coatings, tool life, chip evacuation

1. Introduction

Apart from the benefits related to an improvement in the surface and geometrical integrity of machined components, high-speed cutting (HSC) is considered as one of the key technologies to gain higher productivity and throughput [1]. Especially in drilling, which still represents a time-consuming and, usually, the cycle-time determining machining operation in linked manufacturing [2], any reduction in the cycle time is strongly appreciated. Because of the disadvantageous process characteristics (e.g., enclosed cutting point, slender and weak cutting tools), the high-speed range in drilling already starts at twice the conventional cutting speed [3]. But even in spite of this comparatively moderate increase, high-speed drilling is associated with a notable rise in cutting temperature, amplifying the heat load on the drilling tools and accelerating tool wear [4]. Therefore, mainly drills made of heat and wear resistant tungsten carbide are employed [3].

However, recent research [5] has demonstrated that because of their brittleness, tungsten carbide drills are less suitable for drilling deep holes, so that for this application HSS drills still have to be considered as the first choice. In order to employ these heat-sensitive HSS drills in HSC, the use of appropriate tool coatings is proposed to compensate for the lack in hot-hardness [6,7].

So far, however, research work has predominantly concentrated on fairly large diameter drills drilling holes with a small depth-to-diameter ratio. The aim of this work, therefore, is to investigate in how far state-of-the-art small diameter HSS twist drills can indeed be used for deep-hole drilling with higher cutting speeds. In this context, the effect of the cutting speed on the deep-hole drilling capability, cutting forces, tool life and wear behaviour of selected twist drills is assessed. The results obtained are then compared with those achieved under similar circumstances but with conventional cutting speed, as reported in [8].

2. Experimental Details

The cutting tests were carried out on a vertical CNC milling machine, equipped with a 2.6 kW high-precision spindle with a maximum speed of 40,000 rpm. Plain carbon steel with a carbon content of 0.45% (C45/1.0503) was used as the workpiece material. It was cut into plates of dimensions 75×65×15 mm, which were fixed into a vice mounted on top of a Kistler dynamometer (9271A), see Figure 1.1. This dynamometer was used to measure torque and thrust force while drilling through holes with a depth of 15 mm into the workpiece plates.

Figure 1.1. Experimental setup

All drilling tests were carried out with externally supplied minimum quantity lubrication (18 ml/h), using a fatty acid ester, that contained 20% alcohol in order to improve its cooling capability. All boreholes were pre-centered in order to minimize the effects associated with heavy tool run-outs. The centre holes had a diameter of 1.55 mm and a depth of 3 mm.

The twist drills that were tested are listed in Table 1.1. All of them are of diameter 1.5 mm. They also had a 130° point angle, enlarged chip flutes, a 40° helix angle and a thinned web in order to improve chip evacuation and reduce cutting forces respectively [9,10].

In order to ensure comparability between the different twist drill types, i.e. uncoated as well as coated ones, a common set of cutting parameters was used for all the tests and drills. Based on the recommended conventional cutting speeds for each individual type of twist drill, average values of 26 m/min and 0.26 mm/rev were calculated for the cutting speed and feed rate respectively. The latter resulted in a feed of 143 mm/min. Except for the uncoated drills, these cutting parameters are within ±10% of the values recommended by the drill manufacturers. Since only Co-HSS twist drills were used, the high-speed cutting speed was set to 52 m/min, i.e. twice the conventional speed. The feed rate was kept at 0.026 mm/rev, resulting in a feed of 286 mm/min.

3. Results and Discussion

The torque curves measured while drilling with the conventional cutting speed of 26 m/min are shown in Figure 1.2(a). Doubling the cutting speed to 52 m/min resulted in the torque curves shown in Fig. 1.2(b). Each curve shown in this figure is the average of curves recorded from four twist drills, each of which drilled seven consecutive boreholes. The forces were measured at a tool life of approximately 15 to 20%, thus eliminating any significant effect that the running-in period or extensive tool wear might have had.

At borehole depths less than 3 to 4 times the diameter, effects associated with chip evacuation can almost be neglected, because of the short distance the chips have to travel along the flutes. Frictional forces, e.g. rubbing of the drill's margins along the borehole wall, can also be considered to be comparatively small. Hence, the values given for this depth principally represent the cutting force associated with the drilling process. They can, therefore, provide accurate information whether the increase in cutting speed leads to any change of the cutting force. With increasing borehole depth, however, the chip flow gets increasingly obstructed because of the increased distance the chips have to travel along the chip flutes, which is associated with a rise in frictional force between the tool, chips and workpiece. The torque necessary to overcome this friction is superimposed on the torque required to cut the workpiece material, i.e. to drill the hole. Therefore, the overall torque provides information whether increasing the cutting speed has any effect on the chip disposal process in deep-hole drilling. The thrust force, in contrast, was a very poor indicator for the chip flow, and is therefore not discussed in this paper.

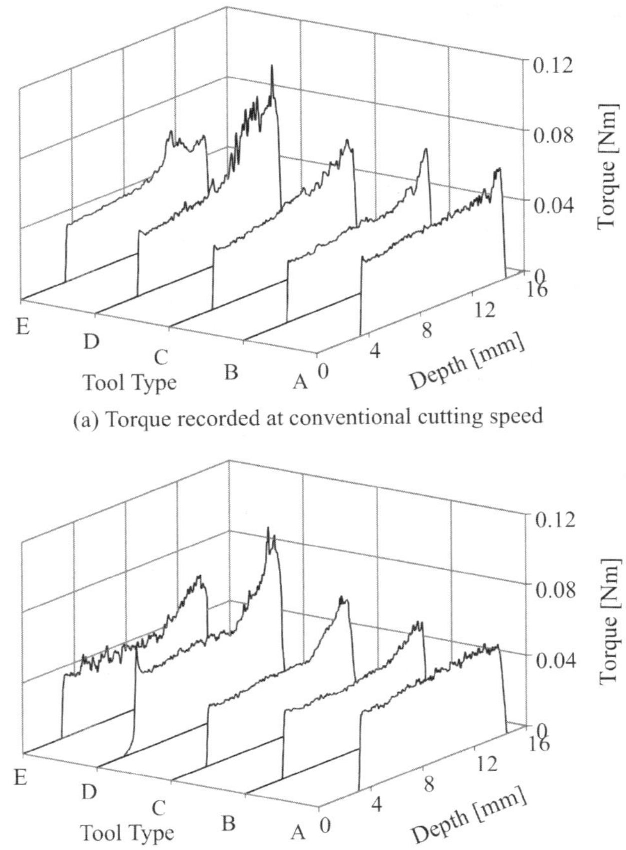

(a) Torque recorded at conventional cutting speed

(b) Torque recorded at higher cutting speed

Figure 1.2. Torque in relation to borehole depth

Generally speaking, the comparatively moderate increase in the torque from the initial value to the maximum value recorded demonstrates that all the tested twist drill types are suitable for drilling holes of 10 times the diameter. However, comparing the drills with each other reveals that the TiCN-coated drills exhibited the worst deep-hole drilling performance, which manifested itself in the strongest increase in torque. This is mainly attributable to the rough surface topography of its chip flutes, which obstructed the chip flow through the drill's flutes. Comparing the surface of the chip flutes of the different drill types employed, see Figure 1.3, clearly shows that the TiCN-coated drills show the by far roughest surface, characterized by a large number of macro-particles, so called droplets. All other drills showed

Table 1.1. Details of the twist drills tested

Tool type	Manufacturer / tool code	Tool coating	Coating thickness	Coating hardness	Coating process / post-coating surface treatment
A	Titex / A1222	uncoated	-	-	-
B	Titex / A1222	TiN	~ 2.4 µm	~ 2300 HV	filtered arc-evaporation / polished
C	Titex / A1249	TiAlN multilayer	~ 2.8 µm	~ 3400 HV	filtered arc-evaporation / polished
D	Guehring / GT100	TiCN	~ 3.5 µm	~ 3000 HV	arc-evaporation
E	Guehring / GT100	TiAlN multilayer	~ 2.9 µm	~ 3300 HV	filtered arc-vaporation / polished

a much smoother surface.

The main reason for this difference in surface roughness is the use of standard arc-evaporation coating technology in case of the TiCN-coated drills, which is prone to droplet formation, compared to the more advanced filtered arc-evaporation, where droplet formation is strongly reduced. In the case of the TiAlN-coated drills, post-coating polishing was applied to further smoothen the coating's surface by removing many of the droplets that still have formed.

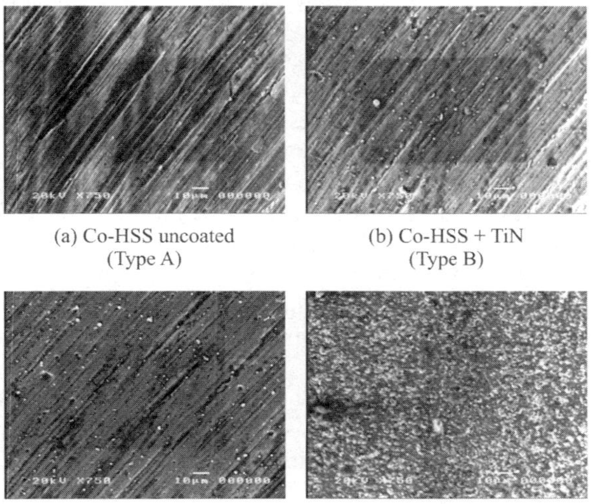

(a) Co-HSS uncoated (Type A)
(b) Co-HSS + TiN (Type B)
(c) Co-HSS + TiAlN (Types C and E)
(d) Co-HSS + TiCN (Type D)

Figure 1.3. Scanning Electron Microscopy (SEM) scans of the chip flute surfaces

Doubling the cutting speed to 52 m/min resulted in no significant change in the torque recorded, neither at initial borehole depth nor the maximum value within each drilling cycle. This indicates that doubling the cutting speed did not cause the chip disposal to deteriorate notably for the twist drill types tested.

Although in general, doubling the cutting speed did not give rise to a change in the torque or a deterioration in the swarf disposal, it has a notable effect on the tool life of all the twist drill types. According to Figures 1.4, which gives the average tool life and the tool life scatter, the tool life decreased significantly at the higher cutting speed. With respect to the uncoated drills of type A, the tool life dropped by 85%. At the conventional cutting speed, the drill tip of the uncoated drill exhibited only very moderate tool wear after drilling 50 boreholes, as can be seen in Figure 1.5. When the cutting speed was doubled, however, the same drill type exhibited strongly abraded cutting edge corners and substantial flank wear after drilling only 25 boreholes. In addition to this, the entire tool tip shows a notable discoloration, which suggests that the drill was subject to an enhanced heat load during cutting. As a consequence, the tool substrate's wear resistance was strongly reduced, making the drill tip more susceptible to abrasive and adhesive wear, manifesting itself in flank and corner wear.

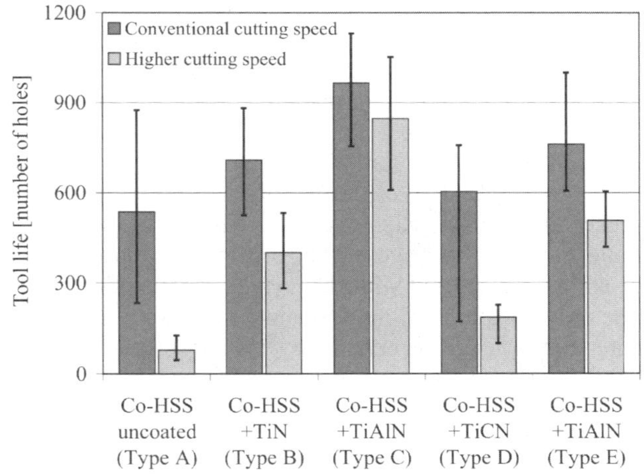

Figure 1.4. Tool lifes recorded with conventional and higher cutting speed

The TiN-coated drills performed considerably better, and reached an average tool life of 401 boreholes. This can be attributed to, firstly, the friction reducing property of the TiN-coating [11,12], resulting in a shorter contact length between tool's rake face and chip and, therefore, a reduced heat flow between chip and tool. Secondly, the coating has a higher wear resistance compared to HSS, which protects the substrate against abrasive wear.

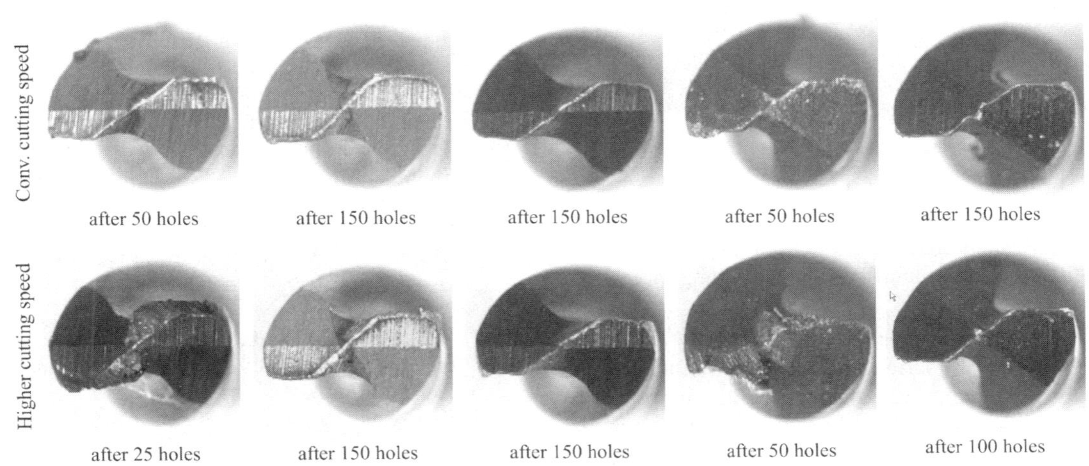

(a) Co-HSS uncoated (Type A)
(b) Co-HSS +TiN (Type B)
(c) Co-HSS+TiAlN (Type C)
(d) Co-HSS+TiCN (Type D)
(e) Co-HSS+TiAlN (Type E)

Figure 1.5. Wear pattern for different twist drill types at 20% of their tool life using conventional and higher cutting speed

However, after drilling 150 boreholes, the coating along the cutting edges and the chisel edge was completely removed. The most likely reason for this accelerated wear progression is the limited hot-hardness of TiN combined with the increased temperature at the cutting point.

Although TiCN has a much higher hardness at room temperature and a similar hot-hardness as TiN, the TiCN-coated drills suffered from their relatively poor chip evacuation capability. Because of the obstructed chip flow, some of the heat from the chips was conducted into the workpiece and tool, thus further increasing the drill tip temperature and accelerating tool wear. The increased cutting temperature during high-speed drilling brought about an increasingly severe thermal load on the tool. As Fig. 1.5 shows, after drilling 50 boreholes, the cutting edges already exhibited extreme flank and corner wear, which was not observed when drilling at the conventional cutting speed.

A considerably better deep-hole drilling capability, i.e. a strongly improved chip evacuation, and a notably higher hot-hardness of the coating caused the TiAlN-coated drills to reach an average tool life of 847 (type C) and 507 boreholes (type E) respectively. At the higher cutting speed, the TiAlN-coated Co-HSS twist drills showed the same moderate tool wear after drilling 150 boreholes as when they were used with the conventional cutting speed. The only notable wear mark is some minor rounding at the outer cutting edge corners, since this part of the tools was operating with the highest cutting speed and was, therefore, subjected to the highest temperature during the drilling process.

Near the end of their tool life, however, all drills showed excessive flank wear and a strong rounding of the cutting edges corners, as the photographs in Figure 1.6 show. Of the three drills depicted, the TiCN-coated one shows the most serious wear. Its cutting edges are almost completely abraded, and its chisel edge is entirely deformed, showing strong discoloration. Since the coating had been completely removed from these parts, the tool substrate below was then directly exposed to the cutting temperature.

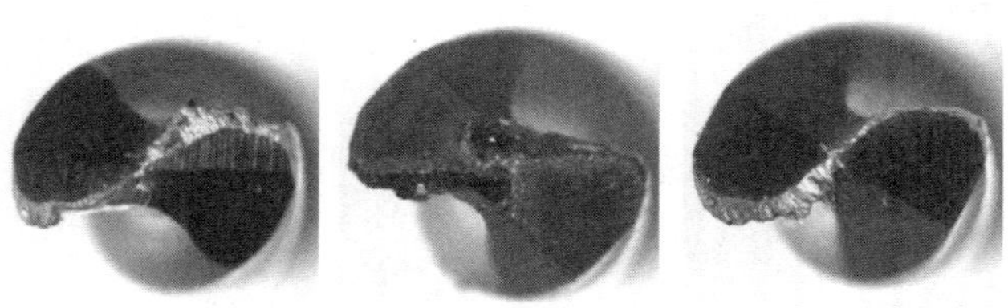

after 600 holes	after 200 holes	after 450 holes
(a) Co-HSS +TiAlN (Type C)	(b) Co-HSS+TiCN (Type D)	(b) Co-HSS+TiAlN (Type E)

Figure 1.6. Wear pattern for different twist drill types at the end of their tool life using conventional and higher cutting speed

The TiAlN-coated drills also show strongly rounded cutting edge corners, serious flank wear, and heavy microwelding. But, unlike the other drills, the unprotected tool substrate along the cutting lips exhibits only a very light discoloration. Since the rake faces still showed an almost intact TiAlN-coating, the tool substrate was still protected from the hot chips moving along the rake face. As a consequence, only a small percentage of the heat stored within the chips was actually conducted into the tool, resulting in the tool tip undergoing a relatively smaller increase in temperature than the other drills.

4. Conclusion

From the results presented in this paper, the following conclusions can be drawn:

- Doubling the cutting speed did not deteriorate the chip evacuation capability of the small diameter twist drills tested.
- By applying a suitable coating, which partly compensates for the lack in hot-hardness of the tool substrate, Co-HSS twist drills can be employed in (deep-hole) drilling using higher cutting speeds.
- One of the key features of these coatings is a smooth surface, so that the chip flow through the drill's flutes does not get obstructed. Hence, the use of coating techniques that produce coatings with a reduced surface roughness (e.g., by reducing or avoiding droplet formation), in combination with the use of post-coating surface treatments (e.g., polishing), appears to be inevitable to provide twist drills that achieve a satisfactory level of performance in deep-hole drilling with higher cutting speeds.

References

[1] Schulz, H., Moriwaki, T., 1992, High-Speed Machining. Ann. CIRP, 41(2), 637-643.

[2] Toenshoff, H.K., Spintig, W., Koenig, W., 1994, Machining of Holes - Developments in Drilling Technology. Ann. CIRP, 43(2), 551-561.

[3] Schulz, H., 1998, Hochgeschwindigkeitsbearbeitung. Hanser, Munich-Berlin.

[4] Ber, A., Goldblatt, M., 1989, The influence of temperature gradient on cutting tool's life. Ann. CIRP, 38(1), 69-73.

[5] Mueller, P., 2000, Hohe Leistung und Prozesssicherheit bei tiefen Bohrungen in Stahl und Guss. Maschinen-Anlagen-Verfahren, 10, 44-45.

[6] Heiler, R., Mueller, P., 1998, Moeglichkeit der HSC-Bearbeitung beim Bohren und Gewindebohren. Maschinenmarkt, 104(52), 18-21.

[7] Cselle, T., Barimani, A., 1995, Today's applications and future developments of coatings for drills and rotating cutting tools. Surf. Coat. Techn., 76-77(2), 712-718.

[8] Heinemann, R., Hinduja, S., Barrow, G., Petuelli, G., 2006, The performance of small diameter twist drills in deep-hole drilling, J. Manuf. Sci. Eng., 128(4), 884-892.

[9] Spur, G., Masuha, J.R., 1981, Drilling with Twist Drills of Different Cross Section Profiles. Ann. CIRP, 30(1), 31-35.

[10] Bhattacharyya, A., Chattopadhyay, A.B., Roy, R., 1981, Chisel-Edge Modification of Small HSS and Carbide Drills for Improved Machinability. Ann. CIRP, 30(1), 21-25.

[11] Wallen, P., Hogmark, S., 1989, Influence of TiN coating on wear of high speed steel at elevated temperature. Wear, 130(1), 123-135.

[12] Klocke, F., Krieg, T., 1999, Coated Tools for Metal Cutting - Features and Applications. Ann. CIRP, 48(2), 515-525.

Comparison of Cutting Strategies for High Productive End Milling

M.S. Uddin[1], A. Matsubara[2], S. Ibaraki[3] and T.Yasuda[4]

[1]Kyoto University, Japan, 6068501, ms_uddin@preceng.mbox.media.kyoto-u.ac.jp
[2]Kyoto University, Japan, 6068501, matsubara@prec.kyoto-u.ac.jp
[3]Kyoto University, Japan, 6068501, ibaraki@prec.kyoto-u.ac.jp
[4]Hitachi Construction Machinery Co.,Ltd, Japan, 300-00131, yasuda06@Hitachi-kenki.co.jp

Abstract. This paper presents a comparison of high-speed milling and heavy milling by using the visual chart approach. To search for a more efficient cutting process for 2.5D parts, an interactive process planning and analyzing method is introduced. Geometries to be machined are decomposed into combinations of machining features. Then, cutting strategies are interpreted into constraints on cutting conditions and cutting performance. Several combinations of machining features are examined based on the evaluation of machining time in the charts.

Keyword: 2.5D parts, machining features, machining time, high-speed milling, heavy milling

1. Introduction

Owing to the development of high-speed machining centers and cutting tools, high-speed cutting has become one of the major trends in many manufacturing applications [1]. In high-speed cutting, small depths of cut and large feed rates are typically selected in accordance with the spindle speed. In contrast, expert operators in job shops often employ a heavy cutting strategy, in which large depths of cut and small feed rates are selected. As end mills have versatility to cut various shapes, a comparison of the cutting strategies is not easy.

Computer-aided process planning systems (CAPP) [2,3,4] have been developed to support the decision making process of human operators. There are two approaches to develop CAPP. One approach is to translate an experienced operator's experiences into intelligent software: production rules, fuzzy logics, and neural networks. These tools are useful, as they can emulate the decision-making process of experts, but the expert's strategies are often implicit and difficult to translate into software. Another approach is to introduce an optimization method to decide process parameters. Although this approach has the potential to give highly optimized solutions, its performance and reliability completely depend on the quality of models and calculation algorithms used in the optimization. Moreover, reasons for selecting the conditions may not be understood intuitively. Wang proposed an economic method to define optimization strategies for single pass end milling operations [5]. In Wang's work, various possible constraints are shown in feed-speed diagrams, which give a deeper understanding of the characteristics of the problem. However, to apply this method to the machining of actual parts, the axial and radial depths of cut have to be also considered.

This paper presents a comparison between high-speed and heavy milling by using a simple process planning method. Higher productive cutting conditions for 2.5 dimensional machine parts are searched. First, a process model to predict the different components of the cutting force, is introduced which is closely related to performance indices. Then, a methodology to compare the machining efficiency using different cutting strategies is described. In this methodology, cutting strategies are interpreted as constraints on cutting conditions and cutting performance. Several combinations of machining features are examined based on evaluation of the machining times. Finally, these results are shown in the form of charts, from which one can visually find a process plan and cutting conditions that result in high productivity.

2. Cutting Force Model

There have been many researches on cutting force models [6]. Among them, a mechanistic model is suitable for CAD/CAM operations, as it is given from the tool geometry, cutting conditions, and cutting coefficients obtained from the experiments. In this work, the model from [7] has been employed as implementation of the algorithm is easy.

Figure 1.1 shows a view of workpiece and end mill. Parameters of the end mill are D: diameter (mm), θ_h: helix angle (rad), N: number of teeth. Cutting conditions are R_d: radial depth of cut (mm), A_d: axial depth of cut (mm), f_t: feed per tooth (mm/tooth), V: cutting speed (m/min), S: spindle speed (min^{-1}). α is the rotation angle of the tooth (rad), α_{st} is the entry angle, and α_{fin} is the exit angle of immersion.

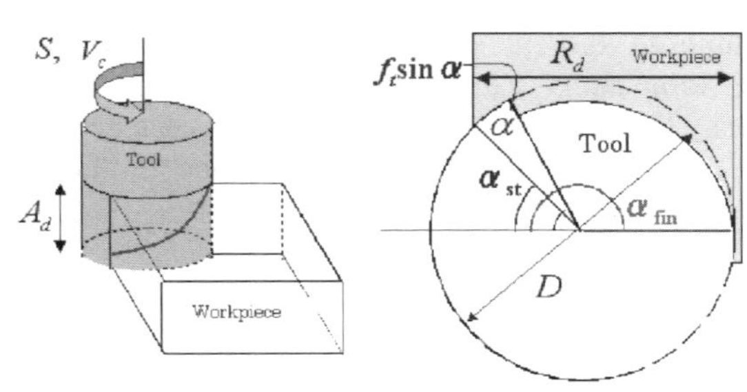

Figure 1.1. Tool and workpiece interference

Figure 1.2 presents the cutting force coefficients, K_t, K_t, and K_t, which are applied at the cutting point in the tangential, radial, and axial directions respectively. These coefficients are used to calculate the feed, normal, and axial components of the cutting forces are described below.

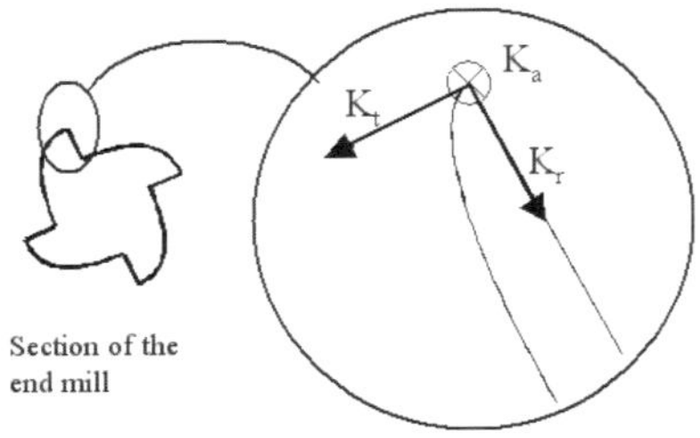

Figure 1.2. Cutting coefficients

$$f_m(\alpha)=\begin{cases}(-K_t\cdot\cos\alpha-K_r\cdot\sin\alpha)\cdot f_t\cdot\sin\alpha & \alpha_{st}\le\alpha\le\alpha_{fin}\\ 0 & otherwise\end{cases}, \quad (1)$$

$$f_s(\alpha)=\begin{cases}(-K_t\cdot\cos\alpha+K_r\cdot\sin\alpha)\cdot f_t\cdot\sin\alpha & \alpha_{st}\le\alpha\le\alpha_{fin}\\ 0 & otherwise\end{cases}, \quad (2)$$

$$f_a(\alpha)=\begin{cases}-K_a\cdot f_t\cdot\sin\alpha & \alpha_{st}\le\alpha\le\alpha_{fin}\\ 0 & otherwise\end{cases}. \quad (3)$$

Cutting forces of one tooth in the feed, normal, and axial directions are obtained as

$$F_{m1}(\theta)=\int_0^{A_d} f_m(\alpha)\,dz, \qquad F_{s1}(\theta)=\int_0^{A_d} f_s(\alpha)\,dz,$$

$$F_{a1}(\theta)=\int_0^{A_d} f_a(\alpha)\,dz \quad (4)$$

where θ is the rotation angle of the end mill. Superposing the cutting forces of all the flutes gives cutting forces applied on the end mill:

$$F_m(\theta)=\sum_{i=1}^{N}F_{mi}\left(\theta-\frac{2\cdot\pi}{N}\cdot(i-1)\right),\quad F_s(\theta)=\sum_{i=1}^{N}F_{si}\left(\theta-\frac{2\cdot\pi}{N}\cdot(i-1)\right),$$

$$F_a(\theta)=\sum_{i=1}^{N}F_{ai}\left(\theta-\frac{2\cdot\pi}{N}\cdot(i-1)\right) \quad (5)$$

where i is the tooth index.

3. Machining Features and Process Planing

3.1 Machinng Features

Process planning involves various operations: recognition of machining features, selection of cutting tools and path patterns, decision of cutting conditions. In this paper, a feature-based process planning is utilised. Figure 1.3 shows examples of machining features and tool paths. As the volume to be cut is decomposed into machining features, tool paths can be parameterized by cutting conditions. There are several combinations of machining features that remove the same volume; one of them is selected based on evaluation criteria.

3.2 Evaluation of Cutting Performance

Cutting performance is usually evaluated on the following criteria: machining time; machining accuracy (geometrical accuracy, dimensional accuracy, surface roughness) and tool life. To compare the performance of different cutting strategies, our system handles these criteria as follows.

Machining time: The machining time is calculated for each combination of machining features by taking possible cutting conditions into account.

Machining accuracy: Machining processes are roughly divided into roughing and finishing processes. Finishing processes use limited cutting conditions to assure surface quality. Since the productivity is the main focus of our system, this work focuses on roughing process, and thus accuracy indices are less prioritised. However, in order not to influence finishing surface, machining errors on surface are examined based on the tool defection model as shown in Figure 1.4. In this figure, a tool is regarded as a simple cantilever, and its deflection is calculated from a predicted cutting force, where y_c is the displacement of the tool centre, θ_c is the deflection angle of tool tip, w is the distributed cutting force (normal cutting forces divided by A_d). Machining errors are evaluated by horizontal and vertical displacements:

$$\delta l_s = y_c + (D/2)(1-\cos\theta_c), \quad (6)$$

$$\delta l_a = (D/2)\sin\theta_c. \quad (7)$$

Tool life: In order to consider tool life as a cutting performance index, a database is necessary for the combination of tool, material, and cutting conditions. Since it is difficult to prepare a prefect database for real applications, the cutting forces have been used as an index to predict the tool life, including the occurrence of tool breakage.

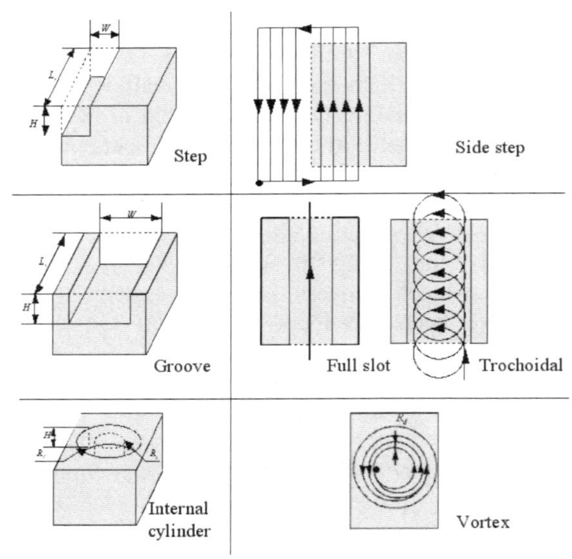

Figure 1.3. Examples of machining features and tool path cycles

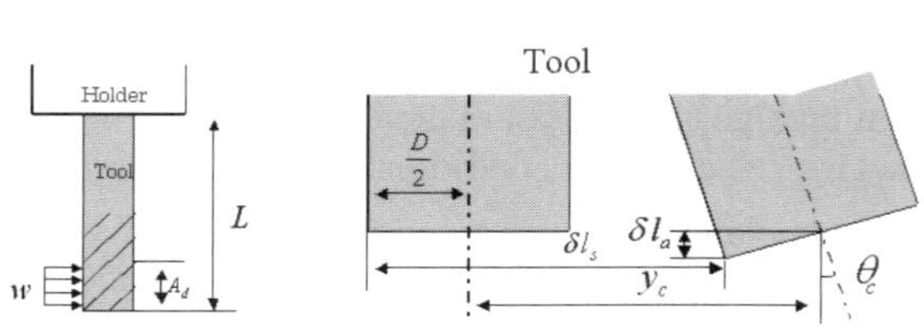

Figure 1.4. Tool deflection model and machining errors

3.3 Interpretation of Cutting Strategies

In the proposed system, each cutting strategy is interpreted as constraints on the cutting conditions or cutting performance. For example, two basic cutting strategies, namely, 1) high speed cutting with smaller depths of cut and higher feed rates, and 2) heavy cutting with larger depths of cut and lower feed rates, are interpreted as constraints as follows:

Constraint 1 (Heavy cutting): In heavy cutting, constraints must be imposed to avoid tool breakage. For example, the cutting force is constrained to be under some allowable maximum value.

Constraint 2 (High-speed cutting with constraints on machining errors): cutting strategies that result in machining errors under a specified value are searched. In such cases, a constraint on the machining error of a side surface is set.

Constraint 3 (High speed cutting): In high-speed machining cases, typically upper limits of depths of cut and cutting forces are constrained.

3.4 Process Planning Procedure

The procedure to determine the cutting conditions is summarized as follows:

(1) The cutting strategy is interpreted into a set of constraints on cutting conditions and performance indices.
(2) Actual constraints on cutting conditions are given.
(3) Constraints on the cutting forces are calculated or given directly.
(4) The machining patterns are prepared by combining defined machining features.
(5) The machining time for each combination of the cutting conditions is calculated and shown in the charts.

As an example of (5), Figure 1.5 shows how to decide on the cutting conditions for machining a feature. This case is for side step milling. Suppose that the constraint is set as, $F_{xy} < 2000$N, where F_{xy} is the resultant cutting force in the feed and normal directions. Possible A_d and R_d are calculated by $A_d = H/m$, $R_d = L/n$, where, m, n are numbers of cutting paths, and H and W are the height and the width of the step, respectively. Then, candidates of cutting conditions that satisfy the constraints are shown as circles. The combination of cutting conditions that minimize the machining time is selected.

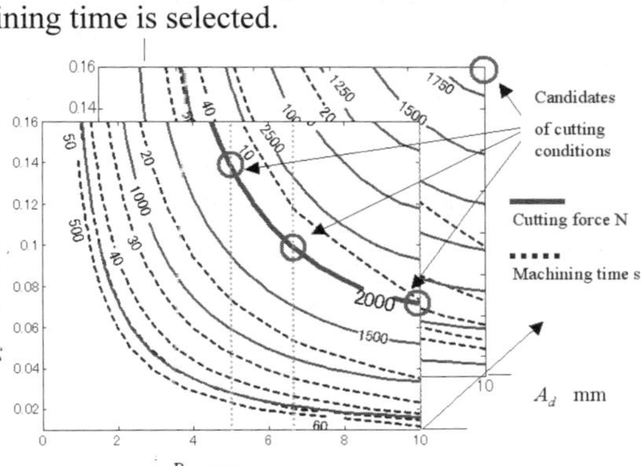

Figure 1.5. Candidates of cutting conditions

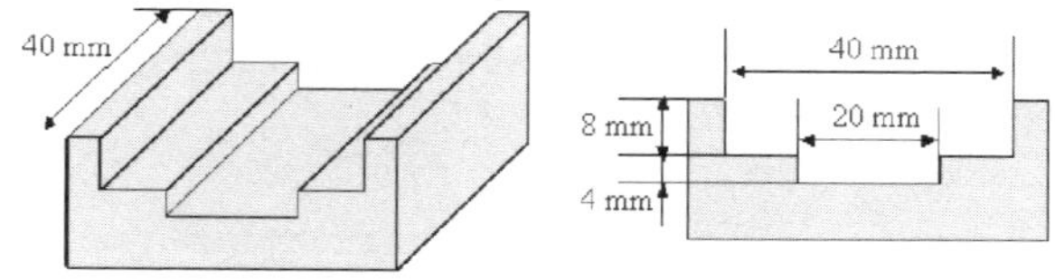

Figure 1.6. Workpiece geometry

Table 1.1. The specifications of the end mill

Material	Sintered carbide
Diameter D	10 mm
Number of teeth N	4
Helix angle θ_h	$2\pi/9$ rad
Young modulus E	534.4×10^9 Pa

Table 1.2. The cutting conditions (fixed values and constraints)

Projection length L	35 mm
Cutting feed rate	$\leq$ 10000 mm/min
Feed rate at air cut	10000 mm/min
Cutting velocity V_c	88.0 m/min
Spindle speed S	2800 min^{-1}
Feed per tooth f_t	0.019 ~ 0.16 mm/tooth
Axial depth of cut A_d	$\leq$ 22 mm
Radial depth of cut R_d	$\leq$ 10mm

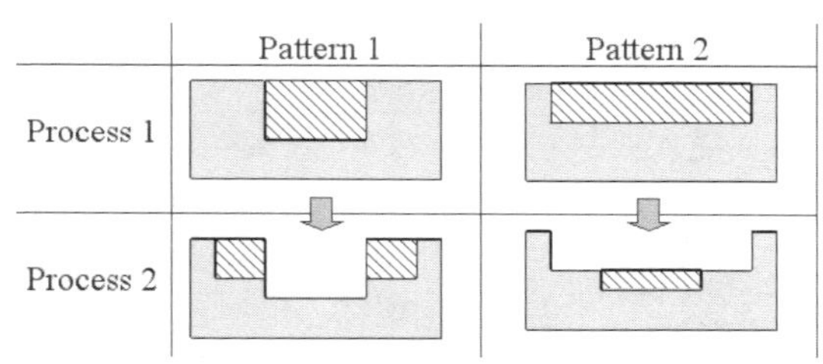

Figure 1.7. Cutting patterns and processes

4. Case Study

This section shows a case study to illustrate the comparison of the cutting strategies using the system. Figure 1.6 shows a workpiece used in the case study. The workpiece material is carbon steel. The specifications of the end mill used in the case study are shown in Table 1.1. Table 1.2 shows cutting conditions and their constraints, which are obtained by considering the machine tool's specifications catalog data of the cutting tool. Possible operation patterns are shown in Figure. 1.7.

Three cutting strategies, as shown in Section 1.2.3, are considered. The combination of cutting conditions that minimizes the machining time is searched. Specifically, constraints are set as follows in each strategy:

Constraint 1 (Heavy cutting): the constraint on cutting force is given as $F_{xy} < 2000$ N.

Constraint 2 (High-speed cutting with constraints on machining errors): the machining error of side surface is constrained as $\delta l_s \leq 40\ \mu m$.

Constraint 3 (High speed cutting): the radial depth of cut and the cutting force are constrained respectively as $R_d < 1$ mm and $F_{xy} < 1000$ N.

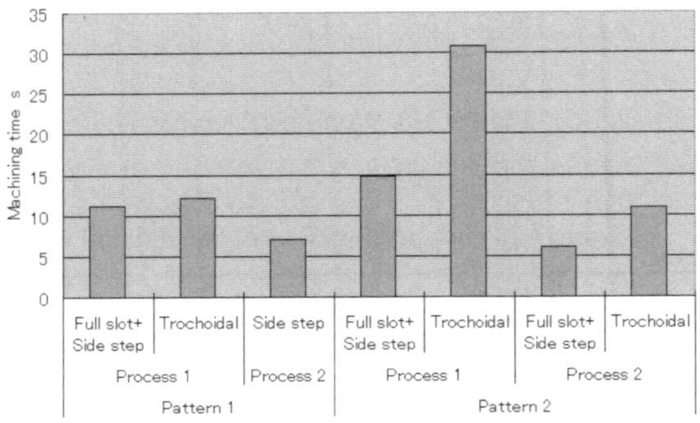

(a) Constraint 1

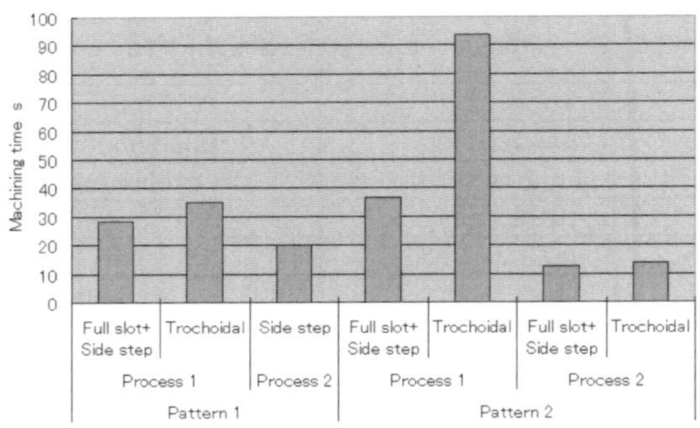

(b) Constraint 2

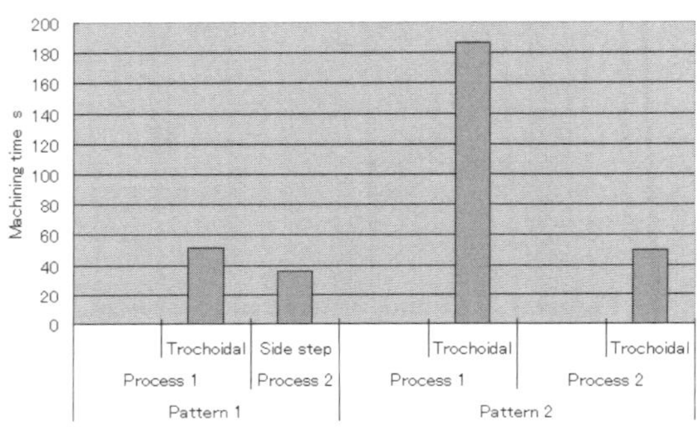

(c) Constraint 3

Figure 1.8. Comparison of machining time

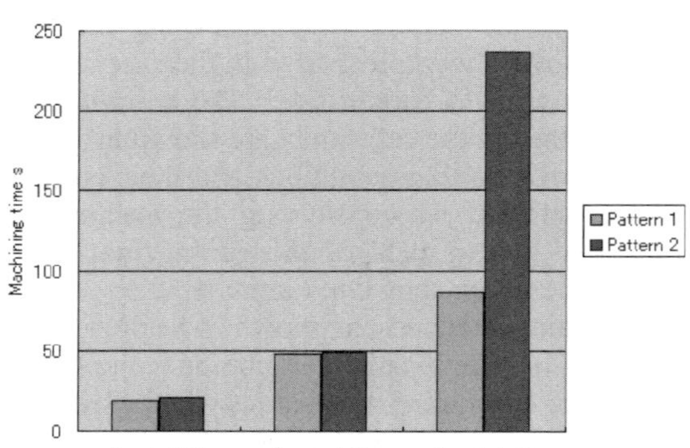

Figure 1.9. Comparison of machining time

Figure 1.8 shows the machining time for each feature and Figure 1.9 summarizes these results. As can be seen in these figures, the machining time gets shorter in the order of constraints 3-2-1. With constraint 3, the machining time in pattern 2 is much larger than that in pattern 1. This is because trochoidal cutting of a slot with a large width takes time. Table 1.3 shows the cutting conditions and calculated cutting forces in each machining pattern 1. In constraint 2, the cutting forces are less than 1000 N, but the machining time is shorter than that in the constraint 3. One reason is that the trochoidal cutting has longer machining time than the full slot and side step cutting. The other reason is that all the cutting conditions are upper bounded by the constraints. One solution to solve this problem is to introduce high-speed tools. In die and mold machining, high-speed tools with smaller depths of cut are often employed for the safety reason. However, since such tools are expensive for machining of medium carbon steel, it is better to use the heavy cut strategy.

5. Conclusion

This paper presents a scheme to compare several machining strategies in end milling process by using visual charts. In this methodology, cutting strategies are interpreted into constraints on cutting conditions and cutting performance. The results show that the constraints on the axial and radial depths of cut are quite important to obtain a high productivity.

References

1 B. P. Erdel, 1998, New Dimension in Manucfactuirng, Hanser Gardner Publications.
2 L. Alting, H. Zhang, 1989, Computer Aided Process Planing: the State-of-art Surevey, International Journal of Production Research, Vol.27, No.4, P553-583.
3 R. Weill, G. Spur and W. Eversheim, 1982, Survey of Computer Aided Process Planning Systems, Annals CIRP, Vol.31, No.2, P539-552.
4 W. Eversheim and J. Schulz, 1985, CIRP technical reports: Survey of Computer Aided Process Planning Systems, Annals CIRP, Vol.34, No.2, P607-614.
5 Wang, 1998, Computer-aided Economical Optimization of End-milling Operations, International Journal of Production Economics, Vol.54, P307-320.
6 F.K.Ehmann, S.G. Kapoor, R.E., DeVor, I. Lazoglu, 1997, Machining Process Modeling: A Review, ASME Journal of Manufacturing Science and Engineering, Vol. 119, P655-663.
7 Y. Altintas, 2000, Manufacturing Automation: Metal Cutting Mechanics, Machine Tool Vibrations, and CNC Design, Cambridge University Press.

10–3

Experimental Investigation of Brittle Material Milling Using Rotary Ultrasonic Machining

K. L. Kuo

National Taipei University of Technology, Taiwan, R.O.C 10608, klkuo@ntut.edu.tw

Abstract. The use of hard brittle materials has become increasingly more extensive. However, it is time-consuming to mill hard brittle materials with traditional metal-cutting techniques that use abrasive wheels. In addition, the tool would suffer excessive wear as well. As the efficacy of using hard brittle materials has increased in recent years, processing such materials for part fabrication has become a challenging problem difficult to solve. However, if ultrasonic energy is applied to the milling process and coupled with the use of hard diamond grits, hard brittle material can be effectively removed.

Rotary ultrasonic machining (RUM) is mostly used in brittle material drilling. Few researchers have focused on the milling application by RUM till now since the milling process is categorized as a horizontal machining mode, whose motion is perpendicular to the longitudinal kinetic energy supplied by the RUM. Besides, the most important hammering action associated with RUM cannot manifest itself, leaving only abrasion action generated by the tool rotation and micro-vibration effective. Therefore, based on this kind of machining process, the easiest and the most effective way to increase the efficiency of the ultrasonic milling is to find out the optimal feed rate and depth of machining.

The present research was aimed to apply the principles and characteristics of RUM for removing hard brittle material and investigating the interactive effect of the feed rate and the depth of machining on surface roughness of the milled hard brittle material. Finally, we attempt to find out the overall efficacy of utilizing the RUM for milling hard brittle material.

Keywords: rotary ultrasonic machining, surface roughness, brittle material.

1. Introduction

The application of ultrasonic energy for material removal was first introduced in 1927 by Wood and Loomis. They applied high frequency (about 70 kHz) onto a glass rod for removing material from the glass substrate while grinding grits were supplied during material processing. They found that it was possible to drill holes in the glass substrate, and the first paper on this research topic was published [1].

The modes of material removal mechanism include hamming, abrasion and hobbing, as depicted in Fig. 1.1(a) [2-4]. As a result, the depth of material excavation was restricted by how deep the slurry can get into the hole in most cases. Prabhakar and his co-workers made use of the fixed grits milling approach for material removal. It is a material removal mode, in which no free rolling of grinding grits is involved; however, the effect of material extraction is enhanced as a result of coupling of the grinding grits and the tool, as shown in Fig. 1.1 (b) [5].

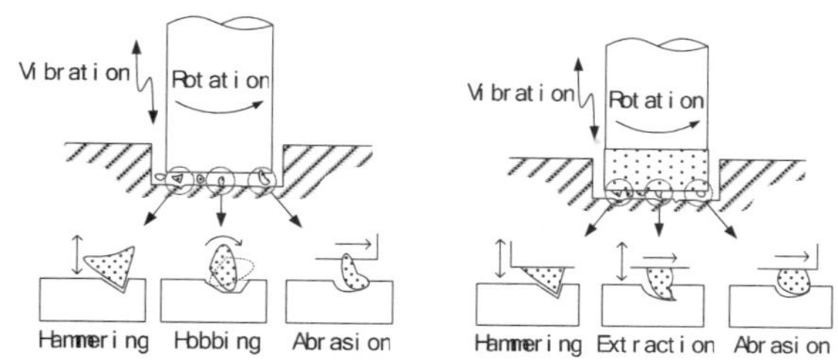

(a) Free grinding grits [2-4] (b) Fixed grinding grits [5]

Figure 1.1. Material removal modes

Under the circumstances, it appears that the easiest and fastest way to increase the efficacy of ultrasonic-assisted milling begins with the improvement of tool design. Pei designed a milling tool, which tapered with a 15° angle with respect to the bottom face of the tool [6].

2. Experimental Parameters and Design

The plane to be milled by RUM is confined to a circular one because of the rotational action associated with the machine. There are two kinds of tool surfaces that make an angle with the plane of the workpiece: one is the conical tool and the other is the globe tool, as depicted in Figure 1.2. Since the shape and the angle associated with the tools are different, the milling depth and the milling efficiency attainable by each respective RUM milling tool would not be the same either.

The vibration force induced by the ultrasonic vibrational energy can be resolved into two components through the tapering angle of the tool: one is hammering force that is perpendicular to the plane of the workpiece, and the other is the grinding force that is parallel to the plane of the workpiece. The two components F_n and F_t relate themselves to the total force F and angle θ defined in Fig. 1.2 as the following two equations:

$$F_n = F\cos\theta \quad (1)$$

$$F_t = F\sin\theta \quad (2)$$

The experiments conducted in the research adopted a globe-shaped tool, which had a conical milling face. The angle stated above will change with the depth of milling h. At the same time, F_n will change accordingly. The relationship between θ and F_n can be determined through radius r of the related globe.

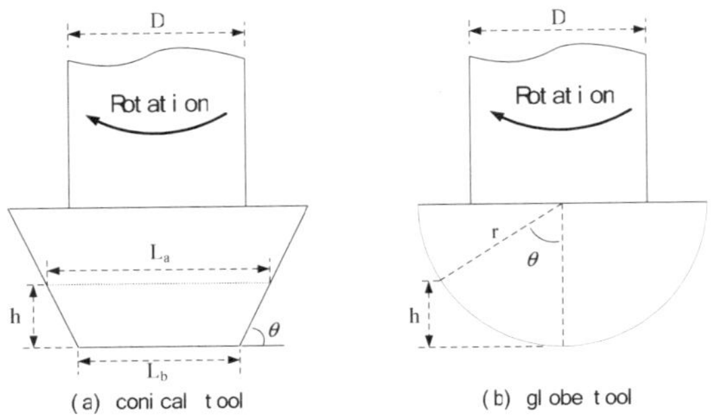

Figure 1.2. Shapes of the conical tool and globe tool

Angle θ can be computed through the following relationship:

$$\cos\theta = \frac{r-h}{r} \tag{3}$$

$$\theta = \cos^{-1}\left(\frac{r-h}{r}\right) \tag{4}$$

Substitution of Eq. 1 into Eq. 3 yields:

$$F_n = F\cos\left(\cos^{-1}\left(\frac{r-h}{r}\right)\right) \tag{5}$$

Rearranging the above expression, we have:

$$F_n = F\left(1-\frac{h}{r}\right) \tag{6}$$

According to Eq. 6, if the depth of milling is not so great, a complete transfer of the ultrasonic energy can be expected. In this case, the material milling efficiency is much greater than that in the conical tool. However, continual increase of milling depth will decrease the value of F_n. As the milling depth approaches radius r, the hammering energy as delivered by force F_n would eventually diminish and become nil. As a result, the milling efficiency would decrease. The design of a globe-shaped tool is simpler; only depth of milling needs to be considered for determining radius r. However, the difference between the cross-sectional areas involved should not be too big. Otherwise, stress concentration would arise, and the transfer of ultrasonic energy would decrease, too.

Ultrasonic milling is to make use of the high-frequency oscillation generated by the piezoelectric ceramic material. Its amplitude of oscillation is about 0.1-0.3 μm, which can be amplified several times through horns. In order to raise the processing capability, greater voltage for producing greater amplitude is recommended. However, improvement as such is rather limited; processing frequency for use has to be set near the resonance frequency of the milling system in order to capitalize on the resonance effect for amplifying the oscillation amplitude and energy. Ten to 100 times of amplification is attainable depending on the distance to the resonance point. As stated above, the experimental setup comprises an automatic frequency compensation system. Working frequency can be automatically adjusted accordingly, according to the feedback of the resonance change so that workpiece surface to ensure an adequate cooling effect.

The material removal rate refers to the material removed per unit time; its computation is made according to the tool shape, length of material milled and milling time taken. As such, disparate material removal rate may result. The globe-shaped tool adopted in this experiment is depicted in Fig. 1.3. The material removal volume may vary in accordance with the depth of milling. Since milling pertains to a horizontal machining process, the most important process parameter is the feed rate. Since the feed rate will determine the overall milling time, a higher the feed will result in a more efficient process.

Figure 1.3. Globe shape milling tool with diamond grits

RUM is a process that makes use of the energy of tiny oscillations to carry on removing the material through milling process. Hence, when milling hard brittle materials, the repeated hammering action generated by the ultrasonic energy can be effectively delivered to the material for its subsequent removal. However, if the same technique is employed to remove a ductile material, the deformation in response to the ductility of the material would absorb the kinetic energy associated with the oscillation, reducing the effect of the hammering action. This leads to a lower efficiency of material removal. Materials suitable for using ultrasonic energy for material removal can be defined as follows:

Brittleness = Shear Stress / Fracture Stress (7)

Hence, ultrasound is suitable for milling hard and brittle materials such as glass and ceramic materials. The purpose of the experiments designed here is to verify the efficiency and practicality of employing ultrasonic milling for removing hard brittle materials.

The material removal rate would increase with feed rate of the tool; however with a concomitant increase in milling resistance. Excessive resistance to milling may coarsen the surface of the workpiece and accelerate tool wear. In order to understand the mutual interactions influencing the above two factors, three levels of feed rate and three levels of milling depth were selected as process parameters to be employed in the material removal experiment. The process parameter matrix designed for the experiment is shown in Table 1.1.

Table 1.1. Process parameters in the ultrasonic milling

Depth f / Milling Feed Rate	H_1 (1 mm)	H_2 (2 mm)	H_3 (3 mm)
V_1 (0.1 mm/sec)	V_1H_1	V_1H_2	V_1H_3
V_2 (0.2 mm/sec)	V_2H_1	V_2H_2	V_2H_3
V_3 (0.3 mm/sec)	V_3H_1	V_3H_2	V_3H_3
Frequency of Tool Oscillation f = 22 kHz, Amplitude A = 25 μm, Rotation Speed = 5000 rpm			

3. Results of Experiment

The plots of surface roughness vs. depth of cut were constructed for entrance and exit areas of the milled lengths, which are shown in Figs. 1.4 and 1.5. A similar trend of roughness vs. depth of cut can be observed in the figures; that is, surface roughness increases with feed rate and depth of cut. However, surface roughness decreases at feed rate V_1, despite the increase in depth of cut from H_1 to H_2. This can be seen from the comparison of tool wear shown in Fig. 1.6. Figure 1.6(a) is the photo of the surface the brand new tool, and Fig. 1.6(b) shows the surface of the same tool after milling with process parameters V_1H_2, where fewer grinding grits are visible due to torn-out of the material that envelops the grits. This has led to a protrusion of the grinding grits, rendering increased cutting angles of the grinding grits. As a result, the surface roughness of the milled surface decreases.

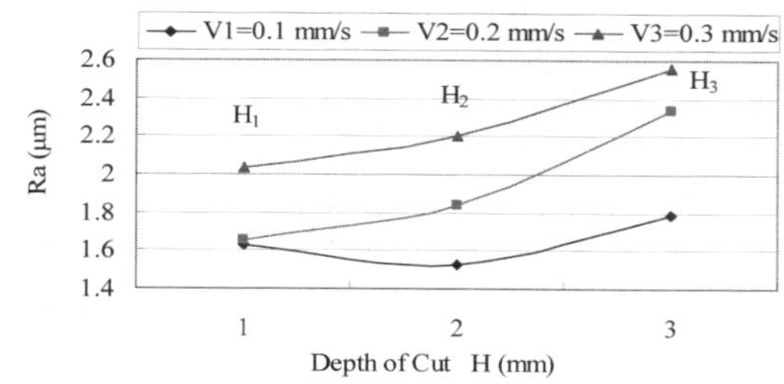

Figure 1.4. Roughness at the entrance area of the milled groove

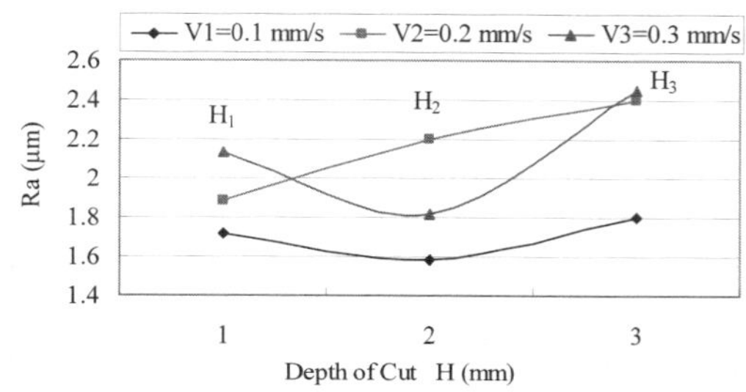

Figure 1.5. Roughness at the exit area of the milled groove

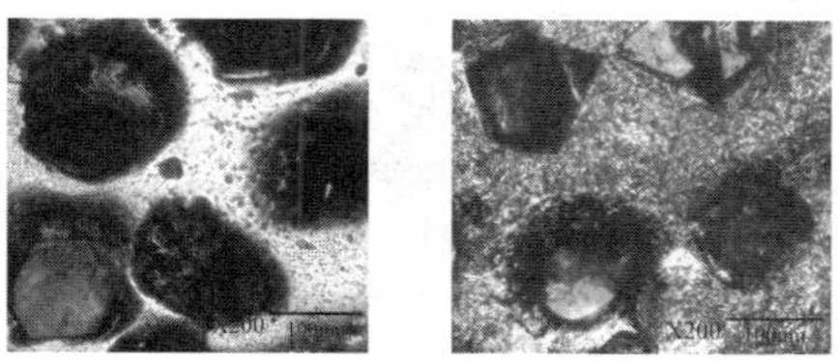

(a) Surface of a new tool (b) Tool surface after milling

Figure 1.6. Comparison between tool surfaces (200X)

Comparison of surface roughness of the milled groove at the exit area reveals that the surface roughness corresponding to process parameters V_1H_2 was smaller than that corresponding to V_1H_1, due to the same reason for the case of the entrance area. Worthy of note is the plot for V_3, which shows an appreciable drop in surface roughness corresponding to depth of cut H_2. This was consistently observed in several experimental runs. Theoretically, the increase in depth of cut and feed rate would roughen the tool's surface quicker. However, no trend as such was observed for the milling condition corresponding to V_3H_2. Nothing in particular was observed when examining Fig. 1.7, except that more grinding grits were found on the surface of the used tool. However, at a 200-time magnification (Fig. 1.8), more protruded grinding grits can be observed here. Because of the disparity in level, focused examination under microscope was not possible, which served as an indirect proof that there existed disparity of two levels involved. Because more grinding grits and more obvious protrusion of the grits were found in the surface of the used tool corresponding to milling condition, V_3H_2, this observation accounts for the obvious decrease in surface roughness of the milled groove.

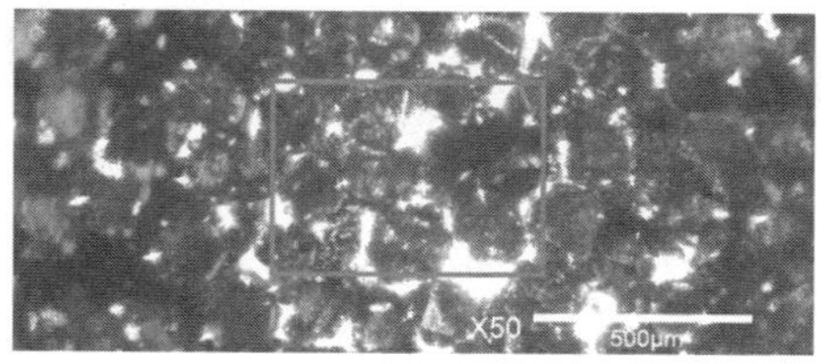

Figure 1.7. Micrograph showing surface of the tool after being used with process parameters V_3H_2

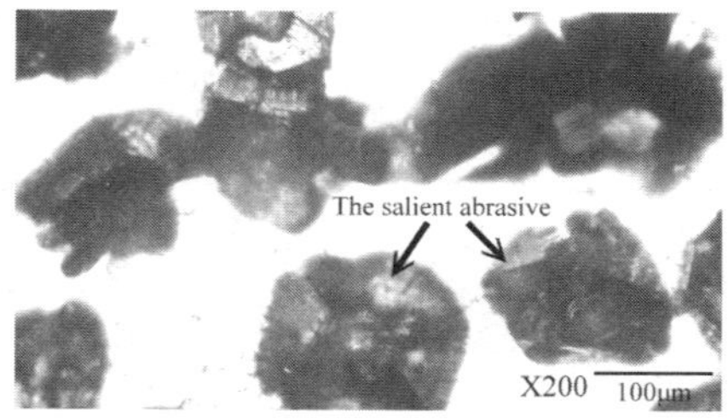

Figure 1.8. Micrograph showing surface of the tool after being used with process parameters V_3H_2

Observations made on the exit areas of the groove milled by the process parameter combinations of V_2H_3 and V_3H_3 reveal no big difference in surface roughness. The considerable drop in surface roughness leads to the conclusion that tool wear influences surface roughness. This agrees with the observation of serious wear at the bottom face and even torn-out of the grinding grits shown in Fig. 1.9. This was presumably caused by selecting inappropriately high values for both depth of cut and feed rate, which led to an excessive resistance to the milling process, and in turn imposed a strong squeezing stress on the grinding grits. As a result, grinding grits were crushed and torn out of the tool. Since the milling tool used with process parameters V_3H_3 has suffered from a stronger wear, more protrusion of the grits was observed. Because there were no grinding grits present on the tool's surface under this condition, the tool was presumably forced to mill (scratch) along the surface of the glass. This is confirmed by the numerous scratch marks generated by the rotational abrasion shown in Fig. 1.10. Hence, detachment of the grinding grits is attributable to coarsening of the milled surface.

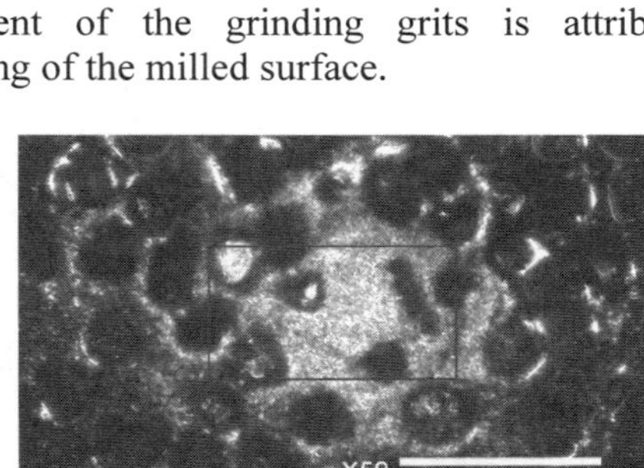

Figure 1 9. Micrograph showing surface of the tool after being used with process parameters V_3H_3

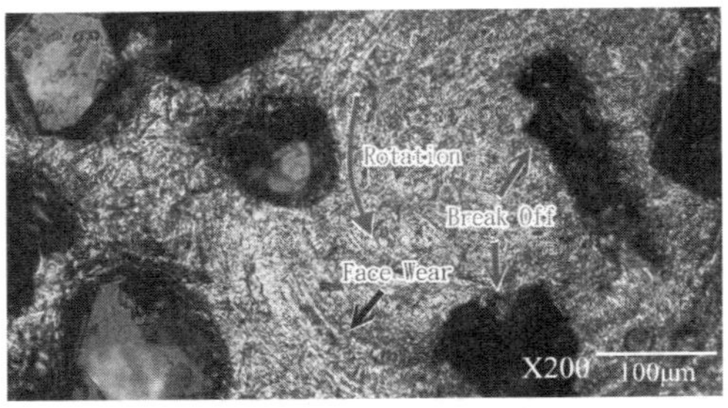

Figure 1.10. The micrograph showing surface of the tool after use at process parameter of V_3H_3

The cause for this will be detailed in the discussion about tool wear. The force taken by the contact surface of the globe tool and the workpiece varies with the depth of cut. Figure 1.11 shows the various regimes of contact between a globe-shaped tool and the workpiece. In this figure, H denotes the depth of cut, and θ is the angle encompassed by the contact pie that corresponds to each depth of cut. As seen in the figure, θ increases with depth of cut H, which in turn affects the conversion of ultrasonic energy to kinetic energy. Figure 1.12 denote the inserts corresponding to a, b and c, respectively.

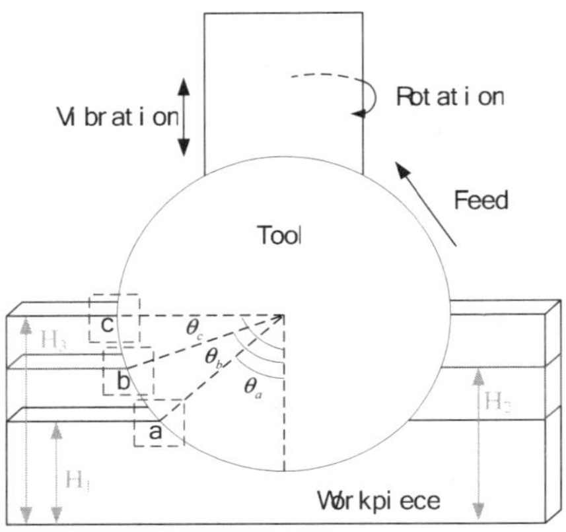

Figure 1.11. The contact regime between tool and workpiece at various depths of cut

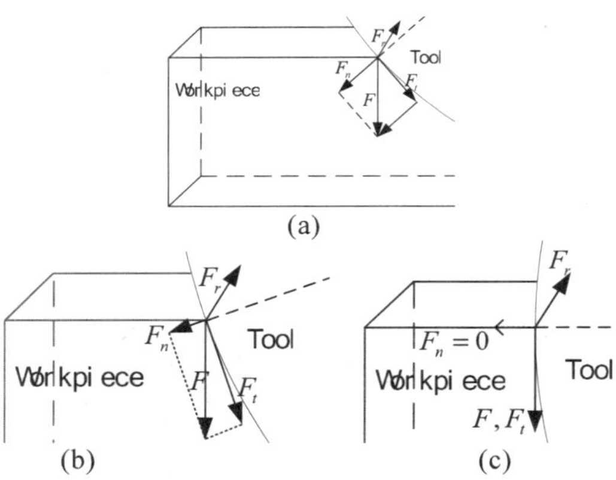

Figure 1.12. Forces of oscillation corresponding to various depths of cut: (a) 1 mm, (b) 2 mm (c) 3 mm

Figure 1.12(a) depicts the groove at a depth of cut of 1 mm, and F_r is the rotational cutting force. F_r is primarily responsible for the formation of blurs in the glass groove, which can be clearly substantiated by examining the photomicrographs presented in Fig. 1.13, where the blurs were formed along the rotational direction. Unbalanced stress arisen from eccentricity of the installed tool may also have contributed to the formation of blurs. In order to reduce the error margin caused by tool installation, eccentricity can be checked during try-out rotation for adjustment before the experiments are carried out.

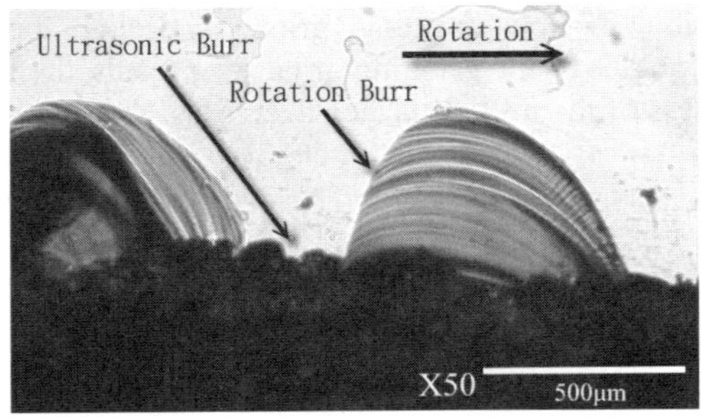

Figure 1.13. Blurs formed when milling was done with ultrasonic-assisted milling

4. Conclusion

1. The resistance to milling increased with feed rate and depth of cut of milling, which in turn causes the increase in tool wear and grit breakage. As this trend reached a certain point, the surface roughness decreased as a result of the decrease in number of grits that had accurate cutting angles and the decrease in grit size.
2. Comparing the interacting influences induced by the milling depth and the feed rate reveals that surface roughness increases with both feed rate and depth of cut at the entrance area of the milled groove.
3. Employing ultrasonic-assisted milling can indeed reduce the processing resistance during milling. In addition, because of the lower processing resistance encountered, this makes it possible to take a faster milling speed for material removal.

References

[1] R.W. Wood and Loomis, Philosophical Magazine, Ser. 7, 1927, pp.417-436.
[2] M. C. Shaw, Ultrasonic grinding, Microtechnic, vol. 10, no. 6, 1956, pp.257-265.
[3] L. D. Rozenberg, V. F. Kazantsev, L. O. Makarov and D. F. Yakhimovich, Ultrasonic cutting, Consultants Bureau, New York, 1964, pp.80-85.
[4] Y. Enomoto, Sliding Fracture of soda-lime glass in liquid environmentals, Journal of Materials Sciences, vol. 16, 1981, pp.3365-3370.
[5] W. R. Tyrrell, Rotary ultrasonic machining, SME Technical Paper, 1970, pp.70-516.
[6] Z. J. Pei and P. M. Ferreira, An experimental investigation of rotary ultrasonic face milling, International Journal of Machine Tools and Manufacture, vol.39, 1999, pp.1327-1344.

10–4

Effect of Edge Beveling on Burr Formation in Face Milling

P.P. Saha[1], D. Das[2], and Santanu Das[3]

[1]Eastern Railway workshop Kanchrapara- 743145, India, p.pratim2007@rediffmail.com
[2]Kalyani Government Engineering College, Kalyani- 741 235, India, dipankar283@yahoo.com
[3]Kalyani Government Engineering College, Kalyani- 741 235, India, sdas_me@rediffmail.com

Abstract. The presence of burrs on the finished workpiece often leads to difficulty in assembly operation, injury to operators, etc., necessitating incorporation of additional processes of deburring. Exit burr formation by milling usually occurs when the cutter moves out of the work piece. One method for minimizing this problem is to prevent the tool from exiting the workpiece while removing material. Burr formation can also be reduced considerably by edge beveling of the workpiece. In the present work, the influence of beveling the exit edge of the workpiece on the burr formation in milling has been experimentally investigated to find out the optimum condition.

Keywords: exit burr formation, milling, edge beveling

1. Introduction

Formation of burrs in milling causes problems in making the finished workpiece. This often necessitates additional operation of deburring. To avoid formation of burr, a number of studies were made [1-14]. Nakayama et al. [1] studied burr formation in various orthogonal metal cutting operations including milling. In milling, the burr formation occurs in the same direction as the primary tool motion and the burr which is three-dimensional appears on the exit surface. Kishimoto et al. [2] and Kitajima et al. [3] did experiments in face milling on burr formation. They analyzed the effects of cutting conditions, lead angle and work piece exit angle on burr size and shape. Chern [4] studied the effect of in-plane exit angle on burr size, and showed the importance of the workpiece geometry and tool geometry on burr formation. Some researchers tried to find out the dependence of burr formation on the stress field in machining [5]. An important aspect of tool and work piece geometry is the exit order of the tool edge because the burr appears near the final exit position of the tool along the work piece edge. Exit here refers specifically to the tool cutting edges moving out of the work piece at an edge while removing material. The exit order of the tool may greatly influence burr formation in terms of burr position on the workpiece in face milling [6].

Pekelharing et al. [7] studied the influence of tool exit on tool life. They found that tool exit leads to negative shearing or foot formation just before the tool exit, causing chip adhesion at low velocities and increased chances of tool failures at high velocities. Ghani and Barrow [8] discovered tool failures and a reduction in tool life when the tool exit occurs under certain conditions. In addition, tool entry and exit conditions are directly related to burr formation in the milling operation. When the tool always enters the workpiece, the edge is usually expected to be burr-free [9]. Optimum tool paths that would maintain favorable cutter tooth entry/exit conditions were also developed [10-11]. Chu and Dornfeld investigated the influence of tool feed direction on the in-plane exit angle and proposed an algorithm for reducing exit burr formation by adjusting tool path orientation [12]. Hassamor et al. adjusted widths of cuts in window-framing tool path planning for eliminating tool exits, but their algorithm is not applicable to the curved workpiece or parts with internal contours [13]. Another work tried to avoid tool exits by adjusting tool positions around the work piece vertices [14].

When burrs remain in the finish material, they lead to assembly failures, sort circuit, injuries to workers or even fatigue failures. All these increase the lead time in production, and hence the production cost. Although there are different deburring processes such as grinding, barrel tumbling, vibratory finishing, centrifugal and spindle finishing, chamfering, abrasive jet matching, water jet cutting, wire brushing, belt sanding, buffing filing, ultrasonic machining etc, the optimum selection of job bevel angles is also reported to have the capacity of reducing the burr formation significantly [15].

The aim of the present work is to study the burr formation mechanism and to investigate the influence of edge beveling conditions of test piece on burr formation in high production face milling of medium carbon steel test pieces. Single tooth uncoated carbide inserts are used to carry out face milling. For all the experiments a milling cutter having diameter 128.6 mm and depth of cut 3 mm were kept constant. Experiments were carried out to find out the optimum range of exit bevel angles of jobs for different machining conditions such that formation of burrs are reduced significantly or eliminated.

2. Mechanism of Burr Formation

The type of burr formed is highly dependent on the in-plane exit angle influencing the mechanism of its formation. Generally five types of burr in face milling are formed: Knife type burr, wave type burr, curl type burr, edge break out, and secondary burr. Knife type burr is created by the pushing out the uncut part near the transition machine surface when the exit angle reaches

150^0. The wave type burr is formed when the exit angle reaches approximately 90^0. Curl type burr is found after machining when the exit angle is less than 45^0. The edge break out is observed when the metal removal rate becomes very high; a rough chamfer and sharp burrs are noted along the tool exit edge. The secondary burr is created when fracture causing separation of the primary burr occurs near the root of the burr [16].

3. Experimental Investigations

Generation of burr mainly depends on the edge beveling of the workpiece. Experiments have been carried out for observing the amount of burr formation corresponding to different edge beveling of workpiece conditions. Details of the experimental set up are given in Table 1.1 and the machining conditions are shown in Table 1.2.

Table 1.1. Experimental set up

Machine Tool	Knee type milling machine, FN 2V Group, HMT Ltd. (India) Speed range (RPM) : 35.5 – 1800 (18 Steps) Longitudinal and Cross feed range (mm/min) : 16 – 1800 (18 Steps) Main Motor Power : 5.5 KW, Feed Motor Power : 1.5 KW
Tool Holder	Cutter diameter : 128.6 mm Holder type (Positive rake) : R/L 265.2-125 ME- 20 AL (Sandvik)
Cutting Tool	SPKN 1203 ED R, Uncoated SM30 (HW)-P30 insert (Sandvik)
Job Material	Medium carbon steel (45 C8), Hardness(BHN) : 175 Composition: C(0.4-0.5%), Si(0.15-0.35%), Mn(0.60-0.90%), S(0.045max), P(0.04%max).
Job Size	100 mm x 70mm x 35mm

Table 1.2. Machining condition

Depth of cut (t): 3 mm (Constant), Number of cutting insert fitted in the cutter: 01 Environment: Dry, In plane exit angle: 120^0 (Constant)			
Experiment Sl. No.	**Cutting velocity, V_c (m/min)**	**Feed, S_z (mm/min)**	**Edge Bevel Angles for each experiment (Degree)**
1	452	40	15, 20, 25, 30.
2	286	25	
3	363	31.5	
4	452	25	
5	286	40	

Table 1.3. Details of Experimental Results

Experiment Sl. No.	**Cutting Velocity, V_c (m/min)**	**Feed, S_z (mm/min)**	**Burr Height (mm) at Edge Bevel Angles (Degree)**			
			15	20	25	30
1	452	40	***	**	*	****
2	286	25	*	****	***	**
3	363	31.5	*	***	****	**
4	452	25	****	*	**	***
5	286	40	*	****	**	***

Note: * - Negligible burr, only can be felt upon touching; ** - Considerable, but not visible; *** - Tiny, but visible; **** - Small but visible.

4. Results and Discussions

Experiments were performed in dry condition on the workpieces with four different exit bevel angles of 15, 20, 25 and 30 degrees at a particular machining condition. At one end of the workpiece, bevel was made for a height of 3 mm. Machining conditions were chosen with three different cutting velocities (high – 452 m/min, medium – 363 m/min and low – 286 m/min and) and three different feeds (high – 40 mm/min, medium – 31.5 mm/min and low – 25 mm/min) following full factorial design of experiments. Details of the experimental observation are given in Table 1.3. At high cutting velocity and feed, minimum burr has been observed at 25^0 edge bevel angle of test piece. With the low cutting velocity and feed, and with medium cutting speed and feed minimum burr was formed at edge bevel angle of 15^0; also with medium cutting speed and feed minimum burr observed at 15^0. With another combination of high cutting velocity and low feed, minimum burr was observed at 20^0 edge bevel angle.

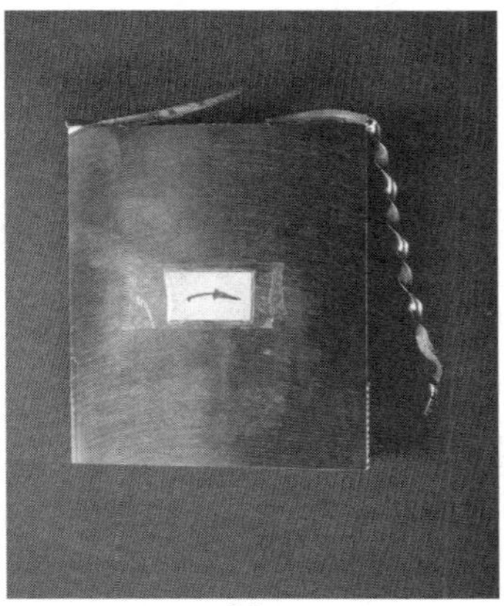

(a)

(b)

Figure 1.1. Photographic view of typical specimens(a) no bevel angle, (b) 15 degree edge bevel of the job

It was found out earlier [17] that when backup material is less or even absent, burr formation is more hence when edge bevel angle is less backup material will be more resulting in less burr formation. Following this reason, similar tendency has been observed at low range of speed in the present case of experimental findings.

Only at high speed conditions with 20 and 25 degrees of bevel angles, minimum burr formation has been observed. This is also expected due to the fact that at high cutting velocity, the plastic straining becomes less resulting in less burr formation; this has also been reported earlier by Avila and Dornfeld [17].

Figure 1.1 (a) shows sizeable amount of burr formation obtained after milling a job with no edge bevel angle (at cutting velocity 452 m/min, feed 40 mm/min), while Figure 1.1 (b) shows no formation of any burr with 15^0 edge bevel angle made on the job. This has been experienced during milling with the cutting velocity of 286 m/min and feed of 40 mm/min. These two photographs clearly indicate the effect of providing edge bevels on the job in controlling the formation of burrs.

Details of the experimental results are shown in Table 1.3., where at different machining conditions and bevel angles, formation of burrs is compared.

5. Conclusions

The following conclusions can be drawn from the experimental investigation carried out:

- It was found that at an optimum value of bevel angle of the workpiece at a machining condition, burr formation becomes minimum and results in good surface quality.
- In most of the cases, at a 15^0 exit bevel angle of the workpiece, minimum burrs have been observed. However, for higher speeds, optimum bevel angle has been formed at 20^0 and 25^0.

References

[1] Nakayama, K., and Arai, M. 1987, Burr Formation in Metal Cutting. Annals of the CIRP, Vol. 36, No. 1, P33-36.

[2] Kishimoto, W., Miyake, T., Yamamoto, A., Yamanaka, K., and Takano, K. 1981, Study of Burr Formation in Face Milling. Bulletin Japan Society of Precision Engineering, Vol. 15, No. 1, P51–52.

[3] Kitajima, K., Miyake, T., Yamamoto, A., Yamanaka, K., and Takazawa, K. 1990, Study of Mechanism and Similarity of Burr Formation in Face Milling and Drilling. Technology Reports of Kansai University, No. 32 March. P1–13.

[4] Chern, G.L. 1993, Analysis of Burr Formation and Breakout. Dissertation for Ph.D, in Mechanical Engineering, University of California at Berkeley, CA, P88.

[5] Iwata, K., Ueda, K., and Ikuda, K. 1982, Study of Mechanism of Burr Formation in Cutting Based on Direct SEM Observation. Journal of the Japan Society for Precision Engineering, Vol. 48, No. 4, P510–515.

[6] Hashimura, M. and Hassa, J. 1999, Effect of In-Plane Exit Angle & Rake Angles On Burr Height And Thickness In Face Milling Operation, Journal of Manufacturing Science and Engineering, February, Vol. 121/13.

[7] Pekrlharing, A.J. 1978, The Exit Failure in Interrupted Cutting, Annals of the CIRP, Vol. 27, No. 1, P5-10.

[8] Ghani, A.K., and Barrow, G. 1985, Tool Failure at Exit During Interrupted Cutting, Annals of the CIRP, Vol. 34, No. 1, P71-74.

[9] Park, I. 1996, Modelling of Burr Formation Process In Metal Cutting, Ph.D. Dissertation, University of California at Berkeley, Department of Mechanical Engineering.

[10] Yong-Sheng Ma. 1995, Tool Path for Face Milling Considering Cutter Tool Entry/ Exit Conditions, Proceedings of the 3rd International Conference on Computer Integrated Manufacturing, Vol. 1, P427-434.

[11] Balasubramanian, N. and Raman, S. 1997, Modeling Geometric Process Variables in Path Planning, Transactions of the NAMRI/SME, Vol. 25, P225-230.

[12] Chu, C.H. and Dornfeld, D.A. 1999, Tool Path Planning for Exit Burr Minimization by Estimating the Tool Length of Primary burrs, submitted for publication, International Journal of Computer Integrated Manufacturing.

[13] Hassamontr, J. 1998, Edge Finishing Planning in Milling, Ph.D. Dissertation, University of California at Berkeley, Department of Mechanical Engineering.

[14] Chu, C.H. and Dornfeld, D.A. 1999, Tool Path Planning for Avoiding Exit Burr, submitted for publication, Journal of Manufacturing Processes.

[15] Sikdar, C. 1993, Effect of cutting edge modification on machining characteristics and performance of coated carbide face milling inserts, Ph.D Dissertation submitted to IIT, Kharagpur, India.

[16] Chern, GWO-Lianq. 2007, Experimental Observation and Analysis of Burr Formation Mechanisms in Face Milling of Aluminium Alloys, International Journal of Machine Tools and Manufacture (Articles in Press).

[17] Avila, M.C., Dornfeld, D.A. 2004, On the Face Milling Burr Formation Mechanisms and Minimization Strategies at high Tool Engagement. Annual Research Reports, University of California at Berkeley.

The Effect of AlCrTiN Coatings on Product Quality in Micro-milling of 45 HRC Hardened H13 Die Steel

A Aramcharoen[1], P T Mativenga[2] and S Yang[3]

[1]Manufacturing & Laser Processing Group, School of Mechanical, Aerospace and Civil Engineering, The University of Manchester, ampara.aramcharoen@postgrad.manchester.ac.uk
[2]Manufacturing & Laser Processing Group, School of Mechanical, Aerospace and Civil Engineering, The University of Manchester, P. Mativenga@manchester.ac.uk
[3]Teer Coatings Ltd, West Stone House, Berry Hill Industrial Estate, Droitwich, Worcestershire, shicai.yang@teercoatings.co.uk

Abstract. The market for micro components is predicted to experience a phenomenal growth in the coming years. Micro milling has been identified as a key technology for micro fabrication especially of micro moulds and dies. However, this demanding application requires the use of advanced micro tooling solutions especially coated tools. In microscale milling; grain size, phase effects and the balance between tool edge radius and undeformed chip thickness, all influence the cutting process. In this paper, high aluminium content, AlCrTiN coating is evaluated in terms of its technological contribution to tool wear reduction, surface quality and burr minimisation in dry milling of H13 tool steel. Thus, the paper presents the scientific contribution of AlCrTiN to the microscale machinability of die steel.

Keywords: micro milling, coatings, tool wear, surface finish, burr formation

1. Introduction

The demand of micro components has increased continuously over the past few years [1]. One of the key processes used for micro component fabrication is mechanical micro machining and mainly micro milling. Current principles of micro milling were derived from conventional/macro milling. However, micro cutting is more challenging as compared to macro scale machining due to the substantial size reduction. Thus, it is doubted whether the problem can be simply solved by scaling down of tooling.

A clear distinct difference between micro and macro mechanical cutting is the size of cutting tool which is less than 1 mm. In addition, the fragile nature of the micro tool [2,3], the ratio of undeformed chip thickness to cutting edge radius [4,5] and the microstructure of workpiece material [6] all bring about the size effect which differentiates micro milling from conventional or macro scale milling.

Premature tool breakage, before any degree of severe wear, and hence unpredictable tool life has been reported when using micro tools [2]. Moreover, in microscale cutting, the ratio of undeformed chip thickness to the cutting edge radius is critical because it has an influence on the effective rake angle. Thus, this will define the trade off between normal cutting and ploughing modes [5]. Unlike macro scale cutting that homogenous materials and classical theory can be assumed, in microscale cutting, the relatively large grain sizes, heterogeneous and multi-phase of workpiece material microstructure are critical in influencing the cutting mechanism, material springback, plucking and surface finish.

The quality of the micro tools is also important, particularly the size of the cutting edge radius. This influences the chip formation and quality of machined components. The size of cutting edge radius is limited by the tool fabrication techniques and the size of carbide grain. Moreover, when the micro tool is coated, the size of edge radius is enlarged and the edge profile could be modified. Experience has also show that the quality of micro tools that is currently available on the market is very variable. Indeed, errors of the micro grinding process in tooling manufacture still manifest in the micro tools sold commercially. It is thus, important at the current technology level that for research purposes all micro tools be subjected to a 100 % inspection before being evaluated.

In addition the grain size and composition of the micro carbides appears to be very critical in influencing premature tool breakage in the micro machining of hardened die materials. Thus, cost and quality issues have to be balanced in the selection of tooling.

Other than the performances of tooling, in terms of wear and fracture resistance, characteristic of machined components are also used to evaluate the micro fabrication process. Indeed, despite all the focus on tool wear the ultimate objective is to deriver a micro product fabricated to the required surface finish, accuracy and geometrical definition without effecting adverse surface integrity defects. In particular in micro components, nano-range surface finish is required by industry, together with close tolerances in geometry and burr free edges. The component quality attributes have been reported as due to the quality of micro tool such as wear progression [7-10]. It is difficult to apply post processing techniques on micro components as they can adversely influence accuracy in dimension and also due to the need to invent new processes for cleaning of micro metal products. Therefore it is highly desirable to fine tune the micro machining process variables and tooling requirements so as to directly achieve the best product quality from manufacture.

It is well established in macro cutting that the use of appropriate coatings significantly improves wear resistance and machined surface quality. For instance, in micro drilling, the use of TiN/AlN multilayer coating on 300 μm drills was reported to lead to the wear performance being improved by 40% as compared to the uncoated drill [11]. However, there is a lack of research evaluating coated tools in micro milling.

Multilayer AlCrTiN is a relatively new coating which has characteristics of high hardness, toughness, thermal stability, good adhesion and exceptional wear resistance. In this paper, the performance of AlCrTiN multilayer coating as compared to uncoated tool is reported for micro milling of die steel material.

2. Experimental Details

The workpiece material used for this investigation was H13 tool steel vacuum hardened to 45 HRC. This material is generally used in the die and mould industry. The evaluated micro tools were 2-flute flat micro end mills of 700 μm in diameter before coating. Some of the tools were then coated with AlCrTiN with two different coating thicknesses of 1.37 and 1.85 μm referred to here as AlCrTiN (LT) and AlCrTiN (HT) respectively. These coating thickness levels are generally lower than the 3-4 μm used in most PVD processes. The idea was to target lower thickness for the micro tool and hence mitigate the negative effects of edge radius increase. The coating process was done using the Teer Coatings Ltd physical vapour deposition (PVD), Closed Field Unbalanced Magnetron Sputtering Ion Plating (CFUBMSIP) technology.

Cutting edge radii of all tools were measured before and after the CFUBMSIP deposition process by using a scanning electron microscope (SEM). The coating process naturally caused an increase of the cutting edge radius of the coated tools. The average edge radius for the uncoated micro tool was 2.25 μm and this increase by approximately 20-48% after coating depending on the coating thicknesses.

Slot milling cutting tests were performed on a Mikron HSM 400 high speed machining centre. Each micro tool was used to undertake 1 meter length of slotting cut. Cutting parameters were kept constant at a spindle speed, N, of 20000 rpm, axial depth of cut, a_p, of 50 μm and chip load, f_z, of 8 μm. These values were established from pilot machining tests with a view to avoid pre-mature tool breakage.

After machining, tool wear was analyzed using an SEM. Workpiece surface roughness and 3D slot profiles were evaluated by the Wyko optical profiler. An optical microscope and laser scanning were used to analyze the burr width and height respectively.

3. Results and Discussion

3.1 Tool wear

Figure 3.1, 3.2 and 3.3 show the conditions of uncoated, AlCrTiN (LT) and AlCrTiN (HT) micro mills respectively before and after 1 meter slotting. Labels (a) in the Figures, display the condition of micro tool before cutting while (b) depicts the damaged condition of the tools after 1 meter cutting length. Figure (c) and (d) illustrate a close-up image of wear patterns on both cutting edges. The images show that edge chipping, wear along the radius of the tool base (radial), wear along the tool diameter circumference (peripheral wear) and wear along the tool axis (axial wear) modes appear on the cutting edges. Comparing Figure 3.1 to 3.2 or 3.3, it is clear that wear progressions on the uncoated tool, Figure 3.1, can be significantly reduced by applying AlCrTiN coating on micro tools (Figure 3.2 and 3.3). In addition, the extent of wear on both cutting edges can be balanced out by the use of coatings. Figure 3.1 shows different wear modes for the cutting edges, whereas the difference is not pronounced for the coated tools. Uniform wear on the cutting edge is desirable as it preserves the micro tool rotational balance and is beneficial for product quality and machining stability.

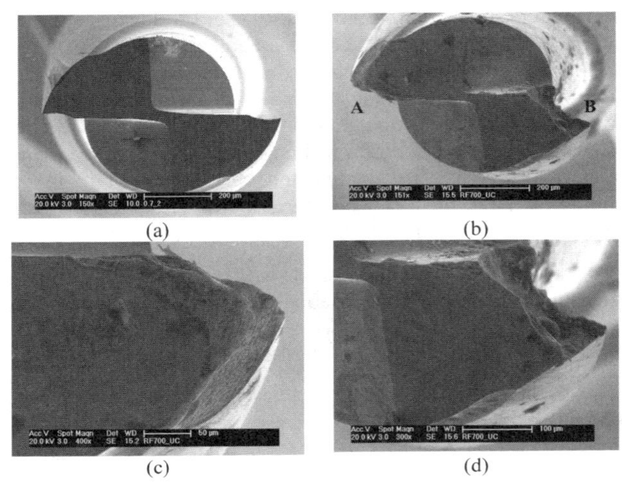

Figure 3.1. Uncoated micro mill

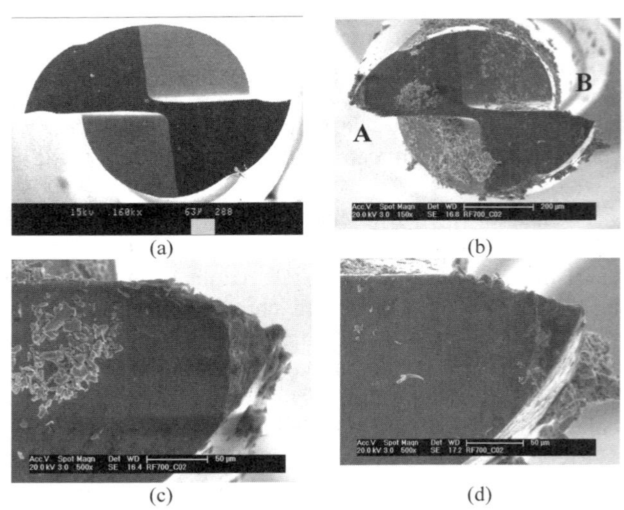

Figure 3.2. AlCrTiN (LT) micro mill

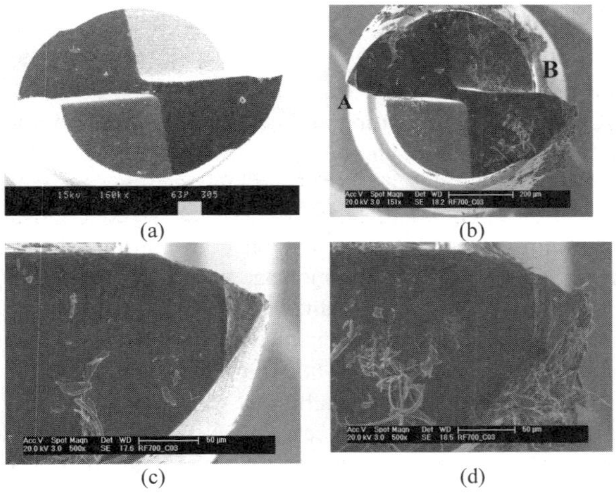

Figure 3.3. AlCrTiN (HT) micro mill

3.2 Surface roughness

Figure 3.4 shows surface finish at the bottom of the groove when slot machining. The use of both low and high coating thickness of AlCrTiN coated micro tools results in a significant improvement in surface finish compared to the uncoated tool. Moreover, the surface finish is in a nano-range when using AlCrTiN coated micro mills. In addition, the surface finish increases with cutting length. This can be attributable to tool wear progression and the enlargement of cutting edge radius which promotes the ploughing effect. For short machining time, there seems to be a no major difference in the surface finish brought about by the coating thickness. After a prolonged cutting time/length the thicker coating appears to generate slightly better surface finish. This strongly suggest that the difference in surface finish for prolonged machining time is attributable to the thicker coating taking a long time to be exhausted in the machining process.

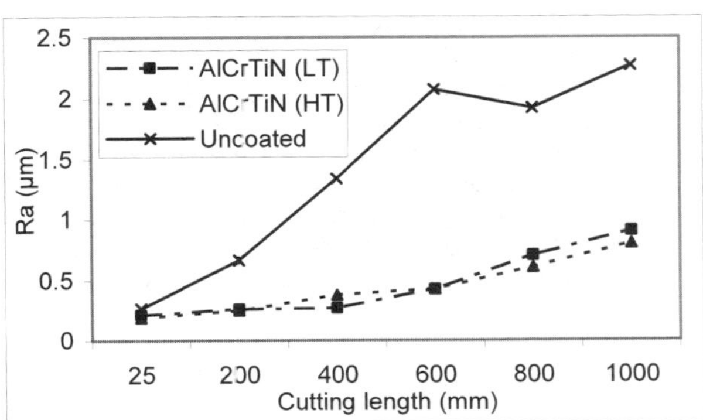

Figure 3.4. Surface finish

3.3 Burr formation

Figure 3.5 shows the top view of the slots at cutting length of 1 meter when using the different micro tools. It can be clearly seen that a large size of top burr is present when using the uncoated tool as compared to AlCrTiN micro tools. Moreover, the size of the top burr for down milling mode is greater than that of up milling mode. This is consistent with established burr theories. This can be because of the stress unload effect and chip tear in down milling mode. There again appears to be no large different in burr formation brought about by the two coating thicknesses. However, similar to the surface finish results above the higher thickness coating is showing a slight burr reduction advantage compared to the low thickness. The mechanisms through which coated tools suppress burr formation are not entirely established. However it is noted that coating of workpiece and re-design of exit angles are active methods of burr formation in macroscale machining. Thus, it is reasonable to assume that if the coating is liberated within the machining cycle it may have the effect of suppressing burr formation, it is further noted that the coating changes the friction coefficient which is assumed to be one of the factors controlling the minimum undeformed chip thickness. The increase in the coating edge radius due to the coating deposition process and its associated increase in the negative rake angle do not seem to override the positive contribution of the coating to burr reduction and surface improvement (as discussed previously).

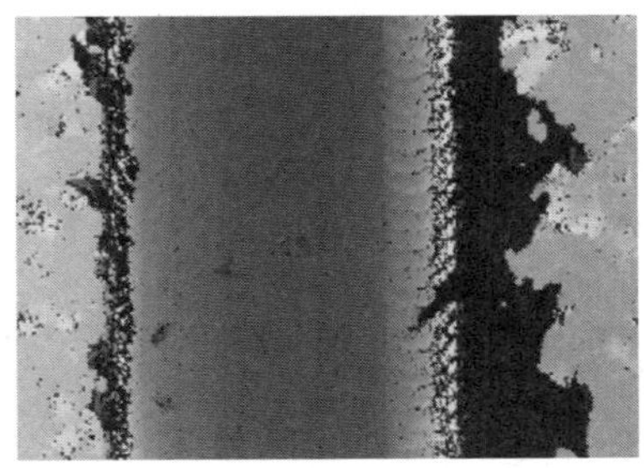

Uncoated

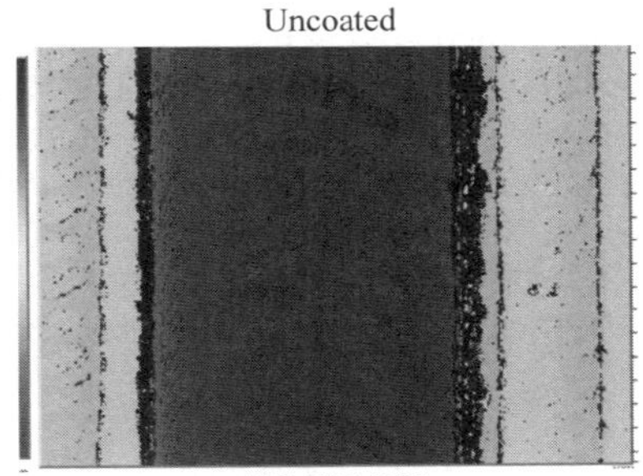

AlCrTiN (LT)

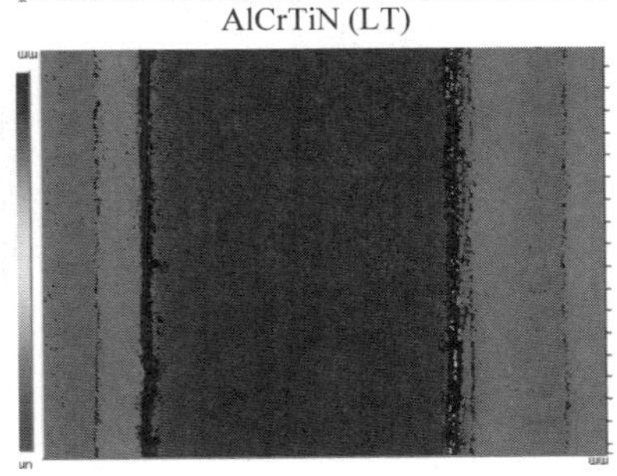

AlCrTiN (HT)

Figure 3.5. Burr formation from Wyko optical profiler

Figure 3.6 and 3.7 show the significant improvement in the size of burr height and burr width respectively when using both AlCrTiN coated micro tools as compared to the uncoated tool. In addition, burr size increase with cutting length which can be because of the increase of cutting edge radius resulting in larger negative rake angle and thus, promoting ploughing.

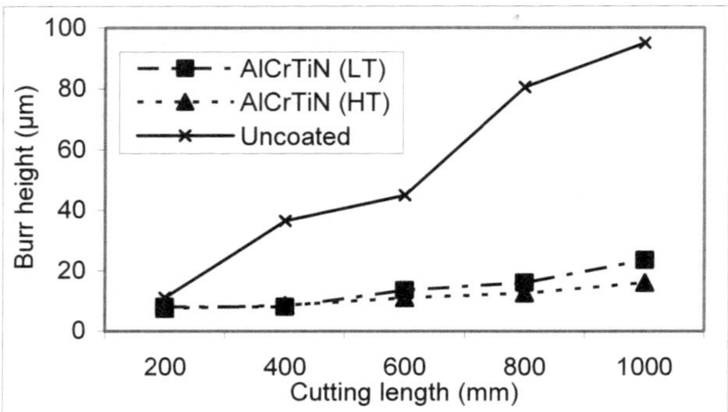

Figure 3.6. Burr height

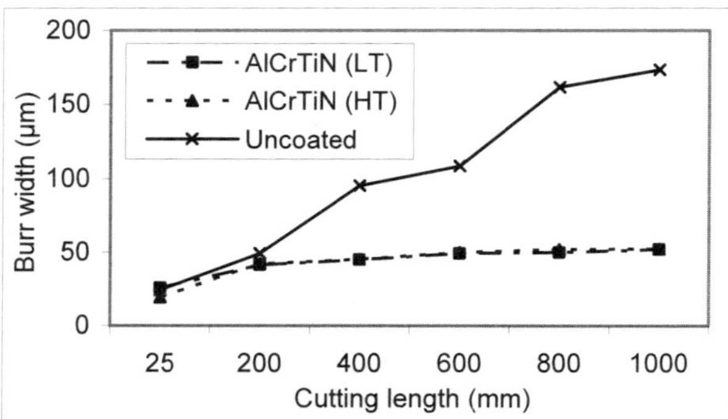

Figure 3.7. Burr width

3.4 Slot profile

Accuracy of micro machining can be indicated by machined dimensional geometry and accuracy. Groove profile is used to evaluate the accuracy in slot machining. Figure 3.8 shows a scan of geometry of the slot after cutting of 1 meter when using uncoated and AlCrTiN coated tools. A significantly reduced error in groove profile can be achieved by the use of AlCrTiN coated mills. This again can be the due to the reduction in wear rate progression.

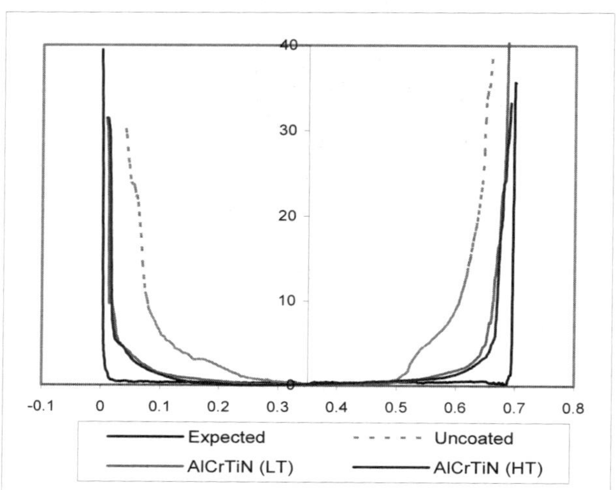

Figure 3.8. Slot profile

4. Conclusions

The aim of this study was to investigate the effect of AlCrTiN (Al rich) coating and coating thickness on micro milling of die steel. The results show that the use of both low and high thickness of AlCrTiN coatings on micro mills are beneficial in terms of tool wear reduction and homogenising the extend wear on both cutting edges. In the case reported here, the increase micro tool edge radius due to the coating process does not override the benefit of the coating protection mechanism. Coating the micro to difference thicknesses does not significantly affect the tool performance for short machining times (this could be finishing operation). However, the thicker coating shown noticeable improvement in surface finish, burr height and slot profile error for prolonged machining times. Overall, significant surface finish improvement, burr size reduction and geometric error reduction are benefits of the AlCrTiN coated micro tools in machining of hardened die material.

Acknowledgement

The authors would like to acknowledge the support of Teer Coatings and Rainford Precision for coating and tooling respectively. We are grateful to Nick Driver for the machining test.

References

[1] Alting L, Kimura F, Hansen HN, Bissacco G, (2003) Micro Engineering. CIRP 52: 635-657

[2] Tansel I, Rodriguez O, Trujillo M, Paz E, Li W, (1998) Micro-end-milling-I. Wear and breakage. International Journal of Machine Tools & Manufacture 38: 1419-1436

[3] Uhlmann E, Schauer K, (2005) Dynamic load and strain analysis for the optimization of micro end mills. CIRP 54: 75-78

[4] Dornfeld D, Min S, Takeuchi Y, (2006) Recent advanced in mechanical micromachining. CIRP 55: 21-24

[5] Bissacco G, Hansen HN, Chiffre LD, (2005) Micromilling of hardened tool steel for mould making applications. Journal of Materials Processing Technology 167: 201-207

[6] Vogler MP, DeVor RE, Kapoor SG, (2004) On the modeling and analysis of machining performance in micro end milling, Part I: Surface generation. Journal of Manufacturing Science and Engineering, Transaction of the ASME 126: 685-694

[7] Takacs M, Vero B, Meszaros I, (2003) Micromilling of metallic materials. Journal of Materials Processing Technology 138: 152-155

[8] Schmidt J, Tritschler H, (2004) Micro cutting of steel. Microsystem Technologies 10: 167-174

[9] Lee K, Dornfeld D A, (2005) Micro-burr formation and minimization through process control. Precision Engineering 29: 246-252

[10] Schaller T, Bohn L, Mayer J and Schubert K, (1999) Microstructure grooves with a width of less than 50 um cut with ground hard metal micro end mills. Precision Engineering 23: 229-235

[11] Yao SH, Su YL, Kao WH, Liu TH, (2005) On the micro-drilling and turning performance of TiN/AlN nano-multilayer films. Materials Science & Engineering A 392: 340-347.

Theoretical Investigation for the Material Removal Rate of Metal Polishing Process

Y.R. Jeng[1], P.Y. Huang[2] and H.J. Tsai[3]

[1]National Chung-Cheng University, Taiwan, 62102, imeyrj@ccu.edu.tw
[2]WuFeng Institute of Technology, Taiwan, 62153, py.huang@mail.wfc.edu.tw
[3]WuFeng Institute of Technology, Taiwan, 62153, hjtsai@mail.wfc.edu.tw

Abstract. The objective of this research is to better understand the mechanisms of material removal in the polishing process. This study is aimed to establish a material removal rate model for metal polishing to predict the result of metal polishing process. This research conducts theoretical study to investigate the material removal rate and its mechanism during the metal polishing process. The present study proposes a material removal rate model for metal polishing based upon a micro-contact model which considers the effects of the abrasive particles located between the polishing interfaces. In addition to such operational parameters as the applied down force, the present study also considers the consumable parameters rarely investigated by previous models, including workpiece surface hardness, abrasive particle size and the standard deviation of abrasive size distribution. The current results show that the material removal rate increases as the average abrasive particle size. In addition, the smaller standard deviation of abrasive particle size distribution can increase the efficiency of metal polishing process. This study also provides physical insights into the interfacial phenomena not discussed by previous models, which ignored the effects of abrasive particles between the polishing interfaces during force balancing.

Keywords: metal polishing, material removal rate, micro-contact mechanics, interfacial phenomena

1. Introduction

Due to the trend for high precision, the requirements for the surface roughness of metal workpiece are becoming stricter. Polishing is the method of choice to planarize the metal surface to a certain tolerable range. In the metal polishing process, workpiece surface roughness is minimized by passing slurry with abrasive particles between the workpiece and the polishing pad [1]. Figure 1.1 shows the schematic view of the metal polishing plateform. The effectiveness of the metal polishing process is determined by a variety of process control parameters, however, due to the complex nature of the inter-relationships between these parameters, control of the polishing process currently remains at an empirical stage. In order to improve the efficiency of the polishing process and the quality of its polishing result, many researchers have explored the effects of the process parameters from a physical perspective and by considering experimental results [2-5].

Xie *et al.* investigated the free abrasive polishing process experimentally and theoretically [6]. Experiments were carried out to understand the effects of particle size, polishing pad and nominal contact pressure on the wear rate and surface roughness of the polished surface. A theoretical model was also developed to predict the relationship between the polishing parameters and the wear rate for the case of hard abrasive particles sandwiched between a soft pad and a workpiece. Chauhan *et al.* observed polished ceramics and metals by microscopic, which indicated that material removal occurs as a consequence of repeated sliding or rolling of the abrasive particles against the workpiece surface [7]. Luo and Dornfeld proposed a solid-solid contact model of the CMP process based on plastic contact over the wafer-abrasive interface and pad-abrasive interface [8]. The fluid effect is attributed to the number of active abrasives, but the effects of slurry particles under hydro-dynamical contact were not investigated.

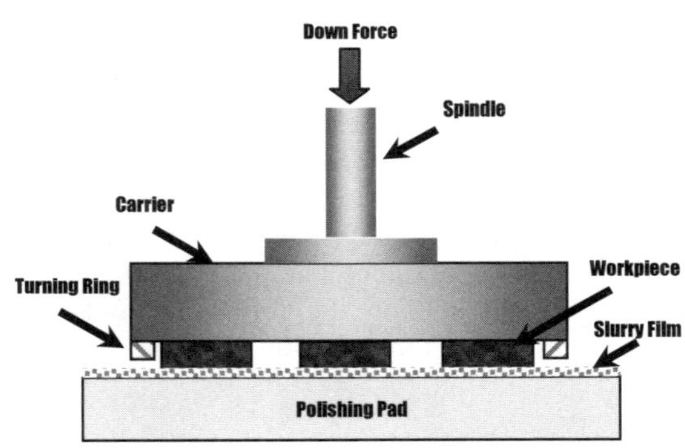

Figure 1.1. The schematic of the metal polishing platform

The current research has recently developed a micro-contact wear model which employs a concept from the work of Jeng *et al.*[9-10], in which the main point is that the particle effects between the polishing interfaces are taken into consideration. The present study adopts this model to investigate the relative influences on the polishing results of the applied down force magnitude, slurry particle size, the standard deviation of abrasive particle size distribution and the workpiece surface hardness. The relationships between the interfacial phenomena and the consumables' characteristics are explored and elucidated.

2. Modeling

Figure 1.2 shows a schematic of contact of the abrasive particles between the polishing interface. To develop the material removal rate model for metal polishing, the following assumptions are made:

1. The direct material removal effect of the polishing pad is far less than that of the abrasives. The material removal occurs primarily as a result of abrasive wear of the surface by the abrasive particles in the slurry.
2. Abrasive particles are spherical in shape, and the size distribution of particles is normal distribution.
3. During the polishing process, the wafer surface and the polishing pad surface are assumed to be parallel to one other and the abrasives are carried into the polishing interface by the slurry flow.
4. The current model assumes the heights of the pad surface asperities to be normally distributed.

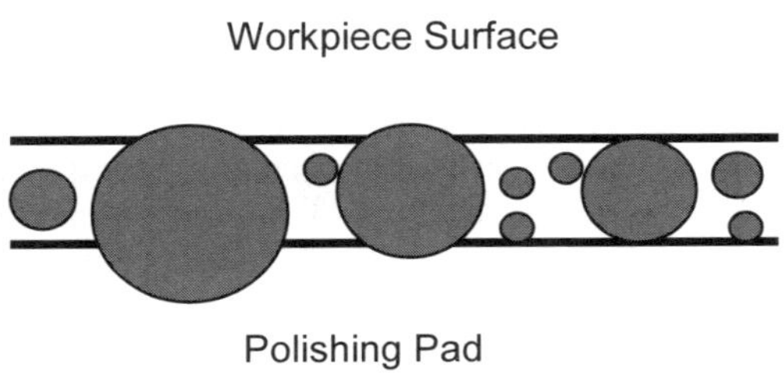

Figure 1.2. Schematic of contact of the abrasive particles between the polishing interface

2.1 Real Contact Area and Contact Force Between the Polishing Interface

The current material removal rate model employs the micro-contact theory proposed by Jeng *et al.* [11] to calculate the real contact area and the asperity contact force between the polishing interface. Substituting the polishing pad and workpiece parameters into the micro-contact model and solving the following equation will give distance d between the polishing interface:

$$\begin{aligned} P_0 = & \frac{4}{3} \cdot E' \cdot \eta \cdot R^{1/2} \cdot \int_d^{d+\omega} (z-d)^{3/2} \cdot \varphi(z) dz \\ & + \eta \cdot \pi \cdot H \cdot \int_{d+\omega_e}^{d+\omega_p} \left[1 - 1\left(1 - k \frac{\ln \omega_p - \ln \omega}{\ln \omega_p - \ln \omega_e} \right) \right] \cdot \left[1 - 2\left(\frac{\omega - \omega_e}{\omega_p - \omega_e} \right)^3 + 3\left(\frac{\omega - \omega_e}{\omega_p - \omega_e} \right)^2 \right] \cdot (z-d) \cdot \varphi(z) dz \\ & + 2 \cdot \pi \cdot \eta \cdot R \cdot H \cdot \int_{d+\omega_p}^{\infty} (z-d) \cdot \varphi(z) dz \end{aligned} \quad (1.1)$$

Furthermore, the real contact area at the polishing interface can be calculated from:

$$\begin{aligned} A_r = & A_n \cdot \eta \cdot \pi \cdot R \cdot \int_d^{d+\omega_e} (z-d) \cdot \varphi(z) dz \\ & + A_n \cdot \eta \cdot \pi \cdot R \cdot \int_{d+\omega_e}^{d+\omega_p} (z-d) \cdot \left[1 - 2\left(\frac{\omega - \omega_e}{\omega_p - \omega_e} \right)^3 + 3\left(\frac{\omega - \omega_e}{\omega_p - \omega_e} \right)^2 \right] \cdot \varphi(z) dz \\ & + 2 \cdot A_n \cdot \eta \cdot \pi \cdot R \cdot \int_{d+\omega_p}^{\infty} (z-d) \cdot \varphi(z) dz \end{aligned} \quad (1.2)$$

In these equations, A_n is the nominal contact area, E' is the equivalent Young's modules for the contact surfaces, η and R are the density and average radius of the asperities, respectively, d is the distance between the rigid flat surface of the wafer and the mean plane of the pad asperities, and $\varphi(z)$ is the distribution function of the asperity heights. Furthermore, H_p is the hardness of the polishing pad material, w, w_e and w_p are the interference, critical interference at the point of initial yielding of the asperity, and critical interference at the point of fully plastic flow, respectively. If w is less than w_e, the deformation of the asperity is classified as elastic. Conversely, if w is greater than w_p, the deformation of the asperity is taken to be fully plastic.

Moreover, the real contact pressure P between the polishing interface and the real contact force F acting on a single effective abrasive can be estimated.

$$P \cdot A = P_0 \cdot A_n \Rightarrow P = P_0 \cdot \frac{A_n}{A} \quad (1.3)$$

$$F = 0.25 \cdot P \cdot \pi \cdot x^2 \quad (1.4)$$

Where x is the diameter of abrasives.

2.2 Material Removal Rate of a Single Effective Abrasive

During the polishing process, the abrasive particles embedded in the real contact area of the polishing interface penetrate into both the polishing pad asperities and the workpiece surface, and this causes these surfaces to deform plastically. As shown in Figure 1.3, the penetration depth of the abrasive on the workpiece $\Delta 1$ and the polishing pad $\Delta 2$ can be determined by the following equations [8].

$$\Delta 1 = \frac{a_1^2}{x} = \frac{2 \cdot F}{\pi \cdot x \cdot H_w} = \frac{\frac{1}{2} \cdot P \cdot x}{H_w} \quad (1.5)$$

$$\Delta 2 = \frac{a_2^2}{x} = \frac{F}{\pi \cdot x \cdot H_p} = \frac{\frac{1}{4} \cdot P \cdot x}{H_p} \quad (1.6)$$

Accordingly, the material removal rate of a single effective abrasive can be got via

$$Vol_{removed\_s} = \Delta 1 \cdot a_1 \cdot V = \frac{\left(\frac{1}{2} \cdot P\right)^{3/2} \cdot x^2}{H_w^{3/2}} \cdot V \quad (1.7)$$

Where V is the relating speed of the polishing interface.

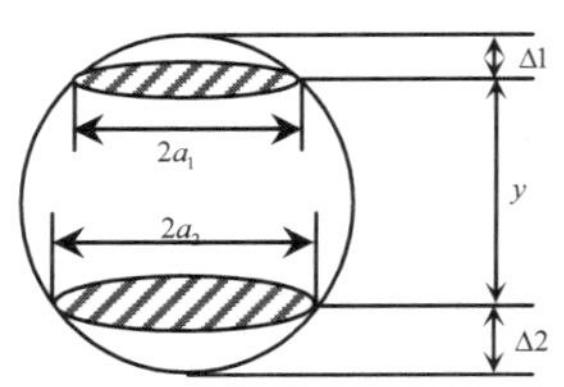

Figure 1.3. Schematic view of effective abrasive particles trapped at the polishing interface

2.3 Number of Effective Abrasives

This study assumes that workpiece surface wear occurs only as the result of the abrasive particles embedded in the workpiece/polishing pad asperity contact area and larger than a particular size, i.e. the remaining particles are assumed to flow freely between the polishing interfaces and hence have no wear effect [14]. In the initial stages of the polishing process, the abrasive particles are distributed uniformly throughout the slurry and are spread over the polishing pad surface. When the wworkpiece contacts the polishing surface, the abrasives in the slurry film become trapped at the top of the polishing pad asperities and are pressed into contact with the workpiece surface.

In the current study, we assume that the abrasive particles distributed within the three times of the root mean square of the pad surface asperities height ($3\sigma_{asperity}$) have chance to be trapped in the real contact area of the polishing interface (A_r). Under these conditions, the number of abrasive particles including in the real contact area can be determined from:

$$n = G \cdot A_r \cdot \left(3 \cdot \sigma_{asperity} - d\right) \qquad (1.8)$$

where A_r is the contact area of the polishing pad asperities and the wafer surface as given by Equation (1.2), d is the separation distance between the wafer and the polishing pad surface after force balancing, and $\sigma_{asperity}$ is the root mean square of the pad surface asperities height.

In Equation (1.8), G is the number of abrasive particles in a cubic meter of the specified slurry and can be estimated by the characteristic of slurry.

The number of abrasive particles trapped in the real contact area including the effective abrasive particles and non-effective particles. From figure 1.3 and 1.4, only the trapped abrasive particles size larger than $x_{\max} - (\Delta 1 + \Delta 2)$ can penetrate into the workpiece surface. Since the abrasive particle size assumed to be normally distributed, the effective abrasive particle number can be estimated by the following equations.

$$\begin{aligned} N &= n \cdot \left(\phi\left(\frac{x_{\max} - x_{avg}}{\sigma} \right) - \phi\left(\frac{(x_{\max} - (\Delta 1 + \Delta 2)) - x_{avg}}{\sigma} \right) \right) \\ &= n \cdot \left(1 - \phi\left(3 - \frac{\frac{1}{4} \cdot P \cdot x \cdot \left(\frac{2}{H_w} + \frac{1}{H_P} \right)}{\sigma} \right) \right) \end{aligned} \qquad (1.9)$$

Where x_{max} and x_{avg} are the largest abrasive particle diameter and the average abrasive particle size respectively.

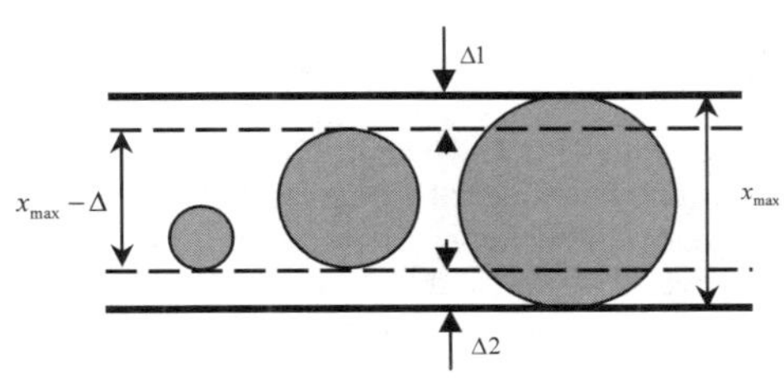

Figure 1.4. Schematic view of effective abrasive particle size

2.4 Material Removal Rate

From equations (1.7) and (1.9), the material removal rate of workpiece surface during polishing process can be got.

$$\begin{aligned} MRR &= \rho_w \cdot N \cdot Vol_{removed\_s} \\ &= \rho_w \cdot G \cdot A_r \cdot (3 \cdot \sigma_{asperity} - d) \cdot \left[1 - \phi\left(3 - \frac{\frac{1}{4} \cdot x_{\max} \cdot \left(\frac{2}{H_w} + \frac{1}{H_P} \right)}{\sigma} \cdot P \right) \right] \cdot \frac{\left(\frac{1}{2} \cdot P\right)^{3/2} \cdot x_{avg-a}^{2}}{H_w^{3/2}} \cdot V \end{aligned} \qquad (1.10)$$

3. Results and Discussions

The current parametric study explores the influence on the material removal rate of operational parameters such as the applied down force and of consumable parameters such as abrasive particle size, the standard deviation of abrasive particle size distribution, and workpiece surface hardness.

Figure 1.5 presents the variation in the material removal rate with applied down force magnitude for various average particle diameters. The simulation results indicate that for a constant average particle diameter, the material removal rate increases with increasing down force magnitude. Furthermore, for a constant magnitude of down force, the material removal rate also increases with increasing average particle diameter. According to the developing model, both larger the abrasive particle size or the applied down force can increase the penetrating depth of the abrasive particle/workpiece contact interface. Consequently, the material removal rate increases. These observations are consistent with the results of previous experimental studies conducted by Xie *et al.* [6]. Equation (1.9) reveals that the number of effective particles is a function of the average particle diameter and the real contact area of the polishing interface. However, the real contact area of the polishing interface increases as the magnitude of the down force increases, and hence the total number of effective particles increases with increasing applied down force or with decreasing average particle diameter. But, the material removal rate in the metal polishing process is influenced more strongly by the penetration depth than by the number of effective particles, and consequently, the material removal rate increases as the average particle size increases.

The effects of the standard deviation of abrasive particle size distribution to the material removal rate are shown in the figure 1.6. According to the theoretical outcomes, the material removal rate increases with the abrasive particle size and decreases with the standard deviation of abrasive particle size distribution. For a constant abrasive particle average diameter, smaller the standard deviation of abrasive particle size distribution will raise the percentage of the contact abrasive particles and increase the effective abrasive particle number. Accordingly, the material removal rate increases. This trend will be more significantly when the average abrasive particle size is larger.

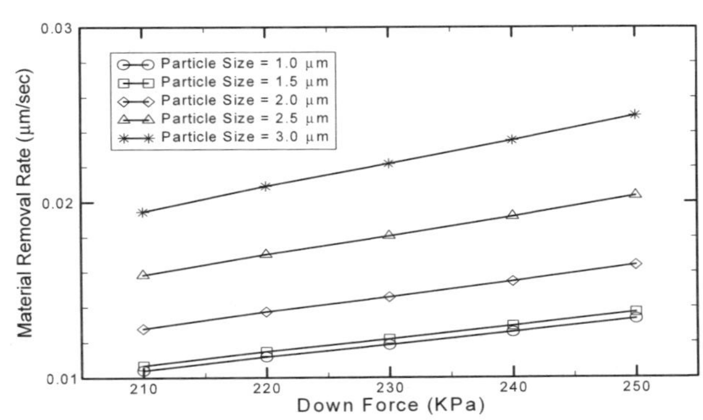

Figure 1.5. Variation in material removal rate with down force for different average particle sizes

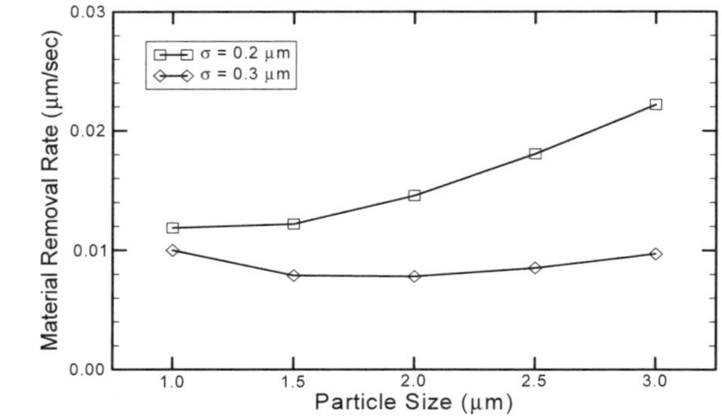

Figure 1.6. Variation in material removal rate with the standard deviation of abrasive particle size distribution for different average particle sizes

Figure 1.7 shows the influence of the workpiece surface hardness on the material removal rate for different magnitudes of down force. The results of figure 1.7 indicate that the material removal rate increases as the magnitude of the applied down force increases, but decreases significantly as the workpiece surface hardness increases. Since these three simulations employ the same polishing pad parameters, the real contact areas of the polishing interfaces are very similar under the same magnitude of applied down force. However, a lower workpiece surface hardness will slightly increase the real contact area. The present theoretical results suggest that the number of effective particles does not contribute greatly to the significant increase in material removal rate observed in the case of a softer workpiece surface. More reasonably, this increase can be attributed to the greater depth of penetration of each effective particle into workpiece surfaces of lower hardness.

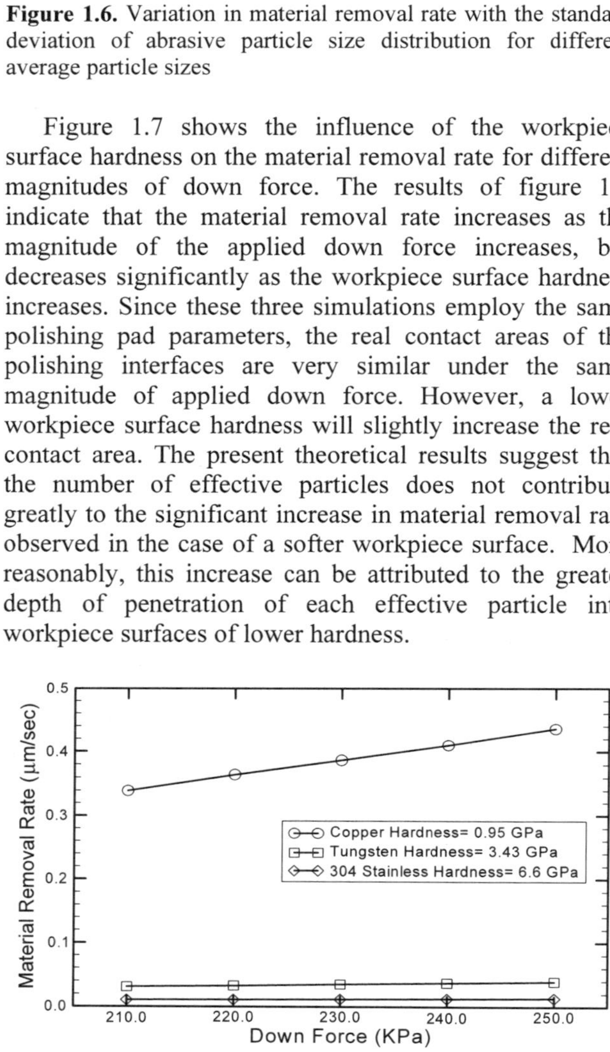

Figure 1.7. Variation in material removal rate with magnitude of down force for different values of workpiece surface hardness

4. Conclusion

This study has presented a material removal rate model which is based on surface asperity contact mechanics and considers abrasive particle effects between the polishing interface. The current model considers the consumable parameters relating to the polishing pad, workpiece surface, and the slurry. A parametric study has been conducted to investigate the relative influences on the polishing results of the applied down force magnitude, slurry particle size, the standard deviation of abrasive particle size distribution and the workpiece surface hardness. The results provide a detailed consideration of the interface phenomena and yield an insight into the physical effects of the operating parameters in the metal polishing process.

Acknowledgements

The authors would like to express gratitude for the support from National Science Council under grant NSC 95-2221-E-274-001 and No. NSC 95-2221-E-274- 008.

References

[1] E.C. Rabinowicz, 1968, "Polishing," Scientific American, Vol. 218, No.6, pp.91-99.
[2] B. Bhushan, 1999, Principles and Applications of Tribology, Wiley & Sons Press.
[3] E. Rabinowicz, L. A. Dunn and P. G. Russell, 1961, "A Study of Abrasive Wear Under Three-body Conditions," Wear, Vol. 4, pp. 345-355.
[4] J. D. Gates, 1998, "Two-Body and Three-Body Abrasion: A Critical Discussion," Wear, Vol. 214, pp. 139-146.
[5] G. B. Stachowiak, G. W. Stachowiak, 2001, "The Effects of Particle Characteristics on Three-Body Abrasive Wear," Wear, Vol. 249, pp. 201-207.
[6] Y. Xie and B. Bhushan, 1996, "Effects of Particle Size, Polishing Pad and Contact Pressure in Free Abrasive Polishing," Wear, Vol. 200, pp. 281-295.
[7] R. Chauhan, Y. Ahn, S. Chandrasekar and T. N. Farris, 1993, "Role of Indentation Fracture in Free Abrasive Machining of Ceramics," Wear, Vol.162-164, pp. 246-257.
[8] J. Luo and D. A. Dornfeld, 2001, "Material Removal Mechanism in Chemical Mechanical Polishing," IEEE Trans. Semiconductor Manufacturing, Vol. 14, No. 2, pp. 112-133.
[9] Y. R. Jeng and P. Y. Huang, 2004, "Impact of Abrasive Particles on the Material Removal Rate in the Chemical-Mechanical Polishing Process: A Micro-Contact Perspective," Joint publication of IEEE and ECS, Electrochemical and Solid-State Letters, Vol. 7, No. 2, pp. G40-G43.
[10] Y. R. Jeng and P. Y. Huang, 2005, "A Material Removal Rate Model Considering Interfacial Micro-Contact Wear Behavior for Chemical Mechanical Polishing," Transactions of the ASME, Journal of Tribology, Vol. 127, No. 1, pp. 190-197.
[11] Y. R. Jeng and P. Y. Wang, 2003, "An Elliptical Microcontact Model Considering Elastic, Elastoplastic and Plastic Deformation," ASME J. Tribology, Vol. 125, pp. 232-240.

Dummy Heat Source Technique for Estimating the Machining Error Caused by Thermal Behavior in Fine Boring

Y.L. Tang[1], H. Sasahara[2], K. Nishimura[3] and T. Watanabe[4]

[1] Tokyo University of Agriculture and Technology, 2-24-16 Nakacho, Koganei, Tokyo, Japan, 184-8588, 50006701207@st.tuat.ac.jp
[2] Tokyo University of Agriculture and Technology, 2-24-16 Nakacho, Koganei, Tokyo, Japan, 184-8588, sasahara@cc.tuat.ac.jp
[3] Nissan Motor Co., LTD., 6-1, Daikoku-cho, Tsrumi-ku, Yokohama-shi, kanagawa, Japan, 230-0053, ki-nishimura@mail.nissan.co.jp
[4] Nissan Motor Co., LTD., 6-1, Daikoku-cho, Tsrumi-ku, Yokohama-shi, kanagawa, Japan, 230-0053, takafumi-w@mail.nissan.co.jp

Abstract. During the process of fine boring of engine cylinder blocks in the automotive industry, a machining error occurs mainly due to mechanical issues and the cylinder's thermal behaviour. This paper introduces a new experimental method to investigate the influence of thermal behavior on machining errors independent of mechanical factors. The temperature distribution in a cylinder during the actual boring process was successfully simulated by heating the cylinder using a YAG Laser. Then, based on the temperature distributions, an FEM model was developed to predict the thermal deformation of the cylinder's diameter during machining. The results showed that the effect of the thermal behavior on machining errors can be investigated. As well, in order to establish parameters of the method, the YAG Laser absorption rate for cast iron (FC250) and the quantity of heat flow into a cylinder during the actual fine boring process were investigated.

Keywords: Fine boring, Machining error, FEM, YAG Laser

1. Introduction

The automotive industry is devoting increasing effort to upgrading the machining accuracy of engine cylinders. Therefore, a kind of high precision manufacturing technology, fine boring, has been applied to enlarge cylinder holes. However, during fine boring, heat caused by metal cutting flows into the cylinder and causes its temperature to rise which in turn leads to a thermal expansion of the cylinder. This will obviously affect the cylindricality of the machined cylinder. Thus, it is important to quantitatively elucidate the correlation between the temperature increase in the cylinder and machining error.

On the other hand, this rise in temperature is also linked to cutting conditions. Temperature rise can be minimized under ideal cutting conditions. Thus, if the correlation between temperature increase and machining error is elucidated, the connection between cutting conditions and machining errors can be clearly estimated, allowing the cutting conditions of the fine boring process to be optimized to achieve high machining accuracy. Moreover, machining errors are affected not only by temperature rise but also by mechanical factors such as elastic deformation or machine vibration [1]. Clarification of the influence of temperature rise on machining errors is also useful to determine whether temperature rise or mechanical factors have a larger influence on machining errors during the boring process. This would provide a clear approach to the optimization of cutting conditions or the refinement of the boring sequence, leading to a higher degree of machining accuracy.

Usually, machining errors are investigated by performing a cutting experiment. However, as mentioned above, besides increases in temperature, machining errors are also caused by other factors during the boring process. Thus, it is very difficult to precisely determine the correlation between temperature increase and machining errors using conventional cutting experiments.

In this study, to investigate the influence of temperature rise and thermal expansion apart from the effects of machining errors, the temperature distribution within the cylinder is simulated by a heating process instead of the actual boring process. A YAG Laser is used as a heat source to heat the cylinder. The quantity of heat transferred into the cylinder causes an increase in temperature and an uneven temperature distribution within the cylinder, mimicking the characteristics of the actual boring process. The temperature distribution causes the thermal expansion of the cylinder. So, in contrast to the actual fine boring, the machining error is affected only by the rise in temperature. Thus, a quantitative correlation between temperature increase and machining error can be obtained. This method is referred to in this paper as the "Dummy Heat Source Technique."

For proving this method's validity, first an actual fine boring process was performed (cutting speed: 650 m/min). A Laser heating experiment was then carried out. The results proved that the temperature distribution of the actual fine boring process can be simulated by heating using a YAG Laser. An FEM analytical model was then developed to predict the thermal deformation of cylinder diameter during the Laser heating experiment. As well, to establish a parameter of the method, the YAG Laser absorption rate of cast iron (FC250) was investigated.

2. Identification of Laser Absorption Rate

When a Laser irradiates a metal surface, a portion of the energy will be transmitted into the metal and a portion of the energy will be reflected off the surface. The ratio of energy absorbed into the workpiece against the energy of the Laser irradiated on the workpiece is defined as follows:

$$\lambda = \frac{\text{Energy absorbed into the workpiec}}{\text{Laser energy irradiated on the workpiece}} \quad (1)$$

The total energy of the Laser irradiated onto the workpiece can be calculated as a multiple of the Laser power (J/s) and the irradiation time (s). On the other hand, the energy absorbed into the workpiece, and alternatively the heat transmitted into workpiece, can be calculated as the temperature difference of the workpiece and the specific heat capacity of the workpiece material using the following relationship:

$$q = \Delta T \cdot M \cdot C \quad (2)$$

Where q is the energy absorbed into the workpiece (J), ΔT is the average temperature difference of the workpiece (ºC), M is the mass of the workpiece (kg), and C is the specific heat capacity of the workpiece (J/kg·K) [2].

In this paper, cast iron (FC250) is used as the material of the cylinder. In order to quantitatively determinate the heat absorbed into the cast iron by heating using a YAG Laser, an identification test of the YAG Laser of the cast iron plate was carried out as shown in Figure 1.1.

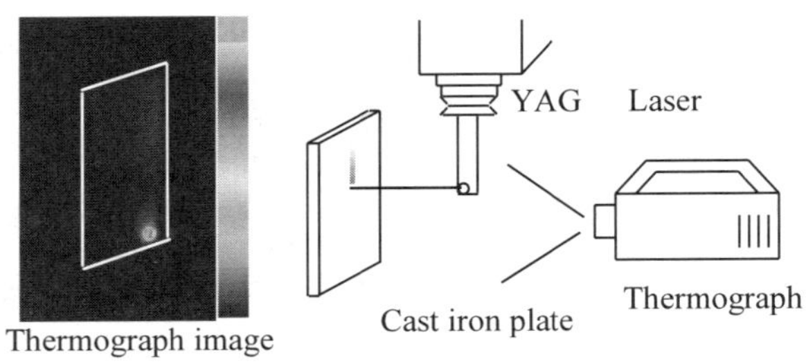

Figure 1.1. YAG Laser absorption experiment

The temperature field on the plate surface was measured using a thermograph camera, and the Laser absorption rate could then be calculated based on the above relationships. The YAG Laser is focused at the surface of the plate giving a beam spot size of 1.0 mm diameter. The Laser-irradiated point then moves at a constant speed along the plate. The thermograph camera (VarioCAM) was positioned at the counter side against the plate surface to measure the temperature field on the plate. By marking the surface of the plate in the thermograph image shown in Figure 1.1 (white line), the average temperature of surface plate was calculated using thermograph software.

In addition, a black coating material (Acheson Aero504) was applied on the surface of the plate, because the absorption rate of the metal surface is relatively low and a sufficient temperature rise couldn't be achieved. The black coating can achieve a higher absorption rate. The experimental conditions are as follows:

- Dimensions of the plate: 90 mm × 45 mm
- Thickness of the plate: 8mm
- Laser power: 60 J/s
- Travel speed: 180, 250, 300 mm/min

The results regarding the measured average temperature rise and the calculated YAG Laser absorption of cast iron (FC250) under various travel speeds are summarized in Table 1.1. The average calculated Laser absorption was then estimated to be that of the Laser absorption of cast iron (FC250): 75%

Table 1.1. Temperature rise and Laser absorption rate

Traveling speed	mm/min	180	250	300
Average temperature rise	ºC	9.9	8.6	7.7
Absorption rate	%	66	76	85

In the above experiments, the difference in temperature between the plate's surface and its inside leads to the estimation error of the absorption rate. The thermograph camera cannot measure the internal temperature. So, in order to investigate the difference between internal temperature and surface temperature, an FEM model identical to the plate used in the experiment was made. FEM temperature analysis was then performed under the same conditions as those used in the identification of the Laser absorption experiment.

Figure 1.2 shows the simulated temperature distribution of the model. The Laser heating point moves from the top of the plate to the bottom. The cross section of the model indicates that the temperature at the irradiating point is high, but a large temperature difference cannot be seen at the upper part where the Laser bean had moved along a couple of seconds earlier.

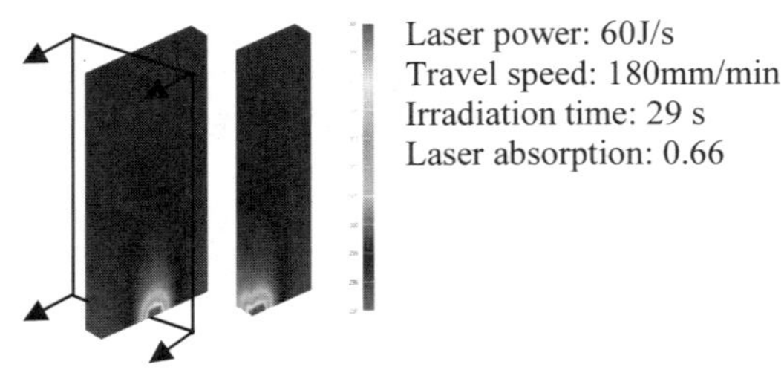

Figure 1.2. Verification of FEM temperature analysis

The average temperature increase in the whole model was calculated and compared with that measured on the plate surface experimentally as shown in Figure 1.3. These comparisons show close agreement between the analytical results and the measurements. These results show the validity of the determined Laser absorption rate.

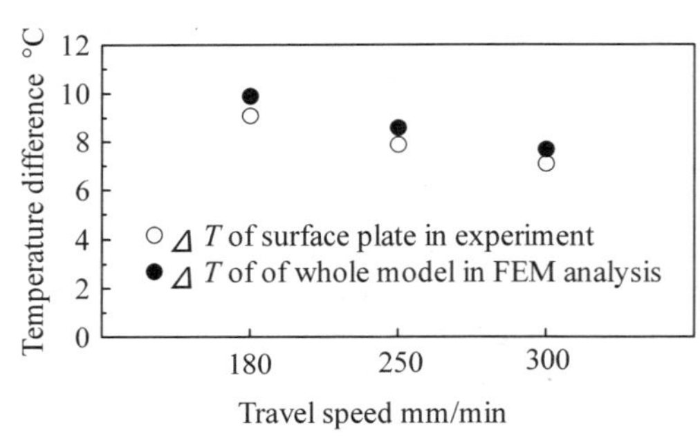

Figure 1.3. Temperature comparison

3. Fine Boring Process Experiment

In order to obtain a sample temperature distribution of the actual fine boring process, a fine boring process experiment was performed under the following cutting conditions:

- Cutting speed: 650 m/min, Feed rate: 0.2 mm/rev
- Depth of cut: 0.325 mm, Cutting fluid: Dry
- Length of the cylinder specimen: 135.0 mm
- Diameter of the cylinder specimen: 83.2 mm
- Thickness of wall: 7.4 mm

The experimental setup is shown in Figure 1.4. To measure the temperature distribution in a cylinder during machining, three pairs of thermocouples (K type) were mounted at three points (A, B, C) at distances of 10, 60, and 120 mm, respectively, from the top of the cylinder. These thermocouples are plugged in a small hole with a diameter of 1.0 mm. The junction of the thermocouples was located at a position 1.0 mm from the machining surface in order to measure the temperature. As an example of the measurement, Figure 1.5 shows the temperature distribution of measured point A.

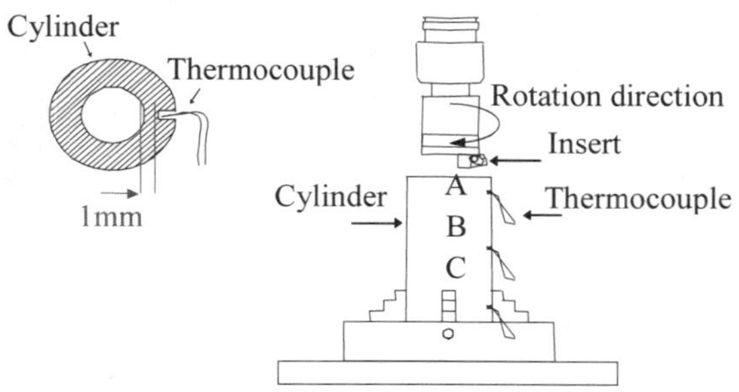

Figure 1.4. Experimental setup of fine boring process

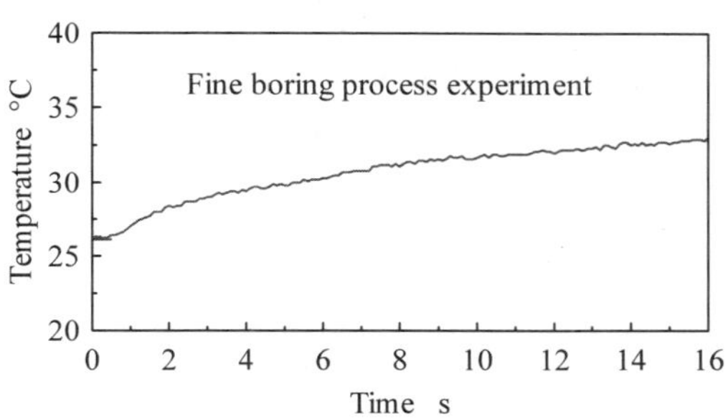

Figure 1.5. Temperature distribution of measured point A

4. Experiment using Dummy Heating Technique on Cylinder

A YAG Laser heating experiment was performed. As a key point of this experiment, the quantity of heat flowing into the cylinder during an actual fine boring process has to be calculated. Thus, sets of cutting experiments and FEM analyses were performed. The heat flow into a workpiece under various cutting conditions was determined in our prvious study [3] in which it was shown that a heat of 4860 J will be transferred into the workpiece under the following cutting conditions: cutting speed 650m/min, feed rate 0.2 mm/rev, depth of cut 0.325 mm, dry machining. Based on the amount of heat and Laser power (maximum power output: 60 J/s), the feed rate of the YAG Laser unit and irradiation time was determined using the following relationship:

q: Heat flows into cylinder = 4860 J
P: Laser power= 60 J/s
λ: Laser absorption rate = 75 %
L: the length of the cylinder specimen = 135 mm

Irradiation time: $t = \frac{q}{P \cdot \lambda} = 108$ s

Feed rate of the Laser unit: $F = \frac{L}{t} \cdot 60 = 75$ mm/min

The experimental setup is shown in Figure 1.6. The rotating Laser unit moves along the axial direction in the cylinder and irradiates the inner wall to heat the cylinder. The rotational speed of the Laser unit is 600 min^{-1}. To compare the heat induced by the Laser with the temperature created during actual fine boring process, three pairs of thermocouples were mounted at three points (A, B, C) 10, 60, and 120 mm, respectively, from the top of the cylinder to measure the temperature distribution.

As an example, Figure 1.7 shows the temperature curve at point A. Comparison this curve with the temperature distribution of the actual fine boring experiment (Figure 1.5) proves that the temperature curve of the fine boring process can be successfully simulated using the dummy heat source technique.

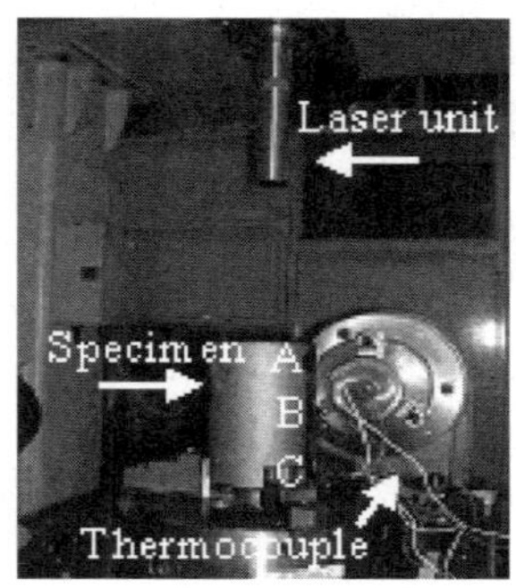

Figure 1.6. Experimental setup of Laser heating experiment

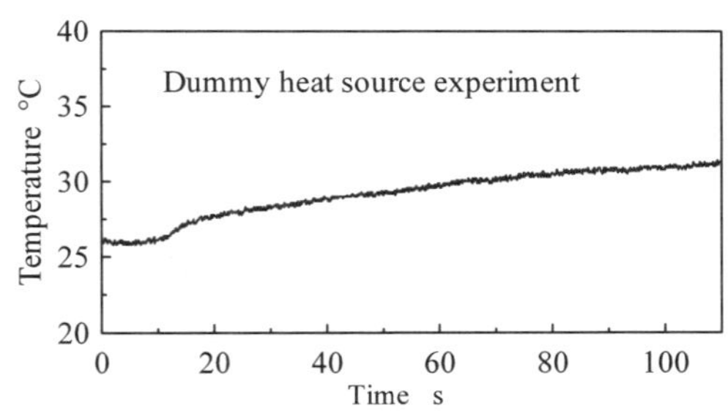

Figure 1.7. Temperature distribution of Laser heating experiment

5. FEM Analysis of Heating and Deformation

It is well known that FEM simulation is a very effective method to investigate the thermal behavior of workpieces during machining. For example, Zhang [4], proposed a model system formulated in the state variable space to describe the thermal behavior of a boring machining system. In this paper, a FEM model was developed to simulate the temperature distribution and thermal expansion of cylinder during a dummy heating experiment. The analytical conditions are the same as those of the dummy heat source experiment.

Figure 1.8 shows the predicted results. It can be seen that the high temperature region moves along the feed direction from the top to the bottom. It should be noted that the temperature at the bottom is higher than that of middle part or top part because the heat is accumulated by heat conduction and the bottom surface of the cylinder is defined as an adiabatic boundary.

The analyzed temperature transition of measurement point A (whose position is shown in Fig. 1.4) is shown in Figure 1.9. Close agreement is seen between the experimental and predicted temperature distributions. Thus, it appears that the thermal expansion of the cylinder during the dummy heat source experiment can be predicted by the FEM analytical model. For example, Figure 1.10 shows the thermal expansion of the diameter of the top of the cylinder during the dummy heating process. It shows that the thermal expansion increases with temperature, finally reaching a constant value.

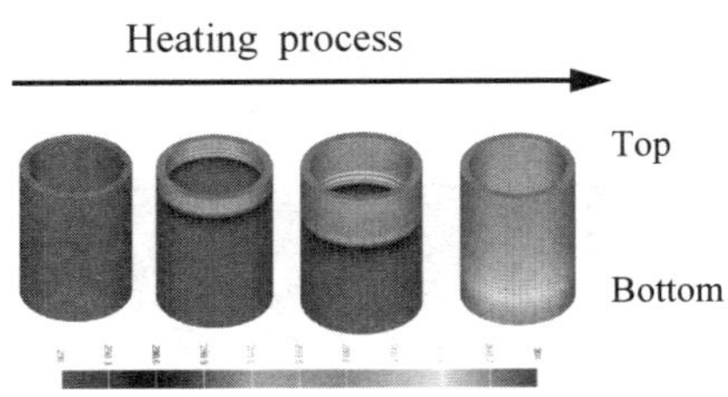

Figure 1.8. FEM simulations Laser heating experiment

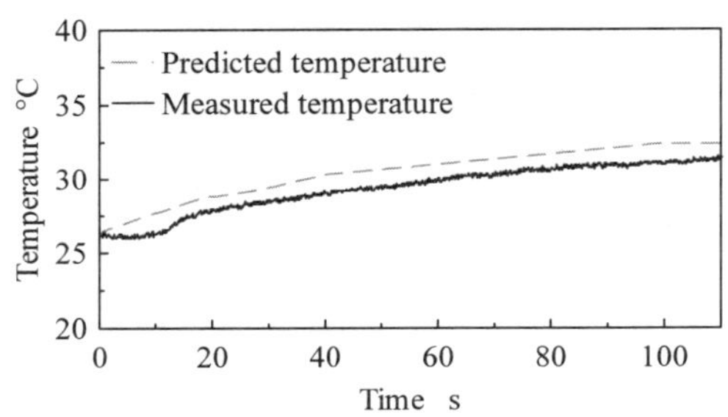

Figure 1.9. Predicted vs. measured temperature

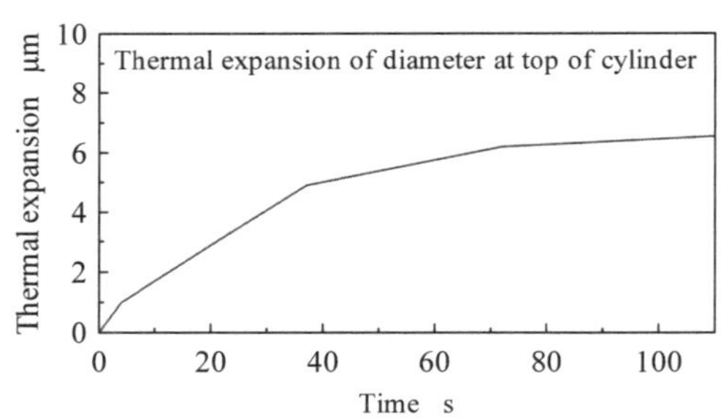

Figure 1.10. Thermal expansion in FEM analysis

The ranges of thermal expansion of the diameters at the top, middle, and bottom of the cylinder during the heating process are shown in Figure 1.11. It can be seen that the range of thermal expansion differs according to the area of the cylinder. The maximum thermal expansion of the cylinder's diameter occurs at the bottom of the cylinder because of the temperature is highest in this area, as shown in Figure 1.11. The thickness of the material removed from the wall will be smaller than that removed under a non-expansion condition during machining, assuming that the actual machining diameter is constant at any regions. After the cylinder is cooled down to the room temperature, the cylinder's diameter shrinks. Then, the machined diameters along the cylinder's axial direction will thus differ according to the cylinder's area. This behavior affects the shape of the cylinder.

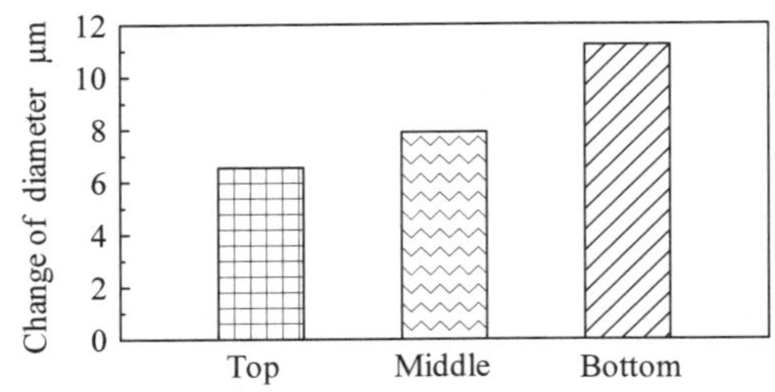

Figure 1.11. Changes of cylinder diameter during heating process

6. Conclusions

The quantitative effects of an increase in temperature on the machining accuracy of a cylinder can be investigated by means of the dummy heat source method. In this paper, in order to demonstrate the method's validity, an actual fine boring process and a Laser dummy heating experiment were conducted. The temperature distributions in the cylinder obtained by the Laser dummy heating experiment were compared with those of the actual boring process. The results proved that the temperature distribution of the actual fine boring process can be substituted by the heating experiment using a YAG Laser. A FEM analytical model was then developed to predict the thermal expansion of cylinder diameter during the dummy heating experiment.

References

[1] Subramani, G., SG Kapoor, RE DeVor, A Model for the Prediction of Bore Cylindricity During Machining, ASME J. Eng. Ind 115 (1993) 15–22.

[2] Woo H.G, Cho H.S, Three-dimensionnal temperature distribution in Laser surface hardening process, Proc Instn Mech Eng. (1999) Vol 213 part B, 695-712.

[3] Y.L. Tang, H. Sasahara, Effect of Machining Strategy on Thermal Expansion in Fine Boring Process, Proc. ICPMT2006, Matsue(Japan), 2006/11, 425-428.

[4] Zheng Y., LI H., Olson W.W., Sutherland J.W., Evaluating Cutting Fluid Effects on Cylinder Boring Surface Errors by Inverse Heat Transfer and Finite Element Methods, Journal of Manufacturing Science and Engineering 122 (2000) 377-383.

An Investigation of the Tool–Chip Interface Temperature and Heat Partition in High-speed Machining of AISI/SAE 4140 Steel with TiN-coated Tool

F. Akbar, P.T. Mativenga and M.A. Sheikh

School of Mechanical, Aerospace and Civil Engineering, The University of Manchester, Sackville Street Building, M60 1QD, Manchester, UK
faraz.akbar@postgrad.manchester.ac.uk

Abstract. Cutting metals at high speeds is associated with high temperatures. Determination of the maximum temperature along the rake face of the cutting tool is therefore of particular importance because of its controlling influence on tool life. This paper reviews the models found in literature for the prediction of the percentage of heat from the secondary deformation zone that flows into the tool in dry high speed turning of AISI/SAE 4140 high strength alloy steel with TiN coated tool. The benchmark involves analytical and experimental based models used for estimating the heat partition coefficient in the tool and the average tool-chip interface temperature. This work is important for quantifying the thermal flux or in selecting and developing appropriate tool coatings.

Keywords: Heat Partition; High Speed Machining; TiN Coating

1. Introduction

Dry machining demands specially engineered cutting materials with extreme hot hardness, improved fracture toughness as well as resistance to abrasion, thermal shock, adhesion, and diffusion. Generally, altering the properties of the tool material by deposition of hard or hard/soft protective coatings can substantially slow down some wear processes, reduce friction and prolong tool life. As a result, currently more than 70% of all cemented carbide inserts feature coatings applied by chemical vapour depostion (CVD), physical vapour deposition (PVD) or a combinations of these deposition techniques [1]. The commercial coatings most widely used in metal cutting are TiC, Al_2O_3 (CVD) technique, TiAlN (PVD), TiN, TiCN (CVD and PVD techniques) [1-3].

In order to make the coating selection more effective much attention should be focused on the role of the properties of the tool materials. Among the available coating systems, titanium-based hard thin films have found the widest acceptance. These tend to improve the wear resistance in many cutting applications [4, 5], by reducing friction, adhesion, and diffusion. PVD TiN coatings are widely used to improve the tribological and mechanical properties of the tool [6, 7]. This paper reports on an investigation of heat partition and the tool-chip interface temperatures for TiN coated carbide tools in high speed machining.

2. Tool-chip interface temperature measurement

The computation of the mean tool-chip temperature is based on dimensional analysis developed by Cook [8]. The rise in mean temperature, $\Delta\theta_t$, at the tool-chip interface is given by equation 1.1.

$$\Delta\theta_t = 0.4\frac{u}{\rho c_p}\left[\frac{V_c t}{\alpha_W}\right]^{1/3} \quad (1.1)$$

where, u is the total specific cutting energy, V_c is the cutting velocity, ρc_p is volumetric specific heat for the workpiece, α_W is the thermal diffusivity of the work material, t is undeformed chip thickness and A_c is the cross-sectional area of the uncut chip. Since the work material properties α_W and ρc_p depend on temperature, this equation is applied iteratively, with new material properties calculated from successive values of $\Delta\theta_t$, until convergence is achieved.

3. Analytical models for prediction of heat partition coefficient

In this study, the prediction of the fraction of heat which flows into the chip is based on the determination of heat partition coefficient (R_{ch}), which defines the percentage of the heat entering the moving chip. The fraction $R_T = 1\text{-}R_{ch}$ therefore defines the heat energy going into the tool which is stationary relative to the heat source. For modelling purposes four different methods have been assessed for the analytical calculation of heat partition coefficient.

3.1 Loewen and Shaw's model

The model developed by Loewen and Shaw [9], for assessing heat partition incorporates modification to the classical solutions of Jaeger's [10] moving band (for the chip) and stationary rectangular (for the tool) heat sources for application to metal cutting. According to this model, the fraction of heat going into the chip R_{ch} can be calculated by equation 1.2.

$$R_{L(ch)} = \frac{\frac{q\, l_{ch}\, A_a}{\lambda_T} - \Delta\theta_S + \theta_o}{\frac{q l_{ch} A_a}{\lambda_T} + \frac{0.377\, q\, l_{ch}}{\lambda_W \sqrt{N_p}}} \tag{1.2}$$

where, λ_W and λ_T are the thermal conductivities of workpiece and tool materials, q is the frictional heat flux, N_p is a non-dimensional parameter, and A_a is the area shape factor which describes the thermal and geometric features of the interface. The non-dimensional parameter N_p is estimated according to equation 1.3a [9],

$$N_p = \frac{V_{ch} \frac{l_{ch}}{2}}{2\alpha_W} \tag{1.3a}$$

and the area shape factor A_a is given by equation 1.3b,

$$A_a = \frac{2}{\pi}\left[\sinh^{-1}\left(\frac{m}{l}\right) + \left(\frac{m}{l}\right)\sinh^{-1}\left(\frac{l}{m}\right) + \frac{1}{3}\left(\frac{m}{l}\right)^2 + \frac{1}{3}\left(\frac{l}{m}\right) - \frac{1}{3}\left\{\left(\frac{l}{m}\right) + \left(\frac{m}{l}\right)\right\}\left\{1 + \left(\frac{m}{l}\right)^2\right\}^{0.5}\right] \tag{1.3b}$$

where, m/l is the aspect ratio of the contact area [11], $m = a_p$ is depth of cut, and l is half length of slider. The rise in the average shear plane temperature, $\Delta\theta_s$, is calculated based on the theory of similarity elaborated for metal cutting by Silin [12]. According to Grzesik [13], the theory of similarity facilitates the calculation of this temperature more accurately than Oxley [14] model. The maximum temperature rise in the shear plane can computed from equation 1.4a [12] :

$$\theta_{s\max} = \frac{\tau_{sh}}{c_p\, \rho\, \tan\phi}\, erf\sqrt{\frac{R_t \tan\phi}{4}} \tag{1.4a}$$

where, τ_{sh} is the shear flow stress, ϕ is the shear angle, and erf is the error function. The non-dimensional thermal number R_t is calculated using the Boothroyd equation (1.4b) [15] as:

$$R_t = \frac{\rho\, c_p V_c\, t}{\lambda_W} \tag{1.4b}$$

The average temperature rise at the shear plane is then modelled using equations 1.5a, 1.5b, and 1.5c [12]:

$$\Delta\theta_s = 0.685(R_t \tan\phi)^{0.07}\,\theta_{s\max} \quad (\text{for } R_t \tan\phi \le 5) \tag{1.5a}$$

$$\Delta\theta_s = 0.620(R_t \tan\phi)^{0.13}\,\theta_{s\max} \quad (\text{for } 5 \le R_t \tan\phi \le 20) \tag{1.5b}$$

$$\Delta\theta_s = 0.820(R_t \tan\phi)^{0.04}\,\theta_{s\max} \quad (\text{for } R_t \tan\phi \ge 20) \tag{1.5c}$$

3.2 Reznikov's model

Reznikov [16] proposed a model for the amount of heat flowing into the chip using the thermal conductivity and the thermal diffusivity values of the tool and workpiece materials according to equation 1.6.

$$R_{R(ch)} = \frac{1}{1 + [(3\lambda_T / 2\lambda_W)\sqrt{(\alpha_W / \alpha_T)}]} \tag{1.6}$$

3.3 Shaw's model

Shaw [11], proposed that the slider is not a perfect insulator and part of 'q' i.e. ($R_{ch}\, q$) will flow into the chip and (1-R_{ch}) q will flow into the tool, and developed the model given by equation 1.7a.

$$R_{SH(ch)} = \frac{1}{1 + [0.754\ (\lambda_T / \lambda_W)/(A_a \sqrt{N_T})]} \tag{1.7a}$$

where, N_T is a dimensionless velocity diffusivity factor, and A_a is the area shape factor evaluated from equation 1.3b. The dimensionless number N_T is given by :

$$N_T = \frac{V_{ch}\, l_{ch}}{2\alpha_W} \tag{1.7b}$$

3.4 Kato and Fujii's model

Kato and Fujii [17] determined the heat partition coefficient for conventional surface grinding based on the values of the heat transmission coefficients (triple products of $c_p\ \lambda\ \rho$) for the workpiece and the grinding wheel materials. Grzesik [18] used a similar approach in metal cutting with multilayer coated tools. By analogy, the heat partition coefficient for given thermal properties of workpiece (W) and tool (T) materials is given by:

$$R_{KF(ch)} = \frac{1}{1 + \sqrt{(c_p\, \rho\, \lambda)_T / (c_p\, \rho\, \lambda)_W}} \tag{1.8}$$

By further manipulating the above heat partition models, the heat going into tool, R_T, can be directly computed, as summarised in Table 1.1. These models are assessed here for heat partition in orthogonal turning using coated tools.

Table 1.1 R_T heat partition models

Model	Equation for the determination of R_T
Loewen and Shaw [9]	$R_{L(T)} = \dfrac{\frac{0.377 q\, l_{ch}}{\lambda_W \sqrt{N_p}} + \Delta\theta_s - \theta_o}{\frac{q\, l_{ch}\, A_a}{\lambda_T} + \frac{0.377 q\, l_{ch}}{\lambda_W \sqrt{N_p}}}$
Reznikov [16]	$R_{R(T)} = \dfrac{[(3\lambda_T / 2\lambda_W)\sqrt{(\alpha_W / \alpha_T)}]}{1 + [(3\lambda_T / 2\lambda_W)\sqrt{(\alpha_W / \alpha_T)}]}$
Shaw [11]	$R_{SH(T)} = \dfrac{[0.754(\lambda_T / \lambda_W)/(A_a \sqrt{N_T})]}{1 + [0.754(\lambda_T / \lambda_W)/(A_a \sqrt{N_T})]}$
Kato and Fujii [17]	$R_{KF(T)} = \dfrac{\sqrt{(c_p \rho \lambda)_T / (c_p \rho \lambda)_W}}{1 + \sqrt{(c_p \rho \lambda)_T / (c_p \rho \lambda)_W}}$

4. Experimental investigations

Turning tests were conducted on a "Dean Smith & Grace 1910T Lathe". A Kistler cutting force dynamometer Model 9121 was used to measure the actual cutting force. Coated inserts of ISO specification TCMW 16T304 6620 with a single layer of TiN coating were used. The inserts were mounted on an ISO specification STGCR/L 2020K 16 tool-holder (screw clamping). The pre-bored workpiece used was AISI/SAE 4140 high tensile alloy steel with an external diameter of 200 mm and 2.5 mm tube thickness. The cutting tests were performed at eight different cutting speeds of 100, 197, 314, 395, 565, 628, 785, and 880 m/min. Feedrate and depth of cut were kept constant at 0.1 mm/rev and 2.5 mm respectively. Temperature was measured using an infrared thermal imager FLIR thermal CAM SC3000.

To determine the emissivity of the cutting tool insert at various temperature levels, the inserts were oven heated to temperatures ranging from 100°C up to 900°C. For example, the thermal emissivity of the TiN coated insert was determined as 0.21 at 700°C.

5. Results and discussion

The stored thermal images were recalled and analysed by using the 'flying spotmeters and the line' (Figure 1.1). Cutting tool temperatures were measured along the tool-chip interface (line LIO1).

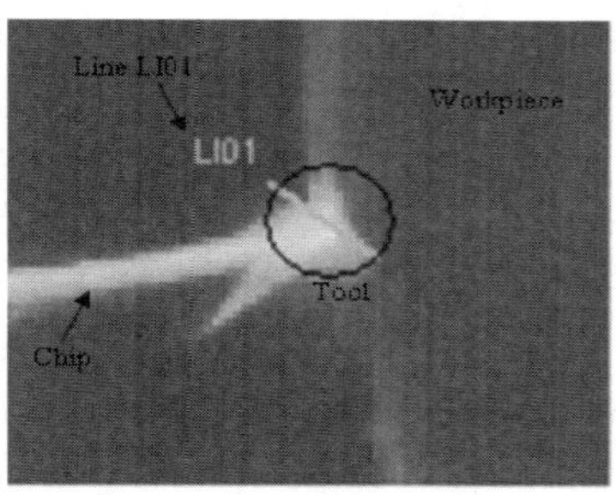

Figure 1.1 Tool-workpiece arrangement through the IR camera

Figure 1.2 presents the tool-chip interface temperature measured with the thermal imaging camera and that predicted by Cook equation (1.1). The data shows a good agreement between the two with only significant deviations occurring in the highest cutting speed region. In addition, the results show an increase of chip interface temperature with cutting speed, as expected.

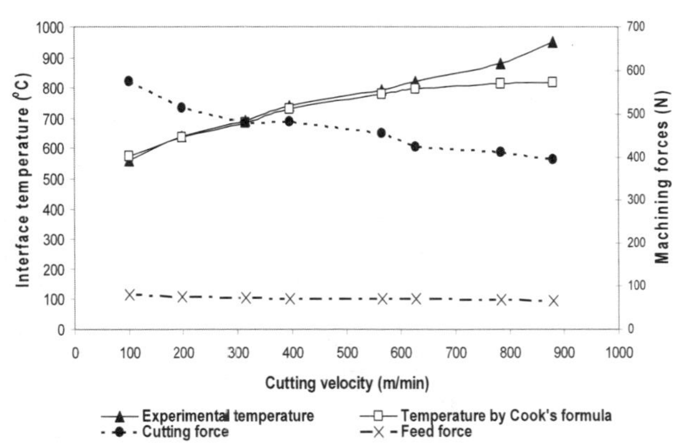

Figure 1.2 Variation of tool-chip interface temperature and machining forces with cutting speed

Also cutting and feed forces for different cutting speeds are shown in Figure 1.2. It is observed that the cutting force decreases appreciably with cutting speed, where as the feed force decreases slightly when the cutting speed is increased. The decrease in cutting force may be explained as due to the reduced chip thickness, which leads to lower contact length (or contact area) and hence higher heat flux and reduced in material shear strength. In this study, the amount of heat resulting from the sliding action of the chip over the cutting insert is characterized by frictional heat flux. The tool-chip contact area is obtained from the optical microscope images of the worn inserts at different cutting speeds. From Figure 1.3 it is observed that the contact area is affected by the cutting speed and frictional heat flux.

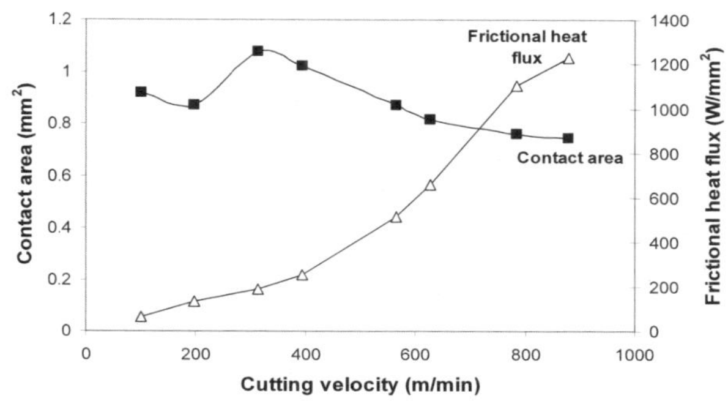

Figure 1.3. Variation of friction heat flux and contact area with cutting speed

It has been reported by Grzesik [18] that the heat partition coefficient varies from 20% to 35% when using multilayer coated tools ($TiC/Al_2O_3/TiN$), and 40% to 50% for uncoated carbide tools in the range of cutting speed of 50 to 210 m/min. The case reported here for a single layer TiN coated tool should produce heat partition coefficient values in between the multilayer coated and uncoated tools, as mentioned above for the conventional cutting speed region. From Figure 1.4, it can be seen that predicted heat partition coefficients by Kato & Fujii $R_{KF(T)}$ and Reznikov $R_{R(T)}$ increase with the cutting speed with the maximum heat going into the tool to be 45% and 55.1% respectively. These two models depend purely on the thermal properties of the tool and workpiece materials, and do not take into account the experimental process parameters like the tool-chip contact length, frictional heat intensity, and the chip velocity. On the other hand, Loewen model $R_{L(T)}$ shows that the heat partition coefficient decreases with an increase in the cutting speed.

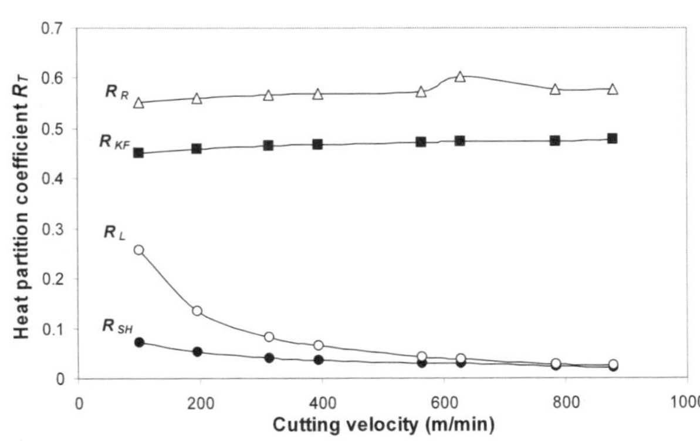

Figure 1.4. Variation of heat partition with cutting speed

From Figure 1.4, it is observed that at cutting speed of 100 m/min heat partition coefficient is 25.8%, and at maximum cutting speed of 880 m/min is 2.5%. In this case, the heat partition coefficient mainly depends on the process variables like cutting speed, depth of cut, tool-chip contact length, and, most importantly, the frictional heat flux. Shaw $R_{SH(T)}$ model predicts the heat partition coefficient to be in between 7.3% and 2.1% which represents the lowest amount of heat going into the tool of all the models.

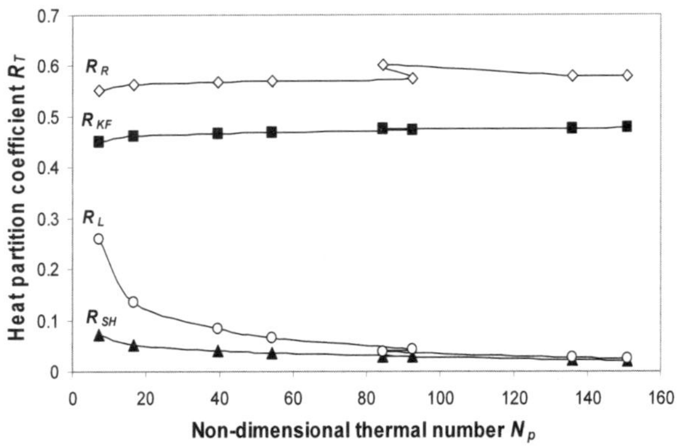

Figure 1.5. Variation of heat partition with thermal number

It has been reported [19] that the heat partition coefficient for a stationary heat source (tool) decreases rapidly with an increase in the non-dimensional thermal number, N_P. For a comparison of this effect, a high Peclet number of $N_P \approx$ 5-20 is characteristic for conventional machining of steel with a carbide tool, and a low value of $N_P \approx 0.5$ is typical for ultra precision machining of aluminum with a single-crystal diamond tool [20]. Figure 1.5 shows the variation of the heat partition coefficient as a function of the thermal number. It can be reasoned, based on Figure 1.5 that the heat partition coefficient $R_{L(T)}$ is in good agreement and decreases distinctly with an increase in the thermal number. In general, a significant reduction of the heat partition coefficient for a stationary heat source (tool) results from a rapid increase in the Peclet number [20]. A similar trend is observed with Shaw model, where $R_{SH(T)}$ decreases with an increase in the thermal number. On the other hand, $R_{K(T)}$ and $R_{R(T)}$ increase with the thermal number.

6. Conclusions

Using Cook's model, it is possible to predict with reasonable accuracy the interface temperature based on the contact characteristics and thermophysical properties of the workpiece and TiN coated tool.

The temperatures of the tool-chip interface increase with cutting speed. Relatively strong influence of the contact characteristics, including the interface temperature, the tool-chip contact area, and the thermal number, on heat partition was observed in this study.

The models developed by Reznikov and Kato & Fujii which are based only on the thermal properties of the tool and workpiece, predict a heat partition increase when the cutting speed is increased. However, models developed by Loewen and Shaw, which contain process variables, predict a heat partition decrease when the cutting speed is increased, and these appear to be closer in magnitude to values reported by Grzesik, for conventional cutting speed. In future studies, the analytical and experimental based predictions will be extended to the finite element prediction of heat partition.

References

[1] Pfouts WR, (2000) Cutting edge coatings. Manuf Eng.125(1):98-107.
[2] Prengel HG, Pfouts WR, Santhanam AT, (1998) State of the art in hard coatings for cutting tools. Surf Coat Technol.102:183-190.
[3] Klocke F, Krieg T, (1999) Coated tools for metal cutting-features and applications. CIRP Ann.48(2):1-11.
[4] König W, (1992) New approaches to characterize cutting tool coatings,. Ann CIRP.41(1):49-52.
[5] Rech J, Djouadi MA, (2001) High speed gear hobbing. Wear.250:45-53.
[6] Erdemir A, Hochman RF, (1988) Surface metallurgical and tribological characteristics of tin-coated bearing steels. Surf. Coat Technol.36:755-763.
[7] Wallen P, Hogmark S, (1989) Influence of TiN coating on wear of high speed steel at elevated temperature. Wear.130:123–135.
[8] Cook N, (1973) Tool wear and tool life. J Eng Ind.95:931-938.
[9] Loewen EG, Shaw MC, (1954) On the analysis of cutting tool temperatures. Transactions of ASME.71:217-231.
[10] Jaeger JC, (1942) Moving sources of heat and the temperature at sliding contacts. Proceedings Royal Society of NSW.76:203-224.
[11] Shaw MC, (1989) Metal cutting principles. Oxford: Clarendon Press.
[12] Silin SS, (1981) Similarity methods in metal cutting. Moscow: Mashinostroenie.
[13] Grzesik W, Nieslony P, (2004) Physics based modelling of interface temperatures in machining with multilayer coated tools at moderate cutting speeds. Int. J. of Mach. T. & Manuf.44:889–901.
[14] Oxley PLB, (1989) Mechanics of machining. An analytical approach to assessing machinability. New York Ellis Horwood.
[15] Boothroyd G, Knight WA, (1989) Fundametals of machining and machine tools. 2nd ed. New York and Basel: Marcel Dekker.
[16] Reznikov AN, (1981) Thermophysical aspects of metal cutting processes. Moscow: Mashinostroenie.
[17] Kato T, Fujii H, (1999) Energy partition in conventional surface grinding. ASME Trans J. Manuf. Sci. Engg.121:393-398.
[18] Grzesik W, Nieslony P, (2003) A computational approach to evaluate temperature and heat partition in machining with multilayer coated tools. Int. J. of Mach. T. & Manuf. 43:1311-1317.
[19] Komanduri R, Hou ZB, (2001) Analysis of heat partition and temperature distribution in sliding systems. Wear.251:925–938.
[20] Komanduri R, Hou ZB, (2001) Thermal modeling of the metal cutting process part ii: Temperature rise distribution due to frictional heat source at the tool-chip interface. International Journal of Mechanical Sciences.43:57-88.

11

Metrology

The Study of Using X-Ray Laminography on Printed-Circuit Board Inspection

H.C. Chen[1], S.C. Lin[2]

[1]Dept. of Power Mechanical Engineering, National Tsing Hua University, Taiwan, 886-3-5715131#33789, g943789@oz.nthu.edu.tw
[2]Dept. of Power Mechanical Engineering, National Tsing Hua University, Taiwan, 886-3-5742920, sclin@pme.nthu.edu.tw

Abstract. This study applies the X-ray Laminography technique to inspect printed circuit boards. The objective is to study effects of some system parameters on the reconstruction quality. The system parameters studied include the kinematical type of the system, the number of synthesized images and the projection angle. The simulation results show that when the translation type is adopted, the translation direction of the X-ray source should not parallel to the major wire direction on other planes. In other words, the direction of wire has to be specified for generating good results when using the translation type. When the rotation type is adopted, the effect of wire orientation on reconstructed image is less significant. The image quality is generally improved by increasing the number of the synthesized images. However, the improvement is less significant when the image number increased. For example, the difference for image reconstructed from 32 frames and 16 frames is less than 1% for the case examined.

Keywords: X-ray, Laminography, printed circuit board, inspection.

1. Introduction

For printed circuit board, visual inspection is frequently adopted to detect for board substrate damage, missing, miss-placed or damaged components, soldering faults and similar quality defects that can be observed by eye. Automatic optical inspection systems are being perfected to perform these inspections automatically in a growing number of installations [1].

However, these automatic optical inspection systems can only detect defects appeared on the surface of the examined object. For interior defects or defects hide behind some other objects, these optical inspection systems became useless. The use of X-ray image provides a way to solve this problem. Traditional transmission X-ray image can generally provide some clues of the defects. In order to provide more reliable and accurate information of the defect, the X-ray computed tomography [2] can be adopted to detect interior defects and provide reconstructed 3-D images.

However, X-ray computed tomography takes much more processing time than traditional transmission X-ray inspection. Even reconstructing one single image for a certain plane, hundreds of images are generally needed to reconstruct a good result. The X-ray laminography technique [3] was another approach to reveal the image on specific slices in an assembly. The laminography technique requires less processing time than the X-ray computed tomography and provide more information than traditional transmission X-ray inspection.

This is achieved by using an X-ray source and detector that move in synchronized motion, while the target being imaged remains stationary. Images are collected continuously throughout the move. All of the images that are obtained are averaged into a final image. This makes object features that are in the focal plane to be prominent in the final image; object features that are out of the focal plane are blurred out, so these features appear only faintly. By looking at a series of laminographs of neighboring planes, it is possible to get a rough idea of the three-dimensional layout of layered circuits.

In this study, it is of interest to optimize the laminography approach for printed circuit board inspection. In the following section, the fundamental theory for X-ray laminography is briefly reviewed. The experimental design and results are then presented and discussed. Finally conclusions were made based on the results.

2. The X-ray laminography

For monochromatic X-ray, the X-ray intensity left after it passing through an object, I_{ray}, can be expressed as [4]

$$I_{ray} = I_0 \exp\left[-\int_{ray} u(x,y,z)ds\right] \qquad (1)$$

where I_0 is the initial intensity of X-ray source; u(x,y,z)is the attenuation coefficient of an object at the location of (x,y,z). X-ray laminography can be used to examine the density of thin slices of an object, in contrast with traditional transmission systems which project the X-ray transmission through a 3-dimensional object onto a 2-dimensional imaging device [5]. Features found with traditional X-ray systems can be measured for width and height, but not thickness (in the z-dimension). Using laminography, the depth of a feature can be found from the acquisition of many X-ray slices.

A laminograph can be created by rotaing and translating an X-ray source in synchronization with a phosphor detector mounted on a movable table. As the table rotates/translates, the detector must be counter-rotated/translated to prevent the image from spinning/moving relative to the camera.

Over the rotation/translation of the detector, only a single plane of the inspected object will maintain a constant position on the detector. Object features above and below this focal plane will rotate/translate around the detector as the table moves. By integrating the images acquired at several locations, the contributions from features not in the focal plane will spread out over a large area, while features at the plane will constantly contribute to the detector intensity.

Kang and Cho [6] derived a geometric projection method to produce cross-sectional images for arbitrary planes using only those images acquired from the focal plane of an object. Their method was adopted in this study.

3. The experimental design

In this study, it is of interest to study effects of some system parameters on the image reconstructed on the focal plane. Table 1.1 summarizes the system parameters simulated. Table 1.2 lists the system parameters and levels studied in this paper.

Table 1.1. System parameters simulated

X-ray type	cone beam
PCB sample	10 mm x 10 mm
Sensor resolution	20 μm/pixel
Source height*	160 mm
Target height**	80 mm

* Source height: Height of source from sensor plane
**Target Height: Height of target plane from sensor plane

Table 1.2. System parameters and levels studied

Factor	Level
Kinematic type	translation, rotation
Number of images	4, 8, 16, 32
Projection angle	45, 54, 63 degree

The system parameters studied include the kinematical type of the system, the number of images taken, and the angle between the X-ray and the examined object.

(1) Kinematical type: The two kinematical types studied are the translation type and the rotation type. It is quite obvious that the rotation type might have a more compact system design than the translation type. However, it is of interest to study which type can result in a better image quality.

(2) Number of synthesized images: This is the number of images collected to reconstruct the image at the focal plane. It is expected that the more images taken the better image results can be achieved. However, the processing time will increases too. So, it is of interest to select a proper amount of synthesized image. Levels tested are 4, 8, 16 and 32 images.

(3) Projection angle: This is the angle between the direction from the X-ray source to the centre point of the object table and the normal direction of the table plane. It is expected that a smaller angle generally reduces system size. For translation type machines, the projection angle is not a constant; therefore, the maximum projection angle is used as the index. Levels tested are 45, 54 and 63 degrees.

In order to study effects of the system parameters mentioned above on the quality of the image reconstructed for the focal plane, a double sided printed circuit image sample was made for these simulation test. The sample was designed based on the latest printed circuit board manufacturing technique. Figure 1.1 shows the circuits on the top layer and bottom layer designed for this study. The wire (white lines) has a width of 200 μm.

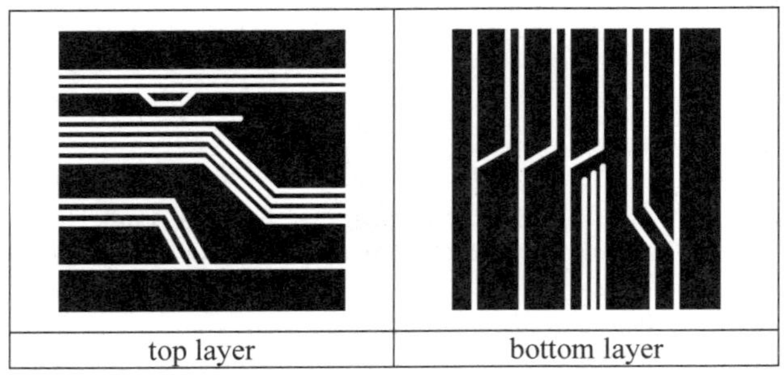

Figure 1.1. The circuit design for the test sample

In order to tell the difference between images, the frequently used correlation coefficient was used as the image similarity index in this test. The correlation coefficient, CC, was determined by the following equations,

$$CC = \frac{1}{m \times n} \frac{\sum_{y=1}^{n}\sum_{x=1}^{m}[f_S(x,y)-\mu_S]\cdot[f_T(x,y)-\mu_T]}{\sqrt{\sigma_S^2\sigma_T^2}} \tag{2}$$

$$\mu = \frac{1}{m \times n}\sum_{y=1}^{n}\sum_{x=1}^{m} f(x,y) \tag{3}$$

$$\sigma^2 = \frac{1}{m \times n}\sum_{y=1}^{n}\sum_{x=1}^{m}[f(x,y)-\mu] \tag{4}$$

where $f_S(x, y)$ and $f_T(x, y)$ are the grey levels at a given image position (x,y) of the referential image and the reconstructed image, respectively;μ_S and μ_T are the mean value of the grey levels of the referential image and the reconstructed image; σ_S ^{2}andσ_T ^{2} are the variance of the grey levels of the referential image and the reconstructed image, respectively.

4. The experimental results and discussions

Figures 1.2 and 1.3 show some reconstructed images for translation type. They show that the reconstructed image on the top layer is quite close to its referential image. However, the reconstructed image on the bottom layer is quite different to the referential image of the bottom layer. It was more like a combination of its referential image and the residual image of the horizontal wire on the top layer. This is due to the major orientation, since the wire on the top layer is horizontal and therefore parallel to the travel direction of X-ray source and sensor, while the major orientation of the wire of the bottom layer is vertical. When reconstructing the bottom layer image, the translation of the X-ray source and sensor were parallel to the wire direction on the top layer, therefore most the horizontal wire images on the top layer are not blurred. In other words, in order to get good reconstruct image the translation direction of the X-ray source should not be parallel to the major wire direction on other planes.

Figure 1.4 shows the interaction effect of the maximum projection angle and number of synthesized image on the image similarity between the tested and the referential images for translation type. It was shown that in general the correlation coefficient increased as the number of synthesized image increased. However, there is not much difference when using 8 or more frames. It was also found that the increase in the maximum projection angle increases the correlation coefficient for top layer, but decreases for the bottom layer. This might be due to the correlation between the spacing between wire and change of projection angle of the X-ray source, so that some wire images taken from one location overlap with other wire images taken from another location.

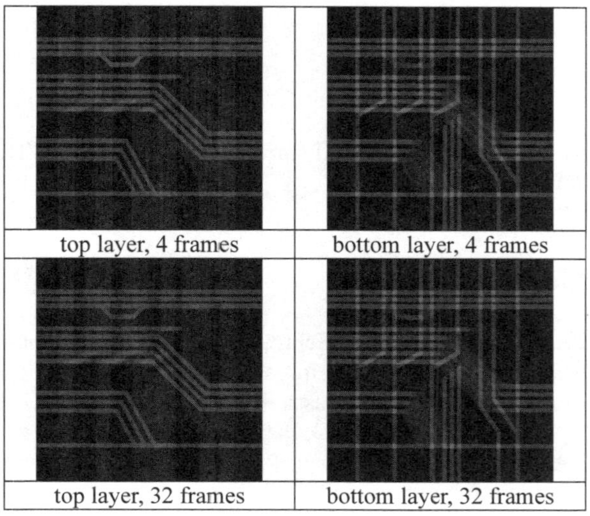

Figure 1.2 The reconstructed image of the focal plane (translation type, with maximum projection angle of 45 degree)

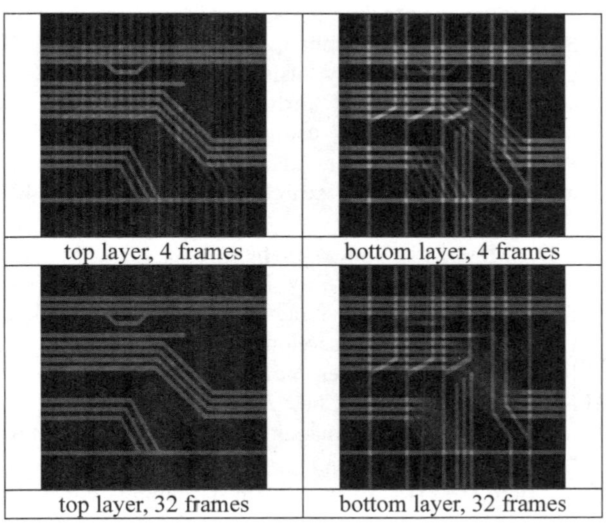

Figure 1.3 The reconstructed image of the focal plane (translation type, with maximum projection angle of 63 degree)

Figures 1.5 and 1.6 show some reconstructed images for rotation type. The reconstructed image of the top layer still had a better result than that of the bottom layer. The reason for this could be due to the wire density of the top layer, which is higher than that of the bottom layer Therefore, there are more residual images left from the top layer when reconstructing the bottom layer image. The image quality generally is improved by increasing the number of the synthesized images. Comparing with the results shown in Figs. 1.2 and 1.3, it was found that the rotational type had a better reconstructed bottom image but a worse top image than the translation type. This is due to the relationship between the movement of the X-ray source and the major direction of the wire in the non-focal plane.

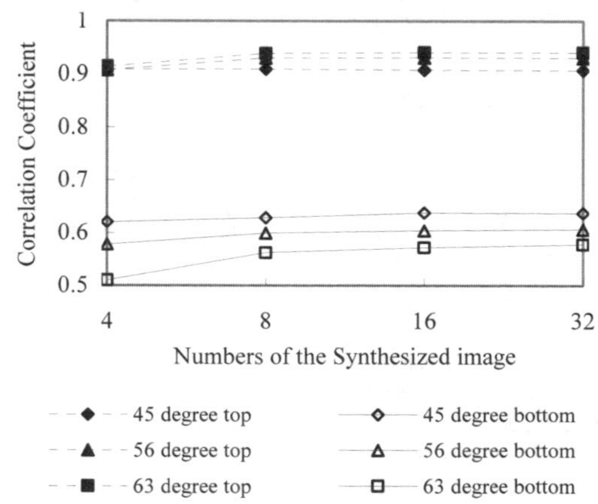

Figure 1.4. Interaction effect of maximum projection angle and image number on image similarity (translation type)

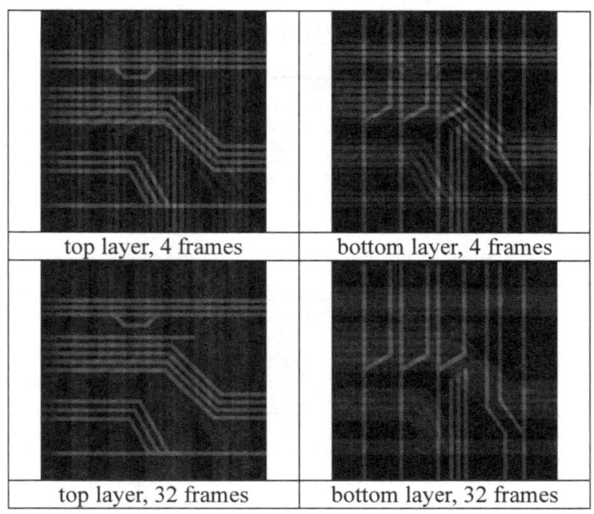

Figure 1.5 The reconstructed image of the focal plane (rotation type, with maximum projection angle of 45 degree)

Figure 1.7 shows the interaction effect of the projection angle and number of synthesized image on the similarity between the tested and the referential images for rotation type. As shown in the figure, the correlation coefficient generally increases as the image number increases for both top and bottom layer at various projection angles. However, it was found that the difference in correlation value is less than 1% when changing from 16 frames to 32 frames. It was indicated that the further increase in image number might not provide a significant improvement in image quality. It was also shown that the increase of projection angle generally improves the image result.

Comparing these results with those shown in Fig. 1.4, it was found that the difference in correlation coefficient for the top and bottom layer is greatly reduced when the rotation type was used. Although the translation type showed better results than the rotation type in reconstructing top layer image, the results with respect to the bottom layer are not acceptable at all. In other words, the direction of wire has to be given in order to achieve a satisfactory result when using the translation type. For rotation type, this specification does not need to be given.

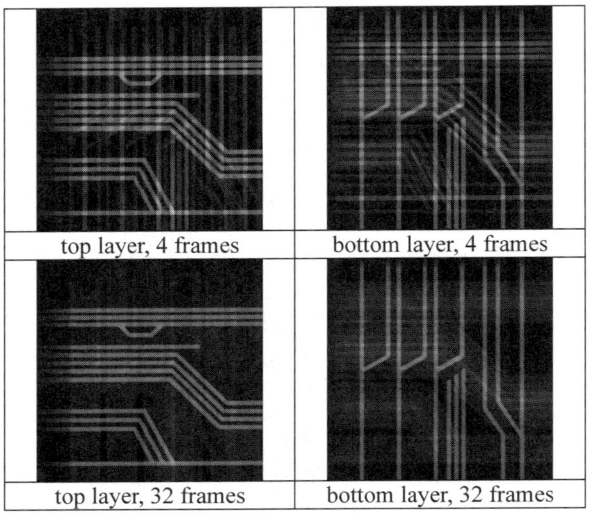

Figure 1.6. The reconstructed image of the focal plane (rotation type, with maximum projection angle of 63 degree)

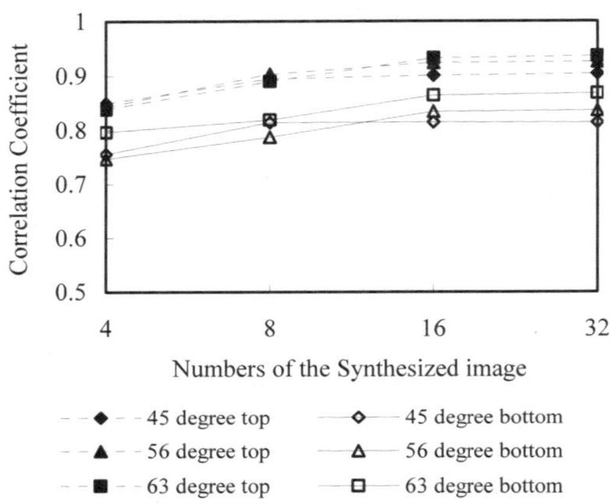

Figure 1.7. Interaction effect of maximum projection angle and image number on image similarity (rotation type)

5. Conclusion

The object of this study was to optimize the X-ray Laminography technique for printed circuit board inspection. The system parameters studied included kinematical type, number of synthesized image and projection angle of the X-ray source. The correlation coefficient is used as the test index. Based on the simulation results, the following conclusions can be made:

(1) When the translation type was adopted, the translation direction of the X-ray source should not be parallel to the major wire direction on non-focal planes. Hence, in order to achieve a satisfactory result when using the translation type, the direction of the wire should be provided. It was also found that the increase in the number of synthesized image generally improve the reconstructed image quality.

(2) When the rotation type was adopted, the increase in number of synthesized images also improved the reconstructed image quality. However, the improvement is less significant when the umber of synthesized image is more than 16. In this test, the difference in correlation coefficient for image reconstructed from 32 frames compared to only 16 frames is less than 1%. It was also found that the difference in correlation coefficient for the top layer and that for the bottom layer is greatly reduced.

Acknowledgements

The authors wish to thank the National Science Council, Taiwan, R.O.C. for their financial support (NSC95-2212-E-007-234).

References

[1] S. Gondrom, S. Schröpfer, Digital computed laminography and tomosynthesis - functional principles and industrial applications, DGZfP Proceedings, BB 67-CD, Paper 11, March 1999.

[2] A.C. Kak and M. Slaney, Principles of Computerized Tomographic Imaging, IEEE Press, 1988.

[3] S. Black, Generating Three Dimensional Models of Solder Joints Using X-ray Laminography, IEEE/CHMT Int'l Electronics Manufacturing Technology Symposium, 1992.

[4] S. Krimmel, J. Stephan, and J. Baumann, 3D computed tomography using a microfocus X-ray source: Analysis of artifact formation in the reconstructed images using simulated as well as experimental projection data, Nuclear Instruments and Methods in Physics Research A 542 (2005) 399–407, February 2005.

[5] T.D. Moore, D.Vanderstraeten, and P.M. Forssell, Three-Dimensional X-Ray Laminography as a Tool for Detection and Characterization of BGA Package Defects, IEEE Transactions on Components and Packaging Technologies, Vol. 25, NO. 2, June 2002.

[6] S. T. Kang, H. S. Cho, A projection method for reconstructing X-ray images of arbitrary cross-section, NDT&E International 32 (1999) 9–20, April 1998.

[7] F. Jian, L. Hongnian, Beam-hardening correction method based on original sinogram for X-CT, Nuclear Instruments and Methods in Physics Research A 556(2006) 379–385, November 2005.

[8] S. Bord, A. Clement, J.C. Lecomte, J.C. Marmeggi, An X-ray tomography facility for I.C. industry at STMicroelectronics Grenoble, ELSEVIER, Microelectronic Engineering 61-62 (2002) 1069-1075.

Realization of Defect Automatic Inspection System for Flexible Printed Circuit (FPC)

Chiu-Hui Chen [1], Chun-Chieh Wang [2], Chun-Yu Lin [3], Yu-Sen Shih [4] and Chung-Fan Tu [5]

[1] Industrial Technology Research Institute, Taiwan, 30011, tracy@itri.org.tw
[2] Industrial Technology Research Institute, Taiwan, 30011, kimiwang@itri.org.tw
[3] Industrial Technology Research Institute, Taiwan, 30011, Aco@itri.org.tw
[4] Industrial Technology Research Institute, Taiwan, 30011, annszu@itri.org.tw
[5] Industrial Technology Research Institute, Taiwan, 30011, CFTu@itri.org.tw

Abstract. Automatic optical inspection equipment in Asia-pacific region market has grown rapidly in recent years. In this paper, an industrial FPC application of automated optical inspection system is presented. An illumination delivers high intensity and uniform lighting. A flat vacuum table applys sufficient suction to overcome panel warping. The processor of the system receives the images and software algorithms to analyze the acquired image. The false alarm rate of the system is less than 10%.

Keywords: FPC, AOI, Affine Transformation, Defect Inspection

1. Introduction

Flexible printing circuit (FPC) is one kind of printing circuit board that is made of flexible material. Due to the characteristic of high flexibility, FPC could connect PCB and parts, which makes the product to fit the requirement of small size. Therefore, FPC are widely used for notebooks, cell phones, LCDs, and digital cameras. Up to now, the inspection of circuit defects is done by human inspection. With electronic devices getting smaller in size and thinner in pitch, and the speed requirement of the on-line process getting more and more critical, human inspection is unable to satisfy the demand of the quality and speed.

Automatic optical inspection (AOI) is used of optics, mechanism, electronic control, and software, which could replace human eye, the software is like the human brain which receives the signals from the eye. Through a specific algorithm, AOI system can inspect the defects of any product. Except the good design of optics and mechanism, and the integration of mechanism and electrical control, AOI system needs to develop an accurate and robust algorithm to inspect the defects [1].

In this paper, we present a technique that is based on machine vision for the defect inspection of FPC. Because the FPC is thin and flexible, the system design must overcome the challenge of lighting and suction to get a better quality of the images, that avoids too heavy computation in the following procedure. The inspection method has to align the inspection image with the reference image, and then carries out image subtraction and morphological processing. In the comparison of the second phase, we get the position of the inspected defects.

2. Algorithm of Defect Inspection

In this paper, we present an effective method based on machine vision for FPC defect inspection. First, the reference image and the inspection image must align to the same position. The border and the inner part of the circuit board generally have fiducial marks for alignment, and these marks may be circle, square, or other specific design shapes. Utilization of the image processing, it's easy to find the correspondent marks in the reference and the inspection images. With affine transformation, the transformed matrix between these two images can be calculated [2]. The kinds of inspected defects of FPC include open, short, pin-hole, over-etch, and under-etch. Line CCD cameras are employed to grab 2D images, which are then compared to the reference image, which is the average of several non-defect images. Figure 1.1. is the flow chart of the defect inspection algorithm. The detail of the defect inspection will be described in the sequence below.

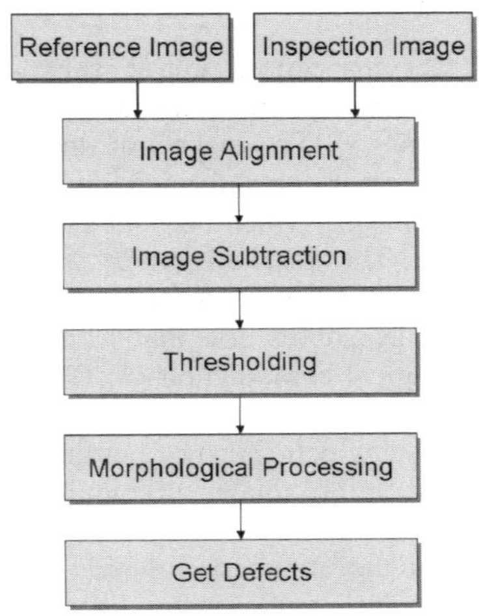

Figure 1.1. Flow chart of defect inspection algorithm

Image alignment is a very important issue in defect inspection. We use affine transformation to align the reference image and inspection image. There are several fiducial marks on the FPC for alignment, and we can also use these marks for inspection. In the reference and inspection image, searching the correspondent alignment marks like circle, and the number of correspondent marks must be more than 3 for accurate alignment. The transformed point (x', y') of the original point (x, y), that is

$$x' = a_1 x + a_2 y + a_3 \tag{1.1}$$

$$y' = a_4 x + a_5 y + a_6 \tag{1.2}$$

$a_1, a_2, ..., a_6$ are the transformation coefficients. Using matrices to describe the relation, that is

$$P_n^{'} = MP_n \tag{1.3}$$

$P_n, P_n^{'}$ are the 3*3 matrices depicting the point correspondences in the reference image and inspection image. M is the affine transformation matrix which is calculated by P_n and $P_n^{'}$.

$$P_n = \begin{bmatrix} x_1 & x_2 & \cdots & x_3 \\ y_1 & y_2 & \cdots & y_3 \\ 1 & 1 & \cdots & 1 \end{bmatrix} \tag{1.4}$$

$$P_n^{'} = \begin{bmatrix} x_1' & x_2' & \cdots & x_3' \\ y_1' & y_2' & \cdots & y_3' \\ 1 & 1 & \cdots & 1 \end{bmatrix} \tag{1.5}$$

$$M = \begin{bmatrix} a_1 & a_2 & a_3 \\ a_4 & a_5 & a_6 \\ 0 & 0 & 1 \end{bmatrix} \tag{1.6}$$

In the defect inspection, image subtraction is the simplest and most useful method that is the first of the techniques applied [3]. The inspection image is grabbed by the line CCD camera, transformed, and compared with the reference image, which is subtracted from the reference image. The residual image after subtraction shows the possible position of defects. Figure 1.2. depicts the result of image subtraction that is the same as the result of XOR logical computation [4]. The result shows the difference between the reference image and the inspection image, which is the possible position of inspected defects. Before image subtraction, one needs to sure that the inspection image is aligned with the reference image, otherwise the subtraction would produce many false defects. Besides accuracy alignment, the uniformity of lighting, and the characteristics of light reflection on the material are important issues. For high accuracy of defect inspection, an additional analysis is required to check whehter the defect is true or false, so that the false defects can be filtered out.

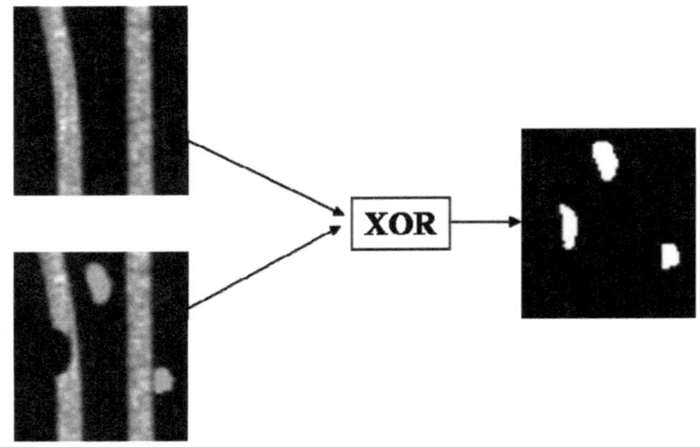

Figure 1.2. Image subtraction is the same as XOR logical computation

After image subtraction, all pixel value of the residual image should be zero if there is no defect in the inspection image. Because of the different acquisition time and the variation in lighting, there are still some counts at the circuit edge after subtraction. If there is a defect, it's residual count would be more than that of the circuit edge. Figure 1.3. is the histogram of the residual image. In our case, the pixels whose values are below 20 are the majority. The residual value between 20 and 40 is mostly at the circuit edge. However, those exceeding 40 are possibly the defects. We can choose a threshold value by the residual histogram, but still have to pay attention that too high a threshold may filter the small defects and those which have less residual value.

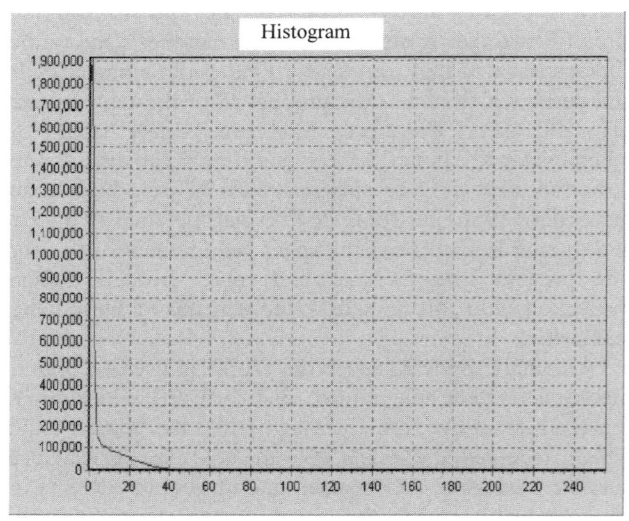

Figure 1.3. Histogram of the residual image

Morphological processing is comprised of several morphological operators. The basic operators are erosion, dilation, opening, and closing operator [5, 6]. Using these operators to analyse the shape and structure of the image has the advantage of high computation speed. Figure 1.4. shows the results of images after erosion and dilation processing. After image subtraction and thresholding, the residual regions (the value more than 255) are defects in theory. There are still false defects, because of noise, the surface of objects, and light reflection. Erosion and dilation processing can remove the small and false defects [7].

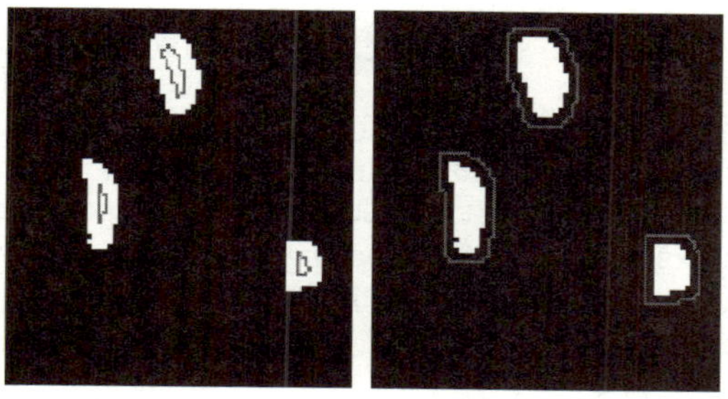

(a) (b)

Figure 1.4. Results of morphological processing, (a) after erosion; (b) after dilation.

After morphological processing, the positions of possible defects are obtained. Because of the glistening from the cover layer of FPC, there are still some false defects that can't be filtered by morphological processing. In order to obtain high accuracy and low false alarms, possible defects need to be checked in a second comparison. In this stage, regions of possible defects will be compared with those of the reference image by calculating the correlation coefficient.

$$r_{xy} = \frac{\sum (x_i - \bar{x})(y_i - \bar{y})}{(n-1)s_x s_y} \qquad (1.7)$$

If there are n data points of X and Y written as x_i and y_i where $i = 1, 2, \ldots, n$. $\bar{x}$ and $\bar{y}$ are the sample means of x_i and y_i, sx and sy are the standard deviations of x_i and y_i. The correlation coefficient r_{xy} can be used to estimate the correlation of X and Y. If X are the defect points of the inspection image and Y are the correspondent points of the reference image, correlation coefficient r_{xy} represents the correlation of the data between inspection image and reference image. Then we set a threshold of correlation coefficient. If the calculated correlation coefficient is more than the setting threshold, the defect is considered a false defect and can be filtered out. After correlation comparison, the false defects are filtered and a higher accuracy is obtained.

3. Results

In this section we will show the result of the FPC inspection system. Figure 1.5. is the FPC inspection system. When the program is started, the CCD camera moves to the initial position. After the reference image and the information of setting are created, placing the inspection FPC on the scaning table, the system scans and inspects the image automatically. Figure 1.6. is the redisual image after image subtraction, thresholding, and morphologicla processing, the white regions express the possible defects. Figure 1.7. shows the final inspection result. The inspected defects are located within the white rectangles. Figure 1.8. shows the defect image grabbed by the review camera and the left window lists defect area and coordinate. Figure 1.9. is the software interface of the system, and shows the result of inspection.

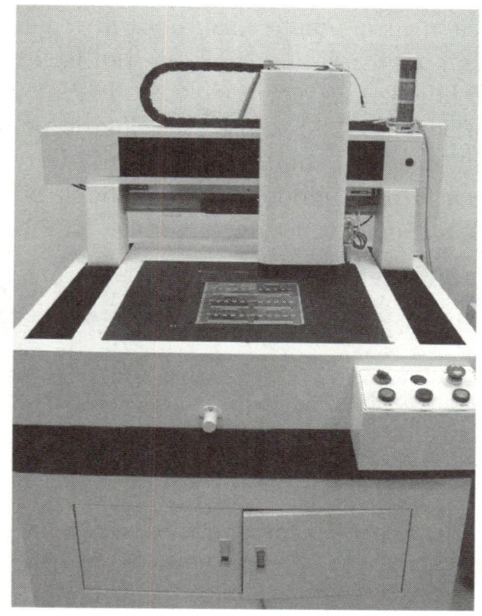

Figure 1.5. FPC inspection system

Figure 1.6. The white regions mean the possible defects, which is obtained after image subtraction, thresholding, and morphological processing.

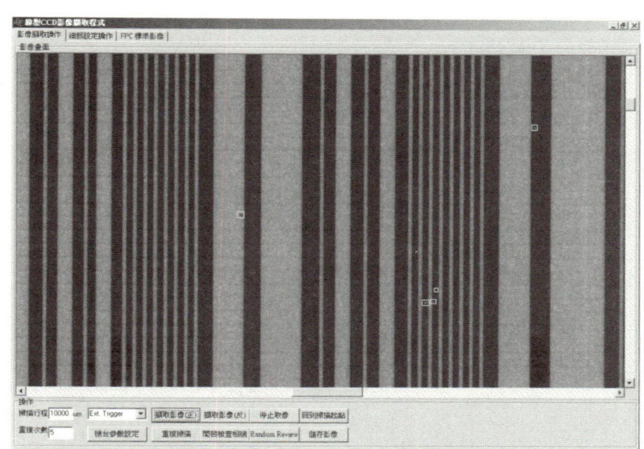

Figure 1.7. Inspection result

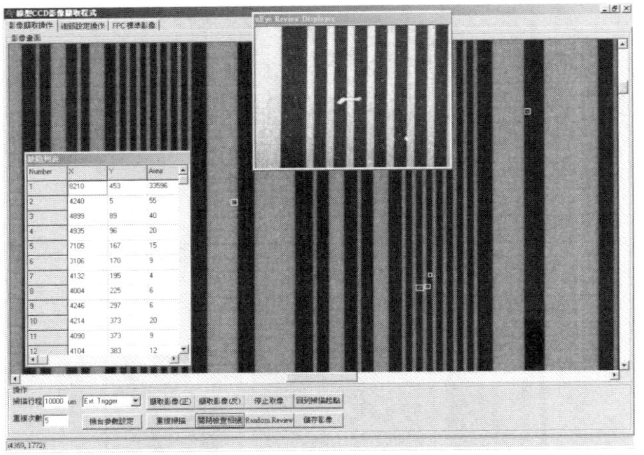

Figure 1.8. The center window shows the defect image grabbed by the review camera; the left one shows the area and coordinate of inspected defects.

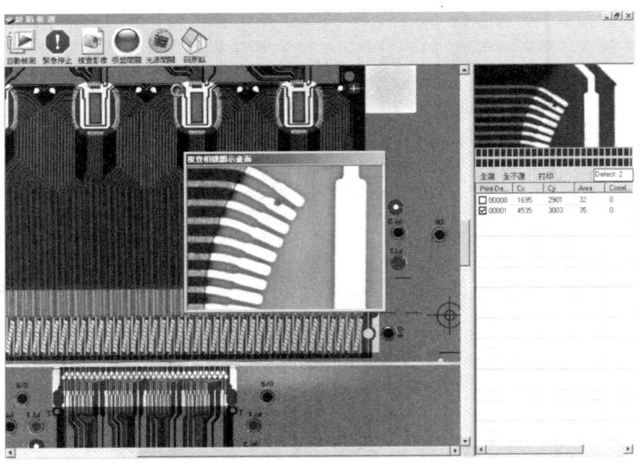

Figure 1.9. Software interface and inspection result

4. Conclusion

In this paper, we presented a technique for FPC defect inspection and realised a defect automatic inspection system. We can effectively detect the out-of-specification defects like line-width, line-spacing, over-etch, under-etch, and pin hole. Because of the flexibility and surface property of FPC, it has to overcome the challenges of suction of FPC and appropriate lighting to get a better quality of the images. In specific design for FPC, the algorithm can inspect the defects more accurately and filter the false defects, which also reduces the review time. The speed of inspection is 5~10 times higher than that of human inspection. The newly developed system does not only achieve the demand of inspection speed, but also achieves a higher accuracy and a lower false alarm rate, and, therefore, it's a better solution for FPC defect inspection.

References

[1] Mogant, M. and F. Ercal, "Automatic PCB inspection algorithms: a survey," Computer Vision and Image Understanding, 63, 287-313, 1996.

[2] Rafael Gonzalez and Richard Woods. "Digital Image Processing." Addison-Wesley, third edition, 1993.

[3] Jon R. Mandeville, "Novel method for analysis of printed circuit images," IBM J. Res. Dev. 29, 1, 73-87, 1985.

[4] Louis J. Galbiati Jr., "Machine Vision and Digital Image Processing Fundamentals," Prentice Hall, 1990.

[5] Milan Sonka, Vaclav Hlavac, "Image processing, analysis, and machine vision," Roger Boyle, PWS publishing, 1999.

[6] Nick Efford, "Digital Image Processing – A Practical Introduction Using Java," Pearson Education, 2000.

[7] Harley R. Myler, Arthur R. Weeks, "Computer Imaging Recipes in C," Prentice Hall, 1993.

Investigation of Auto-focus Technology for Panel Inspection in the Flat-panel Display Industry

C.Y. Chung[1], C.M. Tseng[2], Y.C. Lu[3], K.M. Yeh[4]

[1]Metal Industries Research & Development Centre, Taiwan, 81160, pkr@mail.mirdc.org.tw
[2]Metal Industries Research & Development Centre, Taiwan, 81160, ming@mail.mirdc.org.tw
[3]Metal Industries Research & Development Centre, Taiwan, 81160, ycl@mirdc.org.tw
[4]Metal Industries Research & Development Centre, Taiwan, 81160, yehkm@mirdc.org.tw

Abstract. This study attempts to employ DVD pickup head as a sensor for vertical height measurement in a flat panel display inspection system. Equipped with such a sensor, the system can achieve both high resolution and large depth of field at the same time. In most of the requirements, inspection system not only provides a guide for modifying parameters in manufacturing process, but also position and size of the surface defects. However, as the sample size gets smaller, at least magnification of 50X is necessary so as to generate satisfactory results. Meanwhile, such a high magnification shortens the depth of field to less than 2 μm, which becomes very sensitive to environment disturbances such as mechanical vibrations. We use DVD pickup head as a surface height sensor which provides a coarse range of the height variation of the inspected features such that an evaluation of out-of-focus degree can be achieved. Additionally, the height information gives feedback for auto-focus control loop. In order to incorporate the pickup head into the inspection system, an improved design will be performed to make the size of the sensor more compact physically. The voice coil motor and objective lens will be removed, and the sensor will be arranged such that the measurement position is matched with the area under image inspection. Therefore, the measurement error caused by non-coaxial arrangement can be minimized. In the experiment, the data of surface height has been measured, where approximately the tolerance of 0.3 μm differed from the data measured by a commercial alpha-step instrument. In this study, we have developed a compact, sensitive, low cost and high accurate auto-focus system.

Keywords: flat panel display, insection, DVD pickup head, measuring error, surface profile

1. Introduction

In recent years, the panel inspection technology has continuously advanced and matured along with the development in flat panel displays. A wide variety of inspection and measurement instruments have been developed to date. Among which optical profilometers, including interferometers, triangulation methods, confocal systems, have already been commercialized and put into wide use in the industry. However, as far as the panel inspection industry is concerned, interferometry is only suited for detection of a continuous surface. It is not adequate for a discontinuous surface, which is usually the case for panels after the etching process. If a commercially-available instrument is used, the accuracy is high but has the downside of high costs. To cut down costs, some companies utilize the pattern match method to take continuous images at different distances. Then, computation software calculates and outputs the image with the highest definition. However, the accuracy of this method is limited by the frame rate and algorithm of definition calculation. Hence, a low-cost, simple-structured, high-accuracy auto focus system will be developed in this study to address this issue.

Since the pick-up head has been extensively used in a variety of applications [1-3], it shall be used in this study for scanning the panel surface. When the surface profile information is fed back to the moving mechanism, the position of the camera (or moving carrier) is shifted as well so that auto focusing is achieved. As compared to other sensing methods, such as capacitive or inductive methods, the non-contact optical sensing method developed in this study is better suited for panel inspection. The circuit design, signal acquisition, and surface profilometry of pick-up head have been completed in the preliminary study.

2. 3D Measurement Theory of Pick-up head

2.1 Astigmatic Method

Since astigmatic lens' focal lengths are different in horizontal and vertical directions, two focal planes of different distances exist as collimated light passes through an astigmatic lens. The astigmatic method refers to the way to measure the distance by determining the positions of these two focal planes. As shown in Figure 1.1(a), plane 1 is the tangent focal plane, plane 3 is the sagittal focal plane, and plane 2 is between the two. The spot shape can be determined by placing the four-quadrant photo detector at the image plane. The expression is focus error signal(FES) = (A+C)-(B+D), where A,B,C,D stand for the four quadrants. Figure 1.1(b) shows that when plane1 is the focal plane, a positive voltage is outputted. If plane 3 is the focal plane, a negative voltage is outputted. The FES results in an S-curve [4] as the image plane shifts from a position near the focus to a position away from it. Since the voltage variation between the two ends is linear, i.e. a one-to-one function, the position of image plane can be deduced from this relationship.

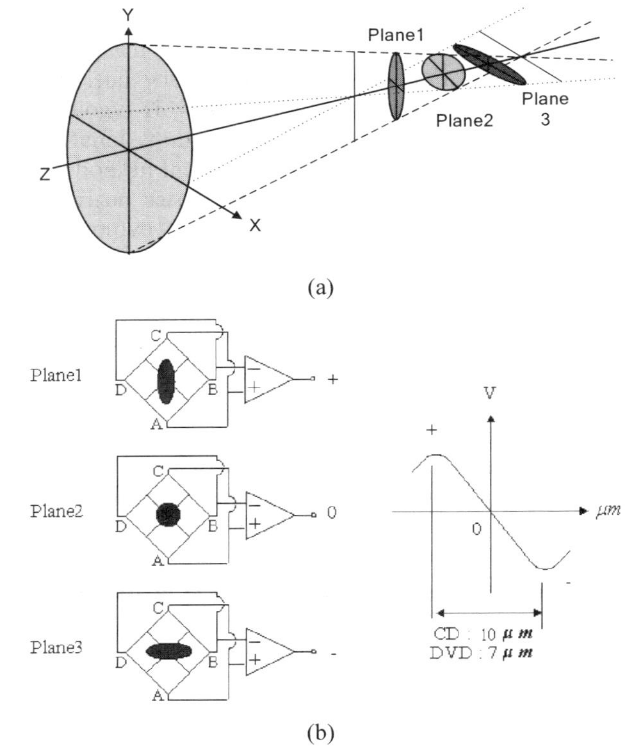

Figure 1.1. Schematic of Astigmatic Method: (a) after collimated light passes through an astigmatic lens, plane1 is the tangent focal plane and plane3 is the sagittal focal plane, (b) Schematic of S-curve [4]

Moreover, for light varying in intensity, four-quadrant photo detector produces S curves with similar shapes but different amplitudes, which means the corresponding functions are different as well. In order to keep the corresponding functions uniform in this study, the FES is normalized, namely, FES/SUM, where SUM(summation) is A+B+C+D. As a result, whenever a variation in light intensity exists, normalization is carried out to preserve the corresponding functions.

2.2 Optical Path of Pick-up head

SONY KHM-210AAA pick-up head [5], as shown in Figure 1.2, is adopted in this study. The optical path is illustrated in Figure 1.3. Inside the head, a laser diode emits a light beam that passes through a beam splitter and a collimator lens, which turns the laser beam into a collimated light beam. This collimated beam is separated into two parts after passing through a lens, where its center consists of concentric grooves tens to hundreds of microns apart. The divided light beams are focused into bi-focal points, which are well-suited for DVD and CD data storage. The reflected light beam traces back the incident path and passes through the collimator lens, the beam splitter, the cylindrical astigmatism lens, and projects onto the four-quadrant photo detector [6]. The detector outputs the focus error signals based on the light distribution in the four quadrants. The disc's defocus distance can then be determined from these signals. The objective of this study is to use the FES to adjust the camera's height so that auto focusing is achieved.

Figure 1.2. Pickup head.

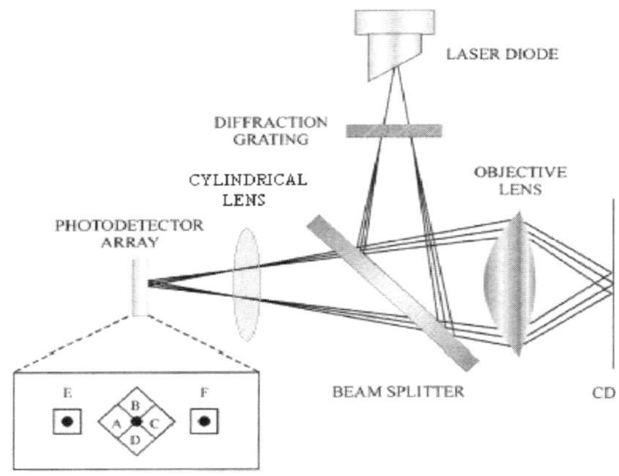

Figure 1.3. Schematic of pick-up head's optical path: consisted of a laser diode, a photodetector array, a beam splitter, a cylindrical lens, etc. [1]

2.3 Circuit Design of Pick-up head

Circuit design mainly deals with processing of four quadrants' signals. In the experiment, turning on the motor at the translation stage produces electromagnetic interference, which in turn influences the signals significantly. Hence, the objective of circuit design is to reduce the interference of exterior noises. The design flowchart is shown in Figure 1.4. Four-quadrant photo detector's signals pass through a terminal board and enter the circuit board. The signals are pre-amplified and are then passed through an active low pass filter for output. Since different DC offsets are required for different types of pick-up heads, a voltage offset circuit is added so that the user can adjust. In doing so, the signals can directly go through an A/D converter and enter the controller for calculation of FES and SUM. Or, they can go together with an analog circuit and be amplified for the second time, filtered, and then passed to the controller. For example, in circuit board 2, analog adding and subtracting are carried out and the result is amplified, filtered, and passed to a PC via an A/D. The advantage is two-fold: the calculation speed is accelerated, and the number of transmission cables is reduced.

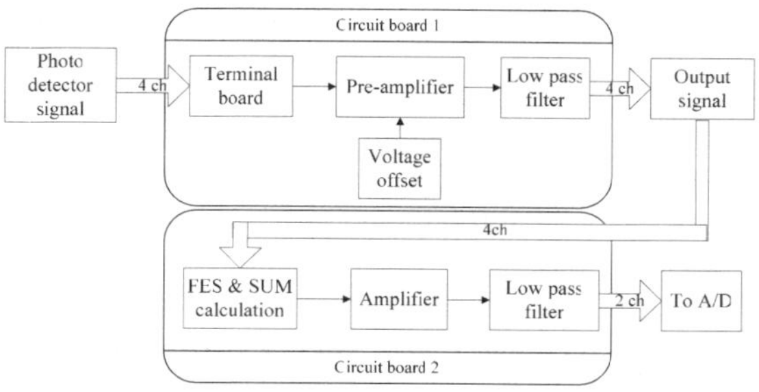

Figure 1.4. Conceptual chart of circuit design: the photo detector signal is the input. It passes a terminal board, a pre-amplifier, a low pass filter, and enters circuit board 2 for calculation. It is outputted after another amplification and filter

3. Experiments and Results

The experimental setup is shown in Figure 1.5. The pick-up head is placed on the Z-axis motor. The object to be inspected is clamped on the XYZ motorized translation stage, where its movement is controlled by a PC. As the Z-axis motor or the translation stage moves, the head picks up analog signals. These signals are then processed and passed to an A/D for conversion into digital signals. They are then analyzed in a PC and the surface profile is reconstructed.

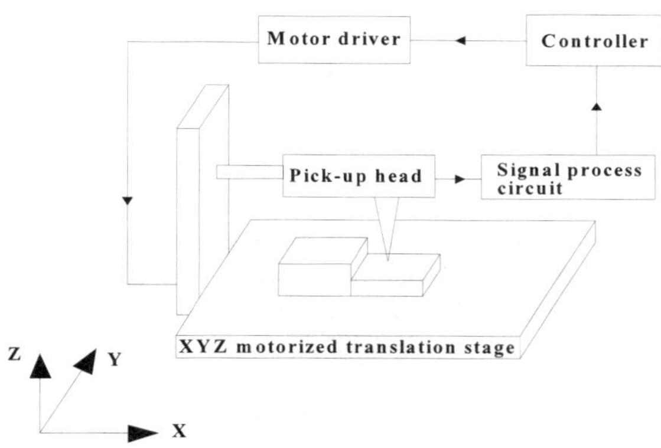

Figure 1.5. System structure: Signals are obtained by the pick-up head, processed, and passed to the controller. The controller outputs commands to drive the motor driver

In our experiment, the first step is to move the Z-axis motor to an appropriate height. Then, it is moved in incremental steps such that the pick-up head's light spot changes from out-of-focus to in-focus and slowly back to out-of-focus again. At the same time, the head's height and voltage are recorded. The purpose of this step is to obtain the S-curve of the head. In the second step, a linear fit, obtained using the least-square method, is applied on the CD's linear range. The linear fit equation will be used for 3D reconstruction. The third step is to activate the scanning process. The fourth step is to reconstruct the 3D profile, Z(x,y), with the voltage, V(x,y), obtained from the scan and the slope(S) of the linear range obtained from the second step, as indicated in Eqn 1.1

$$Z(x,y) = V(x,y)/S \quad (1.1)$$

The voltage curve obtained from the first step is shown in figure 1.6. The upper curve represents the FES and the lower curve represents the SUM. The FES has two linear ranges: one is from CD and the other one is from DVD. The CD and DVD linear ranges are approximately 10μm and 7 μm, respectively. In this study, CD is chosen for its measurement distance and sensitivity. As shown in Figure 1.7, the linear range is approximately 10 μm. In addition, one can see from Figure 1.6 that the reflected light is at its strongest intensity at the CD and DVD ranges of the SUM curve. This coincides with the theory in which an object to be inspected has the strongest reflection energy when in focus. The energy becomes gradually smaller as the object moves away from the focus.

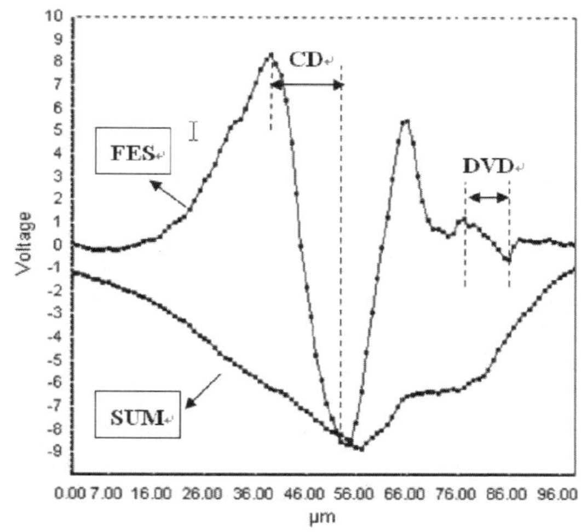

Figure 1.6. FES & SUM voltage vs. position

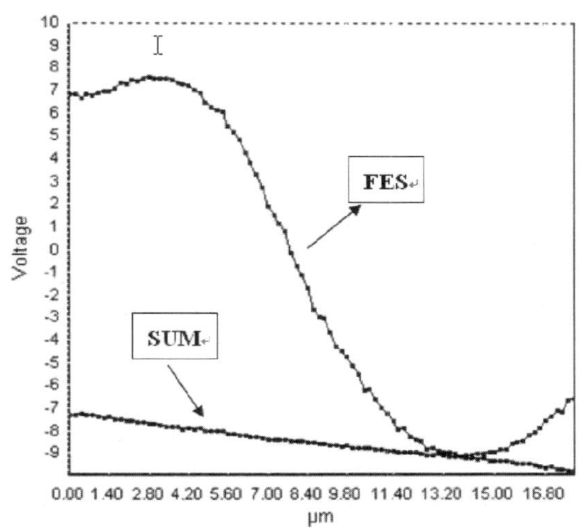

Figure 1.7. FES S-curve & SUM

The objective of the experiment is to detect the light reflected from the object. If the object to be inspected has a low reflectivity, such as a glass substrate, the FES amplitude will become comparatively small. As long as the length of the linear range remains fixed, the slope will become small, which affects the measurement result. Therefore, FES is divided by SUM in order to eliminate the influence from the intensity of reflected light. A positive FES peak value can be expressed as Eqn 1.2 since

(A+C)>>(B+D). Similarly, a negative peak value can be expressed as Eqn 1.3.

$$\frac{FES}{SUM} = \frac{(A+C)-(B+D)}{A+B+C+D} \approx \frac{A+C}{A+C} \approx 1 \quad (1.2)$$

$$\frac{FES}{SUM} = \frac{(A+C)-(B+D)}{A+B+C+D} \approx -\frac{B+D}{B+D} \approx -1 \quad (1.3)$$

As a result, FES/SUM will fall somewhere between 1 and -1 and its slope will remain unchanged regardless of variations in objects' reflectivity. The normalized FES curve is shown in Figure 1.8. It also shows the linear fit line obtained by applying the least-square method on the S-curve's linear range, as mentioned in the second step. The slope of this linear equation can be used for 3D reconstruction in the final step.

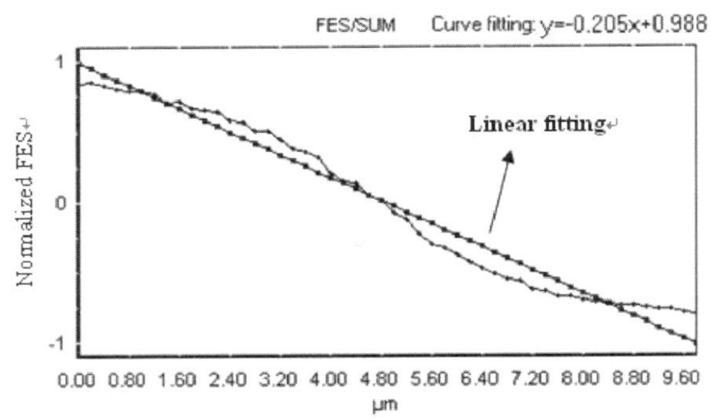

Figure 1.8. Normalized FES & linear fitting line

In the third step, the XYZ translation stage is moved to scan the object. The 3D surface profile is reconstructed by applying the voltage obtained from the scan in Eqn 1.1. Figure 1.9 shows a 2D(x-z) reconstruction plot of an 8 μm step. Its 3D(x-y-z) reconstruction plot is shown in Figure 1.10. The experimental setup used in this study, including a translation stage driven by a three-axis stepping motor, an A/D conversion card, a pick-up head, and a signal processor.

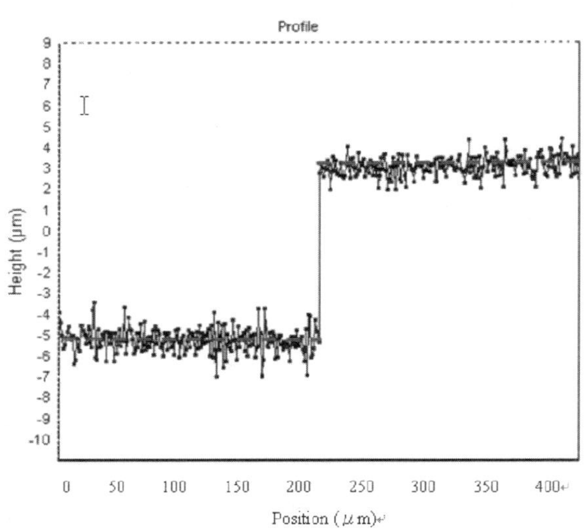

Figure 1.9. Result of line scan

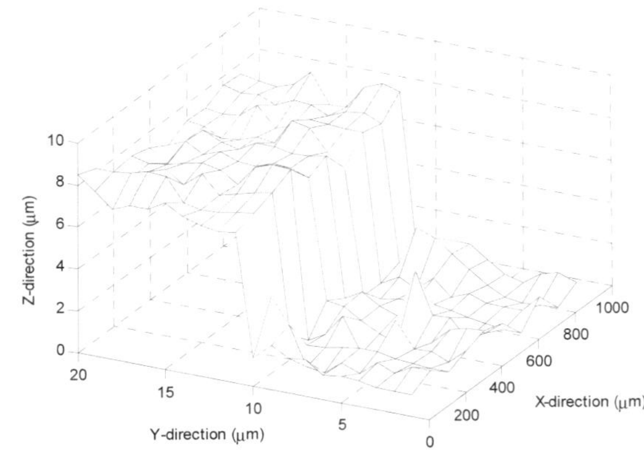

Figure 1.10. Result of area scan

4. Conclusion

In this study, the surface profile of an object has been successfully obtained by a pick-up head. An object with an 8μm step was scanned with an error of approximately lower than 10%, a standard deviation of about 1.41μm, and a sensitivity of 1.6V/μm. A simple, compact, low-cost auto-focus pre-scan system was thus developed. There are three error sources for this system. First, the intensity of light emitted from the laser diode is easily influenced by the temperature. Although it is stabilized by the auto power control (APC) circuit, the stability of long-term testing is not quite good. Secondly, the pick-up head and its circuit are vulnerable to exterior electromagnetic interference. Thirdly, linear-fit equations of higher orders can be considered to reduce fit errors. Subsequent studies will be planned to improve these three issues, with the solution of the third issue as the main aim. Also, on-axis optical path and CCD cameras will be incorporated in the design of a new auto-focus system.

References

[1] A. Bartoli, P. Poggi, F. Quercioli, and B. Tiribilli, (2001) Fast one-dimensional profilometer with a compact disc pickup, Appl. Opt., Vol.40, P1044-1048.

[2] Klaus Ehrmann, Arthur Ho and Klaus Schindhelm, (1998) A 3D optical profilometer using a compact disc reading head, Meas. Sci. Technol., Vol.9, P1259-1265.

[3] Ji-hua Zhang, Lilong Cai, (1997) An auto-focusing measurement system with the piezoelectric translator, IEEE, Vol.2, P213-216.

[4] Chih-Liang Chu, (2006) Investigation of optical inspection focus technology, Technical Report, Dept. of Mechanical Engineering, Southern Taiwan University of Technology, Tainan, Taiwan.

[5] Sony semiconductor product list : Photo Device KHM-210AAA, (1999) Sony Co..

[6] Yung-Chun Ho, (2002) Development of high-density optical storage media, Technical Report, Dept. of Materials Engineering, National Chung Hsing University, Taichung, Taiwan.

Automatic Optical Inspection on Mura Defect of TFT-LCD

S.L. Chen[1], J.W. Jhou[2]

[1] National Cheng-Kung University, No. 1, University Road, Taiwan City 701, Taiwan, slchen@mail.ncku.edu.tw
[2] National Cheng-Kung University, No. 1, University Road, Taiwan City 701, Taiwan

Abstract. Automatic optical inspection plays a key role in inspection of automations and equipments. It can replace manual inspect, reduce the production costs and achieve high speed and high precision with the inspection. Increasing the quality and efficiency of production materially. Mura defect automatic optical inspection system combines with electro-mechanical Systems, image acquisition systems and inspection software. In inspection software, it is to implement polynomial background surface estimation, use the standard which was defined by SEMI and apply auto-focus method to ensure the quality of image. This system has 100% accuracy and spent 45~55 seconds in the experimental stage. There are two options can be selected, in-line or off-line inspection.

Keywords: automatical optical inspection, Mura deffect, TFT LCD

1. Introduction

Recently, Automatic Optical Inspection, or AOI, has been widely used in the automatic production equipment of TFT-LCD, and it has improved the problems of the speed limit of inspection through human eye and the inevitable differences that occur in the precision of inspection from individual to individual with human-eye inspection.

Among the defects inspected with TFT-LCD, Mura means a defect that is larger in size than a pixel. A phenomenon of inhomogenization visible to the human eye can be observed in a region against the region around it.

The Automatic Optical Inspection platform for Mura defects combines a mechanic-electrical system with technologies such as Image Processing & Patten Recognition, which are designed to target the features of Mura defects. In terms of mechanical-electrical system, Chang [1] pointed out that the image acquisition range should be in accordance with the size of the panel to be inspected. With the help of moving machinery, the panels can now be moved to a proper position for image acquisition. Mori and Tanahashi [2] put lightening on both sides of the panel to be inspected as well as the mechanism of rotation and liberation in an attempt to mimic the movement of human-eye inspection. Regarding pattern processing and recognition, to effectively divide Mura defects, Lee and Yoo [3] announced the Polynomial Background Surface Estimation for automatic inspection on Mura defects, while literature [1] improved on the method of dividing the outlier region.

Furthermore, to ensure the clearness of the images, we also discuss the theory of automatic focusing and its application. Poo, Neow and Ang [4] explored eight practical functions for image sharpness, among which the simple operation of sum-modulus-difference method helps speed up the operation. The Robust Focus Measuring Operator announced by Zhang [5] can effectively enhance the speed of alteration in image sharpness function.

1.1 Review on the evaluation of mura

Mura evaluation is based on whether the Object-Region contrariness of the background region and the could-be Mura defects are visible to the human eye. Literature [3] pointed out the relationship between background region and object region.

However, the region of the object region can affect the value of the just-noticeable intensity difference (JND). Literature [6] adopted the SEMU indexical value set up by the Semiconductor Equipment and Material Institute (SEMI) to conduct the evaluation of Mura. The equation is as follows:

$$SEMU = \frac{|I_O - I_b|}{JND} = \frac{|I_O - I_b|}{\left(\frac{1.97}{S^{0.33}} + 0.72\right)} \tag{1}$$

In the equation above, I_o represents the average brightness of the object region, while I_b represents the average brightness of the background. S is the estimated region Mura may occur. Literature [3] redefined I_o and I_b as the following:

$$I_o = \frac{1}{|\Psi_o|} \sum_{p \in \Psi_o} R(p) \quad , and$$

$$I_b = \frac{1}{|\Psi_b|} \sum_{p \in \Psi_b} R(p) \tag{2}$$

where $R(p)$ is the residual of the pixel p in estimated and actual background. Ψ_b is the area of object region.

2. Theory of Polynomial Background Surface Estimation

If we presume I as the image of a window and the location of every pixel as (x,y), we can then present the bivariate polynomial model of likening the image as follows:

$$f_{xy}^{(d)} = \sum_{0 \le m+n \le d} a_{mn} x^m y^n \qquad (3)$$

In the above equation, d represents the order of the polynomial (d = 1, 2, 3, 4, 5, …) and a_{mn} are the various coefficients of the polynomial.

Assuming there are k (k = 3, 6, 10, 15, 21, …) coefficients in a polynomial, there should be k equations in the quest of the various coefficients a_{mn} of the polynomial, and there should be the quantity of k. Literature [1] adopted the Least-Squares Regression Method, LS, and presents it in the form of a matrix $\overline{M}_a^{(d)} \times \overline{V}_x^{(d)} = \overline{V}_b^{(d)}$:

$$\begin{bmatrix} \sum\limits_{x\in I, y\in I} x^0 y^0 & \cdots & \cdots \\ \vdots & \cdots & \cdots \\ \vdots & \sum\limits_{x\in I, y\in I} x^{m+\alpha} y^{n+\beta} & \cdots \\ \vdots & \cdots & \ddots \end{bmatrix}_{k\times k} \times \begin{bmatrix} a_{00} \\ \vdots \\ a_{\alpha\beta} \\ \vdots \end{bmatrix}_{k\times 1} = \begin{bmatrix} \sum\limits_{x\in I, y\in I} I_{xy} x^0 y^0 \\ \vdots \\ \sum\limits_{x\in I, y\in I} I_{xy} x^\alpha y^\beta \\ \vdots \end{bmatrix}_{k\times 1} \qquad (4)$$

We then used inverse matrix to quest coefficient $\overline{V}_x$:

$$\overline{V}_x^{(d)} = \left(\overline{M}_a^{(d)}\right)^{-1} \times \overline{V}_b^{(d)} \qquad (5)$$

Then the various coefficients of the bivariate polynomial can be attained with higher temperature sintering, thus the porous HA is obtained. The operation of Polynomial Background Surface Estimation step is shown in Figure 1.1.

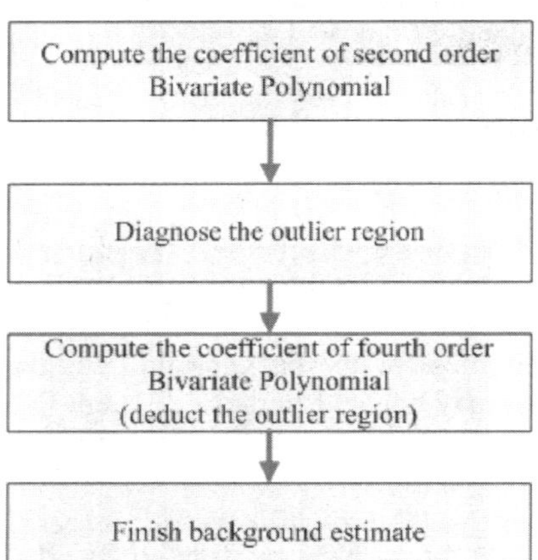

Figure 1.1. The flowchart of polynomial background surface estimation

2.1 Niblack's Threshold Value Division

According to [3], deducted the estimated background image from the original image and then we could attain the image of residuals. We then used Niblack's threshold value division to get the object region to be evaluated.

Niblack's threshold value division mainly uses the average value and the standard deviation of the differences in residuals of all window images against the background to establish a dividing principle that adopts the same threshold value (T) in different window images. This method is robustly in quality.

To realize Niblack's threshold value division, we have to, first, work out average value and standard deviation that belong to the background in the residual image. The operation is as follows:

First, deduct the estimated background from the original image and attain the residuals image:

$$RES_{uv}^{(4)} = I_{uv} (-) f_{uv}^{(4)} \qquad (6)$$

Multiply element by element the bivariate matrix from the residuals image and the outlier region, then deduct the difference of the residuals that belongs to the outlier region:

$$RES'^{(4)}_{uv} = RES_{uv}^{(4)} \circ \Phi_{xy} \qquad (7)$$

Therefore, the calculation of the residuals difference image that belongs to the background as follows:

$$\mu_{-\Phi_{uv}} = \frac{\sum RES'^{(4)}_{uv}}{\sum \Phi_{uv}} \qquad (8)$$

$$\sigma_{-\Phi_{uv}} = \frac{\sum \left(RES'^{(4)}_{uv} (-) \mu_{-\Phi_{uv}}\right)^2}{\sum \Phi_{uv}} \qquad (9)$$

It is then ready for the processing of Niblack's threshold value division. The rule is presented as follows:

$$Z(u,v) = \begin{cases} 0, & \left|RES^{(4)}(u,v) - \mu_{-\Phi_{xy}}\right| / \sigma_{-\Phi_{xy}} > T \\ 1, & \left|RES^{(4)}(u,v) - \mu_{-\Phi_{xy}}\right| / \sigma_{-\Phi_{xy}} \le T \end{cases} \qquad (10)$$

The set up of threshold value (T) in this research will be discussed in latter chapters.

2.2 Analysis and Evaluation Processing

First, we have to filter out the small points in non-object region caused by division, label all object regions, calculate individually on the margins of each object regions, use the information of the margin to calculate on the SEMU indexical value of each object region. The workflow is as Figure 1.2.

After the original image is being processed by the Window processing unit, there exists in the resulting image the noise left by threshold value division as Figure 1.3.

To get rid of the noise and make the object of inspection clear. In this research paper, we use Open Operator [7] in morphology to filter out the noise as Figure 1.4

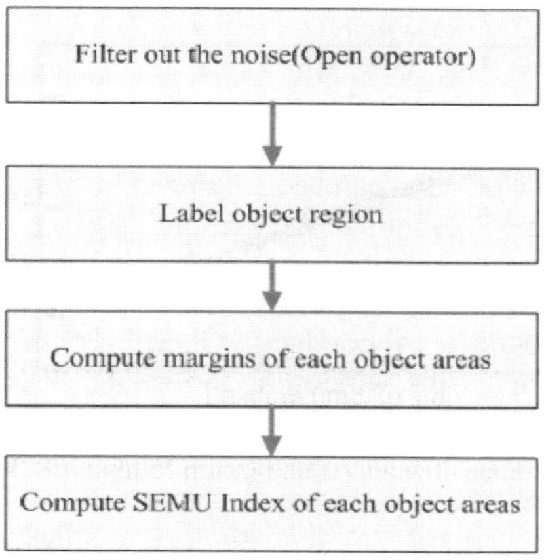

Figure 1.2. Analysis & Evaluation processing flow chart

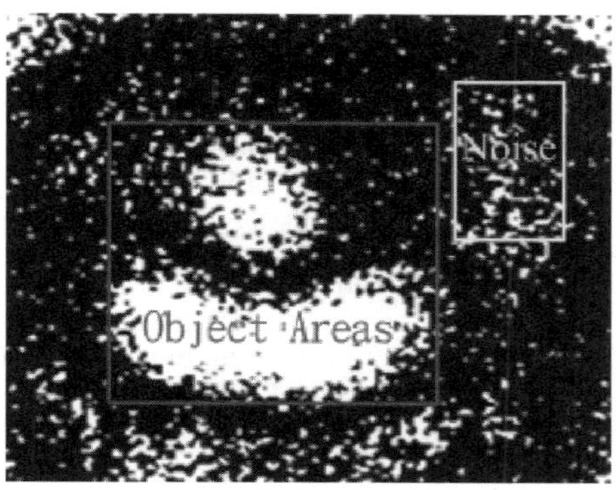

Figure 1.3. Object region and Noise

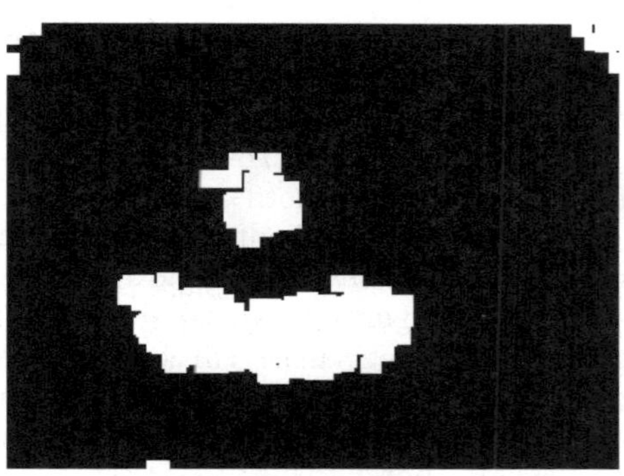

Figure 1.4. Result of opening

3. Experiments and Results

3.1 Effects of calculation Niblack's Threshold Value

Discussing the impact the alteration in Niblack's threshold value division may have on the results of inspection. In this experiment, we chose a cloud-shaped Mura defect as the main observatory object. First, we fixed the parameters. hen we conducted eleven different experiments on threshold values (0.5%, 0.75%, 1%, 1.25%, 1.5%, 1.75%, 2%, 2.25%, 2.5%, 2.75%, 3%). The results are shown in the following Figure 1.5.

From Figure 1.5, we have observed that the outline of the defects in Experiments #5(1.5%), #6(1.75%), and #7(2%) are closer to the defects observed by human eyes. So in this research, the threshold value of Niblack's threshold value division is set in between 1.5~2. We fix it at 1.5 in this research.

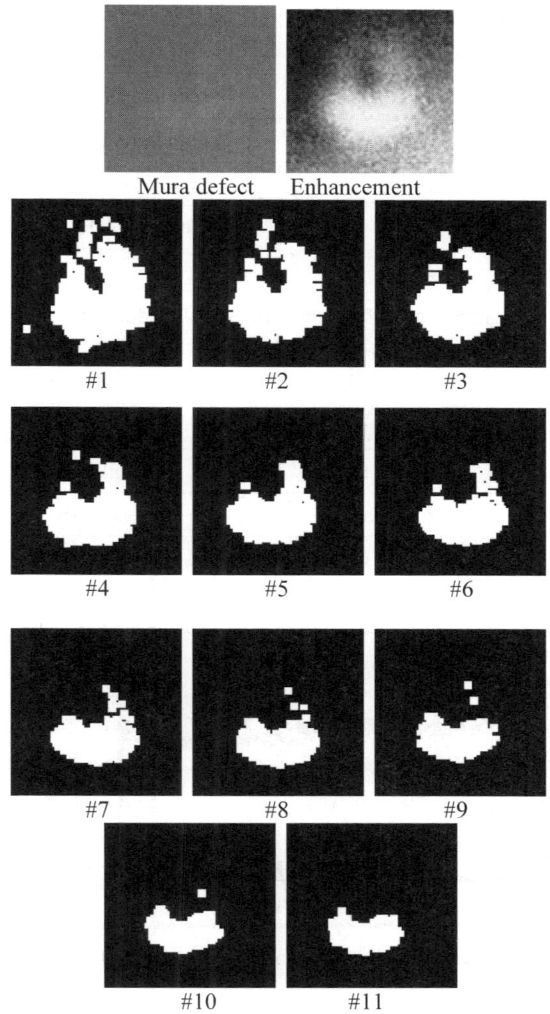

Figure 1.5. Niblack's Threshold segmentation result

3.2 Assessing the Inspection Accuracy and Inspection Speed

Human inspection on the Mura defects is mainly the objective judgment made by the observer. The results acquired are merely the existence of defects and the location of the defects. With automatic inspection, the crucial points for inspection correctness lies in whether the size of object region and its outline equalize the Mura defects observed by human eyes in the evaluation process of Mura defect inspection.

In this research, we chose 7 TFT-LCD 17"images that were actually taken as the images to be used in the accuracy testing. Among them, five of which have each a defect of different shapes, the remaining two are defect-free. For the estimation of the accuracy of the system, we not only studied the correctness of the location of the defects, but we also included the issue where we tried to find out whether the region of the defect and the outline of the defect are similar to those inspected by human eyes. The discussion over the accuracy presented in the testing of the seven images and the results are shown in Table 1.1.

The current inspection accuracy in this research is 100%. The region of the defects inspected and the outline of the defects are similar to the region and the outline of the defects inspected by human eyes. The inspection time needed in each of these testing is measured at between 45~55 seconds.

Table 1.1. Accuracy testing results

Testing Samples	MVI		Kinds of Mura	AOI	
	Num. of Mura	Location of Mura		Num. of Mura	Location of Mura
#1	1	Above Center	Cloudy Mura	4	Above Center
#2	1	Right side	Bias Mura	1	Right side
#3	1	Left side Center Up side	Line Mura	22	Left side Center Up side
#4	1	Top of left	Triangle Mura	1	Top of left
#5	1	down	Region Mura	2	down
#6	0	X	X	0	X
#7	0	X	X	0	X
Num. of Defect by MVI (panel)		5	Num. of Defect by API (panel)	5	Percent of Accuracy 100(%)

4. Conclusion

In this research, we put into practice the Background Surface Estimation method and the quantification and evaluation method of Mura defects in the inspection software for Mura defects. We have drawn the following few conclusions from this research:

1. This research is a combination of image acquisition, movement, and inspection systems. It can automatically provide images for inspection and guarantee a consistent workflow of inspection.
2. With the method of Niblack's Threshold Value Division, the setting of the threshold value concerns greatly the contrariness between the defects and the background which can be inspected by human eyes. We set the threshold value at 5.5, signifying that when the area of the defect is 4624 (Pixels) with a contrariness set at 5, the defects presented happen to be the one that can be recognized by human eyes.
3. In the Polynomial Background Surface Estimation phase, the setting of the size of the window are related to the results of the inspection. It was impossible to conduct inspection when the size setting of the window is too small. Proper size of window can help with the completeness of the inspected region, while increases on the SEMU index value. In this research, we set it at 300×300.
4. With regard to the Niblack's threshold value division, when the value is set at between 1.5~2, the region of the defect divided and its outline are similar to those inspected by human eyes. Therefore, we set the value at 1.5 in this research.

At current stage, the accuracy of inspection in this research is 100%. The area and the outline of the defects inspected are similar to the defects inspected by human eyes. The inspection time is measured at between 45 to 55 seconds.

Acknowledgements

Parts of research results were funded by the National Science council of the Republic of China nuclear Grant NSC 95-2221-E-006-400. This financial support is gratefully acknowledgement.

References

[1] Chang, R.S., 2005, "TFT-LCD Mura Defects Automatic Inspection System using Regression Diagnostics Model", Master's Dissertation, National Cheng Kung University.

[2] Mori, Y., Tanahashi, K., Yoshitake, R., Tamura, T., Moriguchi, K., Yoshizawa, T., Tsuji, S., 2003, "Measurement System and Defect Method of mura in TFT-LCD", IDMC2003, pp.295-298.

[3] Lee, J.Y., Yoo, S.I., 2004, "Automatic Detection of Region-Mura Defect in TFT-LCD", IEICE TRANC. INF. & SYST, pp.3-5.

[4] Neow, P.A., Ang, M.H.Jr, 2001, "*Practical Issues In Pixle-Base Autofocusing For Machine Vision*", IEEE International Conference on Robotics and Automation, ICRA, 3, pp.2791-2796.

[5] Kharitononko, Zhang, X., 2000, "*Digital Focus Detector For Mobile Video Communicators*", IEEE Transactions on Consumer Electronics, 46(1), pp.237-240.

[6] Jia Jia Guo, Bor-Hann Huang, Kerson Wang, Chun-Hsia Chen, 2004. "Development of Mura Generator System", IDMC2004, pp. 441-444.

[7] Gonzalez, R.C., 2002 "Digital Image Processing", 2nd, Prentice Hall.

Development of Simultaneous Confocal Full-field Surface Profilometry for Automatic Optical Inspection (AOI)

L.C. Chen[1], Y.W. Chang[2]

[1] National Taipei University of Technology, Taiwan, 106, lcchen@ntut.edu.tw
[2] National Taipei University of Technology, Taiwan, 106, s4618014@ntut.edu.tw

Abstract. A simultaneous confocal full field 3-D surface profilometer using digital structured fringe projection is presented in the article. Using digital micromirror device (DMD) moiré projection, a digital fringe pattern is developed for lateral scanning with high spatial resolution and measurement efficiency. Four conjugate image sensing modules are configured at four different designated focusing positions, which are controlled by four glass substrates with various thicknesses. A depth-focus response curve can be established by achieving simultaneous vertical scanning of full-field surface profilometry. A standard step height has been measured to attest the measurement accuracy and feasibility of the developed approach. The depth measurement resolution can reach up to 0.1 μm and the maximum measurement error was verified to be within 7% of the overall measuring height size. The measurement efficiency can be significantly improved for on-line automatic optical inspection (AOI).

Keywords: simultaneous confocal full field 3-D surface profilometer, digital structured fringe projection, measurement efficiency,

1. Introduction

The development of a high-speed full-field 3-D surface profilometer using digital Moiré projection is presented in this paper. Optical confocal microscopy for surface profilometry has become important due to its high longitudinal measurable range and excellent vertical resolution. The confocal measurement utilizes the geometrical matching of two conjugate focal points corresponding to both the object surface and the point detector defined by a pinhole aperture. Conventional confocal measurement using laser coherent light not only easily encounters its point-type scanning limitations, but also suffers unacceptable measured errors incurred by its vulnerable scattering problems. To address this, a digital micromirror device (DMD) is deployed as an array of pinholes that may replace the Nipkow disk presently used for lateral scanning in white-light confocal microscopes. The DMD used as a spatial light modulator is an opto-electronic system having a number of independently controlled micromirrors utilized to generate defined spatial modulations of the light field, which is applied as structured light in modern 3-D profilometry. However, this type of approach does not address a key issue encountered in on-line process inspection, in which the vertical scanning time required by the confocal measurement cannot meet extremerely short Tact times required by many automatic optical inspection (AOI) industries.

To minimize the negative impacts mentioned above, an optical confocal measurement system is established to achieve simultaneous vertical scanning of full-field surface profilometry. In the optical system, four conjugate image sensing modules are configured at four different designated focusing positions, which are controlled by four glass substrates with various thicknesses. The DMD chip offers a direct control of the micromirror array by a high-speed parallel interface in making dynamic 3-D surface profilometry feasible under Moiré projection. The optical system can generate and project computer-controlled Moiré fringes onto the object's surface. Along the scanning vertical direction of each measuring point, four image intensities from corresponding image sensing modules can form a focus-to-depth response function, in which the image intensity can reach its maximum when the image of the corresponding DMD pixel is located on the focal plane of the microscope objective. Thus, simultaneous vertical scanning of confocal full-field surface profilometry can be achieved.

Focus operators for 3-D depth detection have been previously studied to measure surface contour information [1-3]. To increase scanning efficiency, several methods have also been developed to fasten the profile measurement. The Nipkow disk as a way to transmit images electrically using a pair of spinning metal disks or the use of a diffractive lens illuminated with a tunable light source has been utilised to improve the confocal scanning speed in lateral space [4]. In recent years, a DMD can be deployed as an array of pinholes that may replace the Nipkow disk presently used for lateral scanning in confocal microscopes [5]. The DMD used as a spatial light modulator is an opto-electronic system having a number of independently controlled micromirrors utilized to generate defined spatial modulations of the light field, which is applied as structured light in modern 3-D profilometry [6]. The DMD chip offers a direct control of the micromirror array by a high-speed parallel interface in making dynamic 3-D surface profilometry feasible under Moiré projection and triangulation measurement principle. However, the required vertical scanning operation still limits the confocal measurement from being considered an effective method for on-line AOI in 3-D surface profilometry. This research thus addresses such an important issue for achieving on-line surface profilometry.

2. Optical design of DMD-based confocal measurement system

In the approach, the digital fringe pattern is first generated by a personal computer, passes through a set of optical lenses, in which a structured-light grid pattern can be produced, and guided into the side port of the stereomicroscope for generation and projection of micro digital fringes onto the surface to be measured, as shown in Figure 1.1. The optical lens unit was composed of the optical lens set of the projection module, a stereomicroscope with an optical objective, glass substrates (GS), non-polarizing beamsplitters (NPB), a reflective mirror and four camera CCDs. To achieve an even light intensity for all the CCDs, the light splitting ratios of the $NPB_{1\ to\ 3}$ (from left to right) are selected as 25%-75%, 33.35%-66.65% and 50%-50%, respectively, in which the incident light intensity for each image sensor can be approximately set to 25%. Meanwhile, to achieve a set of four focusing positions, four glass substrates (GS_1 to GS_4) having various thicknesses are aligned in front of each respective image sensor and are employed to control the image focusing distance. Under such optical arrangement, a series of acquired images with a predefined focus position can be obtained for generation of a focus-depth response curve within one-shot image acquisition. Furthermore, the DMD light projection module can generate and project computer-controlled structured-light patterns onto the object's surface. Along the non-scanning vertical direction of the confocal microscope, four image intensities of each image pixel can be acquired simultaneously.

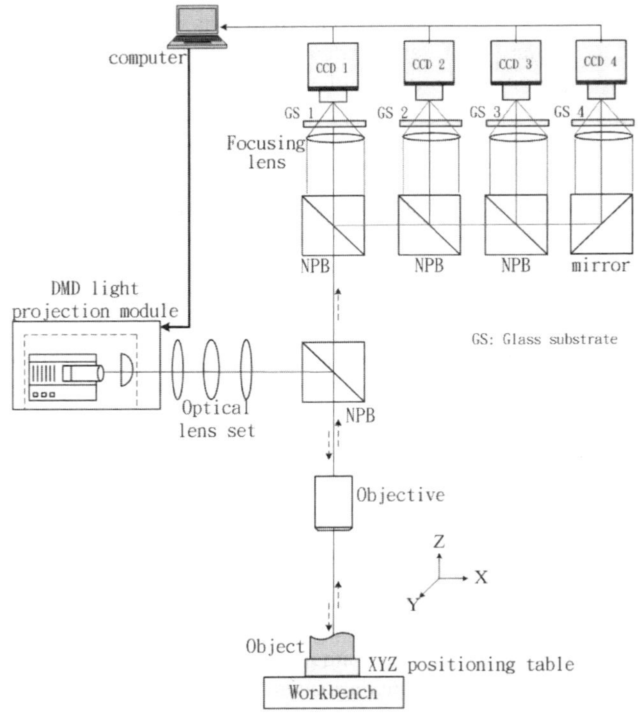

Figure 1.1. The optical layout of the developed measurement system

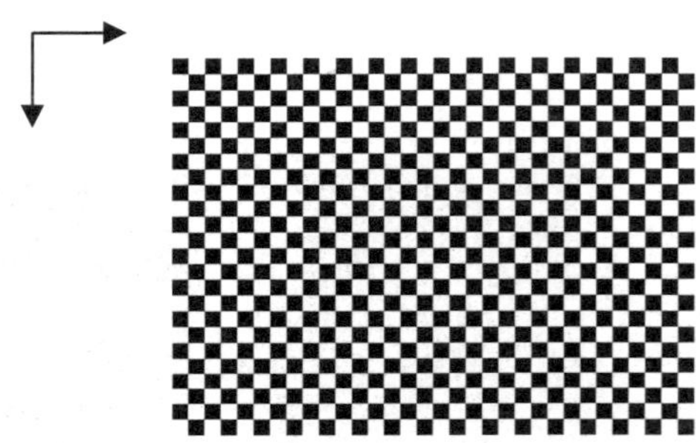

Figure 1.2. Structured light patterns designed for confocal fringe projection

3. Measurement principle

Since DMD is capable of projecting full-field light pattern for transverse confocal scanning, full-field size measurement can be achieved without the need for translating the sample. In the system, DMD is used to project a rectangular light pattern (shown in Figure 1.2) and the designed pattern is projected through a microscopic lens system onto the surface of the specimen. Following this, the light pattern is reflected and imaged onto four CCD cameras from which the light intensity of each pixel-related position is measured and stored for focal measure evaluation. The image intensity of each image coordinate can reach its maximum when the image of the corresponding DMD pixel is locating on the focal plane of the microscope objective. The relation between intensity and displacement along the focal plane can be measured as the focal-depth response curve. Precise 3-D surface contour of the object can be reconstructed when accurate peaking position associated with the focal plane can be defined from its focal-depth response curve.

When the objective lens is designed and corrected for finite conjugate imaging, a thin lens model can be approximately applied to model the optical system as follows:

$$\frac{1}{u_0}+\frac{1}{v_0}=\frac{1}{f} \tag{1}$$

where u_o is the object distance;

v_o is the corresponding image distance when using the GS1;

f is the effective focal length of the objective.

When the CCD_1 is used as a reference sensor and its image focusing is adjusted to be on the focal plane while using the GS_1, the thickness differences between the GS_1 and others can introduce a pre-defined change on each image position for the CCD_2 to CCD_4. The focusing change (Δv) can be modeled using the light refraction principle as follows:

$$\Delta v_i = t_i[\frac{n_g}{\sqrt{n_g^2 - 1}} - 1] \quad (2)$$

where t_i is the thickness variation between the GS_1 and $GS_{1\text{-to-}4}$, respectively.
n_g is the diffractive coefficient of the glass substrate.

By summarizing Eq. (1) and Eq. (2), the corresponding image of the nth CCD camera with respect to an object distance (u_0) can be expressed as Eq. (3):

$$v_n = \frac{f\,u_0}{u_0 - f} + t_i[\frac{n_g}{\sqrt{n_g^2 - 1}} - 1] \quad (3)$$

It is worth noting that the focusing changes for the four image sensors are designed to be equally spanned along the depth of field of the optical system. The depth of the field, D, of an optical system is governed by Eq. (4) as follows:

$$D = \frac{2u_0 A\, f\, c(u_0 - f)}{A^2 f^2 - c^2(u_0 - f)^2} \quad (4)$$

where A is the diameter of aperture;
c is the circle of least confusion;

Therefore, the thickness of $GS_{1\text{-to-}4}$ can be designed by Eq. (5) by deriving it from the formulas (2) and (4).

$$t_i = \frac{i(2u_0 A\, f\, c(u_0 - f))}{N(\frac{n_g}{\sqrt{n_g^2 - 1}} - 1)\left[A^2 f^2 - c^2(u_0 - f)^2\right]} \quad (5)$$

where i is the index number from 1 to 4.
N is the number of CCDs.

4. Experiments and Results

It is important to evaluate if the focal-depth response curve can be successfully generated by the developed system. For this, an industrial sample with ball grid array (BGA) bumps with a diameter of 50 µm and a height of 5 µm was measured and four images, shown in Figure 1.3, were acquired simultaneously from CCD_2 to CCD_4. The focal-depth response curve (shown in Figure 1.4) was calculated and fitted using the sum-modified-Laplacian (SML) operator based on high-pass filtering strategy and curve fitting techniques [7]. The results confirmed that the proposed method is realistic and practical.

Furthermore, to attest the measurement accuracy of the developed measurement approach for 3-D surface profilometry, we conducted an experimental measurement on a Taylor Hobson calibrated step height of 2.370 µm. Figure 1.5(a) shows the image of digital fringes with a rectangular size of 20 µm (H) x 20 µm (V), being projected onto the master piece; Figure 1.5(b) illustrates the 3-D map being reconstructed while the cross section of the profile is shown in Figure 1.5(c). From the analysis of measured errors, it was found that the maximum error of the measurement was within 0.18 µm. It indicates that the maximum measured error, less than 7.0 % of the measured step size, can be achieved by using the proposed method.

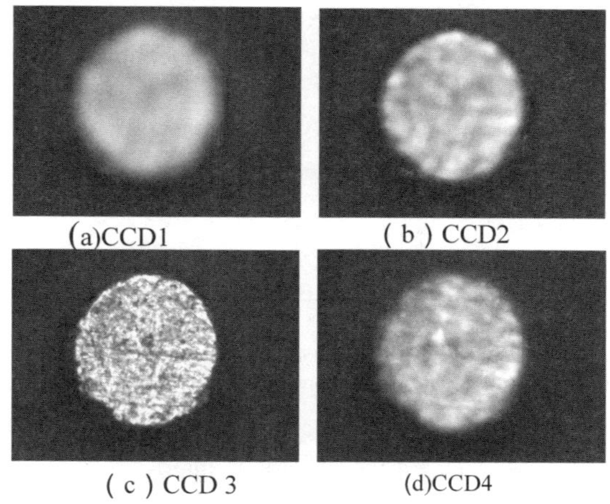

Figure 1.3. Images acquired simultaneously by the developed system when measuring a BGA bump.

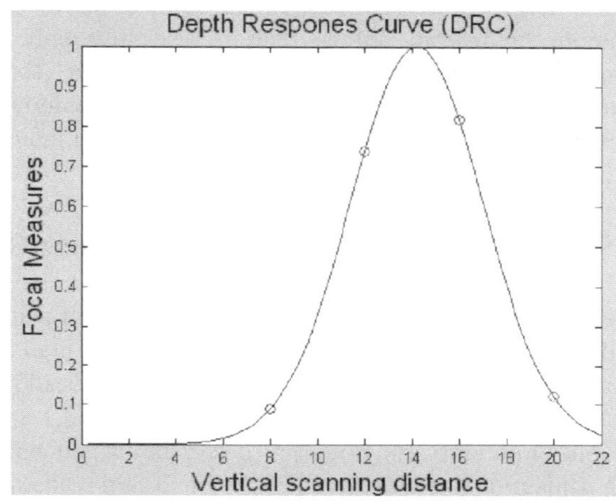

Figure 1.4. Focal-depth response curve being obtained using the sum-modified-Laplacian (SML) operator based on high-pass filtering strategy, where the data points were fitted into a focal response curve through curve fitting.

5. Conclusion

A full-field surface profilometer using an innovative simultaneous confocal approach was successfully developed for on-line 3-D automatic optical inspection. Our approach applied an innovative optical system design to simultaneously obtain the focal-depth response curve without any vertical scanning operation. This enables on-line 3-D profile inspection become feasible for many AOI application. The experimental results also indicate that the maximum measured errors can be still up to 7.0% of the measured step size, which still requires further improvement for achieving more satisfactory measurement accuracy. Furthermore, the measurement depth range currently limited to be within the depth of field (DOF) of the optical system, needs to be further increased for wide industrial inspection application.

(a) standard step height

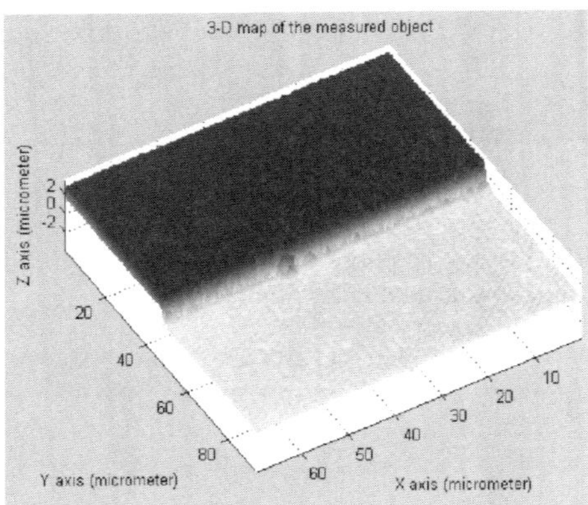

(b) 3-D map

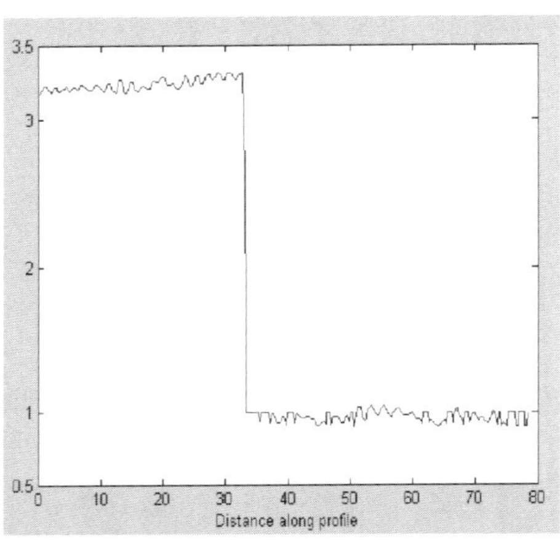

(c) Cross-section profile

Figure 1.5. Measurement results of a calibrated step height of 2.370 μm.

References

[1] S.K. Nayar and Y. Nakagawa 1990, Shape from Focus: An Effective Approach for Rough Surfaces, Proceeding of IEEE Int. Conf. on robotics and automation, P218-225.

[2] E. Krotkov 1987, Focusing, Lecturing Notes, Carnegie Mellon University, Vol. 1.

[3] S. Cha, and etl. 2000, Nontranslational three-dimensional profilometry by chromatic confocal microscopy with dynamically configurable micromirror scanning. Appl. Optics. Vol. 39, No 16, P2605-2613.

[4] T. Tanaami and etl. 2002, High-speed 1-frame /ms scanning confocal microscope with a microlens and Nipkow disks, Appl. Optics., Vol. 41, P4704-4708.

[5] F. Bitte, G. Dussler and Pfeifer, T. 2001, 3D micro-inspection goes DMD, Opt. Laser. Eng. Vol. 36, P.155-167.

[6] R. Höfling, E. Ahl 2004, ALP: universal DMD controller for metrology and testing, Proceedings of SPIE-IS&T Electronic Imaging, SPIE Vol. 5289 P322-329.

[7] L.-C. Chen and W.-J. Kao 2005, Automatic Full-Field 3-D Profilometry Using White Light Confocal Microscopy with DMD-based Fringe Projection, Material Science Forum, Vol. 505-507, P361-366.

Aspheric Lens Decenter Measurement

Jets Chang[1], Bi-Hui Hu[2], Warren Tsai[3], Leon Li[4] and Patrick Chang[5]

[1]Hon-Hai Precision Industrial Corporation, Taipei, Taiwan, jetschang@foxconn.com
[2]Hon-Hai Precision Industrial Corporation, Taipei, Taiwan, Bi-Hui.Hu@foxconn.com
[3]Hon-Hai Precision Industrial Corporation, Taipei, Taiwan, Warren.Tsai@foxconn.com
[4]Hon-Hai Precision Industrial Corporation, Taipei, Taiwan, Leon.Li@foxconn.com
[5]Hon-Hai Precision Industrial Corporation, Taipei, Taiwan, Patrick.Chang@foxconn.com

Abstract. The precision measurement of a lens wedge and decenter is badly needed in the optical testing of the imaging lens production. The three point reference plane measurement method was introduced. We do the evaluation of the wedge and tilt measurement of the aspheric lens by Panasonic UA3P.

Keywords. Aspheric Lens, Optical Imaging Module, 3D Profilometers, Modulation Transfer Function.

1. Introduction

With the development of compact camera modules in mobile devices, the value of small form factor camera has been widely recognized. To replace the traditional imaging optics using a lot of spheric glass lens, the aspheric lens is introduced in the lens module development [1].

An aspheric lens or a sphere is a lens whose surfaces have a profile that is neither a portion of a sphere nor of a circular cylinder. There are several benefits for the aspheric lens in optical systems. The aspheric lens can eliminate spherical aberration and reduce other optical aberrations compared to a simple lens. A single aspheric lens can often replace a much more complex multi-spheric lens system. The resultsing optical system is smaller and lighter, especially good for the compact mobile devices. Besides, it could be make optics cheaper when plastic apheric lens is used. The plastic aspheric lenses can be made by injection molding. The two apsheric surfaces of the plastic lens is set on the mold separates into at least two halves (the core and the cavity) to permit the part to be extracted. Since the two different surfaces of the apsheric lens are set on these two halves, there will be a decenter issue for the ashperes by molding.

Nowadays, as the market demand of small size optical lenses and components, such as pick up lenses, digital camera lenses, cell phone lenses and optical communication elements, increases, the decenter measurement of the aspheres by molding increases as well. If there exists the non-coincidence of optical axis and mechanical axis, it will cause a centering error in the optical system and causes a degradation in optical performance, especially the value of the modulation transfer function (MTF) will decrease.

Instrument for measuring individual surface shapes are known. One of the decenter measurement developed based on one non-destructive method is using a collimator and a rotation stage, however, the rotation stage will introduce the wobbling and the tilting issues during the measurement [2].

Mechanical surface measuring devices such as aspheric profilometers are mainly used in the optical industry. Previously, most contact-type profilometers have performed measurements in two dimensions. These profilometers provide one dimension of accuracy in the aspheric profile measurement of precision optical components. In late 1998, profilometers that perform three-dimension measurements became available from Panasonic Factory Automation, Chicago [3, 4]. They use a single vertically oriented probe that travels in the vertical direction (up and down) as well as in a horizontal plane. Profilometers measure a surface profile on one surface of a lens component at a time and determine the orientation of an optical axis of that surface. However, in order to measure the orientation of two surfaces with respect to one another, the lens component must be turned over so that its second surface can also be measured. The flip-over procedure of the lens measurement produces inaccurate results. The same issue will happen if one uses the interferometer or other measuring device, instead of a profilometer.

2. Theory of Wedge Measurement of an Aspheric Lens by 3D Profilometer

In this section, the definition of wedge and decenter measurement of an aspheric lens is described first. Following is the overview and procedures of measuring the wedge and decenter by using the specific tool.

An aspheric lens surface is usually defined by the location of its vertex, the direction of its optical axis (which is normal to the surface at the vertex) and an aspheric equation including aspheric coefficients describing the surface. The aspheric coefficients are set to zero for a spherical surface. An aspheric lens is usually a lens with two aspheric surfaces. In such aspheric lens component, the relative angular separation between the two optical axes (of its respective surfaces) is referred to as wedge (see Figure 1.1). The shortest distance between the two optical axes is called decenter. The lens center thickness is then defined along the optical axis of the reference surface.

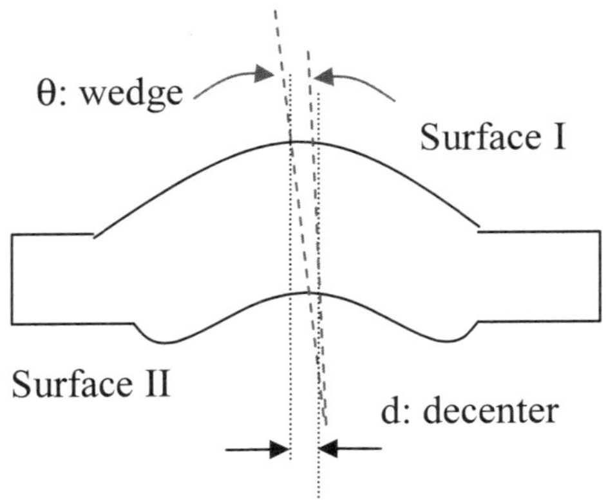

Figure 1.1. Definition of wedge and decenter of an aspheric lens

The theory of 3D Measurement is based on three reference spheres of the specific tooling. The three reference spheres are measured and well-calibrated by standard measurement of the radius of the spheres. Therefore, there will be a first reference plane (P1) marked by the apex of the three reference spheres. The tool is then rotated in Y axis (or X axis). The 2nd reference plane is determined.

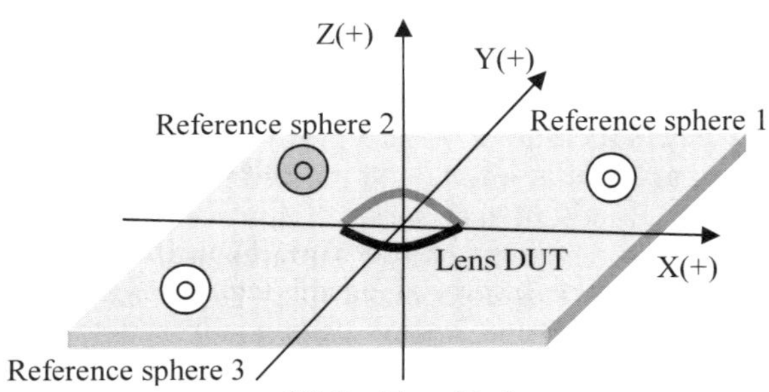

(A) Position side 1

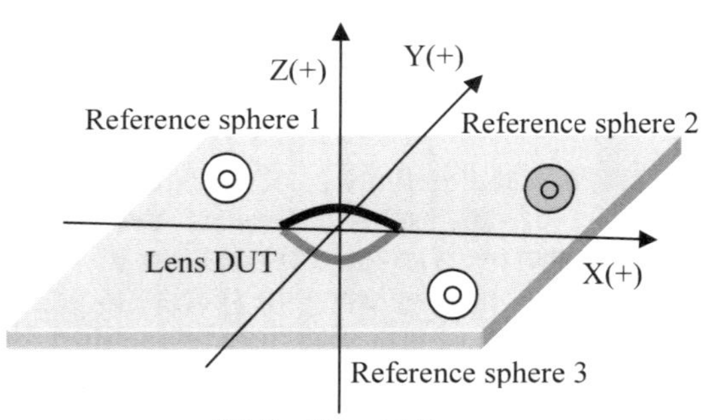

(B) Position side2

Figure 1.2. Configuration of tooling for reference spheres and lens under test

The basic principles of the measurement are as follows:

(1) The center points of the reference spheres are set to be identical and the central coordinates of the reference sphere can be calculated by the 3D measurement system.

(2) The rotation of the apparatus along the Y (or X axis) keeps the same reference plane by measuring the three reference spheres.

(3) The vertexes of the aspheric surfaces are then determined by the 3D measurement system. Hill-climbing and root-mean square error (minimizing R_t) methodology is introduced here to find the coordination of the vertexes of the aspheric surfaces.

(4) Calculate the coordination of the vertex on each surface by the using the vertex coordination of these reference spheres vertexes coordination. There will be two reference planes made in the reference coordination. The decenter can be determined by using the coordination of the vertex of the reference spheres. See Figure 1.3.

θ1 and θ2 are the tilting angles of the normal vectors of the vertexes on each surface with respective to normal vectors of reference planes 1 and 2.

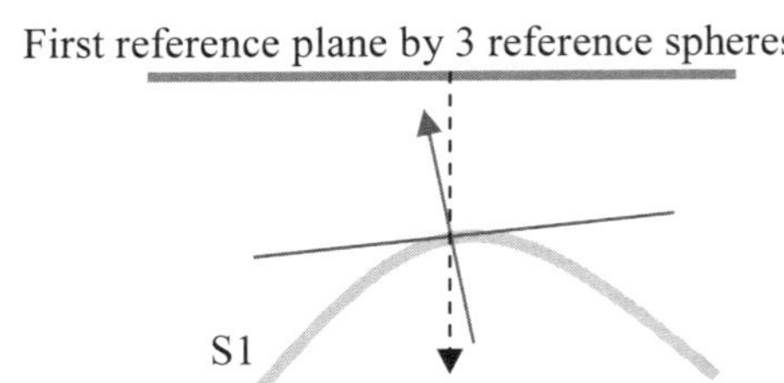

(A) Vertex normal vector of surface 1 with respect to the normal vector of the 1st reference plane

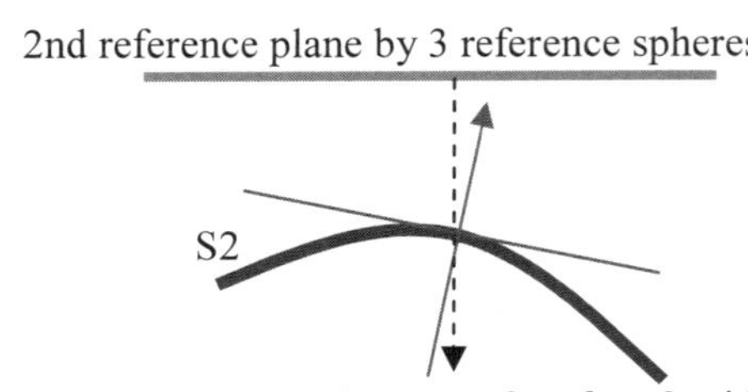

(B) Vertex normal vector of surface 2 with respect to the normal vector of the 2nd reference plane

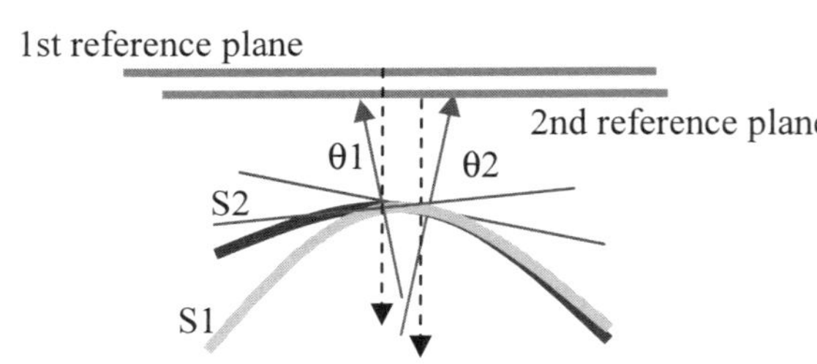

(C)Combination of two reference plane and aspheric vertexes

Figure 1.3. The three basic principles of reference plane measurement for wedge and tilt of a lens (a to c)

According to the measurement procedures listed above, there will be two reference planes and 10 coordination measured by UA3P. Figure 1.4 shows the coordination of the measuring points, respectively.

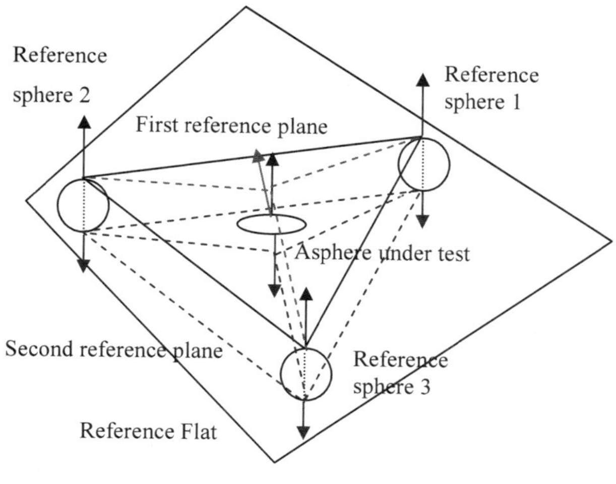

Figure 1.4. Geometrical configuration of aspheric decenter measurement by using three standard spheres and one reference flate.

3. Measurement and Results

3.1 Measurement

The aspheric plastic lenses are measured by this method for the wedge and tilt of the lens. Figure 1.5 shows the measurement results of the same aspheric lens by the three consecutive-injection molding processes. Since the lens mold is controlled by the Sumitomo precision injection molding machine, the positioning of the closing of the lens mold is controlled within 1 micron.

The measurement results of the repeatability of the absolute decenter/wedge of the lenses are listed in Table 1.1. Samples 1 and 2 are in two different cavities of the same mold. We could adjust the position pins of the two separation mold by the results

Table 1.1. Repetibility measurement results of the absolute decenter and wedge of the two lens by the same mold of different cavity.

Results for sample 1	S1	S2	S3	Ave	Std
Abs. Decenter (um)	3.645	3.897	2.967	3.503	0.481
Wedge	0'23"	1'26"	2'9"		
Results for sample 2	T1	T2	T3	Ave	Std
Abs. Decenter (um)	8.05	8.202	8.096	8.116	0.077
Wedge	1'20''	2'05''	0'55''		

3.2 Results Analysis

The variation in the measurements comes mainly from the difference of the reference spheres, the vertex coordination measurement of each reference surfaces including standard spheres and the asphere under test. Hill-climbing with root-mean square error is used for the measurement of the vertex. To minimize the measurement error, the vertex of the asphere surfaces should be within a certain slope around the vertex of the aspheric surfaces because of hill-climbing algorithm. Tilting of the position stage for the standard sphere tooling is also need to be in (almost) parallel with the reference plane of the testing machine.

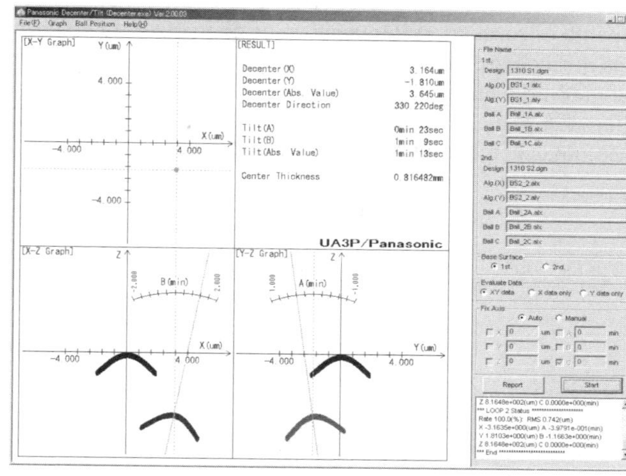

(a)

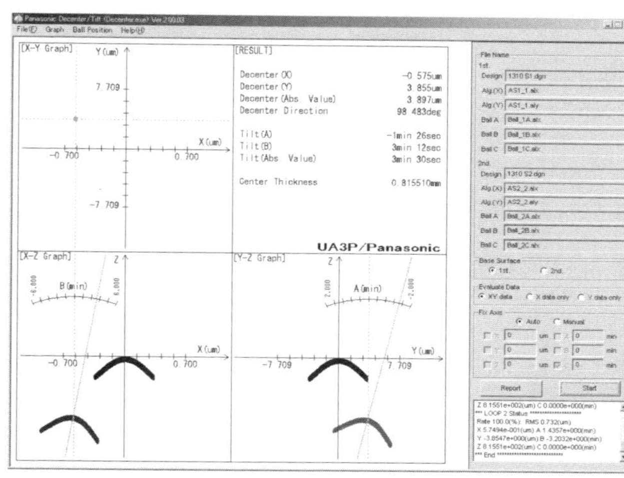

(b)

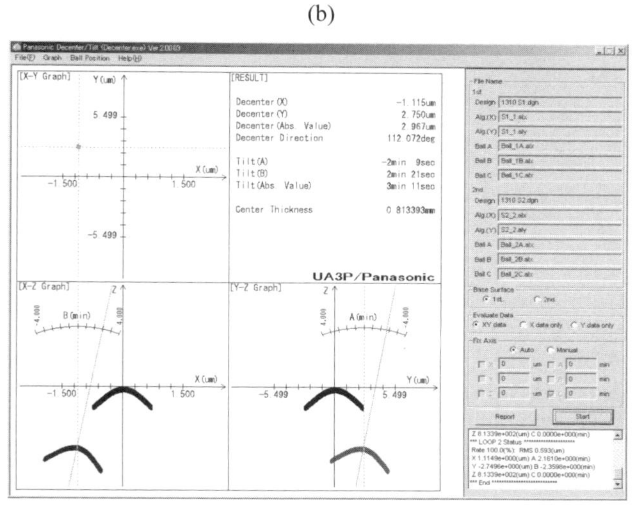

(c)

Figure 1.5. Measurements of the decenter and wedge of an aspheric lens by Panasonic UA3P and the standard spheres tooling

4. Conclusion

The 3D measurement of the wedge and decenter of an apsheric lens demonstrates that this mesurement system is suitable for calibrating the geometric shape of a lens.

The test results show an advantage of the present invention that the apparatus and a method described above is capable of measuring not only the relative orientation of the two surfaces of a typical lens component, but is also capable of measuring the position of diffractive or Fresnel lens surfaces relative to one another. The repeatibility of the measurement is shown in Table 1.1. The tooling using reference plane shows good measurement results for the absolute decenter and wedge of an aspheric lens.

References

[1] Schaub, Michael Patrick, Ph.D. Thesis, Aspheric and diffractive surfaces in one, two and three element lenses, The University of Arizona, 1999.

[2] J. I. Berg, and F. H. Gerhard, United States Patents US 5,831,719, Apr. 12, 1996.

[3] Optical Engineering Technology, http://www.itrc.org.tw/Research/Project/decenter-e.php

[4] J. P. Browen, United States Patents US 6, 072,569, Jun. 6, 2000.

[5] "Front/Rear Face Tilt/Decenter Evaluation Software Instruction Manual," Advanced Production Systems Development Company, Matsushita Electric Industrial Co., Ltd.

11–7

A High-speed Inspection System of Large Area Component

C.Y. Hung[1], C.L. Chang[2], S.C. Lu[3], J.J. Jeng[4], and W.Y. Chen[5]

[1] MSL, Industrial Technology Research Institute, Hsinchu, Taiwan, CY_Hung@itri.org.tw
[2] MSL, Industrial Technology Research Institute, Hsinchu, Taiwan, VincentChang@itri.org.tw
[3] MSL, Industrial Technology Research Institute, Hsinchu, Taiwan, davidlu@itri.org.tw
[4] MSL, Industrial Technology Research Institute, Hsinchu, Taiwan, JJJeng@itri.org.tw
[5] MSL, Industrial Technology Research Institute, Hsinchu, Taiwan, WeiYuChen@itri.org.tw

Abstract. AOI (Automatic optical inspection) systems are becoming more and more important in industrial production lines. Because of the progress of manufacturing technology, the area of TFT-LCD panel components is growing larger, and the requested resolution is getting tinier. Here we are interested in the defect inspection of large area component, such as color filter, TFT array pattern, and substrate glass. The difficulties in its defect inspection are its complex texture and demand for high-speed processing. In this paper, we proposed a high-speed AOI system, including the multi-thread structure and the inspection algorithm. The multi-thread structure can decrease the loading of CPU at host PC. It makes the AOI system becomes simple, flexible, and easier to be managed. The proposed inspection algorithm improves the traditional cell-to-cell comparison method and improves inspection rate. Experimental results show that the proposed AOI system can efficiently detect the defects with high-speed, and it is useful and practical in real inspection line of large area component.

Keywords: high-speed inspection, large area, TFTs

1. Introduction

AOI (Automatic optical inspection) systems are becoming more and more important in industrial production lines. Especially in electronics industry, an attempt is often made to achieve almost 100% quality control of all components and final goods. Compared with manual inspection, the automatic detection of surface defects has the following advantages: high speed, quality assurance, and capable of detecting small size of defects.

Here we are interested in the defect inspection of large area component, such as colour filter, TFT array pattern, substrate glass, etc. The difficulties in its defect inspection are its complex texture and demand for high-speed processing. Because of the progress of manufacturing technology, the area of TFT-LCD panel components is growing larger, and the requested resolution is getting tinier. Even a line-scan CCD (Charge Coupled Device) camera with high resolution is used; several line CCD cameras must be arrayed to finish the inspection in short time. It will also be tough work to calibrate every CCD to the correct position quickly and accurately. Furthermore, high-resolution image data is so enormous that host PC will be over-loading to deal with images grabbed by several CCD cameras at the same time. Mostly multi-CPUs [1] or several PCs [2] in a host PC are used to solve this problem in the past, as shown in Figure 1.1. Too many PCs not only occupy space but are also difficult to be managed; Multi-CPUs in host PC will accelerate the processing speed, but it is inefficient if the computation loading is not divided to every CPU averagely. Furthermore, it did not have a criterion to transfer and process mass image data synchronously, which made the system unbalanced and inefficient. So, we proposed the parallel processing structure working with multiple image processors. The inspection algorithm will be operated in the threads of image processor, and find defects in a short time with parallel processing in multi-thread.

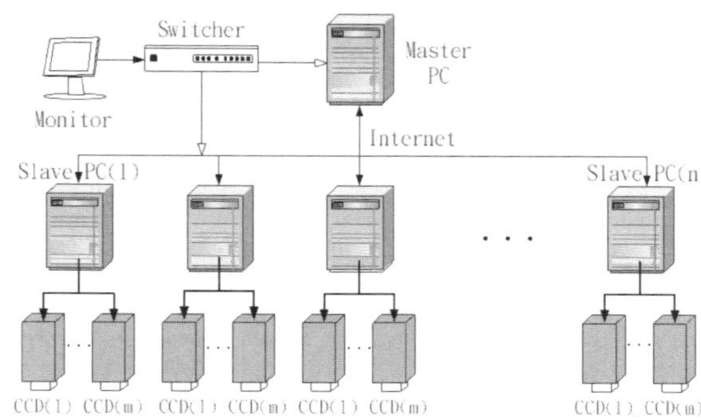

Figure 1.1. The illustration of multi-PC case

Since the characteristic of TFT-LCD images is similar to web materials with texture pattern, it is much valuable to introduce the categories and defect inspection methods of web materials briefly. Web materials can be categorized into two classes: uniform webs and texture webs. Texture webs include materials that are generally textured, such as textiles and wallpaper. In the case of uniform webs, regions that differ from the uniform background are defects. Therefore, simple algorithms, such as thresholding, zero order background tracking, and delta tracker can be used to detect the defects [3,4]. In the case of texture webs, many methods are used to detect the defects such as subtraction [4], template matching, Fourier transform, 1D AR (One-Dimensional Auto Regression) [3,5], fuzzy logic [6], neural network [7], CNN (Cellular Neural Network) [8], and neural-fuzzy network [9], etc. However, almost all of the above methods will take too much time in processing, so it is hard to use them in real industrial application, especially in large area component of TFT-LCD.

Comparing with other methods, the image subtraction method is relatively comprehensible and has the advantage of simplicity and high-speed. For industrial demand of high-speed, most of the commercial products for inspection of TFT-LCD also use cell-to-cell comparison, which basically is a kind of image subtraction technology. However, the traditional subtraction method is suffered from false alarm and defect missing while inspection. In this paper, we proposed a novel cell-to-cell comparison based algorithm to overcome the drawbacks and improve the efficiency of traditional subtraction technology. Experimental results show that the proposed AOI system can precisely detect the defects with high-speed; it is useful and practical in real inspection line of large area component, such as colour filter, TFT array pattern, substrate glass, etc.

The paper is organized as follows. Section 2 describes the structure and algorithm of the proposed AOI system. Section 3 gives experimental setup and results. Finally, conclusions are summarized in the last section.

2. The multi-thread inspection algorithm

The parallel processing structure and inspection algorithm of the proposed AOI system will be described in the following subsections.

2.1 Parallel processing of multi-thread

The proposed inspection system works with the frame grabber with processor and memory, which is called image processor. The processing algorithm will be operated in the image processor threads and inspect defects in a short time with parallel processing of multi-thread. The main inspection procedures and the sub threads are managed by main thread in slave PC; each sub thread works independently and is in charge of the image processing that grabbed by each line CCD camera. Sub thread controls two threads, which manage to grab image and inspection synchronously in the image processor. After inspection, the results will be integrated in the main thread of the slave PC, and be transferred to the master PC for UI (User Interface) output.

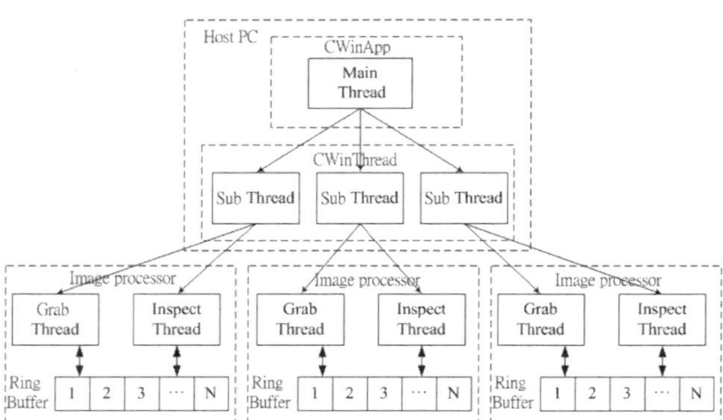

Figure 1.2. The illustratioin of multi-thread structure

The structure of the entire system is shown in Figure 1.2. Each processor and thread is in charge of its position; they work independently and synchronously to make the system efficient and fast. The data flowchart in multi-thread structure is shown in Figure 1.3. We located the grabbed mass image data in ring buffer, and completed the inspection in image processor. So, we just have to transfer the inspection results to the slave PC, which extremely decreases the data transfer loading and processing time.

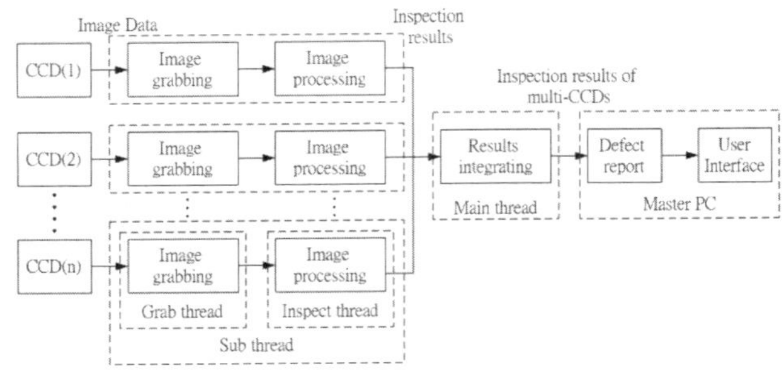

Figure 1.3. The data flowchart in multi-thread structure

The proposed AOI system needs many line CCD cameras to inspect TFT-LCD samples with large area. The proposed method can decrease the loading of CPU at the host PC. Meanwhile, when the system needs more CCD cameras, it only has to increase the image processors and sub threads instead of adding extra more PCs. It makes the AOI system becomes simple, flexible, and easier to be managed.

2.2 The inspection algorithm

The proposed inspection algorithm focuses on the defect inspection in character of regular array pattern, which is a kind of cell-to-cell comparison method. It uses saturate-subtract to compare grey value to above/below or left/right cell pattern, and integrates with OR operation to enhance the contrast of defect image. We called it "saturate-subtract-OR" algorithm described in the following sections.

The traditional cell-to-cell comparison algorithm using the absolute value subtraction and the "AND" operation can detect black and white defects simultaneously. However, it has three disadvantages. First, the absolute-subtract might make negative value, which is surplus information for black defect and caused false alarm. Second, the AND operation took the minimum value that might lead to low contract. Higher contrast would be obtained only when defect had obvious difference between both left and right intervals. Third, it will generate mistakes near the image boundary because of the lost of one side information.

The proposed saturate-subtract-OR algorithm used the saturate value subtraction and the "OR" operation as shown in Figure 1.4.

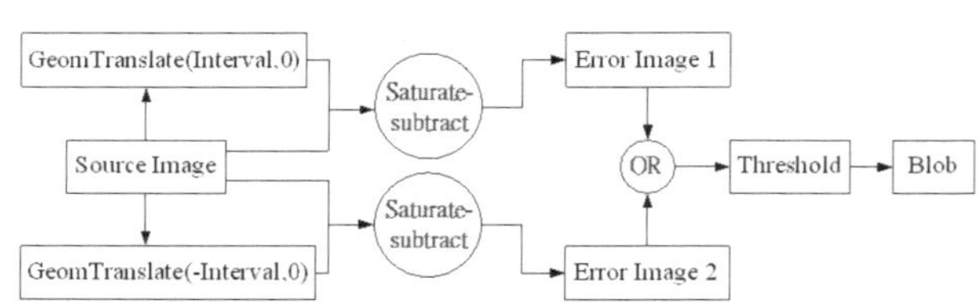

Figure 1.4. The flowchart of saturate-OR algorithm

It has the following advantages: First, with saturate value subtraction, it deletes unnecessary information and eliminates false alarm. Second, with OR operation, higher contrast can be obtained if the defect has obvious difference between either side of left or right interval. The third advantage is the fast computation speed. Compare with the traditional absolute-AND algorithm, the sensitivity to black defects is increased significantly under the condition of equal computation time.

3. Experimental setup and results

The section will describe the system architecture, experimental setup, inspection flowchart, and experimental results, respectively.

3.1 System architecture

As shown in Figure 1.5, fluorescent illumination, line-scan CCD camera array, and PC-based image processors are used in our system. Figure 1.6 shows a photograph of experimental setup.

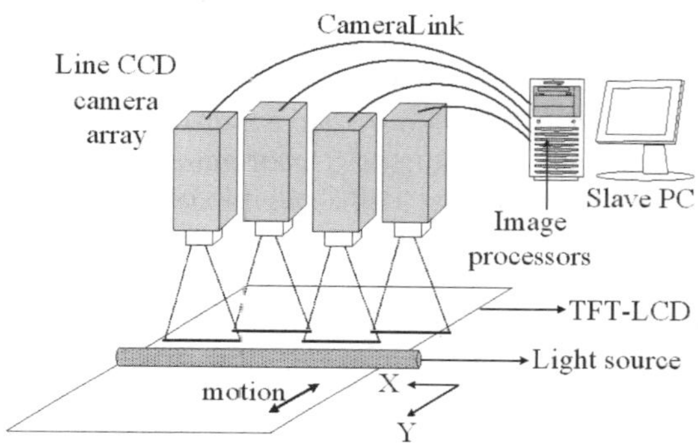

Figure 1.5. System architecture

Figure 1.6. Experimental setup

3.2 Inspection flowchart

The flowchart of defect inspection is shown in Figure 1.7. It consists of image acquisition, image pre-processing, validated region check, defect inspection, defect map output, and defect review. The details of the flowchart of defect inspection are described in the rest of this section.

3.2.1 Image acquisition

Since the proposed system arrayed multiple high-resolution line CCD cameras, the alignment between cameras and stage is necessary to grab non-distortion images. We used our standard glass board to calibrate every CCD to the correct position quickly and accurately. And, because sensor noise and non-uniform illumination will degrade the image quality, the flat field correction is done to obtain better image quality. The system drives stage along y-axis and triggers cameras with preset distance to grab images.

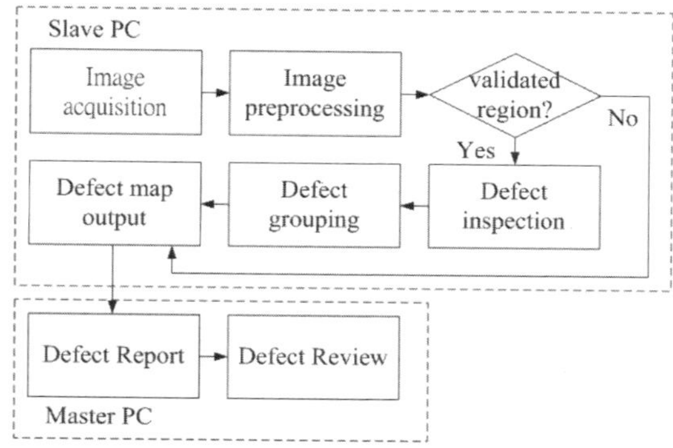

Figure 1.7. The flowchart of defect inspection

3.2.2 Image preprocessing

The task of image pre-processing is to check the validated regions. The grabbed image of TFT-LCD includes validated and invalidated regions. The validated regions consist of regular cells and the invalidated regions generally include some information such as product number and alignment patterns. Before performing defect inspection, the system first chooses the validated regions, which can be determined by horizontal and vertical projections.

3.2.3 Defect inspection

After validated regions have been determined, the defect inspection is performed. It is the major core technology of an AOI system, which will determine the system performance. The algorithm of defect inspection is described as Section II.

3.2.4 Defect grouping and map output

The task of the defect grouping is to group the neighbouring defects as the same group. Then the defect map, such as defect position, size, and shape, is integrated to a specific-format file and transferred to UI of master PC for output.

3.2.5 Defect review

Finally, the user can choose the specific defect, and the system will automatically review it by the area CCD camera with extra-high resolution which is controlled by master PC. Because of the high resolution, it needs accurate alignment to move the area CCD camera to the specific defect position correctly. The high-precision alignment was implemented beforehand by our standard glass board. It helps us to complete the alignment and calibration in a short time.

3.3 Experimental results

Our experimental system, used the proposed multithread structure and inspection algorithm, can process image data of about 80 megabytes per second per CCD. Even if multiple line CCD cameras were installed in one PC simultaneously, the inspection speed remained rapid, and could reach up to 9K lines/sec with 7 μm resolution. We listed some typical defect images and marked the defects with rectangles after the inspection, as shown in Figure 1.8. Compare with traditional algorithm, the defect sensitivity of proposed method is increased significantly under the condition of equal computation time. Experimental results show that most low-contrast defects, such as (d) and (e) in Figure 1.8, could be detected by the proposed inspection algorithm. Compared with the traditional algorithm, the effect of contrast enhancement is shown in Figure 1.9. Adding the comparison interval from two to four sides will enhance the defect contrast as well, but it will take twice the processing time. Table 1.1 shows the comparison of efficiency between two algorithms and two number of comparison sides. Obviously more defects are detected by the proposed saturate-OR algorithm.

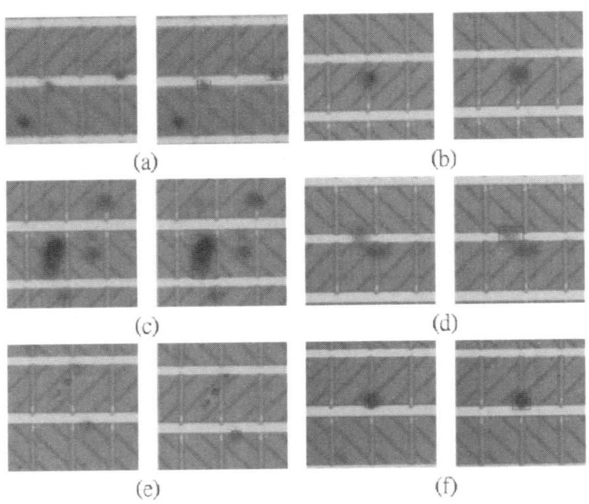

Figure 1.8. The typical defect images and corresponding inspection results

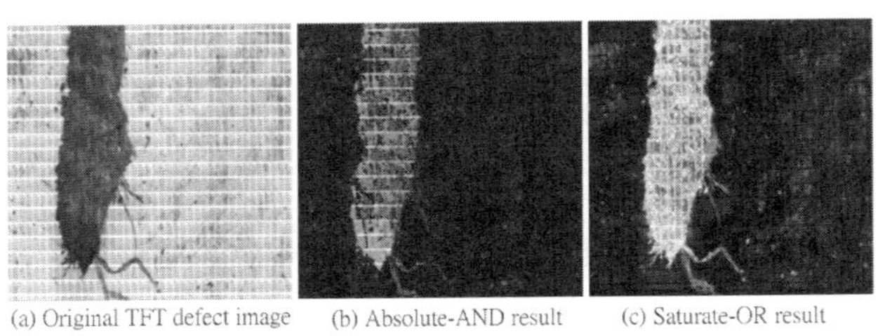

Figure 1.9. The illustration of contract enhancement with saturate-OR algorithm

Table 1.1. The comparison of efficiency between different algorithms

		Left-right interval		Above-below and left-right interval	
Algorithm	Image No.	Defect number	Time (ms)	Defect number	Time (ms)
Absolute AND	1	109	14.285	133	30.18
	2	91	14.25	119	30.118
	3	181	14.274	401	30.139
Saturate OR	1	190	14.299	217	30.186
	2	169	14.284	189	30.146
	3	977	14.297	1312	30.185

4. Conclusions

Currently, MSL (Mechanical and Systems Research Laboratories) is developing AOI equipments, such as PCB (Printed Circuit Board) artwork, colour filter, and TFT array pattern, etc. In this paper, we proposed a high-speed AOI system, including the multi-thread structure and inspection algorithm. The multi-thread structure can decrease the loading of CPU at host PC. It makes the AOI system become simple, flexible, and easier to be managed. The proposed inspection algorithm improves the traditional cell-to-cell comparison method and improves inspection rate. Experimental results show that the proposed AOI system can efficiently detect the defects with high-speed, and it is useful and practical in real inspection line of large area component.

References

[1] J. H. Kim, S. Ahn, J. W. Jeon, and J. E. Byun, Jun. 2001, A high-speed high-resolution vision system for the inspection of TFT LCD, Proceedings. ISIE 2001. IEEE International Symposium, vol. 1, pp. 101-105.

[2] K. Nakashima, 1994, Hybrid Inspection System for LCD Color Filter Panels, 10th International Conference on Instrumentation and Measurement Technology, Hamamatsu, Japan, pp. 689-692.

[3] S. H. Hajimowlana, R. Muscedere, G. A. Jullien, and J. W. Roberts, 1999, An In-Camera Data Stream Processing System for Defect Detection, Real-Time Imaging, vol. 5, pp. 23-34.

[4] T. S. Newman and A. K. Jain, 1995, A Survey of Automated Visual Inspection, Computer Vision and Image Understanding, vol. 61, no. 2, pp. 231–262.

[5] A. Materka and M. Strzelecki, , June 1998, Texture Analysis Methods – A Review, COST B11 report, Brussels.

[6] S. H. Hajimowlana, R. Muscedere, G. A. Jullien, and J. W. Roberts, 2000, Defect Inspection in Web Inspection Using Fuzzy Fusion, IEEE International Symposium. on Circuits and Systems, pp. 718-721.

[7] J. Iivarinen and J. Rauhamaa, 1998, Surface Inspection of Web Using Self-Organizing Map, Proc. of SPIE, vol. 3522, pp. 96-103.

[8] V. Preciado, D. Guinea, R. Montufar, and J. Vicente, 2001, Real-time Inspection of Metal Laminates by means of CNN's, Proc. of SPIE, vol. 4301, no. 39.

[9] C. L. Chang, H. H. Chang, and C. P. Hsu, 10-12 July 2005, "An intelligent defect inspection technique for color filters", Mechatronics, 2005. ICM '05. IEEE International Conference, pp. 933-936

An Air-bearing Displacement Sensor with a Micro-ball Stylus for Profile Measurement of 3D Micro-structured Surfaces

Y. Yoshikawa[1], A. Shibuya, W. Gao[3]

[1]Tohoku University, Japan, 980-8579, yasu-y@nano.mech.tohoku.ac.jp
[2]Tohoku University, Japan, 980-8579, shibuya@nano.mech.tohoku.ac.jp
[3]Tohoku University, Japan, 980-8579, gaowei@cc.mech.tohoku.ac.jp

Abstract. Micro-structured surfaces such as micro-aspheric lenses/lens arrays and micro-gears are getting increasingly important in industries. A lot of micro-structured surfaces have complicated three-dimensional (3D) profiles with short pitches and large amplitudes (high-aspect-ratio). Some of them also have steep surface slopes. It is difficult, however, to use conventional measuring instruments to measure profiles of such 3D micro-structured surfaces. This paper describes a contact-type displacement sensor for profile measurement of 3D micro-structured surfaces. The shaft of the sensor is levitated by air-bearings to reduce the contact force between the sensor and the measured surface. A sharp corn stylus is attached at the end of the sensor shaft. A precision glass micro-ball is glued on the top of the stylus shaft. The diameter of micro-ball is under 50 µm. The displacement sensor is also used to measure the profile of a micro gear with a high-aspect-ratio surface profile.
Keywords: measurement, aspheric profile, air-bearing, displacement sensor, micro-stylus.

1. Introductionn

Micro-aspheric lens are one of the most important components, which are widely used in medical endoscopes and fiber-optic communication systems [1, 2]. The key of these technologies is micro-structured surfaces represented by microlens-arrays and micro-gears. Precision measurement of such surfaces is very important for assurance of the machining accuracy.

Laser microscopes with three-dimensional (3D) measuring abilities can be employed for measurement of micro-structured surfaces such as microlenses and micro-gears. However, laser lights will scatter when the light hits steep surfaces, which makes it difficult to measure micro-structured surfaces with large slope changes.

This paper describes a contact-type displacement sensor for profile measurement of 3D micro-structured surfaces. A sharp corn stylus with a micro-ball as the probe ball, which is called micro-ball stylus, is attached at the end of the sensor shaft. The sensor shaft is levitated by air-bearings. The probe of the micro-ball stylus touches and scans the surface so that the surface profile can be measured. The contact-type displacement sensor is less influenced by disturbances and no special treatment of surface is necessary, which makes the sensor effective in practical use.

The contact-type displacement sensor typically employees a ruby ball as the probe ball for contacting the surface. Ruby balls are hard and have good resistance to wear, resulting in a long life time. However, it is difficult to reduce the diameter of a ruby ball to be smaller than 100 µm, which is necessary for measurement of micro-structured surfaces with short pitches and large amplitudes [3].

On the other hand, a micro-stylus with a micro-ball of 30 µm in diameter has been developed by Hidaka [4]. The ball is formed by melting and rounding the tip of a glass stick. However, it is difficult to control the form accuracy and diameter of the micro-ball. There are also micro-stylus with diamond tips of 2 µm diameter. But it is expensive to manufacture and also difficult to assume the accuracy of the tip [5].

This paper describes the development of a micro-ball stylus with a precision glass micro-ball whose diameter is under 50 µm. The glass ball, which is used as the spacer of crystalline liquid panels, is with low cost and high form accuracy.

The micro-ball stylus is attached to the shaft of a contact-type sensor, which is levitated by air-bearings to reduce the contact force between the sensor and the measured surface [6]. The movement of the sensor shaft is measured by a linear encoder with a resolution of 0.14 nm. The displacement sensor with the micro-ball stylus is used to measure a micro-gear with a high-aspect-ratio surface profile. The pitch of the gear is approximately 130 µm and the height is approximately 65 µm. The maximum surface slope is approximately 70 degrees.

2. Fabrication of the micro-ball stylus

This section describes the fabrication of the micro-ball stylus. The glass micro-ball is glued on the tip of the micro-stylus shaft. The stylus shaft is made from stainless steel. The angle of the shaft top is approximately 50 degrees. A thread is formed on the other end of the stylus shaft, which is used to mount the stylus shaft on the displacement sensor.

A photograph of the system for gluing the micro-ball is shown in Figure 1.1. Two XYZ-stages and two CCD-cameras are employed to glue the micro-ball on the tip of the stylus shaft. The stylus shaft is attached to one of the XYZ-stages. The glass micro-ball is put on the other XYZ-stage. The position of the stylus tip relative to the micro-ball can be easily adjusted by using these XYZ-stages. The micro-ball and the stylus tip are also monitored by the two CCD-cameras from the X- and Y-directions so that the micro-ball can be glued at the center of the stylus tip.

To glue the micro-ball, the adhesive, which is the heat-hardening type, is first attached to the tip of the stylus shaft. Then the tip of the stylus shaft is made contact with the micro-ball by using the XYZ-stage. When the micro-ball touches the adhesive at the tip of the stylus shaft, the micro-ball is attached to the tip of the stylus shaft by the effect of surface tension. The micro-ball can then be glued on the tip of the stylus shaft by heating the adhesive to be hardened. Microscope images of micro-styluses with glass micro-balls of 50 μm and 10 μm diameters are shown in Figures 1.2(a) and (b) respectively.

Micro-styluses with glass tube styluses are also made by the same way. Instead of stainless steels, glass tubes are used as the stylus shaft. The glass tube is stretched to form a stylus with a conical shape by heating. Photographs of micro-styluses with glass tubes are shown in Figure 1.3. Glass micro-balls with diameters of 50 μm and 30 μm are shown in the figure. The angle of the stylus top is approximately 10 degrees and the length is 3 mm.

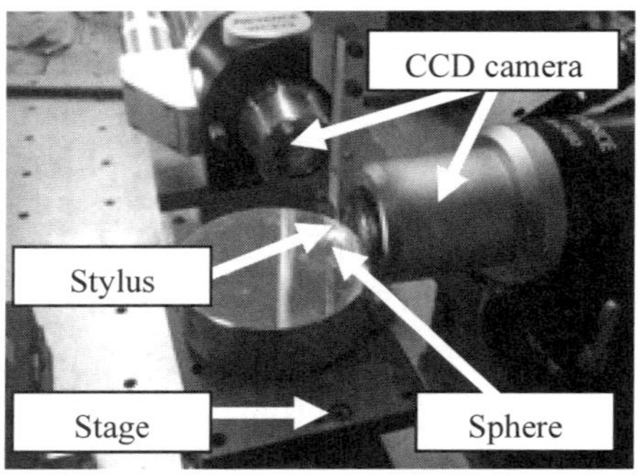

Figure 1.1. Micro-ball gluing system system

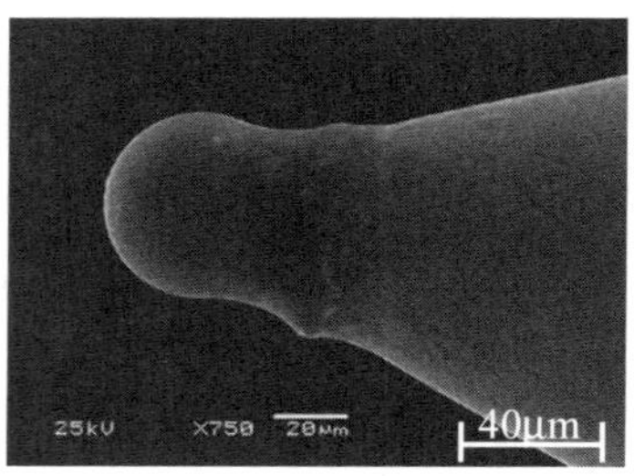

a

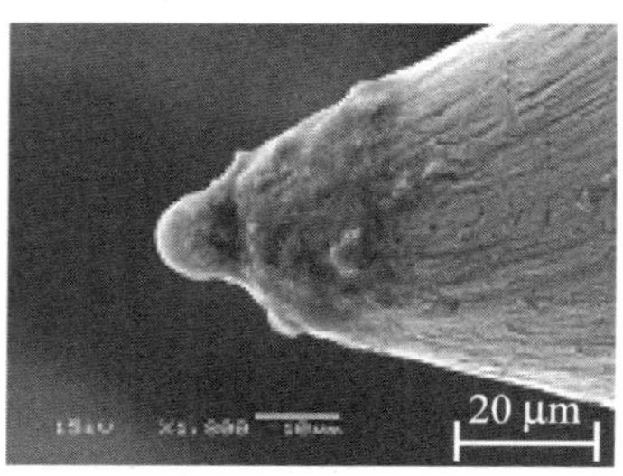

b

Figure 1.2. Microscope images of micro-styluses with stainless shafts.a. ϕ50 μm ball; b. ϕ10 μm ball

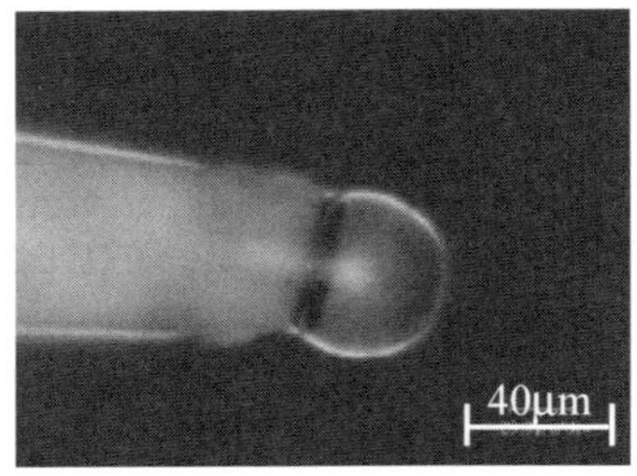

a

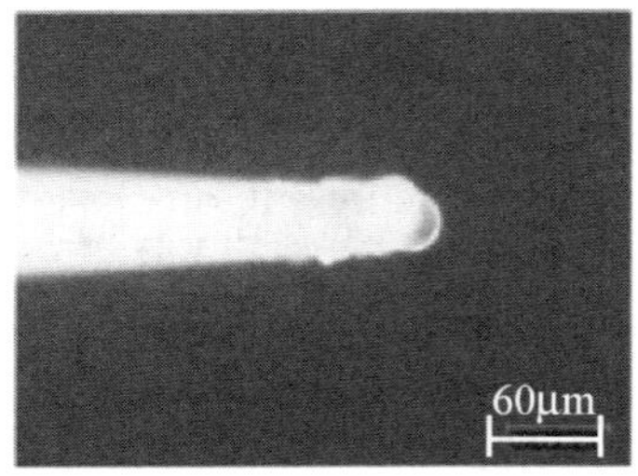

b

Figure 1.3. CCD images of micro-styluses with glass tubes. a. ϕ50 μm ball; b. ϕ30 μm ball

3. Testing of the micro-ball stylus

The strength of the micro-ball stylus is tested by using a strain gauge. The specification of the strain gage is shown in Table 1.1. The micro-ball stylus is inclined 45 degrees and pushed against the strain gauge by moving a manual stage so that an increasing load can be applied to the micro-ball stylus. The micro-ball stylus with the stainless shaft is tested first. The output of the strain gauge, which represents the strength of the micro-ball stylus, is shown in Figure 1.4. From the figure, it can be seen that the micro-ball stylus with the ϕ50 μm ball can tolerate a 40 mN load and that with the ϕ10 μm ball can tolerate a 6 mN load, respectively. It should be noted that the difference in the loading time is caused by the non-constant speed of the manual stage.

The experimental result of the micro-ball stylus with the glass tube shaft is shown in Figure 1.5. From the figure, it can be seen that both the micro-ball styluses with the ϕ50 μm ball and the ϕ10 μm ball can tolerate loads larger than 90 mN, which is limited by the measure range of strain gauge.

Table 1.1. Specification of the strain gage

Manufacturer	KYOWA Electric Instruments Co.
Sensitivity	0.0339 μN/1μV/V
Range (F.S.)	100 mN
Nonlinearity	±0.6% F. S.
Hysteresis	±0.19% F. S.

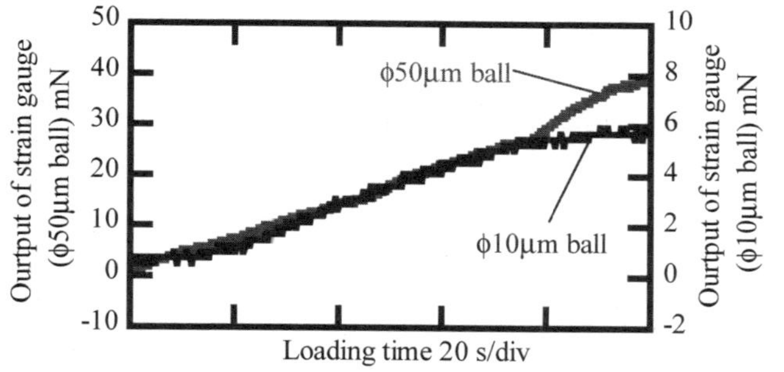

Figure 1.4. Strenght of micro-ball stylus with stainless steel shaft

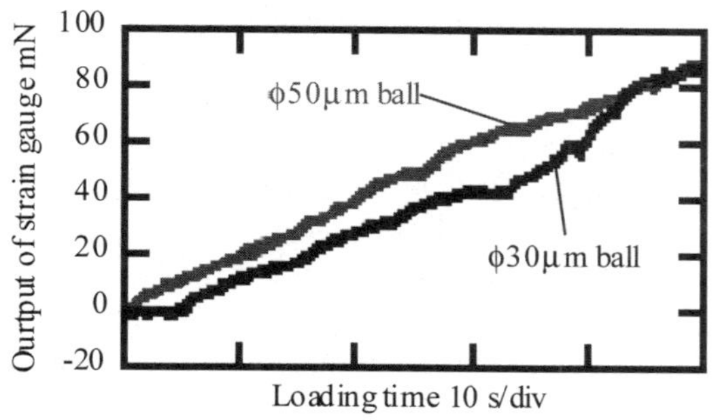

Figure 1.5. Strenght of micro-ball stylus with galass tube shaft

4. Testing of the displacement sensor with the micro-ball stylus

The schematic of the displacement sensor is shown in Figure 1.6. The micro-ball stylus is mounted on an air-bearing displacement sensor. The sensor is based on the inclined self-weight method [7]. The shaft of the sensor, which is set horizontally, is levitated by air-bearings. The measuring force is set by adjusting the tilt of the sensor. The movement of the shaft is measured by a linear encoder, of which the scale is attached to the shaft to avoid the Abbe error. The resolution of the linear encoder is 0.14 nm. The specification of the linear encoder is shown in Table 1.2.

The measuring force of the displacement sensor is evaluated. The displacement sensor is pushed against the strain gage with a measurement range of 0.98 mN. The specification of the strain gage is listed in Table 1.3. Figure 1.7 shows the output of the strain gauge with respect to the displacement of the sensor. It can be seen that the measuring force is less than 0.4 mN within the displacement range of 8 mm.

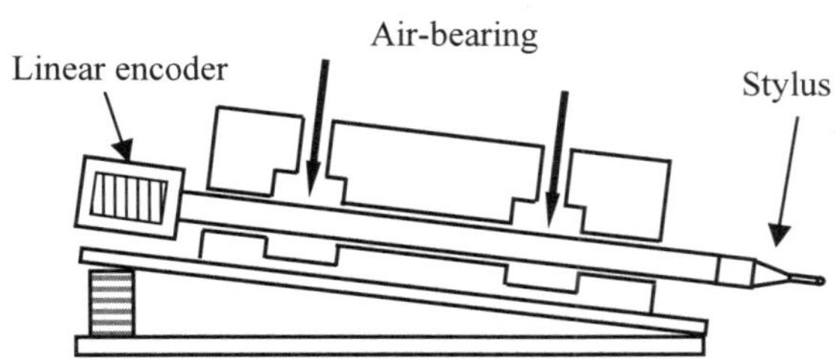

Figure 1.6. Schematic of the air-bearing displacement sensor

Table 1.2. Specification of the linear encoder

Manufacturer	SONY Manufacturing Systems Corporation
Signal period	0.14μm
Interpolation No.	1000
Resolution	0.14nm
Measurement range	8mm

Table 1.3. Specification of the strain gauge

Manufacturer	SSK CO., LTD.
Range (F.S.)	0.98 mN
Nonlinearity	±0.2% F.S.
Repeatability	0.2% F.S.
Hysteresis	±0.2% F.S.

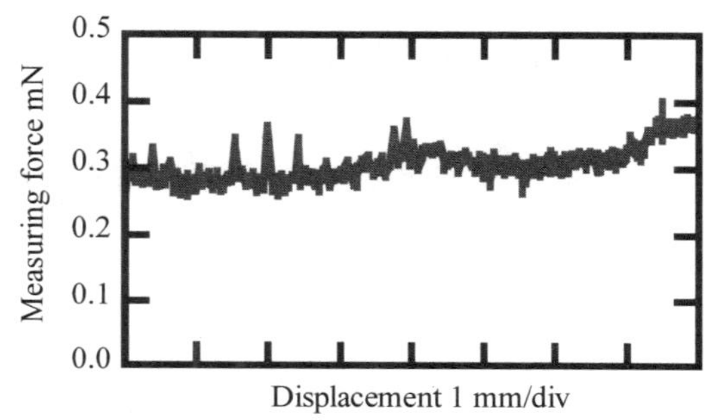

Figure 1.7. Measuring force of the displacement sensor

5. Profile measurement of a micro-gear surface

The surface profile of a micro-gear is measured by the displacement sensor with the micro-ball stylus of glass tube shaft. The diameter of the miro-ball is 50 μm. Figure 1.8 shows a schematic of the measurement system. The displacement sensor is mounted on a linear stage with its axis along the X-axis. The linear stage is used for coarse alignment of the displacement sensor with respect to the micro-gear. The micro-gear is mounted on a spindle with the rototing axis along the Y-axis. When measurement, the displacement sensor is kept stationary and the micro-gear is rotated by the spindle so that the surface of the micro-gear can be scanned by the displacement sensor. The sampling position is determined by the rotary encoder of the spindle.

The stability of the measurement system is tested by keeping the linear stage and the spindle stationary. Figure 1.9 shows the output of the displacement sensor during the stability test. The stability of the measurement system is approximately 20 nm.

A part of the measured surface profile of the micro-gear is shown in Figure 1.10. The horizontal axis shows the rotational angle at which the output of the displacement sensor is acquired by a personal computer. The left vertical axis shows the output of the displacement sensor, and the right vertical axis shows the repeatability of the measurements. It can be seen that the repeatability is approximately ±1 μm. The pitch and height of the micro-gear teeth are measured to be 130 μm and 60 μm, respectively. Figure 1.11 shows measured surface profiles scanned from opposite directions. The first profile is obtained by rotating the micro-gear clockwise and the second one is obtained by rotating the micro-gear counter clockwise. It can be seen from Figure 1.11, the difference between the two profiles is approximately ±10 μm. The difference is caused by the stick-slip movement of the micro-ball of the displacement sensor at the steep areas of the micro-gear surface, which has a maximum slope of approximately 70 degrees.

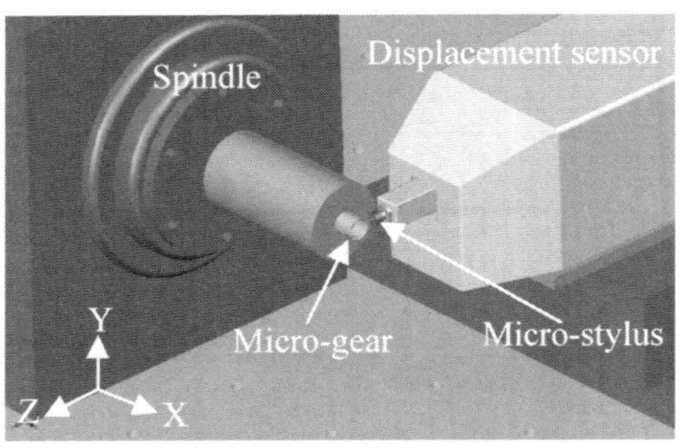

Figure 1.8. Schematic of the measurement system

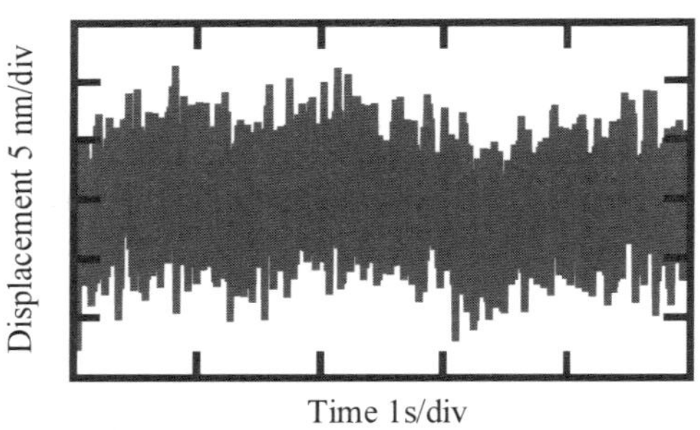

Figure 1.9. Stability of the displacement sensor in the measurement system

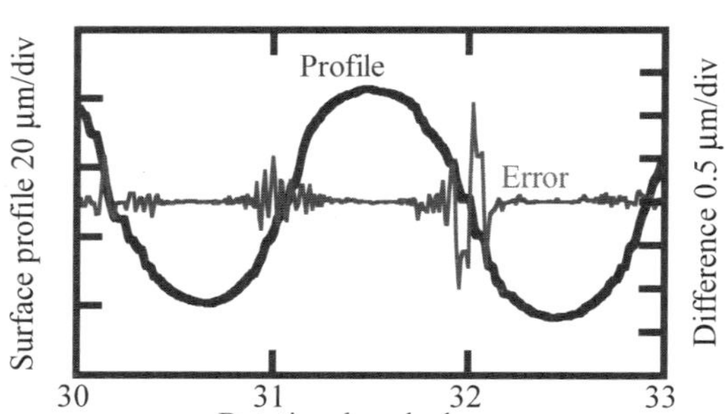

Figure 1.10. Measurement result of the micro-gear surface profile

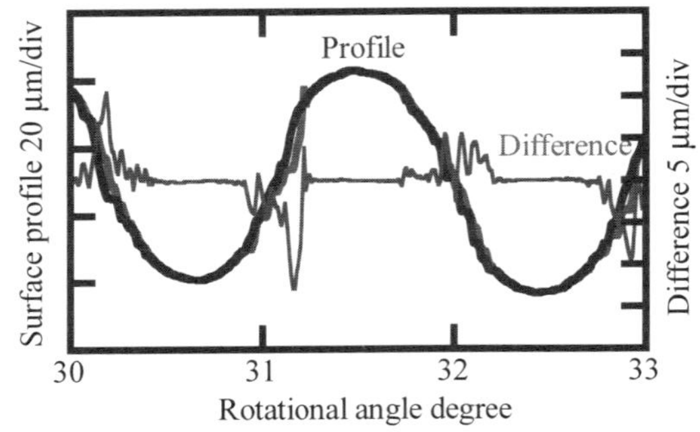

Figure 1.11. Measured surface profiles of the micro-gear from different scanning directions

6. Conclusion

A micro-ball stylus has been developed for profile measurement of high-aspect-ratio micro-structured surfaces. Precision glass micro-balls are employed as the probe ball for an air-bearing contact-type displacement sensor. The diameter of the micro-ball is less than 50 μm. The stylus shaft of the miro-ball stylus has been made by stainless steel and glass tube. It has been confirmed that the micro-ball stylus with the glass tube can tolerate a load larger over 90 mN. The measuring force of the air-bearing displacment sensor is approximately 0.4 mN. The surface profile of a micro-gear with a maximum slope angle of 70 degrees has been measured by the displacement sensor with the micro-ball stylus.

References

[1] Seibel, E.J., Fauver, M. et al. (2003) Micro fabricated optical fiber with microlens that produces large field of-view, video rate, optical beam scanning for microendoscopy applications, Proc. of SPIE, Vol. 4957, P46-55.

[2] W. Gao, A. Shibuya, Y. Yoshikawa, S. Kiyono, and C. H. Park, (2006) Separation of scanning error motions for surface profile measurement of aspheric microlens, Int. J. of Manufacturing Research, in press. Vol. 1, No. 3, P267-282.

[3] A. Weckenmann, et al. (2004) Probing Systems in Dimensional Metrology, Annals of the CIRP Vol. 53, No. 2, P657-684.

[4] K. Hidaka, (2006) Study of a small-sized Ultrasonic Probe, Annals of CIRP, Vol. 55, No. 1, P567-570.

[5] Catalog of Ultrahigh Accurate3D profilometer UA3P, Panasonic Factory Solutions Co., Ltd.

[6] A. Shibuya, W. Gao, Y. Yoshikawa, B. F. Ju, S. Kiyono, (2006) Profile measurement of micro-aspheric surface by an air-bearing stylus with a micro-probe, International Journal of Precision Engineering and Manufacturing, P122-127., in press.

[7] Y. Nagaike, Y. Nakamura, Y. Ito, W. Gao, T. Kuriyagawa, (2006) Ultra-precision On-machine Measurement System for Aspheric Optical Elements, Journal of the Chinese Society of Mechanical Engineers, Vol.27, No.5, P535-540.

In-plane Displacement Measurement with Subnanometer Resolution for Precision Manufacture Platform

C.C. Wu[1], C.C. Hsu[1], J.Y. Lee[2], H.Y. Chen[3] and H.F. Weng[1]

[1]Industrial Technology Research Institute, Hsinchu 300, Taiwan; e-mail: chyanchyi@itri.org.tw
[2]National Central University, Jhongli 320, Taiwan
[3]Machvision Technology Inc., Hsinchu 300, Taiwan

Abstract. A novel method is presented for two-dimensional (2-D) in-plane displacement measurement that is based on the heterodyne interferometry. The method can effectively reduce the the environmental disturbance and allow the measurement of high stability and low uncertainty. According to the experiment results, the subnanometer resolution can be ensured within 250 μm measuring range. The in-plane displacement measurement is also accomplished.

Keywords: in-plane, grating, heterodyne, subnanometer

1. Introduction

Interferometry plays an important role in the precision measurement of displacement for semiconductor technology, nanotechnology, biotechnology, etc. The demand for measuring long-range displacement with nanometric resolution has been increasing with technology advances. In practice, however, the optical interferometer always suffers from noises, which commonly come from atmospheric influences, background vibration, and thermal drift. To achieve higher measurement accuracy, some compensation method has been adopted [1]. Nevertheless, the accuracy still deteriorates with increasing measured length.

The displacement measurement methods can be divided into two categories: homodyne and heterodyne methods. Firstly, for the homodyne method, Dobosz developed a numerical compensation method to correct the nonlinearity of the linear laser encoder, which consists of a laser diode and the reflecting phase grating with 1 μm grating pitch [2]. Salbut proposed five types of the waveguide grating interferometers and 20 nm accuracy of the system was reported using the fringe pattern analysis technology [3]. However, all of the head types used two different gratings and their parallelism might influence the interference signals. Secondly, for the heterodyne method, Wu *et al.* used the synthesis method involved synthesizing higher frequency quadrature signal, mixed with the measurement signals, and added the resultant two signals to the final signal [4]. This final signal had the heterodyne signal form and carried the testing information. This method demonstrated that precise measurement results of the step-motor for a 25 μm stepping movement can be achieved. Wu *et al.* developed dual frequency heterodyne interferometry for nanometer out-plane displacement measurement [5]. The moving grating with constant speed which is induced the Doppler shift frequency in diffracted beams and the heterodyne signal is generated. Hence, stability of the geometric structure of this portion should be a serious concern. The out-plane displacement measurement results of this method possessed the standard deviation of 1.8 nm for the movement range within 0.18 μm. Kao *et al.* proposed a grating interferometer with 2-D grating for in-plane displacement measurement [6]. In Kao's system, a crossed grating with 1.6 μm pitch in both the *x*- and *y*-axis was adopted. An electronic interpolation with a factor of 400 leads to a measurement resolution of 1 nm.

The heterodyne interferometer can provide a high resolution but it always suffers from environmental noise. Even though the grating interferometer can have better immune to these surrounding noises, the measurement resolution is not enough for subnanometer applications. This paper presents a new method for the displacement measurement in precision manufacture platform with subnanometer resolution. According to the optical configuration and heterodyne interferometry technology, it has the merits of high stability to the environmental noises, and high resolution. The experiments of the in-plane displacements are implemented and related results will be detailed.

2. Principles

The proposed optical configuration is shown in Figure 1.1 and the photography of its 3-D optical path is shown in Figure 1.2. For clarity, consider firstly the optical configuration shown in Figure 1.1. A heterodyne laser source HLS directly incidents onto the diffraction grating G and is diffracted into two measurement beams. The measurement beams are guided into ±1st order diffracted beams and the +1st and –1st order diffracted beams propagate two different paths: (1) mirror M_1 → polarization beam splitter PBS → analyzer AN_1 (45°) → detector D_1; and (2) mirror M_2 → polarization beam splitter PBS → analyzer AN_2 (45°) → detector D_2. According to the arrangement of the optical configuration, the interference signal detected by the detector D_1 is from the interference between the *s*-polarization of the +1st order diffracted light and *p*-polarization of the –1st order diffracted beam. Similarly, the interference signal I_{D2} detected by the detector D_2 is from the interference between the *p*-polarization of the +1st order diffracted light and *s*-polarization of the –1st order diffracted beam. Afterwards signals, I_{D1} and I_{D2}, are sent into the lock-in amplifier and the total phase change Φ, which is 4 times magnification of the phase shifting ϕ, can be measured immediately.

Secondly, for 2-D displacement measurement, the 2-D diffraction grating is mounted on the *x-y* stage. Hence, four diffracted beams will be directed along the *x-z* plane and *y-z* plane, respectively. They will be detected by the detectors D_{x1}, D_{x2}, D_{y1}, and D_{y2}, which are used to measure the *x*- and *y*- direction in-plane displacement, respectively. Thus we have the similar signal form as following:

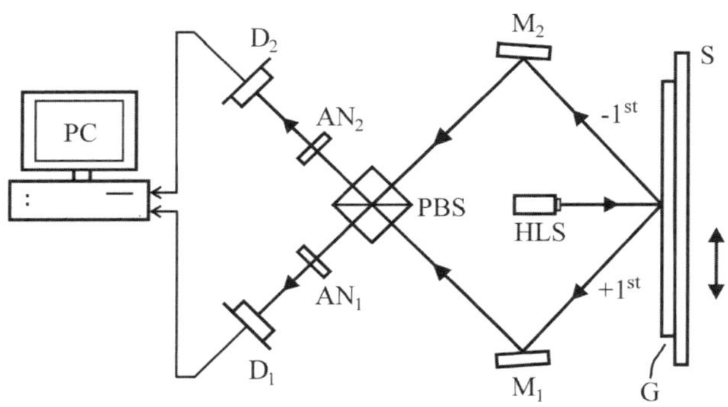

Figure 1.1. Optical configuration for 1-D displacement measurement (2-D optical path only). HLS: heterodyne laser source; M_1 and M_2: mirrors; S: stage; G: diffraction grating; PBS: polarization beam splitter; AN_1 and AN_2: analyzer; D_1 and D_2: detectors.

$$I_{Di1} = \frac{1}{8}[r_{pi(+1)}^2 + r_{si(-1)}^2 + 2r_{pi(+1)}r_{si(-1)}\cos(\omega t + 2\phi_i)], \quad (1.1)$$

$$I_{Di2} = \frac{1}{8}[r_{pi(-1)}^2 + r_{si(+1)}^2 + 2r_{pi(-1)}r_{si(+1)}\cos(\omega t - 2\phi_i)], \quad (1.2)$$

and

$$\phi_i = m\frac{2\pi\Delta_i}{d}, \quad (1.3)$$

where ω is the angular frequency difference between the *p*- and *s*-polarizations of the heterodyne light source; $r_{pi(+1)}$, $r_{pi(-1)}$, $r_{si(+1)}$, and $r_{si(-1)}$ are the reflection coefficients of the *p*- and *s*-polarizations of the $\pm1^{st}$ order diffracted light propagated along *x*- and *y*-direction, respectively, and $i = x, y$. ϕ_i is the phase shifting coming from the 2-D in-plane displacement Δ_i. Keep the fact $\Phi = 2\phi$ in mind. The relation between the in-plane displacement and the corresponding phase change can be expressed as:

$$\Delta_i = \frac{\Phi_i d}{8\pi m}. \quad (1.4)$$

With Eqn (1.4), the in-plane displacements, Δ_i, can be readily determined by the measurement of the corresponding phase changes, Φ_i.

3. Experiments and Results

A He-Ne heterodyne light source, which was linearly polarized at 632.8 nm with frequency stabilized control, was adopted. The 2-D diffraction gratings was mounted on the precision x-y stage (model: 8095, New Focus, Inc.). The stage was driven by the pico-motor™ which was driven by piezo-electric mechanisms. The grating pitch was pre-determined by traceable atomic force microscopy (AFM) and it was 1.1 μm. The reference and measurement signals were received by the detectors and sent into the high speed, multi-channel data acquisition card (model: PCI-6143, National Instruments Corporation). Then the phase change was measured. The phase resolution of this program can be adjusted manually from 0.1° down to 0.001°.

The experiments were measurement by using the 2-D diffraction grating and compared with commercial dynamic calibration system (model: Agilent 5529A, Agilent Company) which were calibrated by the primer standard system in National Measurement Laboratory (NML) of the Center for Measurement Standards (CMS). The sampling-rate of these experiments was kept the same during the whole measurement process and the value was fixed at 128 measured points per second.

The temperature and the atmospheric pressure variations were controlled within 23±2°C and 101.325±0.7 kPa respectively. Two experiments were implemented. One was 1-D displacement measurement along *x*-axis for the design verification; the other was 2-D displacement measurement. 1-D experiment compared the displacements between the PZT and our method. 2-D in-plane displacement experiments were measured simultaneously using our method and Agilent 5529A. The experimental setup was shown in Figure 1.2.

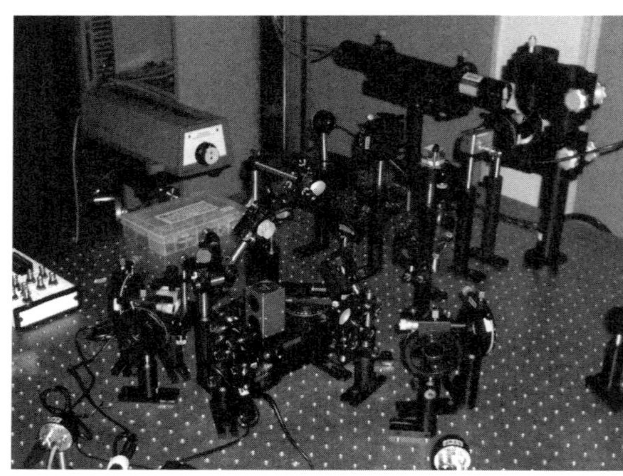

Figure 1.2. Experimental setup.

The measurement range of 1-D experiment was 120 nm. The driving condition of the pico-motor™ was 50 pulses/sec. The short range result was shown in Figure 1.3. The long range result of about 250 μm was shown in Figure 1.4. There was 1.403 μm discrepancy between our method and Agilent 5529A. This might come mainly from the cosine error between two measured axes of Agilent 5529A and our method, and this error could be minimized by careful alignment.

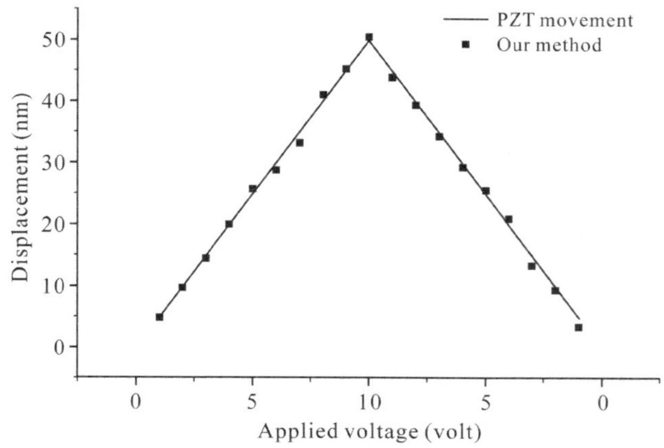

Figure 1.3. 1-D result for 50 nm displacement measurement.

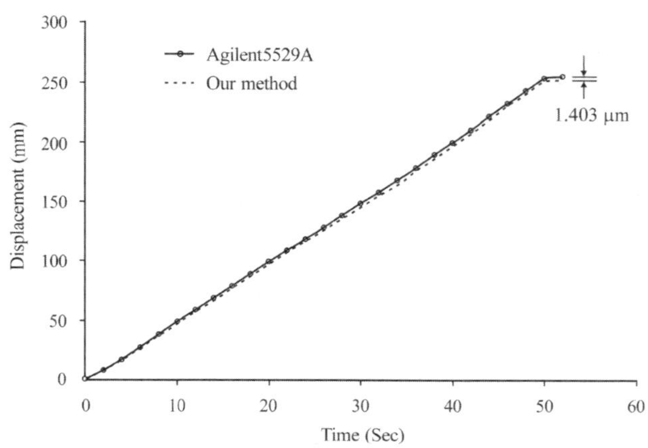

Figure 1.4. 1-D result for 250 μm displacement measurement.

The driving condition of the pico-motor™ was 100 pulses/sec. The 2-D in-plane displacement was measured simultaneously using our method and Agilent 5529A dynamic calibration system. The movement path of the *x*-*y* stage was aligned at 45° with respect to the x-axis, and the total displacements were 120 nm and 140 nm respectively. The measurement results are presented in Figure 1.5 and Figure 1.6. In both figures, our results were closely related to the ones by the commercial measurement system and the differences were 9.5 nm and 6.7 nm for the *x*- and *y*-direction, respectively.

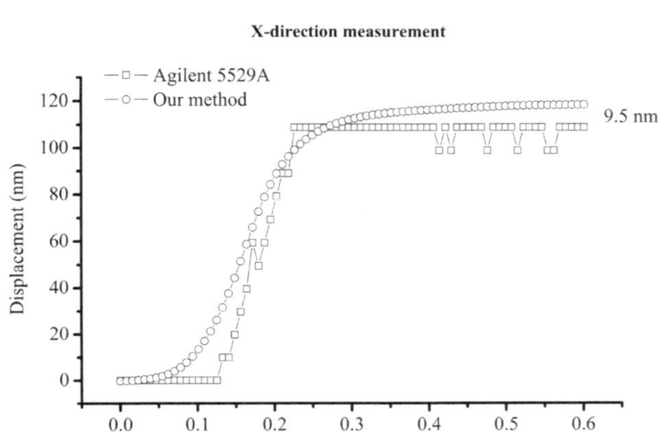

Figure 1.5. 2-D displacement measurement result along *x*-axis.

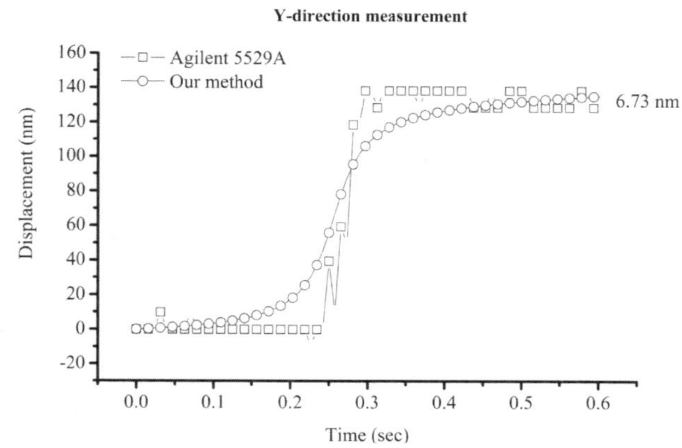

Figure 1.6. 2-D displacement measurement result along *y*-axis.

For the long-range displacement measurement, the difference between our method and Agilent 5529A might be large (Figure 1.4). Because of the coordinate systems of the motorized stage, Agilent apparatus, and the diffracted grating might be not coincidence. The movement directions of these three elements might be different slightly and that will be enlarged as the movement increases. Hence, proper alignment of them will level down the difference of the measurement results by these two measurement methods.

4. Discussion

To test these general noise levels of the system, the stage was held stationary. Two gratings were used to investigate the phase noise. One has a grating pitch of 40 μm and the other 1.6 μm. In this stationary situation, the contributors of the measured phase variation are only low and high frequency phase noises. The measurement results show that the magnitudes of the high and low frequency noises are about 1° and 0.2° within 10 seconds for both grating pitches 40 μm and 1.6 μm. The smaller the grating pitch d, the smaller the phase noise influence in our system. The displacement noises are measured and shown in Figure 1.7.

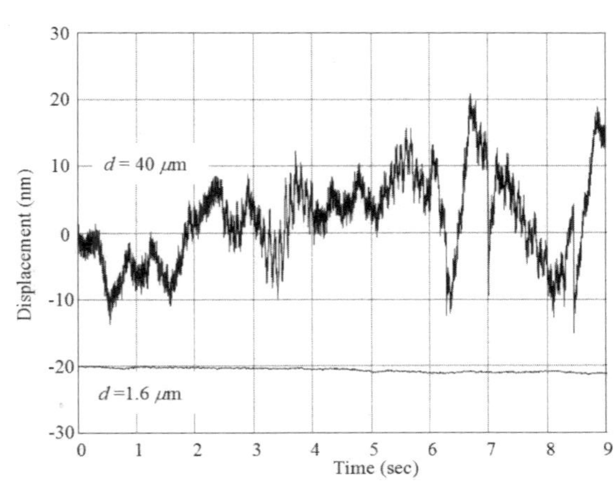

Figure 1.7. The measured displacement noise for different grating pitches, $d = 1.6$ μm and 40μm.

It is obvious that the noise level of grating pitch 1.6 μm is much less than that of grating pitch 40 μm in the displacement measurement using our system. In our system, the grating with 1.6 μm pitch was chosen for displacement measurement. If only high frequency noise is considered, the measurement resolution of our system is estimated to be 0.2 nm. It is well known that increasing the time constant of RC filter of the lock-in amplifier will attenuate the high frequency noise level, and we believe a compact optical configuration will further improve low frequency noise. In addition, by using the 2D grating instead of 1D grating in similar optical configuration, the two dimensional displacement measurements with single interferometer can be accomplished simultaneously.

5. Conclusion

We demonstrate 1-D and 2-D displacement measurement results using a novel heterodyne subnanometric displacement measurement technique. Because of the simple optical configuration and the merit of heterodyne interferometry, our method has some advantages, such as high stability, high resolution, and low uncertainty. There are no significant differences between the results of our method and the commercial system. These lead us to believe that our method can be applied to ultra-precision positioning in controlling the motorized stage movement. We are now undertaking the measurement capacity verification of this technique for the displacement measurement in the picometer scale, and the related results will be demonstrated in the future papers.

Acknowledgement

This study was supported in part by the Ministry of Economic Affairs, Taiwan, R.O.C. under the contract number 5301XS3F10.

References

[1] B. Edlén, The refractive index of air, Metrologia 1966, Vol. 2, P71-80.

[2] M. Dobosz, High-resolution laser linear encoder with numerical error compensation, Opt. Eng. 1999, Vol. 38, P968-973.

[3] L. Salbut, Waveguide grating (moire) microinterferometer for in-plane displacement/strain field investigation, Opt. Eng. 2002, Vol. 41, P626-631.

[4] W.J. Wu, C.K. Lee, and C.T. Hsieh, On the signal processing algorithms for Doppler effect based nanometer positioning systems, Jpn. J. Appl. Phys. 1999, Vol.38, P1725-1729.

[5] G. Wu, Z.J. Cai, and L.J. Zeng, A two-color heterodyne interferometer based on diffraction grating for displacement measurement, Key Engineering Materials 2005, Vol.295/296, P189-194.

[6] C.F. Kao, C.C. Chang, and M.H. Lu, Double-diffraction planar encoder by conjugate optics, *Opt. Eng.* 2005, Vol.44, No.2, 023603.

Width Measurement of Metal Surface Cracks Based on Image Analysis Technique

Li Meng[1], Wang Zhong-yu[2], Shuai Mei[3]

[1]Beihang University, Beijing, China, 100083, 86-10-82339303, limenglee@aspe.buaa.edu.cn
[2]Beihang University, Beijing, China, 100083, 86-10-82339303, mewan@buaa.edu.cn
[3]Beihang University, Beijing, China, 100083, 86-10-82317303, shuaimei@gmail.com

Abstract. Width measurement of cracks has become more important in metal surface inspection. This paper deals with a quantitative image analysis technique for width measurement of cracks in metal surface. Using three procedures including image acquisition, image processing and image analysis, the maximum width of an incident crack can be obtained. Employing a CMOS image sensor, a crack image can be acquired. The processing and analysis of the crack image are accomplished with a self-developed VC++ software running on a microcomputer. Based on Ferret algorithm, the widths of cracks are defined and quantified. The experiment results indicate that the software is convenient and could run quickly, and the method is adaptable. The results obtained can provide highly precise information for the quantification of cracks on metal surface.

Keywords: Quantitative evaluation; Crack width; Image analysis; Ferret algorithm

1. Introduction

There are many reasons which may result in cracks on metal surfaces. The presence of cracks can influence the properties and security of metal-based materials, so width measurement of cracks is significant in the damage diagnosis of metal surface [1].

Traditional manual methods which have low efficiency and big measurement error will be affected by subjective factors and instrument precision. In recent years, many image analysis methods have been put forward to study the efficient measurement and detection of cracks to replace traditional manual method [2-5], unfortunately they mainly focus on image processing, especially the elimination of some disturbing factors such as incoherent illumination and noises. Most of these methods lack quantitative analysis and the width definition of cracks. B. Gérard and J. Marchand proposed an effectual conception of crack width, the application is limited because a torsion coefficient used in calculation and a presumption that the crack is continuous are needed [6]. Another method is counting the transverse number of black pixels after converting the original image of crack into a binary image. Its application is limited by its inadaptability of complicated conditions.

Ferret algorithm is a method usually used in definition of geometrical size of irregular graphics [7]. In Ferret algorithm, the minimum distance between all the two parallels which are tangent with the crack's outline is called Ferret width. Because many cracks on metal surface have irregular shapes, the width of a crack is defined as Ferret width. In this paper, based on image analysis technique, the original image of crack can be pre-processed and quantitatively analysed by the Ferret algorithm.

2. Image Acquisition and Processing

2.1 Image Acquisition

A close-distance optical system consisting of imaging lens, CMOS image sensor, lamp-house and microcomputer is used to acquire the surface image of metal imaged on the plane of the CMOS by the lens. Fig. 1.1 shows the sketch layout of the imaging lens and COMS image sensor. In order to generate images with satisfactory contrast, the inspection area on metal surface should be bright enough. There are four LEDs around the imaging lens to supply adequate illumination. Fig. 1.2 shows the actual lighting setup and optical layout.

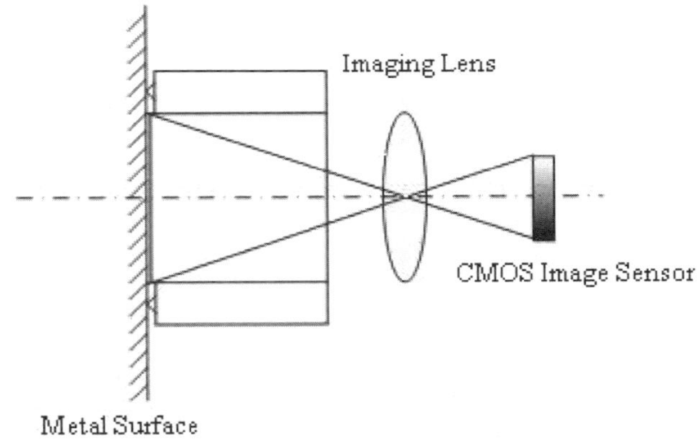

Figure 1.1. Sketch layout of the imaging lens and COMS image sensor

Although the focus of the lens is fixed, the accuracy of the optical system may be influenced because it is not easy to place the plane of the sense organ on the image plane. In order to eliminate the position error on the plane of the sense organ of CMOS, the imaging lens of this optical system is designed as a telecentric object lens.

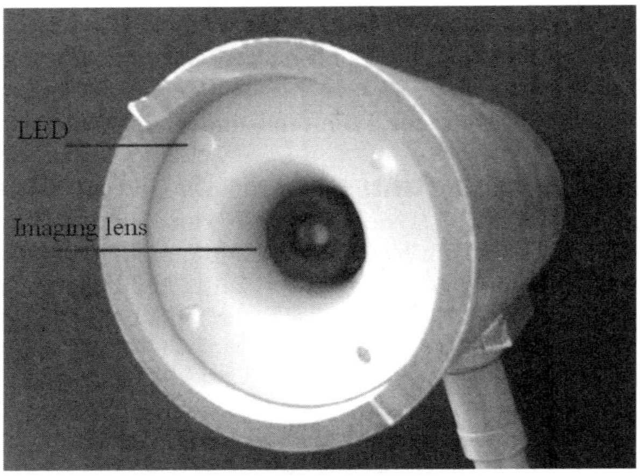

Figure 1.2. Actual lighting setup and optical layout

The image used for subsequent processing is then stored in the microcomputer. The amplificatory multiple and focus of the lens is fixed, so the actual size of metal surface crack is then proportioned to the image size. The system calibration is done by measurement of a standard object. The actual width of the standard object is 1mm, and the pixel width on the image is 40. Through calibration, a coefficient R which presents the ratio of actual width to one pixel width is obtained. In this paper, R is 0.025mm/pixel.

2.2 Image Processing

2.2.1 Graying and Contrasting
Firstly, the original image is grayed into 8-bits grayed image. Then the grayed image must be contrasted to increase the contrast between the crack and the metal surface. After that, the crack's figure is more distinct and the recognition and extraction of its characters are more convenient.

In this paper, the contrasting coefficient is selected as 100, and the grayed image and the contrasted image are shown in Fig.1.3 and Fig. 1.4.

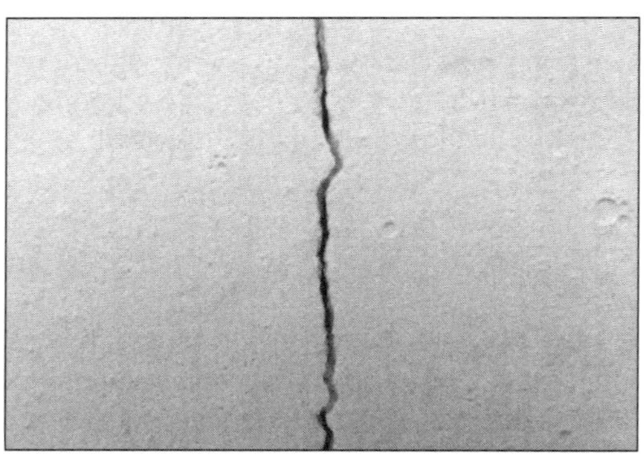

Figure 1.3. The grayed image

Figure 1.4. The contrasted image

2.2.2 Thresholding
Thresholding can make the image processing simpler and it is usually a necessary step before image analysis, character extraction and mode identification. The key of thresholding is the selection of threshold value, because it decides the precision of the extraction of the crack's figure.

Those threshold values obtained manually by an iterative process are not efficient and the precision of the extraction may be influenced by human subjective factors. In this paper, the threshold value k is determined by using the Otsu method. After the decision of k and thresholding, the image is converted into a binary image which has only two gray levels of 255 and 0. The binary image is shown in Fig. 1.5.

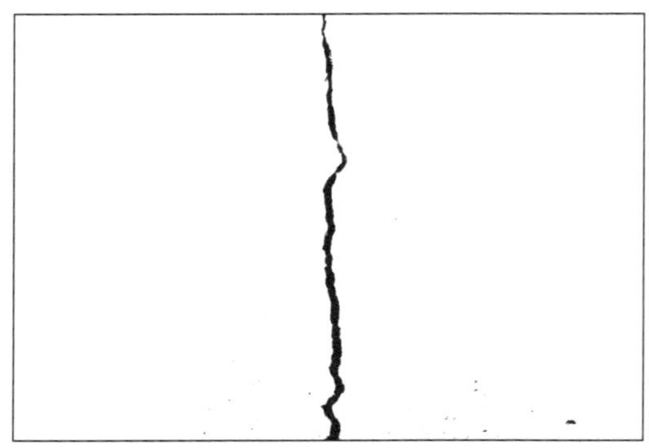

Figure 1.5. The binary image

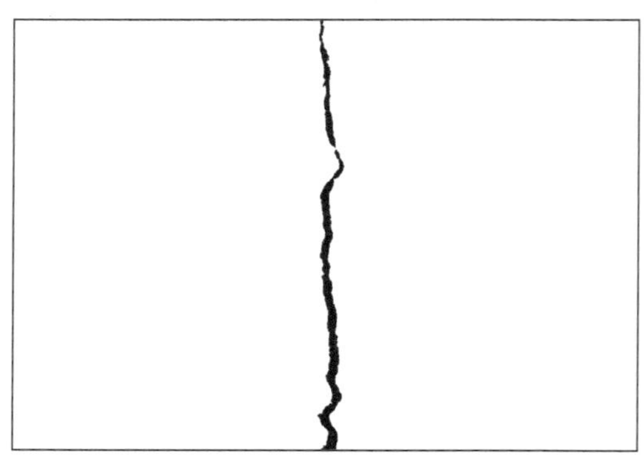

Figure 1.6. The final image

2.2.3 Elimination of Defects
Sometimes there are small individual objects which do not belong to a crack in a binary image. These objects such as isolated points are called defects, which will affect the

correctness of image analysis. An open operation, one of morphological methods, is used to eliminate these defects. The open operation composed of erosion and dilation is a conventional method which can clean out most small defects.

Some relatively bigger defects may be left after the open operation, an area size thresholding method is needed to discriminate farther between the crack and these defects. The area size thresholding method consists of three parts: area labeling, area size measurement and size thresholding in which a threshold value is needed. After this step, the final image is obtained and shown in Fig. 1.6.

3. Width Measurement

The final image is a binary image in which the white areas represent the crack and the black areas represent the background. In order to make the measurement more convenient, some areas without black pixel on the final image are cut out. The final result is shown in Fig. 1.7. Based on Ferret algorithm, the whole image should be divided into several sections before measurement. In this paper, the final image is divided into four sections. The dividing lines should be at the corners where the crack's orientation changes strongly.

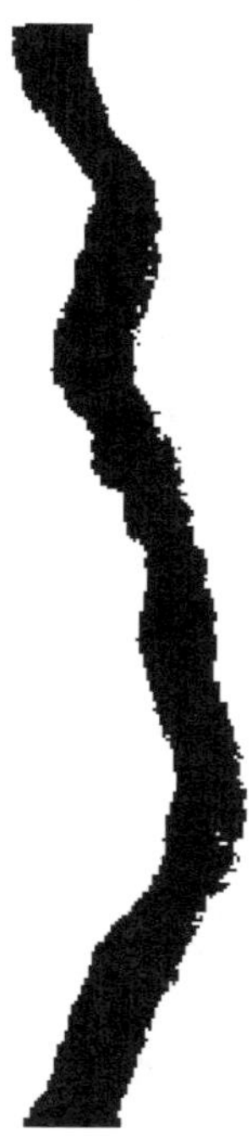

Figure 1.7. Cut image

By VC++ software, the divided image is processed and the four sections are shown in Fig. 1.8. The white areas represent the crack and the black areas represent the background.

Ferret algorithm is applied to each section of the image and every width expressed as pixels of the four sections is obtained. The actual width can be converted by the coefficient R, which is obtained by the calibration done before. If the width of section i is d_i, the actual width of section i W_i can be calculated by $W_i = d_iR$. By comparing widths of each section, the maximum width and the section with the maximum width are obtained. Table 1 shows the analytical results. The orientation is define as the angle which contrarotates from the vertical direction to the the parallels' direction.

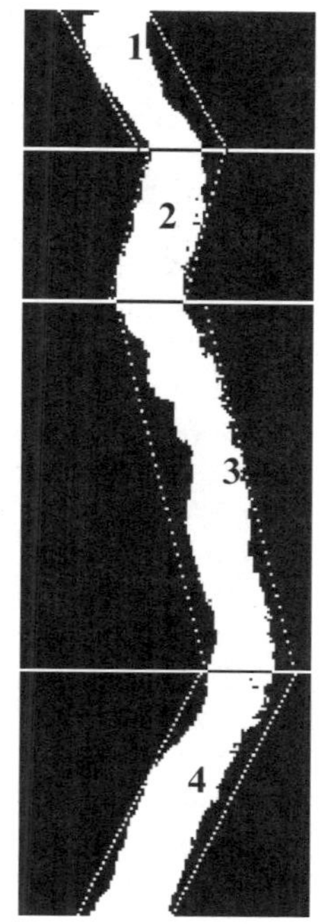

Figure 1.8. Ferret algorithm image

As seen in Table 1.1, the maximum width located in section 3 is 0.780 mm. Because different division of image leads to different result when using Ferret algorithm to analyze cracks, the measurement result is not unique.

Table 1.1. Analytical results

Part Number	Width (Pixel)	Orientation (Degree)	Width (Millimeter)
1	27.84	29.25	0.696
2	26.29	-14.10	0.657
3	31.21	13.79	0.780
4	30.81	-26.12	0.770

4. Conclusion

An automatic technique for width measurement of cracks on metal surface has been developed. Based on image analysis and quantitative calculation, the width of crack has been obtained by Ferret algorithm. The VC++ software used for image processing and analysis is proved convenient and fast. The whole measurement operation can be finished in less than 20 seconds.

Through procedures of graying, contrasting, thresholding and elimination of defects, the crack has been extracted from the original image. Because almost every figure of cracks on metal surface is irregular, Ferret algorithm aiming at quantitative evaluation of irregular graphics is appropriate for this measurement system. The system can provide credible evaluation for inspections of metal surface. In the future, it can be used in steeling industry and mechanical industry.

References

[1] Zou Yiqun, Hou Guicang, Yang Feng. 2004, A Surface Crack Inspection Algorithm Based on Digital Image Processing Techniques, Control & Automation, Vol.20, No.4, P80-81,P105.

[2] Han Yanfang, Shi Pengfei. 2005, Surface Defect Detection Based on Image Processing Techniques, Measurement & Control Technology, Vol.24, No.9, P15-18.

[3] Xu Ke, Sun Haoyang, Chao Lin, et al. 2006, Application of Morphological Filtering to On-line Detection of Surface Crack of Medium and Heavy Steel Plate. Chinese Journal of Scientific Instrument, Vol.27, No.9, P1008-1011.

[4] Stampfl J, Scherer S, Berchthaler M, et al. 1996, Determination of the Fracture Toughness by Automatic Image Processing, International Journal of Fracture, Vol.78, No.1, P35-44.

[5] Ammouche A, Breysse D, Hornain H, et al. 2000, A New Image Analysis Technique for the Quantitative Assessment of Cracks in Cement -based Materials, Cement and Metal Concrete Research, Vol.30, P25-35.

[6] Gérard B, Marchand J. 2000, Influence of Cracking on the Diffusion Properties of Cement-based Materials. - Part I: Influence of Continuous Cracks on the Steady-state Regime, Cement and Metal surface Research, Vol.30, P37–43.

[7] Coster M, Chermant J L, 1985, Précis d'analyse d'images, Presses du CNRS, Paris, France.

A Multifunctional Scanning White-Light Interferometer (SWLI): Construction of 2-D and 3-D Measurements

Jin-Liang Chen, Kevin Kuo, Chi-Hong Tung, Calvin C. Chang

Center for Measurement Standards/ITRI, Taiwan, 30055, JinLiangChen@itri.org.tw
Center for Measurement Standards/ITRI, Taiwan, 30055, kevinkuo@itri.org.tw
Center for Measurement Standards/ITRI, Taiwan, 30055, chtung@itri.org.tw
Center for Measurement Standards/ITRI, Taiwan, 30055, CalvinCChang@itri.org.tw

Abstract. It is necessary and important to find the focal plane in order to locate where the interference fringe is, and to determine the vertical-scan range prior to the scanning operation for an SWLI. A novel scheme is proposed in this paper for simultaneously carrying out the auto-focusing and the real-time 2-D measurement. Conventionally in a SWLI, it always happens that the interference fringe appears across the focal plane, and therefore the in-focus image is blurred by the interference fringe, which makes it difficult to carry out the auto-focusing and the 2-D measurement with the in-focus image. A multi-function SWLI has been developed such that the 2-D and 3-D measurements can be done simultaneously. Although the 3-D profile reconstruction is basic to an SWLI, the real-time 2-D measurement is not trivial. The proposed scheme is based on clarifying the out-of-focus images where interference fringes do not exist. The proposed scheme is applied on some preliminary experiments and it shows that the real-time 2-D measurement can be realized, and moreover, the accuracy of the 3-D profile reconstruction can also be improved.

Keywords: white-light interferometer, auto-focus, 2-D measurement.

1. Introduction

In recent years, SWLI has been widely adopted in the quality control of LCD and semiconductor manufacture procedures, such as surface roughness measurement for wafers, surface profile measurement for micro-optical components, size and height measurement for photo spacers among color filters for LCD display etc. Its vertical resolution may reach far deep to the sub-nanometer class (0.5 nm) [1]. This technology resolves the profile of the specimen by analyzing the interference light intensity on every pixel within the field of view. The height of each pixel on the specimen surface can be calculated based on the vertical scanning position and the interference light intensity for the position of zero optical path difference. The most common method is CPS [2], which looks for the peak position of the coherence envelope of each pixel and collectively profiles the specimen surface. There is another algorithm called PSI, which works on the interference phase and can be used for a more detailed surface with a resolution of 0.1 nm.

SWLI utilizes the low coherence nature of white light for surface profile measurement with wide dynamic range and high resolution. However, the low coherence length of the light source limits the identifiable range of the interference fringes. Usually, in a scanning white-light interferometric measuring system, the low coherence length of a broadband light source confines the interference fringes to within several microns, which poses difficulties for people who are trying to locate the interference fringes or the focal plane manually with high magnification interferometric objectives along the entire vertical range of movement. As regards the production line, automation is definitely necessary. The first step is to find out the range of interference fringes by auto-focusing technique and to carry out vertical scanning along the range. Information about interference fringes inside images can be calculated by computer vision together with auto-focus (AF) algorithm. It would make the searching for interference fringes easier to increase the positioning resolution.

Some studies address improving focuses for normal optical microscopes. However, these AF methods may be influenced by interference fringes and lost their precision [3-5]. At the same time, if the position of focus is different from the position of zero optical path difference, the S/N ratio of the resulting interference fringes will be reduced and thereby the accuracy the profile measurement will deteriorate. Improving the focusing precision of an SWLI has been a key problem for improving the measurement performance.

Sometimes it is beneficial to take 2-D images along with 3-D data for the purpose of lateral measurement. However, images taken by a traditional SWLI are degenerated by interference fringes. A trivial solution would be to switch between interference objective and normal objective by a nosepiece, yet it consumes a lot of time and mechanical motion of the nosepiece may result in dust pollution because of friction. To obtain clear 2-D images without interference fringes, Peter proposed a Fringe-Removal method in 2005 [6], which effectively removes the interference fringes from the 2-D images. The resulting images can be used in 2-D inspection and measurements.

In this study, we establish an SWLI capable of both 2D imaging and 3D profiling. Its 3D measurement speed can reach 10um/sec with a resolution of 0.1nm. It also furnishes 2D images without any interference fringe for transversal size measurement. Besides, two additional functions of auto-focus and self-calibration can be implemented on this SWLI system. These functions increase the contrast of interferograms, reduce S/N ratio, and improve the 3D profile measurement remarkably.

2. The structure of a multi-functional SWLI system

2.1 Scanning white-light interferometer

Figure 1.1 shows the configuration of the Mirau-type SWLI system. It uses a broadband light source of low temporal coherence length. The light is incident from left to right onto the first beam splitter, where it is reflected downward onto the Mirau objective. A reference mirror and a second beam splitter are inside the Mirau objective. The light is split by the second beam splitter into two beams, a reference beam and a measuring beam. The measuring beam goes directly through the second beam splitter onto the surface of the specimen and is reflected back from the surface of the specimen. Then, it travels upward through the Mirau objective again. The reference beam bounces between the second beam splitter and the reference mirror three times and then travels upward, where it meets with the measuring beam, together, they go through the first beam splitter and onto the recording CCD. An actuator is installed on the CCD to adjust its position for proper imaging. In this way, a high contrast interferogram and a clear image without any interference fringe can be obtained.

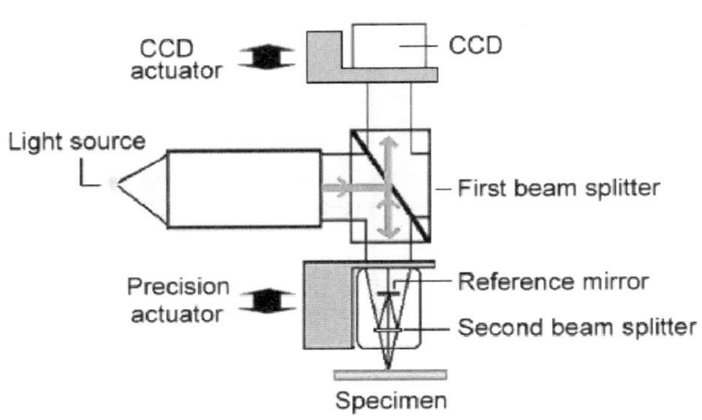

Figure 1.1. A schematic plot of a basic Mirau-type white-light interferometer.

Since the broadband light source has low coherence length in nature, the interference fringes will exist only in the region in which the optical path difference of the two beams is shorter than the coherence length of the light source. The optical path difference between the two beams changes as the precision actuator carries the Mirau objective to scan through a vertical range. The surface profile of the specimen is obtained based on analyzing the peak position of the coherence envelope of each measured pixel. Prior to the operation of vertical scanning, it is necessary to determine the vertical scan range that covers the region in which the interference fringes occur. Conducting this searching and determination process manually would waste much precious time and, which is even worse, it would sacrifice the measurement accuracy. Therefore, it would be nice to use image processing technique to auto-detect the vertical region where the interference fringes take place.

2.2 Coherence peak sensing (CPS)

The processing theory of the coherence peak sensing (CPS) calculates the peak position of the coherence envelope of each measured pixel [7-9]. When a large amount of data is collected for the measurement, the algorithm used in the analysis needs to be as simple as possible for fast data processing. The algorithm of calculating the mass centre of the envelope [7, 9] is well suited for this task since it requires fewer data manipulation. The algorithm can be described as the following equation

$$Z_c = \frac{\sum_{n=1}^{n=N} A_n \times n \times \Delta}{\sum_{n=1}^{n=N} A_n} \tag{1.1}$$

where n is the consecutive sampling number, N is the total number of samples, and Δ is the constant scanning interval between two samples. The weighting function A_n is the square of the interference intensity with DC intensity removed.

Since the interference envelope is related to the power spectrum of the light source [6], which is usually not a perfectly symmetric Gaussian, as shown in Fig. 1.2, the peak position of the envelope is not necessary the position of zero optical path difference. However, if the envelope shape of each measured pixel in the field of view (FOV) is identical, the relative position of the mass centre of the envelope of each pixel will remain unchanged, even though the envelope shape is not a perfectly symmetric Gaussian. By crunching out all the mass-centre positions of all the pixels in the FOV, the surface profile of the specimen can be reconstructed.

2.3 Passive auto-focus algorithm

Since interference occurs at the focal plane of the objective, auto-focus algorithm for the image is helpful in searching for interference fringes on the focal plane. In literature [5], various auto-focus algorithms are compared based on the in-focus images. Among these, the Tenunbaum Gradient (Tenengrad) is best suited for fast focusing and provides very good focusing precision. The equation for the Tenunbaum Gradient is

$$F_{Tenengrad} = \sum_{i=1}^{M} \sum_{j=1}^{N} \left[S_x(i,j)^2 + S_y(i,j)^2 \right] \tag{1.2}$$

where S_x and S_y are obtained by convolution of the Sobel operators in the x-direction and the y-direction. M and N are the dimensions of images.

However, the intensities between the bright and dark fringes may blur the image and reduce the amount of information available about the sharpness of the image. Therefore, image contrast judgment rules should be added. The contrast of interference fringes exceeds the contrast of the background image, so it is utilized to determine the occurrence of the interference fringes. Wherever the interference fringes occur, the pixel intensities of the image would be widely distributed and the wider the distribution, the greater image contrast. Accordingly, the histogram of the pixel intensities of the image is generated. Images with a wide distribution of pixel intensities correspond to images with interference fringes. A threshold for the numerical pixel intensity will be set to eliminate the effect of noise. The equation for the image contrast is

$$F_{histogram} = \max\{g \mid n_g > N_{th}\} - \min\{g \mid n_g > N_{th}\} \quad (1.3)$$

where n_g is the numerical pixel intensity, g, and the noise is filtered using the number threshold N_{th}.

3. Realization of multifunctional SWLI

3.1 Autofocus and automatic calibration

Generally, the positions of zero optical path difference for SWLI are set on the focal plane of the objective for high optical resolution. In the SWLI structure of this study, there are actuators installed at both the objective and the CCD. The actuator at the objective is for automatic focusing. However, the CCD may not be in the right position because of mechanical tolerance or assembly inaccuracy. Under such circumstances, the obtained interference image will be out of focus as shown in Figure 1.2. As a result, the 3D profile by VSI will be affected and its precision will be deteriorated.

Figure 1.2. The blurry interference image before calibration

Before the SWLI calibration, the focuses obtained by AF procedure in the entire travel are shown in Figure 1.3. The x-axis represents the number of images. F_T represents the normalized sharpness of the images. F_H represents the normalized contrast of the images. The interval between images is 100 nm. The result shows that at point H, the peak value of the contrast occurs and so as the interference fringes, and the focal plane should be here. However the peak of the sharpness is located at point T, which means that the corresponding focusing position of the object-image on CCD is different from the position of the zero optical path difference. In this occasion, we need to adjust the position of the CCD until the peak position of the contrast coincides with that of the sharpness. Then, the CCD is in a proper position. As shown in Figure 1.5, the interference image comes with clear quality and high contrast. The S/N ratio of SWLI is improved.

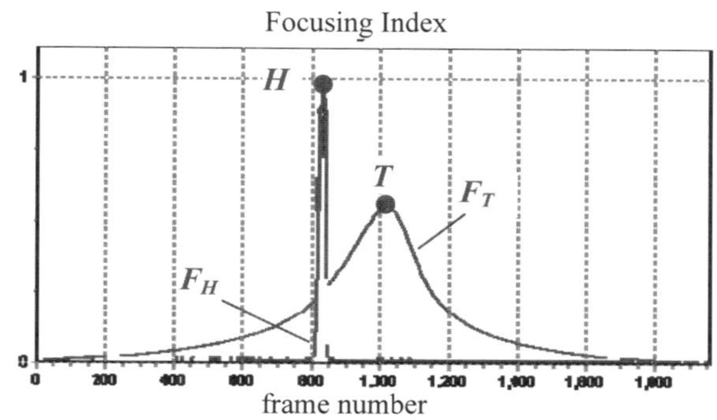

Figure 1.3. The index values for focusing before calibration

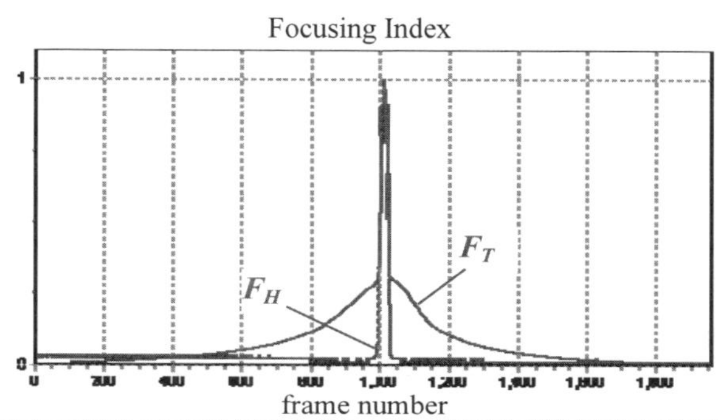

Figure 1.4. The index values for focusing after calibration

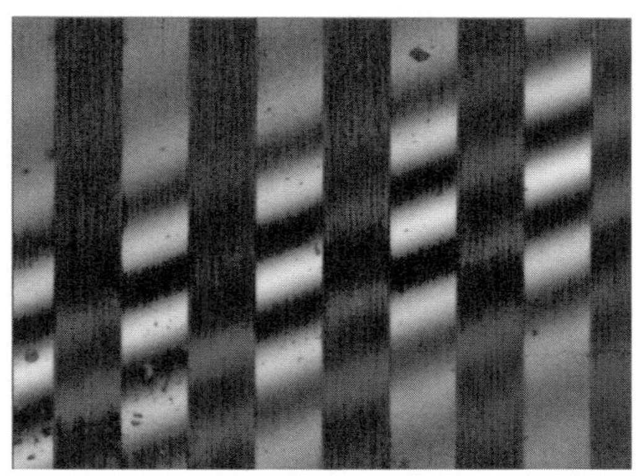

Figure 1.5. The clear interference image after calibration

3.2 Lateral measurement

The movement of CCD and objective can be used to obtain clear images without interference fringe for 2D lateral measurement and defect detection besides SWLI automatic calibration. As shown in Figure 1.3, the span of prominent contrast around point H represents the vertical range of the interference images. Generally, the coherence length of interference is about 2 μm. If the object-surface inclination is taken into account, the vertical range of the interference images will be about 10 μm. Notice also that in the figure, at point T, the image has a maximum sharpness but no interference fringe. It means that this image locates beyond interference range. The reason for the maximum sharpness is that the CCD is at the corresponding focusing plane. As shown in Figure 1.6, the 2D image obtained by this method can be used for 2D lateral measurement. It will be easier to implement defect detection based on clearer 2D images, such as detection of LCD's MURA, short circuit of TFT electrodes, etc.

Figure 1.6. Clear image without interference fringe

3.3 3D Surface Profile Measurement

After adjusting the positions of CCD and objective by AF method, it's possible to obtain high contrast interference images. In this case, good profiling results by ordinary SWLI could easily be obtained as shown in Figure 1.7. Figure 1.8 demonstrates the results of SWLI scanning before calibration for comparison purpose. It shows that if we move the CCD to a proper imaging position, the precision of SWLI 3D morphological measurement will be improved.

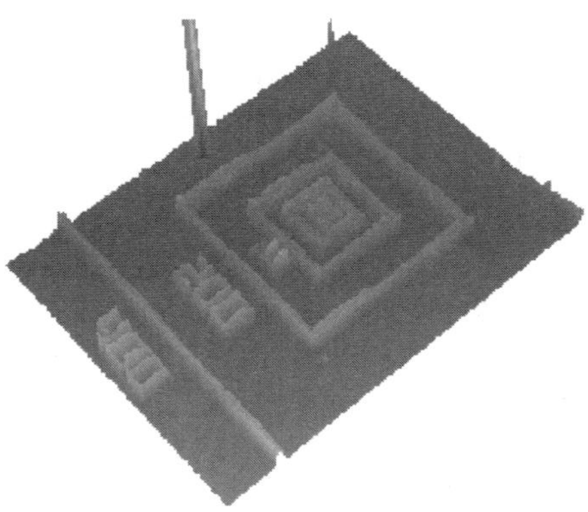

Figure 1.7. Scanning profile after CCD calibration

4. Conclusion

This study proposes an easy structure to implement lateral measurement on top of 3D profile measurement based on SWLI. The advantages include no objective-switching needed and low-cost clearer images with no interference fringe for lateral measurement. Besides, this system also supports automatic calibration. In the illustrative experiment, the SWLI AF based on this method encounters less fringe and installation errors. The position of the CCD and its corresponding images are adjustable. Together with auto-focus algorithm, proper and clear interference images can be obtained. Meanwhile, the precision of focusing and measurement is improved.

Figure 1.8. Scanning profile before CCD calibration

1.5 References

[1] Mehta DS, Saito S, Hinosugi H, Takeda M, Kurokawa T, (2003) Spectral Interference Mirau Microscope with an Acousto-Optic Tunable Filter for Three-Dimensional Surface Profilometry, Applied Optics 42(7): 1296 - 1305

[2] Hua-Chen Hsu, Chi-Hong Tung, Ching-Fen Kao, Calvin C. Chang, (2004) A white-light profiling algorithm adopting the multiwavelength interferometric technique. Proc. of SPIE 5531:244-249

[3] Qingxing Li, Lifen Bai, Shifu Xue, Luyun Chen, (2002) Autofocus system for microscope. Opt. Eng. 41(6):1289-1294

[4] Yu Sun, Stefan Duthaler, Bradley J. Nelson, (2004) Autofocusing in computer microscopy: Selecting the optimal focus algorithm. Microscopy research and technique 65:139-149

[5] Mario A. Bueno-Albarra, Josue Alvarez-Borrego, Leonardo Acho, Maria Christina Chavez-Sanchez, (2004) Fast autofocus algorithm for automated microscopes. Opt. Eng. 44(6):1063601

[6] Peter de Groot, (2005) Generating fringe-free images from phase-shifted interferometry data. Applied Optics 44(33):7062-7069

[7] R. Dandliker, E. Zimmermann, G. Frosio, (1992) Elecroniclly scanned white-light interferometry: a novel noise-resistant signal processing. Optics Letters, 679-681

[8] Paul J. Caber, (1993) Interferometric profiler for rough surfaces. Applied Optics, 32:3438-3441

[9] US Patent 5633715, (1997) Centroid approach for estimating modulation peak in broad-bandwidth interferometry

The Error Analysis of Cross-section Curve with Contact Mode Atomic Force Microscopy Simulation Measurement by Using Trapezoidal Cantilever Based on Molecular Mechanics

Z.C. Lin[1], J.C. Huang[2], C.H. Yeh[3]

[1]National Taiwan University of Science and Technology, Taiwan, zclin@mail.ntust.edu.tw
[2]Tung Nan Institute of Technology, Taiwan, jchuang@mail.me.tnit.edu.tw
[3]National Taiwan University of Science and Technology, Taiwan, M9312101@mail.ntust.edu.tw

Abstract. Based on the molecular mechanics, this paper uses the two-body potential energy function to construct a trapezoidal cantilever nano-scale simulation measurement model of contact mode atomic force microscopy (AFM) under the constant force mode to simulate the measurement the nano-scale standard stair-step sample. This paper investigates the offset distance error occurred during scanning the cross-section curve of the standard step sample. The two probes employed in this study are of different cantilever geometry and probe material (V-shaped Si_3N_4 cantilever and trapezoidal Si cantilever), but have the same tip radius of 3.65 nm. It is found that the probe tip radius and the tip edge oblique angle of the probe influenced the cross-section curve during simulation measurement. Therefore, the factor influencing this cross-section curve appearance is mainly the geometry form of the probe. It can be observed that the cross-section curve under simulation measurement by using the trapezoidal Si cantilever is closer to the cross-section curve of nano-scale standard sample than under simulation measurement by using the V-shaped Si_3N_4 cantilever. It can be known that the probe material can influence the cross-section curve under simulated measurement. The investigation of the error analysis is referential in compensating the error of AFM measurement in the future, and can be used to further enhance the accuracy of AFM scanning and measurement.

Keywords: Atomic force microscopy, Morse potential, Constant force mode, Scan Curve, Cross-section angle

1. Introduction

Currently the tool that is most frequently used to measure nano components or structures is a scanning probe microscopy (SPM). Among various SPMs, atomic force microscopy (AFM) is the commonest one. Invented by Binnig et al. [1] in 1986, AFM is a modified instrument based on the shortcoming of scanning tunneling microscopy that it can only be applied to conductive. Therefore, AFM can be applied not only to the measurement of surface state of conductive materials, but also to the observation of the surface state of non-conductive materials. The process of AFM measurement is easily affected by the shape of the probe and the action force between the probe and the specimen, enabling a certain level of error that occurs on the nano-scale surface topography of the scan result. Therefore, investigations towards the AFM probe tip radius and the shape of the probe tip edge are a very important factor affecting the measurement error.

Therefore, a number of researchers investigated the use of AFM for surface measurement. For example, Villarrubia [2] used the concept of mathematical morphology towards the measurement result and deduced a mathematical expression of the probe's shape of respective data points. Song [3] used SAM technology on Au to joint a test plate with a height of around 0.3 nm and measured it on AFM machine, so as to reconstruct the probe's tip. From the above literature, it is known that most scholars focused on the surface images acquired by measurement to undergo the reconstructing of the probe's size, or the influence of the probe's tip size towards the surface topography and surface roughness. They did not consider the influence of the probe towards the action force on the surface of the test plate.

This paper intends to assemble both the sample and the probe by lattices of atoms, and array a macro appearance and establish a simulated nano-scale measurement model of the constant force mode of contact mode AFM, so as to simulate and analyze the appearance of a sample at stair-step perpendicular height measured by AFM. This paper investigates the offset distance error that occurs during scanning the cross-section of the step standard sample. It analyses the influence of the probe tip radius and the probe tip edge oblique angle on the offset distance and the horizontal height between the scan curve by simulation measurement and the cross-section of nano-scale standard sample.

2. Construction of Measurement Simulation Model by Constant Force Mode AFM

2.1 Physical model of constant force measurement

This paper employs molecular mechanics to simulate the physical model of the AFM probe measurement mechanism. Both the sample and the tip of a trapezoidal cantilever probe are composed of single-crystal silicon atoms according to the ideal lattice arrangement of single-crystal silicon. The height of atomic-scale step sample is

54.31 Å and the scanning distance is 100 Å. The illustration of its simulation model is shown in Figure 1.1.

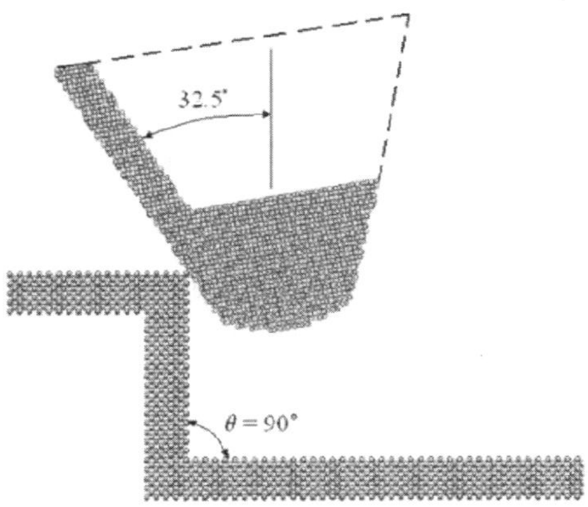

Figure 1.1. Schematic diagram of scanning step simulation model

This paper performs the simulation by using the constant force mode of the contact mode AFM. When the gap between the probe and the sample reaches a certain distance, an action force is created between the trapezoidal cantilever probe and the sample. This action force can produce a rotation of the cantilever. At the measurement point, when the force-induced rotation of the trapezoidal cantilever reaches the fixed deflection angle, this model is immediately regarded in this paper as having reached the preset measured force, and the height at this point is recorded. By scanning horizontally in proper order, the scan curve of the simulation curve can be acquired.

Since the calculation results of Morse potential function already match with the results of many experiments, and the function is suitable for describing the action force among solid atoms, this paper adopts Morse's potential function to conduct the simulation. The definition of Morse's potential function is expressed in equation (1) [4] and the parameters are shown in Table 1.

$$\phi(r) = D \cdot \left\{ e^{-2\cdot\alpha\cdot(r-r_0)} - 2 \cdot e^{-\alpha\cdot(r-r_0)} \right\} \quad (1)$$

To reduce time of calculation in the general use of molecular mechanics, 2.5 times of r_o is selected as the truncated radius (r_c). In order to get a more real situation, this paper selects 3 times of r_o so as to acquire a more accurate solution.

Table 1. Parameters of different Morse Potentials

Parameter	Si-Si[5]	Si-N[6]
D (eV)	3.032	3.88461
$\alpha(10^{10}m^{-1})$	0.7891	2.3266
$r_o\,(10^{-10}m)$	4.208	1.62136

2.2 Calculation of force and deflection for the trapezoidal cantilever of probe

When operating the constant force mode of contact mode AFM, a feedback loop maintains a constant deflection between the trapezoidal cantilever and the sample by vertically moving the scanner at each data point (x,y) to maintain a "setpoint" deflection. Thus, this paper simplifies the setpoint deflection as a fixed rotation θ_{final} mode.

This paper simplifies the action force in n direction as F_n, and the torque of cantilever as M. Substitute them in the equation for acquisition of deflection angle θ. Then the deflection angle θ_{force} under this receiving action force can be determined.

Figure 1.2 is an illustration of receiving action force of a certain atom at the tip. Here, $\overrightarrow{F_i}$ is the action force of the atom of the sample towards the ith atom at the tip through two-body potential energy. In the figure, ψ is the fixed oblique angle of AFM, generally being 10°. If there are m molecules at the tip, the moment on the cantilever the of probe caused by all the atoms at the tip is indicated in equation (2). Thus, $\overrightarrow{M_y}$ is the moment in y-axis direction in equation (3). The receiving forces of the probe's cantilever in x and z directions are shown in Figure 1.3.

$$\overrightarrow{M} = \sum_{i=1}^{m} \overrightarrow{M_i} = \sum_{i=1}^{m} (\vec{r_i} \times \vec{F_i}) \quad (2)$$

$$\overrightarrow{M_y} = -\sum_{i=1}^{m} \begin{vmatrix} r_{ix} & r_{iz} \\ F_{ix} & F_{iz} \end{vmatrix} \vec{j} \quad (3)$$

When the distance between the probe and the sample reaches a certain distance, the action force in normal direction and action force in tangential direction will be created between the probe and the sample.

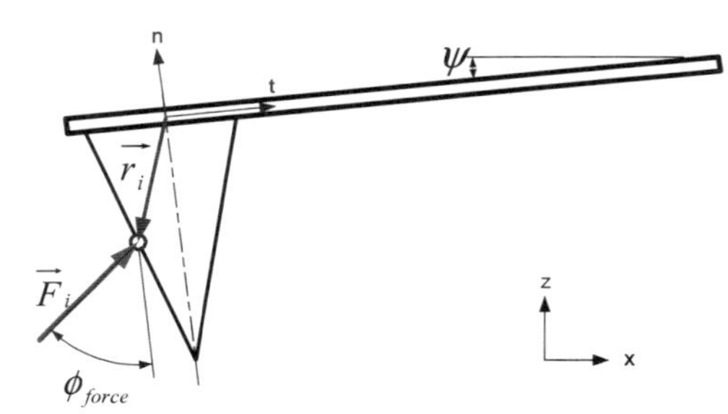

Figure 1.2. Illustration of receiving action force of a certain atom at the tip [7]

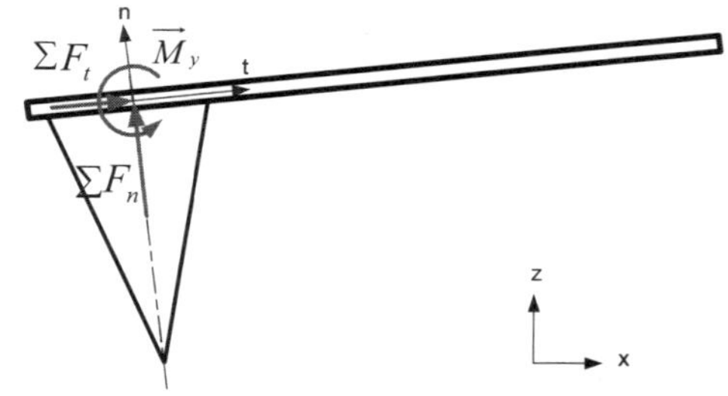

Figure 1.3. Illustration of the receiving action force of cantilever of probe [7]

Since the contact mode AFM atomic action force is generally at nanoNewton (nN) scale and the general action force in the normal direction is about 2-20 nN [8,9].

Within this paper we assume that the action force of the probe's cantilever in the normal trapezoidal direction is 20 nN, which is used for calculating the fixed bending value (deflection angle θ_{final}) created by the probe during measurement. The characteristics and size of the probe are specified in Figures 1.4 and 1.5.

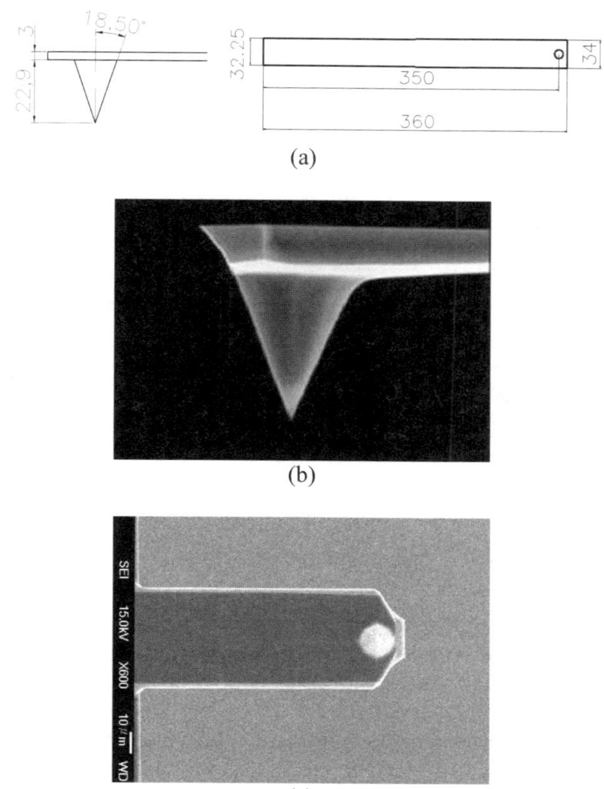

Figure 1.4. Tip of the probe: (a) size chart with units in μm; (b) and (c) pictures taken by SEM

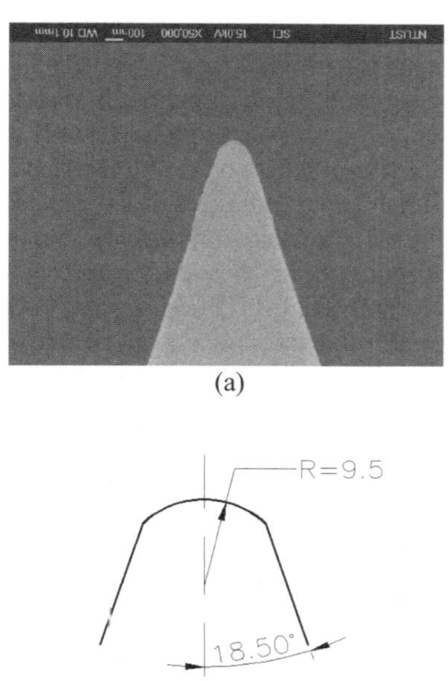

Figure 1.5. Point of the probe: (a) size chart with units in μm; (b) picture taken by SEM

This paper has simplified the trapezoidal cantilever, as shown in Figure 1.6. The external force is P, the external bending moment is M and the cantilever weight is W. ion order to determine the deflection angle of the trapezoidal shape cantilever probe at the length L_f, equation (4) is used.

$$\theta(\ell_f)= \frac{12\ell}{Eb\alpha t^3}\left\{ P(\ell_f-\gamma)[\ln(\gamma-\ell_f)-\ln(\gamma)-\ell_f]+M[\ln(\gamma-\ell_f)-\ln(\gamma)] -\frac{W}{2}\left\{(\ell-\gamma)^2[\ln(\gamma-\ell_f)-\ln(\gamma)]+2(\ell-\gamma)(-\ell)+\frac{1}{2}(\ell^2-2\gamma\ell)\right\}\right\} \quad (4)$$

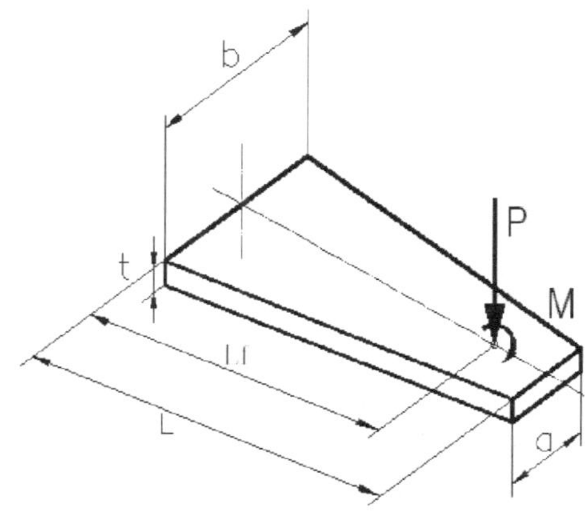

Figure 1.6. Illustration of trapezoidal shape cantilever

3. Verification of the AFM Measurement Simulation Model

In [10], the half-part oblique angle of the V-shaped cantilever probe tip is 22.5°, and the probe tip radius is 36.5 nm. According to the standard lattice arrangement of Si_3N_4, N atoms and Si atoms form the probe model, whereas Si atoms form the sample. The fixed deflection angle, 7.7092×10^{-5} (rad), is used to simulate the scanning of the nano-scale step standard sample. The simulated scan curve in [10] has been compared with the real AFM scan curve, so as to prove the correctness of the simulation measurement model in.

In order to reduce the time required for the simulation, the probe tip radius (36.5 nm) of the V-shaped cantilever is shrunk by a factor of 10 to 3.65 nm. It is substituted in the simulation model in [10] for AFM simulation measurement, so as to acquire the simulated scan curve of the nano-scale standard step sample. The fixed deflection angle, 7.7092×10^{-5} (rad), is used to simulate the scanning of the nano-scale step standard sample. The simulation results indicated in [10] are illustrated in Figure 1.7 and Figure 1.8.

This paper adopts the trapezoidal cantilever probe, with its probe tip radius being 3.65 nm and the half-part oblique angle of the probe tip being 22.5° to simulate the scanning of the nano-scale step standard sample. The fixed deflection angle $2.88*10^{-5}$ (rad) is used to simulate the scanning of the nano-scale step standard sample. The simulated size is shown in Figure 1.9, and the simulation result is shown in Figure 1.10.

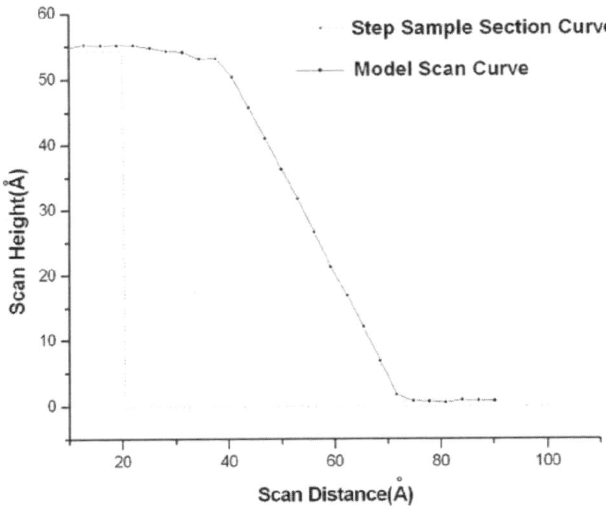

Figure 1.7. Simulation result of the scan curve of V-shaped cantilever [10]

To compare the difference between Figure 1.7 and Figure 1.10, a new plot combining these two figures is shown in Figure 1.11. It can be found that both simulated scanning results are close to each other. Therefore, it is proved to be correct for this study to construct the simulation measurement model at the contact force mode AFM.

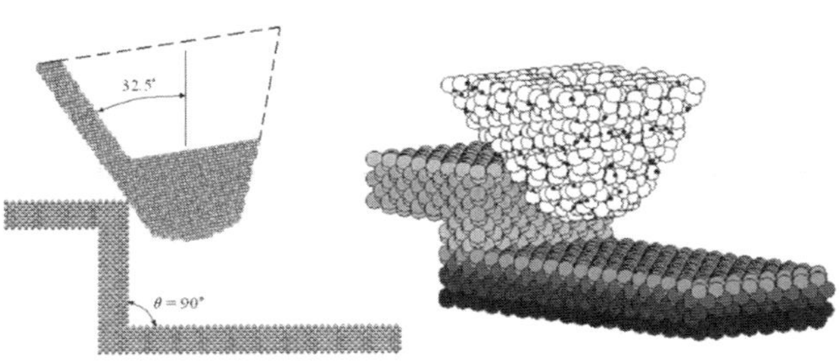

Figure 1.8. Schematic diagram of simulation of the V-shaped cantilever [10]

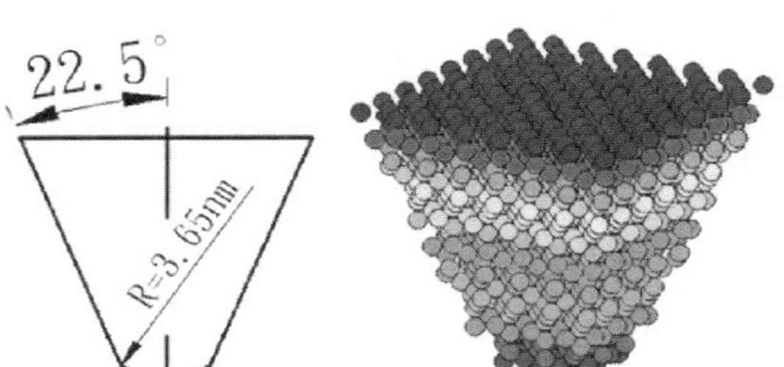

Figure 1.9. Size chart and schematic diagram of the trapezoidal probe

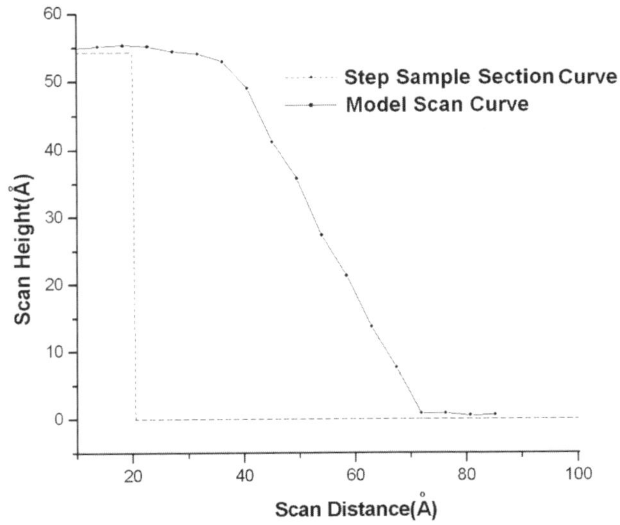

Figure 1.10. Simulated scan curve chart of trapezoidal probe with tip radius of 3.65mm

4. Effect of the Geometric Shape and Material of Probe for Scan Curve

The scan curve in [10] is a simulated scan by the V-shaped (Si_3N_4) cantilever towards the single crystal silicon sample. In the study, the probe is a single-crystal silicon trapezoidal cantilever, and the sample is single-crystal silicon. Comparing it with the simulated scan curve presented in [10], as shown in Figure 1.11, it is divided into 3 parts, indicating the differences between the trapezoidal probe and the V-shaped probe.

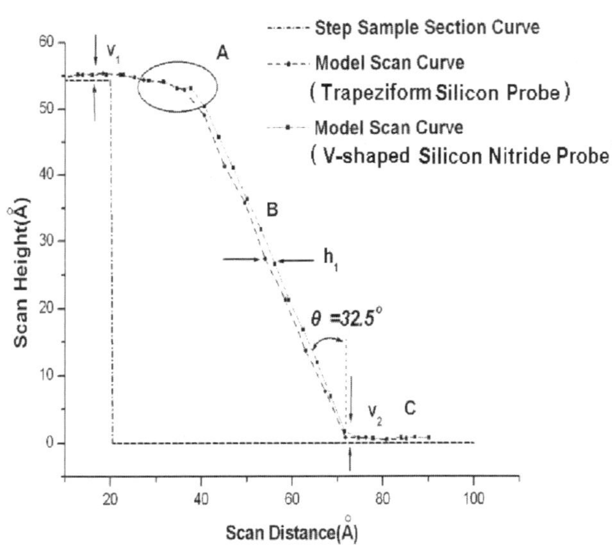

Figure 1.11. Comparison between the scan curve of trapezoidal probe and the scan curve of the V-shaped probe of literature [10]

Part A:

Comparing the scan curve simulated by the trapezoidal cantilever probe (Si-Si) with the scan curve simulated by the V-shaped cantilever probe (Si-N) [10], it is close to the surface profile of the nano-scale standard sample in Part A. The perpendicular height difference between the two simulated curves is about V_1=0.1Å. From [5] and [6], it is known that the binding energy of Si-N is 3.88461 and the binding energy of Si-Si is 3.032, implying that D_{Si-N}□D_{Si-Si}. The material parameter value of Si-N is 2.3266, and the material parameter value of Si-Si is 0.7891, implying that α_{si-n}□α_{si-si}. The r_o of Si-N is 1.62136 and the balance distance value of Si-Si is 4.208, implying that $r_{o\ si-n}$□$r_{o\ si-si}$. From the above data, it is known that the Morse potential action force between the Si-N atoms is greater than the Morse potential action force between the Si-Si atoms.

Part B:

In this zone, what affects the simulated scan curve most is mainly the oblique-side effect of the probe. The different Morse potential action forces of the probes mainly cause the deviated distance h1. It is observed from the comparison that the simulated scan curve (Si) is closer to the real sample than the simulated scan curve (Si-N).

Part C:

The scan curves at this zone are mainly extending those of Part B. Therefore, the simulated scan curve is mainly affected by the probe tip edge oblique angle.

5. Analysis of sensitivity

5.1 Effects of trapezoidal cantilever probe tip on the scanned profile of the atomic-scale step

Regarding the simulation of number of atoms of the trapezoidal tip radius, when the probe tip radii are 10.5 Å, 9.5 Å and 8.5 Å. The study uses the probes with cantilever probe tip radii at 10.5Å, 9.5 Å and 8.5 Å and their half-part oblique angle θ=18.5°, then the curve as shown in Figure 1.12 can be acquired. From Figure 1.12, the scanned profile of atomic-scale step sample by the probes of different tip radii can be seen. In this figure, it can be found that when the probe tip radius is smaller, it is easier to scan the actual cross-section. The results are consistent with the traditional cognition. It means that the sharper the probe, the more real effects can be acquired.

2 Effects of the oblique side angle of the trapezoidal cantilever probe tip for the scan profile of the atomic-scale step sample

The study uses the probes with the half-part oblique side angles θ= 13.5°, θ= 18.5° and θ= 23.5°of cantilever probe tip round angle (R = 9.5 Å), then the curve as shown in Figure 1.13 can be acquired. It can be observed in Figure 1.13 that if the oblique side angle of the probe is large, the scan curve moves outwards to the lateral sides of the step sample. As a result, it can be concluded that the geometric shape of the probe affects the simulated scan curve to a significant degree.

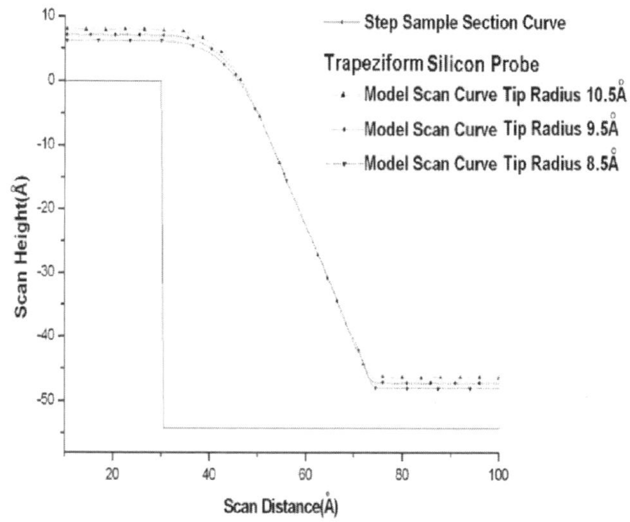

Figure 1.12. Scanned profile of atomic-scale step sample by the probes of different tip radii

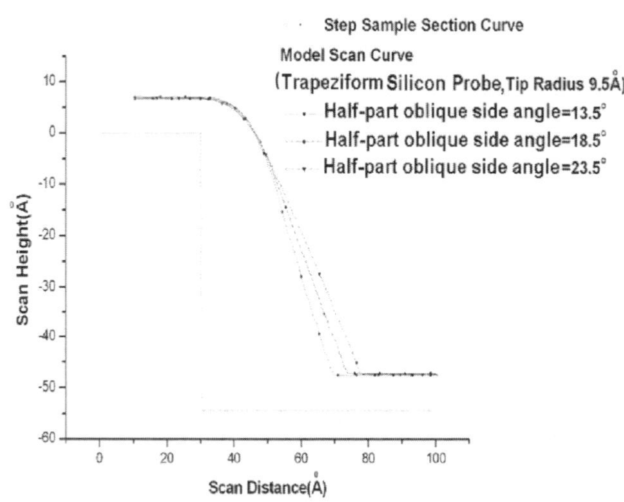

Figure 1.13. Profile of atomic-scale step sample scanned by probes with different oblique side angles

6. Conclusion

This paper investigates the offset distance error that occurs during scanning the cross-section of the step standard sample. The two probes with different cantilevers and probe materials (V-shaped Si_3N_4 cantilever and trapezoidal Si cantilever), both of which have the same tip radius of 3.65nm, are employed in this study. The study investigates the influence of the probe tip radius and the probe tip edge oblique angle on the offset distance and the horizontal height between the scan curve by simulation measurement and the cross-section of nano-scale standard sample.

It is found that the probe tip radius and the probe tip edge oblique angle influence the cross-section curve during simulation measurement. It can be known that probe material can influence the cross-section curve under simulation measurement.

References

[1] Binning G. and C.F. Quate, 1986, Phys. Rev. Lett. Vol.56, P930-933.
[2] Villarubia, J.S. 1994, Surface Science, 321, P 287–300.
[3] Song Xu, A. Nabil, and Gang Yu. Liu, 2001, American vacuum society, P175-176.
[4] Liu, Shih-Che, 2003, Using Molecular Mechanics Simulation on Nano-Scale Atomic Force Microscopy Measurement and Probe Sensitivity Analysis, Master Thesis, Department of Mechanical Engineering, National Taiwan University of Science and Technology, Taipei, Taiwan.
[5] Michael Rieth, 2003, Nano-engineering in science and technology: an introduction to the world of nano-design, World Scientific.
[6] David L. Martin, Donald L. Thompson and Lionel M. Raff, 1986, Journal of Chemical Physics, Vol.84, P 4426-4428.
[7] Marian C.M., M. Gastreich and J.D. Gale, 2000, Phys. Rev. B, Vol.62, P 3117-3124.
[8] Sergei N. Magonov, and Myung-Hwan Whangbo, 1996, Surface analysis with STM and AFM: Experimental and Theoretical Aspects of Image Analysis, VCH Publishers.
[9] Meyer E., R.M. Overrney, K. Dransfeld, and T. Gyalog, 1998, Nanoscience-Friction and rheology on the nanometer scale, World Scientific.
[10] Zeng, Bo-Jhan, 2004, Investigation of Simulation on Atomic Force Microscopy Measurement by Using Molecular Mechanics and Forecasting Actual Cross-Section Appearance, Master Thesis, Department of Mechanical Engineering, National Taiwan University of Science and Technology, Taipei, Taiwan.

11–13

Molecular Dynamics Simulation of AFM/FFM Surface Observation: Influence of Probe Tip Shape

J. Shimizu[1], L. Zhou[2] and H. Eda[3]

[1]Ibaraki University, Hitachi, Japan, 3168511, jshimizu@mx.ibaraki.ac.jp
[2]Ibaraki University, Hitachi, Japan, 3168511, lbzhou@mx.ibaraki.ac.jp
[3]Ibaraki University, Hitachi, Japan, 3168511, eda@mx.ibaraki.ac.jp

Abstract. Several molecular dynamics simulations have been performed to clarify dynamically the contact mechanism between specimen surface and probe tip in the surface observation by an atomic force microscope (AFM) or friction force microscope (FFM). In the simulation, a three-dimensional simulation model is proposed where the specimen and the probe are assumed to consist of monocrystalline copper and rigid diamond or a carbon atom, respectively. The effect of cantilever stiffness of the AFM/FFM is also taken into consideration. The surface observation process is simulated on a well-defined Cu{100} surface. As a result, the influence of the probe tip shape on the force images was evaluated, and it was proven that the surface atomic images could be observed even when a flat tip probe was used due to the periodicity of the specimen surface.

Keywords: AFM, molecular dynamics, stick-slip, friction, cantilever, two-dimensional friction phenomenon, simulation

1. Introduction

Many surface analyses are now performed on the atomic scale using surface measurement instruments such as the scanning tunneling microscope (STM) and the atomic force microscope (AFM). In particular, the AFM is widely used in the fields of advanced technology, because it can be utilized for any material and possesses atomic-scale resolution even in air. The atomic-scale stick-slip phenomenon was described for the first time by Mate et al. [1] thorough the AFM or friction force microscope (FFM) surface measurement. It is a unique friction process phenomenon at an atomic level. Recently, other many researchers have attempted to clarify the atomic-scale friction mechanism using an AFM or FFM, and some useful insights have been reported [2, 3].

However, there remain unresolved problems. These include whether a surface image obtained through an AFM surface measurement is expressed by a specimen surface atom or periodicity of the specimen surface and the behaviour of the specimen atoms at the contact point. Komiyama et al. [4] have simulated AFM surface image with and without point defects using the Morse potential function and have investigated the possible resolution limits of AFMs. However, they did not take into account the cantilever and the specimen stiffness. Sasaki et al. [5] and Holscher et al. [6] have analysed image pattern of FFMs using a three-dimensional model where the cantilever stiffness was taken into consideration, although no account was taken of specimen stiffness. Many other researchers have also simulated the friction processes between a slider and a specimen surface or between two solid surfaces using the molecular dynamics (MD) or other theoretical model [7-15]. As a result, various interesting phenomena have been reported, such as the atomic-scale stick-slip phenomenon, the superlubricity, the origin of the friction, the friction anisotropy and so on. In particular, Ohzono et al. [16] simulated the molecular stick-slip motion and the friction anisotropy observed experimentally between an AFM tip and an ordered monolayer of n-alkane chains which tilt in one of six equivalent stable directions using a slider with a single atomic scale protrusion, connected to an external force control unit via three orthogonal springs, and commented about the friction anisotropy. Our group has simulated the atomic-scale stick-slip phenomenon using a simple model where the effect of the cantilever stiffness of the AFM is taken into consideration [17], and has reported some worthwhile insights, such as the relationship between the atomic force images and behavior of the probe tip [18].

In this paper, to clarify dynamically the contact mechanism between specimen surface and probe tip in surface observation using an AFM/FFM, in particular to clarify the influence of the probe tip shape on atomic force images, several MD simulations have been performed.

2. MD Simulation model

The specimen and the probe are assumed to consist of monocrystalline copper and rigid diamond, respectively. The scanning direction of the probe is taken as the x axis and direction perpendicular to the sliding direction is taken as the y axis. The x-y plane corresponds to the specimen surface. The direction perpendicular to the specimen surface is taken as the z direction. These are chosen according to a right-hand coordinate system. An existing Morse potential [19] is applied for the description of the interaction between a pair of copper atoms. Since the actual potential between a copper atom and a carbon atom pair has not been clarified, the Morse potential proposed by Inamura et al. [20] is applied here. A total of 723 copper atoms are considered in the simulations.

In order to consider the influence of spring constant of the AFM cantilever, a model as shown in Fig. 1.1 is proposed. The probe tip and the supports A, B and C are connected by means of springs in the x, y and z directions, their spring constants being k_x, k_y and k_z, respectively. Support A moves at a constant sliding speed V, and supports B and C move at the same sliding speed as that of the probe tip, keeping a constant distance between it and the specimen surface. In the friction process, spring forces are exerted on the probe tip. By taking the mass of the probe tip to be M, its position in the x direction at time t to be $x(t)$, its relative displacements to the support in the y and z directions to be $y(t)$ and $z(t)$, respectively, the force exerted on it in each direction to be $F_x(t)$, $F_y(t)$ and $F_z(t)$, respectively, the following equations of motion are obtained:

$$M\ddot{x} = F_x(t) + k_x\{Vt - x(t)\} \quad (1.1)$$

$$M\ddot{y} = F_y(t) - k_y y(t) \quad (1.2)$$

$$M\ddot{z} = F_z(t) - k_z z(t) \quad (1.3)$$

where $F_x(t)$, $F_y(t)$ and $F_z(t)$ are obtained through the MD calculations. Leap-frog method is applied to the integral calculations of Eqs (1.1)-(1.3) and Newton's equations of motion (MD calculations).

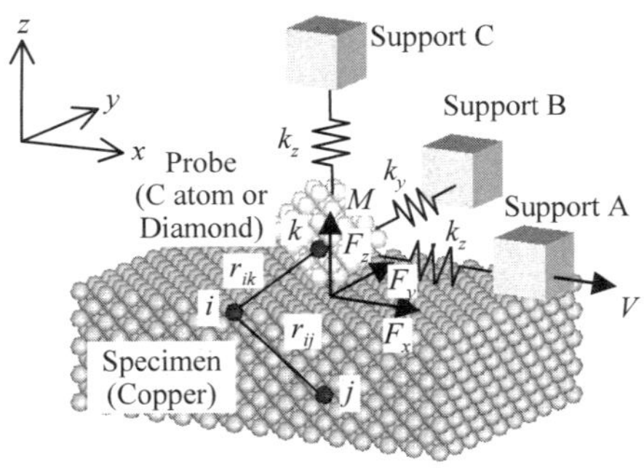

Figure 1.1. MD simulation model

3. Ideal Atomic Force Images

Ideal reactive force images are obtained by displaying the reactive forces acting on the probe assuming the stiffness of the spring to be infinite and the probe and the specimen to be perfect rigid bodies. Here and in the simulations, the x, y and z axes correspond to the [100], [010] and [001] directions of copper, respectively, and the (001) surface of copper corresponds to the specimen surface, well defined at an atomic level. The scan positions or directions of the probe are set as shown in Fig. 1.2 ($a \times 3a$). In Fig. 1.2, (1) and (2) denote the cases where the probe slides just on top of the surface atoms and at the position shifted by $a/10$ (a = 0.36 nm is lattice constant of copper) from top of the most surface atoms, respectively. In order to investigate the influence of the probe tip shape on force images, a single carbon atom, a diamond lattice and a four diamond lattices have been selected as the probes, as shown in Fig. 1.3.

Figs. 1.4 to 1.6 show the ideal reactive force images using a carbon atom probe, a diamond lattice probe and a four diamond lattices probe, respectively. The calculation of the reactive force acting on the probe is carried out using the interatomic potential equal to the MD simulation. Considering the correspondence with the simulation, the friction force is considered positive when the spring force affects the probe in the negative direction of the x-axis ($-F_x$); the lateral and normal forces are positive when the spring forces affect the probe in the positive direction of the y and z axes (F_y, F_z), respectively.

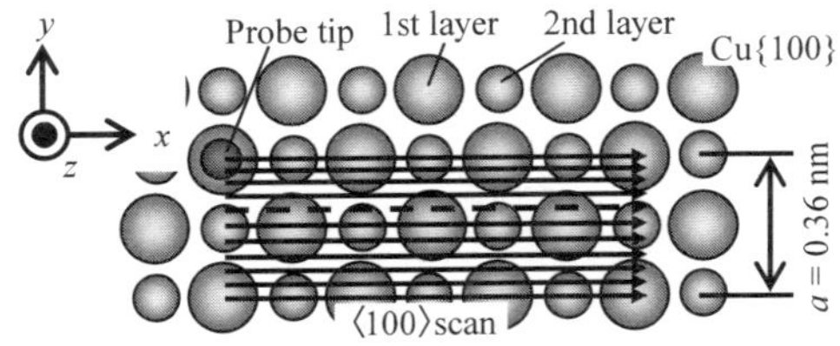

Figure 1.2. Scanning position and direction

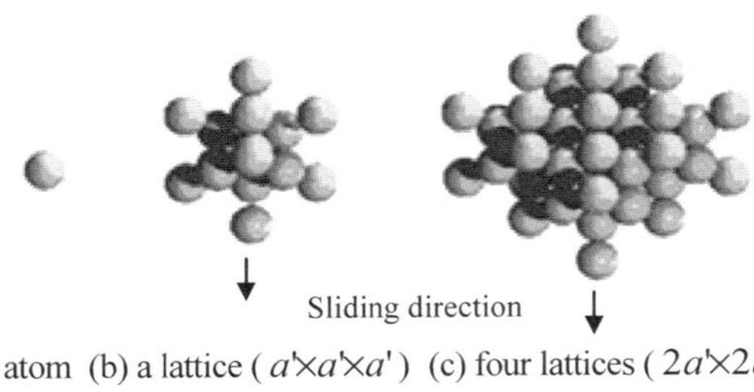

Figure 1.3. Shapes of the probe (a' = 0.357 nm is the lattice constant of diamond)

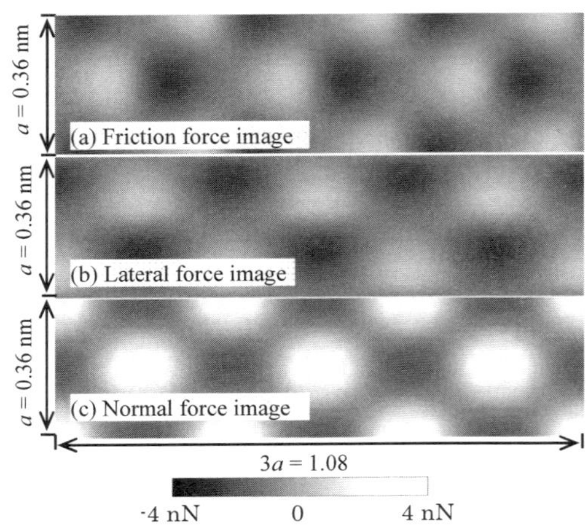

Figure 1.4. Ideal force images by a single carbon atom probe

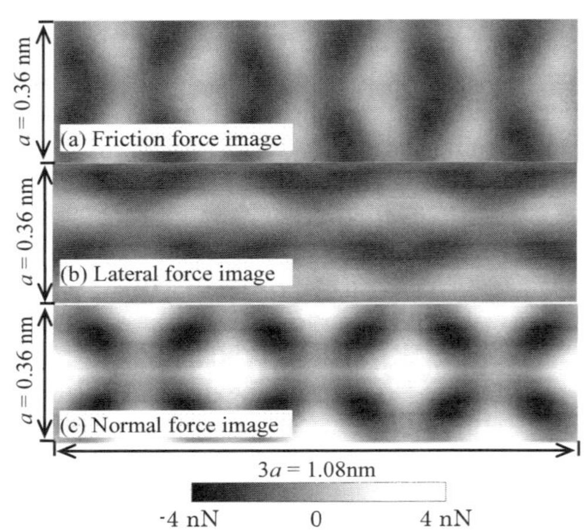

Figure 1.5. Ideal force images by a diamond lattice probe

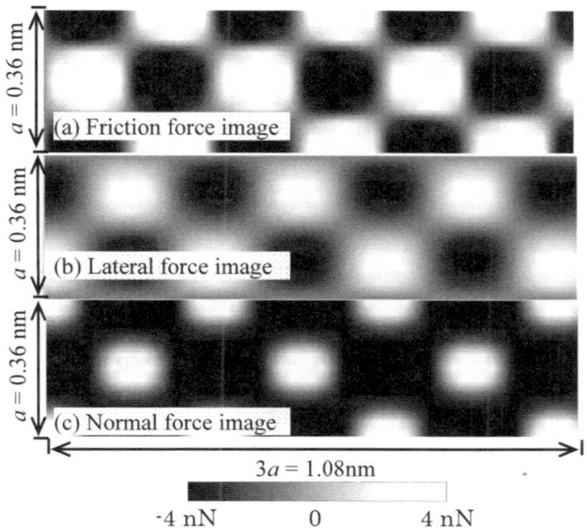

Figure 1.6. Ideal force images by a four diamond lattices probe

The difference between Figs. 1.4, 1.5 and 1.6 is remarkable. In Figs. 1.4 and 1.6, the reactive force images of each specimen surface atom can be clearly expressed, and the position of each specimen surface atom can be clearly seen. However, in Fig. 1.5, the pattern of reactive force images of each atom is complicated. However, the specimen surface atom interval can be obtained from the periodicity of the images.

4. MD Simulation Results and Discussion

Table 1.1 gives the MD simulation conditions applied here. To reduce the noise due to the lattice vibration on the reactive forces, the initial temperature is set at absolute zero. The cantilever stiffness is set as for an actual AFM cantilever.

Table 1.1. MD simulation conditions

Initial temperature	0 K
Mass of probe M	(a) 0.2×10^{-26} kg (an atom) (b) 3.6×10^{-26} kg (18 atoms) (c) 2.7×10^{-25} kg (54 atoms)
Sliding speed V	5 m/s
Spring constant k_x, k_y, k_z	(a),(b) 5 N/m, (c) 10 N/m

4.1 The case of a single carbon atom probe

Figure 1.7 shows the snapshot of atomic arrays and the position of the probe in every 7.5 ps using a single carbon atom probe in the case of the broken line scan in Fig. 1.2. Figure 1.8 shows the reactive force (spring force) images by using a single carbon atom probe.

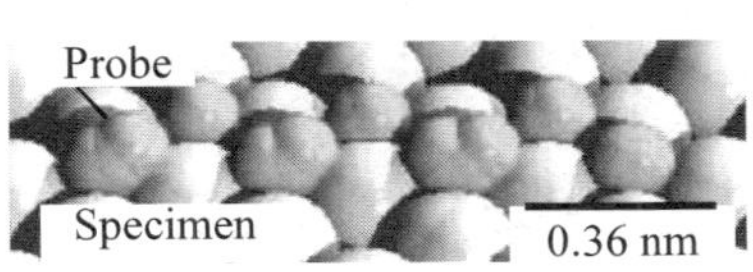

Figure 1.7. Snapshot of atomic arrays and position of the probe every 7.5 ps using a single carbon atom probe in the case of the broken line scan in Fig. 1.2

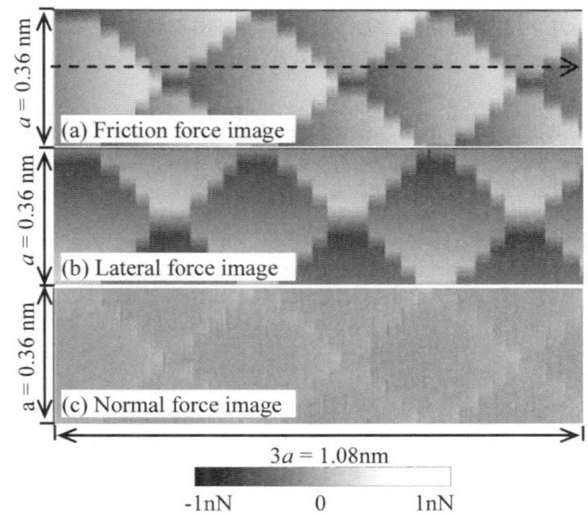

Figure 1.8. Force images using a single carbon atom probe

In Fig. 1.7, the generation of wraparound of the probe tip and the two-dimensional friction process are confirmed even in a one-dimensional sliding. Such a two-dimensional atomic-scale stick-slip [2] can be understood by comparing Fig. 1.7 and Fig. 1.8. The atomic-scale stick-slip phenomenon, which is an elementary process of the friction, can be fundamentally explained according to the model of Tomlinson [21].

4.2 The case of a single diamond lattice probe

Figure 1.9 shows the snapshot of atomic arrays and position of the probe centre atom every 7.5 ps using a single diamond lattice probe in the case of the broken line scan in Fig. 1.2. Figure 1.10 shows the reactive force images by using a single diamond lattice probe.

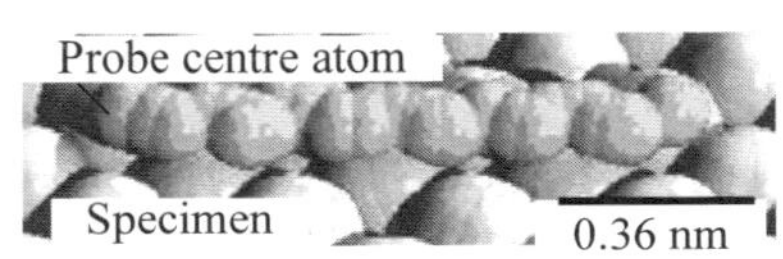

Figure 1.9. Snapshot of atomic arrays and position of the probe centre atom every 7.5 ps using a single lattice probe in the case of the broken line scan in Fig. 1.2

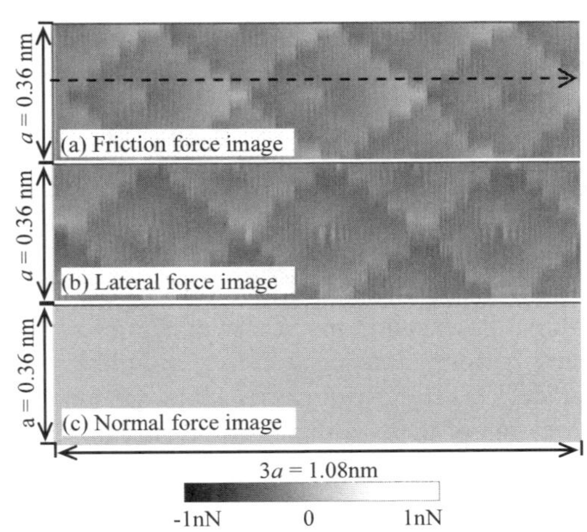

Figure 1.10. Force images using a single lattice probe

The two-dimensional atomic-scale stick-slip phenomenon of Fig. 1.9 is weakened compared to that of

Fig. 1.7. Therefore, the contrast of the atomic force images seems to be weakened in Fig. 1.10.

4.3 The case of a four diamond lattices probe

Figure 1.11 shows the snapshot of atomic arrays and position of the probe centre atom every 7.5 ps using a four diamond lattices probe in the case of the broken line scan in Fig. 1.2. Figure 1.12 shows the reactive force images by using a four diamond lattices probe.

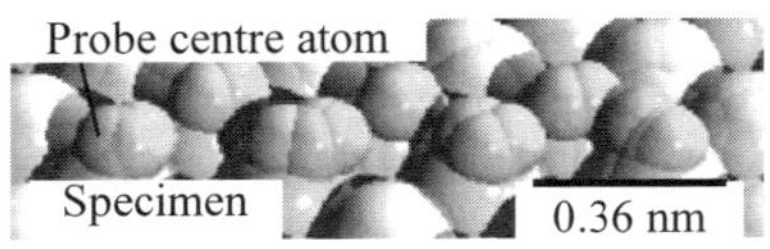

Figure 1.11. Snapshot of atomic arrays and position of the probe centre atom every 7.5 ps using a four lattices probe in the case of the broken line scan in Fig. 1.2

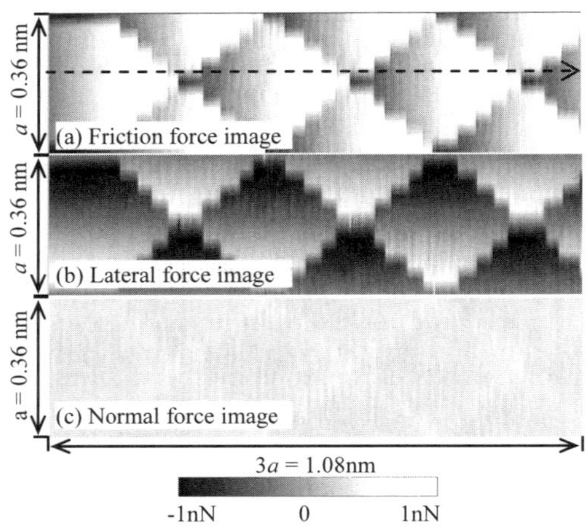

Figure 1.12. Force images using a four lattices probe

In Fig. 1.11, the generation of the two-dimensional atomic-scale stick-slip can be observed similar to Fig. 1.7 although the stick points are on the tops of the specimen surface atoms. From Fig. 1.12, clear surface atomic force images can be confirmed.

As a result, it was proven that the surface atomic images could be observed even when a flat tip probe was used due to the periodicity of the specimen surface.

5. Conclusion

To clarify the contact mechanism in AFM/FFM surface observation, especially the influence of the probe tip shape, three-dimensional molecular dynamics simulations have been performed using a simple cantilever stiffness effect model. The following conclusions are obtained: (1) The generation of the two-dimensional atomic-scale stick-slip phenomenon is important for the expression of the accurate surface atomic images. (2) The surface atomic images can be observed even when a flat tip probe is used due to the periodicity of the specimen surface.

References

[1] Mate CM, McClelland GM, Erlandsson R, Chiang S, (1987) Atomic-Scale Friction of a Tungsten Tip on a Graphite Surface. Phys. Rev. Lett. 59:1942-1945.

[2] Fujisawa S, Sugawara Y, Ito S, Mishima S, Okada T, Morita S (1993) The Two-Dimensional Stick-Slip Phenomenon with Atomic Resolution. Nanotechnology 4:138-142.

[3] Morita S, Fujisawa S, Sugawara Y (1996) Spatially Quantized Friction with a Lattice Periodicity. Surf Sci Reports 23:1-41.

[4] Komiyama M, Tsujimichi K, Ohkubo S, Tazawa K, Kubo M, Miyamoto A (1995) Simulation of Atomic Force Microscopy Image Variation along the Surface Normal: Presence of Possible Resolution Limit in the Attractive Force Range. J. Jpn Appl. Phys. 34:L789-792.

[5] Sasaki N, Tsukada M, Fujisawa S, Sugawara Y, Morita S, Kobayashi K (1997) Analysis of Frictional-force Image Patterns of a Graphite Surface. J. Vac. Sci. Technol. B15: 1479-1482.

[6] Holscher H, Schwarz UD, Zworner O, Wiesendanger R (1998) Consequences of the Stick-Slip Movement for the Scanning Force Microscopy Imaging of Graphite. Phys. Rev. B57:241-2481.

[7] Landman U, Luedtke WD, Ribrarsky MW (1989) Stractural and Dynamical Consequences of Interactions in Interfacial Systems. J. Vac. Sci. Technol. A7:2829-2839.

[8] Shinjo K, Hirano M (1993) Dynamics of friction: Superlubric state. Surf. Sci. 283:473-478.

[9] Cieplak M, Smith ED, Robbins MO (1994) Molecular origins of friction: The force on adsorbed layers. Science 265:1209-1212.

[10] He G, Muser M, Robbins MO (1999) Adsorbed Layers and the Origin of Static Friction. Science 284:1650-1652.

[11] Harrison JA, Stuart SJ, Brenner DW (1999) Atomic-Scale Simulations of Tribological and Related Phenomena. Bhushan B, editor. Handbook of Micro/NanoTribology, CRC Press 525-594.

[12] Sorensen MR, Jacobsen KW (1996) Stoltze P. Simulation of Atomic-Scale Sliding Friction. Phys. Rev. B53:2101-2113.

[13] Buldum A, Ciraci S, Batra IP (1998) Contact, Nanoindentation, and Sliding Friction. Phys. Rev. B57:2468-2476.

[14] Lan H, Zhang S (2004) Molecular Dynamics Simulation on Nanotribological Properties of Molecular Deposition Film during the Scan Process. Trib. Int. 37:661-665.

[15] Ni B, Sinnott SB (2001) Tribological properties of carbon nanotube bundles predicted from atomistic simulations. Surf. Sci. 487:87-96.

[16] Ohzono T, Fujihira M (2000) Effect of an Atomic Scale Protrusion on a Tip Surface on Molecular Stick-Slip Motion and Friction Anisotropy in Friction Force Microscopy. Jpn. J. Appl. Phys. 39:6029-6034.

[17] Shimizu J, Eda H, Yoritsune M, Ohmura E (1998) Molecular Dynamics Simulation of Friction on the Atomic Scale. Nanotechnology 9:118-123.

[18] Shimizu J, Eda H, Zhou L (2002) Molecular Dynamics Simulation of the Contact Process in AFM Surface Observations. Tribotest 9:101-115.

[19] Girifalco LA, Weizer VG (1959) Application of the Morse Potential Function to Cubic Metals. Phys. Rev. 114:687-690.

[20] Inamura T, Takezawa N (1992) Atomic-Scale Cutting in a Computer Using Crystal Models of Copper and Diamond. Annals of the CIRP 41:21-24.

[21] Tomlinson GA. (1929) A Molecular Theory of Friction. Phil. Mag. 7:905-39.

Measurement of Mechanical Properties of Boron Nanowire Using Nano-manipulation System

C.H. Lin[1], M. Chang[2], X. Li[3] and J.R. Deka[4]

1Chung Yuan Christian University, Chung Li-320, Taiwan, ROC, ming@cycu.edu.tw
2Chung Yuan Christian University, Chung Li-320, Taiwan, ROC, ming @cycu.edu.tw
3University of South Carolina, Columbia, SC 29208, USA, LIXIAO@engr.sc.edu
4Chung Yuan Christian University, Chung Li-320, Taiwan, ROC, juti@cycu.edu.tw

Abstract. Nanotubes and nanowires have shown great potential in a broad range of applications from nanocomposite, nanoelectronics to nanoelectromechanical systems. The small dimensions of such materials also impose a tremendous challenge for experimental studies of their intrinsic mechanical, electrical and other device related properties. The present research has developed four degrees of freedom nano mechanical manipulation system for use inside a scanning electron microscope (SEM). This manipulation system is used to monitor three dimensional nano manipulation processes and study the mechanical response of crystalline boron nanowire (BNW). Crystalline boron nanowire is synthesized by vapor liquid solid (VLS) technique. The electron beam induced deposition (EBID) method is used to clamp boron the nanowire to the AFM tip inside the SEM vacuum chamber. The mechanical property of the BNW is measured by applying continuously increasing load on it. SEM image can analyze the deformation and buckling behavior of the boron nano wire. Transmission electron microscopy (TEM) and atomic force microscope (AFM) were used to measure the cross-sectional area of nanowire. The Young's modulus of the boron nanowire has been measured to be 117± 4.1 GPa.

Keywords: Boron nanowire, Young's modulus, Nano manipulation, Electron beam induced deposition

1. Introduction

One-dimensional nanostructure has developed a great technological interest because of their potential fundamental characteristics and practical applications in areas such as material science, chemistry, physics and engineering [1,2].

Boron has the most varied polymorphism of any of the elements in the periodic table. It belongs to the III A family, one spot left to carbon in the periodic table. Boron has a very high melting point of 2300°C, its coefficient of elasticity is approximately 380-400 GPa [3], and it possesses a hardness close to diamond. Boron fibers are produced from the halogenated boron, return to original boron, and adhere to the metal fiber or chemical fiber continuously in an evaporator. Crystalline boron nanowires are a semiconductor and exhibit electrical properties consistent with those of the elemental boron [4]. Several methods were used in the past decade to access mechanical properties of BNW. The mechanical properties of boron fibers have been intensively studied by tensile testing method which gives Young's modulus value from 390-440 GPa [5]. The Young's modulus of BNW has been measured to be approximately 72.7±3.9 GPa by nanoindenter method [6]. The average Young's modulus of boron nanowire measured with resonance method and tensile testing is found to be approximately 320 GPa [7].

In the present study, compression experiments are carried out on vertically grown BNW on silicon substrate with Vapor Liquid Solid (VLS) technique [8]. For Young's modulus measurement of the prepared NW, a nano-manipulator is setup inside a scanning electron microscope (SEM). The nanomanipulator has three axis linear motions and single axis rotational probe motion provided by picomotors. The manipulation system is designed and constructed for easy handling, structuring and mechanical characterization of nanoscale objects inside SEM. The Young's modulus of BNW measured with the compression technique is 117±4.1GPa.

2. Synthesis of Boron NanoWire

VLS technique is employed to synthesize boron nanowire (BNW) on silicon substrate. In the VLS process, the nucleation and an anisotropic crystal growth is achieved with generation of solid/liquid alloy interfaces using metal catalyst. A metal catalyst Au is used to generate a solid/liquid alloy interface for nucleation of the nanowire. Silicon substrates are coated with a 3 nm thick Au film using sputtering technology. The nanowire growth is initiated after the solid saturation is reached in the solid/liquid alloy. Scanning electron microscopy (SEM) and transmission electron microscopy (TEM) are applied to characterize the morphology and microstructure of the BNW. Figure 1.1(a) shows an SEM image of VLS grown BNW on silicon substrate. The length of the prepared BNW is observed to be approximately 10-25 μm and its diameter is 80-120 nm. A selected portion of a single BNW under transmission electron microscope (TEM) is shown in Figure 1.1 (b). The micrograph shows that the crystalline BNW is covered with an oxide layer that forms when crystalline NWs are exposed to air after post annealing.

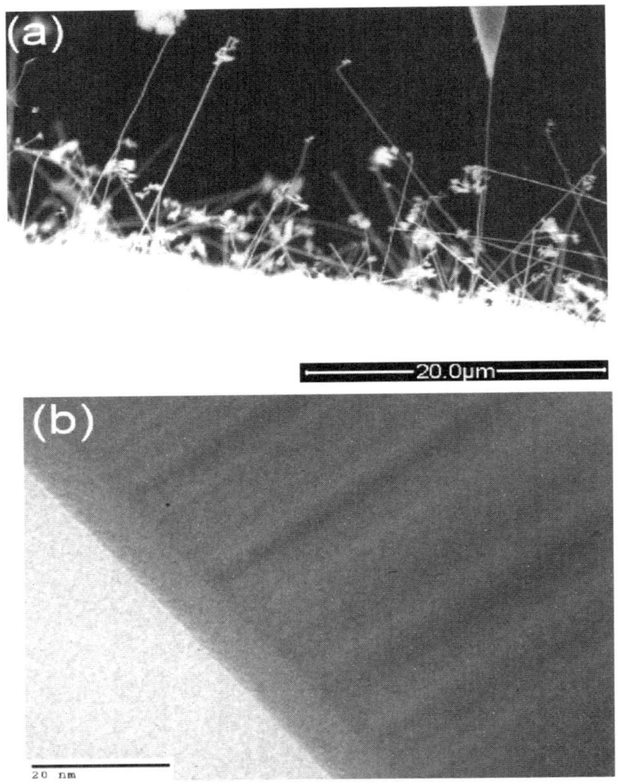

Figure 1.1. (a) SEM image of VLS grown crystalline boron nanowires on Si substrate; (b) TEM image of a single NW, which confirms the existence of oxide layer

3. Experiments and Results

3.1 Instrumentation

The coefficient of elasticity measurements of BNW are performed using the technique used for three dimensional manipulation of carbon nanotubes under a scanning electron microscope [9]. The nanomanipulator is designed and fabricated inside the vacuum cavity of an ESEM FEI Quanta 200's scanning electron microscope to place NW. The resolution of the microscope is specified as 3.5 nm at 30 kV. In order to prevent the crashing of sensor and objective lens in the SEM vacuum cavity, the x-y and z-θ motions are grouped into two low profile opposing stage sub-modules attached symmetrically on the SEM platform around the axis of the electron beam. The schematic design of the nanomanipulator is shown in Figure 1.2.

Coarse three axis linear motions up to ±5 mm and single axis 360° rotational probe motion are provided by sigma stage driven picomotors (8301-V and 8341-V, New Focus). An integral x-y stage guides the motion parallel to the x-y plane of the SEM stage, and a separate z axis stage is used for the motion along the SEM beam axis. The rotational motion normal to the beam is accomplished using a picomotor rotating actuator mounted on top of the Z stage. This actuator is attached mechanically to the AFM tip. Three straight picomotors are used to provide linear motion and another picomotor is used to support the rotating tip. This SEM stage nanomanipulator occupies approximately $10^3 mm^3$. The overall system size of the nanomanipulator is, length ≈100 mm, width ≈50 mm and height ≈ 25mm. The electrical control signals for the picomotors are conducted through the isolated vacuum feedthrough passing through the SEM main chamber wall. The translation stages share a common electrical circuit, which is often grounded, and the rotating tip and sample holder on top of the Z-stage and X-Y stages, respectively are all connected through the vacuum feedthrough so that articulating elements can be electrically addressed separately outside the vacuum. The maximum traveling range of the picomotors is 12.5 mm, with a single step size of picomotor 30 nm. A piezotube (PZT) of size 3mm×3mm×2mm is placed on top of the Z stage for continuous motions when the linear stage reaches the desired position for manipulation operation. By the precise voltage control, a spatial resolution of 0.1 nm for the PZT can be achieved. The variation of sample positions are observed by the real time images of SEM.

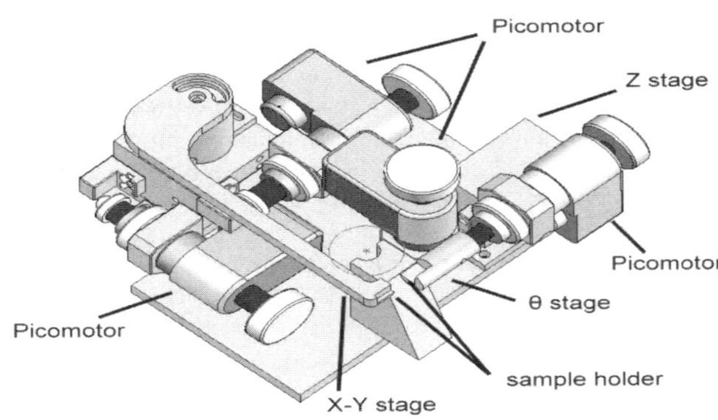

Figure 1.2. Schematic diagram of the nanomanipulator inside

3.2 Determination of Coefficient of Elasticity of Cantilever Probe

Two AFM tips are used to hold the BNW on the nano manipulator stage inside the SEM. One AFM tip with soft cantilever [CSG11, NT-MDT] is attached to the rotating actuator and another AFM tip with rigid cantilever [NSG 20] is fixed on the X-Y stage and placed opposite to the rotating stage of the manipulator. The AFM tip with soft cantilever is used as a sensor for the determination of buckling force of NW via its deformation of cantilever and its coefficient of elasticity.

The elasticity modulus of the cantilever probe is determined using resonance phenomenon. An AC signal from a function generator is applied to the piezo actuator attached to the cantilever probe. As a result, the cantilever probe is excited with the frequency of the AC signal. By tuning the frequency of the AC signal, the cantilevered AFM tip is excited in such a way that maximum amplitude is achieved when the frequency of the signal matches the natural resonance frequency of the cantilever. SEM images are analyzed to extract resonance frequency of the cantilever. The scanning rate of the SEM image is 2.05 Hz which is shown in Figure 1.3.

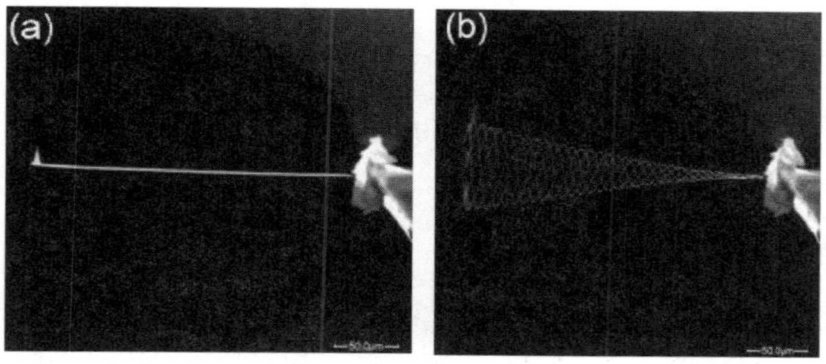

Figure 1.3. AFM probe at (a) static condition; (b) resonance condition

From the measurement, it is observed that the cantilevered AFM tip resonates at 16.588 KHz. From the calibrated data (NT-MDT) the coefficient of elasticity of the soft cantilever probe is computed to be 0.06 N/m. The elasticity coefficient of rigid cantilever is determined as 48 N/m.

3.3 Mechanical characterization of Boron Nano Wire

The boron nanowire source is placed on the sample stage of the nanomanipulator. By the relative movement of the atomic force microscope probe on the z-θ stage and the sample on x-y stage, the AFM tip is brought into mechanical contact with the top of a single nanowire. When the tip is brought close enough to the BNW, it sticks to the AFM tip through Van der Waals forces. To ensure a rigid connection between the BWN and the AFM tip, amorphous carbon is deposited on the joint by electron beam irradiation deposition (EBID) technique for 15 minutes. This carbon originates from the contaminants present in the sample surface and the residual gases inside the SEM chamber.

As mentioned, two AFM tips are used to hold BNW on the nano manipulator stage. The rigid cantilever is gradually moved upwards by actuating the piezo actuator. Since the maximum deflection of the soft cantilever is less than five percent of the cantilever's length, the applied force (F) on the NW is obtained by measuring the cantilever deflection (d) and multiplying it by the cantilever elasticity coefficient (k) [10]. The rigid AFM cantilever (NSG20) requires 5 steps for 100 nm deflection and it serves as the force sensing element. The nanowire is compressed by increasing the axial compressive load (P) continuously. SEM images are recorded to analyze the compressive loading process and find out the critical compressive load (P_{cr}).

When the NW is in stable equilibrium it is straight. In the moving process of the cantilever, the NW remains straight or is slightly bent till the NW is in neutral equilibrium. The NW is in unstable equilibrium when the applied axial compressive load is greater than the critical load (P_{cr}) and buckles under the slightest disturbance. Figure 1.4 shows buckling of NW due to applied load. Figure 1.5 shows the load displacement curve of the BNW. The buckling instability is observed in the flat region of the curve. The critical compressive load is found to be 0.258 μN.

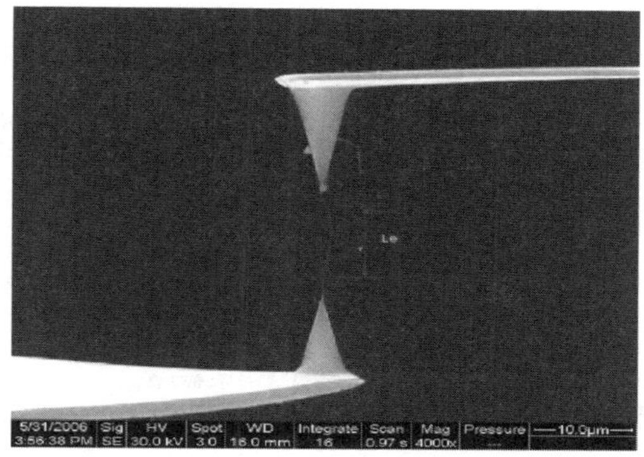

Figure 1.4. SEM image showing buckling of NW attached between two AFM cantilevered tips

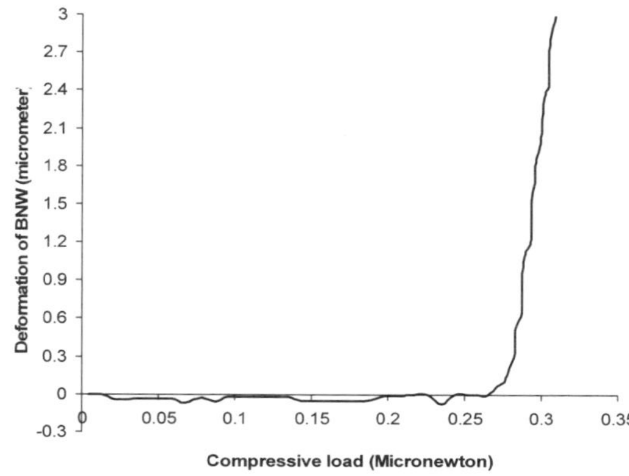

Figure 1.5. Compressive load versus deformation of

The critical load for an ideal elastic column of length L is given by

$$P_{cr} = \frac{\pi^2 EI}{(L_e)^2} , \qquad (1)$$

where E is Young's modulus, I is the Moment of Inertia and L_e is the effective length of the column and equal to 0.7L as shown in Figure 1.4.

The Young's modulus of the BNW is computed by rearranging the formula. The moment of inertia of the nanowire is calculated by considering the NW as an elliptical shape. Two axes of the elliptical shaped nanowire are measured with TEM and AFM micrograph. A mesh copper TEM grid is used to place nanowire inside the TEM which is shown in Figure 1.6(a). Figure 1.6(b) shows the image of a BNW with a width of 153 nm. The thickness of the oxide layer is 6 nm. The vertical distance of the NW is determined by AFM image scanning of the BNW. The AFM tip is moved to the midpoint of the NW. The cross sectional analysis of the NW is presented in figure 1.7 and height of the NW is determined to be 115 nm. The details of the parameters considered for the determination of Young's modulus of the two BNWs with the above mentioned technique are listed in Table 1.1. The Young's modulus is computed to be approximately 117±4.1 GPa.

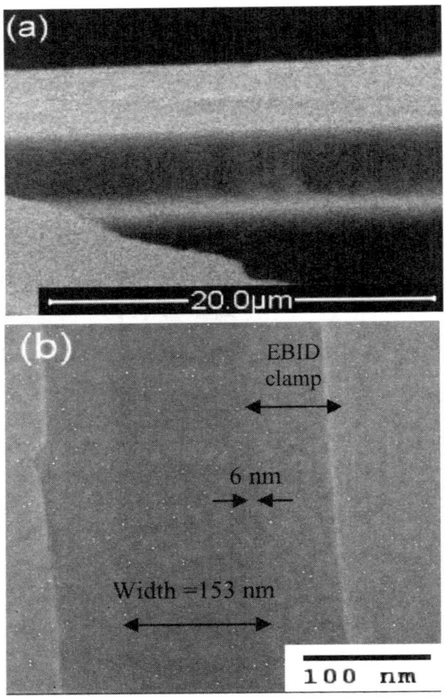

Figure 1.6. (a) BNW on copper net; (b) TEM image of BNW

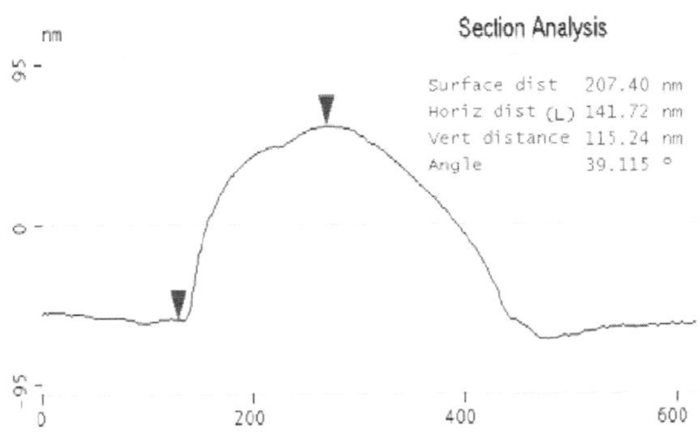

Figure 1.7. AFM analysis of BNW

Table 1.1: Young's modulus of two BNWs

Nano wire	1	2
Critical load (µN)	0.2586	0.2434
Length (µm)	13.563	12.863
Width (µm)	0.153	0.223
Height (µm)	0.115	0.098
MI (μm^4)	1.94×10^{-5}	1.75×10^{-5}
Young's modulus (GPa)	121.10	114.29

4. Conclusion

The Young's modulus of crystalline BNWs is studied with buckling method using a nanomanipulator inside a SEM. The NWs diameter is measured with TEM and AFM analysis. The coefficient of elasticity of crystalline boron nanowire measured with this nanomanipulator shows a high elasticity coefficient. The Young's modulus of the NW is in the range of between 73 ~ 105 GPa, depending on different cross sections.

This investigation confirms the capability of the nanomanipulator inside the SEM as a constructive device for measuring mechanical properties of nano scale materials. The nano scale material can be held in the three dimensional space from SEM real-time images without sacrificing positional resolution. This free-space three-dimensional manipulation is more advantageous than the two dimensional AFM manipulation. The AFM manipulation is not in real time and restricted to the sample surface whereas the nanomanipulator inside SEM is not limited by such type of restriction.

Acknowledgements

The authors gratefully acknowledge the support of the Center-of-Excellence Program on Membrane Technology, the Ministry of Education, Taiwan (project number 956049303), and the National Science Council of Taiwan, the Republic of China (project number NSC 95-2221-E-033-038).

References

[1] Chen LH, Jin S, 2003, Packaging of Nanostructured Microelectromechanical Systems Microtriode Devices, Journal of Electronic Material, Vol 32, No. 12, P1360-1365.

[2] Fukuda T, Arai F. et al. 2003. Assembly of Nanodevices with Carbon nanotubes through Nanorobotic Manipulations, Proc IEEE Vol. 91, No. 11, P1803- 1818.

[3] Tavadze F. N., Lominadze J. V., et al. 1981, The Effect of Impurities on the Mechanical Properties of Zone-Melted Boron, J. Less-Common Met., Vol. 82, P95-97.

[4] Wang D. W., Lu J. G., et al., 2003, Electrical Transport in Boron Nanowires, Applied Physics Letter, Vol. 83, No.25, P 5280- 282.

[5] Tally CP. 1959, Mechanical Properties of Glassy Boron, Jour. Applied Physics, Vol. 30, No.7, P 1114-1115.

[6] Xu Z.H.., Li X., 2006, Sample Size Effect on Nanoindentation of Micro/nanostructures, Acta Materiala, Vol.54, No.6, P 1699- 1703.

[7] Ding W., Calabri L. et al. 2006, Mechanics of Crystalline Boron Nanowires, Composites Science and Technology, Vol. 66, P 1112-1124.

[8] Otten C. J., Lourie O. R., et al., 2002, Crystalline Boron Nanowire, Jour. of American Chemical Society, Vol. 124, No. 17, P4564-4565.

[9] Yu M. F., Dyer M. J., et al. 1999, Three-Dimensional Manipulation of Carbon Nanotubes Under a Scanning Electron Microscope, Nanotechnology, Vol.10, No.3, P 244- 252.

[10] Ruoff R. S., Qian D., et al., 2003, Mechanical Properties of Carbon Nanotubes: Theoretical Predictions and Experimental Measurements, Comptes Rendus. Physique, Vol. 4, No.9, P 993-1008.

12

Precision Machine Design

12–1

Modelling and Simulation Considerations of a Micro-chuck Mechanism

Y.M. Chen[1], T.Y. Wu[2], S.H Tu[3], C.P. Chiang[4]

[1]Intelligent System Engineering Division, Industrial Technology Research Institute (ITRI), Taiwan, higherself@itri.org.tw
[2]Renewable Energy Division, Industrial Technology Research Institute (ITRI), Taiwan, jeffrey@itri.org.tw
[3]Renewable Energy Division, Industrial Technology Research Institute (ITRI), Taiwan, shunhsu@itri.org.tw
[4]Renewable Energy Division, Industrial Technology Research Institute (ITRI), Taiwan, cp.cshiang@itri.org.tw

Abstract. This paper addresses modelling and simulation considerations of chuck (friction grip) mechanism for high speed rotary cutting instrument shank (dental driller). The chuck (needle gripper) mechanism is embedded within a micro rotor-shaft casing. The shaft casing is a thin shell structure with dimension of 8 mm in length, 1.6 mm diameter and 0.5 mm shell thickness. High precision manufacturing is one of the dominant requirements in producing this dental equipment component. In the design for manufacturing case of such a micro mechanism, modelling and simulation become essential to ensure the precision requirements and the required chuck gripping force are satisfied. For this reason, accuracy of design analysis is vital. This work performs and compares stress analysis of micro-chuck gripping mechanism via both linear and non-linear theories. In addition, modelling and simulating the actual shank-gripping and releasing processes are performed by taking both plastic theory and contact mechanics into considerations. As a result, residual strain which degenerates the gripping force can be predicted. Moreover, the combination of non-linear finite element analysis and solid mechanics theory offers a promising environment for analyzing micro-chuck mechanism with precision manufacturing requirements.

Keywords: chuck, shank, dental driller, gripping force

1. Introduction

High speed rotary cutting instrument, often refered to as a dental drill, is one of the essential tools to a dentist. Typical dental drills operate at 200,000 to 400,000 revolutions per minute [1,2], the source of rotation power is based on converting highly compressed fluid into mechanical shaft work via a micro turbine. The shaft casing is a thin shell structure with dimensions of 8 mm in length (see Figure 0.1), 1.6 mm diameter and 0.5 mm shell thickness. The chuck (needle gripper) mechanism which is used to hold the dental needle (shank), is embedded within a micro rotor-shaft casing.

There are two chuck mechanisms commonly employed in a dental driller, namely taper and friction chucks (see Figure 0.2). The taper chuck mechanism design utilises a geometric taper section to produce the gripping force on the shank. The advantage of such a design is that the degeneration of the resulting gripping force is less likely to occur while in service. However, precision manufacturing has been one of the crucial steps so as to balance the dental drill; an unbalanced mechanism can cause the dental needle to displace from the shaft axis and results in irreversible imprecise cutting. As a result, damage could be significant since this dental tool is designed to operate at high speeds.

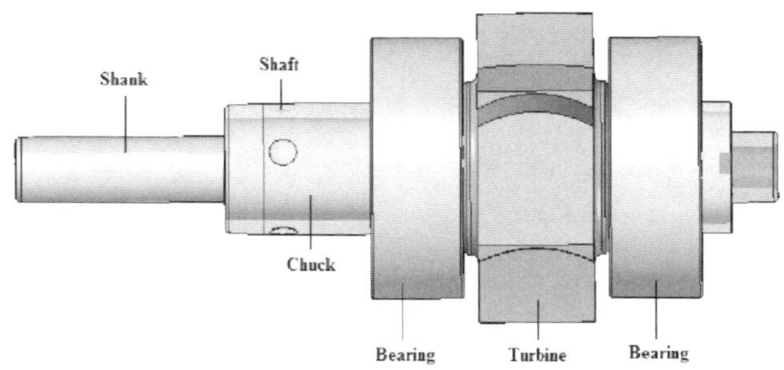

Figure 0.1. Dental driller shaft components

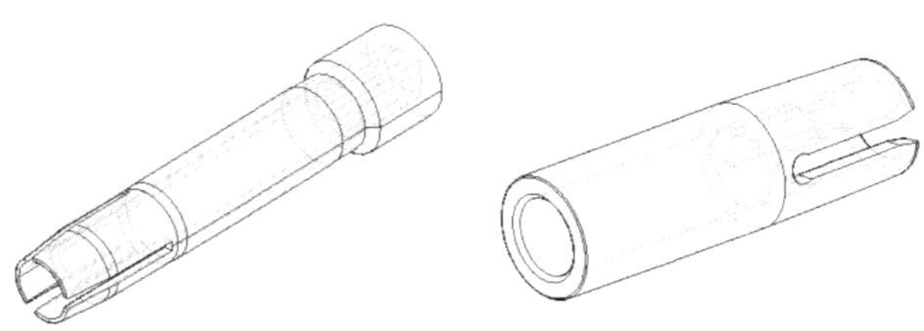

Taper Chuck | Friction Chuck

Figure 0.2. Chuck Mechanisms

The friction chuck mechanism is based on holding the dental drill with a circular section. The advantage of this design is that the shank can be easily balanced with the shaft axis. But the disadvantage of this design is that the design gripping force tends to degenerate if manufacturing tolerances are not probably monitored and controlled, especially when dealing with micro scale components.

This work investigates a friction chuck mechanism. In the design for manufacturing scenario, understanding the reasons for the degeneration of the gripping force is vital to designers. By taking the gripping force into consideration, the designer can ensure its operational efficiency and extend its service life.

2. Friction Chuck Mechanism

The friction chuck mechanism is a thin-shell cylindrical-like structure with two sections of different radii (see

Figure 0.3), the front section is with a larger radius *R1*, and the rear is smaller in section with radius *R2*. The dimension is 8 mm in length, 1.6 mm inner shell diameter and the shell thickness is 0.5 mm. The outer shell is welded to the shaft casing and their interface is at the front section of the friction chuck mechanism. The rear section is smaller so as to provide adequate clearance between the rear section of the friction chuck mechanism and the shaft casing. The difference between R1 and R2 is defined in this paper as *Friction Contact Thickness* (FCT). It should be pointed out here that design of the gripping force is directly related to the amount of FCT i.e. the gripping force increases with FCT.

$$R1 - R2 = FCT$$

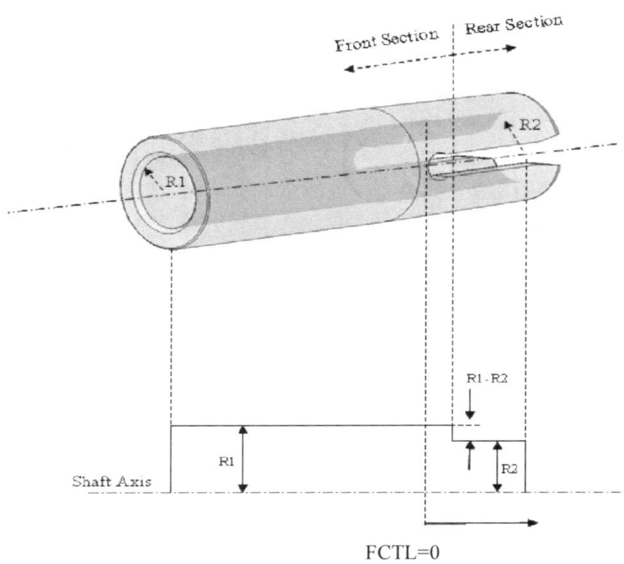

Figure 0.3. Friction chuck mechanism

The dental needle (shank) radius is the same as R1, i.e., the radius of the front end of the friction chuck mechanism. When the shank is inserted into the friction chuck mechanism, the front end has the function of holding the shank in alignment with the shaft axis, the rear section however, due to difference in section area, would be forced to expand. This expansion of the rear section must be tolerated by the clearance exist between the shaft casing. The expanded rear section thus provides gripping force to the shank. The magnitude of the gripping force in this case is thus equal to the product of the contact normal force and the friction coefficient. This is often referred to as Coulomb friction force, in this case, it is also the gripping force generated by the friction chuck mechanism. Another design parameter that was considered during the gripping force design is the location of the FCT. In this paper, this design parameter is referred to as the *Friction Contact Thickness Location* (FCTL) and it is measured from the location of the hole as shown in Figure 0.3.

According the International Organization for Standardization (ISO) specification, one class of dental needles needs to have a radius designed and manufactured between 0.795 mm and 0.8 mm, i.e. with a manufacturing tolerance of 5 µm. For this type of dental needle, the required needle gripping force must be designed to be greater than 2.2 kilogram force (Kgf) and torque greater than 1.6 Kgf-mm. To attain the listed specifications, design of friction chuck mechanism becomes a crucial task. To ensure the required gripping force is maintained at the specified level during the service life by considering over thousands of uses of this dental equipment and lowering the cost on precision manufacturing. One of the important considerations in designing friction chuck mechanisms for manufacturing is that consideration of the manufacturing tolerances must be taken into account, especially when the dimensions of this dental equipment is micro-scale based, slight variations to the millimetre could alter the structure behaviours and subsequently fail to attain the design requirements.

In this study, commercial finite element package ABAQUS was employed. Finite element models of both the friction chuck mechanism and a shank were constructed for simulation purpose to study the actual process of inserting a shank into a chuck. The geometries were constructed using CAD and imported into ABAQUS for finite element analysis. Due to symmetry, only a quarter of the geometry was analyzed. In this modelling, the following assumptions have been made: (a) the material of friction chuck mechanism has been assumed to behave plastically and hence plastic strain is allowed and set to 0.2% offset; (b) the point of contact between the shank and chuck, i.e., the line separating the front and rear section shown in

Figure 0.3 initially set to zero displacement, but allows micro displacement to compensate for the expansion behaviour of the rear section, and (c) the surface of shank is defined as primary contact interface and the inner surface of chuck as the second contact interface. It should be pointed out here that material properties of both shank and chuck are as follows: E = 20000 kg/mm^2, ν = 0.28, σ_y = 133.6 kg/mm^2 and friction coefficient of 0.1.

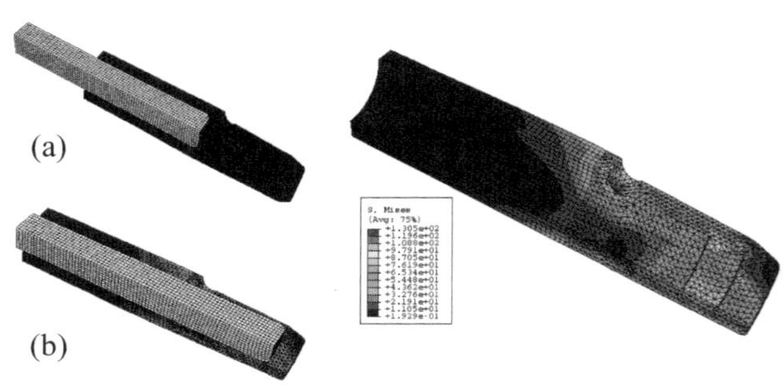

Figure 0.4. Simulation of Shank and Friction Chuck

Figure 1.4 depicts the simulation of inserting a shank into the chuck; Fig.1.4(a) shows the initial position where shank position is still on the left of the line where FCTL = 0, and Fig.1.4(b) shows the shank has caused the rear section of the chuck to expand, thus generating gripping force onto the shank. The simulation result can be considered as a good indicator of the validity of the constructed finite element model. With this confidence in the finite element model, the relationship between chuck gripping force and various parameters was investigated and the results described in the following sections.

2.1 Variation of FCTL vs. Gripping Force

In this section, the relationship between the FCTL and gripping force is investigated. Based on ISO standard, it is assumed that the radius of the shank to be 0.8 mm. Hence in this analysis of the friction chuck mechanism, the radius, R1, is defined as 0.8 mm and by presetting R2 to be 0.75 mm, FCT is equal to 0.05 mm. Several values of FCTL are studied as shown in Figure 0.5. It should be pointed out here that due to double symmetry in the cylindrical-like structure, the analysis is based one quarter of a cylinder-like structure hence the magnitude of the gripping force needs to be scaled by a factor of four in order to reflect the full-scale result.

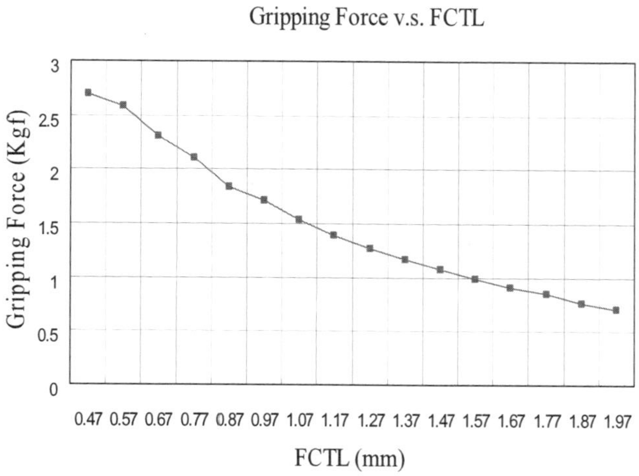

Figure 0.5. Relationship between gripping force and FCTL.

ISO standards require the friction chuck mechanism to be designed with a gripping force of not less than 2.2 kgf, one fourth of 2.2 kgf is 0.55 kgf. Figure 0.5 shows that the relationship between gripping force and FCTL. It should be clear that the gripping force reduces with FCTL shift further backwards. At the cut-off line, i.e., gripping force is 0.55 kgf, the FCTL is greater than 2 mm away from line marked FCTL=0 (see Figure 0.5). In other words, when FCTL is at 2 mm, the magnitude of the gripping force can still meet the ISO specification.

From a design point of view, setting the FCTL as far back as possible is a desirable consideration because sufficient length of the front section (see Figure 0.3) is needed to hold the shank steady in aligning with the shaft axis. However, setting the FCTL backwards indicates that the length of the friction chuck needs to be longer which incurs additional weight and cost. For this reason, ineffective materials can be removed from friction chuck. However, removing unwanted materials from the friction chuck belongs to the problem of shape optimisation, the shape of holes or notch needs to be carefully investigated to avoid stress concentration and at the same time, ensuring that the required gripping force is above the acceptable level.

Another problem is that most designs of dental drills would need to consider is the degradation of gripping force due to both the residual stress and plastic strain. Consider a very common case when dentists, for surgical reason, need to replace different kind of drills. Over the service life of dental hand-piece, replacement of the drill needs to be taken into the consideration when designing the chuck mechanism to ensure that, due to frequent insertion of the shank, the gripping force does not degenerate below the acceptable limit.

2.2 Residual Stress vs. Gripping Force

In order to study the effect of residual stress on the degeneration of the gripping force, circular holes, with radius of 0.3 mm, were cut out next to the original hole. The purpose is to investigate the residual stress distrubution and also study the effect of removing additional weight from the chuck component on the residual stress distribution.

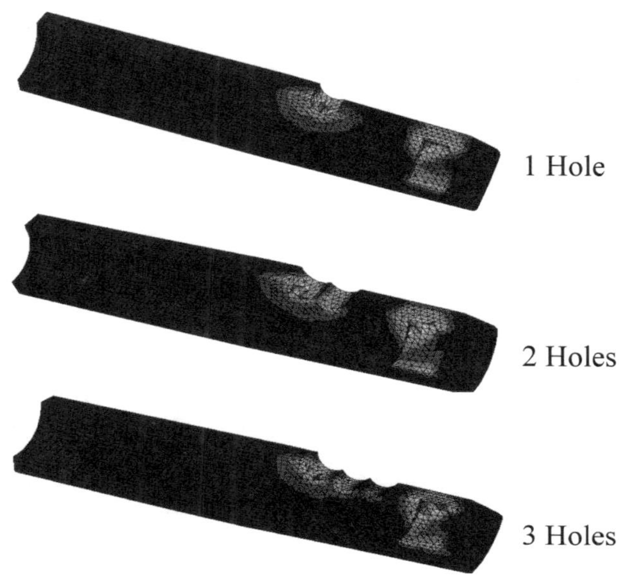

Figure 0.6. Residual stress distribution vs. notches

Figure 0.6 depicts the residual stress distribution of the friction chuck mechanism versus number of holes. It can seem that the distribution of residual stress tends to concentrate on the first hole; the second and third hole do not seem to alter the distribution much. This simply indicates that additional weight can be removed from the friction chuck component subject to the condition that the degeneration of gripping force resulted from residual stress is within the acceptable limit.

To further explore this possibility, the degeneration behaviour of the gripping force and the frequency of shank insertion are studied. In this study, it is not the magnitude of the actual gripping force, but rather the trend of degeneration. Since the actual magnitude of gripping force can be designed with carefully tuning various parameters, such as the FCT, FCTL and perhaps also considering removing additional material, shape and/or topology of the holes etc. In other words, the magnitude of the gripping force would be meaningful if an optimisation considers all parameters. For example, one could attempt to define an optimisation with the objective function of minimizing the gripping force with the constraint that the minimum gripping force has to be above 2.2 kgf (or 2.2 kgf multiplied by a safety factor). With this objective function, the design variables will permit FCT and FCTL to be varied. In addition, one could also consider geometry optimisation of the cut-outs sections to further reduce the structural weight. Alternatively, one could treat this as a geometry shape optimisation problem for the rear cut-outs section and define objective function aimed to minimise the weight of the friction chuck, subject to the constraint that the gripping force, FCT and FCTL are within their prescribed limits. Only when all the design variables are considered and optimised, meaningful results will be obtained.

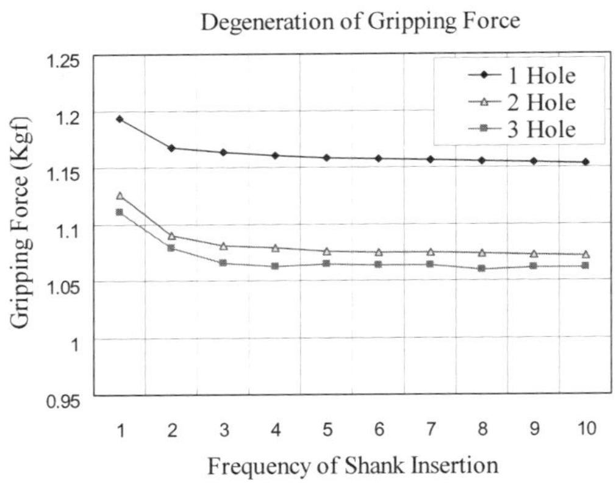

Figure 0.7. Gripping force degeneration behaviour

Figure 0.7 shows how the gripping force degenerates with the frequency of insertion. The behaviour is similar for all three different geometry cut-outs profiles. The degeneration of the gripping force reaches a steady value after the third insertion. With only one hole in the rear section, the percentage of gripping force degeneration after the third insertion is just under 3%. Whereas with two and three holes, degeneration of the gripping force after the third insertion is just under 6%. Clearly, removing more materials tends to increase the degeneration rate of gripping force.

3. Concluding Remarks

The design and modelling of friction-chuck mechanism has been considered in this work. Commercial finite element package ABAQUS has been employed to study the interaction between shank and chuck. The ISO specification indicates the required gripping force for friction-chuck mechanism needs to be greater than 2.2 kgf. In this work, the relationship between chuck gripping force and the FCTL has been investigated and it has been noticed that the gripping force degenerates with an increase in FCTL. In addition, to simulate frequent replacement of dental drill during product service life, a study on the degeneration of gripping force behaviour versus frequency of drill replacement has been performed. It was found the after the third insertion of shank, the degeneration of gripping force reaches a steady state value. However, if one considers a minimum weight design, it should be noticed that the behaviour on removing more material would tend to increase the degeneration rate of the gripping force. Care must be taken into consideration while designing friction-chuck mechanisms. In this work, several behaviours of the friction chuck have been studied and the task of refining the design parameters such as the FCT, FCTL, minimum weight design potentials, and notch shape design have been suggested for further work using an optimization approach.

Acknowledgement

The financial support from the National Science Council (NSC) of Taiwan to the 35th International MATADOR conference is gratefully acknowledged.

References

[1] Dyson, J. E., and Darvell, B. W., (1993), The development of the dental high speed air turbine hand piece Part 1, Aust. J., 38(1), 49-58.
[2] Dyson, J. E., and Darvell, B. W., (1993), The development of the dental high speed air turbine hand piece Part 2, Aust. J., 38(1), 131-144.
[3] International Organization for Standardization, (1985) Dental rotary instrument – number coding system-part 1: General Characteristics, International Standard, ISO 6360-1.
[4] International Organization for Standardization, (1986) Dental rotary instrument - part 2: shape and specific characteristics, International Standard, ISO 6360-2.
[5] International Organization for Standardization, (1985) Dental rotary instrument – shanks, International Standard, ISO 1797.
[6] International Organization for Standardization, (1986) Dental rotary instrument - part 1: steel and carbide burs, International Standard, ISO 3823-2.
[7] ABAQUS Version 6.6 Documentation.

12–2

Prediction of Machining Accuracy of 5-Axis Machine Tools with Kinematic Errors

M. Sharif Uddin[1], S. Ibaraki[2], A. Matsubara[3], and T. Matsushita[4]

[1]Kyoto University, Kyoto, Japan, 606-8501, ms_uddin@preceng.mbox.media.kyoto-u.ac.jp
[2]Kyoto University, Kyoto, Japan, 606-8501, ibaraki@prec.kyoto-u.ac.jp
[3]Kyoto University, Kyoto, Japan, 606-8501, matsubara@prec.kyoto-u.ac.jp
[4]Okuma Corp., Aichi, Japan 480-0193, te-matsushita@gmx.okuma.co.jp

Abstract. Kinematic errors due to geometric inaccuracies in 5-axis machine tools cause deviations in tool positions and orientation from commanded values, which consequently affect geometric accuracy of the machined surface. The present research work studies the prediction of machining accuracy of a 5-axis machine tool with its kinematic errors. First, kinematic errors associated with linear and rotary axes of a 5-axis machine tool with tilting rotary table types, are identified by a DBB method. By using an error model of the machine tool, erroneous tool position and orientations are computed. Then, machining error with respect to the nominal geometry is predicted and evaluated. With the aim to improve the geometric accuracy of a machined surface, an error compensation for tool position and orientation is also presented. As an example, the machining of a tilted taper cone by using a straight end mill, as described in the standard NAS979, is considered in a case study to experimentally verify the prediction and the compensation of machining errors in 5-axis machining.

Keywords: 5-axis machining, kinematic errors, geometric inaccuracy, DBB, error compensation

1. Introduction

With a tremendously increasing need for machined components with geometric complexity and high dimensional accuracy, 5-axis machine tools are extensively used in the manufacturing of dies and molds, and aerospace parts. Because of having its versatile functionalities, a 5-axis mache tool offers notable benefits including increased material removal rate, enhanced surface accuracy, and reduced effective machining time. However, it is well known that kinematic errors due to geometric inaccuracies of structural components in a 5-axis machine tool can cause significant errors on the tool position and the orientation with respect to workpiece, hence leading to a geometric inaccuracy of machined surface in actual cutting operation.

In the past, numerous research works has been carried out to realize kinematic errors, and to simulate and improve motion accuracy of 5-axis machine tools. These works attempted to mainly build up an error model with kinematic errors, and then to predict the motion accuracy of the machine tool (e.g. [1]). There are also many works reported on the identification of kinematic errors on a 5-axis machine tool. Among them, the schemes using the telescopic double ball bar (DBB) measuring device have been widely accepted (e.g. [2]). More recently, a 3D probe ball device [3] is also used to identify some link errors of a 5-axis machine tool.

While the research work reviewed above has focused on the measurement of kinematic errors and the evaluation of motion accuracy of machine tools, no or little work is available in the literature, focusing on the prediction of machining accuracy with kinematic errors in 5-axis machine tools. Unlike conventional 3-axis machines, it is very difficult to intuitively understand the effect of kinematic errors on a 5-axis machine on the machining accuracy, since the motion of a 5-axis machine is generally quite complex. In the industry of 5-axis machine tool builders, the NAS979 standard [4], the machining of a tilted taper cone, is widely known and is often conducted to investigate the machining accuracy of a 5-axis machine tool. However, it is generally quite difficult to diagnose error sources of the machine from error profiles of the machined taper cone. As a basis to establish such a diagnosis methodology from machining results, and to implement a compensation scheme of the machine's motion errors, it is important to understand the effect of kinematic errors on the machining error.

The present research work proposes a simulator of machining errors in 5-axis machining with the effects of kinematic errors due to geometric inaccuracies of 5-axis machine tools. Three squareness errors associated with linear axes and eight kinematic errors associated with rotary axes are evaluated and considered in the present study. At first, these total eleven kinematic errors for a 5-axis machine tool of tilting rotary table type are practically identified by the telescopic DBB measuring device. Then, machining errors with respect to the nominal geometry are predicted and evaluated to realize the effects of kinematic errors. In order to enhance geometric accuracy of machined surface, an error compensation for tool position and orientation is also presented. As an example, the machining of a tilted taper cone by using a straight end mill, as described in the standard NAS979, is considered in case studies to experimentally verify the prediction and the compensation of machining errors in 5-axis machining.

2. Kinematic Errors in 5-axis Machine Tools

2.1 Definition of Kinematic Errors

In this study, a 5-axis machine tool with a tiltling rotary table is considerd as the target. Figure 1.1 shows the basic configuration of the machine tool. The machine contains three linear axes (**X**, **Y**, **Z**) for generating linear motions in *X*, *Y*, and *Z* directions and two rotary axes (**A**,**C**) for generating rotary motions on the tilting-rotary table about *X* and *Z* axes respectively.

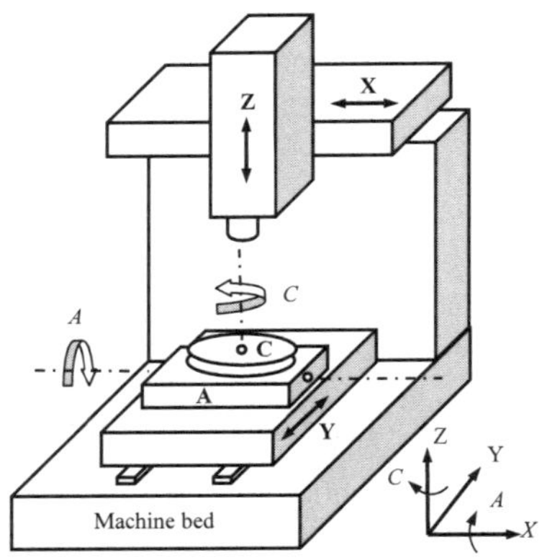

Figure 1.1. 5-axis machine tool with tilting rotary table

Inasaki et al. [5] have pointed out that there are 11 kinematic errors associated linear and rotary axes in this type of 5-axis machine tool; namely eight kinematic errors (α_{AY}, β_{AY}, γ_{AY}, β_{CA}, δx_{AY}, δy_{AY}, δz_{AY}, δy_{CA}) associated with rotary axes and three kinematic errors (γ_{XY}, α_{YZ}, β_{ZX}) associated with linear axes. The basic definitions of these kinematic errors are as follows; α_{AY}: the angular error of **A**-axis with respect to **Y**-axis about *X*-axis. Similarly, β_{AY} and γ_{AY} are angular errors of **A**-axis about *Y* and *Z*-axes respectively. β_{CA}: is the angular error of **C**-axis with respect to **A**-axis about *X* axis. δx_{AY}, δy_{AY}, and δz_{AY} are linear shifts of **A**-axis from **Y**-axis in *X*, *Y* and *Z* directions respectively. δy_{CA} is the linear shift of **C**-axis with respect to **A**-axis in *Y* direction. γ_{XY}, α_{YZ}, and β_{ZX} are defined as the perpendicularity or squareness errors of three linear axes (**XYZ**) on *X-Y*, *Y-Z*, and *Z-X* planes respectively. These eleven kinematic errors will be considered in error modeling of the machine tool and then, effects of them on machining accuracy will be evaluated as well.

2.2 Identification of Kinematic Errors

In this study, a DBB measurement method is applied to identify the above kinematic errors considered. The DBB method already specified in ISO230-1 has recently been applied to the measurement of linear and angular kinematic errors existing in 5-axis machine tools. A number of DBB tests with specific measurement patterns proposed by Tsutsumi and Saito [6] are conducted to identify kinematic errors. The details of the measurement system and procedures for estimating the kinematic errors can be found in [6].

3. Error Modeling of 5-axis Machine Tool

According to the coordinate system and kinematic chain of the target machine tool as shown in Fig. 1.1, an error model is established. A brief description of the error model is as follows:

Consider an initial coordinate of the tool tip, $p_t^0 = [0001]^T$, and for a command tool path trajectory the position $(\hat{X}_c \hat{Y}_c \hat{Z}_c)$ and orientation $(\hat{A}_c \hat{C}_c)$ of the tool tip in reference coordinate. The position of the tool tip with kinematic errors associated with a liner axes, ${}^r p_t$ can then be obtained by:

$${}^r p_t = {}^r T_t p_t^0$$

where, ${}^r T_t$ is a 4x4 erroneous homogeneous transformation matrix (HTM) due to kinematic errors, and can be expressed as:

$${}^r T_t = D^1(\hat{X}_c) D^6(\gamma_{XY}) D^2(\hat{Y}_c) D^4(\alpha_{YZ}) D^5(\beta_{ZX}) D^3(\hat{Z}_c)$$

where, $D^{1\text{-}6}(*)$ represent the 4x4 HTMs for purely linear and rotary motions. The left-side superscript *r* denotes a vector in the reference coordinate, while right-side subscripts *t* and *c* indicate the tool tip's and command tool tip's locations respectively.

Then, by taking into account 8 kinematic errors associated with rotary axes, the erroneous position of the tool tip in workpiece coordinate, ${}^w p_t$ can be obtained as:

$${}^w p_t = ({}^r T_w)^{-1} ({}^r p_t)$$

where, ${}^r T_w$ is a 4x4 erroneous HTM due to kinematic errors, and can be expressed as:

$${}^r T_w = {}^r T_a {}^a T_c$$

The HTMs, ${}^a T_c$ and ${}^r T_a$ can be obtained as:

$${}^a T_c = D^2(\delta y_{CA}) D^5(\beta_{CA}) D^6(\hat{C}_C)$$

$${}^r T_a = D^1(\delta x_{AY}) D^2(\delta y_{AY}) D^3(\delta z_{AY}) D^4(\alpha_{AY}) D^5(\beta_{AY}) D^6(\gamma_{AY}) D^4(\hat{A}_C)$$

Hence, the matrix ${}^w p_t$ contains tool positioning errors in workpiece coordinate and will generate erroneous tool position due to existing kinematic errors in a 5-axis machine tool.

4. Computation of Machining Errors

By using the above error model, once the erroneous tool position and orientation for a tool path trajectory is given, the next step is to compute machining error on workpiece surface. In this study, as an example, 5-axis machining of a tilted taper cone (NAS979) is assumed to be carried. Hence a simple and easy procedure to compute machining errors with respect to nominal geometry of the taper cone workpiece surface is presented. The detailed algorithm for computation of machining errors is given as follows:

(1) Given a set of points on a nominal surface of workpiece, $W_i (i = 1, \ldots, N_w)$, let's define the vectors normal to those points of the surface, $\vec{v}_{w,i} (i = 1, \ldots, N_w)$ as can be shown in Fig. 1.2.

(2) Let's consider a plane made by the tool axis vector, $\vec{v}_{t,j}$ and the tool motion direction vector, $\vec{M}_{t,j}$ for two successive points on the tool path trajectory,

$p_j^w (j = 1,...,N_p)$. Then the plane is shifted toward the nominal workpiece surface by an amount of tool radius, r (see also Fig. 1.2).

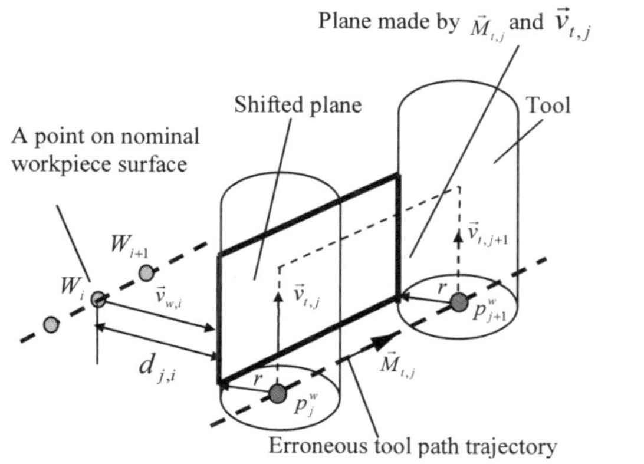

Figure 1.2. Algorithm for computation of machining error

(3) For a point, p_j^w on the tool path trajectory, calculate the distance, $d_{j,i}$ between the shifted plane and a point, W_i on the nominal workpiece surface along the normal vector, $\vec{v}_{w,i}$.

(4) For a point, p_j^w on the tool path trajectory and all the points on the nominal workpiece surface, $W_i (i = 1,...,N_w)$, similarly calculate the distances, $d_{j,i}$. Then find the minimum distance, $\min d_{j,i}$. This is the machining error on the nominal workpiece surface for a point, p_j^w on the tool path trajectory.

(5) Similarly for all the points on the tool path trajectory, $p_j^w (j = 1,...,N_p)$, repeat the steps (2)-(4). This will lead to the whole computation of the machining errors on the nominal workpiece surface.

5. Case Studies

In this work, two case studies are demonstrated to evaluate the effects of kinematic errors and hence, to justify the prediction of machining errors in 5-axis machining. As an example, the machining of a tilted taper cone workpiece as specified in NAS979 is considered in this study. Figure 1.3 shows the 5-axis machining configuration and different parameters of the taper cone workpiece. In Fig. 1.3, D, θ, and φ are defined as the diameter of tool path trajectory in workpiece coordinate, taper angle, and tilt angle about X-axis of the taper cone respectively, while (C_x, C_y, C_z) is the center location of workpiece on the work-table. H_t and H_b are heights of the taper cone and base cylinder.

5.1 Simulation and Experimental Results

The machining test is conducted by using a commercial 5-axis machine tool with the configuration shown in Fig. 1.1. First, kinematic errors of this machine are experimentally identified by using the methodology presented in Section 1.2.2. Table 1.1 summarizes the identified kinematic errors.

For case study I, machining errors on the bottom and top surface of the taper cone are simulated with the following machining conditions: D=210mm, θ=30°, φ=15°, (C_x,C_y,C_z)=(0,-100mm,53mm), H_t=20mm, and H_b=30mm. Then, using the above machining parameters of the taper cone, actual cutting test with a straight end mill is carried out. Aluminum alloy, a soft material, is used as the taper cone workpiece. Spindle speed of 5500min^{-1}, feed rate of 1000 mm/min, and tool radius of 10 mm are used in cutting test. Conservative feed rate and soft workpiece material are chosen to reduce effects of cutting process (e.g. cutting force, tool deflection, etc) on machining accuracy. After cutting test, surface measurements taken by a roundness measuring device (Talycenta1000) are carried out to measure machining errors on bottom and top surfaces of the machined taper cone workpiece surface.

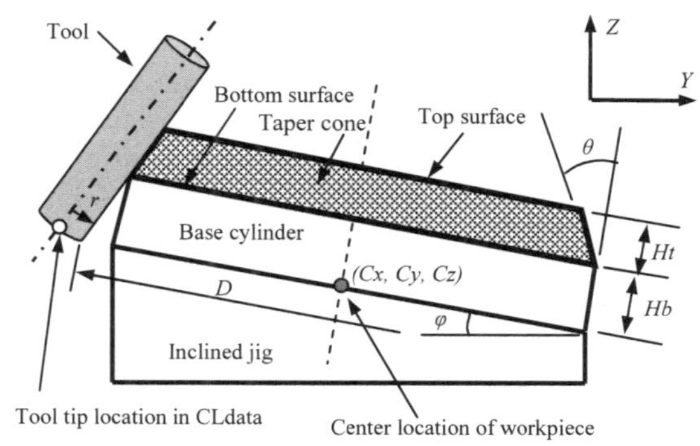

Figure 1.3. Machining configuration and parameters of taper cone

Table 1.1. Identified kinematic errors by DBB tests

8 kinematic errors with rotary axes		**3 squareness errors with linear axes**	
Error parameter	value	Error parameter	value
α_{AY}	+0.0001°	γ_{XY}	-0.0034°
β_{AY}	-0.0071°	α_{YZ}	+0.0054°
γ_{AY}	-0.0081°	β_{ZX}	+0.0037°
β_{CA}	+0.006°		
δx_{AY}	+0.10 μm		
δy_{AY}	- 4.0 μm		
δz_{AY}	+2.8 μm		
δy_{CA}	+3.4 μm		

Figure 1.4 shows simulated and measured machined surface trajectories for bottom and top surface of the taper cone with respect to its nominal surface. It is seen from Fig. 1.4 that kinematic errors in 5-axis machine tool have significant effects on machining accuracy. Also, simulated surface trajectories well agree with measured surface trajectories.

Case study II is carried out at another set of parameters of the taper cone and center location of workpiece, which are as follows: D=206.4mm, θ=30°, φ=75°, (C_x,C_y,C_z) = (0,-103mm,93mm), H_t=20mm, and H_b=30mm. Values of feedrate and tool radius are used as same as those in first case study but spindle speed of 5000min^{-1}. Figure 1.5 compares simulated and measured machined surface trajectories with respect to nominal surface of the taper cone. In this case, simulated results are also found to have a good match with measured ones. Further, a comparison of simulated and measured circularity errors for both case studies is summarized in Table 1.2.

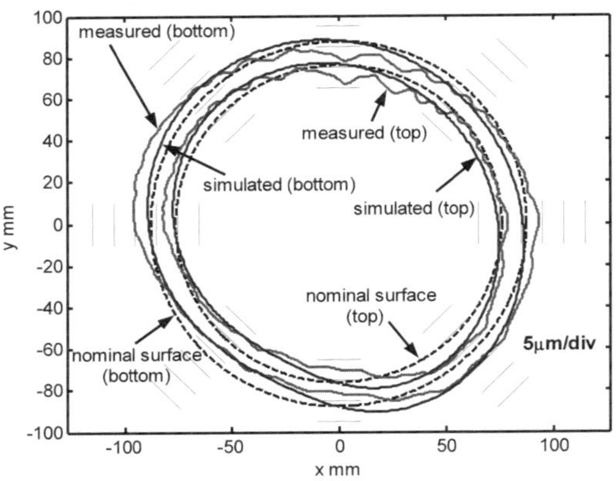

Figure 1.4. Simulated and measured machined surface trajectories for case study I

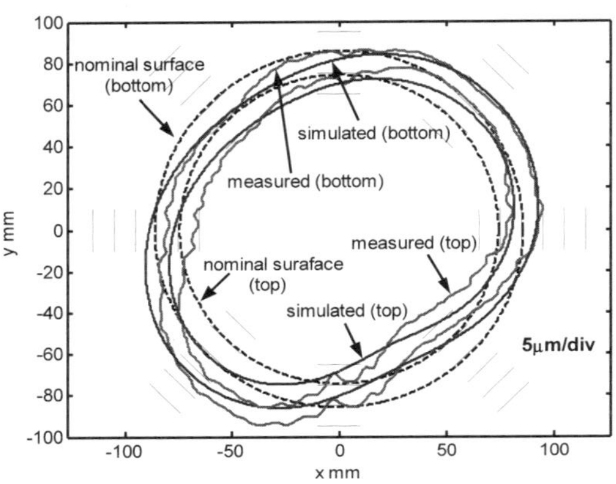

Figure 1.5. Simulated and measured machined surface trajectories for case study II

Table 1.2. Summary of circularity errors for both case studies

Circularity errors	Case study I		Case study II	
	simulated	measured	simulated	measured
Bottom (μm)	7.2	9.4	13.3	17.3
Top (μm)	6.7	8.8	13.0	19.5

5.2 Error Compensation

With an aim to improve machining accuracy, an error compensation for tool position and orientation is carried out. First, an erroneous tool path due to kinematic errors is calculated. Error values in tool position and orientation between erroneous and nominal tool paths are obtained. Then, a compensated tool path is calculated by simply canceling error values of tool position and orientation from its nominal tool path. In this work, the presented error compensation is demonstrated for the case study II. Cutting test for error compensation is carried out using the same machining condition in case study II. Figure 1.6 shows compensated machined surface trajectories obtained from experiments. It is found that, by error compensation, circularity errors are reduced from 17.3 μm to 10.1 μm at bottom and 19.5 μm to 11 μm at top surfaces of the machined taper cone, and hence machining accuracy is improved significantly.

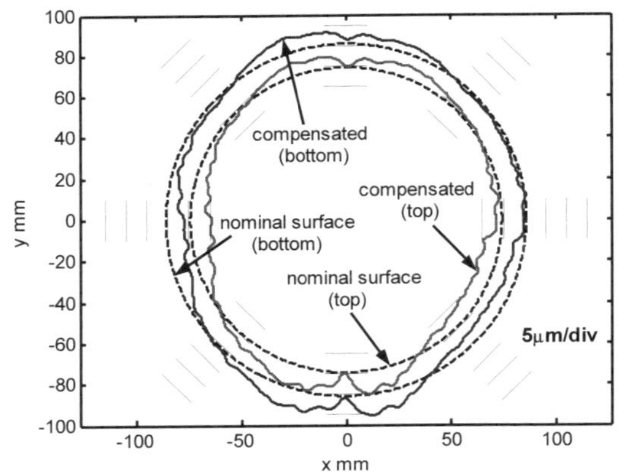

Figure 1.6. Compensated machined surface trajectories for case study II

6. Conclusion

In this paper, machining accuracy of a 5-axis machine tool with kinematic errors is predicted and evaluated. By using an error model with kinematic errors, machining errors on a standard tilted taper cone (NAS979) workpiece is simulated. Case studies with cutting experiments on a target 5-axis machine tool verify the effectiveness of prediction of machining accuracy in 5-axis machining. An error compensation for tool position and orientation is demonstrated. Experiments for error compensation show a significant improvement in machining accuracy in terms of reduction of circularity errors of the machined surface.

References

[1] S.H. Suh et al. 1998, Error Modeling and Measurement for the Rotary Table of Five-Axis Machine Tools, Int. J. of Adv. Manuf. Tech., Vol.14, P656-663

[2] S. Sakamoto, I. Inasaki, 1994, Identification of Alignment Errors in Five-Axis Machining Centers, Trans. of JSME, Vol.60(C-575),P2475-2483.

[3] W.T. Lei, Y.Y. Hsu, 2002, Accuracy Test of Five-Axis CNC Machine Tool with 3D Probe-Ball. Part II: Errors Estimation, Int. J. of Mach. Tools and Manuf., Vol.42, P1163–1170.

[4] National Aerospace Standard, Uniform Cutting Tests-NAS (National Aerospace Standard) Series: Metal Cutting Equipment Specifications, National Standards Association, Washington DC, USA.

[5] I. Inasaki, K. Et al. 1997, Shaper Generation Theory of Machine Tools – Its Basis and Applications, Yokendo, Tokyo, P95-103 (in Japanese).

[6] M. Tsutsumi, A. Saito, 2003, Identification and Compensation of Systematic Deviations Particular to 5-Axis Machining Centers, Int. J. of Mach. Tools & Manuf., Vol.43, P771-780.

Active Aerostatic Bearings for Ultraprecision Applications

H. Mizumoto[1], S. Arii[2], Y. Yabuta[3], Y. Kami[4] and Y. Tazoe[5]

[1] Tottori University, Koyama-Minami Tottori, Japan, mizu@ike.tottori-u.ac.jp
[2] Tottori University, Koyama-Minami Tottori, Japan, arii@ike.tottori-u.ac.jp
[3] Tottori University, Koyama-Minami Tottori, Japan, yabuta@ike.tottori-u.ac.jp
[4] Nachi-Fujikoshi, Okake Namerikawa Toyama, Japan, ykami@po.hitwave.or.jp
[5] Nachi-Fujikoshi, Okake Namerikawa Toyama, Japan, ytazoe@po.hitwave.or.jp

Abstract. For ultraprecision applications, two types of active control restrictors for aerostatic bearings are evaluated. The first restrictor is called an aerostatically controlled restrictor (ACR), a bearing-pressure-feedback type. The change in the pressure on the bearing surface directly controls the restriction gap, and the stiffness of the aerostatic bearing with the ACR can be infinite. The second restrictor is called an active inherent restrictor (AIR), a bearing-gap-feedback type. The change in the air-gap on the bearing surface is detected by an electric sensor, and then an electric actuator controls the restriction gap. By using the AIR, infinite bearing stiffness and nanometer-order of rotational accuracy of the spindle can be realized. Positioning with sub-nanometer order of resolution can also be executed by an aerostatic guideway with the AIR.

Keywords: active control, aerostatic bearing, air spindle, guideway, machine tool, positioning, rotational accuracy

1. Introduction

Aerostatic lubrication is one of essential technologies in the field of precision engineering and nanotechnology. Spindles and tables of ultraprecision machine tools and measurting machines are supported and guided by aerostatic bearings and guideways. However, owing to the limitation of the air supply pressure available, the bearing stiffness is relatively low, and the rotational accuracy of the aerostatic bearing can not be high enough for ultraprecision applications.

For improving the performance of aerostatic bearings, we proposed several active control systems [1-5]. In the present paper, we compare and reevaluate two of these active control systems. One system employs a bearing-pressure-feedback type active restrictor called an aerostatically controlled restrictor (ACR) [1,2], and another system employs a bearing-gap-feedback type called an active inherent restrictor (AIR) [4]. Load characteristic and motion accuracy of the aerostatic bearings with these active restrictors are analyzed from the ultraprecision viewpoint. The analyses show that the infinite bearing stiffness can be realized by both active control systems. However, nanometer order of rotational accuracy and the positioning with sub-nanometer order of resolution can only be executed by active aerostatic bearings with the AIR, the active restrictor of a bearing-gap-feedback type.

2. Actice aerostatic bearing with aerostatically controlled restrictor (ACR)

Figure 1.1 shows the active aerostatic thrust bearing using an air-film restrictor [2]. Its thrust bearing surface A is divided into two regions, and the inner region with large air gap acts as the air-film restrictor. Pressurized air is supplied to the inner region of A via the air inlet D. Air pressure drops slowly on the inner region, and quickly on the outer region of the bearing surface A. For the active control, the restriction ring (6) whose side is used as the inner region of the bearing surface is elastically supported by another aerostatic bearing B (control-bearing) incorporated. The restriction ring is moved by the applied load to the spindle and the air gap on the inner region varies, while the air gap on the outer region can be constant. Thus, the increasing and decreasing of the applied load is compensated by the change in the pressure distribution on the bearing surface. The restriction ring, the key-element of this mechanism, is supported by the aerostatic control-bearing B, therefore, this active control system of the bearing-pressure-feedback type is called an aerostatically controlled restrictor (ACR).

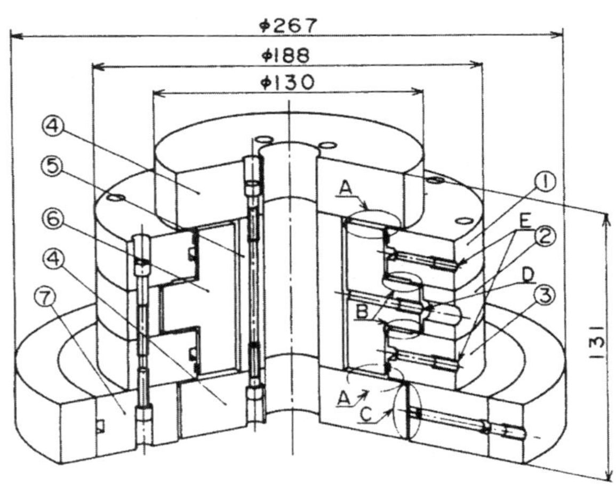

Figure 1.1. Active aerostatic thrust bearing with aerostatically controlled restrictor [2], 1: outer ring 1,2: outer ring 2, 3: outer ring 3, 4: thrust plates, 5: spindle, 6: restriction ring, 7: base, A: main thrust bearing, B: control bearing, C: radial bearing, D: air inlet for main bearing, E: air inlet for control bearing

For several supply pressures p_p* to the control-bearing, the load characteristic of the active thrust bearing is shown in

Fig. 1.2. When the supply pressure p_P* is high, the restriction ring is supported rigidly and moves little for the load increment. Therefore, the thrust bearing behaves as an ordinary passive aerostatic bearing. As p_P* is decreased and the mobility of the restriction ring increases, the spindle displacement decreases and the bearing stiffness increases by the effect of the active control. At the optimum value, p_P* = 0.34MPa, the spindle does not displace for load fluctuation and the bearing stiffness is infinite. Further decrement of p_P* makes the bearing stiffness negative, i.e. the load for the downward yields the upward movement of the spindle.

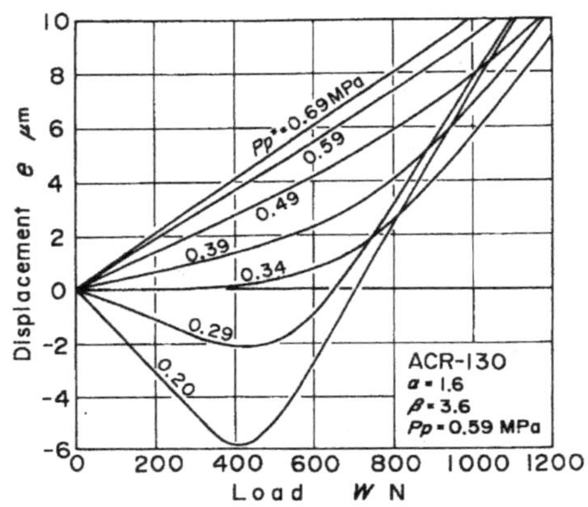

Figure 1.2. Load characteristic of active aerostatic thrust bearing with aerostatically controlled restrictor [2]

As shown in Fig. 1.2, the active control restrictor of the bearing-pressure-feedback type can make the bearing stiffness infinite. However, the infinite bearing stiffness does not guarantee the high rotational accuracy of the spindle. In the ultraprecision applications, the rotational accuracy of the spindle is crucial. The active restrictor of the bearing-pressure-feedback type only controls the bearing pressure, and does not control the spindle position directly. Therefore, the application of restrictor of the bearing-pressure-feedback type for the ultraprecision machine is restricted.

3. Actice aerostatic bearing with active inherent restrictor (AIR)

An air bearing spindle with active control system of the bearing-gap-feedback type [4,5] is shown Fig. 1.3. Spindle is suppored by an opposed-pad thrust bearing and a radial bearing. On the upper thrust bearing surface, eight restrictor units (AIR units) are embedded, while there is no restrictor on the lower bearing surface. Such structure reduces the influence of geometrical error of the lower thrust bearing on rotational accuracy because the stiffness of the lower bearing is zero. Figure 1.4 shows the AIR unit. The unit has an air outlet hole of 0.3 mm in diameter at the center of the resin. When the unit is embedded in the bearing surface, pressurized air is supplied from the backside of the unit. Passing through the hole in the actuator, the air blows into the bearing surface.

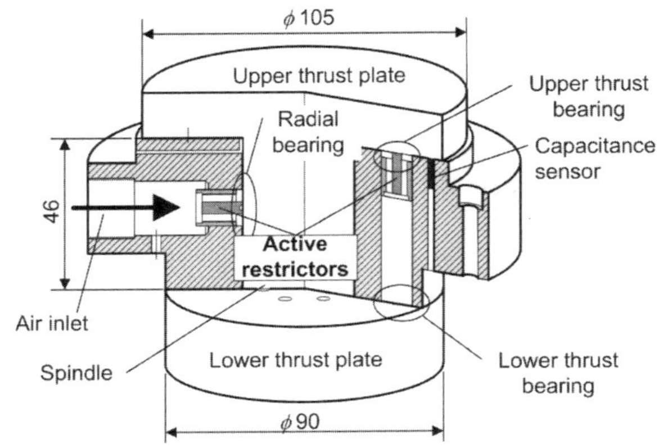

Figure 1.3. Air bearing spindle with active inherent restrictor

The principle of operating mechanism of the active inherent restrictor (AIR) is shown schematically in Fig. 1.5. Inherent restrictor is a kind of an orifice, and the effective orifice area is given by $\pi \cdot d \cdot h_d$, where d is the diameter of the inlet hole and h_d is the restriction gap on the bearing surface. The piezoelectric actuator (PZT) moves the elastic resin up and down, and the restriction gap h_d, i.e. the orifice area can be changed. According to the displacement of the spindle (change in the bearing gap h) caused by an additional load U_L, the restriction gap h_d is increased by controlling the length of the piezoelectric actuator, and the bearing pressure Po increases to compensate the additional load without changing the bearing gap h. Consequently, the stiffness and the rotational accuracy of the spindle can be improved.

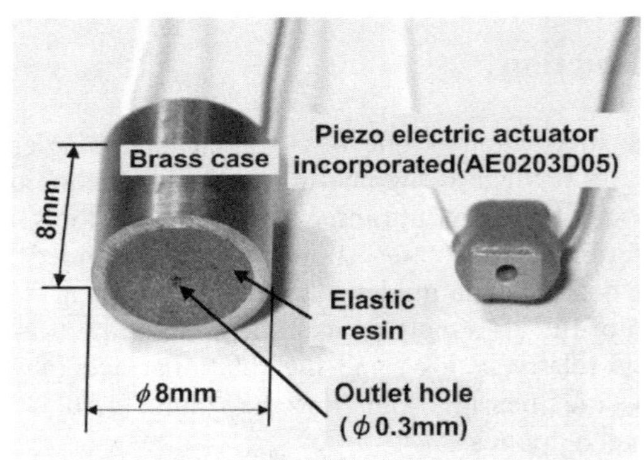

Figure 1.4. Unit of active inherent restrictor and piezoelectric actuator incorporated

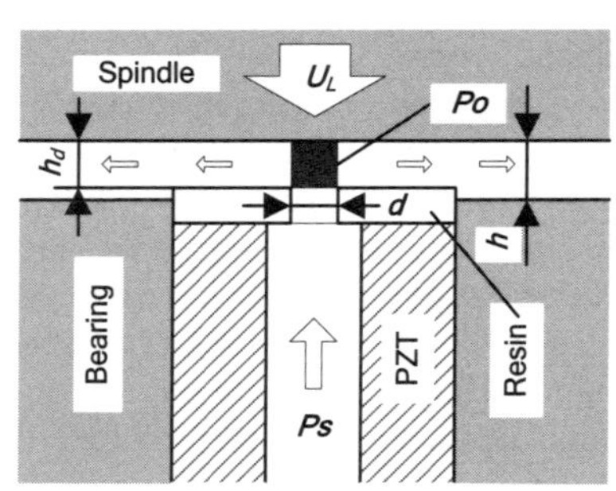

Figure 1.5. Operating principle of active inherent restrictor

The control system for the air-bearing spindle is shown in Fig. 1.6. In the thrust direction, vibration of the upper thrust plate is detected by a capacitance sensor embedded in the bearing surface. Detected analogue vibration is converted into digital data and input to a microcomputer. The computer calculates the control voltage for the piezoelectric actuator by using a PI algorithm. The calculated result is converted into analogue voltage and supplied to the AIR units via a power amplifier. In the radial direction, the vibration is measured at a reference ball on the thrust plate. By using the same data processing equipment as in the thrust direction, the AIR units are controlled.

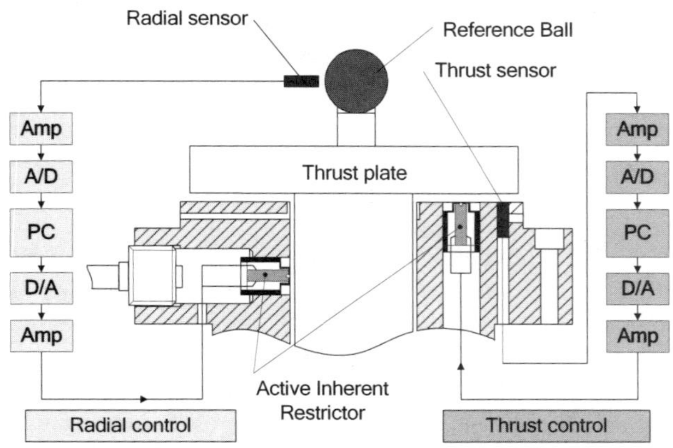

Figure 1.6. Control system for air bearing spindle with active inherent restrictors

The effect of the AIR control on the rotational accuracy in thrust direction is shown in Fig. 1.7, where N is the rotational speed. Without active control, the amplitude of the spindle vibration measured on the thrust bearing surface is about 0.1 μm. This is an unacceptable level of vibration for ultraprecision applications. After the active control starts, the amplitude of the vibration is reduced to less than 10 nm. In the radial direction, the amplitude of the vibration under control is also decreased to less than 10nm. Thus the rotational accuracy of the air bearing spindle can be improved to the nanometer order by the active control system using the AIR. In the bearing-gap-feedback type, the vibration at the measuring point can be controlled. Therefore, this type of active control system has the ability to reduce the vibration at the cutting point of the machine tool or the measuring point of the measuring machine. Such ability of the active control bearing is useful for ultraprecision applications.

4. Active aerostatic guideway with active inherent restrictor (AIR)

The application of the AIR for the aerostatic guideway is shown in Fig. 1.8. This is the cross sectional view of an ultraprecision tri-mode positioning system proposed by the author [7]. A rectangular machine table is guided by an aerostatic guideway and positioned under three positioning modes, namely, coarse, fine and ultra-fine mode. The basic coarse mode positioning employs a twist-roller friction drive (TFD) [8] driven by an AC servomotor. The TFD and the table are connected by an aerostatic coupling that reduces the vibration generated in the TFD. In the coarse mode, full stroke of the machine-table travel (40 mm) can be controlled with nanometer order of resolution. For the fine mode positioning, a fine adjustment using piezoelectric actuators incorporated into a motor holder drives the TFD.

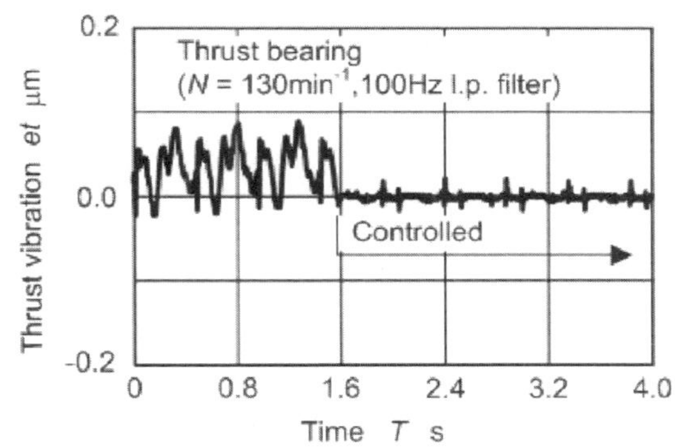

Figure 1.7. Effect of active control on thrust vibration of air bearing spindle with AIR

The ultra-fine mode of the positioning system employs an aerostatic guideway with the AIR. This active aerostatic guideway not only guides the machine table, but also controls the air gap on the bearing surface. The machine table consists of four plates surrounding the rectangular guide rail. Inside of the plates of the machine table, there are the vertical and horizontal bearing surfaces where the AIRs units are embedded. Control system for the active aerostatic guideway is almost the same as the control system for the active aerostatic bearing shown in Fig. 1.6. Lateral displacement of the table is detected by a fiber-optic sensor. The detected analogue displacement of the table is converted into digital data by a 16-bit AD converter and input to a microcomputer. The control

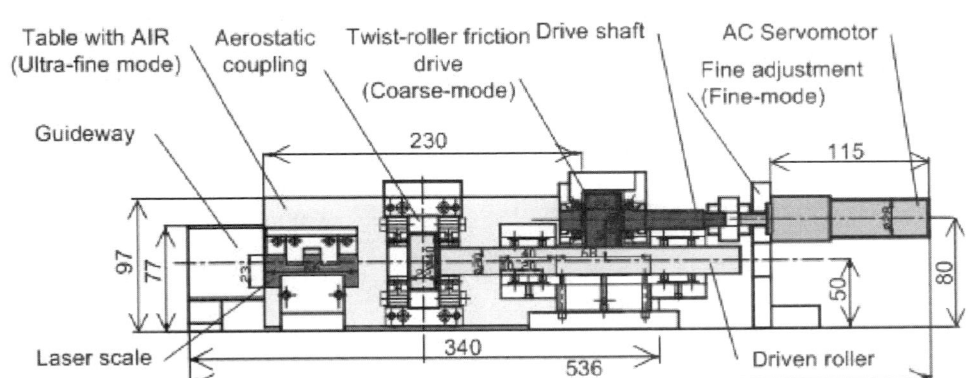

Figure 1.8. Tri-mode ultraprecision positioning system using active aerostatic guideway with AIR

signal for the actuator in the AIR is calculated by using the PI algorithm, and then converted into analogue voltage by a 16-bit DA converter. The converted analogue voltage is fed to the AIR units via a power amplifier.

The vibration of the machine table in the horizontal direction is shown in Fig. 1.9. Without control (see Fig. 1.9 upper), the straightness of the table motion is bout 20nm. This level of straightness is fairly good but not enough for ultraprecision applications. For realizing nanometer and sub-nanometer order of positioning, higher motion accuracy of the table should be needed. By using the control system for the guideway, the AIR embedded in the guideway controls the bearing gap according to the table fluctuation detected by the sensor. The effect of the active control on the table motion is shown in Fig. 1.9 lower. It can be seen that the vibration of the table is in the order of nanometer and the straightness of the table motion is less than 1nm. Thus, the AIR is effective for improving the performance of aerostatic equipments.

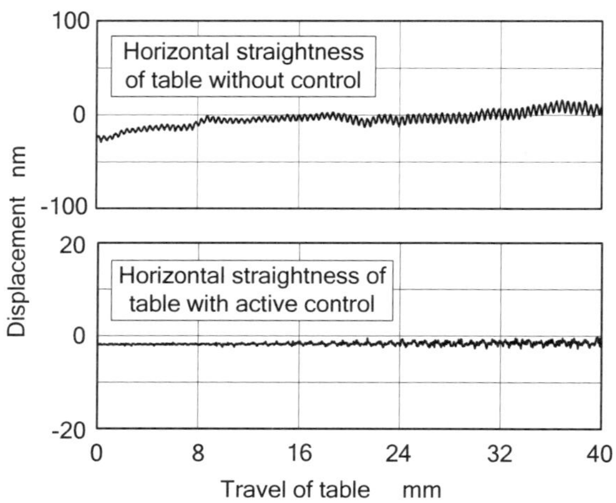

Figure 1.9. Straightness of motion of machine table

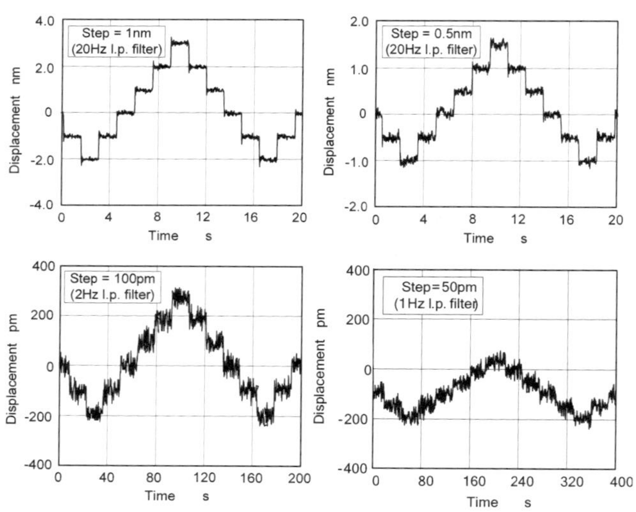

Figure 1.10. Ultraprecision positioning using active aerostatic guideway with AIR

By some minor modifications of the control program for the AIR, the active aerostatic guideway functions as the ultra-fine mode positioning device. The results of step positioning of various step widths (from 1 nm to 50 pm) are shown in Fig. 1.10, where same width of step is repeated five times in one direction, then in the reverse direction. It is shown that the steps with the width of 1 nm, 0.5 nm and 100 pm are clearly resolved. With considerable noise, however, the steps of 50 pm can also be recognized. It has been reported that the positioning resolution of another ultraprecision positioning system using the active aerostatic guideway is 10 pm [6]. Therefore, the positioning resolution using the active aerostatic guideway can be in the order of 10 pm.

5. Conclusion

From the viewpoint of ultraprecision application, the effect of the active control of aerostatic bearing is analysed. It is relatively easy to improve the bearing stiffness by the active control. However, the improvement for the rotational accuracy of the bearing for ultraprecision application can only be realized by the bearing-gap-feedback type active restrictor. The analyses show that the active aerostatic bearings will be an essential tool in the next-generation of ultraprecision engineering.

References

[1] H. Mizumoto, et al, 1989, An Aerostatically Controlled restrictor for Obtaining an Infinite Stiffness Aerostatic Thrust Bearing, Bull. of Japan Society For Precision Engineering, Vol.23, No.3, pp211-216.

[2] H. Mizumoto, et al, 1990, Zero-compliance aerostatic bearing for an ultra-precision machine, Precision Engineering, Vol.12, No.2, pp75-80.

[3] H. Mizumoto, et al, 1992, An Infinite-Stiffness Aerostatic Bearing with an Exhaust-Control Restrictor, Bull. of Japan Society for Precision Engineering, Vol.26, No.3, pp225-230.

[4] H. Mizumoto, et al, 1996, Active inherent restrictor for air-bearing spindles, Precision Engineering, Vol.19, No.2/3, pp141-147.

[5] H. Mizumoto, et al., 2001, Rotational accuracy and positioning resolution of an air-bearing spindle with active inherent restrictor, Proceedings of 10th Int. Conf. of Precision Engineering, Vol.1, pp.119-122.

[6] H. Mizumoto, et al., 2001, Picometer positioning system using aerostatic guideway as motion-reduction mechanism, Proceedings of 16th American Soc. for Precision Engineering annual meeting, Vol.1, pp.119-122.

[7] H. Mizumoto, et al., 2005, A Two-axis Tri-mode Ultraprecision Positioning System for Picometer Positioning, Proceedings of 5th Int. Conf. of the euspen, Vol.2, pp.495-498.

[8] H. Mizumoto, et al, 1996, Picometer positioning system using twist-roller friction drive, Precision Engineering, Vol.19, No.2/3, pp141-147.

Jacobian Inversion Analysis and Extraction of Dexterous Workspace in PKMs

A. Khalid, S. Mekid

School of Mechanical, Aerospace & Civil Engineering, The University of Manchester, Manchester M60 1QD, UK

Abstract. The emerging interest in the use of parallel kinematic machines for machining applications requires extensive research to determine proper workspace and dexterity when designing PKMs. Theoretically, the jacobian matrix is used to calculate all the performance parameters in PKMs. The condition number plays the role of an observer calculated for every point of the workspace in order to identify the location's reachability and dexterity. The paper discusses issues related to the Jacobian inversion to determine the dexterity space. A program has been made to analyse the extraction of dexterity from workspace with two examples to illustrate this. Issues concerning dexterity and the jacobian inversion are also discussed.

Keywords: Manipulator workspace, Dexterity, SPS system.

1. Introduction

Parallel kinematic machines use different types of joints in their structure. The joints and links work in combination to form a closed loop kinematic chain. The combination of a number of closed loop chains form a PKM system. The joints are spherical, prismatic, universal and revolute which play a determinant role in the final shape of the workspace. Each type of joint causes a unique effect on the system's overall mobility.

Grubler's equation [1] is considered to calculate the degree of freedom or mobility of the system. It gives only the mobility of the system in the directions where major motion is carried out. However, minute motions in other directions have not been counted. In this paper, two examples are observed for the calculation of dexterity and workspace which will show that the manipulator's movement in constraint directions are also accounted.

2. Analytical Modelling-General Formulation

The modelling of a PKM system will require a general modelling method. Eq. (1) refers to the common base platform designed in PKM systems. The number of limbs can be ranged from three to six depending upon the stiffness requirement of the system. In this study, three limbs are proposed in the upcoming examples. Three limbs form an equilateral triangle and 120° apart from each other. The projected joint locations a_i (Fig.1.1) location with respect to the platform center can be given as follows (Fig. 1.1 and 1.2).

$$\begin{bmatrix} a_{1x} \\ a_{1y} \\ a_{1z} \end{bmatrix} = \begin{bmatrix} r_p \\ 0 \\ 0 \end{bmatrix}, \begin{bmatrix} a_{2x} \\ a_{2y} \\ a_{2z} \end{bmatrix} = \begin{bmatrix} r_p c_\alpha \\ r_p s_\alpha \\ 0 \end{bmatrix} \text{ and } \begin{bmatrix} a_{3x} \\ a_{3y} \\ a_{3z} \end{bmatrix} = \begin{bmatrix} r_p c_\beta \\ r_p s_\beta \\ 0 \end{bmatrix} \quad (1)$$

The upper moving platform has a radius r_p. The rotation representation of the moving platform is based on Euler method [2] which gives the transformation rotation matrix of the upper platform with respect to the global reference system (GCS).

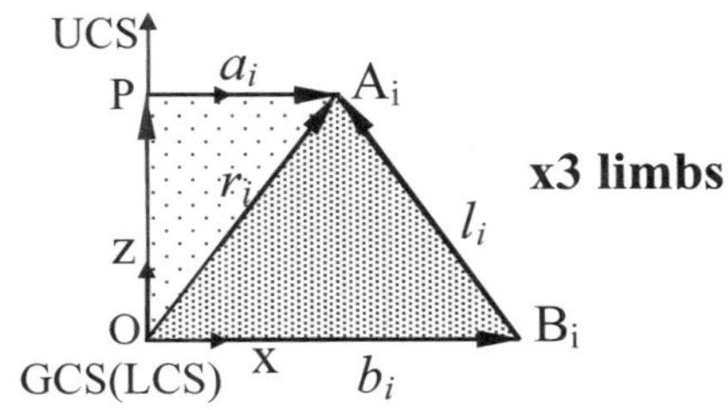

Figure 1.1. Forward (ΔOPA_i) & inverse kinematics (ΔOB_iA_i): Transformation from lower (global) to upper coordinate system.

The global coordinates for each upper joint A_i (Fig.1.1), can be calculated using the basic transformation in Eq.(2).

$$\begin{bmatrix} a_{ix} \\ a_{iy} \\ a_{iz} \\ 1 \end{bmatrix}_{GCS} = \begin{bmatrix} u_x & v_x & w_x & x \\ u_y & v_y & w_y & y \\ u_z & v_z & w_z & z \\ 0 & 0 & 0 & 1 \end{bmatrix}_L^U \begin{bmatrix} a_{ix} \\ a_{iy} \\ a_{iz} \\ 1 \end{bmatrix}_{UCS} \quad (2)$$

The three components of the upper joints in terms of global reference can be written in the following way:

$$\begin{aligned} r_{ix} &= x + u_x a_{ix} + v_x a_{iy} + w_x a_{iz} \\ r_{iy} &= y + u_y a_{ix} + v_y a_{iy} + w_y a_{iz} \\ r_{iz} &= z + u_z a_{ix} + v_z a_{iy} + w_z a_{iz} \end{aligned} \quad (3)$$

The type of lower platform joints e.g. spherical, will determine the constraints to be put on the system. By using the constraints, equations can be derived for the forward kinematic solution. The vector loop diagram locating the joint comprises the lower platform, limb and upper joint location and can be written as:

$$\vec{r}_i = \vec{b}_i + \vec{l}_i \quad (4)$$

The actuator can either be placed in b_i or l_i depending on the design requirement. The module of the actuator vector can be found using the square of length. If l_i is chosen to be the actuator then the modulus length of ith leg can be written as:

$$l_i = \pm\sqrt{r_i^2 + b_i^2 - 2 r_i b_i} \quad (5)$$

The manipulator's Jacobian provides the link between the actuators and the chosen points on the end effector. From the Jacobian information, velocities, acceleration, stiffness and workspace singularities can be analysed. The end effector can move in all the six directions described by:

$$X=[x \quad y \quad z \quad \psi \quad \theta \quad \varphi]^T \tag{6}$$

The manipulator has three prismatic actuators b_i:

$$q=[b_1 \quad b_2 \quad b_3]^T \tag{7}$$

Then, the kinematic constraints imposed by the limbs can be written in the jacobian form

$$f(X,q)=0 \tag{8}$$

where f is an n-dimensional implicit function of q and X. Differentiating with respect to time, a relationship can be obtained between the input joint rates and the end-effector output velocity as

$$J_x \dot{X} = J_q \dot{q} \tag{9}$$

The overall Jacobian matrix J can be written as

$$J \dot{X} = \dot{q} \quad \text{where } J = J_q^{-1} J_x$$

The Jacobian formed here has a very important role to play. The inversion of the Jacobian depends also upon the final shape of the matrix. To deal with these issues, two case-studies are considered next.

3. PRS and SPS systems

3.1 PRS system

The upper and lower platform local coordinates of joints with respect to point 'P' and 'O' are given by eq. (10) and eq. (11), respectively (Figure 1.1.2). The Base could have an optional inclination angle γ.

$$\begin{bmatrix} a_{1x} \\ a_{1y} \\ a_{1z} \end{bmatrix} = \begin{bmatrix} r_p \\ 0 \\ 0 \end{bmatrix}, \begin{bmatrix} a_{2x} \\ a_{2y} \\ a_{2z} \end{bmatrix} = \begin{bmatrix} r_p c_\alpha \\ r_p s_\alpha \\ 0 \end{bmatrix} \text{ and } \begin{bmatrix} a_{3x} \\ a_{3y} \\ a_{3z} \end{bmatrix} = \begin{bmatrix} r_p c_\beta \\ r_p s_\beta \\ 0 \end{bmatrix} \tag{10}$$

$$\begin{bmatrix} b_{1x} \\ b_{1y} \\ b_{1z} \end{bmatrix} = \begin{bmatrix} b_1 c_\gamma \\ 0 \\ b_1 s_\gamma \end{bmatrix}, \begin{bmatrix} b_{2x} \\ b_{2y} \\ b_{2z} \end{bmatrix} = \begin{bmatrix} b_2 c_\gamma c_\alpha \\ b_2 c_\gamma s_\alpha \\ b_2 s_\gamma \end{bmatrix} \text{ and } \begin{bmatrix} b_{3x} \\ b_{3y} \\ b_{3z} \end{bmatrix} = \begin{bmatrix} b_3 c_\gamma c_\beta \\ b_3 c_\gamma s_\beta \\ b_3 s_\gamma \end{bmatrix} \tag{11}$$

Then, the constraints are defined through the use of revolute joint on the lower end of the limb. Forward kinematic solution is established with the help of the following three constraints.

$$r_{1y} = y + u_y r_p = 0 \tag{12}$$

$$r_{2y} = r_{2x} \tan(\alpha) \tag{13}$$

$$r_{3y} = r_{3x} \tan(\beta) \tag{14}$$

The parameters x, y and ϕ are dependent [2]. The inverse kinematic solution in terms of actuator values could be determined by dot multiplying eq. (4) with itself to yield the actuation expressions needed for the Jacobian:

$$b_1 = r_{1x}\cos(\gamma) + r_{1z}\sin(\gamma) \pm \sqrt{(r_{1x}\cos(\gamma) + r_{1z}\sin(\gamma))^2 - (r_{1x}^2 + r_{1y}^2 + r_{1z}^2 - l_1^2)} \tag{15}$$

$$b_2 = r_{2x}\cos(\gamma)\cos(\alpha) + r_{2y}\cos(\gamma)\sin(\alpha) + r_{2z}\sin(\gamma) \pm \sqrt{(r_{2x}\cos(\gamma)\cos(\alpha) + r_{2y}\cos(\gamma)\sin(\alpha) + r_{2z}\sin(\gamma))^2 - (r_{2x}^2 + r_{2y}^2 + r_{2z}^2 - l_2^2)} \tag{16}$$

$$b_3 = r_{3x}\cos(\gamma)\cos(\beta) + r_{3y}\cos(\gamma)\sin(\beta) + r_{3z}\sin(\gamma) \pm \sqrt{(r_{3x}\cos(\gamma)\cos(\beta) + r_{3y}\cos(\gamma)\sin(\beta) + r_{3z}\sin(\gamma))^2 - (r_{3x}^2 + r_{3y}^2 + r_{3z}^2 - l_3^2)} \tag{17}$$

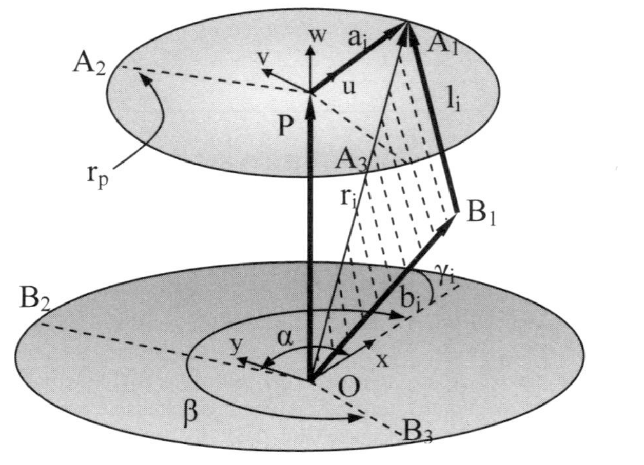

Figure 1.2. Vector model of the inclined PRS manipulator.

Finally, the Jacobian for the 3-PRS system is presented below:

$$J = \begin{bmatrix} \frac{s_{l1x}}{F_1} & \frac{s_{l1y}}{F_1} & \frac{s_{l1z}}{F_1} & 0 & 0 & 0 & 0 & 0 & 0 \\ 0 & 0 & 0 & \frac{s_{l2x}}{F_2} & \frac{s_{l2y}}{F_2} & \frac{s_{l2z}}{F_2} & 0 & 0 & 0 \\ 0 & 0 & 0 & 0 & 0 & 0 & \frac{s_{l3x}}{F_3} & \frac{s_{l3y}}{F_3} & \frac{s_{l3z}}{F_3} \end{bmatrix} \tag{18}$$

where $F_1 = s_{b1x}s_{l1x} + s_{b1y}s_{l1y} + s_{b1z}s_{l1z}$, $F_2 = s_{b2x}s_{l2x} + s_{b2y}s_{l2y} + s_{b2z}s_{l2z}$, $F_3 = s_{b3x}s_{l3x} + s_{b3y}s_{l3y} + s_{b3z}s_{l3z}$

s_{bi} and s_{li} are unit vectors along the actuator and the leg l_i respectively. s_{bi} are readily available from eq. (11) and s_{li} can be defined as follows:

$$\begin{bmatrix} s_{lix} \\ s_{liy} \\ s_{liz} \end{bmatrix} = \begin{bmatrix} (r_{ix} - s_{bix})/l_i \\ (r_{iy} - s_{biy})/l_i \\ (r_{iz} - s_{biz})/l_i \end{bmatrix} \tag{19}$$

The inversion of the Jacobian is discussed later.

3.2 SPS system

The upper platform coordinates are similar to those written in eq. (10) while the lower platform is supposed to be twice the size of the upper platform.

From Figure 1., the closed loop vector diagram gives the following equation which can be represented further in terms of r_i as follows

$$\overline{OP} + \overline{PA_i} = \overline{OB_i} + \overline{A_iB_i} \tag{20}$$

The actuator $\overline{A_iB_i}$ is determined as:

$$\overline{A_iB_i} = p + {}^B R_A \, {}^A a_i - b_i \tag{21}$$

The length of the ith actuator is obtained by taking the dot product of the vector $\overline{A_iB_i}$ with itself.

$$d_i^2 = [p + {}^B R_A \, {}^A a_i - b_i]^T [p + {}^B R_A \, {}^A a_i - b_i] \quad \text{for } i = 1,2,3 \tag{22}$$

where d_i denotes the length of the ith actuator. Extending eq. (22) yields

$$d_i^2 = p^T p + [{}^A a_i]^T [{}^A a_i] + b_i^T b_i + 2p^T [{}^B R_A \, {}^A a_i] - 2p^T b_i - 2[{}^B R_A \, {}^A a_i]^T b_i \tag{23}$$

By taking the square root of eq. (23), the inverse kinematics solution of three limbs can be found straight forwardly.

$$d_i = \pm\sqrt{p^T p + [{}^A a_i]^T [{}^A a_i] + b_i^T b_i + 2p^T \left[{}^B R_A {}^A a_i\right] - 2p^T b_i - 2\left[{}^B R_A {}^A a_i\right]^T b_i}$$

for i = 1,2,3 (24)

There are two possible solutions for limb lengths at each given location of the moving platform. The negative limb length is physically not feasible. The overall Jacobian is found for this example as follows.

$$J = \begin{bmatrix} s_{1x} & s_{1y} & s_{1z} & (a_{1y}s_{1z} - a_{1z}s_{1y}) & (a_{1z}s_{1x} - a_{1x}s_{1z}) & (a_{1x}s_{1y} - a_{1y}s_{1x}) \\ s_{2x} & s_{2y} & s_{2z} & (a_{2y}s_{2z} - a_{2z}s_{2y}) & (a_{2z}s_{2x} - a_{2x}s_{2z}) & (a_{2x}s_{2y} - a_{2y}s_{2x}) \\ s_{3x} & s_{3y} & s_{3z} & (a_{3y}s_{3z} - a_{3z}s_{3y}) & (a_{3z}s_{3x} - a_{3x}s_{3z}) & (a_{3x}s_{3y} - a_{3y}s_{3x}) \end{bmatrix} \quad (25)$$

Three equations are available from eq. (24) and three equations are formulated from direct kinematic solution. Six other equation are used from the orthogonal conditions of the rotation matrix given in [6]. A total of 12 equations are solved for 12 parameters, i.e three actuator values and nine variables u_x, u_y, u_z, v_x, v_y, v_z , w_x, w_y and w_z defined in eq. (2). It will result in numerous solutions because all the equations are of the second order and 2^{12} solutions are expected. Raghavan [4] has applied the continuation method and found that 6-SPS platform has 40 solutions. However, it is possible that some of them are not feasible.

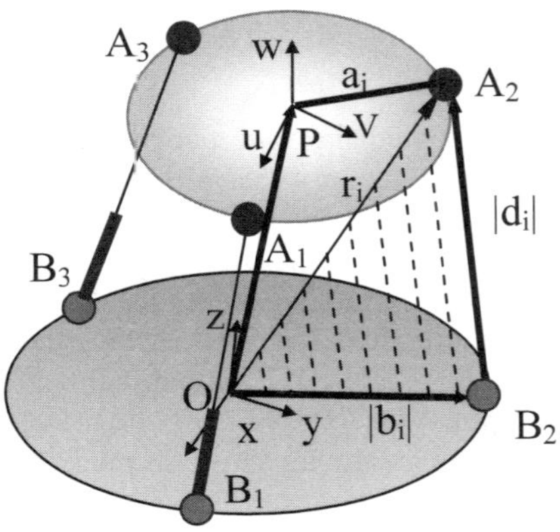

Figure 1.3. Mechanism architecture and the vector loop diagram for SPS system.

4. Inversion Analysis

The PKM workspace is determined with the inversion of the Jacobian. The dexterity is defined within the workspace as a superior capability of the PKM to achieve full orientations in 3D space. Some issues in the determination of the workspace and the corresponding dexterity are discussed.

Type of Matrix: The Jacobians found in eqs.(18) & (25) are 3x9 and 3x6, respectively. The matrix inversion of these conventional Jacobians is an issue. The determinant is easily found for a square matrix. The conventional method, shown previously, will usually give a rectangular (over or under determined) matrix which needs further transformation to allow inversion. The Screw method explained in [6] gives by default square jacobians that are easy to inverse by definition. A rectangular matrix needs further transformation or decomposition to turn it into a square one. This leads to incorrect results if the inversion is not properly made. A pseudo technique is suggested.

Methods of Inversion: The matrix decomposition can be used for the purpose. There are a few methods available to decompose $m \times n$ matrices. As both cases at hand are under-determined problems, therefore, pseudo inverse based on singular value decomposition is then a convenient method. The method factorises the matrix J where $m < n$ and when the rank of the matrix J is not equal to the number of columns as required by this method. The idea is to turn a single rectangular matrix into its factors of square matrices. The square matrices are then used to take their own inverses and get multiplied again in the reverse order. The whole workspace of the manipulator can be checked for the matrix inversion. Pseudo-inverse method provides a partial solution for the inverse of rectangular matrices. The J matrix is decomposed as

$$J = U \times S \times V^T \quad (26)$$

where V is n×n orthogonal matrix consists of the eigenvectors of $J^T J$. U is $m \times m$ orthogonal matrix and consists of the eigenvectors of JJ^T. S is the $m \times n$ diagonal matrix consists of the square roots of the eigenvalues of JJ^T arranged in descending order. For eq.(26), the pseudo-inverse matrix is written as:

$$(J^{-1})_{pseudo} = V \times (S^{-1})_{pseudo} \times U^T \quad (27)$$

This technique will help in the determination of singularities within the workspace but this is not sufficient as the inversion depends also on the condition number defined next. The manipulator performance (i.e.kinematic manipulability) within the workspace in introduced by Eq. 28 [5].

$$m(J) = \sqrt{\det(JJ^T)} \quad (28)$$

Effect of the condition number: The workspace regarding the accessible orientations of the moving platform is considered as the dexterity of the manipulator. The inversion of the Jacobian may lead to an amplification of the motion explained by the condition number $k(J)$.

$$k(J) = \|J\|_2 . \|J^{-1}\|_2 \quad (29)$$

where $\|.\|$ is a 2-norm of a matrix and is defined as the square root of the ratio of the largest to the smallest eigenvalues of the Jacobian matrix. The locations where the condition number reaches a value of 1, have their dexterity at maximum. A matrix is said to be well conditioned when k is small, and, ill conditioned when k is large. A well conditioned Jacobian matrix gives a more uniform set of possible velocities. Actually, the minimum value of $k(J)$ is 1. If 0 is found then the location is singular [3]. The value of the condition number increases near the workspace boundaries where the dexterity becomes very low. The dexterous workspace is a subset of the total reachable workspace.

To find out the reachable and dexterous workspace, the condition given in eq. (30) is required to be implemented in the dexterity searching program. The maximum condition number is given as the input and the program start search from the defined reference point in a certain direction. When it reaches to a maximum value in that direction, the program stops. Then, it restart searching from the reference point in another direction.

$$cond(J) \le cond(J)_{\max} \quad (30)$$

5. Desterity and workspace results

The dexterity characteristic for each type of platform has been computed. The key dimensions of the platforms are l_i=1m and r_p = 1m, arbitrarily chosen. Figure 1.4 shows the dexterity computed for PRS platform using conventional forward kinematic (a) and pseudo-inversion of the conventional Jacobian (b) with low condition number i.e. ≤1.5. The pseudo inversion gives almost similar results compared to conventional method. However, small discrepancies could be seen at the envelop in Fig. 1.4(b). The Screw Jacobian obtained from the screw method [6] is a square matrix for which the dexterity is shown in Fig. 1.4(c).

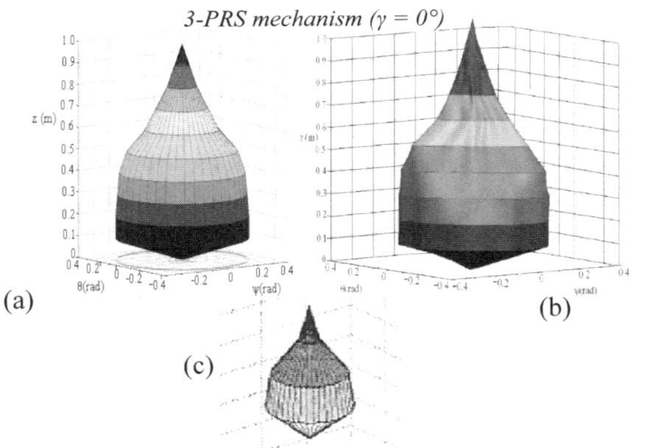

Figure 1.4. (a) Dexterity using conventional Jacobian at cond(J) ≤ 1.5 (γ = 0°) (b) Dexterity obtained for the cond(J) ≤ 1.5 by the pseudo-inverse of conventional Jacobian. (c) Dexterity obtained with the screw method (*not to scale*).

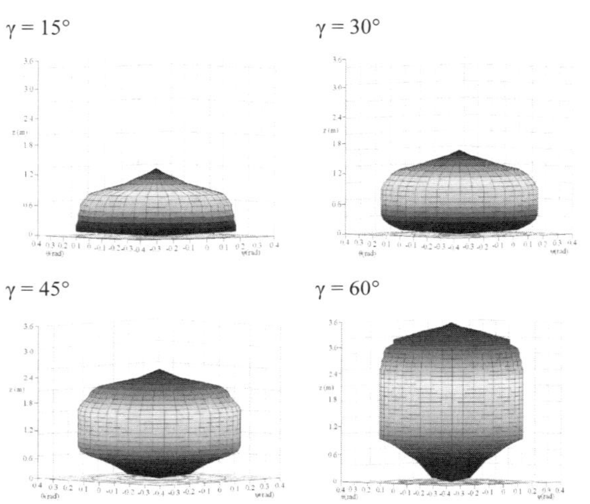

Figure 1.5. Dexterity workspaces using conventional Jacobian at condition number ≤ 1.5 for the 3-PRS mechanism at 4 different inclination angles.

Figure 1.5 shows an extension of the dexterity depending on the angle of the chosen posture (actuator angle γ). The information required by the designer is the workspace generated by the PKM and its corresponding dexterous workspace. This subset workspace will secure a better dexterity for the cutting tool for example. Figure 1.6 and 1.7 show the workspace and its corresponding dexterous workspace.

The ratio of the dexterous and real workspace is better in SPS than in PRS platforms. The use of spherical joints induces larger workspace compared to PRS. For SPS system, one of the solutions is picked up randomly to formulate the workspace and dexterity in Figure 1.7.

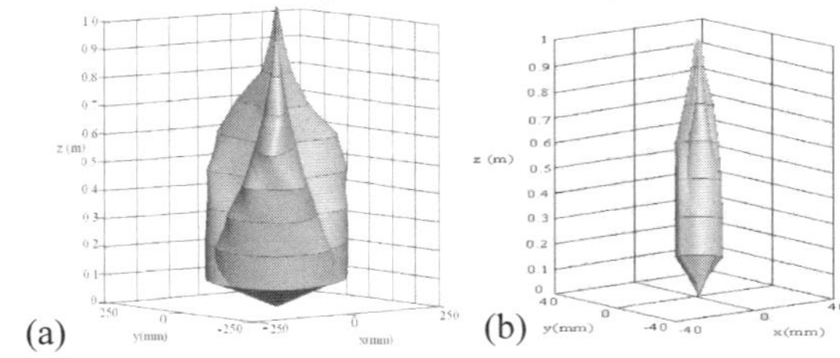

Figure 1.6 (a) Real workspace (3-PRS) (b) Dexterous workspace.

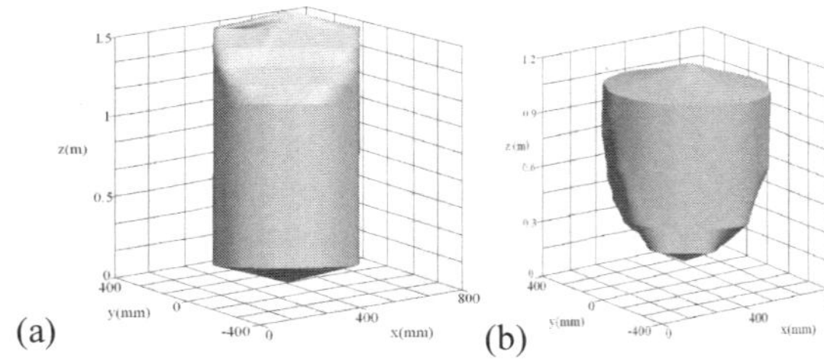

Figure 1.7. (a) Real workspace (3-SPS) (b) Dexterous workspace

6. Conclusion

The work presented in this paper has discussed the extraction of the dexterous workspace considered as a subset of the whole workspace. A technique of pseudo inversion has been applied to non-squared jacobians where it is difficult to find solutions, and, compared to forward kinematic for dexterity. Two conditions to be observed in this type of computation: the condition number that must be kept lower (≤1.5), and, preventing unfeasable solutions among those that may be found with SPS as the latter platform exhibits highly non-linear equations. The first obtained results reveal that the SPS has the workspace and the dexterous space larger compared to PRS platform as expected due to its high degree of freedom. The next step is to validate experimentally these results with measurements of the end effector reachability and dexterity. More legs will be added to investigate its influence on the stiffness to find a trade-off between dexterous workspace and axes stiffness.

References

[1] Gogu Grigore, 2005, Mobility of mechanisms: a critical review, Mech. and Machine Theory, Vol. 40, P1068-1097.

[2] J. A. Carretero, M. A. Nahon et al., 2000, Kinematic Analysis and Optimization of a New Three Degree of Freedom Spatial Parallel Manipulator, ASME Journal of Mech. Design, Vol. 122, No. 1, P17-24.

[3] Merlet J. P., 2006, Jacobian, Manipulability, Condition Number and Accuracy of Parallel Robots, ASME Journal of Mechanical Design, Vol. 128, P199-206.

[4] Raghavan M., 1993, The Stewart Platform of General Geometry has 40 configurations, ASME Journal of Mechanical Design, Vol. 115, P277-282.

[5] T.Yoshikawa, 1985, Manipulability of Robotic Mechanisms, Intern. Journal of Robotics Research, Vol. 4, No. 2, P3-9.

[6] Tsai Lung Wen, 1999, Robot Analysis: The Mechanics of Serial and Parallel Manipulators, John Wiley & Sons, New York.

The Structure Design of a Micro-precision CMM with Abbé Principle

K.C. Fan[1], C.L. Liu[2], P. T. Wu[3], Y. C. Chen[4] and W. L. Wang[5]

[1]National Taiwan University, Taiwan, 10627, fan@ntu.edu.tw
[2]National Taiwan University, Taiwan, 10627, clliu@ntu.edu.tw
[3]National Taiwan University, Taiwan, 10627, r93522638@ntu.edu.tw
[4]National Taiwan University, Taiwan, 10627, d92522028@ntu.edu.tw
[5]Hefei University of Technology, Heifei, China, 230009, w_weili@163.com

Abstract. This paper presents the design considerations of a micro precision CMM. The structure of this CMM consists of the bridge, the Z-spindle and the XY stage. A novel bridge designed in a pagoda shape will be proposed and analyzed to prove its superior stiffness with force balanced condition. A high precision spindle design with counterweight and thermal error suppression are considered. The design of novel a coplanar XY-stage that observes the law of Abbé Principle is particularly proposed. Constructing these three structural modules the developed small CMM could meet the Abbé principle and provide high geometrical accuracy of the structure. Designed CMM has measurement lengths in X: 25 mm, Y: 25 mm and Z: 10 mm. Driven by the ultrasonic nanomotor and fed back by the hologram scale, each axis can achieve 1 nm resolution.

Keywords: Micro-CMM, Abbé Principle, Co-planarXY stage, Pagoda brideg, Structural design.

1. Introduction

Coordinate Measuring Machines (CMM) are versatile instruments for three-dimensional measurement of almost all kinds of workpieces. They are widely used in industry for precision measurement with general axial resolution from 0.1 to1 µm, accuracy from 1 to 5 µm, measuring range from several centimeters to meters, and contact probe diameter from 0.5 mm to 5mm. Except those ultra-high precision CMMs that are equipped with laser interferometers as position sensors, common CMMs are based on the readings of linear scales that are offset from the probe moving axis. In other words, its structural design has inherent Abbé errors [1]. The Abbé principle is called the first principle in metrology by scholars. Since Abbé first proposed this principle in 1890, it has been referenced by machine tool and measuring instrument societies to the accuracy design of the structures. The definition of Abbé principle is: in displacement measurement the reference should be in line with the displacement to be measured or on its extension. In Bryan's generalized interpretation, if the Abbé principle is not possible in the system design, either the slideway that transfers the displacement must be free of angular motion or the angular motion data must be obtained to compensate for the Abbé error by software. Following the Abbé principle, Bryan's principle extended the concept to straightness measurement [1] and Zhang extended it to other geometrical measurements [2].

Nowadays, many existing machine tools and coordinate measuring machings (CMM) still cannot comply with Abbé principle, except some special cases such as the "Ultimat" CMM by Lawrence Livermore Lab. [3], the Nanomeasuring machine (NMM) by TU Ilmenau [4], and others.

Many meso to micro scaled parts have been able to be produced by new manufacturing technologies in recent years, such as MEMS, LIGA, micro machining, etc. The required dimensions are in the range of mm to µm, resolution needed is from 1 µm to 10 nanometers, and accuracy demanded is from several micrometers down to about 50 nanometers. To face the inspection demand of these micro parts, the conventional CMM is no longer available to meet the requirements. The development of small CMMs has become an interesting topic in recent years. Collaborative research between National Taiwan University and Hefei University of Technology started the micro/nano-CMM project in 2001. Miniaturizing of the CMM structure with nano-motion and nano-scale techniques has been developed in sequence, but the Abbé principle was partially implemented [5, 6].

This paper presents the design considerations of a novel micro precision CMM that has force-balanced structure to yield high stiffness and conforms to the Abbé principle in 3D space. Designed CMM has measurement lengths in X: 25 mm, Y: 25 mm and Z: 10 mm. Driven by the ultrasonic nanomotor and fed back by the hologram scale, each axis can achieve 1nm resolution. The bridge design and the Co-planar XY stage design are described in the following.

2. The Pagoda Bridge Design

Rectangular types of bridges are always employed in precision CMM structures for mounting the Z-ram and probe. The deformation at the center of the bridge is very critical because of the concentrated load from the spindle and the generated driving force, which will react to the bridge. Although its static deflection caused by the total weight of the bridge itself and the ram does not influence the measuring accuracy, the generated driving force, evenif comparatively small, will induce dynamic deformation of the bridge up to submicron level. In order to meet the high precision requirement in nanometer measurement, the stiffness of the bridge has to be redesigned.

Figure 1.1 shows two types of bridges: rectangular shape and circular shape. Assuming they have the same width (2R), height (R), thickness, and carring the same ram load (P), from the structural analysis the maximum staticx deflection at the bridge center will be:

For rectangular bridge: $\delta_{y\max 1} = 0.55\dfrac{PR^3}{EI}$

For circular bridge: $\delta_{y\max 2} = 0.24\dfrac{PR^3}{EI}$

It is apparent that the circular shaped bridge has higher stiffness.

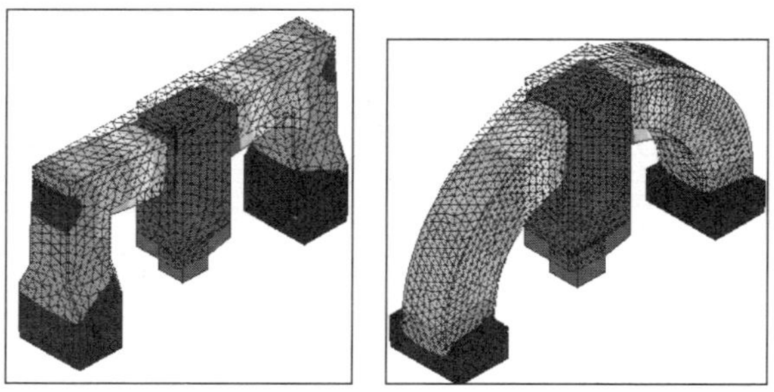

Figure 1.1. (a) Rectangular bridge, (b) Circular bridge

Again, if we consider the bridge is designed in an arch shape of the same dimension, as shown in Figure 1.2 (a), it must be even stiffer than the circular bridge as most of the bridges and tunnels in civil construction are of similar shape. Moreover, the arch shape can further be optimized for the minimum deflection design with respect to the parameters of radius R and angles of α and β, as shown in Fig. 1.2 (b). The objective function can be formulated by:

$$\begin{cases} Min \quad f(X) \\ X = [x_1, x_2, x_3] = [\alpha, \beta, R] \\ s.t. \quad 0^\circ \le \alpha \le 45^\circ \\ \quad 55^\circ \le \beta \le 90^\circ \\ \quad R_{\min} \le R \le R_{\max} \\ \quad 0 < \max(\sigma) \le 25Mpa \end{cases}$$

f(X) denotes the calculated deflection at the bridge center with design variables (R, α, β) and certain given dimension and stress constraints. The analysis can be done by finite element analysis in association with optimization technique. Figure 1.3 (a) shows the optimized arch shape.

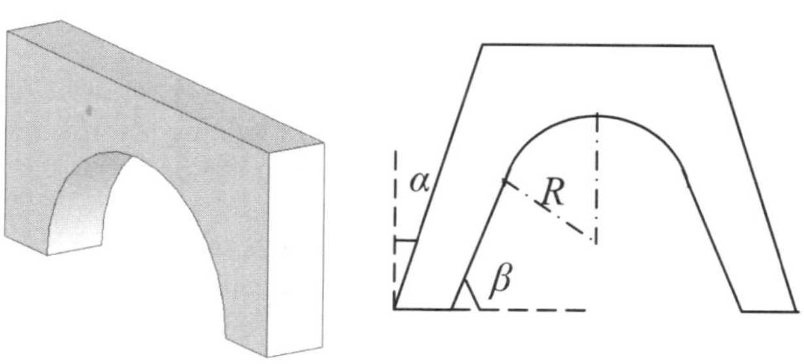

Figure 1.2. (a) Arch bridge, (b) Optimized arch bridge

From the above analysis it is seen that the conventional rectangular bridge is not the best design. The tapered arch bridge should possess the highest stiffness in the CMM structure. Moreover, the best CMM structure should be force balanced in all directions. The tapered arch shape provides force balance only in one plane. Therefore, this study proposes the pagoda bridge, which is symmetrically constructed from the optimized tapered arch shape. This pagoda bridge has a force-balanced structure in all directions, as shown in Fig. 1.3 (b).

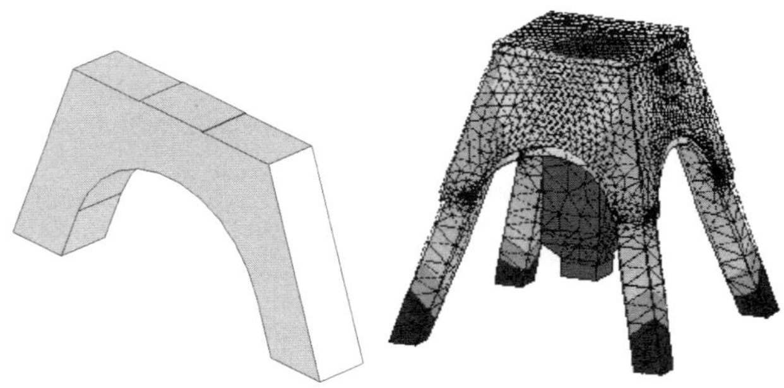

Figure 1.3. (a) Optimized arch bridge, (b) Pagoda bridge

In order to compare the stiffness of different bridge shapes with the required micro-CMM, we design the same width of 220 mm, height of 110 mm, thickness of 60 mm, supporting pad of 70 mm×100 mm×40 mm, and under the same ram weight of 29.4 N. The bridge material is granite with a density of 2660 (Kg/m^3), Young's modulus of 60 GPa and Poisson's ratio 0.3. From FEM analysis by AnSYS 7.0 software the center deflection of each type of bridge are compared in Table 1.1. It can be seen that the pagoda bridge has the highest stiffness among all and the force balance structure. The dynamic driving force generated by the actuator (such as the motor) will be superimposed on the bridge due to its reaction force. Although this driving force is small compared with the static load, it will inevitably generate a dynamic deformation of the bridge. For instance, a force of 1 N will incur 12.3 nm deflection at the rectangular bridge center and 3.9 nm at the pagoda bridge center. This will cause the same amount of measuring error of the micro-CMM. The pagoda bridge certainly has the minimum influence.

Table 1.1. Comparison of Stiffness

Bridge type	Deflection due to self-weight	Deflection due to self- weight and ram load
Rectangular	0.156 μm	0.362 μm
Circular	0.102 μm	0.174 μm
Arch	0.090 μm	0.154 μm
Optimum arch	0.060 μm	0.131 μm
Pagoda	0.060 μm	0.115 μm

3. Abbé Principle in 3D Coordinate

The famous Abbé Principle only indicates the rule in 1D motion. For a CMM, it is in 3D coordinate. This study thus proposes the extended Abbé Principle for a 3-coordinate motion machine:

1. Each merasuring axis should be in line with its corresponding motion axis, and
2. All three measuring axes should interset at the functional point of the machine.

This concept can be realized by Figure 1.4, where the intersecting point of three measuring axes will be the cutting point of a machine tool or the measuring point of a CMM. Obviously, all current configurations of side-mounted linear scales in machine tool and CMM designs have inherent Abbé errors. This is why the volumetric error analysis by 21 geometrical error terms is so important for the error compensation. Since for all the vertical types of 3-coordinate machines the vertical ram has its measuring axis in line with its motion axis, the only condition to meet with the 3D Abbé Principle relies on the XY plane motion.

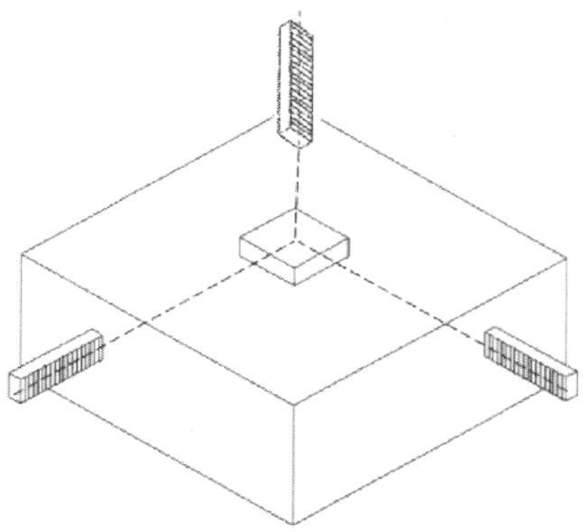

Figure 1.4. Abbé principle for 3D coordinate

4. The Co-planar XY Stage

A conventional XY stage is normally stacked up by two linear stages. Not only the position-dependent cross-talk between the axes cannot be avoided, the Abbé error of the lower stage is large due to the large Abbé offset.

A novel co-planar Abbé free XY stage is thus proposed in this study. The moving table slides along a common base plane which is precision ground to less than 1 μm flatness and isolated from the base plate of the stage. This is the essence of co-plane motion. Two push-pull bars move the table in X and Y directions via two linear slides respectively, as shown in Figure 1.5 (a). Each bar is driven by an ultrasonic motor through a linear slide, which is supported by a heavy block fixed to the base. A hologram grating scale is then mounted on the top of the bar along the axial direction. The motion of the bar can be detected by an optical reader mounted on the base plate. Figure 1.5 (b) shows the entire configuration of the developed Co-planar stage. The stroke of each axis is 25 mm. Since it has been proved the minimum step of the motor and the resolution of the optical reader (called Linear Diffraction Grating Interferometer, LDGI) can all reach to 1 nm [5, 6], the designed co-planar stage can have 25 mm stroke and 1nm resolution in each axis. Moreover, the intersection of two scales' axis is right to the functional point of the XY stage, it observes the law of Abbé principle in 2D space. An additional feature of this design is the hollow common base plane that allows a sensor to be embedded in to measure the straightness error of the moving table at the functional point, as shown in Figure 1.6. This agrees with the Bryan principle [1]. In addition, the squareness of X and Y motions can be assured by the smooth motion of two linear slides.

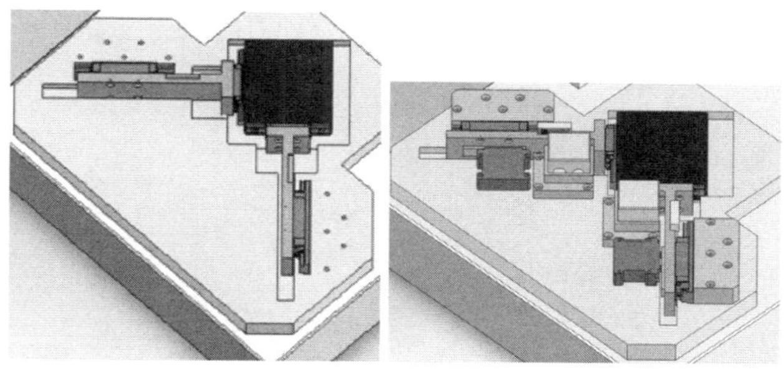

Figure 1.5. (a) Push-pull co-planar motion mechanism, (b) Adding supporting block, ultrasonic motor, hologram scale and the optical reader

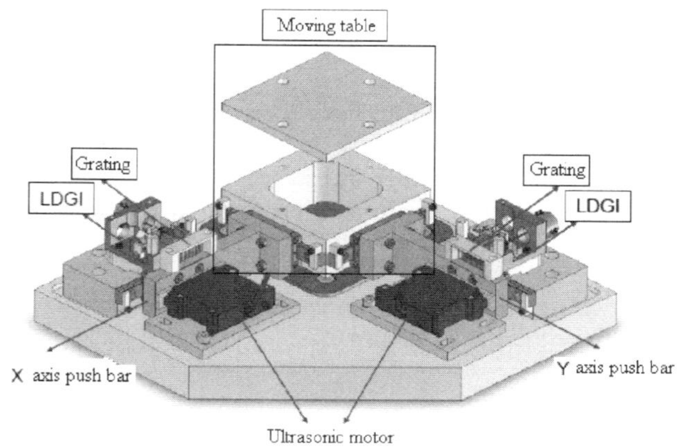

Figure 1.6. Complete structure of the co-planar XY stage

5. The Ram Design

Due to the symmetrical geometry of the pagoda bridge, the ram is mounted in the center of the bridge through the top plate, and with the geometry as symmetrical as possible. Figure 1.7 shows the structure of the ram of which the moving part is driven by the same ultrasonic motor from one side and a LDGI (displacement sensor) is fixed on the other side to balance the weight. A hologram grating scale is mounted on the moving part in line with the ram axis. The other two sides of the moving block are mounted on two linear slides. The end of the ram can chuck a high speed spindle for machining purpose, or a probe for measurement purpose. A coaxial counterweight is also designed to balance the driving force.

6. Assembly of Micro-CMM

Each of the key modules of the CMM structure has been explained on respective design concept from the precision engineering point of view. The assembly of the CMM is illustrated in Figure 1.8. It can be seen that the bridge geometry is symmetrical and the ram is in the center of the bridge. The XY stage has the co-planar feature with its centre of mass as low as possible. The extension lines of three linear scales coincide at the functional point (cutter or probe) at all commanded positions of X, Y, and Z directions. The straightness error of the moving table can be detected by an embedded sensor in line with the functional point.

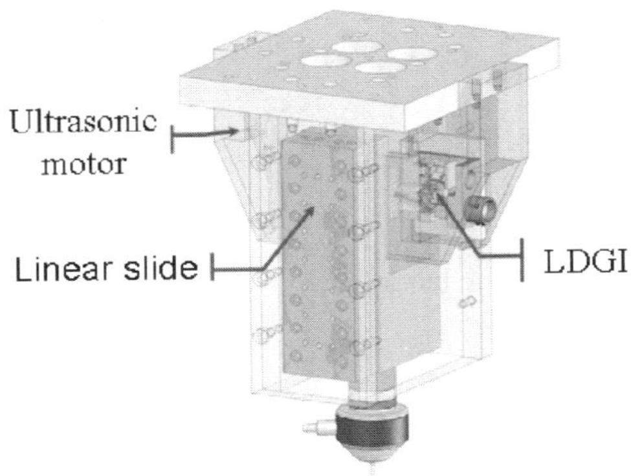

Figure 1.7. The structure of the ram

The machine is under the stage of assembly and fine adjustment. The whole dimension of the machine is only about 350 mm×350 mm×400 mm, with strokes of 25 mm×25 mm×10 mm, and resolution to 1nm. The next step will to study the motion control of the machine so as to achieve the CNC function.

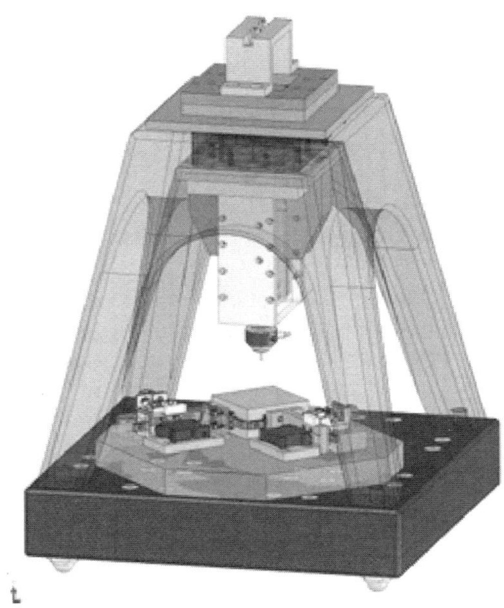

Figure 1.8. Structure of the developed micro-CMM

7. Conclusion

A novel micro-CMM structure has been designed with respect to the key modules, namely the pagoda bridge, the co-planar XY stage and the ram. All are designed based on the concept of precision engineering in terms of high stiffness, force balance, Abbé's principle and Bryan's principle. We admit that the height change of the workpiece still has Abbé offset. As it is for the dimensional measurement and profile cutting of 3D micro parts, the actual profile height will be only a few mini-meters to micrometers. The Abbé error is deemed very small. The next step will to study the motion control of the machine so as to achieve the CNC function.

Acknowledgement

The main part of this research is sponsored by the National Science Council of Taiwan under contract number NSC-94-2212-E002-001. The bridge design is collaborated with Hefei University of Technology sponsored by the National Science Foundation of China (50420120134).

References

[1] Bryan J. B. 1979, The Abb é Principle Revisit: An Updated Interpretation, Precision Engineering, P129-132.

[2] Zhang G. X 1989, A Study of the Abbé Principle and Abb é Errors , Annals of CIRP, Vol.38, No.1, P525-529.

[3] Bryan J. B. and Carter D. L. 1979, Design of a new Error-corrected Co-ordinate Measuring Machine, Precision Engineering, P125-128.

[4] Jäger J., Manske E., et al, 2002, Nanopositioning and -Measuring Technique, 7th International Symposium on Laser Metrology, Novosibirsk, Russia, Vol. 4900, P755-762.

[5] Fan K. C., Fei Y. T. et al, 2006, Development of a low cost Micro-CMM for 3D Micro/nano Measurements," Measurement Science & Technology, Vol. 17, P524-532.

[6] Fan K. C., Fei Y. T. et al, 2007, Study of a Non-contact Type Micro-CMM with Arch-bridge and Nanopositioning Stage, Robotics and Computer-aided Manufacturing, Vol. 23, Issue 6, P.276-284.

12–6

Evaluation of Size Effect on Micro-machine-tools Design for Microfactory

M. Yamanaka[1], S. Hirotomi[2] and K. Inoue[3]

[1]Tohoku University, 6-6-01, Aramaki-Aoba, Sendai 980-8579, Japan, yamanaka@elm.mech.tohoku.ac.jp
[2]Shimano Inc, Japan
[3]Tohoku University, Japan

Abstract. A micro lathe with a machine base size of 150×100 mm (postcard size) was developed by the authors. It has a good cutting performance for brass test workpieces. To know the developed machine is most suitable size for machining the given workpiece size, the error functions, which affect the cutting performance, are examined how to change to the machine tool size. The calculation method and a simple size effect of error functions are proposed and their usefulness is examined by some calculation results.

Keywords: Microfactory, Machine tools, Lathe, Machine design, Size effect

1. Introduction

Ordinary production systems use enlarged machine tools aiming at high rigidity regardless of the size of the parts produced. The concept of microfactory is to use small machine tools corresponding to the miniaturization of workpieces. This is useful for saving resources, space, and energy. Some prototypes were developed to realize this concept [1]. Okazaki developed an NC micro lathe which is 32x28x30 mm in size [1]. Park developed a miniaturized 3-axis milling machine of 200×300×200 mm [2]. The authors are developing micro lathes aiming at the practical use in mass production systems of small/micro parts for optics, IT and electronics, until now [3]. The machine with a base size of 150×100 mm (postcard size) has been developed and the relation between machine characteristics and cutting performance are evaluated to clarify advantage/disadvantage of the developed lathe [4].

To design micro machine tools, it is important to properly understand the merits and demerits of downsizing. Mishima proposed a design tools [5] combined with the form-shaping theory of machine tools [6] and Taguchi method [7]. It offers a simplified method to consider the deformation of machine tool structures and the component errors. The developed micro lathe has a good cutting performance for the test workpiece. However, it is obscure that the developed machine is most suitable size for machining the given workpiece size. It is impossible to make many sizes of machine tools and evaluate their performances.

In this paper, the outline of developed micro lathe is introduced, and the influence of accuracies according to the machine size to the cutting performance is examined in a simulation. The calculation method considering the size effect on micro machine tools design is discussed with the calculation results.

2. Micro Lathe

The maximum size of target workpiece was assumed to be ϕ5 ×10 mm and the lathe was designed. The size of the machine base was decided to be of A6 size (150×100 mm) for its foot print area considering the degree of miniaturization, its practicality, assembly and, in particular, a sufficient rigidity. Each device of the machine was designed with this restriction. The developed micro lathe MTS 2 (product of Nano Corporation, Japan) is shown in Fig. 1.1. X and Z-axis tables, which use the same modules, are guided by cross-roller ways, and driven by a ϕ2-triangle screw thread and an AC 15-mm in square servomotor. The size of the table is about 50×80 mm (excluding motor). The weight including the motor is 585 g. The headstock is fixed on the X-axis table and the tool post is mounted on the Z-axis table.

Figure 1.1. Appearance of developed micro lathe MTS2 (Nano Corporation, Japan, base size: 150x100 mm)

A micro motor of 16 mm diameter is used to drive the work spindle in the headstock from the balance with the table. A rotary speed of more than 10000 min^{-1} is necessary to obtain a cutting speed of about 150 m/min for small-diameter workpieces. Because of this, a direct drive is preferred, but a reducer is required to drive the workpiece because the motor torque of a ϕ16 mm class motor is extremely small (only a few mNm), which is generally not enough to obtain a sufficient cutting power. The rotary speed is secured as much as possible by selecting a high-speed motor. The planetary roller reducer using traction drives is adopted for this work spindle. The reduction ratio is 4. The size and weight of the headstock is 75×40×30 mm, its weight is 230 g (include motor).

3. Calculation of Error in Consideration of Size Effect

There are many error functions, which affect the cutting performance. Here, only four error functions, namely the assembly error, the motion error, the deformation by cutting force and the thermal deformation, are considered, and how they change to the machine tool size is examined. Then we propose the calculation method of error in consideration of size effect. The model with dimensions of above-mentioned micro lathe and those of enlarged/reduced are used to calculate by using the form-shaping theory. The deviation from the target position of the tool point is obtained and the proposed method is evaluated.

3.1 Calculation method

The above-mentioned micro lathe is modeled as shown in Fig. 1.2. The model consists of eight elements. The numbers from 0 to 7 are given to the reference point in coordinate transformation on each element. Though the bed is actually one solid component, it is divided into two parts virtually to consider the influence of the bed's rigidity. The target position in the tool point is apart from the point 0 to h in Z-direction and R in X-direction, respectively.

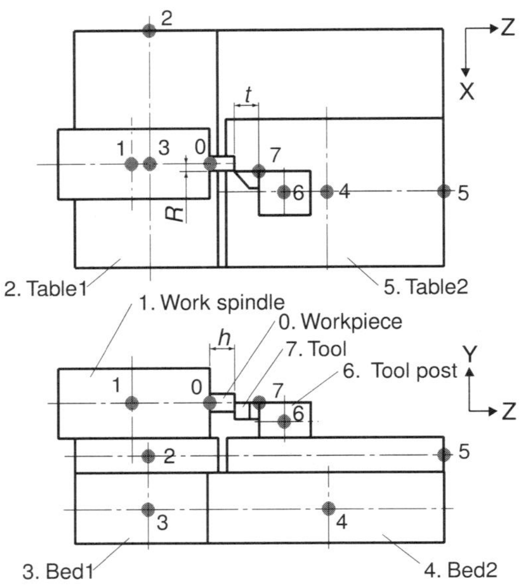

Figure 1.2. Micro lathe model for calculation

An orthogonal coordinate system S_i corresponding to an element i is defined. The transformation from S_i to S_{i+1} is represented by the coordinate transformation, which is represented by the homogeneous transformation matrices A_i. Vector $\mathbf{r}_w$ represents the relative displacement between the workpiece and the tool, and vector $\mathbf{r}_T$ represents the position of the tool point. The relation between $\mathbf{r}_w$ and $\mathbf{r}_T$ is given by Eq. (1), and $\mathbf{r}_w$ is defined as the form-shaping function.

$$\mathbf{r}_w = A_0 A_1 A_2 A_3 A_4 A_5 A_6 \mathbf{r}_T \tag{1}$$

The error between element i and i+1 is represented by another homogeneous transformation matrix ε_i as follows.

$$\varepsilon_i = \begin{pmatrix} 1 & -\gamma_i & \beta_i & \delta_{xi} \\ \gamma_i & 1 & -\alpha_i & \delta_{yi} \\ -\beta_i & \alpha_i & 1 & \delta_{zi} \\ 0 & 0 & 0 & 1 \end{pmatrix} \tag{2}$$

The form-shaping function considering the error is defined as $\mathbf{r'}_w$ expressed by

$$\mathbf{r}'_w = \varepsilon_0 A_0 \varepsilon_1 A_1 \varepsilon_3 A_2 \varepsilon_3 A_3 \varepsilon_4 A_4 \varepsilon_5 A_5 \varepsilon_6 A_6 \mathbf{r}_T \tag{3}$$

The form-shaping error function $\mathbf{r}_e$, which expresses the error as a quantitative deviation from the target position, is defined as Eq. (4).

$$\mathbf{r}_e = \mathbf{r}'_w - \mathbf{r}_w = \begin{pmatrix} \Delta x \\ \Delta y \\ \Delta z \end{pmatrix} \tag{4}$$

The error between elements of the micro lathe is represented by 6 components, which are translational and rotational motions along the X, Y and Z axes, as δ_x, δ_y, δ_z, α, β, and γ, respectively.

The limited four error functions may affect between elements shown in Fig. 1.3. The assembly error appears between all elements except between the bed 1 and the bed 2, which are divided into two parts. The motion error appears between the elements that move relatively during cutting. Because force and heat propagate through an element, the deformations by cutting force and heat appear between all elements. Moreover, as shown in Table 1, it was analyzed and classified which error components are affected by above-mentioned four error functions. For example, the motion error between the workpiece and the work spindle depends on the runout of the work spindle. Because the workpiece rotates around the Z-axis, the translational error in X and Y except Z-axes are considered. The assembly error can not define the direction to appear, and is assumed to appear in all 6 components. The deformations by cutting force and heat are assumed to appear in 3 components of translational error.

3.2 Size effect of error

Table 2 shows to what extent the amount of error of each item in Table 1 changes when the size of the machine tool is changed n times, which is examined by other reports [8, 9] and the calculation. It is thought that the difficulty of assembly is constant regardless of the machine size. Therefore, the size effect of the assembly error is 1. The motion error depends on the accuracies of guides in a slide table or bearings in a work spindle. Here, it is considered that the machine element of same absolute accuracy can be selected regardless of the size, and the size effect of the motion error is assumed to be 1 as well as the assembly error. The cutting force is divided in 3 directions, namely principle, the thrust and the feed force. A simple bend and

Table 1.1. Error functions affected on each error component and actual values of error in n= 1

	α [deg]	β [deg]	γ [deg]	δ_x [μm]		δ_y [μm]		δ_z [μm]	
ε_0 (0-1)	A: ±0.05	A: ±0.05	A: ±0.05	A: ±1.0 M: ±0.5	C: −6.15 T: −0.8	A: ±1.0 M: ±0.5	C: 1.34 T: 0.8	A: ±1.0	C: −3.13 T: −0.8
ε_1 (1-2)	A: ±0.05	A: ±0.05	A: ±0.05	A: ±1.0 M: ±1.0	C: −2.15 T: 0.5	A: ±1.0 M: ±0.5	C: 0.46 T: 0.5	A: ±1.0 M: ±0.5	C: −1.08 T: −0.5
ε_2 (2-3)	A: ±0.05	A: ±0.05	A: ±0.05	A: ±1.0	C: −2.07 T: −0.2	A: ±1.0	C: 0.44 T: 0.2	A: ±1.0	C: −1.04 T: −0.2
ε_3 (3-4)					C: −1.02 T: 0.1		C: 0.22 T: 0.1		C: −0.51 T: −0.1
ε_4 (4-5)	A: ±0.05	A: ±0.05	A: ±0.05	A: ±1.0	C: −2.07 T: 0.2	A: ±1.0	C: 0.44 T: 0.2	A: ±1.0	C: −1.04 T: 0.2
ε_5 (5-6)	A: ±0.05	A: ±0.05	A: ±0.05	A: ±1.0 M: ±0.5	C: −3.02 T: 0.4	A: ±1.0 M: ±0.5	C: 0.65 T: 0.4	A: ±1.0 M: ±1.0	C: −1.51 T: 0.4
ε_6 (6-7)	A: ±0.05	A: ±0.05	A: ±0.05	A: ±1.0	C: −3.25 T: 0.6	A: ±1.0	C: 0.70 T: 0.6	A: ±1.0	C: −1.63 T: 0.6

A: Assembly error, M: Motion error, C: Deformation by cutting force, T: Thermal deformation

compression of beam according to the direction of applied force on each element is considered, the size effect of deformation by cutting force is assumed to be $1/n$ and $1/n^2$, respectively. If the temperature increases, generally a material expands. The deformation δ_t of a member having a length μ can be calculated using the formula

$$\delta_t = \alpha_t \Delta T \ell \tag{5}$$

where, α_t is the linear coefficient of thermal expansion and ΔT is the temperature rise of the member. Therefore, the size effect of thermal deformation is considered as n.

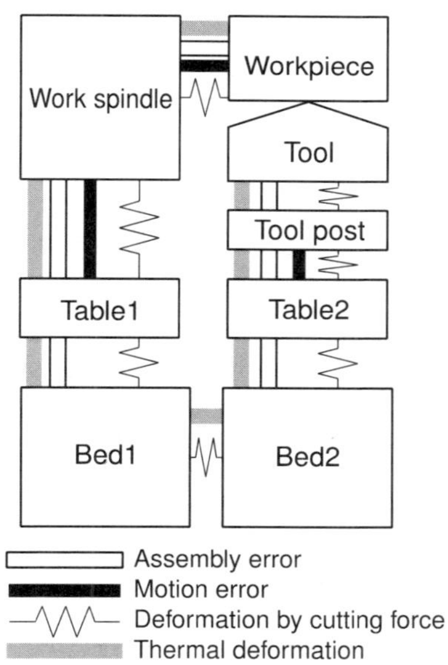

Figure 1.3. Generation of error between elments of micro lathe

Table 1.2. Size effect of error functions

Error function	**Size effect**	
Assembly error	1	
Motion error	1	
Deformation by cutting force	Bending: $1/n$	Compression: $1/n^2$
Thermal deformation	n	

3.3 Result and discussion

The dimensions of each element of the model in Fig. 1.2 are required to calculate the error. The real dimensions of the developed micro lathe shown in Fig. 1.1, and these are used as that of a model for n= 1. Then the model is enlarged n times and the error in Eq. (4) is calculated by the method mentioned above using the size effect of errors shown in Table 1.2. And how the error changes according to n is examined changing n from 0.5 to 10. However, the dimensions of workpiece and tool are assumed to be constant regardless of n, where R is 2.5 mm, h is 10 mm and t is 8 mm. In this method, actual values of error are required for the calculation. Hence, the magnitude of each error for n= 1 is decided as shown in Table 1.1. The assembly and the motion errors were decided in consideration of actual micro lathe. Both errors can take a positive or negative value, and are expressed with the sign of + or -. The deformations by cutting force and heat are obtained by FEM. The deformation of workpiece and tool were not considered. The element 7 is fixed and the magnitude of deformation of each element is obtained. The tangent force applied to a workpiece is calculated by the rated torque of the motor of the work spindle, and it is assumed to be the principal cutting force. The ratio among 3-divided cutting forces is decided by the reference [8]. The deformation of each element when the atmosphere rises by 20 K is calculated as the thermal deformation. Because the target is a cylindrical component, the errors ΔR in cutting radius and Δz in longitudinal direction are defined as the evaluation function of error. R' and R are the actual and instructed cutting radii, respectively. ΔR is given as follows.

$$\Delta R = R' - R \qquad R'^2 = (R + \Delta x)^2 + \Delta y^2 \tag{6}$$

The assembly and motion errors take positive or negative values in Table 1.1, and the number of those items is 44. If the case of plus/minus in each item is combined, there are 2^{44} patterns to calculate. This combination is so large that the orthogonal table in Taguchi method [7] is applied and the calculation is held in small combinations. Here, the orthogonal table of L64 is used to obtain $\mathbf{r}_e$ by Eq. (4) and the evaluation functions are calculated in 64 cases.

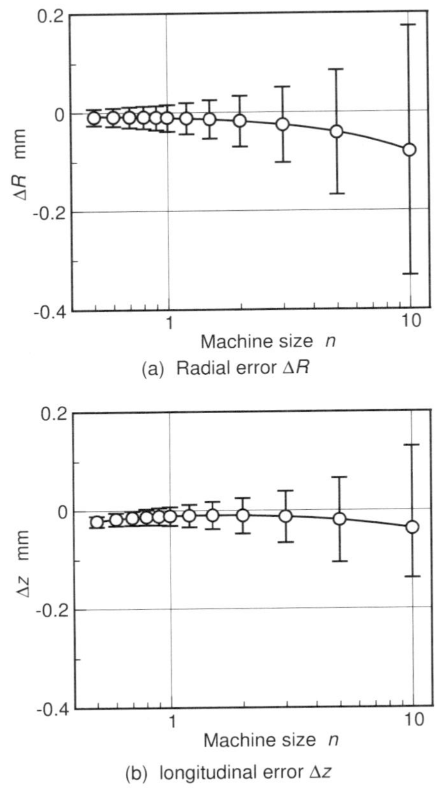

Figure 1.4. Relation between evaluation functions and machine size

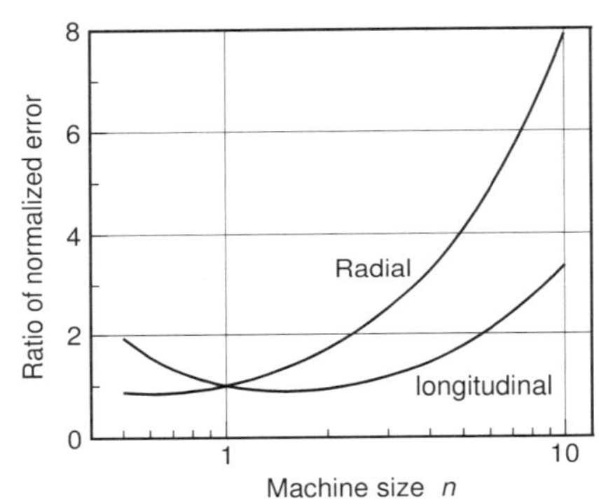

Figure 1.5. Tendency of change of normalized evaluation functions

The relation between the machine size n and the average of evaluation function is shown in Fig. 1.4. The max/min values are also plotted in the figure. The magnitude of ΔR increases as n increases. The difference between the maximum and the minimum values increases too. The assembly error and the thermal deformation may be the reason for this. Though the magnitude of assembly error does not depend on n, as shown in Table 1.2, the movement in coordinate transformation becomes large as n increases, so that r_e becomes large. The thermal deformation increases because of its size effect. As for Δz, there is the minimum value near $n = 1.5$. Δz increases as well as ΔR when n increases more than it. When n becomes small, Δz increases slightly by the size effect of the deformation by cutting force. This suggests that an optimum size of this kind of machine exists for a given size of workpiece. Here, R is set as 2.5 so that $R >> \Delta x$, Δy and ΔR is almost constant from Eq. (6) though n becomes small.

Each mean value for n is normalized by dividing by that for n= 1 to know the tendency of change of evaluation function to n. The result is shown in Fig. 1.5. ΔR and Δz become 8 and 3.4 times as large as at n= 10. It can be said that the sensitivity of ΔR to the machine size is higher. The calculation result by the proposed method changes greatly by the values of error and the size effect shown in Tables 1.1 and 1.2. In this study, though, the size effect of error functions was simplified, more examination is necessary to improve the accuracy of calculation.

4. Conclusions

The influence of accuracies according to machine size on cutting performance is examined. The conclusions may be summarized as follows:

1. The error function, which affects the cutting performance, is examined. Then the calculation method of error in considerlation of size effect is proposed applying the form-shaping theory.
2. The calculation model is built by using the dimensions of the actual micro lathe. The relation between machine size and error is obtained. The assembly error and the thermal deformation have the strongest influence.

References

[1] Okazaki, Y., et al., (2004) Microfactory: Concept, history and developments. Trans. ASME, J. Manufacturing Science and Engineering 126: 837-844

[2] Park, J. K., et al., (2006) A precision meso scale machine tools with air bearings for microfactory. Proc 5th Int. Workshop on Microfactories, Besancon, France (CD-ROM)

[3] Iijima, D, et al., (2004) Micro Turning System 3 (MTS3): A practical CNC lathe for microfactories. Proc. 4th Int. Workshop on Microfactories, Shanghai, China 1: 50-55

[4] Yamanaka, M, et al., (2006), Relation between mechanical characteristic and cutting performance of micro lathe. Proc 5th Int. Workshop on Microfactories, Besancon, France (CD-ROM)

[5] Mishima, N., (2003) Design of a Miniature Manufactuirng System for Micro-fabrication. Proc 10th Annual Conf. for Concurrent Engineering, Modeira, Pirtugal 1129-1135

[6] Reshtov, D. N., Portman, V. T., (1988) Accuracy of machine tools. ASME Press, New York

[7] Taguchi, G., Konishi, S., (1994) Quality engineering series. ASI Press

[8] Sugita, T., et al., (1984) Fundamentals of cutting (in Japanese). Kyoritsu Publish.

13

Process and Machine Tool Control

13–1

Closed-loop Sensor System for Automated Manufacturing Machines

A.A. Shaik[1], G. Bright[2] and W. L. Xu[3]

[1]Council for Scientific and Industrial Research, Material Science and Manufacturing, South Africa, ashaik@csir.co.za
[2]University of Kwa-Zulu Natal, School of Mechanical Engineering, South Africa, brightg@ukzn.ac.za
[3]Massey University, Institute of Technology and Engineering, New Zealand, w.l.xu@masey.ac.nz

Abstract. Many automated multi-axis machines operate under 'open loop' control. The exact position in space of the end effecter or tool head is not sensed directly. The software controlling the robots or automated manufacturing machines make a calculated estimate of where the tool-head should be. This is often achieved by monitoring sensors on axes that track linear translation and rotations of shafts or gears. For low precision applications this system is appropriate. However, positional errors often occur. This method may not be suitable for high precision robots and automated machine tools.
There exists a need for a sensor system that is capable of acquiring the exact spatial coordinates of the tool point or end effector directly. The aim of this research project was to design a low cost sensor system that would precisely locate the tool points' spatial coordinates directly by 'closed loop' control. This aided in the reduction of errors encountered in 'open loop' control. The system was capable of seamless integration with existing techniques for motion control. The sensor system was able to locate the tool head in 2D space. With simple additions and modifications it rendered itself applicable to 3Dspace location of the tool head. It was modular, sufficiently robust and error immune such that it would work in almost any environment.

Keywords: Automation, Manufacturing, Closed Loop Control

1. Introduction

The need for a sensor system that provides direct feedback of the end effector's spatial coordinates is essential for precise assembly and machining. The objective of this research project was to design a parallel robot that would have an integrated sensor system. The sensor system should be low cost and be capable of locating the tool points' spatial coordinates directly or aid in reducing the errors encountered in the 'open loop' control. The system should be capable of seamless integration with existing techniques for motion control and should be able to locate the tool head in 2D space. With simple modifications it should render itself applicable to location in 3D space. It should be modular, sufficiently robust and error immune to work in almost any environment.

There are numerous technologies available for the purpose of locating objects in space. These are the Global Positioning System (GPS – with 10 m accuracy, a Differential GPS solution by [1] reduces the errors to less than 5 cm); proximity location through Bluetooth, WiFi and Cellular networks (position is deduced by knowing the cell with which a device is communicating) and triangulation systems (using lasers, ultrasound, etc.).

In addition there are imaging methods used to determine position. [2] Discusses a low cost solution employing a camera and LCD (Liquid Crystal Display) screen to locate an object's coordinates in 2D with high accuracy. VSLAM (Visual Simultaneous Location and Mapping) systems used in mobile robotics use cameras to build a map of its surroundings for navigation purposes. There are also image processing techniques that can determine movement from consecutive frames provided by a camera. These images also provide information regarding the surroundings and are completely passive. The accuracy and resolution however depend on how far from the object the camera is and the resolution of the camera itself.

All these technologies are suitable for locating large objects in a relatively large space, for instance locating people in buildings [3]. In industry, however, robotic arms used for assembly, welding or spray-painting move in a confined space and the position of the end effector must be determined with millimeter to sub-millimeter resolution.

For manufacturing purposes the resolution of computer numerically controlled (CNC) machines range from micrometers to nanometers. For such purposes Laser interferometer technologies are sometimes used. These position sensing systems have excellent resolution as well as accuracy and are used in manufacturing environments for IC (Integrated Circuit) design, prototyping and manufacturing. An optical heterodyne interferometer designed at NASA's Jet Propulsion Laboratory can measure linear displacements with an error of 20 pm (Pico-meters, 10^{-12}m).

Grid encoders offer another solution; grids made by OPTRA have a coverage range up to 380 mm × 380 mm. Their accuracy and repeatability lie in the same range as Laser Interferometers (i.e. micro- to nanometer resolution). Both Grid encoders and Laser Interferometers are expensive technologies, usually costing hundreds of thousands to millions of dollars. Furthermore, they can not measure absolute position directly; they both use fringe patterns (due to light wave interference) to measure relative displacement. The integration of these displacement measurements coupled with the knowledge of the end-effectors' initial position yields the current position.

2. Proposed Sensor Feedback System

The problem of locating the end effector of a robot in real world space is first reduced to finding its position in a 2D plane with regard to a point reference. The general problem of location in 3D space is solved by attaching two 2D planes at right angles. With such an arrangement 2 axes coincide and if the reference point of each plane coincides, the result is a 3 axis sensor system for position location. This paper attempts to document a solution of finding the end effector in a 2D plane.

After consideration of the available physical quantities (ultrasound, infrared light, radio waves, etc.) used when locating objects, it was decided that a laser would be most suitable for the reasons that follow. A laser light sensor can be conditioned to provide a digital output. Comparatively analogue sensors require digitization for use in digital systems. Analogue signals are compromised by atmospheric effects, temperature, humidity and unshielded noise from surrounding machinery. Triangulation utilizing radio, ultrasound or infrared waves is not suitable as multiple reflections from surrounding surfaces cause interference and provide unreliable results. They also require modulation and demodulation to distinguish the signals generated from any that can be created by the environment. The proposed sensor concept utilizes a direct approach, with a laser attached to the end effector and a sensor grid (a grid of laser sensors equally spaced in rows and columns) mounted directly above it. This is a natural choice as the coherent nature of laser light makes finding the end effector in 2D space easy if the laser beam remains perpendicular to the sensor plane at all times. The end effectors' location is the same as the sensor which is stimulated (in a 2D plane, depth has no meaning). This perpendicular constraint can be enforced by the use of tilt sensors that can tell the orientation of an object, corrections can then be made to keep the end effector perpendicular. Also the inherent nature of the robot can be sufficient to force this condition. The Flex-Picker is one such example as its arms force the end-effector to remain parallel to its base at all times.

It must be stressed that this sensor system requires only bit (1 or 0) information for each sensor. Each sensor is either stimulated or not stimulated. This makes data processing and transfer far simpler and makes control easier. The resolution is limited to the spacing between sensors. If the spot light is smaller than the spacing between sensors, these will represent a dead zone where beam tracking will be lost completely. The laser light detectors are phototransistors with a Darlington configuration. Current fabrication techniques can accommodate millions of transistors on a sliver of silicon. These fabrication methods can be used to construct a detector screen with an exceptional and practical resolution. Resolution affects data output, a greater resolution implies more data per unit area (more sensors).

A hybrid type system would involve a sensor grid with a comparatively lower resolution. Each sensor provides a checkpoint. Knowing the exact spatial distance between these detectors provides the controller a means to limit the errors incurred. Instead of accumulating errors from one extremity to the next, errors only exist between successive detectors

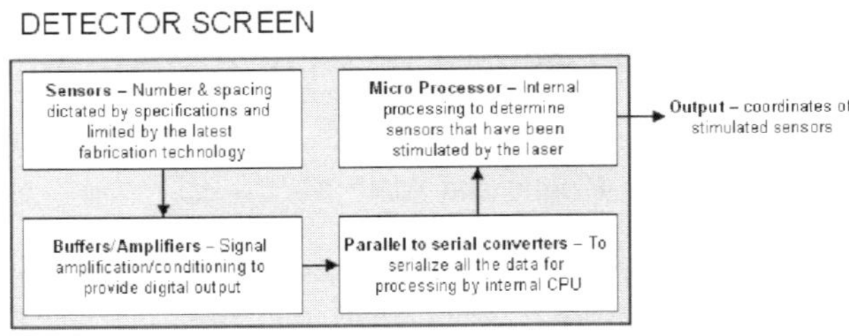

Figure 1.1. Block Diagram of intended sensor system.

3. Mechatronic Design

The mechatronic design consists of 3 parts the mechanical, electronic and software components. As stated in [4], the term mechatronics is used for the integration of microprocessor control systems, electrical systems and mechanical systems. The mechanical structure was designed to test the electronic hardware and software control, the purpose of which is to validate the proposition. The mechanical design was also undertaken to explore parallel mechanisms. The design is documented to deliver a simple prototype. The specification on resolution was relaxed as it was stated that fabrication techniques can produce a screen with a realistic resolution. More emphasis was placed on creating a cheap system that could reduce errors.

Mechanical Structure:

The mechanical structure is based on a Flex-Picker pick and place parallel kinematics industrial robot. It is a scaled adaptation. The design consists of 4 articulated arms; 4 servo motors; a plate end effector with attached laser; ball-cup joints and a mounting frame. The entire mechanical structure is 600mm in length, 400mm wide and 500mm high. Figures 1.2 and 1.3 illustrate the parts and the assembly.

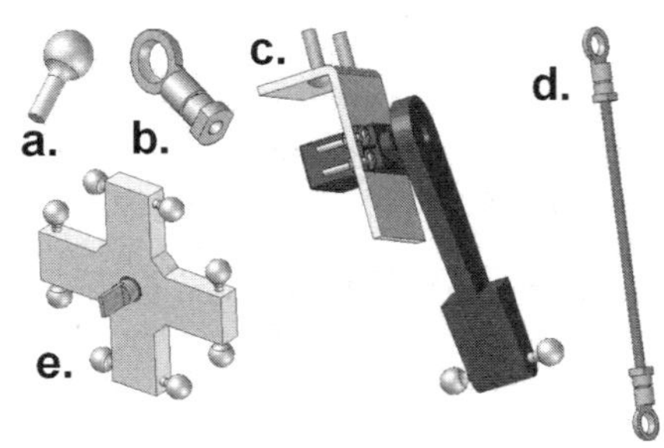

Figure 1.2. Significant mechanical parts that position the laser: (a) Ball from bearing, (b) Cup from modified bearing socket, (c) Servo motor with upper arm attached and mounting bracket, (d) Lower arm component, (e) Laser and laser mounting

It must be noted that the lower arm components are held together via 2 springs (not shown), one just below the 'elbow' and the other just above the 'wrist' for each forearm. The ball cup joints give a large degree of freedom. These were made from ball in socket bearings. The upper arms swing from side to side whereas the lower arms can move up, down, left and right and can even rotate about the 'elbow' by sequencing pairs of its basic motion (induced by rotating pairs of servos). The laser can move about a section of space, which is roughly a hemisphere below the sensitivity area (the square cut-out on the servo mounting frame, Figure 1.3 (a). The guidelines for parallel mechanism design were followed as provided in [5].

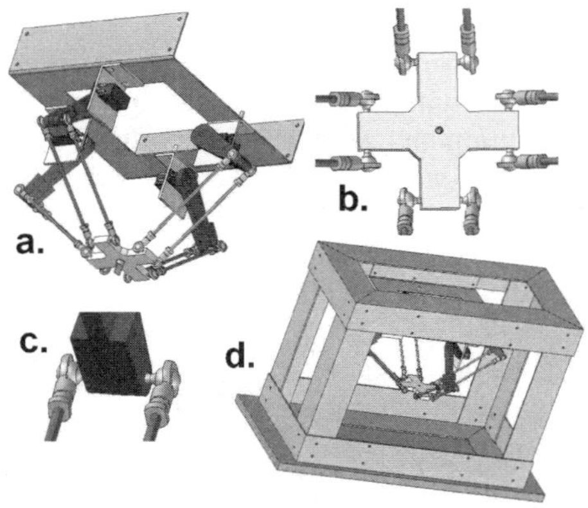

Figure 1.3. Assembly of all components resulting in the final structure: (a) All servos mounted with upper arm and lower arm attached to laser, (b) Lower arm attachment to laser end effector, (c) Articulated arm 'elbow' joint, (d) Complete Assembly

Electronic Hardware:

The system designed is a hybrid type as mentioned in section 1.2. The resolution is low due to the 5 mm diameter of the LED type package of the laser sensor. Also the PCB tracks routing sensors to buffers and data converters occupy significant space, which could not be avoided. To improve resolution a smaller package should be used, preferably surface mount, however these could not be sourced at the time this screen was made. Each sensor grid consists of 64 BP103 phototransistors arranged on an 8x8 grid. The resolution (distance between the centres of 2 successive sensors) is 18 mm on both the rows and columns.

This phototransistor has a daylight filter to prevent wrongful stimulation by ambient light. It is sensitive to light wavelengths in the range 600–900 nm. A 650 nm key-ring laser is being used as a sensor stimulant. This is a cheap and effective solution. Although the output power of this laser is less than 1 mW, it is sufficient to turn the phototransistor on. The sensed signal has to be buffered/amplified to ensure that the voltage level output from the LPT133 is within the proper digital range (0 – 0.8 V for a logic 0 and 3.5 – 5 V for a logic 1), according to [6] for proper electronic design. For this purpose each sensor on a column is passed to a transistor driver within a ULN2803, which consists of 8 transistor drivers. Eight driver chips are used, one for each column. The outputs from each ULN2803 are fed to a parallel to serial data converter, the 74LS166 to serialize the data for transfer to a PC. The 8 output serial lines from the data converters are fed to an ATMEL ATmega8515 microcontroller. There are eight bytes of data and each bit represents one sensor's current state. The controller is used to transfer the 8 bytes of data to the PC via its USART transceiver and the PC's RS232 serial port. It also controls the 74LS166 data converters and the servo motors. The microcontroller enables the data converters and clocks the data out of each of them.

Software:

The software has two parts to it, i.e. the microcontroller code and the user interface.

Microcontroller Code:

There are 4 parts to the microcontroller code, i.e. receiving and interpreting commands from the PC; sensor data acquisition; data transfer and servo rotation.

Receiving and Interpreting Commands – There are commands for activation, sensor data acquisition, data transfer and servo rotation. These tell the controller when to run the corresponding routines.

Sensor Data Acquisition - This routine enables/disables; clears and clocks the data out the parallel to serial converters. See [7] for a complete description of how this is achieved.

Data Transfer - Once the 'Sensor Data Acquisition' routine completes, the 8 bytes of data await transfer. The serial port of the PC works with the ASCII character set. Each data byte representing a number in the range 0-255 must be sent out the microcontroller's serial port via 3 ASCII characters (one character for each of the hundreds, tens and units digits) representing numbers (0-9 which in ASCII is 0x30-0x39 in hexadecimal notation).

In total there are 32 character bytes (4 character bytes per data byte, including byte completion character) transferred from controller to PC. Using a Baud Rate of 9 600 (bits per second) and including a parity check bit (9 bits per byte) this takes 0.03 s to complete.

Servo Rotation - As there are 4 servo motors, 4 pulse-width modulated signals have to be generated. Incoming commands indicate a particular servo and the length of its pulse-width. The pulse-width value will be received in 3 bytes. In total there are 16 character bytes transferred from PC to controller (4 bytes per servo) to position the laser as required. The time taken to do this is 0.015 s (9 600 Baud). Timers within the controller will ensure that the PWM signals comply with the desired range of 1-2 ms.

User Interface- The user interface provides a visual display of the data received, 64 colored circles represent the 64 phototransistors. The display routine searches through each data byte (columns) for low bits (rows) and changes the color of the corresponding circle (blue when not stimulated, red when stimulated). The interface also allows the user 2 options for control; either via a mouse or a selection grid. With the mouse the user can manually control the laser and move it anywhere within its mechanical constraints. When the selection grid is active the laser will position itself, first finding a reference point and then move along rows and columns. It will pass over all sensors which have been selected by the user.

Control Design Overview:

The sensor grid provides direct feedback of the position of the laser/tool head at discrete points in space. The purpose of the grid is to reset accumulated errors in the positioning control system. The coordinates of the sensors are stored in the controlling software. When errors are accumulated and need to be reset the end effector is moved to sensor that is closest to the end effector. Once it is positioned and the sensor is not stimulated it knows that there are errors in its positioning and it then spirals out to locate the sensor and reset the error.

4. Performance/Operation

The novelty of this sensor system lies with the fact that it can determine the position of the end effector directly, provided that the constraints mentioned are strictly adhered to. It does not determine position from calculation as in laser triangulation systems and it does not integrate displacements as in laser interferometers.

The novelty of the mechanical design is with respect to the 4 arms that attach to the end effector. The arrangement and configuration has the advantage of easier control, a simplified kinematics model and added stiffness.

Sensor Grid:

The sensor grid operates as intended after having integrated all the functions described on the microcontroller. The project is still in the testing phase and is currently being integrated into the system. Much work is still needed to ensure the screen works properly with all data conversion and transfer routines. The second version of the screen is currently being designed to improve resolution and system repeatability.

Mechanical Structure:

The most important aspect of the mechanical structure is the articulated arms. A simulation model was created to simulate and animate its movement, to ensure that it complies with the requirements of the design. The results are shown in Figure 1.4. A motion generator is attached to each servo head and is set to follow a harmonic function. The end effectors' motion spanned the entire range of the sensitivity area. Most importantly the laser mounting remained parallel to the sensor grid PCB (sensor plane) throughout its motion, this ensures that the laser beam is always perpendicular to a detector. The singularities for this mechanism occur when the arms are completely extended or folded, when the upper and lower legs are co-linear. This situation was avoided in the simulation and can be avoided in reality.

5. Conclusion

The objective of creating a parallel robot with an integrated sensor system has been met. The mechanical system works with the control system designed. The only issue with the mechanical aspect is that the ball joints. They do not provide sufficient positioning accuracy and repeatability. The use of high performance ball joints or compliant joints in a future version of the robot will eliminate this problem.

The sensor system is of low cost and provides direct feedback. Resolution is currently the only problem, but as mentioned fabrication techniques can resolve this issue. This system is modular and facilitates inclusion in existing systems. 3D location is possible with two of these sensor grids placed at right angles to each other. The sensor grid is also robust and is not affected by ambient light.

A fair comparison with best practices of industry can only be made with a sensor system that has roughly the same resolution and accuracy. This project sought to test the theory that such a system could be created, and if it would be useful. As such it has accomplished its purpose, but a fabricated screen could provide proper quantitative performance measures.

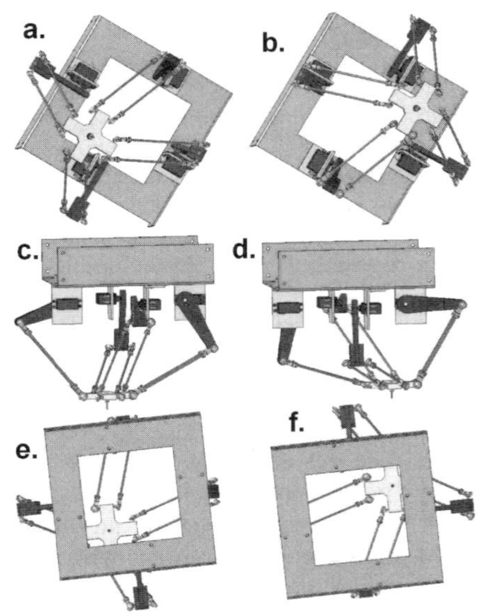

Figure 1.4. Simulation of laser tool head movement: a and b Bottom views, c and d Side views, e and f Top views

References

[1] Willgoss, R., V. Rosenfeld and J. Billingsley 2003, High Precision GPS Guidance of Mobile Robots.

[2] Ziegert, J. 2005, Active Vision/Display Sensors for Precision Positioning. University of Florida, USA.

[3] Dempsey, M, 2003, Indoor Positioning Systems in Healthcare, a Basic Overview of Technologies. Radianse, Inc.

[4] Bolton, W. 2003, Mechatronics, Electronic Control Systems in Mechanical and Electrical Engineering, Third Edition.

[5] Creamer, R. H. 1976, Machine Design, Second Edition.

[6] Sedra, A. S. and K. C. Smith, 1998, Microelectronic Circuits, Fourth Edition.

[7] Shaik A.A, Prof. G. Bright and Prof. W. L. Xu, 2006, Modular Sensor System For Flexi-Picker And Multi-Axis Automated Machines

Friction Compensator for Feed Drive Systems Consisting of Ball Screw and Linear Ball Guide

R. Sato[1] and M. Tsutsumi[2]

[1]Tokyo University of Agriculture and Technology, 2-24-16, Naka-cho, Koganei, Japan, 184-8588, ryuta@cc.tuat.ac.jp
[2]Tokyo University of Agriculture and Technology, 2-24-16, Naka-cho, Koganei, Japan, 184-8588, tsutsumi@cc.tuat.ac.jp

Abstract. Quadrant glitches are often observed on the circular trajectories of NC machine tools. In this study, a friction compensator that can compensate for the friction characteristics in a microscopic region is proposed. In the compensator, friction forces and torques are modeled using a simple mathematical model, and the estimated friction torque around the motor axis is added to the torque command. In order to evaluate the effectiveness of the proposed friction compensator, the compensator was applied to an actual feed drive system and its mathematical model. As the results of the experiments and simulations, it was confirmed that the proposed friction compensator can eliminate the quadrant glitches effectively, even if the radius and feed rate change.

Keywords: NC machine tool, Feed drive system, Circular motion, Quadrant glitch, Friction model, Friction compensator

1. Introduction

The accuracy of NC machine tools is one of the most important aspects of fabricating mechanical parts, molds and dies. Since the trajectory of the cutting tool relative to the workpiece defines the machined surface, highly accurate motion is always required. In NC machine tools, generally, the relative motions are generated by the motion of feed drive systems consisting of a servomotor, a ball screw, and linear ball guides. Hence, the motion accuracy of feed drive systems is almost equal to the accuracy of NC machine tools. Circular motion tests are usually applied to evaluate the accuracy of motion of the feed drive system. For circular trajectories, large quadrant glitches are often observed as the motion changes across the x and y quadrants. It is well known that this phenomenon is caused by the friction forces acting on the feed drive mechanism.

In order to eliminate the quadrant glitches, various research studies have been carried out. Tung et al. investigated machining errors due to friction and proposed a compensation method based on repetitive control [1]. Erkorkmaz and Altintas proposed a feed forward friction compensator based on a friction model as a function of sliding velocity [2, 3]. Tan et al. applied a dual-relay feed back approach toward modeling and compensation of frictional effects [4]. In addition, Mei et al. simulated the quadrant glitches by a mathematical model with friction and proposed a compensation method based on the simulation results [5, 6].

On the other hand, it is known that the pre-sliding friction characteristics influence the quadrant glitches. From this point of view, Park et al. modeled the pre-sliding friction as a time function and proposed a friction compensator [7]. Suzuki et al. also proposed a friction compensator based on a pre-sliding friction model consisting of a spring and damper [8].

However, although many kinds of methods for friction compensation have been proposed, as mentioned above, no method can eliminate the quadrant glitches of circular motion under various radii and feed rates. In this study, a compensator that can compensate for the pre-sliding friction characteristics and transient response of the servo amplifier is proposed. In order to clarify the effectiveness of the proposed compensator, circular motion tests under various radii and feed rates were carried out. From the results of the experiments and simulations, it was confirmed that the proposed friction compensator can eliminate the quadrant glitches effectively, even if the radius and feed rate change.

2. Experimental Apparatus

The X-Y table system shown in Fig. 1.1 was used to evaluate the proposed method. This system consists of two feed drive systems (one each in the X- and Y-axis directions). The mechanism consists of an AC servo motor and amplifier, a ball screw and linear ball guides. The AC servo motors and ball screws are directly connected by a coupling. The two AC servo motors used were identical.

The whole system was controlled by a personal computer with a DS1103 (dSPACE GmbH) DSP board. The position in each axis was detected by two linear encoders with a resolution of 0.1 μm. The rotational angle of the servo motors was detected by the attached rotary encoders of the motors with a resolution of 10,000 pulses/rev. Both the position and rotational angle were digitally measured and recorded through counter ports of the DSP board.

Feed drive systems generally have both velocity and position control loops. In the experimental system, the positions were controlled based on measured positions. The rotational velocities were controlled based on the rotational velocities, which were numerically calculated from the measured rotational angles. These controls were carried out on both the X and Y axes.

Since the focus of this study was the dynamics of the system, all experimental results shown in this paper were measured using linear encoders. The measured data from the linear encoders indicate the dynamic characteristics of each axis, except geometrical errors such as the straightness. Motor torque curves were also measured using analog monitor outputs from the servo amplifiers and AD converter ports of the DSP board.

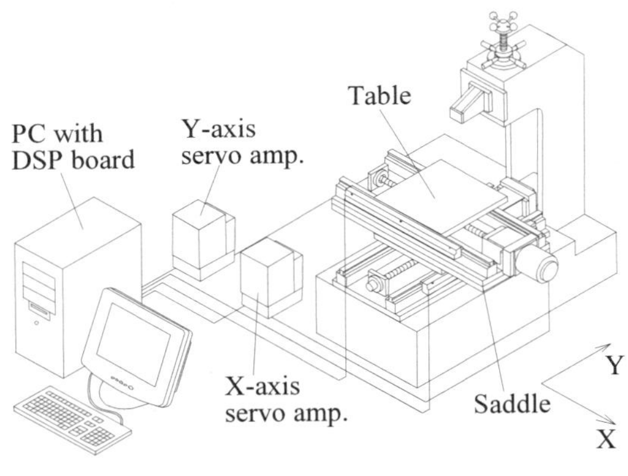

Figure 1.1. Experimental apparatus

3. Friction Characteristics of Feed Drive Mechanism in the Microscopic Region

Friction exhibits significantly different characteristics during motion direction changes, and this attribute influences the generation of quadrant glitches. In order to investigate the friction during motion direction changes, the motor torque and table displacement were measured under sine wave motion. The amplitude of the sine wave motion was 50 μm with a frequency of 0.05 Hz.

Figure 1.2(a) shows the measured result of the relationship between the table displacement and motor torque. The motor torque incorporates the inertial force, viscous friction force, and nonlinear friction force. In this case, however, the velocity and acceleration are quite small. Hence, the measured result shown in Fig. 1.2(a) represents the nonlinear friction characteristics.

As can be seen from this figure, the friction force is almost proportional to the displacement in the microscopic region. This characteristic is known as the nonlinear spring characteristic. Cuttino et al. and Fukada et al. investigated the nonlinear spring characteristic of ball screws [9, 10], and Futami et al. investigated this characteristic in linear ball guides [11].

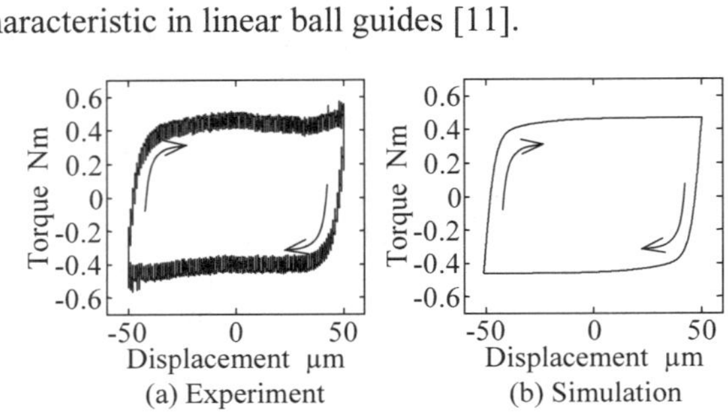

Figure 1.2. Relationship between table displacement and motor torque in the microscopic region

In this study, therefore, a simple model for the nonlinear spring friction is proposed in Eqn. (1.1). In this equation, f_c is the Coulomb friction torque around motor rotor [Nm], x' is the table displacement from the motion direction changing point [m], and a is the inclination coefficient of the friction torque.

$$f = f_c(2\tanh(a \cdot |x'|) - 1)\,\mathrm{sgn}(\dot{x}) \qquad (1.1)$$

Figure 1.2(b) shows the simulated result of the relationship between the table displacement and friction torque. As shown in this figure, the nonlinear spring characteristic is well expressed by Eqn. (1.1).

Since it can be assumed that the velocity is quite small when the motion direction changes, in this study the relationship between velocity and friction force, such as the Stribeck effect, is not introduced into the model.

4. Proposal for the Friction Compensator

The friction compensator shown in Fig. 1.3 is proposed based on the proposed friction model mentioned above. Figure 1.3(a) shows a block diagram of the controller and servo amplifier, while Fig. 1.3(b) shows the detail of the proposed friction compensator.

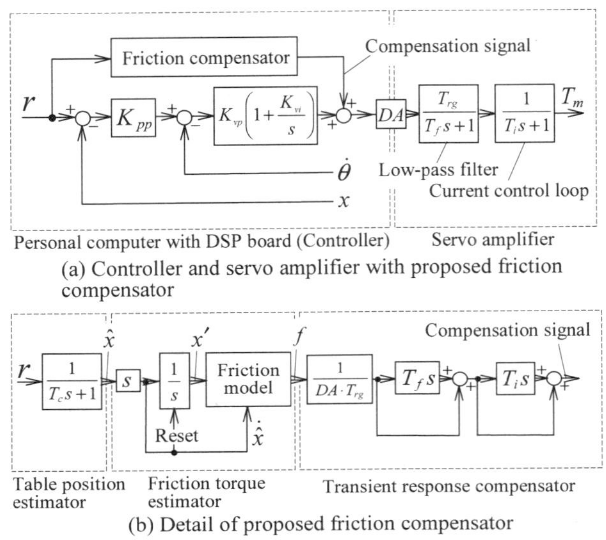

Figure 1.3. Block diagrams of proposed friction compensator

The proposed friction compensator generates a compensation signal from the positional command r [m], and the compensation signal is added to the torque command signal generated from the velocity controller, as shown in Fig. 1.3(a). Where K_{pp} is the proportional gain of the positional loop, K_{vp} is the proportional gain of the velocity loop, K_{vi} is the integral gain of the velocity controller, DA is the DA converter gain [V], T_{rg} is the torque command gain [Nm/V], T_f is the time constant of the low-pass filter [s], T_i is the time constant of the current control loop [s], and T_m is the output torque of motor [Nm].

The torque command signal from the controllers is input into the amplifier. The amplifier controls the current of the motor armature in order to generate the desired torque. In this study, the current control loop is modeled as a first-order system, as shown in Fig. 1.3(a). The amplifier also has a low-pass filter.

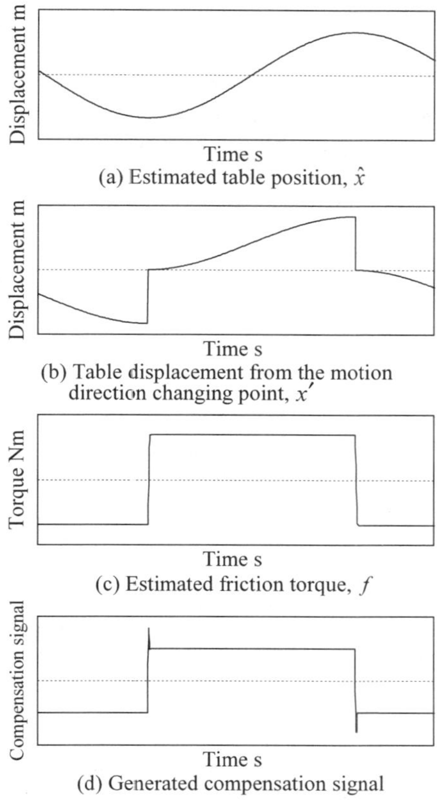

Figure 1.4. The signal patterns in the proposed friction compensator

If the output torque of the motor, T_m [Nm], is the same as the friction torque, the friction torque will be counterbalanced. From this point of view, the friction compensator was developed aimed to make the motor torque to same as the friction torque changing.

Figure 1.3(b) shows the block diagram of the friction compensator that can make the motor torque the same as the friction torque. The compensator consists of three components: the table position estimator, the friction torque estimator, and the transient response compensator. Figure 1.4 provides examples of each signal in the proposed compensator.

In the friction compensator, first, the actual table position is estimated by the table position estimator (Fig. 1.4(a)). In this study, the whole feed drive system was simply modeled as a first-order system which has the time constant T_c [s]. The time constant is almost the same as the inverse number of the position loop gain, G_p [1/s].

Friction torque f [Nm] can be obtained based on the estimated table position, $\hat{x}$ [m]. In order to obtain the friction torque by Eqn. (1.1), the table displacement from the motion direction changing point x' [m] and the derivative of the estimated table position $\dot{\hat{x}}$ [m/s] are required. The displacement can be obtained through the following steps:

Step 1: Differentiate the estimated table position.
Step 2: Integrate the derivative of the table position.
Step 3: Reset the integrator when the sign of the derivative changes.

As a result, the displacement from the motion direction changing points (Fig. 1.4(b)) and the derivative of the table position can be obtained simultaneously. Therefore, it becomes possible to estimate the friction torque by the proposed friction model (Eqn. (1.1)). The parameters of the friction model can be identified by matching the experimental and simulation results, as shown in Fig. 1.2.

The motor torque T_m should be the same as the estimated friction torque, f (Fig. 1.4(c)). However, in actual systems, there are transient delays in the electrical circuit of the servo amplifiers. The delays can be compensated by the inverse transfer functions, as shown in Fig. 1.3(b), where DA, T_{rg}, and T_f are known parameters. Although the time constant T_i is unknown, it can be identified through the step response tests [12].

As described above, the proposed friction compensator can be easily achieved, and the parameters also can be determined systematically.

5. Simulation and Experimental Results

Figure 1.5 shows the simulation results of circular motions. The simulations were carried out using the mathematical model developed by the authors [12][13]. It was previously confirmed that the model can simulate actual behaviors accurately.

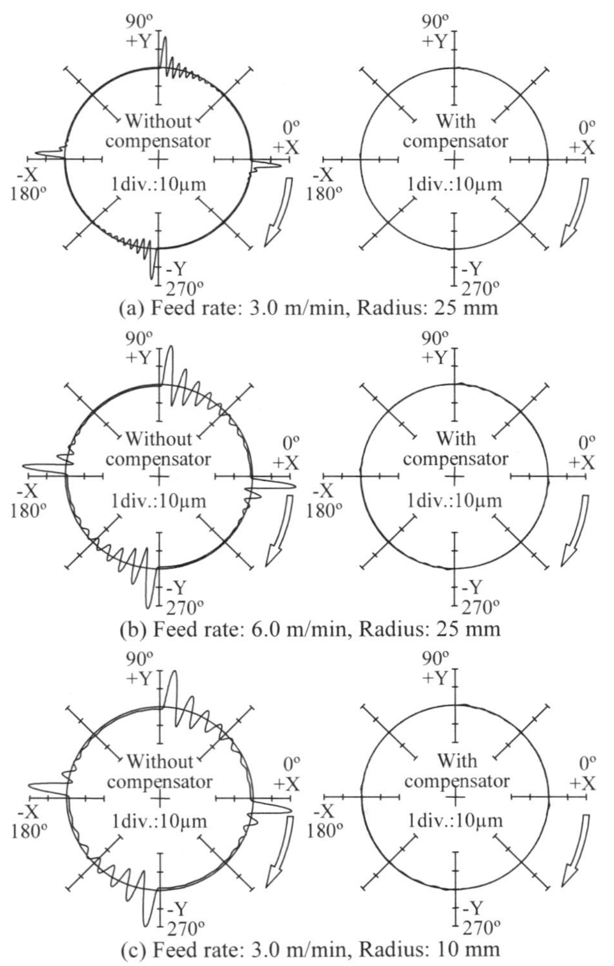

Figure 1.5. Simulation results of circular trajectories

As you can see from the figures, the quadrant glitches are completely eliminated, although the heights and shapes of the glitches change depending on the feed rate and radius of the circular motion.

The experimental results of the circular motions are shown in Fig. 1.6. The experiments were carried out by using the experimental apparatus shown in Fig. 1.1. It can be seen from both the simulation and experimental results that the quadrant glitches can be effectively eliminated by the proposed friction compensator, even if the radius and feed rate change. Of course, the parameters of the friction compensator were not changed in all simulations and experiments.

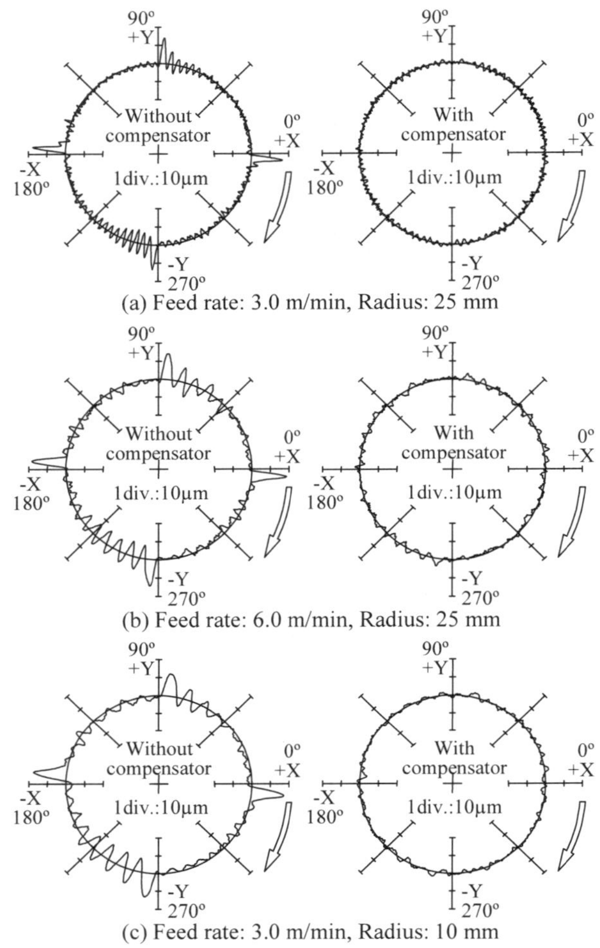

Figure 1.6. Experimental results of circular trajectories

6. Conclusion

In this study, a friction compensator based on a nonlinear spring friction model was proposed. The effectiveness of the friction compensator was evaluated by simulations and experiments of circular motions. The conclusions can be summarized as follows:

1. The nonlinear spring characteristic of friction can be expressed by the proposed friction model.
2. The proposed friction compensator can be easily realised, and the parameters can be determined systematically.
3. The quadrant glitches can be effectively eliminated by the proposed friction compensator, even if the radius and feed rate change.

For practical use, the effectiveness of the proposed friction compensator will be confirmed in other feed drive systems. Furthermore, a friction compensator for feed drive systems consisting of sliding guideways should also be developed.

References

[1] Tung, E.D., Anwar, G. and Tomizuka, M., 1993, Low Velocity Friction Compensation and Feedforward Solution Based on Repetitive Control, Transactions of ASME, Journal of Dynamic Systems, Measurement, and Control, June, Vol. 115, pp.279-284.

[2] Erkorkmaz, K. and Altintas, Y., 2001, High Speed CNC System Design. Part II: Modeling and Identification of Feed Drices, International Journal of Machine Tools & Manufacture, Vol. 41, pp.1487-1509.

[3] Erkorkmaz, K. and Altintas, Y., 2001, High Speed CNC System Design. Part III: High Speed Tracking and Contouring Control of Feed Drives, International Journal of Machine Tools & Manufacture, Vol. 41, pp.1637-1658.

[4] Tan, K.K., Lee, T.H., Huang, S.N. and Jiang, X., 2001, Friction Modeling and Adaptive Compensation Using a Relay Feedback Approach, IEEE Transactions on Industrial Electronics, Vol. 48, No. 1, pp.169-176.

[5] Mei, X., Tsutsumi, M., Tamazaki, T. and Sun, N., 2001, Study of Friction Error for a High-Speed High-Precision Table, International Journal of Machine Tools & Manufacture, Vol. 41, pp.1405-1415.

[6] Mei, X., Tsutsumi, M., Tao, T. and Sun, N., 2004, Study on the Compensation on Error by Stick-Slip for a High-Precision Table, International Journal of Machine Tools & Manufacture, Vol. 44, pp.503-510.

[7] Park, E.-C., Lim, H. and Choi, C.-H., 2003, Position Control of X-Y Table at Velocity Reversal Using Presliding Friction Characteristics, IEEE Transactions on Control Systems Technology, Vol. 11, No.1, pp.24-31.

[8] Suzuki, Y., Matsubara, A., Kakino, Y. and Tsutsui, K., 2003, A Stick Motion Compensation System with a Dynamic Model, Proceedings of the International Conference on Leading Edge Manufacturing in 21st Century, Niigata, Japan, pp.525-528.

[9] Cuttino, J.F., Dow, T.A. and Knight, B.F., 1997, Analytical and Experimental Identification of Nonlinearities in a Single-Nut, Preloaded Ball Screw; Transactions of the ASME, Journal of Mechanical Design, Vol. 119, pp.15-19.

[10] Fukada, S., Naruse, S. and Matsumoto, T., 2000, Studies on Microscopic Behavior of Ball Screw (1st Report), Journal of the Japan Society for Precision Engineering, Vol. 66, No. 7, pp.1070-1075.

[11] Futami, S. and Furutani, A., 1991, Nanometer Positioning Using AC Linear Motor and Rolling Guide (2nd Report), Journal of the Japan Society for Precision Engineering, Vol. 57, No. 10, P1808-1813.

[12] Sato, R. and Tsutsumi, M., 2005, Modeling and Controller Tuning Techniques for Feed Drive Systems. In Proc. of the ASME, Dynamic Systems and Control Division, Part A, DSC-Vol.74-1, pp.669-679.

[13] Sato, R. and Tsutsumi, M., 2006, Dynamic Behavior of Feed Drive Systems around Quadrant Changes in Circular Motion. In Proc. of the CIRP 2nd International Conference on High Performance Cutting (CD-ROM).

Radial Basis Function Sliding-mode Control for the Laser Alloying Process

Hung-Yi Chen[1] and Shiuh-Jer Huang[2]

[1]Mingchi University of Technology, Taipei, Taiwan, 243, hychen@mail.mit.edu.tw
[2]National Taipei University of Technology, Taipei, Taiwan, 106, hd3601@ntut.edu.tw

Abstract. Since the Laser alloying process system has nonlinear coupling and time varying dynamic characteristics, it is very difficult to establish an accurate process model for designing model-based controller. Hence, an adaptive radial basis function sliding-mode control is proposed by combining the advantages of adaptive neural network and sliding-mode control strategies. It has on-line learning ability to deal with the system's nonlinear and time-varying uncertainties by adjusting the control parameters. Two adaptive radial basis function sliding-mode controllers are designed for tuning the Laser power and the traverse velocity simultaneously to tackle the absorptivtiy and geometrical variations of the work pieces. The simulation results show that the good surface lapping performance is achieved by using this intelligent control strategy.

Keywords: Laser alloying, radial basis function, sliding-mode controller, on-line learning.

1. Introduction

It is well known that Laser surface treatments have been widely used for improving the mechanical and chemical properties of metal parts. The results of Laser alloy process are determined by the Laser beam power density and beam velocity during the Laser alloying process. Since the Laser surface process has high sensitivity to the disturbance, the surface treatment results may have significant variation even under the same operation conditions. The desired temperature cycle and temperature distribution of the work piece may be perturbed by the process disturbance, which may induce the unacceptable quality results.

Some of the process parameters, such as the Laser beam velocity and the Laser power, are observable and controllable, but some of them, such as the focus and dynamic positioning error and melt pool depth, can not be monitored. The major process disturbances in Laser alloying process are absorptivity variation and work piece geometrical change [1]. Hence, it is necessary to design a good feedback control system to achieve reproducible results with constant alloyed surface properties. Romer et al. [2] chose the Laser power as a command signal to suppress the effects of the absorptivity disturbance, and regulated the traverse velocity to compensate for the negative effect of geometrical disturbance. They employed PID and LQG controllers to control this multi-input and multi-output system based on an identified deviation model. Good control performance was obtained by using two PID controllers with mode-switching characteristic for Laser power and beam velocity, respectively. Therefore, it is still a single-input and single-output control strategy as each time only one controller was activated by model-switching.

Since the dynamic characteristic of this surface alloying process is very complicated, it is difficult to establish an appropriate dynamic model for a model-based controller design. Hence, a model-free control strategy is employed to solve this problem. Shieh et al. [3] employed a fuzzy logic control method to develop two fuzzy controllers for regulating the Laser power and the traverse speed simultaneously to achieve the desired temperature distribution on the alloying surface. However, the design of a traditional fuzzy controller fully depends on an expert, or the experience of an operator. A time consuming adjusting process is required to achieve the specified control performance. Here, an adaptive radial basis function (RBF) sliding-mode controller is developed to tackle the variation and disturbance of this Laser alloying control system. The proposed control strategy is based on a RBF structure combined with sliding-mode control scheme. An adaptive rule is employed to on-line adjust the weights of RBF by using the reaching condition of a specified sliding surface. Two adaptive RBF sliding-mode controllers are designed for regulating the Laser power control and traverse speed simultaneously to achieve the desired temperature distribution on the alloying surface. The control performance is evaluated based on numerical simulation results. For simulating the absorptivity and geometrical change of the work piece, step-wise disturbances are introduced to verify the robustness of the proposed controller.

2. Laser Alloying Process Model for Simulation Purpose

Since the Laser surface alloying process has complicated and nonlinear dynamic behavior and all the process parameters are interacted during the Laser alloying process, a linearized deviation model for a certain nominal operation conditions was identified based on the experimental results of real Laser alloying using a pseudo-random-binary-signal input signal [1]. It will be used to evaluate the dynamic performance of this adaptive RBF sliding-mode controller based on numerical simulation. The Laser power and traverse velocity are considered as inputs to system dynamic model identification process, and the melt pool area $S_{m,T}$ and the temperature T_l behind the melt pool are considered as the process outputs. The linearized deviation model of the Laser surface alloying process is defined as:

$$\begin{bmatrix} \Delta S_{m,T} \\ \Delta T_i \end{bmatrix} = G(q)\begin{bmatrix} \Delta P_L \\ \Delta v \end{bmatrix} = \begin{bmatrix} G_{11}(q) & G_{12}(q) \\ G_{21}(q) & G_{22}(q) \end{bmatrix}\begin{bmatrix} \Delta P_L \\ \Delta v \end{bmatrix} \quad (1.1)$$

where *G(q)* is the matrix transfer function and G_{ij} represents an auto-regressive model, relating the j-th input to the i-th output. The following transfer function model was established as [1]:

$$G_{11}(q) = \frac{0.15q^3 - 0.017q^2 + 0.034q + 0.0049}{q^4 - 0.76q^3 + 0.20q^2 + 0.23q - 0.13}$$
$$G_{21}(q) = \frac{-0.045q^3 - 0.1908q^2 + 0.4935q - 0.020}{q^4 - 0.76q^3 + 0.20q^2 + 0.23q - 0.13}$$
$$G_{12}(q) = \frac{0.67q^3 - 2.62q^2 + 0.37q - 0.78}{q^4 - 0.76q^3 + 0.20q^2 + 0.23q - 0.13}$$
$$G_{22}(q) = \frac{2.11q^3 - 1.52q^2 + 8.31q - 9.49}{q^4 - 0.76q^3 + 0.20q^2 + 0.23q - 0.13} \quad (1.2)$$

The nominal operation conditions were set as $P_{L0} = 1250W$, $v_0 = 50mm/s$, $S_{m,T0} = 1850^\circ C$, $T_{i0} = 1640^\circ C$, $f_s = 100Hz$ for this linearised discrete time model.

Different disturbances should be introduced into the system outputs for co-relating the absorptivity or dimension changes of the work piece to evaluate the controller robustness. Fig. 1.1 shows the introducing disturbances of the melt pool area and the temperature behind the melt pool, which is representing the absorptivity and geometrical defects, respectively.

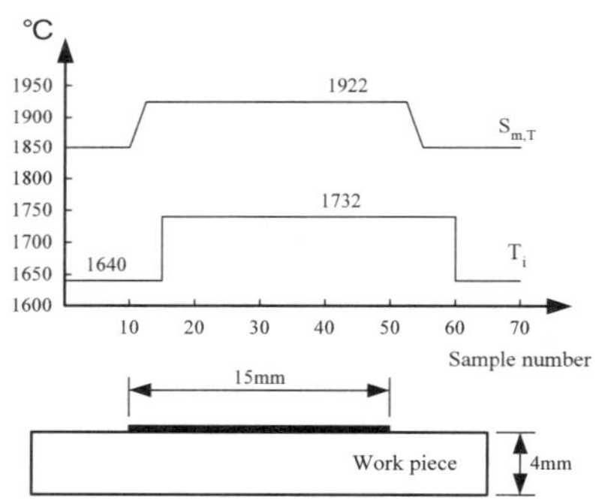

(a) Absorptivity disturbance

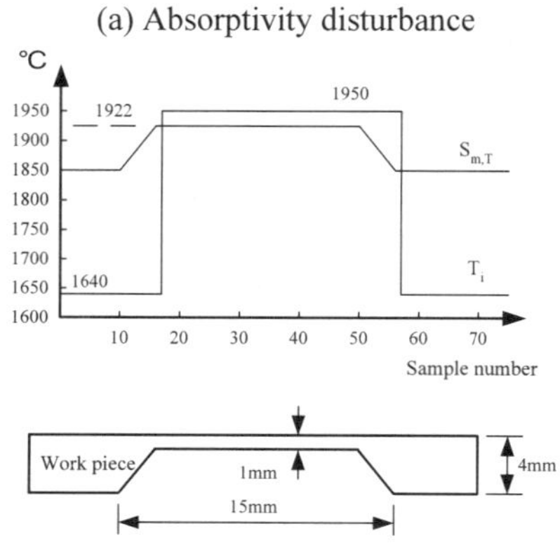

(b) Geometrical disturbance

Figure 1.1. The commencing disturbances of temperature $S_{m,T}$ and T_i.

3. RBF Sliding-Mode Controller

The sliding-mode control theory has been widely employed to control non-linear systems, especially the systems have model uncertainty and external disturbance. Here the sliding surface variable, *s*, is employed as the controlled variable for designing a neural network (NN) controller. Consider a second-order system with e and $\dot{e}$ as the phase plane variables, the sliding surface on the phase plane can be defined as:

$$s = \left(\frac{d}{dt} + \lambda\right)e = \dot{e} + \lambda e \quad (1.3)$$

where s is the sliding variable, e is the tracking error and λ is a strictly positive constant. This sliding variable, *s*, will be used as the input signal for establishing a NN model to calculate the control law. Recently, the RBFNN [4,5] has been widely used to model and approximate non-linear functions. Here a RBFNN is employed to model the relationship between the sliding surface variable, *s*, and the system control law, *u*. The Gaussian function is used as the activation function of each neuron in the hidden layer. The excitation values of these Gaussian functions are distances between the input values of sliding variables, *s*, and the central positions of Gaussian functions.

Here, a RBFNN-based controller is proposed by combining an adaptive rule and the technique of sliding-mode control. The control input of the RBFNN control is

$$u = \sum_{j=1}^{n} w_j \exp\left(-\frac{\|s - c_j\|^2}{\sigma_j^2}\right) \quad (1.4)$$

where n is the number of hidden layer neurons, w_j is the weighting between input layer neurons and hidden layer neurons, and c_j and σ_j are the spread factor and central position of the Gaussian function, respectively. The sliding surface variable, *s*, is specified as the input value of the RBFNN and an adaptive rule is introduced to adjust the weightings between hidden and output layers neurons.

Based on the Lyapunov theorem, the sliding surface reaching condition is $s\dot{s} < 0$. If a control input u can be chosen to satisfy this reaching condition, the control system will converge to the origin of the phase plane. Since a RBFNN is employed to approximate the non-linear mapping between the sliding input variable and the control law, the weightings of the RBFNN should be regulated based on the reaching condition, $s\dot{s} < 0$. The adaptive rule is derived from the steep descent rule to minimize the value of $s\dot{s}$ with respect to w_j. Then the updated equation of the weighting parameters is

$$\dot{w}_j = -\Gamma \frac{\partial s(t)\dot{s}(t)}{\partial w_j(t)} \quad (1.5)$$

where Γ is the adaptive rate parameter. Based on the chain rule, the above equation can be rewritten as

$$\dot{w}_j = -\Gamma \frac{\partial s(t)\dot{s}(t)}{\partial u(t)} \frac{\partial u(t)}{\partial w_j(t)} = \Gamma b s(t) \frac{\partial u(t)}{\partial w_j(t)}$$

$$= \gamma s(t)\exp\left(-\frac{\|s-c_j\|}{\sigma_j^2}\right) = \gamma s(t)\phi_j(s) \tag{1.6}$$

where the adaptive rate parameter, Γ, and the system input parameter, b(t), are combined as a learning rate parameter, γ. Then, the weightings between hidden and output layers neurons can be on-line adjusted to achieve the learning ability of RBFNN. Those weighting values are the most important parameters in this control scheme which determine the ratio of each radial basis function in the control law. The control block diagram of this RBF sliding-mode controller is shown in Fig. 1.2.

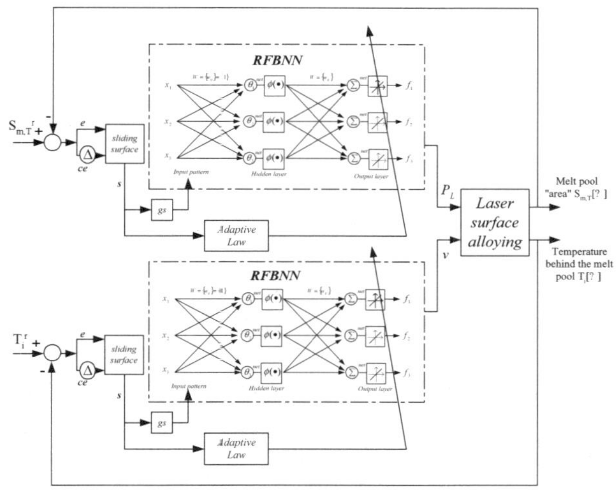

Figure 1.2 The adaptive radial basis function sliding-mode control block diagram.

4. Computer Simulation Results

Two simulation cases are executed in order to evaluate the dynamic performance of the proposed RBF sliding-mode controller. In order to verify the availability of the designed controllers in practical implementation, different step-wise disturbances are introduced into the system outputs to simulate the corresponding absorptivity or geometrical dimension changes of the work piece during the Laser alloying process. The sampling frequency was selected as 100 Hz to match with the reference dynamic model and the response speed limitation of the practical spectral and ratio pyrometers for temperature profile measurement during the control process. The parameter λ was set as 150. The selected value of the learning speed parameter γ was set as 0.28 for the melt pool area $S_{m,T}$ and 0.00055 for the temperature T_i behind the melt pool.

Case A: The work piece has absorptivity change.

When the surface of the work piece was coated with 3 μm graphite layer over a width 15 mm, the approximate stepwise disturbances 72 and 92°C are introduced to the output variables $S_{m,T}$ and T_i, respectively, in the case of absorptivity change. The melt pool area, $S_{m,T}$, output response and the Laser power adjustment of this control system are shown in Fig. 1.3 (a) and (b), respectively. The temperature behind the melt pool and the transverse velocity change are shown in Fig. 1.3 (c) and (d), respectively. The Laser power is decreased to about 1000 W to counteract the disturbance. It takes about 17 steps (170 ms) for the melt pool area error to reduce to less than 20°C. At the end of the absorptivity disturbance, the melt pool area drops below its reference value, then the Laser power is increased to 1255 W. The deviation of the melt pool area from its reference value during this disturbance is less than 2°C except at the instants of disturbance commencing and removing. The maximum deviation of the temperature T_i is less than 10°C except at the instants of disturbance commencing. The transverse speed is reduced to 49.5 mm/sec and is restored to 50.3 mm/sec after the end of the disturbance. Both the Laser power and transverse velocity variations are within the working range of this system. The Laser power is taking care of the major role of temperature control.

Case B: The work piece has geometrical change.

When the thickness of the work piece was reduced from 4 mm to 1 mm over a width 15 mm, the approximate stepwise disturbances 72 and 310°C are introduced into the output variables $S_{m,T}$ and T_i, respectively, in the case of geometrical change. The melt pool area output response and the Laser power of this control system are shown in Fig. 1.4 (a) and (b), respectively. The temperature behind the melt pool and the transverse velocity are shown in Fig. 1.4 (c) and (d), respectively. The Laser power is decreased to about 930 W corresponding to this disturbance. Simultaneously, the beam velocity is decreased to about 44.7 mm/sec. At the end of the geometrical disturbance, the melt pool area drops below its reference value, then the Laser power is increased to 1252 W. The maximum deviation of the melt pool area from its reference value is less than 4°C except at the instants of disturbance commencing. The temperature error of T_i is reduced to less than 20°C in 16 steps (160 ms) except at the instants of disturbance introducing. Both the Laser power and transverse velocity variations are within the working range of this system.

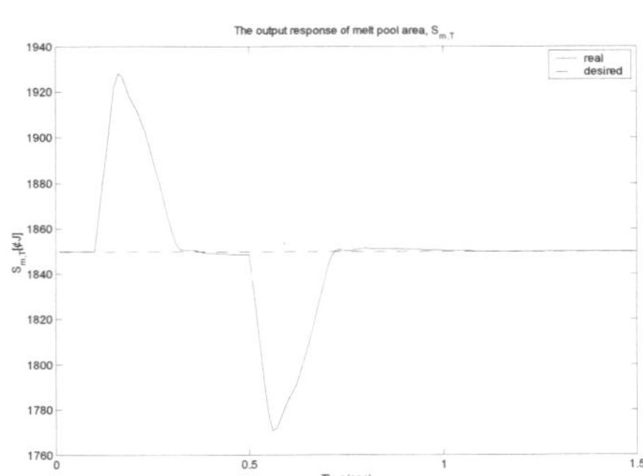

(a) output response of melt pool area, $S_{m,T}$

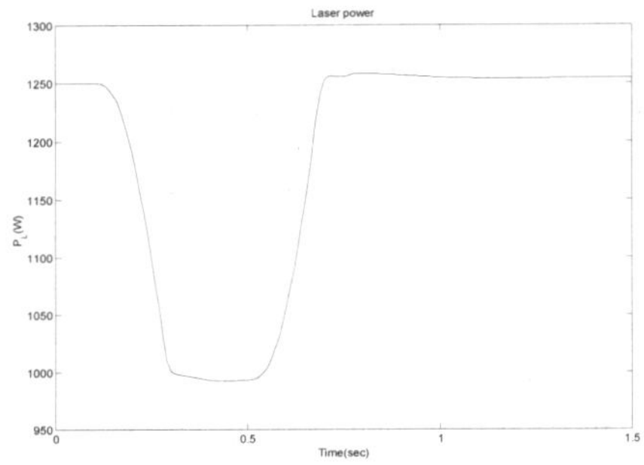

(b) Laser power control input P_L

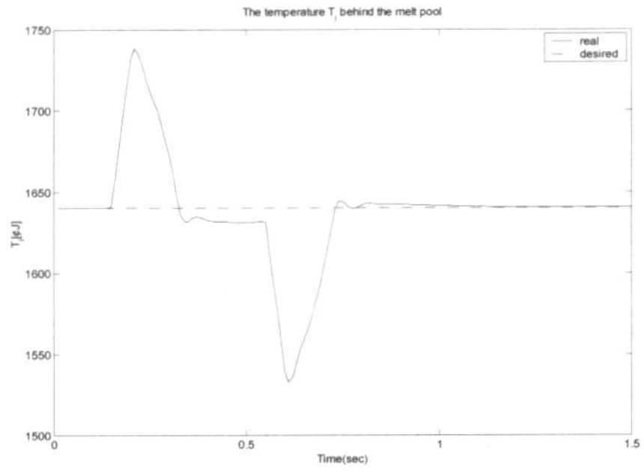

(c) temperature T_i behind melt pool

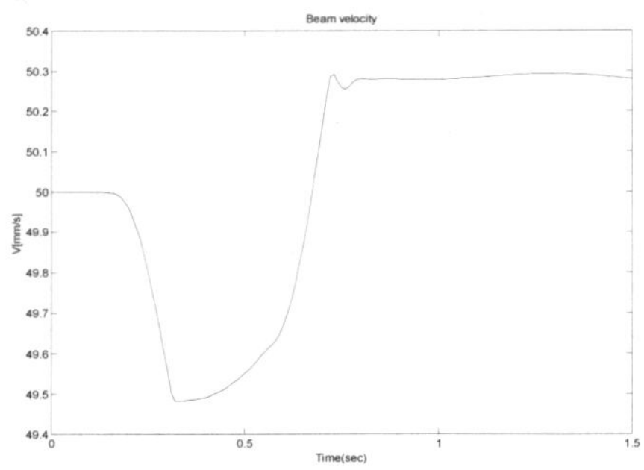

(d) traverse velocity V

Figure 1.3. Work piece with absorptivity disturbance.

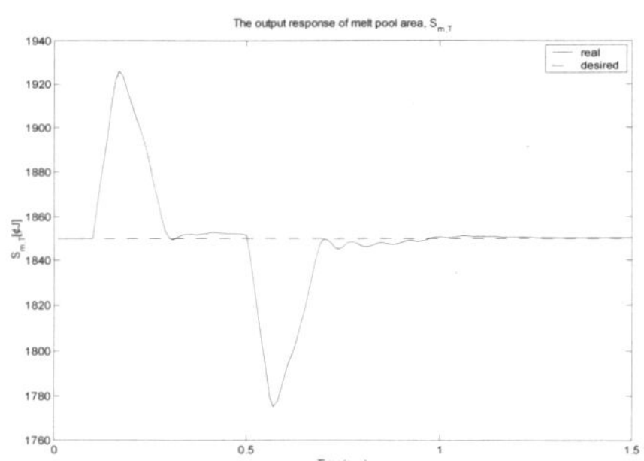

(a) output response of melt pool area, $S_{m,T}$

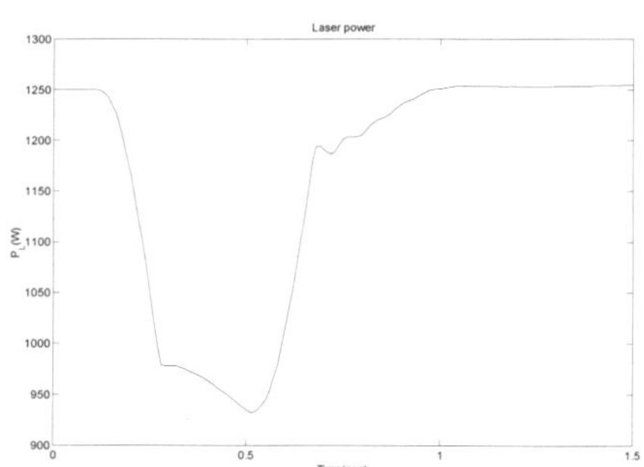

(b) Laser power control input P_L

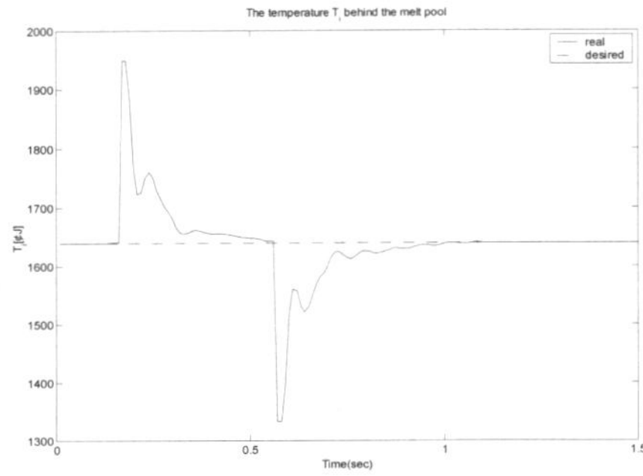

(c) the temperature T_i behind the melt pool

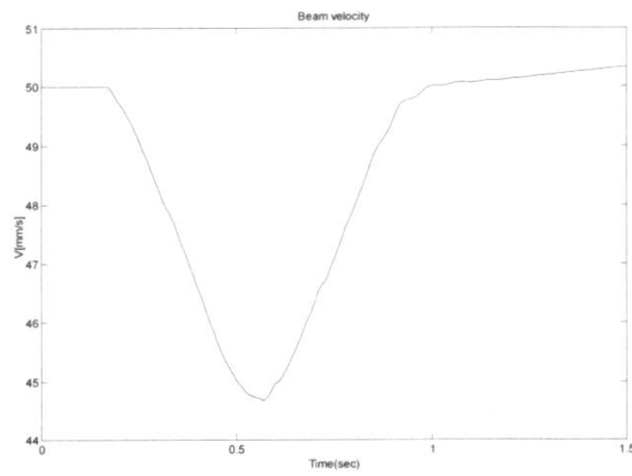

(d) traverse velocity V

Figure 1.4. Work piece with geometrical disturbance.

5. Conclusion

The Laser surface alloying process has nonlinear coupling and time varying behaviour. It is difficult to design a model-based controller for this control system. An adaptive RBF sliding-mode controller was employed to form a closed loop control for this Laser surface alloying process. Both simulation cases have shown that the proposed RBF sliding-mode control strategy can compensate the practical operation disturbances of the Laser surface alloying process and monitor the Laser alloying process successfully and effectively.

Acknowledgements

This research was supported by the National Science Council of the Republic of China, with contract NSC95-2221-E-131-005.

References

[1] Romer, G. R. B. E., 1998, Modelling and control of Laser surface treatment, Ph.D. thesis, University of Twente, Dept. of Mechanical Engineering, Enschede, the Netherlands.
[2] Romer, G. R. B. E., Meijer, J. and Aarts, R. G. K. M., 1999, Multivariable control of Laser alloying of Ti6Al4V, In Proceedings of ICALEO'99, San Diego, USA, Nov. 15-18.
[3] Shieh, S. W., Huang, S. J., and Li, L., 2001, Fuzzy logic control for the Ti6Al4V Laser alloying process, The International Journal of Advanced Manufacturing Technology, Vol.18, P247-253.
[4] Sanner R. M. and Slotine, J. –J. E , 1992, Gaussian network for direct adaptive control, IEEE Trans. Neural Networks, Vol.3, P837-863.
[5] Hwang C. L., 1999, Neural-network-based variable structure control of electrohydraulic servosystems subject to huge uncertainties without persistent excitatin, IEEE/ASME Trans.Mechatronics, Vol.4, No. 1, P50-59.

Real-Time Tracking and Stereo Vision-Based Control of a Goldfish-Catching System

C.C. Ho[1], C.L. Shih[2]

[1]National Taiwan University of Science and Technology, Taipei, Taiwan, 106, burt.ho@gmail.com
[2]National Taiwan University of Science and Technology, Taipei, Taiwan, 106, shihcl@mail.ntust.edu.tw

Abstract. Designing a real-time visual tracking system to catch goldfishes is a very complex task because a fish's precise relative 3D positions inside a water tank are difficult to determine. In this paper, we present a new visual servo control system using stereo vision to acquire the 3D positions of goldfishes. The continuously adaptive mean shift (CAMSHIFT) algorithm and optical flow vision tracking methods are applied to provide feedback of a tracked fish's real-time position and motion trajectory in a water tank at a high frame-rate. Two low-cost, well calibrated cameras are used to track the goldfishes and a 5-axis robot manipulator is used to catch a specified goldfish. The designed visual tracking and servo system is less sensitive to lighting influences and thus performs more efficiently. Experiments of the proposed method yielded very good results as the real-time 3D vision successfully tracked two fishes and guided the manipulator to catching one of them in a net attached to its end-effector.

Keywords: fish catching, stereo vision, visual servoing, CAMSHIFT

1. Introduction

The basis of all different approaches in visually tracking moving objects and eventually catching them is called visual servoing – the control loop using visual feedbacks to maintain the position of a robot manipulator moving toward a specified target [1]. A commonly used system to process the visual feedback is the machine vision measuring system. The advantage of the machine vision measuring system is its ability to perform measuring without contact, when contact measurements are not allowed by the environment. Machine vision systems can precisely measure a number of variables such as length, angle, position, orientation, color, etc. However, the biggest problem of the machine vision system is its over-sensitiveness to the noises in presented images, e.g. disturbances caused by unwanted objects in the background/foreground and non-ideal lighting conditions. A visual tracking system must be capable of recording image features in real-time. The execution of feature-extracting and template-pattern-matched computing while accounting for all these disturbances occupies too much of the machines' memory for them to perform motion tracking time-efficiently, and hence limiting their real-time control performances [2,3]. Furthermore, the required image-capturing equipments are too costly.

In order to track the trajectory and compute the positions of moving goldfishes in real-time, researchers have resorted to different methods that process the visual feedbacks at lower costs, computation wise and time wise. A method featuring statistical analysis is designed by Ukigaya et al. [4]. They proposed a model that recognizes the color distribution patterns of moving objects in video sequences. Based on their analysis of the active visual patterns in a series of images of a swimming goldfish, they built a color distribution pattern model, which also serves as a probabilistic state transition mechanism. Another approach, proposed by Suzuki and Minami [5,6], utilizes genetic algorithm (GA) to search the fish feature. Raw grayscale images are processed via GA to track the swimming fishes. This approach demonstrates real-time recognition of the fishes' 2D position and the visual servoing of a hand-eye robot used to catch one of the fish. However, these tracking-catching schemes are only for tracking objects in 2D.

In this paper, a hybrid control scheme based on CAMSHIFT and optical flow algorithm is applied to track the trajectory and to compute the 3D position of swimming goldfishes in real time, using a stereo vision system with two CMOS cameras and a robot to catch the specified goldfish. The proposed tracking algorithm is extremely tolerant to lighting influences and highly efficient in computing compared with previous discussed template-pattern-matching algorithms. CAMSHIFT is a modified version of the mean shift algorithm [7] and was originally developed by Bradski [8] to detect and track faces in a video stream. Similar to the mean shift algorithm, which ascends a gradient of a probability distribution of colors to find the nearest dominant mode, CAMSHIFT tracks the mode of color probability distribution across the frames of a video sequence. The tracking performance is further improved by optical flow method to overcome the objects overlapping problem.

2. Stereo Visual Servo System

Image capturing and processing are the two major challenges in a stereo visual tracking system. The main goals of these respective processes are to provide visual feedback, and to identify the objects we wanted to track (in this case goldfishes) and determine their 3D positions. The video-signal-capturing process is conducted by the stereo sensors in the stereo tracking system with two low cost CMOS cameras with pixel resolutions of 320 x 240. The synchronizing concurrent video frames are then transmitted via USB to a PC in the system buffered memory, and finally displayed in the capture windows using the windows driver model (WDM) functions. The windows driver model can reach a high video frame rate with the two CMOS cameras operating the frame rate of 30-image-per-second; both capable of providing real-time visual feedback. The visual servo system disscussed in this paper is shown in Figure 1.1. Figure 1.2 shows the block diagram of the visual servo system which consists of an image-based system calibration unit, a hybrid CAMSHIFT tracking unit, a depth from binocular stereo unit, a inverse kinematics unit, and a robot arm position control unit.

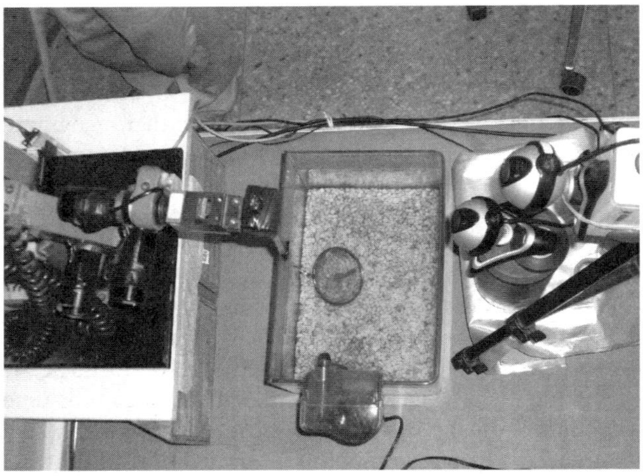

Figure 1.1. The visual servo system to catch the goldfishes.

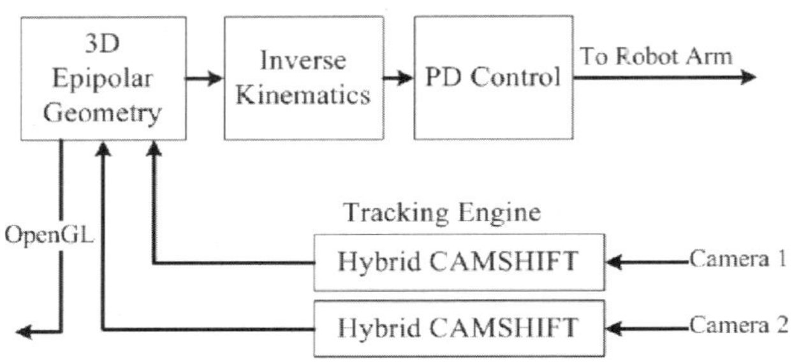

Figure 1.2. The system block diagram of the proposed visual servo system.

2.1 Calibration of Image-based System

Camera calibration and robot hand-eye transformation are the two fundamental steps needed to be taken before any reliable image processing and visual servo can be performed. To obtain the 3D position of an object from a 2D image, the relationship between the coordinates of a set of points in the 3D space, the tank, and the coordinates of their corresponding points on the 2D plain, the retrieved image, must be found out via the basic characteristics of cameras, generally grouped into intrinsic and extrinsic parameters. Through camera calibration one can obtain the intrinsic and extrinsic parameters. The intrinsic parameters, which are independent of a camera's position, describe its focal length (f_x f_y), principal point (C_x, C_y), and distortion coefficients. The extrinsic parameters, which offer information for the transformation between the coordinate systems of the camera's image and the concrete world, include a 3D translation vector t and a rotation matrix R. For stereo cameras, the relative extrinsic parameters between the cameras also have to be obtained. The combination of a camera's intrinsic and extrinsic parameters presents the full set of data needed to locate the corresponding position of a point in an image in the outside world. References [9] and [10] provide a complete review of the fundamentals and techniques of camera-calibration.

In this paper, a static pattern calibration is applied to obtain two sets of parameters of the cameras and empower automatic edge detection. The key in this process is the usage of a flat checkerboard whose squares have predetermined dimensions. The intrinsic calibration routine is performed after several views of the board with known geometry pattern, using points on the model plane and their projection onto digital images. The extrinsic calibration also implements the checker-patterned board. The extrinsic parameters can be calculated linearly using an array containing the location of points in the digital images and another array containing the location of their counterparts in the real world.

2.2 Hybrid CAMSHIFT Algorithm

CAMSHIFT algorithm is a non-parametric technique that can track a specified target's 2D position efficiently across a series of images. Optical flow describes the motion of feature points between successive images in a video stream. The hybrid tracking algorithm can be summarized in the following:

1. A region of interest (ROI) window is selected to be the source image for color probability distribution computation.
2. An initial search window is centered at the data point position.
3. The color distribution of the region centered at the mean shift search window is calculated. The mean location of the discrete probability image is found within the search window using moments: given that $I(x,y)$ is the pixel value (the intensity of the discrete probability image) at (x, y) within the search window, one can compute the zeroth moment by:

$$M_{00} = \sum_x \sum_y I(x, y) \tag{1.1}$$

and find the first moment for x and y by

$$M_{10} = \sum_x \sum_y xI(x, y) \tag{1.2}$$

$$M_{01} = \sum_x \sum_y yI(x, y) \tag{1.3}$$

Then the mean search window's location can be calculated as followed:

$$x_c = \frac{M_{10}}{M_{00}} \tag{1.4}$$

$$y_c = \frac{M_{01}}{M_{00}} \tag{1.5}$$

4. Iterate mean shift algorithm until the centroid of the probability image in which mean shift algorithm converges is found. Store the zeroth moment (the distribution area) and the mean location.
5. Search the centroid location of neighboring fishes within twice the area of the mean search window. Optical flow tracking is applied when more than one fish lies inside this range. Then, go to Step 3.
6. The size of the search windows is set as a function of the zeroth moment found in Step 4 and the center of the search window is placed at the mean location for the following frame. Repeat Step 3 and Step 4.

When tracking the 2D position of a colored object, in our case a goldfish, CAMSHIFT operates on a color probability distribution image derived from color histograms. The center and size of the region representing the goldfish is computed and used as settings for the search window on the following frame of the video sequence, in agreement with the steps mentioned above. Figure 1.3 shows two digital pictures of two goldfishes. The pictures are taken by each of two cameras that has been recognized and processed within 33[ms] video rates by applying CAMSHIFT algorithm. As shown in Figure 1.3, only limited regions surrounding the current CAMSHIFT window, which includes images of two goldfishes, are transformed into a discrete probability image. This tends to result in a large reduction in computational costs. Figure 1.4 shows the problem of partial overlapping which causes the CAMSHIFT tracking failed. As the color probability distribution of specified fish is fixed, the optical flow technique can be applied to track the specified location of discrete image. The neighboring fishes overlapping problem is then improved by applying the optical flow method.

Figure 1.3. Two tracked goldfish within 33 [ms] video rate in both cameras to produce two stereo pair respectively.

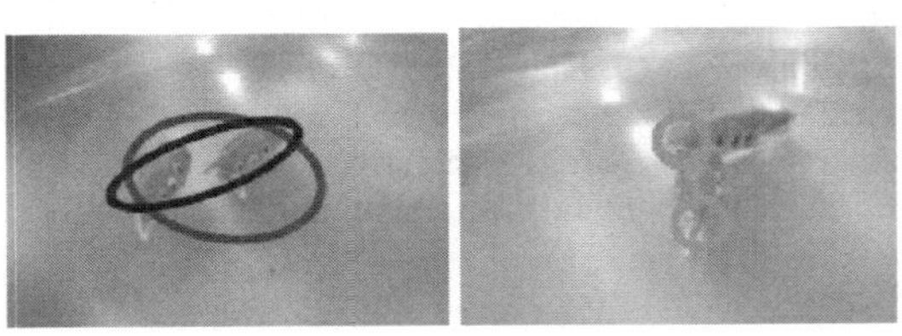

Figure 1.4. The neighboring fishes overlapping causes CAMSHIFT to fail (left), but tracking improved by applying the optical flow method (right).

2.3 Depth calculated through Binocular Stereo

The purpose of using two cameras instead of one to track a goldfish becomes clear at this point: to provide data of the goldfish's position in a third dimension. The calculation of depth via binocular stereo is a common way to capture the complete 3D information of a scene [11]. First, feature points in the images obtained by the first and second camera create a set of stereo pairs. Each pair produces two rays, one for each point, that would intersect in the actual 3D world. This approach transforms the coordinate systems of the two cameras into absolute coordinates. The real-time 3D positions of the two goldfishes, tracked with the above mentioned 3D point measuring method, are rendered using the Open Graphics Library (OpenGL for short) [12].

2.4 Hand-Eye Coordinate Transformation

In the final phases of visual servoing, sending visual feedbacks to control a robot, the 3D coordinates of the tracked object in the stereo vision reference frame must be translated into coordinates in the manipulated robot's reference frame. This procedure is called the hand-eye coordinate transformation. In order to obtain necessary information between the hand/eye coordinates, the manipulator moved to four non-co-plane spots while being tracked by the binocular stereo system with hybrid CAMSHIFT algorithm. A strong source of light, here a LED, is attached to the hand of a manipulator to make the tracking easier. According to the obtained homogenous transform matrix, the stereo vision system can use the 3D position information captured by two cameras to control the motion of the end-effector.

3. Experiments and Results

Results of experiments implementing the proposed process derived in previous sections are presented in this section. The intrinsic parameter calibration is achieved by calculating the projections of a grid with 9 × 7 frames. For each camera, the calibration routine was performed 30 times, with the results shown in Table 1.1. The positions and orientations of the stereo cameras in respect to the real world's coordinate system could be obtained through the extrinsic calibration routines shown in Section 1.2.1. After the extrinsic calibration, the two cameras' 3D positions were verified at (538, -51, 745) and (578, -48, 619) respectively, both positioned at the top of the water tank for the best view of the targeted the goldfish. The calibration routines are followed by the determination of the machine's hand-eye transformation matrix, a process shown in Section 1.2.4.

The image processor used in our system is a 3.2 GHzPentiumIV/512MB-RAM PC running on Windows XP with two Logitech QuickCamPro 4000 cameras connected via USB. The resolutions of the images captured by the two cameras are set at 320 × 240 pixels.

The size of the water tank is 300 × 200 × 180 mm deep and the diameter of circular net used to catch the goldfish, which is set at the end of the manipulator, is 80 mm. The manipulator used in the experimental system is a 5-axis manipulator, Micro-robot 88-5. The position control system of the Micro-robot 88-5 system is an independent joint PD control system. The experiment is started by choosing a fish for the system to track, which can be done with the mouse. Once the initial search window is set, the system automatically began tracking the selected objects in 2D applying the hybrid CAMSHIFT algorithm described in Section 1.2.2 and further computing the target's depth with binocular stereo vision. In Section 1.2.3, the 3D position and orientation is hence fully retrieving the targeted goldfish in real-time. We have recorded the moving path of the tracked goldfish in this experiment using the calibrated stereo camera system in a sequence of images. After these steps the manipulator is set to strike at the tracked goldfish's predicted 3D position in extremely short future. Then the inverse kinematics is computed by the PC and the joint angle commands are transmitted from the PC to the Micro-robot 88-5 controller via RS-232 interface and updated at the rate of 10 Hz.

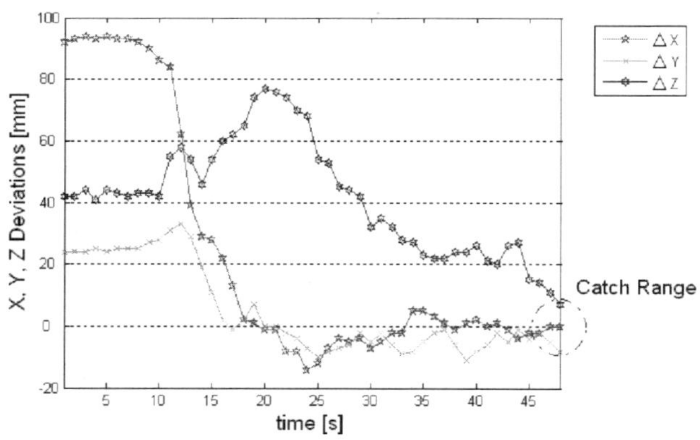

Figure 1.5. Position deviations between the goldfish and the net.

Table 1.1. Intrinic parameter calibration result

Camera	f_x	f_y	C_x	C_y
CAM 1	404	406	165	130
CAM 2	410	410	156	124

Unit: Pixel

The manipulator catches the fish by pulling up the net, which was initially below the water surface in the tank, and approaching the target continuously under position commands. Figure 1.5 shows the deviations along the *X-Y-Z* directions in the real world coordinate system. As shown, the deviations decreased in the region lying within the net's diameter.

The system is also tested by several external disturbances, such as changes in lighting conditions and partial occlusions of the tracking cameras' vision, and the results show that the tracking system is hardly affected. With promising experimental results, the implementation of the goldfish tracking system described in this work has been proven robust successful.

4. Conclusion

In this paper, stereo vision tracking of a fish and the visual servoing for catching it is achieved by applying a computational efficient hybrid CAMSHIFT algorithm using two low-cost, calibrated USB cameras, which can operate in real-time at a high speed. The proposed visual tracking system could also be extensively used for simultaneous recognition and tracking of multiple objects that are not occluded by each other. The stereo vision based servo system has been proven capable of working in non-ideal real-world conditions under disturbances like poor lighting conditions, partial occlusion of camera vision, object rotations, and scale variations in experiments.

References

[1] R. Kelly. 1996, Robust asymptotically stable visual servoing of planar robots, IEEE Trans. Robot. Autom., Vol.12, No.5, P759-766.

[2] Luigi di Stefano, Stefano Mattoccia, Martino Mola. 2003, An efficient algorithm for exhaustive template matching based on normalized cross correlation, Proc. 12th Int. Image Analysis and Processing (ICIAP), P322-327.

[3] Sumi, K., Hashimoto, M., and Okuda, H.. 1995, Three-level broad-edge matching based real-time robot vision, Proc. of International Conference on Robotics and Automation, P1416-1422.

[4] Atsuo Ukigaya, Tatsuyuki Kawamura, Masatsugu Kidode. 2005, Statistical analysis of distribution patterns for moving objects - skill acquisition to focus a target region for a catch-robot, Proceedings of JSAI.

[5] H. Suzuki and M. Minami. 2002, Fish catching using gazing-GA visual servo - verification of robustness for lighting condition varieties, Proc. Sice Annu. Conf.

[6] H. Suzuki and M. Minami. 2005, Visual servo to catch fish using global/local GA search, IEEE/ASME Trans. on Mechatronics, Vol.10, No.3, P352-357.

[7] Y. Cheng. 1995, Mean shift, mode seeking, and clustering, IEEE Trans. Pattern Anal. Machine Intell., Vol.17, P790-799.

[8] G.R. Bradski. 1998, Real time face and object tracking as a component of a perceptual user interface, Proc. IEEE Workshop Applications of Computer Vision, P214-219.

[9] M.-C. Villa-Uriol, G. Chaudhary, F. Kuester, T. C. Hutchinson, and N. Bagherzadeh. 2004, Extracting 3d from 2d: selection basis for camera calibration, Proc. IASTED Computer Graphics and Imaging (CGIM), Kauai, HI, P315-321.

[10] Hutchinson, T. C., Kuester, F., Doerr, K.-U., and Lim, D. H. 2006, Optimal hardware and software design of an image-based system for capturing dynamic movements, IEEE Transactions on Instrumentation and Measurement, Vol.55, No.1, P164-175.

[11] Ramesh Jain , Rangachar Kasturi , Brian G. Schunck. 1995, Machine Vision, McGraw-Hill, Inc., New York, NY,

[12] Richard S. Wright, Jr. Michael Sweet. 2000, OpenGL SuperBible Second Edition, Waite Group PRESS.

13–5

Nano-positioning Drive with Piezoelectric Actuator Integrated into Support Bearing Unit of Ball Screw

A. Matsubara[1], T. Fujita[2], D. Kono[3], N.Tanaka[4] and Y.Watanabe[5]

[1]Kyoto University, Japan, 6068501, matsubara@prec.kyoto-u.ac.jp
[2]Kyoto University, Japan, 6068501, tomoya.fujita@t2003.mbox.media.kyoto-u.ac.jp
[3]Kyoto University, Japan, 6068501, Daisuke.Kono@t02.mbox.media.kyoto-u.ac.jp
[4]NSK Ltd, Japan, 2518501, tanaka-nob@nsk.com
[5]NSK Ltd, Japan, 2518501, watanabe-yas@nsk.com

Abstract. This paper presents a new fine positioning device with a piezoelectric actuator integrated into the support-bearing unit of a ball screw. By moving the outer race of the support bearing, the piezoelectric actuator gives a preload to the bearings, which translates the screw shaft and table. Since the rotation of the screw shaft gives coarse motion to the table, the whole system can perform the positioning of the table in a long stroke with a high resolution. On the developed system, open loop and closed loop tests are carried out to evaluate its positioning performance. It was found that the piezoelectric actuator translates the table in the order of nanometers in the open loop test and the designed closed-loop control system can control the table position with a resolution of 20 nanometres.

Keywords: Dual stage, Precision positioning, Piezoelectric actuator, Ball screw drive, Closed loop control

1. Introduction

In recent years, the demand for high precision machine tools has been increasing in manufacturing areas such as die and mold, optical parts, and high precision parts for MEMS. For such high precision machines, positioning systems with a resolution in the sub-micrometer range and a stroke of several hundred millimetres are necessary.

Linear motors have been extending its application especially in the ultra-precision positioning, as they are frictionless and have good dynamic response. They are suitable for nano-positioning drives [1], but it is difficult to obtain high load capacity in compact design. In contrast, ball screw drives provide a higher load capacity and stiffness. It is, however, more difficult to achieve the positioning accuracy in the sub-micrometer range by using conventional designs.

Dual (double-stage) positioning systems have been developed to achieve positioning in a long stroke with a high resolution. Typical approach is to stack a fine positioning stage with piezoelectric actuators on a coarse motion stage driven by a ball screw [2-6]. In the design of the fine table, flexible elements such as elastic hinges are used so that the piezoelectric actuator can move the table in the desired direction. As the piezoelectric actuator needs preload, a compressing spring element must be installed in the same structure. These flexible elements, however, do not assure axial stiffness against pulling forces. Moreover, it is difficult to enhance the table's stiffness against the angular moment caused by cutting forces applied on the tool.

Chen built a piezoelectric actuator with double nuts so that it can control the preload force of the nut and position of the table [7]. Although this approach can prevent external moments from being directly applied on the actuator, the integration of the actuator and the tuning of the preload is not easy. Otsuka developed a small stage that has a piezoelectric actuator between the support-bearing unit and the base unit [8]. In this table, 1 nm positioning resolution was achieved by using a high accurate linear encoder, but the stiffness in pulling direction of the piezoelectric actuator is not assured.

To cope with these problems, one of the authors developed a new fine positioning device, in which a piezoelectric actuator is integrated into the support-bearing unit of a ball screw [9]. The piezoelectric actuator provides a fine motion to the table and a preload to balls in the support bearing. The preloaded bearing system provides higher table stiffness with regard to pulling forces to the piezoelectric actuator. This paper presents the motion characteristics and performance of the fine positioning device. To experimentally investigate them, several motion tests are carried out with the developed device in the open loop. Then, simple positioning controllers are designed and examined to find out the resolution of the system.

2. Nano Positioning Drive with Piezoelectric Actuator

2.1 Drive Mechanism

Figure 1.1 shows the mechanism of the fine positioning system. The system employs the same components as a conventional ball screw drive except for the support-bearing unit. The support-bearing unit consists of a pair of angular contact ball bearings in DF arrangement and the preload components. In the combined bearings, there is a clearance between the outer races, while a spacer is inserted between the inner races. A piezoelectric actuator is built between the end cap (preload cap) of the unit and the spacer that contacts the outer race.

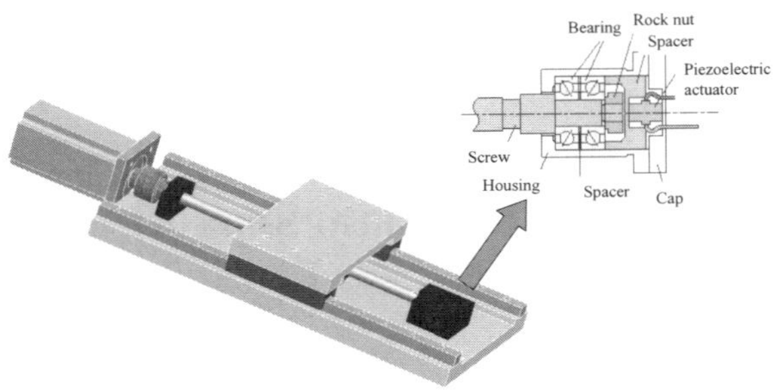

Figure 1.1. The fine positioning mechanism

When a voltage is applied to the piezoelectric actuator, it expands and moves the outer race of the bearing on the end-cap side. Assuming that the outer race on the opposite side does not move and the elastic deformation of the bearing balls absorbs gap elimination of clearance, the screw shaft moves in axial direction. Then, the table moves due to the axial motion of the screw shaft. If the screw shaft motion is perfectly converted into a table motion, the table moves by the distance half of the eliminated gap.

2.2 Developed System

We designed the positioning system for the Z axis for a high accurate machine. Tables 1.1 and 1.2 show the specifications of the ball screw drive and the piezoelectric drive used for the developed system. Main components of the ball screw drive (coarse drive) are a high precision ball screw, an AC servo motor, support bearing units, and a flexible coupling. In the fine motion, the nominal stroke of the table is 10 μm, which is half of the stroke of the piezoelectric actuator itself. The table is guided by linear ball guideways.

Table 1.1. The specification of the ball screw drive

Travel distance		200 mm
Table mass		5 kg
Ball screw	Diameter	12 mm
	Lead	5 mm
Rated output of the servo motor		200 W
Resolution of the rotary encoder		261144 pulse/rev
Nominal resolution of positioning		19 nm

Table 1.2. The specification of the piezoelectric drive system

Piezo electric actuator	Displacement	20 μ m
	Size	5 x 5 x 20 mm
	Expansion/Input voltage	μ m/ V
	Load capacity	1800 N
Amplifier	Gain	x 50
	Dynamic response	1 KHz

To control the coarse table, we use an updated CNC controller (MELDAS MAGIC 68, Mitsubishi electric), whose resolutions of command trajectories and position feedback signals are one nanometer. This controller employs conventional three cascade feedback loops: motor current PI-loop, velocity PI-loop, and position P-loop. We use the semi-closed loop control. Compared to the nominal resolution of 19 nm, the actual resolution was found to be 200 nm from the positioning test.

For the control of the fine motion, we use a PC with a 16-bit DA board. The command voltage is transferred from the PC to the piezo-amplifier through the DA board.

3. Open Loop Tests of Fine Positioning

In order to investigate the motion characteristics of the fine positioning device, several motion tests are carried out. First, we checked the positioning accuracy and yaw motion of the table. Figure 1.2 shows the experimental setup. The dual positioning table is placed on a vibration-isolated table and the table motion is measured at 3 points on one side of the table (P_1, P_2, and P_3 shown in Fig. 1.2). A capacitance-type sensor with the resolution of 1 nm is used for the measurement of the table displacement. The measured displacement signals are filtered with a low pass filter and captured with an A/D board. The cutoff frequency of the low pass filter is 10 Hz and the sampling frequency is 100 Hz.

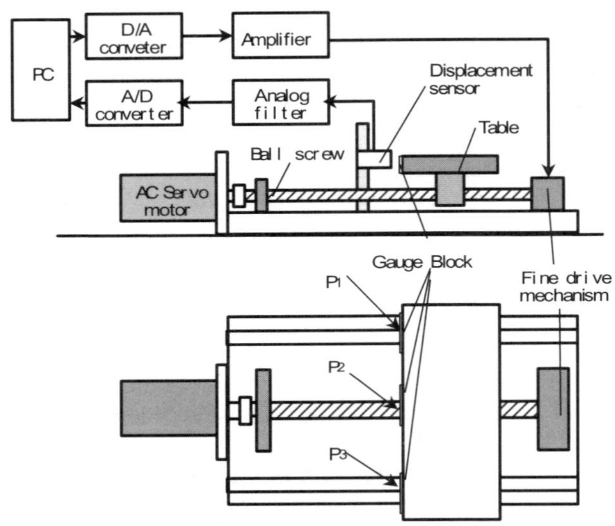

Figure 1.2. Experimental setup

A step-shaped reference position trajectory is given to the amplifier and table displacements are measured. The reference position is incrementally increased and then decreased by 3 steps. The holding time is 0.5 s for each step. The input voltages that correspond to the widths of 50, 20, 10, and 5 nm are calculated and commanded. Each step test is repeated 5 times. Figure 1.3 shows examples of the measured displacement at the table's center (P_2).

Steps can be clearly observed in 50 nm step and 20 nm step tests. Although significant noise components exist in 10 nm step and 5 nm step tests, steps can be still observed. In repeated tests, the variance of step sizes in uni-directional positioning was 8 nm, but step sizes are larger when the table is driven forward than backward. As test results show an existence of the nonlinearity, a sine wave with variable amplitude was given to the amplifier and the table displacement is measured. Figure 1.4 shows the relationship between the command voltage and measured displacement. As shown in this figure, the hysteresis characteristic exists in the drive mechanism. Same tests are carried out with different frequencies of the sine wave. Then, the hysteresis was found to be repeatable and position-dependent.

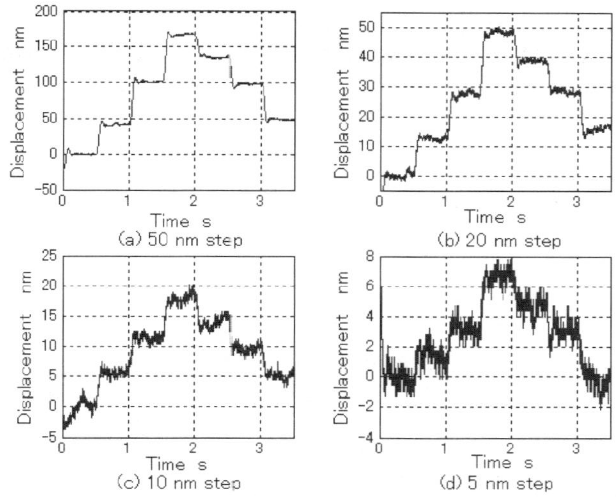

Figure 1.3. Measured step responses of table at P_2

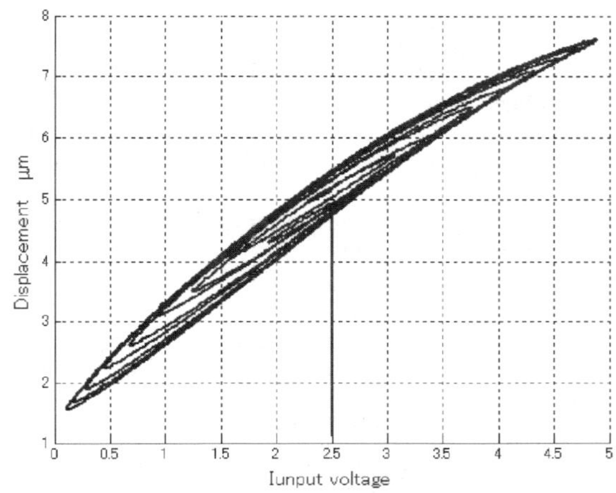

Figure 1.4. Relationship between command voltage and measured displacement

Based on the measured hysteresis, the step command is modified so that each step size of the table motion becomes the same with increasing and decreasing steps. The table displacements are measured at P_1, P_2, and P_3 simultaneously, and shown in Figure 1.5. The displacements at P_1 - P_3, which are overlapped, are almost the same, which implies that yaw motion of the table is negligibly small.

4. Closed Loop Tests

Due to the nonlinear relationship between the command voltage and the table position, it is difficult to achieve fine positioning by open loop control. So, the table position is controlled by closed loop scheme. A linear encoder with a resolution of 10 nm is attached to the table and used for closed loop control. The measured table position is fed back to the controller (PC) through a 32bit counter board. Three control schemes, P-control, PI-control, and feedfoward (FF) +PI control, are examined for fine positioning. Figure 1.6 shows the block diagram of a FF+PI controller. The controller parameters are as follows: K_p is a position loop gain; K_{vi} is an integral gain; K_f is a feedforward gain. A PI controller can be obtained by setting $K_f = 0$; a P controller by setting $K_f = 0$ and $K_{vi} = 0$. The control command is sent to the piezo amplifier through the 16bit D/A board.

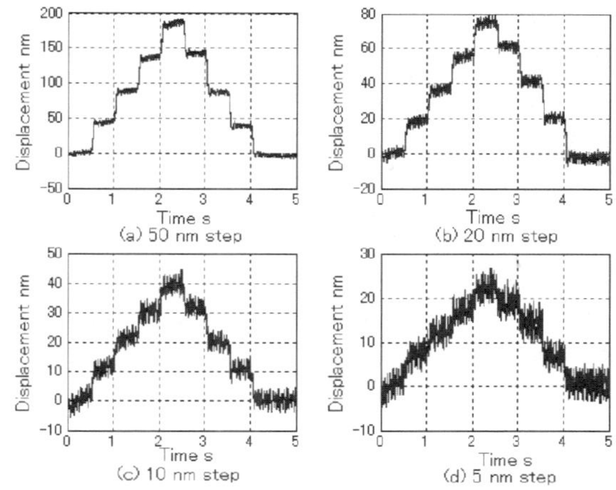

Figure 1.5. Measured step responses of table at P_1, P_2, and P_3

Figure 1.7 shows the experimental setup. In order to test the positioning accuracy of a vertical axis, the positioning system is vertically attached to the column. The position accuracy is evaluated by measuring the displacement at the table and the tool tip. A dummy tool is attached to the table and the displacement of the tool tip is measured by using a capacitance-type sensor.

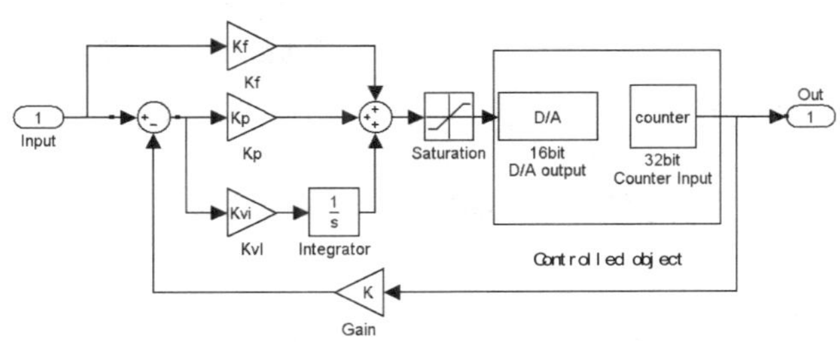

Figure 1.6. Block diagram of feedforward +PI control

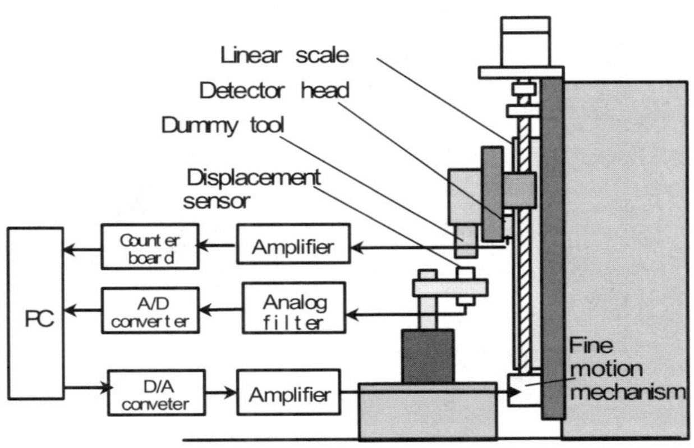

Figure 1.7. Experimental setup

First, the performance of each controller is compared with respect to the steady state error, undershoot, and overshoot of the step response. Finally, the FF+PI control was found to be the best among three controllers. Next, the controller gains are tuned so that the step responses satisfy the following criteria: the settling time is 10 ms; the steady state error is less than 20 nm. After the tuning, the step response test with various step sizes is carried out.

Figure 1.8 shows the measured table displacement (feedback signals from linear encoder). Steps can be observed in the table's response to input step sizes of 50 nm and 100 nm. However, it is difficult to observe step shapes in the table responses to input step sizes of 10 nm and 20 nm. This is because fluctuations have a range of 20 nm. Figure 1.9 shows the tool tip displacement. Steps can be observed in the tool tip response to input step sizes of 50 nm and 100 nm. Even though fluctuations are significant, steps can be seen in the response to the input step size of 20 nm. Although the response to the input step size of 10 nm fluctuates significantly, the table motion can still be recognized. The displacements measured at the table and tool tip show a good degree of agreement, which implies that pitch motion is very small in this positioning system.

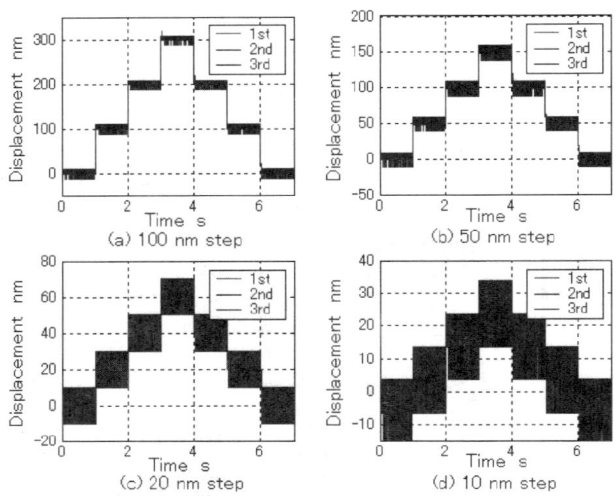

Figure 1.8. Step response of the table

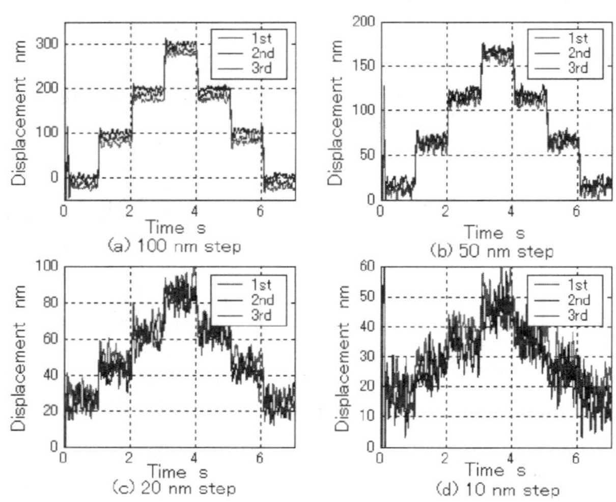

Figure 1.9. Step response of the tool tip

5. Conclusion

A fine positioning device, which is a ball screw drive with a piezoelectric actuator in the support-bearing unit, has been developed and several tests were carried out to evaluate its positioning performance. In the open loop test, table motions with a nanometer order resolution were observed with small yaw motions. A feedforward + PI closed loop controller was designed and fine pointing tests were carried out. The controller was able position the table and tool tip with a resolution of 20 nm, which is twice as much as the resolution of the feedback sensor.

References

[1] S. Futami, A. Furutani, S. Yoshida, 1990, Nanometer Positioning and its Micro-dynamics, Nanotechnology, Vol.1, P31-37.

[2] S. Moriyama, T. Harada, A. Takanashi, 1984, Precision X-Y Stage with a Piezo-driven Fine-table, Bulletin of the Japan Society of Precision Engineering, Vol.50, No.4, 718-723 (in Japanese).

[3] C. W. Lee, S. W. Kim, 1997, An Ultraprecision Stage for Alignment of Wafers in Advanced Microlithography, Precision Engineering, Vol.21, P335-345.

[4] H. J. Pahk, D. S. Lee, J. H. Park, 2001, Ultra Precision Positioning System for Servo Motor Piezo Actuator using the Dual Servo Loop and Digital Filter Implementation, International Journal of Machine Tools & Manufacture, Vol. 41, P51-63.

[5] A. Woronko, J. Huang, et al. 2003, Piezoelectric Tool Actuator for Precision Machining on Conventional CNC Turning Centers, Precision Engineering, Vol.27, P335-345.

[6] T. Elfizy, G. M. Bone, M. A. Elbestawi, 2005, Design and Control of a Dual-stage Feed Drive, International Journal of Machine Tools & Manufacture, Vol. 45, P153-156.

[7] J. S. Chen, I. C. Dwang, 2000, A Ballscrew Drive Mechanism with Piezo-electric Nut for Preload and Motion Control, International Journal of Machine Tools & Manufacture, Vol.40, P513-526.

[8] J. Otsuka, S. Ichikawa, Y. Yamaguchi, 2003, Development of Small Ultraprecision Displacement Small Ultraprecision Positioning Device with 1 nm Resolution, Journal of the Japan Society of Precision Engineering. Vol. 69, No. 10, P1428-1433 (in Japanese).

[9] N. Tanaka, 2006, Development of Nano Positioner, NSK Technical Journal, No. 680. P29-35 (in Japanese).

Real-time Monitoring of a Pulsed Power for Reactive Magnetron Sputtering Using a LabVIEW System

F.L. Wen[1], C.H. Wen[2], I. Hsu[3], H.S. Wu[4] and H.J. Wen[5]

[1]St. John's University, Taipei, Taiwan, 25135, ericwen@mail.sju.edu.tw
[2]National Chiao Tung University, Hsinchu, Taiwan, 30010, hsien@onlinenet.com.tw
[3]Ta Hwa Institute of Technology, Hsinchu, Taiwan, 30740, hsuichien@yahoo.com.tw
[4]Yuan Ze University, Taoyuan, Taiwan, 32003, cehswu@saturn.yzu.edu.tw
[5]St. John's University, Taipei, Taiwan, 25135, wenhj@mail.sju.edu.tw

Abstract. Due to a pulsed power offering the instant energy for extremely high plasma density, the purpose of this study is to develop the real-time monitoring system for the modern pulsed plasma coating by the LabVIEW technique. One selected exampl of the thin-film formation was testified based upon various N_2 gas flows for depositing ZrN_x film on the substrate of a p-type (100) silicon wafer through pulsed-DC reactive magnetron sputtering. The results indicate that the specific pulsed parameters affect the crystallized status on the ZrN_x film, in which a pulsed power was monitored by a LabVIEW system. Also, the characteristics of ZrN films in crystal orientations and grain sizes have directly relationship to various N_2 flow rates.
Keywords: Reactive sputtering, pulsed magnetron coating, zirconium-nitride film, layer characterization

1. Introduction

It is important to fabricate functional thin film with the specific surface structure and properties. Due to the wide applications of thin film engineering, they have the various stabilities in thermal and chemical situation and also demonstrate high hardness, anti-corrosion, and magnetic and optical features. The interesting feature involving the wide variation of electrical or semiconductor resistivity depends upon the chemical composition ratio of contents in compound films. There are two types of preparations for thin films, such as CVD and PVD, and also the PVD method is recovered as more popular processing in recent decades [1-8]. An innovative concept for thin film manufacturing by applying a GPIB card combined with the commercial available LabVIEW software tool has been proposed, based on a Windows computer to construct a monitoring platform. LabVIEW, a powerful graphical programming language developed by National Instruments, is easy to learn and can be used to develop a monitoring instrument rapidly. Once, this instrument needs a modification without requiring any new equipment. Users can create a whole set of laboratory-instrument control functions using LabVIEW but be less cost. Through a GPIB, an instrument-control card, LabVIEW enables a computer to easily retrieve the external signals, such as a real-time voltage and current signals, to proceed with analysis, monitoring and control. The current study is toward application in an on-line real-time monitoring system for sputtering systems by GPIB cards integrating with the LabVIEW software.

In order to investigate the influence of nitrogen flow rate on characteristics of zirconium-nitride film (ZrN_x) through pulsed magnetron coating on the substrate of a p-type (100) silicon wafer, the real-time monitoring concept is introduced to the processing technique by various N_2 flow rates through a pulsed-DC reactive magnetron sputtering with the unipolar-pulsing configuration. Through analysis of experimental results, one of selected processes as a sample for the pulsed power implementation is demonstrated by X-ray diffraction analysis.

2. Configuration of a Pulsed Plasma System

A schematic diagram of pulsed plasma system for preparation of thin films in this work is shown in Figure 1. The size of the stainless-steel reactor is 450×280×250 mm. The base pressure is evacuated to less 5 x 10^{-4} torr by a turbo-molecular pumping system. In the reactor, two disc electrodes of 2" in the diameter are placed parallel to each other, apart by 5 cm. Argon gas is used as a working gas and nitrogenous gas is a reactive gas for nitride films. Pulsed plasma is controlled around the substrate by the magnetron mechanism from a pulsed power with power of 300 W, the pulsed duration of 5-50 μ s, the frequency range of DC-50 kHz and the pulse repletion rate of 1-50 kHz. For the repetition rate of more than about 0.1 kHz, the after-glow plasma by pulsed power remains near the substrate during the off-time of the voltage pulse arrangement.

Plasma Generation using a Pulsed Power [7]

The plasma discharge models as a quasi-steady state in three-dimensional space in the reactor, and is assumed at a constant forward speed. A schematic sketch of the pulsed-plasma system is considered in this study as shown in Figure 1. The device consists of two coaxial electrodes (E1 and E2) separated by a hat-shaped insulator at one end, while the other end is open. The space between the electrodes is filled with the fed Argon gas. Both banks of big capacitors (C1 and C2) are charged by DC power supplies (DC1 and DC2) when the switch S1 and switch S2 are opened. Via the high power transistors: T1, T1*, and T2, T2* switching when the switch S1 and S2 are closed, these IGBT devices facilitate easy ignition of the pulsed plasma. During the ON and OFF times, the diagonal IGBTs are switched and opened continually (T1 and T1*) or (T2 and T2*). The ON and OFF times are freely adjustable range from the μ-seconds to seconds. The DC source can be used in the voltage, current or power controlled mode. Thus, the control configuration of a pulsing system is constructed to offer the various operating modes such as unipolar, bipolar, and DC mode.

Due to pulse technology delivering more power at the same energy level, the SPIK, the pulsed DC power controller, has a quick arc recognition and arc shutoff function which sensitivity can be regulated via the adjustable arc level, thus preventing the substrate surface from eventual burns caused by arc discharges. It also avoids the separation of macro particles into the substrate surface. By using of the fast Fourier transformation (FFT) it can be shown that the Dirac delta function contains the largest number of frequency components. The concept of power impulse is introduced to perform the state of the plasma focus as described,

$$v_k(t-a)=\begin{cases}\dfrac{1}{k} & \text{if} \quad a\le t\le a+k \\ 0 & \text{otherwise}\end{cases}, \qquad (1)$$

where $1/k$ is the voltage of amplitude acting from $t = a$ to $t = a + k$, and k is positive and small. In electrical power, the integral of a electromotive forces $v(t)$ acting on the equivalent capacitance circuit (RC circuit) of plasma system over a time interval $a \le t \le a+k$ is call the impulse of the electromotive force or the impulse power.

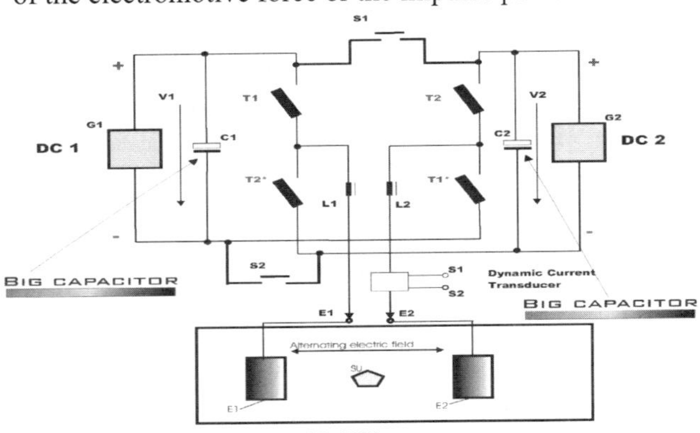

Figure 1. Illustrative diagram for plasma generation concept of a pulse power [7]

A coaxial line is used as an inductive storage device for the capacitor banks C1 and C2, which are discharged in order to deliver the required instantly high current. The high power density in the plasma focus causes the increasing peak power by more than a factor of 10~20 times due to electrostatic fluctuations. The inductively stored energy is coupled to the plasma very effectively. Ion and electron beams are generated by the plasma focus which exceed power densities.

Development of the On-line Monitoring System

The real-time instrumentation-monitoring-interface technique is based on the LabVIEW on-line data acquisition functions to transmit signals for monitoring a pulsed power. For this purpose, we use the graphic user interface (GUI) language to perform the monitoring system and to integrate some stand-alone measurement instruments, such as an oscilloscope, an arbitrary function generator, a current probe, and a high voltage differential probe, to reach a single PC-based control environment. Equipped with the instrument control and a connection to current and voltage signals, the GUI system can display pulsed parameters, such as pulsed duration, frequency, and the pulse repletion rate in voltage and current signals. The personal computer adopts Windows 2000 operation system and a low-cost NI-GPIB control card. A personal computer was used to control a SFG-830 arbitrary function generator (AFG) via the NI-GPIB card to send out the pre-designed or desired pulsed waveforms of various frequencies to a pulsed controller SPIK 2000. The computer would then receive, via the data acquisition from an oscilloscope communication interface, the output voltage and current signals of the pulsed power. The block diagram of a system structure is shown in Figure 2.

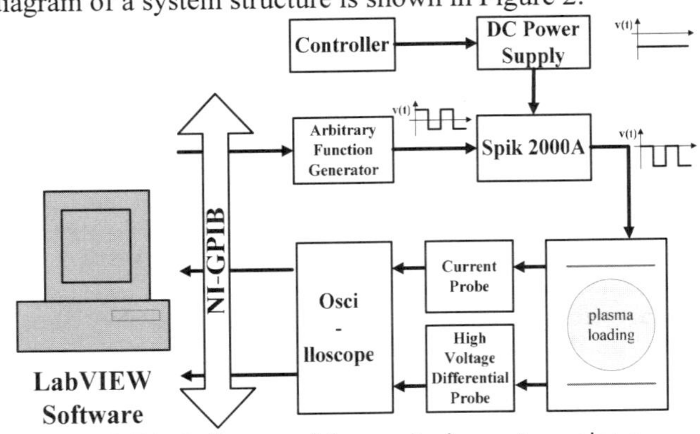

Figure 2. Block diagram of the monitoring system using a LabVIEW system for pulsed plasma loading

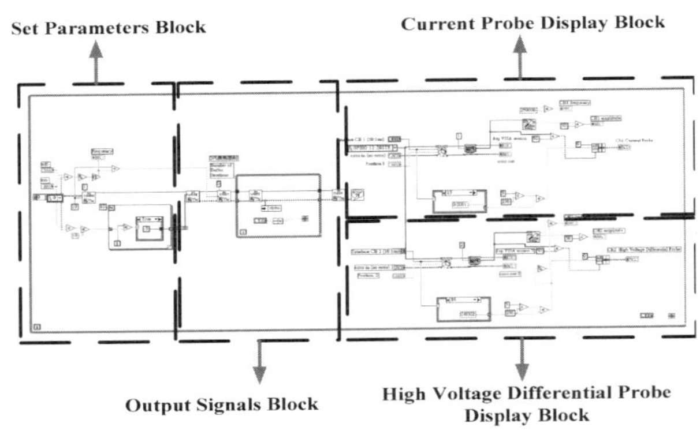

Figure 3. LabVIEW codes' blocks

The LabVIEW codes' blocks approximately comprise three graphic programming zones, as shown in Figure 3(a), namely (1) the parameter setting, (2) the signal acquisition and parameter acquiring, and (3) the pulsed configuration displays. These zones are interconnected one to another through software signal lines. The parameter-setting zone is used to construct the pulsed configuration, such as frequencies, duty cycle, on-off time intervals or repletion rate. The SFG-830 AFG received the desired pulsed-waveforms from the output signal block of LabVIEW programming. These signals drove a pulsed controller to produce a series match pulsed power to the pair of magnetron frames to form pulsed plasma. The signal acquisition and the pulsed waveform demonstrate were written into the current and voltage display blocks.

3. Selected Example for Thin Film Coating and Pulsed Parameter Arrangement

The silicon wafers were chosen as the substrate materials. The substrates were 4 inches in diameter and 0.525 mm in thickness of P-type (100) silicon wafers. The wafers were cut into a quarter, and then the specimens were ultrasonically cleaned in acetone and 2-propanol progressively for 5 minutes, respectively. And then they were washed by distilled water and air-dried. The experimental specimens were loaded into the chamber of a pulsed-DC reactive magnetron sputter for deposition.

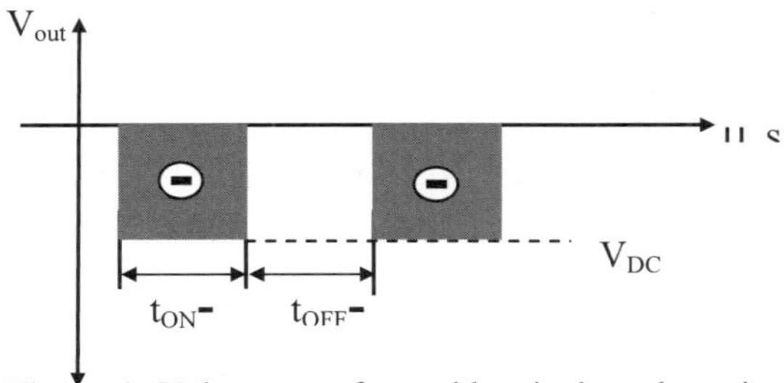

Figure 4. Voltage waveform with unipolar pulse trains

During this whole processing, the working gas (Ar) and reactive gas (N_2) were introduced into the chamber and mass flow controllers were used to regulate both gas flows. The flow rate of Argon gas was set at 25 sccm and that of Nitrogen gas was operated from 0.5 to 3.0 sccm. The pulsed-DC power was operated at 320 W using a constant-power mode. The distance from target to substrate was adjusted to 50 mm. The symmetric unipolar or bipolar pulse power could be employed to generate the plasma, but it was only by using of one DC-power supply for deposited processing. Therefore, there is the same level of voltage, current and power since DC1 is equal to DC2 from the same DC power supply. For improving sputtering deposition of zirconium particles on the silicon substrate, it is desirable to increase the plasma density and thus achieve more ions to be used for surface interaction during the growth of thin films. The plasma density increases with increased applied power to a magnetron sputter source. However, using a DC source, increasing the power leads to the generation of heat and eventually the destruction of the silicon substrate. Therefore, in this study by a pulsed power, the average power and heat generation can be controlled or be arranged. The symmetric unipolar power arrangement could be easily implemented by a SPIK controller, which offers the least average power and less heat generation but keeps the same energy consumptions. One example of a unipolar-pulsed power waveform for voltage is controlled by SPIK as shown in Figure 4. The technique gives an extremely high plasma density and the degree of ionization for the sputtered particles. We have experimented that the high-power-peak pulsed technique instantly generates a substantial amount of ionized metal vapor causing an enhanced sputtering. General considerations for a pulse train in the duty cycle is defined by the duration of the pulsed on-time t_{on},

$$\tau_{dutycycle}(\%) = \frac{t_{ON}}{T}, \quad (2)$$

where τ represents a duty cycle (%), and T is the period time of a completed cycle of pulse trains. However, when considering unipolar or bipolar pulsed modes, there are different parameter arrangements, specifically pulse duration and duty cycle. Other parameters will be modified significantly compared to a non-pulsed plasma system. Normally, in our unipolar pulsed plasma mode, only the negative pulsed on-time t_{on} is counted into the working duration for sputtering, and the duty cycle is redefined as,

$$\tau_{uniploar\ dutycycle}(\%) = \frac{t_{ON}^{-}}{T}, \quad (3)$$

where τ represents a duty cycle (%) for unipolar modes, and T ($= t_{ON}^{-} + t_{OFF}^{-}$) is the period time of a completed cycle of pulse trains. For a pulse train closed to rectangular shape one would use the full width at half maximum (FWHM) as the relevant on-time. For the sinusoidal-like pulsed trains, the duty cycle concept is unable to well definition. If the duty cycle calculation is applicable, the average power can be easily determined from peak values of Dirac's delta function as,

$$P = \frac{\tau}{k}, \quad (4)$$

where P stands for power quantities (W). If the pulse rise and fall times are not much shorter than the constant portion of a pulse train or without constant portions, the average power will be [8],

$$\overline{P} = \frac{\int_{t_1}^{t_2} p(t)dt}{t_2 - t_1}, \quad (5)$$

where $p(t)$ stands for instant power (W), and (t_1, t_2) is the time interval offered an average power over a plasma system. Hence, the amplitude of currents can be larger orders of magnitude, leading to the higher instant power and particle energies, higher degree of ionization in the plasma system, and other advantages.

4. Experimental Results and Discussion

Since thin film properties are highly correlated to the material and processing, the appropriate method to investigate the properties of the ZrN_x films is very important. For the thin film with crystal ambiguity, the power XRD instrument in less glance incident angle (GIAXRD) shall be proper equipment to identify the crystal orientation on surface microstructure. In this a study, using the thin-film X-ray diffractometer, MXP18 GIAXRD at the Instrument Center of National Tsing Hua University, Hsinchu, Taiwan, the crystal structure of deposited ZrN films was observed. Lower diffractive angles have less penetrated depths by X-ray resulting that stronger diffractive signals can be detected. The Cu Kα line at 1.54056 Å was used as the source for diffraction. The operating power was 40 kV × 150 mA. For the $\theta/2\theta$ scans, a scan rate of 2 deg/min was used, and the 2θ scan range was from 20 to 60 degrees.

Through GIAXRD analysis as shown in Figure 5, the intensity peak of (111) crystal orientation of ZrN on the silicon substrate is appeared in the spectra at the flow rate of 0.5 sccm. It sustains the (111) crystal orientation till the flow rate of 3.0 sccm and holds higher

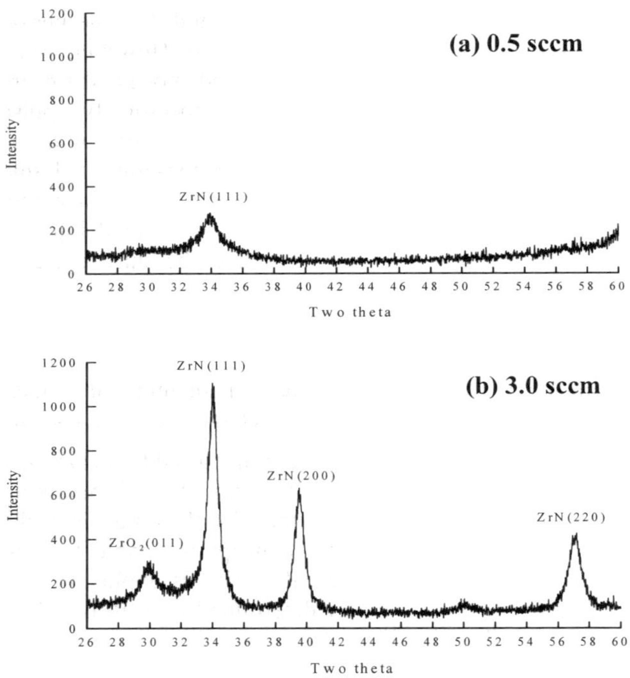

Figure 5. GIAXRD spectra based on various N_2 flow rates

intensity peaks as the increased N_2 gas flows. Due to the tendency of reduced strain energy for the crystal structures, the crystal orientation of (111) has the more obvious guiding which belongs to the preferred orientation for ZrN films. Also, the orientations of (220) and (200) on the ZrN film appear intensity peaks in spectra when N_2 flow rate at 3.0 sccm, as shown in Figure 5(b). Their intensities are proportional to the reactive gas flows. The result is inducted from enough pulsed electrical power to strike out the Zr ions with higher energies. After Zr ions were deposited on the substrate surface, these ions were easier movement (diffusion) and were accumulated at the stable location with lower potential energies. They have regular phases and highly preferred orientation.

In addition, the more flow rate of active gas has the stronger peak intensities of ZrN crystallizing orientation. It is proof of the helpful factor for crystal formation while there are more nitrogen atoms in plasma chamber. However, from the spectra in Figure 5 shown of ZrO_2 (011) at N_2 flow rate of 3.0 sccm, one possible reason for the existence of oxides results from more nitrogen molecules absorbing too much energies to the compound of nitrides, especially in the plasma sputtering with a constant pulsed power.

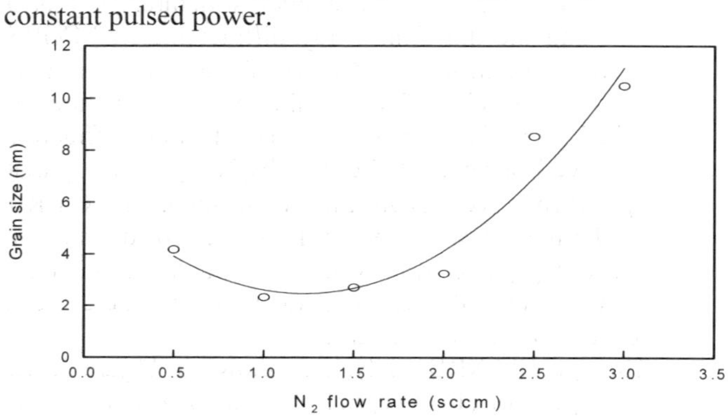

Figure 6. Estimated grain size by the GIAXRD technique with various N_2 flow rates

According to the estimation of Scherrer equation as (6), the least grain size of 2.3 nm at 1.0 sccm and largest grain size of 10.5 nm at 3 sccm.

$$GrainSize = \frac{0.9\lambda}{B \times \cos\theta}, \qquad (6)$$

where λ is the wavelength of X rays, B stands for the full width at half maximum (FWHM) of intensity peaks, and θ is appeared angle of intensity peaks. The effect of N_2 flow rate on the grain size of the ZrN film surface is illustrated as Figure 6. If N_2 flow rate is less than 2.0 sccm, there is almost no modification on the surface grain sizes. The flow rates will remain in effect on grain size after more N_2 gas is added. One proper explanation for the phenomena is the surface effect on a silicon substrate. Since the larger grain size of ZrN film is closed to the substrate, more growth rates of thin films are inherent in the microstructure's crystal orientation of silicon substrates.

5. Conclusion

The study demonstrates the configuration of a real-time monitoring pulsed power system for reactive magnetron sputtering using a LabVIEW platform. One selected example, with the special parameter arrangement and pulsed trains, has also shown that the grain size on the zirconium-nitride film is modified by various N_2 flow rates. The GIAXRD results indicate that the reactive gas highly affects on crystal orientations based on the observation of intensity peaks at the spectra. There are many subjects left and needed to be studied such as efficiency of energy transfer into plasma loading, system response to a different pulsed train input, and real-time turning. The monitoring system integrated with a pulsed power for plasma sputtering of thin films has the great potential application in advanced thin film manufacturing.

Acknowledgment

Partial financial support for this work was provided by the National Science Council of Taiwan government, under the contract NSC 94-2622-E-129-007-CC3.

References

[1] Braithwaite, N S. J. 2000, Introduction to gas discharges, Plasma Sources Sci. Technol., Vol.9, P517-527.

[2] Conrads H., Schmidt M. 2000, Plasma generation and plasma source, Plasma Source Sci. Technol., Vol.9, P441-454.

[3] Bunshah R. F., Deshpandey C. V. 1985, Plasma assisted physical vapor deposition processes: A review, J. Vac. Sci. Technol. A, Vol.3, P553-560.

[4] Chapman B. 1980, Glow discharge process, John Wiley & Sons, Canada, New York.

[5] M. Konuma. 1992, Film Deposition by Plasma Techniques, Springer-Verlag, Berlin, Germany.

[6] Wen F.L., Lo Y.-L., et al. 2006, A Pulse DC Plasma Deposited Resistor Process, Mater. Sci. Forum, Vol.505-507, Pt2, P865-870.

[7] G. N. Mark. 2003, DC-Pulse Power Supplies, MELEC, Ottersweier, Germany.

[8] Andre Anders. 2003, Fundamentals of Pulsed Plasmas for Materials Processing, ICMCTF Invited Talk, P1-30.

Windows CE-based Intelligent Remote Monitoring System

Shang-Liang Chen[1], Ting-Hao Lin

[1]National Cheng-Kung University, Taiwan, Institute of Manufacturing Engineering, slchen@mail.ncku.edu.tw

Abstract. An intelligent remote monitoring system based on Windows CE OS is developed in this research. It includes an embedded OS, a remote monitoring system, a Bluetooth interface and an intelligent agent. With the characteristics of the embedded system, the developed monitoring system requires less resources than general high-end monitoring system. The system has hard real-time (1 ms) function for monitoring the disorder of machine. In addition, Bluetooth wireless technology plays an important role for remote message communication and connection between different neighbourhood machines. For monitoring, several intelligent agents are designed to judge if there is any anomaly of machines. The intelligent agents can also search and locate the position of the equipment engineer. Then, the monitoring messages will be transmitted to the equipment engineer.

Keywords: intelligent remote monitoring system, Windows CE, Bluetooth interface,

1. Introduction

Remote site engineers can realize the working condition of a machine and give necessary maintenance immediately. When the machine shows any anomaly, the remote monitoring system transmits an alarm message through the internal communication interface. However, the anomaly signal light on a machine is used to represent the anomaly of the machine currently. Therefore, the anomaly needs to be found by engineers. A Bluetooth wireless technology with searching function is designed and developed in this research to immediately transmit the alarm messages to the PDA carried by the equipment engineers. Therefore, they can quickly know which machine is disordered and deal with the problems right away. Fig. 1.1 is a schematic diagram of a Windows CE based intelligent remote monitoring/control system developed in this research.

The current PC-based monitoring system used in industry is mainly based on a Windows 2000/XP operating system (OS). Windows 2000/XP systems are easy to be used and highly compatible to end-users. However, they are also very complicate with many application programs included. The system's real time performance is also limited. Applying the system to the industrial applications, many functions of the 2000/XP system are not necessary or need to be modified. For example, industrial controllers should have the hard real-time function, but the function does not exist in a Windows 2000/XP system. Therefore, a third party solution is usually used to meet these requirements, which increases the costs. To solve this problem, an embedded OS is rapidly developed in recent years. The embedded OS can support the requirements from different hardware platforms. This embedded system is also excellent in the integration between software and hardware.

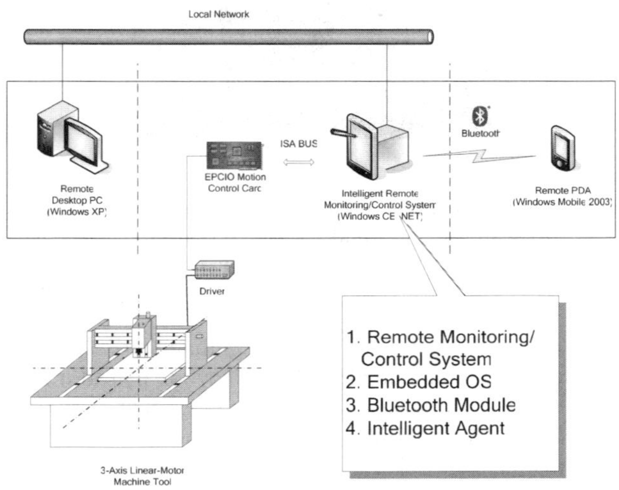

Figure 1.1. Windows CE based monitoring system

In this paper, Windows CE.NET was selected as the OS. It contains complete characteristics of an embedded system. For example, hard real-time function, complete support for Bluetooth, IEEE802.11 wireless network, compatible with Visual Studio, etc. Windows CE.NET OS provides advanced real-time functions, such as 256 priority levels for executing thread and supports for nested interrupts. The nested interrupts, if necessary, allow lower priority interrupts to be serviced immediately within a higher priority interrupt cycle, instead of having them to wait until the end of a higher priority ISR.

In this paper, two steps are adopted to develop the embedded intelligent remote monitoring/control system. Fig. 1.2 shows the detailed procedures of these two steps, which are described as follows:

The first step is the design of the embedded OS:

1. Planning the related software and hardware of an embedded OS for achieving the desired efficiency.

2. Selecting the CPU used in the system.
3. Selecting the peripheral interfaces of the system.
4. Selecting the internal service modules of the system.

The second step is to build up an intelligent remote monitoring/control system:

1. Planning the network modules of the remote monitoring/control system.
2. Designing HMIs (Human Machine Interfaces) of the remote monitoring/control system.
3. Selecting the communication interface and uploading the Bluetooth device driver.
4. Designing the communication functions.
5. Defining the functions of the intelligent agent of the system.
6. Constructing the modules of the intelligent agents defined in the above.

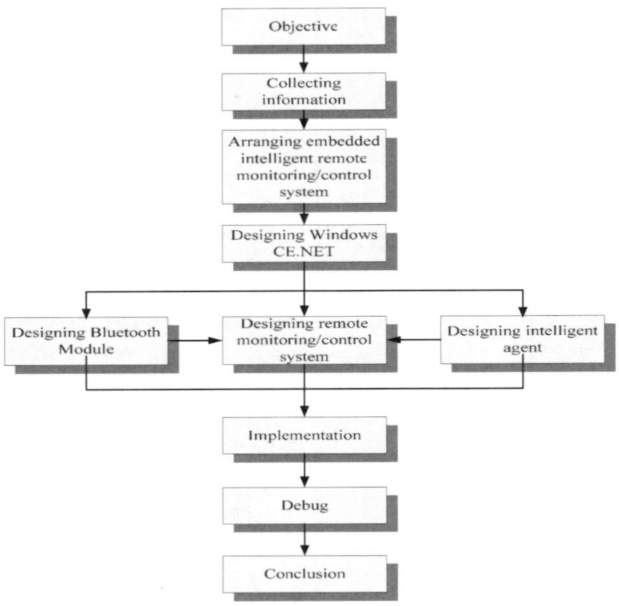

Figure 1.2. Procedures of research methodology

2. Design and development

For designing a Windows CE.NET OS, two parts are described in this paper (see Fig. 1.3). The first part is the systematic level design. In this part, all kinds of featured modules operated by Windows CE are arranged and integrated into the system. The second part is hardware structural design. The hardware structural design is decided by its applications. After the systematic level design and hardware structural design, the created Windows CE.NET OS will be tested.

2.1 Systematic Level Design

1. Plan the required specifications of applications.
2. Choose the CPU structure and build the board support packages (BSPs).
3. According to the required real-time function, design the kernel interrupt clock and interrupt vector table.
4. Design the driver features of peripheral interfaces (network, USB, Bluetooth, etc.).
5. Design the software module's features and upload them into the system.
6. Set the related registry of the system (i.e. platform.reg, command.reg, etc.).
7. Integrate, compile, and debug the embedded OS.
8. Download the compiled NK.bin files through RS232 and perform the starting test.
9. Recompile the release version. Download the release version to the storage devices of the development board. Build up software development kit (SDK) for developing the application programs.

2.2 Hardware Structural Design

1. Design the required specifications for the applications.
2. Choose a suitable CPU and development board.
3. Choose peripheral devices including storage devices, RAM size, motion cards, data acquisition cards, USB interfaces, Bluetooth module, etc.
4. Setup the whole hardware structure. Download the obtained NK.bin files of Windows CE.NET and perform the test.

2.3 Structure and function design

The basic structure of a remote monitoring systematic contains the function of data acquisition, monitoring, and remote communication. Fig. 1.3 shows the remote monitoring system of a new generation automatic factory using Bluetooth wireless communication technology. To achieve these functions, three modules are included in this system. They are Windows CE real-time OS, Bluetooth communication interface for message receiving and sending, and the intelligent agent module for sending alert messages.

There are:

(a) Web-based connection and sign-in:
With TCP/IP protocol, the remote monitoring/control system is started and data can be transmitted to the web page of the remote internet explorer via the internet. Objects provided by ASP.NET and Windows CE.NET are adopted as main functions of connection. Remote users need to sign-in their ID and password to ensure the safety of system.

(b) Web-based system status and data transmitting:
After the connection of remote network, users can realize the motion status of the current machine via the internet. One of the advantages of using Web-based data transmitting method is that the used OS platform can communicate with most kinds of OS that support standard HTML language.

(c) Record and store the data:

With encoder, AD/DA converters and Digital I/O provided by motion control card, motion position, velocity, the related data from machine can be collected every 1 ms. The embedded OS can provide the timers for accomplishing the command.

(d) Display the motion path:
The system can read the position data and display a 2D motion path diagram immediately.

(e) File transmitting:
With the FTP protocol, the system can transmit extra monitoring or executing commands to the machine from the remote site. Windows CE.NET kernel provides related functions for setting up FTP server.

(f) Dialogue window of Real-time Text Message:
When the Bluetooth wireless connection is on, both the host-site and client-site have the right to online and real-time send text messages. Text can be input from the dialogue window and sent by pressing the 'Send Message' button.

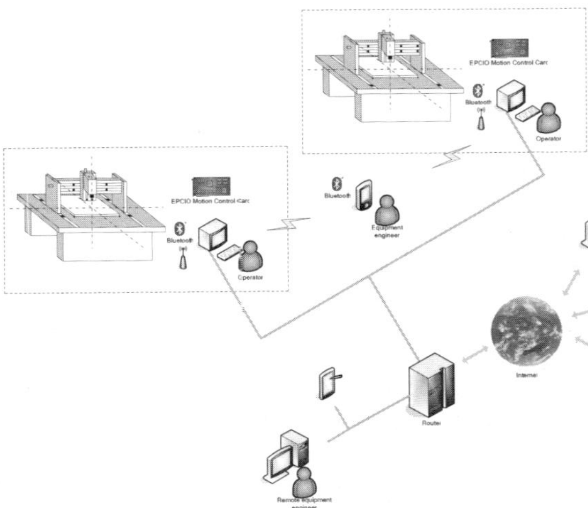

Figure 1.3. Remote monitoring/control system using Bluetooth wireless communication technology

2.4 Rules of Intelligent Agents

In this paper, the concept of Intelligent Agent is adopted in the remote monitoring/control system. Monitor Agent (to judge the monitoring data) and Alert Agent (to send alert messages to objects when there is something anomalous) are two types of intelligent agents. Using the definition of intelligent agent (such as: belief, goal, and action), the expected function of the monitoring system can be implemented. Using the automation feature of intelligent agents is helpful for data monitoring. Using the decentralization feature of intelligent agents is also helpful to find out the location of the equipment engineers and to communicate with other machines

2.5 Monitor Agent

Monitor Agents are used to store log files of the machine, to catch the knowledge the agent needs, and to judge if there is any anomaly by means of pattern comparison. There are two kinds of pattern comparisons. The first one is that the programmer designs the alert message features in advance and adds them to the system. The alert message rules will be started when the monitored data matches the alert message features. The second one is to record the monitored data all the time. If there is any anomaly, the frequency of anomaly will be recorded and an alert message will be sent to the operator after the anomaly number is accumulated to a preset number. The machine can also automatically judge if there is anything wrong and stores its feature. The system will inform the operators in advance. In this paper, the built belief, goal, and the rule of action of the Monitor Agent are executed in the remote monitoring/control system. For the anomalies judgment, here, the anomalies of each axis are judged by the position, speed, torque, servo on signal, driver alarm, etc.

3. Results and discussions

3.1 Verification of the Intelligent Remote Monitoring

The developed six functions are verified here. For data storage function, the Data_Thread is built to store data. The default path can be set to store data into inner disk on module (DOM) or USB flash disk. The cycle time is 1 ms for reading and recording. The recorded data can be read to understand if there is any mistake before anomaly occurs.

For verification of the web-based remote data display function, the developed service of web servers will be uploaded after Windows CE.NET OS is on. After the intelligent remote monitoring/control system is started, two procedures will be executed: one is to create the sign-in pages and renew the default homepage of the original web server. The second is to create the monitoring page and upload it to the web servers. The monitoring page is hyper-linked with the homepage. The monitoring data will be updated on the pages. The real-time remote data display is accomplished in this research.

For renewing data on the pages, one Web_Thread is mainly in charge of reading data of machines and displaying data on the remote pages. In addition, HTML and ASP language are used to display data on the monitor pages and stored into the web servers. Monitoring data can be real-time displayed and renewed on the remote site. After the above two procedures are accomplished, the IP addresses of the remote monitoring system are input from the remote site on different OS platforms. With the verification of passwords, the status of machines and the data can be displayed after sign-in.

For the motion control function, the verification is executed by Draw_Thread. The position, speed, and the figure of 2D motion path are drawn by reading the data of each axis of machine

The experiment of intelligent agents has two parts. One is the Monitor Agent test and the other is the Alert Agent test. Monitor Agent is used to judge if there is any anomaly by the monitoring rule with the monitored data.

The experiment uses an output simulated motion command. Assumes the command is corresponded to the anomalous condition of the monitoring rule. The experiment is to test if the Monitor Agent can judge the anomalous condition correctly by its own knowledge database.

The simulated motion commands can be categorised in two types: position monitoring rule and speed monitoring rule. After choosing two simulated anomalous statuses, the DEMO button is pressed. The screen can display the current position (red lines) and the speed (blue lines). Fig. 1.4 and is the screen showing the starting of the intelligent remote monitoring system and judging anomalies of machines by Monitor Agent.

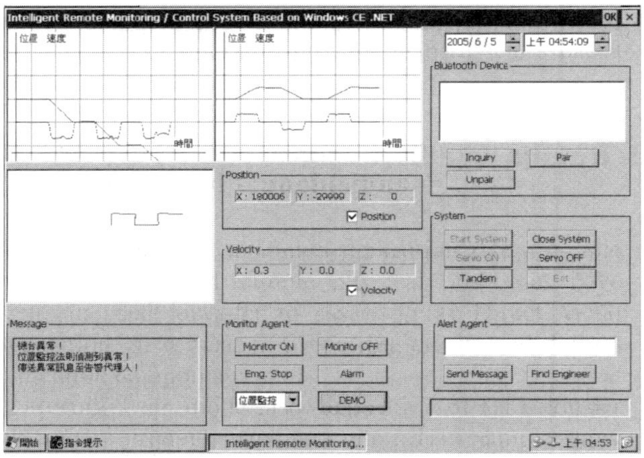

Figure 1.4. Monitor Agent detects position anomalous

4. Conclusions

The complete structure of the embedded intelligent remote monitoring system is successfully designed and implemented in this research. The system includes Windows CE.NET OS, Bluetooth Module, and intelligent agent. The developed intelligent agent integrates the Bluetooth interface to make the system have the wireless transmission capability. The developed Windows CE.NET-based intelligent remote monitoring/control system has the ability of 1 ms real-time performance, data storage, web-based remote data displayed, anomalies judgment, remote Alert, and etc. The results are summarized as follows:

(a) The web-server constructed by the HTTPD service of Windows CE.NET OS can successfully transmit the monitored data and display on the remote site.
(b) The Monitor Agent of the intelligent agents can correctly judge the default anomaly of machine and successfully send anomalous messages to the Alert Agents. The Alert Agents will further send a message to the remote PDA of the equipment engineers.

References

[1] J. Lee, “Machine performance monitoring and proactive maintenance in computer-integrated manufacturing: review and perspective.” , Int. J. Comput. Integrated Manuf., 1995

[2] R. A. Errath and J. J. Dominguez, “Remote drive condition monitoring” , in Proc. Conf. Rec. 1999 IEEE-IAS/PCA Cement Industry Tech. Conf., Roanoke, VA, Apr. 11–15, 1999

[3] P. W. Tsc, “Web and Virtual Instrument Based Machine Remote Sensing Monitoring and Fault Diagnostic System”, ASME 2001 Design Engineering Technical Conf. and Computers and information in Engineering Conf. Pittsburgh PA, September 9-12, 2001

[4] J. T. Huang, C. K. Chen, W. L. Lee, “Dynamic Case-Based Reasoning for Remote Diagnosis and Maintenance System,” 2002 IEEE/ASME International Conference on Advanced Manufacturing Technologies and Education in the 21st Century, Aug. 11-14, 2002

14

Rapid Prototyping

14–1

Digital Additive Manufacturing: From Rapid Prototyping to Rapid Manufacturing

K.K.B. Hon

University of Liverpool, UK, hon@liv.ac.uk

Abstract. The emergence of stereolithography in 1998 is a milestone for an entirely new class of layer-based manufacturing processes. This new manufacturing approach which allows direct digital manufacturing from CAD to a physical part, is the most flexible and versatile form of fabrication process. This paper provides an overview on the development of digital additive manufacturing processes, commonly known as rapid prototyping processes, and chart its progress and challenges towards the ultimate goal of rapid manufacturing with supporting case examples.

Keywords: stereolithography, digital additive manufacturing, rapid prototyping

1. Introduction

Manufacturing is crucial to creation of wealth and provision of quality of life. Manufacturing covers numerous aspects from systems design and organisation, technology and logistics, operational planning and control. The study of manufacturing technology is usually classified into conventional and non-conventional processes. The former grouping of processes are in existence before 1950 and the latter ones generally refer to those adopted by industry after 1950 and they are typified by EDM, ECM, USM, LBM, etc.

In 1988, an entirely new approach of additive manufacturing process based on the fabrication of layer by layer principle was demonstrated in the form of stereolithography (SL). In the early days, SL had poor dimensional accuracy and its application was restricted to 'touch and feel' models. With improved technology and accuracy, it became an accepted method for prototyping and the term rapid prototyping was accepted universally. In the early nineties, other forms of layer-manufacturing processes were also developed and numerous studies were undertaken to compare various processes in this new group of manufacturing technology [1-2]. In mid-Nineties, RP processes were firmly established as an indispensable part for rapid product development.

Numerous terms were coined to describe this group of layer-based processes, for instance, freeform fabrication, layered-manufacturing, material ingress manufacturing, etc. [3]. However, the term Rapid Prototyping is the one most widely accepted [4-5]. RP can be defined as the direct fabrication of a physical part from a 3D model by additive manufacturing. As there are no tool setting, workpiece fixturing and tool wear compensation requirements, it is the most flexible and direct form of digital manufacturing. In fact, 3D Systems has switched recently to the term 'Additive Digital Manufacturing' instead of 'Rapid Prototyping'.

Prototyping is generally restricted to the point of pre-production but it does not mean that such prototypes could not be functional. As RP evolves into Rapid Manufacturing (RM) which provides more space for application and exploitation, a more robust definition for RM is required and quite a few have been proposed, as for instance, Wohler [5]. A modified and more robust definition based on Rudgley is: Rapid Manufacturing is the manufacture of end-use functional products using layer-based additive manufacturing techniques [6].

Rapid Manufacturing is an emerging subject and the first international conference solely dedicated to this topic was only initiated this year. As the root of Rapid Manufacturing is Rapid Prototyping, this paper aims to provide an overview of both technologies, current developments as well as future challenges.

2. Classification of Digital Additive Processes (DAM)

A systematic classification system for DAM processes is based on the four distinctive forms of raw materials. As shown in Figure 1.1, there are four main forms of materials, i.e., liquid, powder, solid or semi-solid and sheet material. The diagram shows some of the most popular processes as well as a number of development stage systems such as Selective Laser Melting (SLM) and Rapid Freeze Prototyping (RFP). Presently, there are far more system manufacturers for liquid, powder and solid based system as they are more versatile and flexible.

3. Characteristics of Digital Additive Manufacturing

The operating principles of commercially available DAM systems are now well-documented [4-5]. Reviewing DAM technologies as a whole, a number of generic characteristics could be identified that are unique when compared with conventional model-making processes by machining.

- Part building can be made directly from CAD data resulting in a much shorter process chain compared with traditional subtractive process
- Absence of tooling requirement contributes to

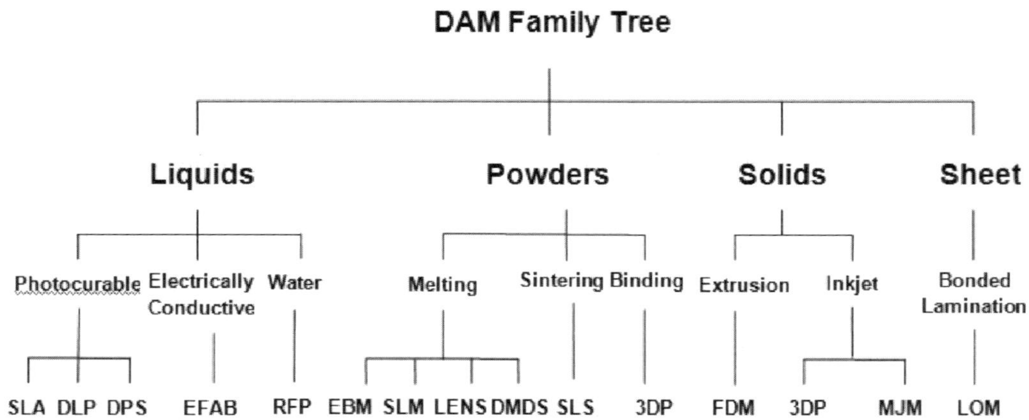

Figure 1.1. Digital Additive Manufacturing family tree with representative processes

significant lead time reduction.

- The layer-by-layer building mechanism allows great geometrical freedom in making parts which were previously impossible.
- DAM processes are generally highly automated which allow unattended overnight part building, increase machine utilization and reduce unit cost.
- The layer manufacturing approach has a beneficial effect of rendering the building time insensitive to the complexity of part geometry.
- Each DAM process operates for a very narrow range of materials.
- Some DAM processes allow functional gradient or multi-materials.
- Components build by DAM process could have highly anisotropic material properties.

Combining all such characteristics, it is obvious to conclude that DAM is the most flexible digital manufacturing technology available. New technical possibilities such as conformal cooling channels in plastic injection tooling, surface features for optimised heat transfer are becoming economically viable.

4. Present Trend in Digital Additive Manufacturing

The most popular DAM system is FDM which has sold over 4000 units to date. Its market acceptance is due to affordable machine costs, ability to produce functional end-use parts, ease of installation and operation and low maintenance requirements. This will satisfy a large group of product development needs where accuracy is less critical.

Another system which is gaining rapid industrial application is the 3D printer. The starch-like material used in a 3D printer is the cheapest per unit volume. This encourages the designer to experiment with more design variants before a final design is chosen. This direction is moving towards low-cost RP/RM systems.

The quest for precision is a continuous driver for DAM. Presently, it is possible to have a layer thickness of the order of 17 μm. It is obvious that the smaller the layer thickness, the less is the staircase effect. For tooling and master pattern applications, high precision DAM process is indispensable.

A definite trend on materials is the advances on stronger materials for functional applications. For instance, nano powders are added to SL epoxy resin to enhance mechanical properties [7], production of composite materials by SLS [8]. There is also development of more medical grade materials suitable for implantation. Other trends include faster systems, for instance, spiral growth manufacturing, micro-scale systems [9-10].

5. Transition to Rapid Manufacturing

Rapid manufacturing is a misdemeanour as it was originally meant to be a new catch phrase for marketing purposes. The use of rapid manufacturing technology does not guarantee that it is the quickest way of producing a part. RP and RM could not compete with conventional mass production methods when the part geometry is not complex. For instance, the following exemplary processes are much faster than layer-based rapid manufacturing processes:

- Injection moulding of plastic parts
- Blanking of sheet metal components
- High speed machining of dies and moulds
- Extrusion of bars and rods

In addition, the production of simple geometrical shapes such as cylinders, holes, flats, and tapers, standard CNC machining is also much faster.

The exploitation of the easiness for making complex parts offers new opportunities for producing high value-added parts especially if they are functional. This characteristics forms one the foundation principles for the growth of rapid manufacturing.

It is now increasingly evident that rapid manufacturing fulfils a number of niche applications for making unique, one-of-a-kind products [11]. The main drivers for such new opportunities are due to:

- Emergent needs for mass customisation.
- Demand for personalised healthcare products.
- Opportunities for low-volume, high value added products.
- Time-critical requirement for parts with complex

geometry.

Based on such evolving trends and the increasing deployment of rapid manufacturing, a growing number of rapid manufacturing case studies have been reported, as described below:

5.1 Teeth straightening

The traditional method of tooth alignment by braces does not offer visual appeal. With the use of transparent stereolithography teeth aligner, it has no adverse effect on appearance and this method is gaining acceptance especially in USA as shown in Figure 1.2 [12].

Figure 1.2. Application of Stereolithography for teeth-straightening [12]

5.2. Hearing aid

One of the most successful applications for rapid manufacturing is hearing aids [13]. This is a prime example for mass personalisation as each human ear is unique. It has been reported that over 1.5 million hearing aid shells have been produced to date by Siemens, using 3 RM methods, i.e., SLS, SLA and DLP [14].

5.3. Other applications

Successful RM applications have been reported for jewellery, luxury car, formula 1 racing car, tissue engineering scaffolds and others. New applications for microsystems have also been reported. In fact, there are no restrictions on the applications of RM technology [15].

6. Challenges and Application Issues

6.1 Machine design

The design of rapid prototyping machines has not changed fundamentally since their inception in the early Nineties. Issues on robustness, reliability, build time still need further improvements. The 'material addition' per unit time remains behind 'material removal' for established processes such as high speed machining, HEDG where a specific removal rate of 1000 $mm^3/mm \times sec$ could be achieved for steel.

6.2 Machine cost

High machine costs were an obstacle for DAM on market penetration in the early 90s. In recent years, a number of low cost machines have become available in the market at around £13.5k (US$25k), for instance, FDM, 3D Printing. On the other hand, expensive systems such a Laser Engineered Netshape System (LENS) could cost as much as £750k (US$1.4 million). In the operation of machine, not only the machine cost but also the consumables, operator and ancillary costs have to be taken into consideration

6.3 Materials cost

One of the impediments of DAM technology is the relatively high cost of materials. This is especially true for SL systems as not only the unit volume cost is high but also a vat of photo curable resin has to be maintained. This adds to the operating costs significantly. For RP to progress to wider applications, lower costs of materials and ultimately lower operating costs will be a worthy target.

6.4 Dimensional accuracy

Most RPM processes involve phase transformation which leads to volumetric changes and residual stresses. Some forms of geometrical distortions are inevitable and numerous investigations have been reported [16-17]. In the case of SLA, the shrinkage is caused by density change from liquid phase to solid phase from 1.13 gm/cm^3 to 1.35 gm/cm^3 for Somos 9110 epoxy resin. Even with the most advanced hatch pattern, curling remains a process characteristic as shown in Figure 1.3 [18].

6.5 Support materials removal

Support for part building is a characteristic feature in RP for liquid or extrusion based processes. Supports serve to connect the part to the build-platform, to support overhang building, to create a flat down-facing surface and to reduce the risk of delamination and deformation. For powder based RP methods, there is no need for support. As for FDM, new water-soluble support materials have greatly simplified the downstream removal process.

6.5 Part size

The largest part that could be built in a SL vat is 1500 x 750 x 500 mm^3. Larger parts could only be constructed by joining two parts together by epoxy resin, mechanical methods or by design. The ability to build large parts is constrained by optics as well as the prohibitive costs of holding a large vat of epoxy resin. For economic reasons, RPM equipment based on point-of-use materials dispenser such as FDM will be a more viable solution for large parts.

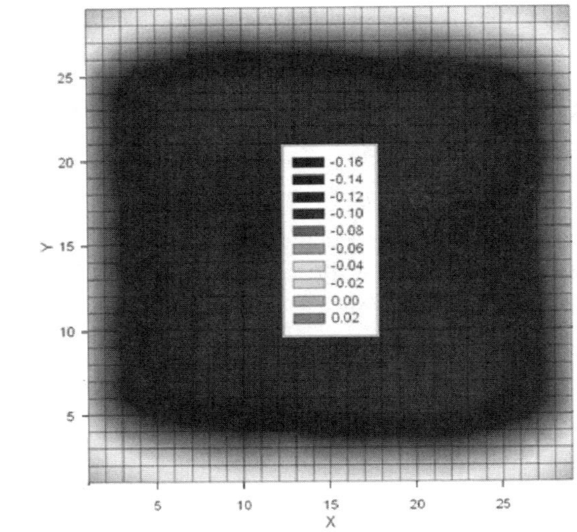

(a)

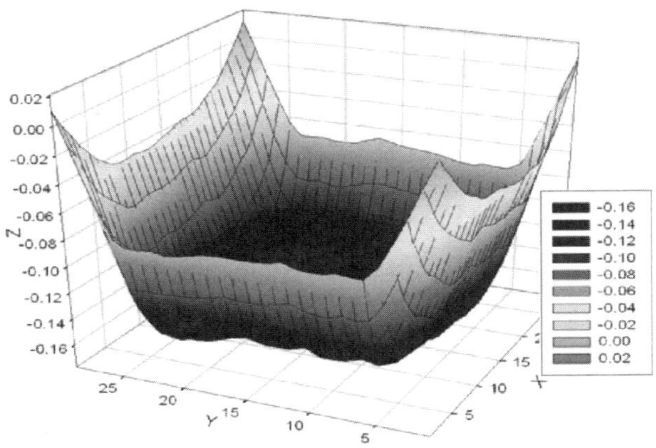

(b)

Figure 1.3. (a) Contour graph and (b) 3D surface profile of part built with Bisector scanning pattern (mm)

7. Conclusions

The history of RP and RM is short when compared with conventional metal cutting processes. The attractions of this new group of layered manufacturing processes are versatility, flexibility and simplicity. Tremendous improvements on processes, materials, speed and accuracy have been made since the pioneering stereolithography process first appeared in the market 18 years ago. Shakeout of technology such as LOM and SGC have been witnessed as a result of market force. With sustained worldwide efforts on research and development, advances and developments in RP and RM will further broaden their applications as cost-effective solutions to customised, small and medium volume production.

References

[1] Juster NP and T.H.C. Childs THC (1994), A comparison of rapid prototyping processes, Proc. 3rd European Conf. Rapid Prototyping and Manufacturing, 35-52.

[2] Chua CK, Chou SM and Wong TS, (1998) A study of the state-of-the-art rapid prototyping technologies, International Journal of Advanced Manufacturing Technology 14/2:146-152

[3] Kruth JP (1991) Material Incress Manufacturing by Rapid Prototyping Techniques, Annals of the CIRP, 40/2:603-614

[4] Jacobs PF, (1992) Rapid Prototyping and Manufacturing: Fundamentals of Stereolithography, SME

[5] Wohler T, (2006) Wohler Report 2006, Wohler Associates Inc.

[6] Rudgeley M, (2001) Rapid manufacturing-the revolution is beginning, Proc. uRapid 2001, 441-444

[7] Anon, (2006) Somos NanoForm 15120 data sheet

[8] Hauser C, Sutcliffe CS, Egan M and Fox P, (2005) Spiral Growth Manufacturing (SGM) – A Continuous Additive Manufacturing Technology for Processing Metal Powder by Selective Laser Melting, Proc. Solid Freeform Fabrication, 1-12

[9] Hon KKB and T.J. Gill TJ, (2003) Selective laser sintering of SiC/Polyamide composites, Annals of the CIRP, 52/1:173-176

[10] Huang S, Heywood MI, Young RCD, Farsari M and Chatwin CR, (1998) Systems control for a micro-stereolithography prototype, Microprocessors and Microsystems, 22/2:67-77

[11] Pham DT and Dimov SS, (2001) Rapid Manufacturing: The Technologies and Applications of Rapid Prototyping and Rapid Tooling, Springer-Verlag

[12] www.invisalign.com

[13] www.materialise.com/eGazette_200602/magicsRM_ENG.html

[14] Martin R, (2006), Application of Rapid manufacturing for the production of individualised hearing instruments, Proc. 1st Int. Conf. Rapid Manufacturing, 154-167

[15] Juster NP and T.H.C. Childs THC (1994) Linear and geometric accuracies from layer manufacturing, Annals of the CIRP, 43/1:163-6

[16] Onuh S and Hon KKB, (1998) Application of the Taguchi method and new hatch styles for quality improvement in stereolithography, Proceeding of the IMechE, Part B- Journal of Engineering Manufacture, 212/B6:461-472

[17] Levy GN, Schindel R and Kruth JP, (2003) Rapid Manufacturing and Rapid Tooling with Layer Manufacturing (LM) Technologies, State of the Art and Future Perspectives, Annals of the CIRP, 52/2:589-610

[18] Hon KKB, Han C and Edwardson SP, (2006) Investigations on New Scanning Pattern for Stereolithography, Annals of the CIRP, 55/1:217-220

Effects of Mold Pattern with Anti-adhesive Treatment in Nanoimprint Lithography

T. L. Chang*, F. Y. Chang, H. Y. Lin, W. L. Lai, J. H. Tsai, and S. H. Chang

Mechanical and Systems Research Laboratories, Industrial Technology Research Institute, Hsinchu, Taiwan 310
*Tel: +886-3-5915843. Fax: +886-3-5826104. E-mail: *tlchang@itri.org.tw*.

Abstract. A study of a top-down technique of nanoimprint lithography (NIL) for developing a novel bottom-up approach of chemical vapor deposition to generate anti-adhesive coatings between Ni mold stamps and polymethyl methacrylate (PMMA) substrate is presented in this work. A polybenzoxazine ((6,6'-bis(2,3-dihydro-3-methyl-4H-1,3-benzoxazinyl))) molecule self-assembled monolayer (PBO-SAM) regarded as an anti-adhesive coating agent demonstrates that non-fluorine-containing compounds can improve the nanoimprint process in Ni/PMMA substrates. In this experiment, the nanostructure-based Ni stamps and imprinted PMMA molds are fabricated using an electron-beam lithograph (EBL) and a homemade nanoimprint equipment, respectively. The diameter of pillar patterns obtained is 200 nm with 400 nm pitch on Ni stamp surface. Based on the hydrophobic PBO-SAM surface in this forming condition, the results of Ni mold stamps indicate over 90% improvement in controlling quality and quantity.

Keywords: Nanoimprint lithography, Ni Stamp, PBO-SAM, Anti-Adhesive Layer

1. Introduction

Nanoimprint lithography (NIL) is one of the attractive methods employed as a powerful tool for making nanometer-size patterns [1,2]. Rapid development of NEMS/MEMS technologies is required for optoelectronic devices, quantum computing devices, biosensors and the electronics devices. The conventional photolithograph may not be the appropriate method for some tasks [3]. X-ray, ion-beam, and electron-beam lithography, are not suitable for fabricating large-area patterns. NIL provides flexibilities in terms of material, pattern size, structures, and substrate topography [4]. Nowadays, the NIL fabrication method receives great attention that can open new doors for interdisciplinary nanoscale research and commercial products because of its combination of high patterning resolution at low cost and high throughput. However, several application issues have to be solved before this lithography technology is mature enough for industrial scale processes. The imprinted mold process is often carried out at high temperature (>100°C above the glass transition temperature of a polymer) and high pressure (>100 bar) that are obviously undesirable. The thermal cycle of heating and cooling processes can cause distortion of the mold and the imprinted substrate. One particular issue is the control of anti-adhesive layers treatment between the stamp and polymer to preventmechanical failure that can affect the imprint quality and the stamp lifetime. Schift *et al.* demonstrated the use of fluorinated trichlorosilane as the anti-adhesive coatings on silicon at a microscale level [5]. Park *et al.* also used a fluorinated silane agent to achieve a better anti-adhesive coating process for Ni mold stamp [6]. However, few attempts have so far been made at a non-fluoro material in NIL process of anti-adhesive coating treatment for Ni stamps. For sustainability of the environment, there is the need to employ non-fluoro materials. Also due to Ni-based materials are soft, more effective roller to roller (R2R) nanoimprint techniques can be developed through the use of these materials. In this study, PBO-SAM is developed as anti-adhesive coating agent between Ni stamps and PMMA substrates in order to improve nanofabrication techniques, namely, NIL.

2. Experiment

2.1 Preparation of polybenzoxazines

Polybenzoxazines were prepared by reacting 4,4'-isopropylidenediphenol (bisphenol-A, BA-m), formadldehyde and methylamine.. The first step in the synthesis process was to obtain benzoxazine monomers from a mixture of formaldehyde/dioxane (0.4 mole/80 mL) and methylamine/dioxane (0.2 mole/20 mL) at 10°C for 10 minutes in a jar. After stirring at 500 rpm, the solution of bisphenol-A /dioxane (0.1 mole/100 mL) can be added to achieve chemical reactions for 6 hours. Then the solution was processed to obtain a viscid liquid in 200 mL diethyl ether. To remove solutions of formadldehyde and methylamine that were not used in the reaction, 2 N NaOH aqueous solution (20 mL) and distilled water (20 mL) were added by washing 5 times. Finally, benzoxazine precursors are obtained after vaporizing diethyl ether. By heating the benzoxazine precursors at 140°C for 1 hour, BA-m polybenzoxazines can be obtained.

2.2 Fabrication of nanopatterns by NIL process

4-inch P-type Si(100) wafers are used in this study. To prepare SiO_2-based Ni (atomic weight 5.87 g/mole) substrate, there was a sequential deposition of Ti (5 nm), and SiO_2 (20 nm), followed by O_2-plasma treatment. A Ti interlayer was used to enhance the adhesion between the Ni substrate and SiO_2 layer. After cleaning by using acetone, isopropanol, and deionized water, the sample was spin-coated by a photoresist (ZEP520A-7, Nippon Zeon Co., Ltd.). The master mold was fabricated using a Crestec CABL-8210 electron-beam direct write tool (30 keV, 100 pA) with reactive ion etching (RIE) of Ni film in Figure 1.1(A). Next, the imprinted nanopatterns can be employed via hot embossing lithography (HEL) that is performed by the homemade equipment shown in Figure 1.1(B). The fabrication of nanopatterns utilizing HEL procedure with PBO-SAM coating method is schematically shown Figure 1.1(C). A 200 nm thin PMMA film (molecular weight 15 kg/mole) was spin coated on a SiO_2 substrate, followed by baking on a hot plate for 30 minutes at 160°C. And then the PBO-SAM coatings as an anti-adhesive agent were deposited by chemical vapor deposition (CVD) process. The master was replicated by embossing on the PBO-SAM/PMMA substrate film for 10 minutes at 150°C and 50 bar. Finally, PBO-SAM/PMMA layer that remained on the bottom of embossed nanostructure was removed by RIE treatment. After each imprint of stamps and substrates were fabricated, a microscope was used to observe them, and the water contact angle (CA) was measured to determine their wetting and adhesive properties.

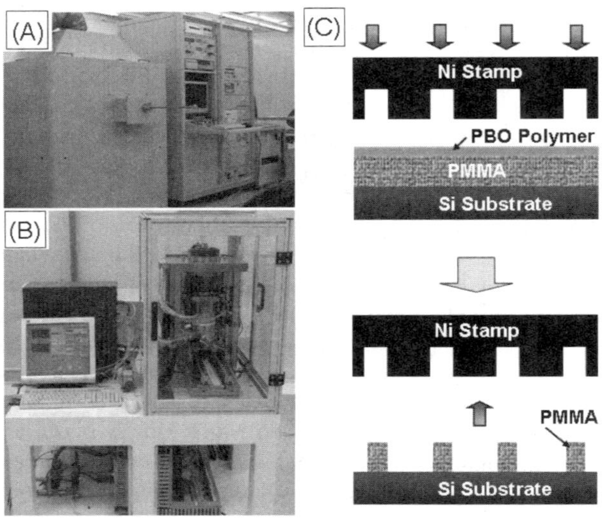

Figure 1.1. (A) EBL equipment for fabricated mold stamp. (B) HEL equipment for nanoimprint pattern with the computer controlled electronics. (C) Schematic diagram of nanostructures using NIL process

3. Results and Discussion

In this study, the property of polybenzoxazines is investigated by a Fourier Transform Infrared Spectrometer (FTIR, model: FTS-40, MASS, USA). As shown in Figure 1.2., characteristic peaks occur in the absorption spectrum of polybenzoxazine solution at 1470 cm^{-1}, 1655 cm^{-1}, and 2900 cm^{-1} indicating vibrational modes of molecular bonds. In addition, strong hydrogen-bonding interactions between the polymer chains which act as chemical cross-links were observed in the structure of polybenzoxazines ,. Here, the polybenzoxazines consists mainly of OH…N and O-…H+N intramolecular hydrogen bonds. This material is like polytetrafluoroethylene (Teflon) that is a low surface-free-energy material as it combines water repellence with other desirable properties.

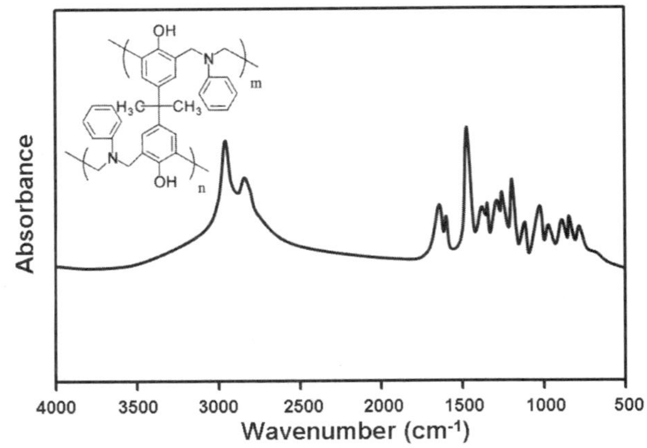

Figure 1.2. FTIR absorption spectrum of polybenzoxazines

The master Ni nanopatterns including a diameter of 200 nm and a height of 340 nm with a 400 nm period that can be characterized using scanning electron microscopy (FE-SEM: JEOL, JSM-6500F) are shown in Figure 1.3. Herein if the nanopatterns did not have PBO-SAM coatings, adhesion problems occur leading to bad replication of nanostructures. The parts of pillar nanopatterns without PBO-SAM coatings can be broken as shown in Figure 1.4(A). The other parts of pillar molds without PBO-SAM coatings can be difficult to peel off from the PMMA substrate due to the hydrophilic/hydrophobic properties of the material as shown in Figure 1.4(B). FE-SEM images of the Ni stamp show some surface areas are defective. Theses fractures of Ni mold indicates that mechanical failure are caused by interfacial adhesion between the imprinted MMA substrate and Ni mold stamp.

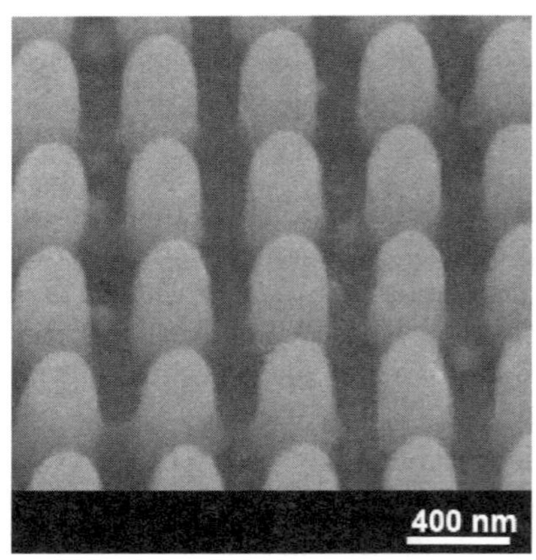

Figure 1.3. FE-SEM micrograph of a Ni mold stamp with 200 nm diameter pillars and a period of 400 nm

In this experiment, polybenzoxazines can be regarded as a novel material based on their unique properties that can be considered a discovery of a non-flourine, and low surface free energy to solve anti-adhesive treatment between PMMA substrate and Ni stamp. Moreover, the hydrogen bonding between polybenzoxazine polymer chains plays a significant role in promoting some attractive properties. Here, the solution of polybenzoxazine is prepared by blending THF solution. Based on the chip immersing 0.1 wt% polybenzoxazine solution, the Ni samples with polymer molecules could improve the anti-adhesive properties in NIL process.

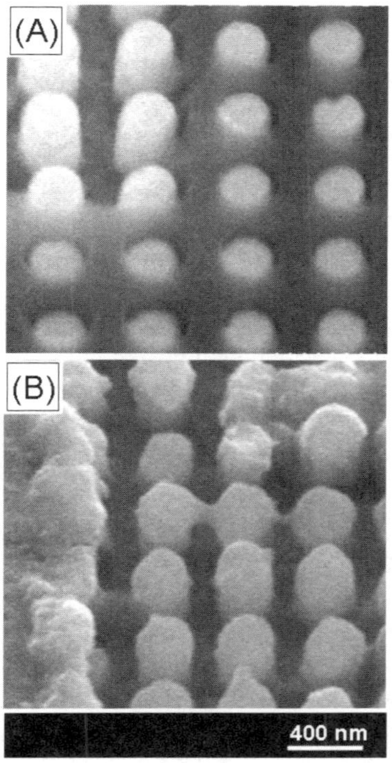

Figure 1.4. FE-SEM micrographs of Ni-imprinted stamps without anti-adhesive treatment

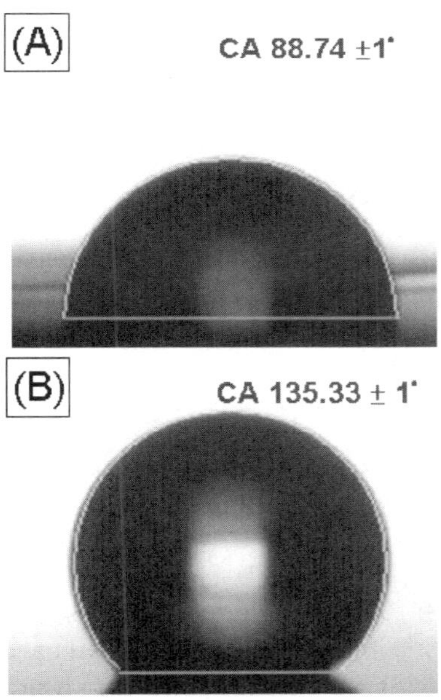

Figure 1.5. Contact angles of water drops on (A) a PMMA polymer film surface, (B) a smooth PBO-SAM coating film surface.

The water CA measurements give a good idea of the density of hydrophilic groups at the surface of the substrates that shows a marked constant in wetting behavior. After chemical cleaning with polybenzoxazines and acetone, the measured results of water CA measurements for Ni stamp on PMMA substrate surface utilizing comparison between studies before and after PBO-SAM coating treatment, as shown in Figure 1.5(A), and Figure 1.5(B), are 88.74±1°, and 135.33±1° receptively. The contact angles of the samples used in this research are measured at 25°C by injecting a 10-μL liquid drop with a goniometer (model: GH-100, Kruss, Germany) interfaced with an image-capture software. The measured results indicate that a Ni stamp with PBO-SAM coating treatment has low surface free energy ($\gamma = 16.43$ mJ/m^2). This surface free energy is even lower than Teflon ($\gamma = 1.03$ mJ/m^2), nevertheless, despit the fact that the polybenzoxazine polymer material is cheaper and easier to prepare before an anti-adhesive experiment process.

Ni nanopattern molds with PBO-SAM coatings have good and uniform nanopatterns that can be verified by FE-SEM image in Figure 1.6(A). Besides, the mold is characterized by atomic force microscopy (AFM, model: Autoprobe M5, Park Scientific Instruments) measurement in Figure 1.6(B) and 1.6(C). A 3-D AFM image of a 2×2 μm^2 PBO-SAM coating square was obtained via the NIL process. The image section clearly reveals an excellent homogeneity and fidelity of nanopatterns on this area.

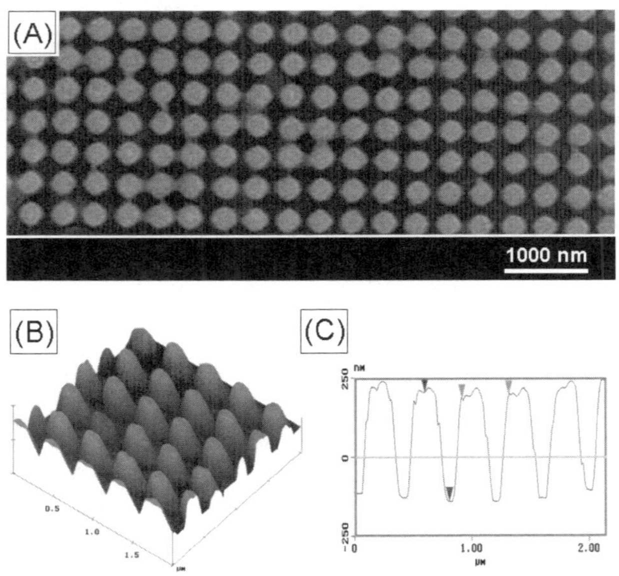

Figure 1.6. (A) FE-SEM micrograph of Ni-imprinted stamp with anti-adhesive treatment process. (B) AFM image of master mold and (C) is its sectional profile view

4. Conclusion

In conclusion, a new chemical coating process without using fluorinated silane molecule material treatment has been successfully developed for using Ni stamp to imprint PMMA substrate. Besides, the measurement of water contact angle can be found to play an important role for quality after using a material coating treatment. Furthermore, this study successfully demonstrates PBO-SAM coatings can solve adhesion problems in the NIL process thereby, provides the capability and flexibility necessary for future optimization and development.

References

[1] M. D. Austin, H. X. Ge, W. Wu, M. T. Li, Z. N. Yu, D. Wasserman, S. A. Lyon, and S. Y. Chou., 2002, Ultrafast and Direct Imprint of Nanostructures in Silicon, Nature., Vol.417, 835-837.

[2] Chou, S. Y., C. Keimel, and J. Gu., 2004, Fabrication of 5 nm Linewidth and 14 nm Pitch Features by Nanoimprint Lithography, Applied Phys. Lett., Vol.84, 5299-5301.

[3] M. Agirregabiria, F. J. Blanco, J. Berganzo, M. T. Arroyo, A. Fullaondo, K. Mayora, and J. M. Ruano-López, 2005, Fabrication of SU-8 Multilayer Microstructures based on Successive CMOS Compatible Adhesive Bonding and Releasing Steps, Lab Chip, Vol.5, 5545 - 552.

[4] W. Hu, E. K. F. Yim, R. M. Reano, K. W. Leong, and S. W. Pang, 2005, Effects of Nanoimprinted Patterns in Tissue-Culture Polystyrene on Cell Behavior, J. Vac. Sci. Technol. B, Vol. 84, 2984-2989.

[5] H. Schift, L. J. Heyderman, C. Padeste, and J. Gobrecht, 2002, Chemical Nano-Patterning Using Hot Embossing Lithography, Microelectron.Eng., Vol.423, 61-62.

[6] S. Park, H. Schift, C. Padeste, B. Schnyder, R. Kötz, and J. Gobrecht, 2004, Anti-Adhesive Layers on Nickel Stamps for Nanoimprint Lithography, Microelectron.Eng., Vol.73-74, 196-201.

Component Repair Using Laser Direct Metal Deposition

W. Wang, A. J. Pinkerton, L. M. Wee and L. Li

Laser Processing Research Centre, School of Mechanical, Aerospace & Civil Engineering, The University of Manchester, Sackville Street, Manchester M60 1QD, UK
wei.wang-5@postgrad.manchester.ac.uk, andrew.pinkerton@manchester.ac.uk, l.wee@manchester.ac.uk, lin.li@manchester.ac.uk

Abstract. Laser Direct Metal Deposition (LDMD) is investigated as a method for repairing cracks and defects in metallic components. An experimental investigation has simulated flawed components in the form of plates of H13 hot work tool steel with a square slot milled into the upper surface of each. H13 powder was deposited into the grooves using the LDMD method with a 1.5 kW diode laser. Different combinations of deposition parameter were tested and each sample analysed in terms of mass deposition rate, deposition microstructure, evidence of porosity and mechanical properties, including microhardness and tensile strength. In this work the results are presented and interpreted using statistical techniques to allow the overall quality of the repair process to be related to some of the key LDMD input process variables. Results show that the LDMD method can produce high quality repairs for internal defects; however, the range of values obtained in each test also highlights the sensitivity of the process to correct selection of deposition parameters.

Keywords Diode Laser; H13 steel; Direct Metal Deposition; Repair

1. Introduction

During the life of a component it may be subject to local impacts, thermal cycles and/or stresses or other testing conditions liable to create local defects or cracks. For example, cracks can be initiated in welds by mismatched creep properties, in foundry goods due to residual thermal stresses and in turbine shafts and blades due to high cycle fatigue. Cracks are common initiators of failures that cause high value components to be discarded as useless.

Repair and refurbishment are increasingly common practices, seeking to maximize unit lifetime, availability and profitability. There are significant environmental benefits to repair rather than replacement, including saving new material, avoiding the energy and emission needs for refining and transporting the material and reduction of waste. The development of ultrasonic techniques means that internal cracks can now be detected within a component before failure occurs, which further advances the potential of the repair method.

Historically, fusion welding has been used for part repair. Methods include TIG and Plasma-Arc Welding (PAW). However, these methods produce a lot of heat and cause high residual stresses, which result in distortion and heat related effects in the base metal. Electron beam welding (EB) can avoid the use of TIG and gives a more precise heat flux but it needs a vacuum environment and incurs more cost. Because of the high energy density available using a laser beam and the high geometric flexibility the Laser Direct Metal Deposition (LDMD) method is well suited to repair applications. The method involves creating a moving melt pool on the material to be repaired and blowing powder into it using an inert gas stream. The added material increases the size of the pool which rapidly solidifies into a raised track on the surface when the laser moves on. LDMD for surface repair has been widely researched [1,2]. It has been aided by the development of ever more powerful and cheaper lasers and is now used in both civil and military applications [3].

Grum and Slabe [4] showed that the powder laser surfacing method could produce favourable, compressive residual stresses near the surface of a component and the same authors reported a study on the repair of narrow and comparatively deep surface cracks in 12% Ni hot-working maraging tool steel using a laser surface melting method [5]. Wang et al [6] used a dual beam laser arrangement for repairing surface cracks in Ni-based superalloy components. The method was designed for cracks of 0.1 – 0.3 mm width. Song et al. [7] studied the rebuilding of V-grooves of up to 20 mm in depth in medium carbon steel with stainless steel using multi-layer deposition with a CO_2 laser. However as different materials were used that could not strictly be classed as a like-for-like repair technique.

This paper researches a way to repair interior cracks by milling a narrow groove down to the affected area to remove the crack and surrounding material and then refilling the volume with the same metal using LDMD. A diode laser is used since the LDMD process is more energy efficient with a laser of this type rather than with Nd:YAG or CO_2 lasers [8]. Laser power, powder mass flow rate and traverse speed are varied and each sample is analysed in terms of mass deposition rate, deposition microstructure, evidence of porosity and mechanical properties, including microhardness and tensile strength.

2. Experimental procedure

Experiments were carried out on specimens made of the chromium hot work tool steel H13. This is a steel that is designed to meet the highest quality standards in the die casting industry, and to excel in the most demanding hot work tooling applications. The chemical composition of H13 is given in Table 1.1.

Table 1.1. Chemical composition, % by weight, of H13 tool steel [9]

C	Cr	Ni	Si	Mn	Mo	V	S
0.4	5.3	–	1	0.3	1.4	1	0.005

When delivered, the steel was in the soft annealed state. For the deposition, substrate specimens with a size of 24 x 16 x 12 mm were prepared. The volume to be repaired in the specimen was prepared by milling a square groove 3.3 mm wide and 6.0 mm deep. Prior to the experiment the upper surfaces of the substrates were grit blasted and then degreased using ethanol.

The experimental set-up used is shown in Figure 1.1 and comprises a Laserline LDL 160–1500 Diode laser with a maximum power of 1500 W and a dual wavelength output of 808 nm and 940 nm (infrared). The beam was directed onto the surface using a lens with a focal length of 300 mm such that the upper surface of the substrate lay in the imaging plane. The laser spot size at the sample surface was experimentally measured by exposing an infrared detector card with a sensitivity of 1.75 x 10^{-9} Wmm^{-2}, to the pilot beam of the laser. It was found to have a rectangular beam shape with size equal to 3.5 mm (slow axis) x 2.5 mm (fast axis). During the experiment the work piece was moved parallel to the slow axis of the beam.

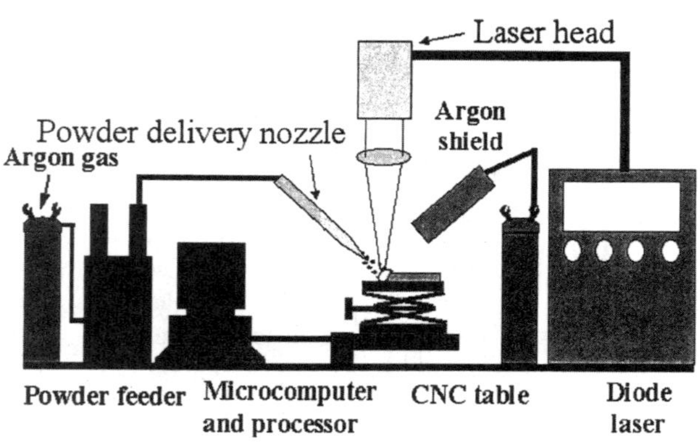

Figure 1.1. The experimental set up (modified from [10])

During the deposition process, the diode laser beam was kept stationary and the work piece was moved with a CNC x-y table. The laser was switched on at the start of the block and off at the end. Prior to each experiment and between tracks the height of a sample was adjusted in the Z-direction using a manual adjustment mechanism on the worktable. Argon was used to convey H13 powder with a particle size of 53-150μm from a FST PF-2/2 disk type powder feeder manufactured by Flame Spray Technologies Ltd. The melt pool was also encompassed in a 0.67 litres/sec low velocity flow of argon from a lateral nozzle with an internal diameter of 30 mm, which acted as a shield gas to reduce surface oxidation during deposition.

A matrix of experiments was performed using different laser powers, P, i.e. P_1 = 900 W, P_2 = 1200 W, and P_3 = 1500 W; different powder mass flow rates, F, i.e. F_1 = 0.237 gs^{-1}, F_2 = 0.472 gs^{-1}, F_3 = 0.71 gs^{-1} and different traverse speeds, V, i.e. V_1 = 1 mms^{-1}, V_2 = 3 mms^{-1}, V_3 = 6 mms^{-1}. As many deposition runs as necessary to completely fill a machined groove were made at each parameter combination.

The repair quality was initially evaluated by visual observation. The samples were sectioned, ground and polished. Some of the polished specimen surfaces were etched with 3% nital (10ml nitric acid, 90ml methanol) and the morphology and microstructures were examined using optical microscopy and a Philips XL32 ESEM-FG. Others were tested for hardness in and around the repaired area using a Buehler MicroMet digital microindenation hardness tester. Some repair deposits were performed specifically for tensile strength testing by using longer substrate pieces. After repair the sides of these pieces were ground away to remove edge effects and the underside of the specimen ground to give a total depth of 7.7 mm (groove depth plus 1.2 mm). Any defects in the base of the groove were thus preserved but the test mainly tested the strength of the repair rather than the substrate. A gauge length of 25 mm was evaluated at a strain rate of 3 mm/min using an Instron 4507 tensile/compressive test rig.

3. Results and Discussion

3.1 Mass Deposition Rate

Mass deposition rate is quantified as the number of runs taken to completely fill the machined groove to at least the level of the original surface. Each run is defined as a movement of the deposition head from one side of the sample to the other and then back to the original position which corresponds to two tracks.

For the matrix of experiments described in section 1.2 it took between 1 and 12 runs to fill the machined area. The number of tracks required increased at lower powder mass flow rates, higher speeds and lower powers. Previous work [11] has indicated that these primary process variables can often be combined to give two input parameters. These are line energy, x_1 (defined as laser power divided by traverse speed, so $x_1 = P/V$) and line mass, x_2 (defined as powder mass flow rate divided by traverse speed, so $x_2 = F/V$). The number of tracks required was found to show good correlation to the reciprocal of the line mass, indicating that this gave a good indication of the depth filled in a run. Figure 1.2 illustrates the relationship found. A relationship of the same general type was found for line energy but the correlation was considerably poorer.

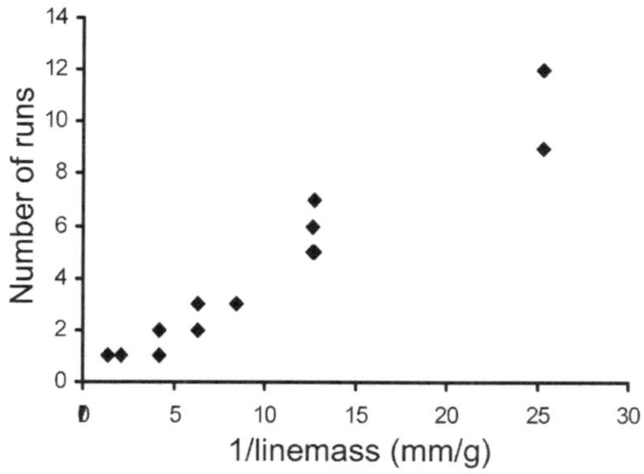

Figure 1.2. Variations in the number of runs required to fill the machined groove with line mass

3.2 Microstructure

Figure 1.3 shows micrographs of cross-sections through selected samples prepared at a line mass of 0.237 gmm^{-1}. Samples typically show a dark central region surrounded by a lighter area before the colour becomes uniform. Closer examination of the outer (unaffected) region from any of the samples allows it to be identified as ferrite with a homogeneous distribution of carbides. Figure 1.4 shows a higher magnification optical micrograph (a) and an SEM image (b) of the deposition area and fusion zone respectively from a typical sample.

The microstructure in the rebuilt zone is fine, free of microporosity and cracks. Grain direction on moving outwards is radial and the grains are highly elongated, indicating directional cooling. This reflects the 'self-quenching' nature of the LDMD process and the fact that the majority of the absorbed heat from the laser is rapidly conducted to the substrate. There is a distinct fusion bonding interface between the added material and the substrate.

The central, dark region seen in Figures 1.3(a) and 1.3(c) and Figure 1.4(b) is tempered martensite. The cooling rates typically experienced in LDMD are rapid,. Previous work has shown that surface deposition of single tracks of H13 produces martensitic structures and in normal heat treatment H13 can be fully hardened by an air cool [12]. The image in Figure 1.5(b) is taken in the lower region of the machined region so the material has been reheated by the subsequent passes to deposit further tracks. Some primary carbides are also visible.

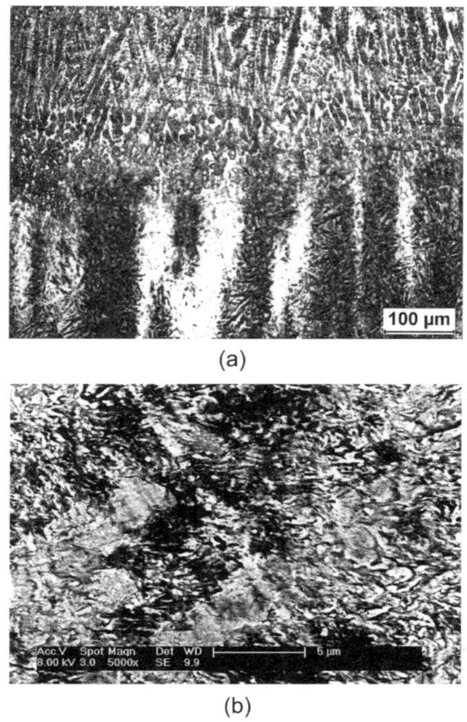

Figure 1.4. Microstructure in a transverse plane of the sample prepared at line energy of 1200 Jmm^{-1} and line mass of 0.237 gmm^{-1}; (a) optical micrograph of fusion zone, (b) SEM of deposited material.

The surrounding, light regions are areas that have been heated above the critical temperature only once. In samples that required only one pass this region would have been created on the return of the laser and deposition head. The material is thus untempered martensite, which is light etching. Sample 1.3(b) was created with only one run so the martensite in the centre of the deposited region has not been fully tempered by subsequent heat cycles.

The size of the martensitic, heat affected zone (HAZ) is quantified in Figure 1.5. It indicates that the width of the HAZ increases with line energy but unexpectedly the depth remains approximately constant at all line energies.

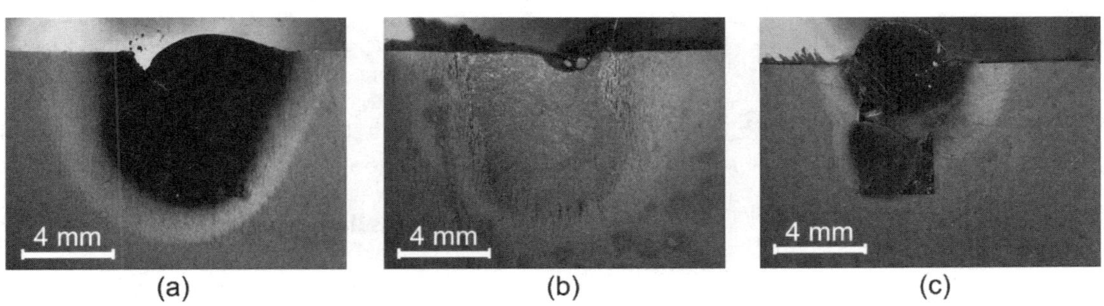

Figure 1.3. Optical micrographs of microstructure and build characteristics in a transverse plane from LDMD samples of H13 steel prepared at line masses of 0.237 gmm^{-1} and line energies of, from left to right, 1500, 1200 and 900 Jmm^{-1}.Microhardness

This can perhaps be explained by the fact that higher line energy increases the HAZ near to where the laser beam is absorbed but also leads to a greater amount of material being deposited. This moves the source of the heat higher in the groove for subsequent passes of the laser, reducing any further heating effect near the base of the groove.

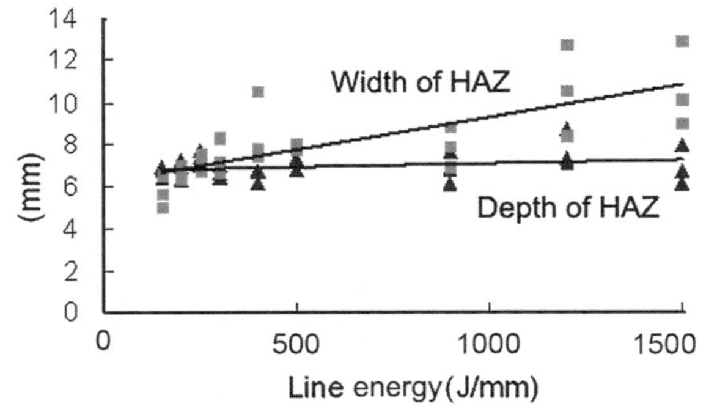

Figure 1.5. Variations in the width and depth of the heat affected zone with line energy

3.3 Microhardness

Figure 1.6 shows microhardness profiles in transverse cross-sections through two samples after multi-pass deposition. These profiles are typical of all the samples.

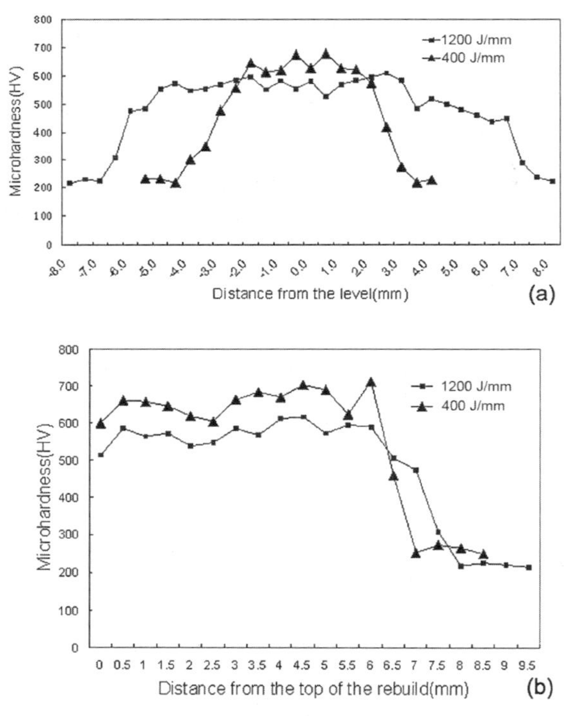

Figure 1.6. Microhardness distribution along the rebuilt cladding layers of samples at two line energies

The microhardness within or close to the bottom of the groove vertically and within or close to the sides of the groove horizontally reach approximately 700 HV, which can be contrasted with values of approximately 200 HV in the original substrate and with the normal working hardness of the material, which is 400 – 600 HV [13p146]. These hardness values are further evidence of the martensitic nature of the structure in these areas. Comparison with Figure 1.3 indicates that the limit of the hardened zone in both horizontal and vertical directions matches the HAZ identified visually. The HAZ is wider for the sample prepared at higher line energy (Figure 1.6(a)) however the peak hardness is lower. This can be explained by the fact that the sample prepared at the higher line energy required one run while the sample prepared at the lower line energy required nine. As H13 is a highly alloyed tool steel (see Table 1.1), it is known to exhibit secondary hardening.

3.4 Porosity

The rebuilt zone is in all cases composed of two or more deposited layers. Factors that could lead to flawed bonding and porosity include inconsistent line energy, misplaced tracks, an oxide layer preventing or hindering fusion and inconsistent laser power resulting in inadequate energy input. The degree of porosity was quantified by analysis of optical micrographs of the type shown in Figure 1.4 using the Photoshop 8.0 software package. Porosity did not show clear trends with any single input variables but was clearly highly dependent on deposit shape. It was most often apparent in the corners or at the sides of the machined area between adjacent tracks. From analysis of the data the best correlation could be obtained with powder mass flow rate divided by laser power (also equal to line mass divided by line energy). This relationship is illustrated in Figure 1.7. It is interesting that this gives a result in grams per Joules, or in other words indicates that the more energy that is provided for each unit mass of added material the lower the porosity.

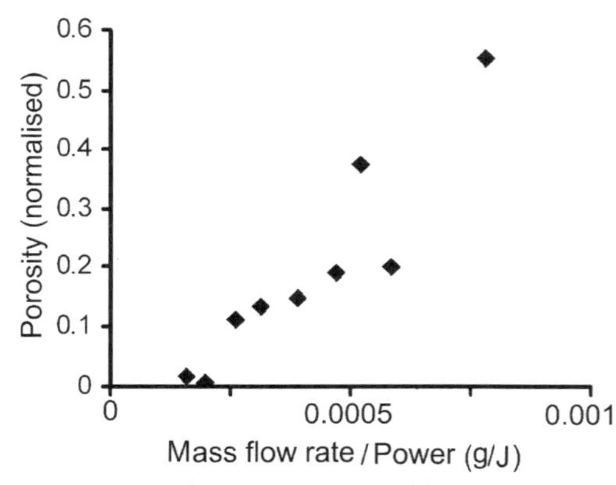

Figure 1.7. Porosity in the deposited sample at different values of the parameter laser power / mass flow rate

3.5 Tensile Strength

Longitudinal tensile testing was carried out on two disparate samples prepared at high and low line masses and energies. Sample A was produced at line energy of 250 Jmm^{-1} and line mass of 0.0395 gmm^{-1} using 9 runs. Sample B was produced at line energy of 1200 Jmm^{-1} and line mass of 0.3338 gmm^{-1} using 1 run. The microstructures of samples A and B are shown in Figure 1.3(a) and 1.3(b) respectively.

The results are shown in Figure 1.9. Specimen A failed under 428 MPa stress at 4.6 mm elongation. Specimen B failed under 309 MPa stress at 3.2 mm elongation. These values are much lower than both the ultimate tensile strength of soft annealed H13, which is 670 MPa, and that of tempered H13 (500 HV), which is 1.7 GPa [12]. In fact, Figure 1.8 indicates that any extension was purely elastic and the yield point was not reached before brittle failure occurred.

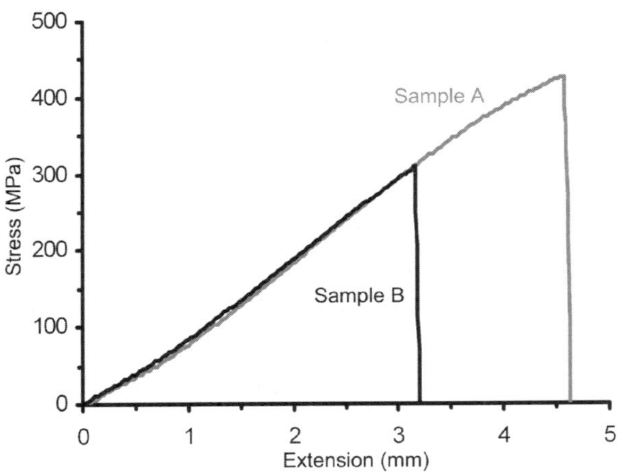

Figure 1.8. Stress-extension curves for LDMD samples A (250 Jmm^{-1}, 0.0395 gmm^{-2}) and B (1200 Jmm^{-1}, 0. 3338 gmm^{-2})

Examination of the failed specimens indicated that in both cases failure had been in or around the fusion zone, near the original groove wall and that there appeared to be incompletely bonded areas on these faces. This was not identified in the micrographs in Figure 1.3, however these micrographs did examine only one transverse plane. The laser beam did not impinge directly onto the side walls so creation of a fusion bond would have relied solely on the heat conducted from molten or previously deposited powder. Clearly this was not sufficient to establish a full strength bond in all areas. The more passes the laser made over the groove the more opportunity there would have been for reheating and formation of a fusion bond, which goes some of the way to explaining why specimen A (9 passes) was stronger than sample B (1 pass).

The micrographs shown in Figure 1.3 show another possible reason for the premature failure of sample B. Small cracks can be seen around the fusion area in this sample. These could be due to stresses around the repair zone, caused by thermal contraction during the rapid cooling phase. Stress relief is normally recommended for H13 after hardening [13].

4. Process Model

From the above consideration, the quality of the final repair appears to be determined by multiple, inter-related parameters; the deposition rate and number of passes, in particular seem to have secondary effects on other properties. To quantify these links a multivariate statistical model is developed [14].

Line energy and line mass are taken as the independent (exogenous) variables and number of tracks, porosity, HAZ depth and HAZ width as the dependent (endogenous) variables. All data were normalized with respect to their average values before use in the evaluation of the matrix which was performed using MatLab (http://www.mathworks.com). The correlation matrix is symmetric so only the upper part is presented in Table 1.2.

Table 1.2. Correlation coefficients between line energy, line mass, porosity and HAZ dimaensions in the final deposit

	Line energy	Line mass	Number of runs	Porosity	HAZ Depth	HAZ Width
	χ_1	χ_2	y_1	y_2	y_3	y_4
χ_1	1	0.78	-0.55	0.27	0.21	0.72
χ_2		1	-0.57	0.68	0.13	0.38
y_1			1	-0.39	0.20	-0.21
y_2				1	-0.02	-0.01
y_3					1	0.73
y_4						1

The table confirms the significant relationship between number of runs and line mass shown in Figure 1.2, between HAZ width and line energy shown in Figure 1.5 and between porosity and line mass shown in Figure 1.7. Other relationships highlighted as significant are those between porosity and number of runs (negative correlation) and between HAZ width and both line mass and HAZ depth (positive correlations).

The same normalized data and regression analysis was used to develop empirical expressions for the combined effect of the process parameters on the deposition repair quality. First and second order main and interaction effects were accounted for in the regression model, which has the general form:

$$Q = a_0 + a_1x_1 + a_2x_2 + a_3x_1^2 + a_4x_2^2 + a_5x_1x_2$$

Here Q is the laser deposition quality parameter (the number of tracks, porosity, width or depth of the HAZ), x_1 is line energy and x_2 is line mass. The regression model coefficients and the adequacy of fits as suggested by R^2 are given in Table 1.3.

Table 1.3. Regression coefficients for Equation (5), relating major sample characteristics to the process parameters of line energy and line mass

Coefficient	Porosity	Number of runs	HAZ Depth	HAZ Width
a_0	0.003	0.58	0.54	0.31
a_1	-0.94	0.79	1.31	1.47
a_2	2.08	-2.35	-0.005	-0.16
a_3	1.01	-1.00	-0.94	-0.73
a_4	0.22	1.10	-0.20	-0.14
a_5	-2.12	1.02	-0.05	-0.09
R^2	0.87	0.92	0.86	0.98

Once expressed in this form these relations can be conveniently manipulated in algebraic or graphical form for selection of optimum processing parameters based on the priorities of a user.

5. Summary and Conclusions

Process parameters and mass deposition rates, sample microstructures and sample physical properties have been analysed in order to assess Laser Direct Metal Deposition as a means of repairing cracks and defects in metallic components.

Results from diode laser deposition of H13 powder into square grooves machined into a H13 sample show that this can be achieved with partial success given the correct seletion of parameters. However, the lack of a full strength fusion bond over the full area of the vertical faces of the groove reduces the tensile strength of the repaired component.

The parameter relationships for the process have been linked via a statistical model. From the analyses, by correlating process parameters and from this model, it can also be surmised:

- The groove is filled vertically at a rate proportional to the line mass; or in other words, the number of runs required to fill a groove is linearly proportional to the reciprical of the line mass
- In cross section, the microstructure formed is a central tempered martensite zone surrounded by a untempered matensite periphery. The width of the zone increases with line energy.
- The final porosity in the repair increases approximately linearly with the parameter of (mass flow rate / power).

The method is promising, but careful selection of parameters is important and the method has some weaknesses, principally poor strength at the sidewalls due to incomplete fusion bonding that need to be addressed.

Acknowledgements

The authors would like to thank technical staff at the University of Manchester for their assistance, particularly Author Sumner and Dave Mortimer for machining/grinding and tensile testing of samples respectively.

References

[1] Islam, M. U., Xue, L., et al., 2001, Process for manufacturing or repairing turbine engine or compressor components, US.

[2] Krause, S., 2001, An advanced repair technique: Laser powder build-up welding: Sulzer Technical Review, Vol. 83, No. 4, P 4-6.

[3] Plourde, R., 2004, Laser-Based Repair System Reclaims High Value Military Components., P 5p.

[4] Grum, J., and Slabe, J. M., 2003, A comparison of tool-repair methods using CO_2 laser surfacing and arc surfacing: Applied Surface Science, Vol. 208-209, P 424-431.

[5] Grum, J., and Slabe, J. M., 2006, Effect of laser-remelting of surface cracks on microstructure and residual stresses in 12Ni maraging steel: Applied Surface Science, Vol. 252, No. 13, P 4486-4492.

[6] Wang, Z.-K., Ye, H.-Q., et al., 2001, Laser repairing surface crack of Ni-based superalloy components: Transactions of Nonferrous Metals Society of China (English Edition), Vol. 11, No. 4, P 572-575.

[7] Song, J. L., Deng, Q. L., et al., 2006, Rebuilding of metal components with laser cladding forming: Applied Surface Science, Vol. 252, No. 22, P 7934-7940.

[8] Pinkerton, A. J., 2004, A Mathematical and Experimental Investigation of Thin Wall Laser Direct Metal Deposition, PhD Thesis, UMIST.

[9] American Society for Metals (ASM International), ed., 1990, Metals Handbook, v. 1: Properties and Selection: Irons, Steels and High Performance Alloys: Materials Park Ohio USA, ASM International.

[10] Syed, W. U. H., 2006, Combined Wire and Powder Deposition for Laser Direct Metal Additive Manufacturing, PhD Thesis, The University of Manchester.

[11] Steen, W. M., 2003, Laser Material Processing: London, Springer-Verlag.

[12] Payson, P., 1962, The Metallurgy of Tool Steels: New York, JohnWiley & Sons Inc.

[13] Wilson, R., 1975, Metallurgy and Heat Treatment of Tool Steels: London, McGraw-Hill.

[14] Krzanowski, W. J., 1998, Principles of Multivariate Analysis: A User's Perspective, Clarendon Press.

14–4

Selective Laser Sintering of High-density Alumina Ceramic Parts

Z. H. Liu[1], J. J. Nolte[2], J.I. Packard[3], G. Hilmas[4], F. Dogan[5] and M. C. Leu[6]

[1]University of Missouri-Rolla, USA, 65409, liuzhehui@yahoo.com
[2]University of Missouri-Rolla, USA, 65409, jjn7td@umr.edu
[3]University of Missouri-Rolla, USA, 65409, jiprzf@umr.edu
[4]University of Missouri-Rolla, USA, 65409, ghilmas@umr.edu
[5]University of Missouri-Rolla, USA, 65409, doganf@umr.edu
[6]University of Missouri-Rolla, USA, 65409, mleu@umr.edu

Abstract. Manufacturing high temperature ceramic parts with Solid Freeform Fabrication (SFF) technology has been of great research interest lately. Described in this paper is a process based on Selective Laser Sintering (SLS) to fabricate 3-dimensional alumina green parts layer-by-layer, followed by binder burnout and sintering to obtain high-density ceramic parts. This SLS process has been developed to manufacture 3D parts including airplane models, letter bars, etc. from mixtures of alumina powder and stearic acid binder. These mixtures were generated by mixing alumina powder with stearic acid in a ball mill. The average particle size of the alumina particles is 0.26 µm. The SLS process parameters were tuned such that the green parts can be built with reasonably sharp geometrical features and that the sintered parts have good mechanical and physical properties. Proper binder burnout conditions were determined in order that the resultant parts after the binder burnout process have no visible cracks. Alumina ceramic parts were obtained by firing the green parts at 1600°C. The alumina ceramic bars achieved 255 MPa in flexural strength, and the bar relative density achieved is 88%. These results are discussed.

Keywords: Selective Laser Sintering, 3D geometry, alumina ceramic parts, process parameter tuning

1. Introduction

High temperature ceramics have high stiffness and inertness to chemicals and good functionality at high temperatures. However, high temperature ceramics are difficult or impossible to melt and cast into a final shape. Therefore, the conventional ways of creating forms, for instance by forging, casting and extrusion, cannot be used to make parts of complex goemetry with these materials. Moreover, high temperature ceramics are hard and difficult to machine.

Selective laser sintering (SLS) is a promising way for manufacturing high temperature ceramic components having 3D geometries [1]. Nelson *et al.* [2] used SLS to fabricate silicon carbide/polymethyl methacrylate (PMMA) green parts. Melting took place only in the PMMA phase which acted as a glue for the silicon carbide powder. However, a real ceramic part, by burning out the binder from the green part followed by sintering at high temperature, has not yet been attainable. Subramanian *et al.* [3, 4] performed SLS on an alumina/polymer binder system followed by post-process infiltration in an attempt to improve the green part density. However, the ceramic components obtained still exhibited a high porosity (~50%) and a low strength of only 8 MPa for alumina powder with an average particle size of 15 µm. The strength of the green samples fabricated using alumina with an average particle size of 2 µm was too low to handle. Recently, Bourell *et al.* [5] modified the fabrication process of silicon carbide ceramic parts by using phenolic resin as a binder and post-process infiltration. However, defects such as overfilling and warping occurred during high temperature processing.

It is well known that the sintering temperature of a ceramic decreases and its strength increases with a decrease in ceramic powder size due to an increase of surface free energy and thus an increase in the driving force for sintering [6]. As an example, the sintering temperature for alumina decreases from 1675 to 1500°C with a decrease in particle size from 4.8 to 1.4 µm [7].

The work described in the present paper has developed a novel process for SLS of ceramic components by coating an organic binder on fine ceramic (alumina) powder. SLS is used to fabricate green ceramic parts which subsequently undergo binder burnout and sinter in a furnace at high temperature to produce ceramic components for testing and analysis. The total volumetric shrinkage from green parts to sintered ceramic parts is quantitatively investigated. The relationship among SLS operating parameters, binder burnout conditions, and the structure and mechanical properties of the alumina ceramics are discussed.

2. Experimental

A calcined alumina (Al_2O_3) powder with a purity of 99.8% Al_2O_3, surface area 8.9 m^2/g, and particle size of 0.4 µm [8] (grade A-16SG, was purchased from Almatis Inc. was used in the study. Stearic acid ($CH_3(CH_2)_{16}CO_2H$, Cat. No. AC17449-0025, Fisher Scientific) was used as the organic binder. 500 g of stearic acid was directly mixed with 2360 g of alumina powder at room temperature in a ball mill for two hours. The powder mixture was sieved through a 35mesh sieve prior to loading into the SLS machine.

The SLS machine used in the current study was a SinterStation 2000 (manufactured by DTM Corporation, which was acquired by 3D Systems in 2001). It uses a high-power carbon dioxide laser (wave length = 10.6 μm) to fuse small particles of polymer powders into a part representing a desired 3D object. The sintering chamber utilized flowing nitrogen gas to maintain an inert gas environment. The laser power values of the SLS machine were calibrated using an Ophir power receiver (SN 145602) and a Nova 2 laser power meter with a resolution of ± 0.1 Watt.

Binder burnout was performed in air in a muffle furnace (Isotemp, model 550-126). Burnout occurred by heating the green samples from room temperature to 600°C for several days and then cooling at a rate of 10°C/min to room temperature. Parts after binder burnout exhibited some strength that allowed for careful transport from one place to another. The parts were sintered at 1600°C for 2hours and cooled slowly to room temperature in a box furnace capable of reaching temperatures as high as 1700°C (Lindberg, model 51644). The heating and cooling rates were 5 and 10°C/min, respectively.

The particle size and distribution of the alumina powder was measured using a laser scattering particle analyzer (Beckman Coulter, model: LS 13320). A thermogravimetric analyzer (TGA) (Shimadzu, model TGA-50) was used to characterize the thermal behavior of stearic acid coated alumina powder in air. The heating rate was 10°C/min. The density of a green bar was calculated by specimen weight and volume. The volume of a green bar was estimated by measuring the length, height and width of the bar at the center of each surface. The density of a dry ceramic bar was measured according to the Archimedes method. Ceramic test bars were polished to a standard size (45 mm×4 mm×3 mm) for measuring flexural strength according to ASTM Standard C1161. To obtain a statistical analysis of the mechanical strength of the composites, 5 test bars were prepared and tested with a screw-drive mechanical testing machine (Instron, model 5881).

3. Results and Discussion

3.1 Green Parts

Several key SLS operating parameters had to be tuned to create a green part of sufficient quality during the study. The majority of the energy for fusing the organic binder is provided by the heaters located directly over the powder bed. The powder temperature in the part build area should be slightly lower than the melting point of the organic binder. The temperature in the areas of the two powder cartridges was held between room temperature and the temperature in the part build area during SLS processing. The additional energy for melting the binder was subsequently applied by the laser. The laser power, beam spacing, laser scan speed, scan count, etc., are major laser parameters to be optimized for melting the organic binder. The part build piston was set to move down in increments of 100-200μm, while the cartridge piston moved up in increments of 150-400 μm. The roller speed was fixed in the range 0.1-0.3 m/sec. The temperature in the part build area was set to 60°C, below the melting point of stearic acid (67-69°C). The powder in the powder cartridge areas was heated to 38°C to prevent curling of the samples caused by the temperature difference between the fused green part and the fresh powder from the powder cartridges. This temperature was also low enough to allow the material to spread smoothly. If the powder temperature was set too high in the powder cartridge areas, the powder clumped together and did not spread evenly during the SLS processing.

The energy density (e) of the laser applied on the powder bed can be defined by [2]:

$$e = \frac{P}{\delta \upsilon_b} \tag{1}$$

where P is the laser power, δ the laser beam spacing and υ_b the laser scan speed. P, δ, and υ_b are adjustable laser sintering parameters.

Based on equation 1, the energy density increases with an increase in laser power and with a decrease in laser beam spacing or laser scan speed. This parameter has a large effect on the density and bend strength of green bars. Both relative density and bend strength of silicon carbide/PMMA green parts were found to increase with energy density [2]. The increase in the bend strength of alumina/PMMA green bars versus the increase of laser energy density is bell-shaped [3, 4]. Excessively high or low energy densities both result in a reduction in the strength of green bars.

Typical values of SLS operating parameters for melting stearic acid blended with fine alumina powder at a volume fraction of 0.5 are given in Table 1.1. These parameters have been used to successfully fabricate green parts. The organic binder is burned in-situ when too much laser power, e.g. 12 W, is applied. On the other hand, the organic binder cannot be melted if the laser power is too low (below 3 W).

Table 1.1. SLS operating parameters used in this work

Operating parameter	Value
Laser power	4.0, 4.8, 5.5 and 6.2watts
Laser scan count	1
Laser scan speed	889mm/sec
Laser beam spacing	127μm
Temperature	60°C in part build area and 38°C in powder cartridge area

SLS parts produced from metallic powder/polymer composites [1] and inorganic powder/polymer composites [2-5] have been shown to contain large amounts of porosity. For instance, the relative density of many porous parts is only about 50%. The relative density of green alumina/stearic acid bars in the current study was investigated. Assuming the mixture of alumina with stearic acid is fully dense, the theoretical density of the mixture is calculated using:

$$\rho_m = \varphi_1 \rho_1 + \varphi_2 \rho_2 \qquad (2)$$

where ρ is density, φ is volume fraction; subscripts m, 1 and 2 denote mixture, alumina and stearic acid, respectively. Taking ρ_1=3.986 g/cm^3 [6], ρ_2=0.84 g/cm^3 [9], φ_1=0.5, φ_2=0.5, a value of ρ_m=2.41 g/cm^3 is obtained for the theoretical density.

In general, the relative densities (45-49%) of these bars are low. In other words, these bars were highly porous. High porosity of green parts leads to poor mechanical properties, high shrinkage of parts after binder burnout, and a decrease in the flexural strength of the ceramic parts after sintering.

3.2 Parts after Binder Burnout

The TGA results for a typical alumina/stearic acid powder mixture containing 51vol.% stearic acid is shown in Fig. 1.1. The onset of weight loss occurred at approximately 150°C, with weight loss being complete at approximately 400°C.

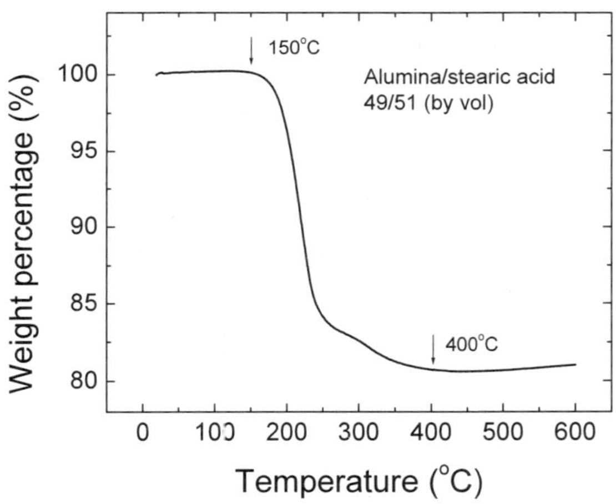

Figure 1.1. TGA analysis of stearic acid/alumina mix containing 51 vol. % stearic acid

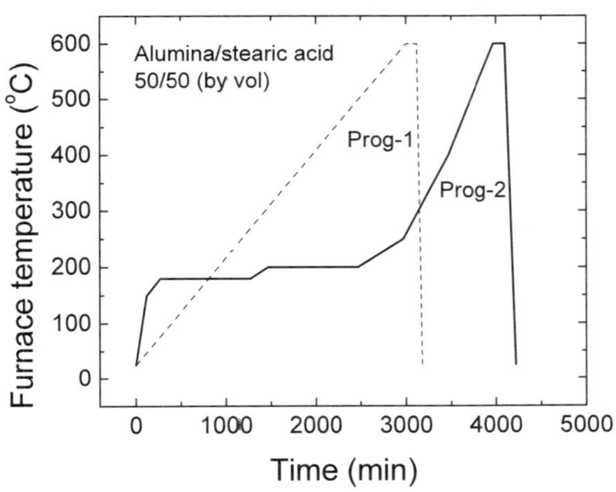

Figure 1.2. Furnace temperature vs. time profiles for two different binder burnout programs

The binder burnout used in the current study for the alumina/stearic acid parts is shown in Fig. 1.2. The initial binder burnout profile (Prog-1) consisted of increasing the temperature from room temperature to 600°C at a rate of 0.2°C/min, holding at 600°C for 120 minutes, and allowing the samples to cool to room temperature within 120minutes. However, new binder burnout profile (Prog-2) was added in later stages of the study which allowed the samples to stay at lower burnout temperatures for considerably longer times, resulting in a much slower, steadier binder burnout procedure. Prog-1 resulted in 11 out of 17 bars cracking. Prog-2 on the other hand produced bars without visible cracking.

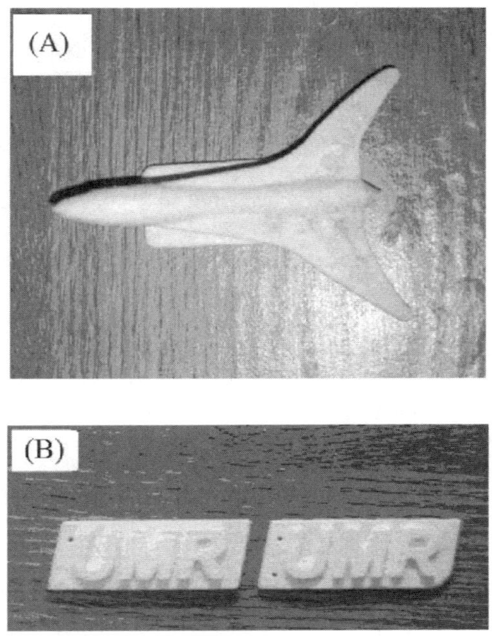

Figure 1.3. Alumina ceramic model (A) and letter bar (B)

3.3 Ceramic Parts

Flexure bars, letter bars and an airplane model were fabricated by the SLS process, underwent binder burnout, and were sintered at 1600°C to obtain ceramic parts. Some of the obtained alumina ceramic parts are shown in Fig. 1.3. The relative density of the alumina ceramic bars whose green bars were processed with a laser power of 5.5watts was measured to be ~88%. The average four-point bend strength of the ceramic bars was 255±17 MPa. This strength is much higher than 8MPa reported by Subramanian et al. [2, 3]. The average particle size of the starting alumina powder used in the current study is 0.26μm measured in this work, which is much smaller than those (2 and 15μm) used by Subramanian et al. The fine alumina particles create a large surface area and strong driving force for sintering. However, the flexural strength of these bars is still lower than that (> 337 MPa [7]) of alumina ceramic bars whose green parts are produced by a high-pressure pressing operation. It is well known that the strength of ceramics decreases substantially with increasing porosity [10]. The porosity of alumina ceramic bars is believed to account for the low flexural strength.

4. Conclusions

Alumina ceramic parts having 3D geometries were successfully fabricated by Selective Laser Sintering. SLS was used to create green parts of alumina/stearic acid powder, where stearic acid acted as a binder. The powder for SLS processing was prepared by mixing alumina powder with stearic acid powder in a ball mill and sieving the mixed powder. It was found that the SLS parameters have to be properly set in order to fabricate green parts with enough mechanical strength to undergo binder burnout and sintering. Laser energy density plays an important role in the SLS process. A high laser power (too high in energy density) may burn the binder while a low laser power may not supply sufficient energy to melt the binder. The porosity in green alumina parts produced by the SLS process was measured to be between 45 and 50%.

Alumina ceramic parts were obtained by sintering parts at 1600°C after binder burnout. Flexure bars produced by this process achieved a relative density of 88% and average flexural strength of 255 MPa. This strength is much higher than the strengths of other ceramic material systems produced by SLS as reported in the technical literature, indicating a significant improvement in our ability to produce advanced structural ceramics by the described SLS technique.

Acknowledgements

This project was supported by the Center for Aerospace Manufacturing Technologies (CAMT) at the University of Missouri-Rolla. CAMT is funded by the Air Force Research Laboratory under contract FA8650-04-C-5704. Mike Hayes and David Dietrich of the Boeing Company have provided technical guidance and support for the project.

References

[1] Venuvinod PK, Ma WY, (2004) Rapid prototyping, Kluwer Academic Publishers, Boston.

[2] Nelson JC, Vail NK, Barlow JW, Beaman JJ, Bourell DL, Marcus HL, (1995) Selective laser sintering of polymer-coated silicon carbide, Ind. Eng. Chem. Res. 34:1641.

[3] Subramanian K, Zong G, Vail N, Barlow J, Marcus H, (1993) Selective laser sintering of Al_2O_3, Proc. 4th Solid Freeform:350

[4] Subramanian K, Vail N, Barlow J, Marcus H, (1995) Selective laser sintering of alumina with polymer binders, Rapid Prototyping J. 1:24.

[5] Bourell DL, Beaman JJ, Stevinson B, (2006) Support-free infiltration of selective laser sintered no-metallic performs, Proc. 2006 NSF Design, Service, and Manufacturing Grants and Research Conference, St. Louis, Missouri, USA.

[6] Rahaman MN, Chapter 7, (2003) Appendix E, Ceramic processing and sintering, Second edition, Taylor & Francis.

[7] King AG, Chapter 10, (2002) Ceramic technology and processing, Noyes Publications.

[8] Data sheet of reactive alumina for ceramic applications from Almatis.

[9] https://www1.fishersci.com/Coupon?cid=1334&gid=193537.

[10] Rice RW, (1996) Grain size and porosity dependence of ceramic fracture energy and toughness at 22°C, J. Mater. Sci. 31:1969.

15

Robotics

15–1

Design of High Sensitive Quadruped Robot with Modular Assembly

P.S. Pa[1] and C.M. Wu[2]

[1]National Taipei University of Education, Taiwan, (R.O.C.), myhow@seed.net.tw
[2]National Taipei University of Education, Taiwan, (R.O.C.), cokenmine@yahoo.com.tw

Abstract. The fast development of technology generates new sensitive robots with increased intelligence and more human interaction. Consequently, the demand for robots gradually increases. This research focuses on a new quadruped robot with multiple functions, high sensitivity and modular assembly. Compared with a biped robot, the quadruped can move just as well as the biped but more stably. Besides, under additional loads, the four feet of the quadruped robot reinforce its carrying ability and reliability, which helps it carry more objects and enhances functionality. Based on different requirements and demands, the highly sensitive quadruped robot provides an expandable interface to add different sensing components, like body sensors, carbon dioxide sensors, humidity sensors, ultrasonic sensors, and so on. Based on the real situation, the user can choose the necessary sensing components. In addition, when combined with a wireless communication module or independent 1.2G Radio Frequency CCD wireless image transmission system, the user can control the robot remotely and instantly. The design helps the quadruped robot expand its applications. By assembling and disassembling modules and changing different sensing components, the highly sensitive quadruped robot can be used for tasks like manufacturing, mine-exploring, emergency rescue, security, and home care. Moreover, the remote control function of the robot will increase interaction with human beings and become highly involved in people's lives.

Keywords: Highly Sensitive, Modular, Quadruped robot, Sensing Component, Interaction.

1. Introduction

The research of the quadruped robot combines different disciplines such as mechanism, mechanics of machinery, electronic engineering, control engineering, biological engineering, and robotics. The major research content includes: the design of the leg mechanism, gait planning, walking track, and balance & control theories. In 1977, Gollidary & Hemani adopted Langrangian's dynamics theory to deduce and linearize the mathematic model of the kneeless biped robot to analyze its stability, manipulability, and observability [1]. The robot has no body but with a high pelvis that connects the arms. It is a biped robot with "loaded" 12 DOF, including: 3 DOF in the hip, 1 DOF in the knee, and 2 DOF in the ankle. There are 5 linkages with 4 DOF from the front view, and 7 linkages with 6 DOF from the side view. The weight of two feet is only around 5 ~ 10% of the total weight. If the load of the arms has little impact on the balance, and the proportion of the weight of the two arms to the total weight is small, the robot will be able to up and down the stairs, walk on a slope with less than 10% degree, turn, and step back [2].

A walking track for robots to follow is necessary before successfully driving the robot to walk or move something from one place to another place. There are several ways to generate the walking track. One is through observing the qait of real human [3] another by instant calculation. In 1970, Vukobratovic et al. adopted the numerical methods to calculate the dynamic walking track of the biped robot [4]. Kato applied the same method to generate the dynamic walking track for his biped robot, WL-10RD [5]. However, it takes a long time to calculate the reference track when the robot is walking, and it is hard to adjust to different surfaces. Unless the CPU can calculate faster or the algorithm is simplified, the numerical method still has the problem of slow calculation. Other methods to generate the walking track are based on the least power [6]. Modern commercial robots are equipped with hybrid spin and slip junctures, connected by the arm linkage or the wrist parts of the robot's arm. The spin juncture controls the movement of two linkages in a specific degree, and the slip juncture controls only the line movement between the two linkages [7]. The analysis and control of the robot's arm requires the development of the analysis and control method. An arm with multiple juncture points is impacted by many force interactions and external environment, and it requires a more complex analysis. Researches of Paul about the homogenous transformation matrix method and the coordinate transformation in the analysis of the robot's kinematics are useful references [8].

The research uses program language to design all necessary walking movements of a quadruped robot. After editing, the research loads all movements in the internal memory of the robot. In addition, the research combines a reliable and low cost circuit with programs to decrease the development barrier of the quadruped robot and encourage further research. The research also develops a connecting interface for external controls to load different sensors into the robot. Moreover, the 1.2G wireless image transmission system is installed in the robot to provide the user with real-time monitoring function. Finally, the robot loads the wireless communication module, and the result is completed. The development of the modules helps the robot to alter its function flexibly and reduce the development cost effectively. Later, the design and implementation of new modules for specific robots' movements will shorten the special robot's development process and reduce the robot's cost hugely.

2. Collecting the relative researches and data of the quadruped robot

(1) Collect the theories and knowledge about how the quadruped robot walks.
(2) Develop a single chip (microprocessor) with expansible design for controlling all movements of the quadruped robot.
(3) Implement the full functions of a quadruped robot.
(4) Use the computer program language to edit all walking movements of the quadruped robot, including step forward, step backward, turn left, turn right, sit down, and stand up automatically after falling down, and save in the memory of the robot.
(5) Install a 1.2 G radio frequency CCD wireless image transmission system and deploy the full vision system in the quadruped robot to offer real-time monitoring functions.
(6) Integrate the wireless communication module (2.4G RF wireless communication module or GSM module) into the quadruped robot for real-time remote controls.
(7) Load different sensors in the quadruped robot for controlling remotely and returning by the wireless module.

3. Research Method

(1) Document Analysis

Collect all kinds of literature and data related to walking robots to sort out what makes up a walking robot and the associated theories.

(2) Theoretical Analysis

Before actual implementation, conduct theoretical analysis on the motion of walking robots. This includes analyzing the robot's center of gravity, the direction the motors are running in, the angles and rotation speeds. Feasibility is first established in theory before practical implementation.

(3) Block Theory

All functions of the quadruped robot in the research are divided into five modules for implementation and testing. When all functions are working, modules are connected together. This method not only simplifies the development process but also debugs easily. Besides, it greatly improves the robot's production. The following describes the five modules:
The controller module - control interface and circuit design.
(a) The robot's body module - motors and mechanism.
(b) The vision system module - 1.2G GF CCD image transmission.
(c) The wireless communication module - 2.4G RF communication module or GSM module
(d) Sensors: infrared sensor, CO_2 sensor, temperature and humidity sensor, etc.

(4) Test Theory

The method is for developing different walking movements of the quadruped robot and saving the movements in the memory. The user controls the robot remotely by the wireless communication module and watches remote real-time images from the vision system. Finally, sensors are loaded in the robot to collect the external data and report to the user by the communication module.

4. Research Process

(1) Hardware Circuit Design and Implementation

The hardware design and implementation includes total circuit design, controller implementation, integration of the communication interface in the computer side, memory planning, and total mechanism design. Besides, an interface is developed for external connection and wireless communication. The research emphasizes on the circuit design of the robot but without consideration of balance calculation on walking. The hardware circuit in the research is mainly for motor controls.

(2) Computer Program Design

In terms of the computer program design, Visual Basic is used as the computer program language and the platform of the Human-Machine Interface (HMI). The program is a tool to edit all walking movements of the robot, such as step forward, step backward, turn left, turn right, stand up automatically after falling down.

(3) Final Assembly of the Quadruped Robot

The quadruped robot is composed of the controller, the motors, the connectors, the sensors, and the battery. Of all components, the motors are the biggest part of the whole robot. Every joint of the robot is controlled by a motor. The research adopts the Korean AI motors as the key joint of the robot. The motor, with a maximum mobility angle up to 333 degrees, meets the specification requirement of the robot in the research. The connectors between motors are composed of the motor's original accessories, the simple aluminium chips, and the copper pillars for supporting and connecting different motors. The controller has a single chip microprocessor core for motor control. Figure 1.1 shows the diagram of AI motor control setup.

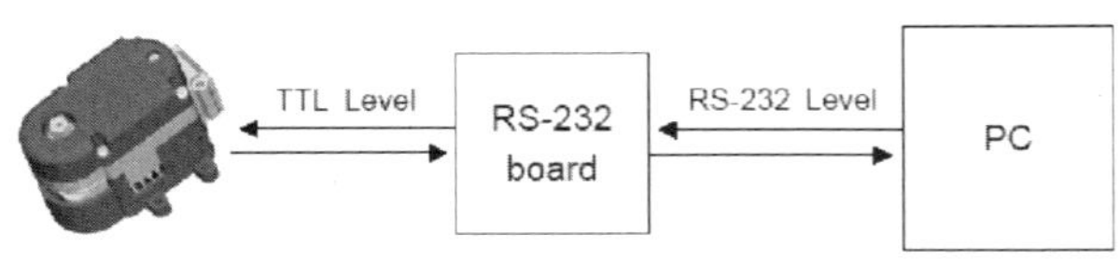

Figure 1.1 Diagram of AI Motor Control Setup

(4) Key Illustration of the Interface Program

Before moving the AI motor, a 4-byte control code must be sent in advance. The code is composed of the start code, the motor speed and the serial code, the position code, and the end code. The length of every code is 1 byte.

Table 1.1 Speed and Motor ID Number

Speed			Motor ID Number				
7	6	5	4	3	2	1	0

Start code=0xFFH
The following figure shows the format of the motor speed and the serial code

Speed: 0(max) ~ 4(min) – There are 5 different speed options.

Motor series: 0~30 – Up to 31 AI motors for control.

Position code: 0 - 254 (position)

End code: (motor speed and serial code XOR position code) and 0x7F

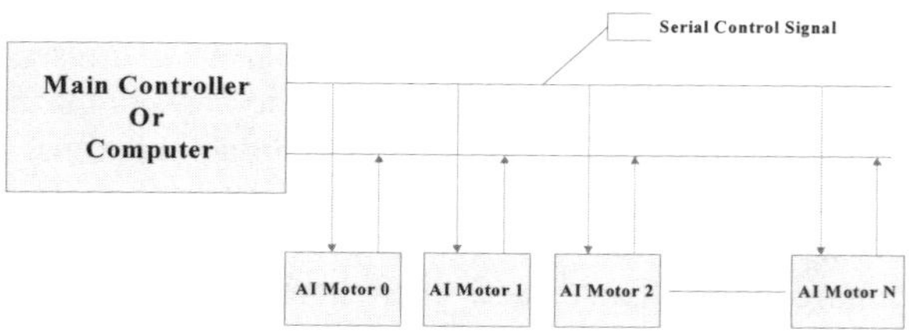

Figure 1.2. Demonstration of the Serial Control

(5) System Infrastructure of the Full System of the Quadruped Robot

The control side (PC or single chip) is able to control up to 31 AI motors simultaneously. By the serial transmission, the control side drives the motor to move to the target angle. By data transmission between motors, the control side is able to control and connect motors.

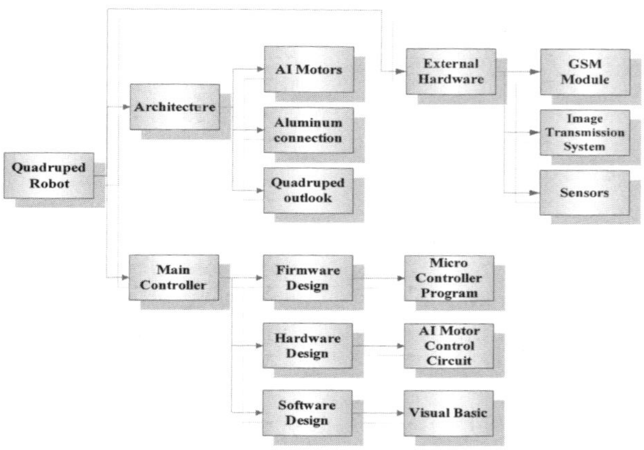

Figure 1.3. System Infrastructure of the Full System of the Quadruped Robot

The design of the quadruped robot in the research has three parts. The first part is the design of the infrastructure that includes the connection of 14 AI motors, the motor connection components, and the figure of the quadruped robot. The second part is about the design of the controller, including firmware design (the program controls in the microcontroller by Keil C language), hardware design (simple control circuit), and software design (HMI by Visual Basic language). The final part is the integration of the external hardware, including the dual half-duplex wireless communication module, 1.2G wireless image transmission system and sensors. When the three parts are integrated, the prototype of the quadruped robot with modules is completed.

(6) Design and Implementation of the Controller Hardware Circuit of the Quadruped Robot

The hardware design integrates the microcontroller (the AT89S52 chip of ATMEL), the memory (the command storage of the quadruped robot's movements), Power Supply Module (provides the power needed by the motor and circuitry), the AI motor control interface (the motor control by the serial transmission), the wireless communication interface (2.4G RF wireless communication module), the connection pins for accepting the data from the external sensors, the communication interface of the PC side for connecting the PC and downloading movements and commands, and the rotary switch for changing different control mode. By combining the above circuit, the control function of the robot is completed.

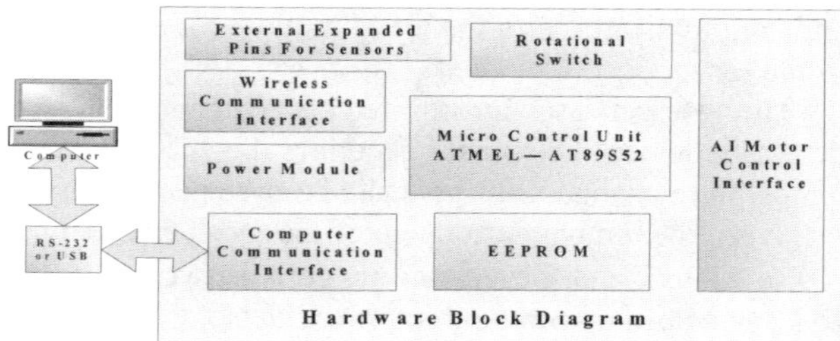

Figure 1.4. Design and Implementation of the Controller Hardware Circuit of the Quadruped Robot

(7) Body Module of the Robot

Fourteen Korean AI motors are used in the research for the main power of the quadruped robot. There is one motor in the head and the neck separately. Each of the four feet has three motors equally. By connecting the motor's attachments, the single aluminum chips, and the copper pillars, the prototype of the quadruped robot is done. After loading the previous hardware circuit and programming all movements (step forward, step backward, turn left, turn right, stand up again when falling down) including and keeping the movements in the memory, the robot is able to complete all movements under the mechanism range.

Figure 1.5. Body Module of the Robot

(8) Vision System Module of the Robot

The external 1.2G wireless image transmission module is for transmitting images to the PC side of the user by the USB converter. The user can even trigger the recording function to save the target image in the AVI (MPEG-4) image format.

(9) Wireless Communication Module of the Robot

The 2.4G RF wireless communication module used in the research has the nRF2401 Single Chip 2.4GHz Radio Transceiver and adopts the half-duplex communication for dual control and data return. As long as the 2.4G wireless communication module is starting, the power light of the transmitter is on. When any control key is pressed, the Tx light will be on for data transmission. At this time, use the control button to drive the quadruped robot's movements. When the sensor receives the returning data, the Rx light turns on and shows the status of the sensor. When no control key is pressed, the transmitter and receiver keep connection for communication to offer the dual/half-duplex function. With the GSM communication module, the user can use a mobile phone to control the robot remotely. When the user presses the buttons of the mobile phone and transmits the signal to the GSM module, the microcontroller of the robot decodes the signal and has the robot move. The controller then sends the AT command and communicates with the GSM module for status report of the sensors.

The research uses an infrared receiver module as the detecting device. The sensor generates the charge from the temperature change, and it is called a pyroelectric infrared receiver. The temperature range of the receiver is between -10 to +50°C, and the voltage range is between 3 to15 V DC. By amplifying the output of the detector and passing through the voltage comparator circuit, the receiver can detect the human body. The sensor receives the infrared rays radiated from all hot objects, including the human body. When no hot object movement is detected, the output of the sensor is 0 V. When a hot object is detected, the output of the sensor is 5 V. The robot detects hot objects by analyzing the output of the sensor, and it returns the result to the user by the 2.4G RF wireless communication module. The research develops the expandable connecting circuit. When the user connects different sensors to the control board, 5 V will be output to the control board when something is detected. The controller returns the signal immediately to the user for real-time reporting.

(10) Quadruped Robot Module

The upper part of the above figure shows the lens of the 1.2G RF wireless image transmission module. From the lens, the quadruped robot dog can deliver the captured image to the user side. By controlling the movements of the motors in the neck and the head, the robot can move the position of the lens and lock on the watching image. The infrared sensor is located in the milky-white half dome on the top of the motor of the neck and is a human body detector. The controller has two parts. The upper part is the 2.4G RF wireless communication module and is for receiving commands from the user and returning the detected signals. The lower part is the control module of the robot and is for all movement control. Under the controller is the lithium battery with 8.4 V voltage and 2000 mAh battery capacity. The battery fully supports the power requirement of the robot.

Figure 1.6. Quadruped Robot Module

5. Conclusion

The main design concept of the module of the sensitive quadruped robot is based on the fixed hardware specifications with combination of different sensors to meet the specific requirements under different situations. Therefore, no new robot is developed for the specific requirement. In addition, the resolution decreases the development cost and time largely. Following the sensor format of the robot and redesigning the sensor, the user can change the firmware of the robot in a short time. By changing the module, the quadruped robot in the research can expand its functions for detecting, mine exploring, rescuing, surveillance, and even home care. Its remote control function increases the interaction between robot and people and improves people's lives greatly. The concept of the firmware change in the research would be one of the main directions in robot development. The concept not only decreases the development cost but also makes multi-function robots with low price possible. It will help the development and spread of robots in future.

Acknowledgement

The study is supported by National Science Council, contract 95-2411-H-152-014- and 95-2622-E-152-001-CC3.

References

[1] Golliday, C.L., and Hemami, H. 1977, An Approach Analyzing Biped Locomotion Dynamics and Designing Robot Locomotion Control, IEEE Transactions on Automation Control, Vol. 22, No. 6, P963-972.

[2] Sadain, P., Rostami, M., Thomas, E., and Bessonnet, G. 1999, Biped Robots: Correlations between Technological Design and Dynamic Behaviour, Control Engineering Practice, Vol. 7, P401-411.

[3] Ambarish Goswami, 1999, Postural Stability of Biped Robots and the Foot-Rotation Indicator (FRI) Point, The International Journal of Robotics Research Vol.18, P523-533.

[4] Frank, A.A. 1970, An Approach to the Dynamic Analysis and Synthesis of Biped Locomotion Machines, Med. Biol. Eng., Vol. 8, P465-476.

[5] McGeer, T. 1998, Passive Walking with Knees, IEEE International Conference on Robotics and Automation, Vol. 3, P1640-1645.

[6] Shih, C.L., Zhu, Y., Gruver, W.A. 1991, Optimization of the biped robot trajectory, Decision Aiding for Complex Systems, Conference Proceedings, Vol.2, P899 – 903.

[7] Shih, C.L., Chi, C.T., and Lee, W.Y. 2000, Walking Experiment of a Biped Robot with Individual DC Motor Current and Position Control," The Fourth International Power Electronics Conference, Tokyo, Japan, P1249-1254.

[8] Jones, Bryan A. 2006, Practical Kinematics for Real-Time Implementation of Continuum Robots, IEEE Transactions on Robotics, Vol. 22/6, P1087-1099.

Development of a Mobile Platform for Service Robots

H. K. Huang[1], K. S. Tseng[2], Y. M. Chu[3] and C. Y. Liu[4]

[1]Industrial Technology Research Institute, Hsinchu, Taiwan 310, R.O.C., harry.huang@itri.org.tw
[2]Industrial Technology Research Institute, Hsinchu, Taiwan 310, R.O.C., seabook@itri.org.tw
[3]Industrial Technology Research Institute, Hsinchu, Taiwan 310, R.O.C., SimonChu@itri.org.tw
[4]Industrial Technology Research Institute, Hsinchu, Taiwan 310, R.O.C., CILiu@ itri.org.tw

Abstract. The development of a mobile platform is presented in this paper, which consists of system overview, motion control, obstacle avoidance and implementation of the system. The hardware platform is equipped with two driven wheels (differential drive) including shaft encoders for motion tracking. The system also contained several infra-red (IR) sensors for obstacle detection and avoidance. A camera was equipped on the system to record and send the images/video of the environment back to the central control room. The control software of the system is based on a Digital Signal Processor (DSP) and Windows CE platform. The DSP is the main motion control unit for the system. It is responsible for the movement of the system, environmental sensing and decision making. Windows CE is used for the function of wireless networking and communication with the central control room. Finally, several experiments were conducted to verify the current capability of the mobile robot platform. From the results of experiments, the mobile platform shows satisfactory performance in motion control, positioning, patrolling and monitoring the environment.

Keywords: mobile robot, hardware/software platform, motion control, collision avoidance

1. Introduction

According to the forecast of the Japan Robot Association (JARA) on service robots, robots used by non-manufacturing industry will have a market worth around 17 billion US dollars in 2010, and this figure will reach around 52 billion US dollars in 2025. In addition, from the survey conducted by International Federation of Robotics (IFR) and United Nations Economic Commission for Europe (UNECE), the number of accumulated installations of service robots will research 6.7 million, which is worth 9.7 billion US dollars. These figures show that the human-robot coexisting world is gradually realized, and thus the interaction between service robots and humans will become much closer in the foreseeable future. In addition, with the recent advancement in robotics and sensing technologies, more and more research has been conducted in the field of service robots.

A guard system is reported on for buildings which utilize an autonomous mobile robot [1]. The guard robot is a wheel-type autonomous robot which moves along a planned path, always alert for anything unusual from moving objects, fire or leaking water. The robot is equipped with cameras and while patrolling, it sends images back to the monitoring station. After it finishes patrolling, the robot can automatically return to its homebase and dock in a battery recharging station.

The intelligent automation laboratory of National Chung Cheng University, Taiwan [2] also developed a security robot. This robot has five sensory subsystems: fire detection, intruder detection, motor diagnosis, obstacle detection, and environment detection. The client user can supervise these sensory signals through the Internet. Both desktop PCs and personal digital assistants (PDAs) can be used as the user interface for remote surveillance and control system. These security robot systems can increase the security of homes and offices.

Schraft et al. [3] have developed a robot mobile platform for assistance and entertainment purpose in a museum. The platform has the capability to execute the tasks to welcome visitors, offer them exhibition-related information and entertain them.

An autonomous mobile robot system was reported by Jensen et al. [4] for allowing complex collaborative interaction with non-experienced users while giving a pre-programmed tour of a public exposition. Siegwart et al. [5] developed a mobile robot, Robox, which is designed for operation in a mass exhibition. The robot is an autonomous mobile platform with capabilities such as multi-modal interaction and global localization/obstacle avoidance based on multiple Gaussian hypotheses.

A patrol robot system for hospitals has been developed [6]. The system consists of an autonomous mobile robot that can judge the condition of patients while moving around sickrooms, and a patrol information terminal to allow observation of the patrol conditions from the nurse center.

Burgard et al. [7] have developed mobile robots, which can be deployed as interactive agents in populated environments such as museum exhibitions or trade shows. The mobile robots can be tele-operated over the Internet and provides remote access to distant users. Trahanias et al. [8] have proposed a number of techniques that further improved Web-operated mobile robots. These techniques include effective map-building capabilities, a method for obstacle avoidance based on a combination of range and visual information, and advanced Web and onboard robot interfaces.

Thus, it motivates us to develop a mobile platform for service robots. The development of the system is presented in this paper. The paper is organized as follows.

First, the overview of the system is presented. After that, the motion control of the system is explained. Then, the section of obstacle avoidance is presented. Following that, two experiments were conducted for the robot motion control. Finally, some future developments are discussed for the improvement of the current capability of the system.

2. System Overview

The hardware of the mobile platform is equipped with two driven wheels (differential drive) including shaft encoders for motion tracking (see Figure 1.1). The system was equipped with a camera to record and send images/videos of the environment back to the central control room. The control software for the system is based on a Digital Signal Processor (DSP) and Windows CE platform. The DSP is the main motion control unit of the system. It is responsible for the movement of the system, environmental sensing and decision making. Windows CE is used for the function of wireless networking and communication with the central control room.

The system has two driving wheels at left and right of the body centre. Encoders are mounted at the motors for driving wheels to realize odometery system. A controller estimates the robot current position based on odometry and controls the driving wheels to move to the given position. UBOT adopted TI DSP2812 as its CPU. DSP communicates with FPGA through the Serial Peripheral Interface (SPI). FPGA is responsible for control of sensors, such as ultrasonic sensors and encoders. DSP is used for control of algorithms and communication with the host computer.

The system adopted RS-232 as the communication interface between DSP and Windows CE. It divided the commands into Command ID and Message ID. In Windows CE, the system uses .NET Compact Framework with C/C++ as the development tool. The main programs were coded by using .NET, and functions required to be coded as COM objects were generated by using C/C++. As shown in Figure 1.2, the software architecture of the UBOT is divided as three parts: robot platform, central control server, and monitoring program [9]. These three parts are connected via the 802.11x wireless network. The Windows CE provides the functions of the robot through the Web Service. The central control server uses the Web Service for giving commands to the robot, and it is also used for sending back some related information.

The central control server contains a .NET Remoting service program and an interface of the Web Service. The monitoring programs can be saved in several kinds of mobile devices, such as PDAs, Smart Phones or Notebooks etc. The monitoring programs can give commands via the Web Service of the Server. Then, the Web Service in the Server will send the commands to the Remoting service program through the Remoting mechanism.

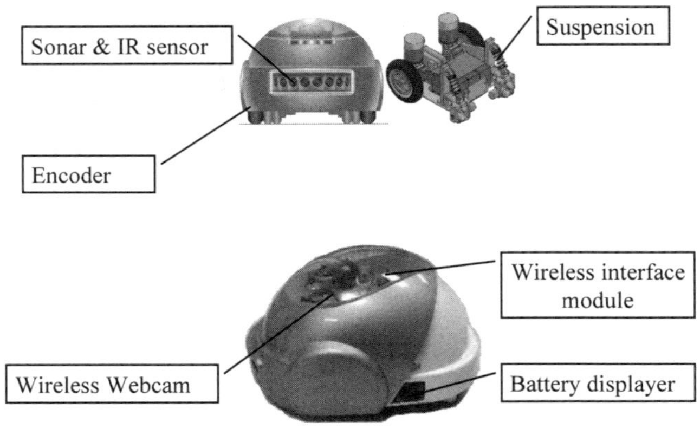

Figure 1.1. Hardware the mobile platform

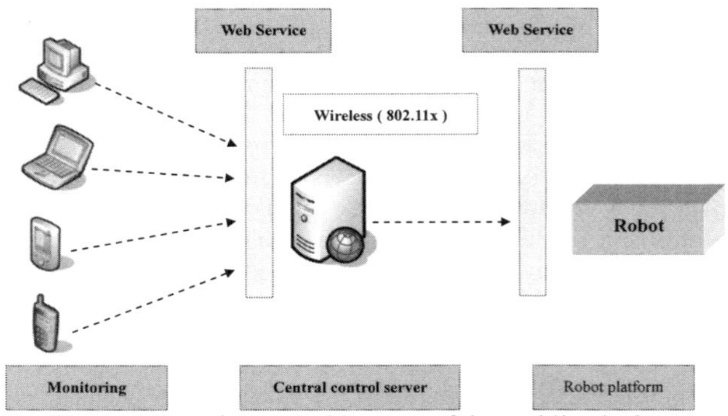

Figure 1.2. Software architecture of the mobile platform

3. Motion Control

We consider our mobile platform system with n generalized coordinates $\mathbf{x}$, subject to m constraints, and its dynamics equation can be described by [10]

$$M(x)\ddot{x} + V(x, \dot{x})\dot{x} = B(x)\tau + A^{T}(x)\lambda \qquad (1.1)$$

where T is the input vector, λ is the vector of constraint forces, M(x) is an $n \times n$ symmetric and positive definite matrix, $V(x, \dot{x})$ is the $n \times n$ centripetal and Coriolis matrix, B(x) is the input transformation matrix, and A(x) is the matrix associated with the constraints.

The system uses an encoder and an electrical compass as its feedback sensors to implement the fixed-point motion control. Figure 1.3 shows the control diagram of the system. It is a position feedback control. The input data (q) are velocity of both wheels and the output data (L) are moving distance and rotated angle generated from the encoder and the electrical compass. With the matrix transformation (J), the robot position (X_r) can be obtained. The robot position error (X_e) is modified by the PI controller and then sent back to the robot plant. Thus, a closed-loop control is completed.

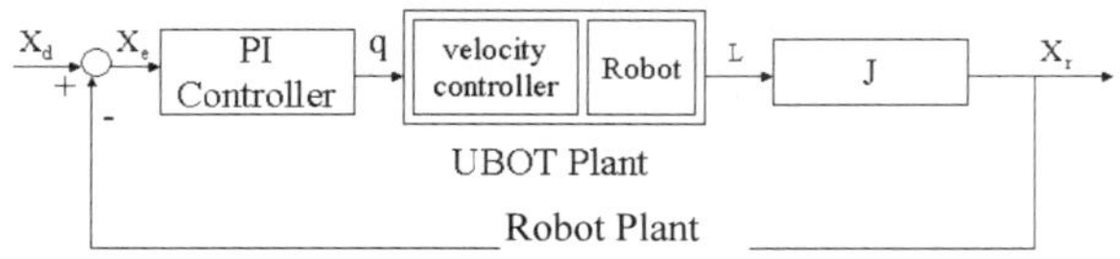

Figure 1.3. Control diagram of the mobile platform

The kinematic equations of the mobile platform are derived as below [11]

$$q = \begin{bmatrix} V_L \\ V_R \end{bmatrix}, \quad L = \begin{bmatrix} l \\ l \\ \theta \end{bmatrix}, \quad J = \begin{bmatrix} \cos\theta & 0 & 0 \\ 0 & \sin\theta & 0 \\ 0 & 0 & 1 \end{bmatrix},$$

$$X = \begin{bmatrix} x \\ y \\ \theta \end{bmatrix}, \quad X_e = \begin{bmatrix} x_d - x_r \\ y_d - y_r \\ \theta_d - \theta_r \end{bmatrix}, \quad q = K \cdot X_e$$

$$K = \begin{bmatrix} K_{11} & K_{12} & K_{13} \\ K_{21} & K_{22} & K_{23} \end{bmatrix} \qquad (1.2)$$

where
V_L: input voltage of the left wheel
V_R: input voltage of the right wheel
l: the measured distance
θ: rotated angle
x: coordinate in the X axis
y: coordinate in the Y axis
X_d: desired position
X_r: real position
K: parameters of the controller

4. Obstacle Detection

The mobile platform needs to have the capability to detect obstacles. Thus, it required some environmental information obtained from the sensors, preferably non-contact sensors, and, moreover, than it can make decisions based on the input coming from these sensors. Among all non-contact sensors, infrared (IR), Laser and ultrasonic sensors are most commonly used. Ultrasonic sensors have the advantage of a wide sensing area. Since an ultrasonic sensor has a cone-shaped scanning range, the user can use several sensors to achieve 360 degrees of coverage. However, this also means that, if the installed position of the sensor is close to the ground, the sensor would be disturbed by the reflected waves from the ground, which would result in a low performance of the ultrasonic sensors.

Compared to the ultrasonic sensors, the IR and Laser sensors have a smaller sensing coverage. To increase the coverage of these sensors, the user is required to install a large number of sensors. This will increase the total costs significantly. Thus, it is proposed in this research to apply both IR and ultrasonic sensors for our mobile platform system. The effective sensing range for the IR sensor is 0.3 m, and the range for the ultrasonic sensor is from 0.5~1.0 m with the angle coverage from 26° to 22°. The system contained six infra-red (IR) sensors and six ultrasonic sensors for obstacle detection and avoidance.

5. Implementation

The experimental setup is shown in Figure 1.4. The setup consists a position measuement system made by PhaseSpace Inc. The system contains 8 cameras (see Figure 1.4). They are used to track the position of LEDs (see Figure 1.5) attached to the mobile robot platform. The accuracy of the measurement system can reach 0.1 cm in a 5 m x 5 m area. Two experiments has been conducted in verify the performance of the proposed mobile platform. In the first experiment, the robot turned itsef into a specified orientation (0°, 45°, 90°, 135° or 180°) and then moved in that direction for a distance of 100cm at the speed of 10 m/min (see Figure 1.6). The actual moving distance of the robot was obtained by the position measuremet system. The test result is shown in Table 1.1. It was found that the maximum position error is around 2.7 cm at the orientation of 45°. For the second experiment, the robot autonomously moved from position (0, 0) through positions (20,20), (30, -20), (-10, -30), and then went back to (0,0). It was found that the mobile robot can execute the fixed-point motion control satisfactorily.

Figure 1.4. Experimental setup

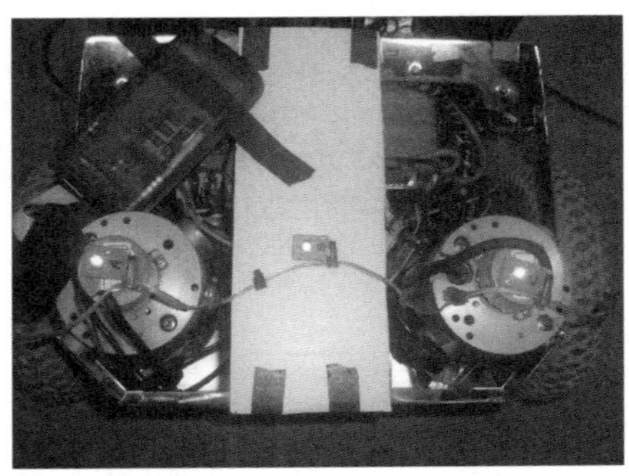

Figure 1.5. LEDs attached on the mobile platform

6. Conclusion

This paper presents the development of a mobile robot system. Its hardware platform is equipped with two driven wheels including shaft encoders for motion tracking. Its control software for the system is based on a DSP and

Windows CE platform. There are some areas considered important for enhancing the capability of our mobile robot system, and they are currently under development for future studies.

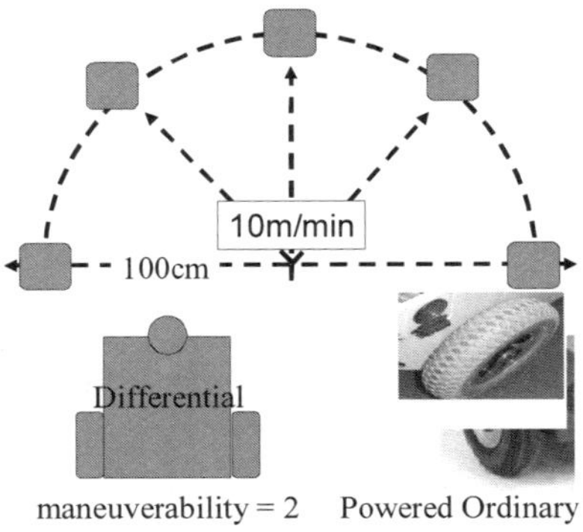

Figure 1.6. Testing paths for positioning errors of the mobile platfiorm

Table 1.1. The result of the positioning test of the mobile platform

Orientation	Actual moving distance (cm)	Positioning errors(cm)
0°	99.2	0.8
45°	97.3	2.7
90°	100	0.0
135°	98.8	1.2
180°	97.5	2.5

These areas include simultaneous localization and mapping (SLAM), autonomous navigation, and distributed control architecture. Without the capability of autonomous navigation, the robot can only move in a fixed area, and this is not sufficient for completing a detailed patrolling task. In addition, regarding to distributed control architecture, due to the recent progress on DSP technologies, the parallel or multi-layer DSP can be applied to speed up the data processing of the robot. Such developments as described above will undoubtedly lead to a more advanced mobile robot system.

References

[1] Shimosasa, Y., Kanemoto, J., Hakamada, K., Horii, H., Ariki, T., Sugawaea, Y., Kojio, F., Kimura, and Yuta, S., 2000, "Some results of the test operation of a security service system with autonomous guard", Proceedings of 26th Annual Conference of IEEE Industrial Electronics Society, vol. 1, pp.405-409.

[2] Luo R. C., Shen S. H., Su K. L., and Chang T. C. 2002, "Remote mobile security system using personal digital assistant (PDA) for security robot", Proceedings of IEEE/ASME International Conference of Advanced Manufacturing Technologies Education in 21st Century

[3] Schraft R. D., Graf B., Traub A. and John D. 2001, "A mobile robot platform for assistance and entertainment", Industrial robot, vol. 28, no. 1, pp.29-34

[4] Jensen B., Froidevaux G., Greppin X, Lorotte A, Mayor L, Meisser M., Ramel G., Siegwart R. 2002, The interactive autonomous mobile system RoboX, IEEE International Conference on Intelligent Robots and Systems, vol. 2, pp. 1221-1227

[5] Siegwart R; Arras K O.; Bouabdallah S; Burnier D; Froidevaux G; Greppin X; Jensen B; Lorotte A; Mayor L; Meisser M; Philippsen R; Piguet R; Ramel G; Terrien G; Tomatis N 2003, Robox at Expo.02: A large-scale installation of personal robots, Robotics and Autonomous Systems, vol. 42, no. 3-4, pp. 203-222

[6] Sato K., Ishii M.; Madokoro H. 2003, Testing and evaluation of a patrol robot system for hospitals, Electronics and Communications in Japan, Part III: Fundamental Electronic Science (English translation of Denshi Tsushin Gakkai Ronbunshi), vol. 86, no. 12, pp.14-26

[7] Burgard, W; Trahanias, P; Hahnel, D; Moors, M; Schulz, D; Baltzakis, H; Argyros, A 2003, Tele-Presence in Populated Exhibitions Through Web-Operated Mobile Robots, Autonomous Robots, vol. 15, no. 3, pp. 299-316

[8] Trahanias P., Burgard W., Argyros A., Hahnel D., Baltzakis H., Pfaff P., Stachniss, C. 2005, TOURBOT and WebFAIR: Web-operated mobile robots for tele-presence in populated exhibitions, Robotics & Automation Magazine, IEEE, vol.12 , no. 2, pp. 77 – 89

[9] Tsai J. Y. and Yu H. H. 2005, "Small intelligent robot", Journal of Mechatronic Industry, vol. 265, pp. 146-153 (in Chinese)

[10] Fukao T., Nakagawa H., Adachi N. 2000, Adaptive tracking control of a nonholonomic mobile, IEEE Transactions on Robotics and Automation, vol. 16, no. 5, pp.609-615

[11] Siegwart R. and Nourbakhsh I. R. 2004, Introduction to Autonomous Mobile Robots, MIT.

15–3

Study of an Industrial Multi-Agent Framework with an Application to Multi-robots Fire-rescue Simulation

Y-M. Chen[1], C-I. Chu[2], Y-P. Lee[3], T-C. Chou[4]

[1]Intelligent System Engineering Division, Industrial Technology Research Institute (ITRI), Taiwan, higherself@itri.org.tw
[2]Intelligent System Engineering Division, Industrial Technology Research Institute (ITRI), Taiwan, achilleschu@itri.org.tw
[3]Intelligent System Engineering Division, Industrial Technology Research Institute (ITRI), Taiwan, erinlee@itri.org.tw
[4]Intelligent System Engineering Division, Industrial Technology Research Institute (ITRI), Taiwan, zcchou@itri.org.tw

Abstract. This paper presents the authors' experience in implementing a multi-agent system created for communication and coordination purposes among mobile robots. The complete system described here is an industrial application focused and aimed at integrating robots, environment and user agents via an agent platform. The work bridges the gap between robots and a JAVA based agent platform. A scenario based multi-robot fire-rescue simulation is used as a study example. The scenario can be rewritten to suit different industrial requirements. The multi-agent framework and its control hierarchy is discussed, path planning with object-avoidance issue is addressed. The actual simulation demonstrates feasibility of this type of agent based system integration. Future work would focus on integration of environment and user agents.

Keywords: multiagent, robots, scenario, coordination

1. Introduction

Robot technology has been an active area of research for many years. Increasingly, relevant technologies such as artificial intelligence, vision, navigation, human-robot interaction, communication, image based identification and multiagent systems, have reached a higher peak??. Incorporation of these technologies has considerably increased the applicability of robots in providing services in the real-world. The growing trend of integrating robotic systems and place robots into real-world applications is an emerging robot industry.

Research into integration of service robots has been a topic of interest to many researchers leading to suggestions of a range of future application of robots. As a result, different types of robots have been proposed such as medical operation robots, construction robots, security and surveillance robots, human companion robots, soccer robots, robot pets etc. System requirement of each type of robot varies according to the characteristics in their specific field of application. For example, surveillance robots may emphasize on vision and navigation functions, thus leading to the integration of an Omni-directional camera and wheels. medical operation robots would focus on the aspect of precision and hence the fine tuning between simulator and actuator is crucial. Soccer robots lay emphasis on coordination. This is where a multiagent system plays a significant role. Multiagent systems (*MAS*) focus on systems in which various intelligent agents working together with each other. Agents can share a common goal, as in the case of soccer robots where a team of robots work together with the objective of scoring goals. Generally, the agents are considered to be autonomous entities such as software programs or robots. Their interactions can be either cooperative or selfish [1]. Two types of multiagent systems have been identified: centralised and distributed systems. A multi-tasking robot is based on centralised multiagent system framework because cooperative agents are integrated within a single host. Soccer robots are a good example of a distributed multiagent system [2] which consists of numerous robots, each of whichcan be considered as an agent within the framework.

Both industry and military have shown research interests in Multi-Agent Systemsbecause theyare promising as innovative in conceptualizing, scheming, and implementing software systems. This promise is principally attractive for creating software that operates in environments that are open and distributed. For example, the perspective of military organisations on multiagent systems seems to be directed by an interest in the integration of a *MAS* composed of air-force, navy, army and also space task-forces, such as surveillance and remote-sensing satellites. The National Institute of Standards and Technology (*NIST*) have proposed a *4D/RCS* methodology in which the so-called reference model architecture is the core. The reference model architecture is characterized by a generic node that is applied to all the hierarchical control levels. *4D/RCS* prescribes a hierarchical control principle that decomposes high level commands into actions that employ physical actuators and sensors [1]. On the other hand, an industrial perspective of *MASs* can be a team of robots collaboratively working together to complete a task or offering services. For example, surveillance robots can be coordinated to replace security guards. In addition, intelligent buildings that security in homes can be viewed as a *MAS* with mostly stationary agents. Following this trend, it can be deduced that future industrial demands fromrobot technology would not be satisfied with single multifunctional robots whichcarriy outspecific tasks, when a coordinated robot team can offer multiples of what such single robots can provide. For this reason, the driving force behind studying *MAS* becomes obvious. This work describes a study and implementation of an industrial multiagent system: a multi-robot coordination system and its application to a fire rescue simulation.

2. Multi-Agent Framework

The multiagent system described here is based on the idea of linking three categories of agents together via an agent platform. The three categories of agents are an environment agent, a robotic agent and user agent. This idea is not new and has been used to resolve coordination problems using a combination of robots and an environment [15]. However, this idea was perceived by the authors from an industrial point of view with an interest of exploring its future industrial values. It is believed that a variety of industrial based agent systems can be derived from implementation of this concept.

The environment agents are defined as a group of stationary agents in a prescribed environment which can be a hospital, office, meeting room, factory, home, museum, library. The agents within these environments can be light, electric doors and windows, air-conditioning units, surveillance cameras, security systems, microphones, etc. The idea is to redesign existing appliances and electronic equipments so that they enable two-way communication with a main agent platform via either wired or wireless internet network [5]. Once this communication route is bridged, controls and coordination of these agents can be scripted in a desired scenario fashion based on customers' preferences. This allows numerousautonomous environment entities to be coordinated in a way that is tailored to the needs.

Robotic agents are basically a group of robots. An agent platform can act as a coordinator via wired or wireless networks. Robots can be of types such as security robots, surveillance robots, robot pet, etc. Each kind of robot entering an environment must first register with the agent platform. This registration process is similar to the so-called *plug & play* behaviour where robotic agents report to the agent platform the services theycan provide and then authorise the agent platform to dominate its controls. The purpose of allowing an agent platform to dominate is toprovide coordination from the agent platform and ensure cooperative operation with other agents. The robotic agent group is different from the environment agent in the way that most robots possess movable platforms, which enables navigation within the environment. The combination of environment and the robotic agents thus enable a wide-variety of services to be provided.

The user agent group consists of devices that enables the user to communicate with an agent platform. These devices can be a cell phone, personal digital assistant, computer, microphone, etc. While the agent platform is capable of coordinating various agents according to a user's specification, the interaction between an agent platform and users still needs to be bridged. This allows users to constantly monitor the status of an *MAS* and also alter system preferences when needed.

The control hierarchy of our industrial based *MAS* consists of five main layers, namely hardware execution, function control, behaviour, autonomous and user control layers (see Figure 0.1)

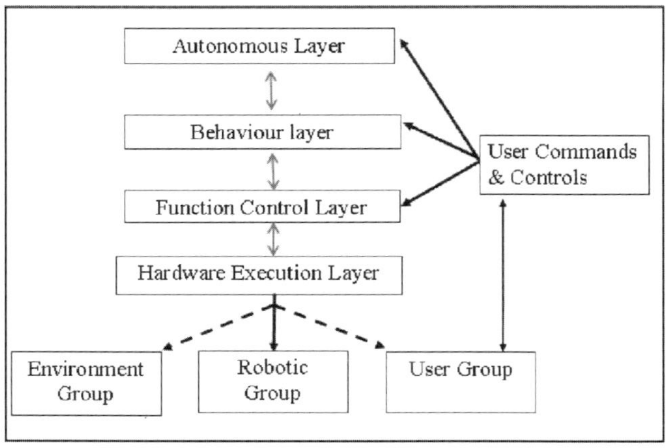

Figure 0.1. Control hierarchy

The hardware execution layer removes hardware and operating system dependencies from the application. All that is required is a configuration file that describes each hardware platform. The configuration files configure the software modules to seamlessly run on a variety of agents. The function control layer defines the control format of each agent. For example, most environment agents utilise *on* and *off* types of control whereas robotics agents cooperatively manifest a variety of behaviourssuch as navigating to a specific location, voice communications, lifting, turning, and jumping behaviours. The behaviour layer contains a list of behaviours and its availability depends on the availability of both the hardware execution layer and function control layer. Behaviours can be chained together to form behaviour networks. The autonomous layer executes a series of behaviour algorithms or behaviour networks in a predefined time schedule.

Over the past three decades, researchers in the field of artificial intelligence have tried to develop rules and algorithms that simulate the human thinking process. This type of research is reported to have met with extreme difficultiesbecause human reasoning is largely based on experience. The process of learning starts from birth and accumulates through life with theexperience being irreplaceable and relatively difficult to duplicate in digital forms. Although numerous attempts to develop a digital human brain has failed, several useful techniques such as decision trees, machine learning algorithms, neural networks, genetic programming, evolutionary programming, genetic algorithms, optimization were developed and evolved from those failed attempts over the past few decades. These techniques have been diversely applied over many disciplines and have been demonstrated to yield impressive results.

Agents are specifically designed to provide service. Based on service requirements, hardware is integrated to perform necessary tasks and exhibit desired behaviours. On the other hand, software codes are programmed in the way in which services are provided. The desired services or behaviours often require inputs from human experience. In other words, the agent is expected to do the right thing at the right time when dealing with a specific situation. The right reaction is often judged and measured by how a typical human-being would react in the same situation. This is one of the reasons artificial intelligence is crucial in the field of robotics and also multiagent systems. The so-called expert systems within the field of artificial intelligence have been particularly useful for service applications because specific services and situations can be considered and pre-programmed, so as to assist agents face all possible circumstances.

Agent coordination is an important aspect in multiagent systems. Coordination strategy has been an active research topic in the field of robotics and controls [6]. Typical coordination problems studied are based on robots exploring an uncertain environment [7-13]. However, the industrial application of these algorithms is somehow limited because real world coordination of robots for navigation is more of a map-based procedure. The map marks all the barricade locations in a fixed environment. The robot can then navigate with path-planning based on the features defined in the map. However, this only works if the pre-memorized terrain has not been altered. In most real-world applications, the map is constructed via supervised remote controls or manually. In other words, the exercise domain needs to be pre-memorised. Thus, avoiding the needs of applying algorithms that randomly explore or navigate within an unknown field. However, some researchers have utilised these randomly exploring algorithms to construct environment maps. Typical industrial applications are cleaning [14-16] and mowing robots [17] which are expected to perform their jobs without supervised intervention.

3. Multi-Agent Fire Rescue Scenario

A simple fire rescue scenario has been designed and treated like a test bed for an implementation of a *Multi Agent System*. The scenario location is located in the robotic laboratory within the Mechanical and System Research Laboratory of the Industrial Technology Research Institute (*ITRI*), which is roughly a 10 m x 5 m area. Three robots were chosen to play a part in this fire-rescue scenario. The robots have not yet been integrated with any fire-rescue equipments at this stage because the objective of the simulation is aimed at demonstrating multi-robot coordination via a Multi-Agent Platform (*MAP*).

The scenario starts by assuming a fire alarm is triggered by environment sensors, which report to the *MAP* that a location marked *X* needs assistance in putting out fire. Following this, the *MAP* evaluates the necessary rescue action. In order to make the scenario more realistic, it is assumed the environment sensor is able to report the intensity of the fire to the agent platform. In addition, the intensity of the fire is characterized on three levels as low, medium and high. After receiving the request for assistance message from the environment sensor, the *MAP* decided to send one robot (assuming it is *R1*) to location *X* to help put out the low intensity fire. While *R1* reaches location *X* and putsthe fire off, another request for assistance message will be sent, at the same time, by the environment sensor requesting for assistance at location *Y*. Again, this is reported as another low intensity fire incident. The *MAP* then needs to decide which robot to send to location *Y*. Assuming at this stage that the *MAP* sends *R1* from location *X* to location *Y* and sends *R2* to location *X*. While *R1* and *R2* are both in their preferred places, the last low intensity fire alarm will be reported by the environment sensor, requesting for assistance at location *Z*. At this stage, assuming the *MAP* sends *R1* from location *Y* to location *Z*, *R2* from location *X* to location *Y* and finally the last robot *R3* to location *X*. This concludes the first half of the scenario. At this stage, fire locations *X*, *Y* and *Z* should be occupied by robots *R3*, *R2* & *R1*, respectively.

The second half of the fire-rescue scenario starts by deactivating the fire alarm signal at location *X* (indicating the fire in location *X* has been put out) and switch the fire intensity level at location *Y* from low to medium. It is then expected that the *MAP* makes the correct decision by sending one robot to support *R2* at location *Y* (assume the *MAP* sends *R3*). Once *R3* has reached its assigned destination, the fire alarm signal at location *Y* is then deactivated and at the same time, the fire level at location *Z* is adjusted from low to high. When this happens, the *MAP* is expected to send the two robots at location *Y* to location *Z*. This fire rescue scenario is completed by deactivating fire alarm signal and expects the three robots to return to their original locations together.

4. Path Planning

Path planning is an active research field within the field of robotics. The robot path planning problem usually has an object avoidance objective in mind. This is to say, when a robot is given a destination coordinate, navigation towards the requested position should be performed to obtain a plan of an intermediate path which avoids any obstacles, and, most importantly, minimises the distance from the starting point to the destination point.

The approaches proposed for these types of problems include: the Rapidly-Exploring Random Tree (*RRT*) approach [18] which is a random search algorithm designed for handling differential constraints; Voronoi diagram and Dijkstra's algorithm [12] based approach for finding the shortest path between any two arbitrary points in a prescribed environment and path planning within an unstructured environment [20]. Other algorithms focus on self-exploration of an unknown terrain. But generally, from an industrial point of view, most object avoidance path planning applications would be map based. This is true now because robots today are specifically designed to provide service within a prescribed environment. Outside this predefined scope, its programmed intelligence may have difficulties in adaptation and, at the same time, having offering quality services. For this reason, map-based or domain restricted path planning strategies tend to receive more industrial attentions.

For the multi-robot fire rescue simulation presented in this paper, the Voronoi diagram-based approach has been adopted for defining environmental conditions and Dijkstra's algorithm for calculating the shortest path between two points.

5. Voronoi Diagram & Dijkstra's Algorithm

Voronoi diagrams have been utilized in dealing with robot path planning problems. Blaer [12] represents the floor layout using a two dimensional map. The obstacles are represented using dense polygons before a Voronoi diagram is constructed. The Voronoi diagram drawnbased on the predefined dense polygonal points tend to partition the space into numerous smaller cells. Voronoi edges that are closer to the geometry entities are then removed and the remaining Voronoi edges are potentially the robot's path. The remaining Voronoi edges are made up of many Voronoi vertices from which the shortest distance based on the map definition between any two points can be searched using the Dijkstra's algorithm (see Figure 0.2).

With this approach, all possible paths resulted from Voronoi edges. The way in which Voronoi edges are generated directly influences the map created. The usual Voronoi diagram for a discrete set of given points only contains edges that are straight line segments, while the results from other Voronoi algorithms can contain parabolic arcs as well as straight lines. One may argue that the parabolic arc is a better path due to its smooth curvature.But the parabolic edges generated aroundobstacles require more complex calculations to generate Voronoi. The authors believe straight line Voronoi edges offer an acceptable quality to uniformly partition a prescribed domain space. For a less dense domain space, the Voronoi diagram may partition the given space with both straight lines and parabolic arcs with dense Voronoi edges generated near the predefined obstacles and coarse edges elsewhere. Coarse Voronoi edges simply mean that possible paths are limited due to the coarse resolution of the Voronoi diagram. In addition, most applications of the Voronoi diagram for path planning suggest thatr obstacles within the path domain should be defined with dense polygonal points. In this way, due to irregular shapes of the layouts, the Voronoi diagram drawn does not have a the quality required for the navigated space. In other words, the optimum path cannot be found if the map is poorly defined. For this reason, in a two dimensional space, it is suggested, in this paper, to cover the navigation region with closely-spaced grid points so that the quality of navigating the domain is improved. In addition, obstacles can be simply be represented with dense points rather than a specified geometry entity such as a polygon.

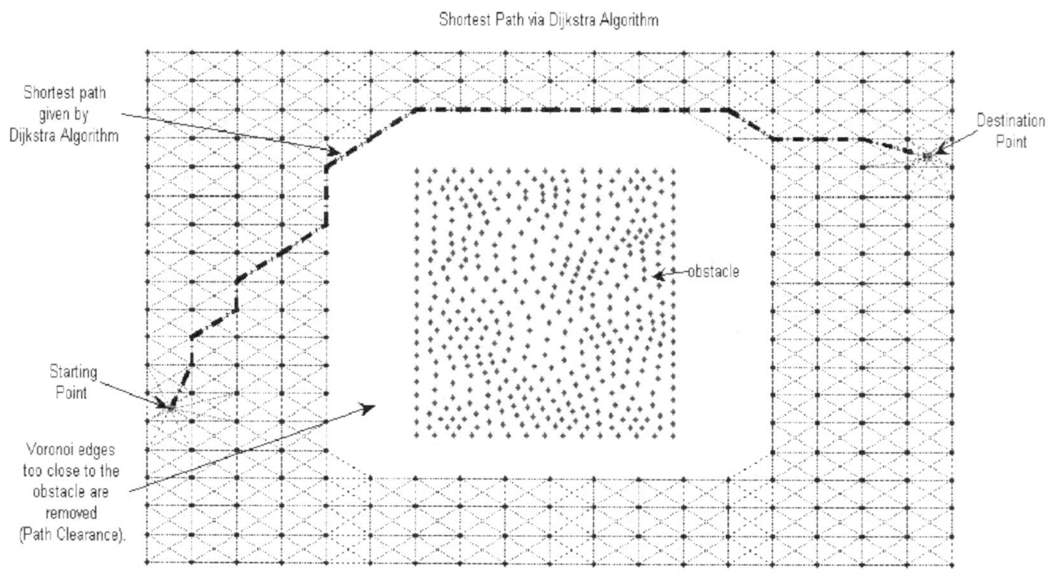

Figure 0.2. Illustration of path Planning

6. Experimental Discussion

The implementation of the prescribed scenario requires integration of three major components: hardware, software and controls. Hardware in this case consists of three robots with navigation and communication functions The Software employed is an agent platform where coordination among various agents is hosted while control is achieved through an algorithm coordinating robots to implement this scenario. Java Agent Development (*JADE*) software framework is adopted as the base framework for our multi-agent work. *JADE* is entirely written in *JAVA* language and complies with the Foundation of Intelligent Physical Agent (*FIPA*) specifications. The control component has been simplified in this case due to two reasons. The existing *FIPA* specifications embedded in *JADE* already defines anagent communication format which can be simply applied without re-inventing the wheel, and the level of difficulty involved in simulating this multi-robot fire rescue scenario is not high. The agent communication is via *RS232* wireless internet network and *RS232* commands are used to pass control messages between agents.

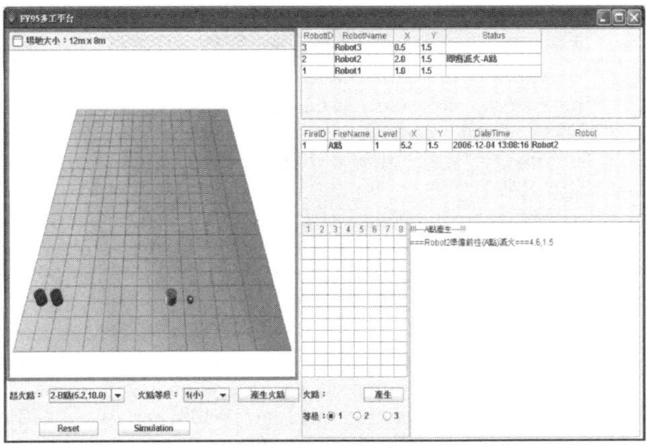

Figure 0.3. JADE based agent control platform

Figure 0.4 illustrates the actual communication process between the agent platform and robots. The environment sensor on the top left corner indicates the signals emitted from the fire alarm. Once the signal is initiated, based on the predefined scenario, the agent platform understands that assistance from robots are required. The agent platform then queries the current coordinates of agents (both agents 1 and 2). Agents 1 and 2 are related to actual robots 1 and 2, respectively. The agents then communicate with the actual robots via wireless field using *RS232* commands. The gathered coordinates are then sent back to the agent platform for further evaluation. The dotted line shown in Figure 0.4 thus represents the separating line between virtual and real space and it in this space that the wireless internet field exists.

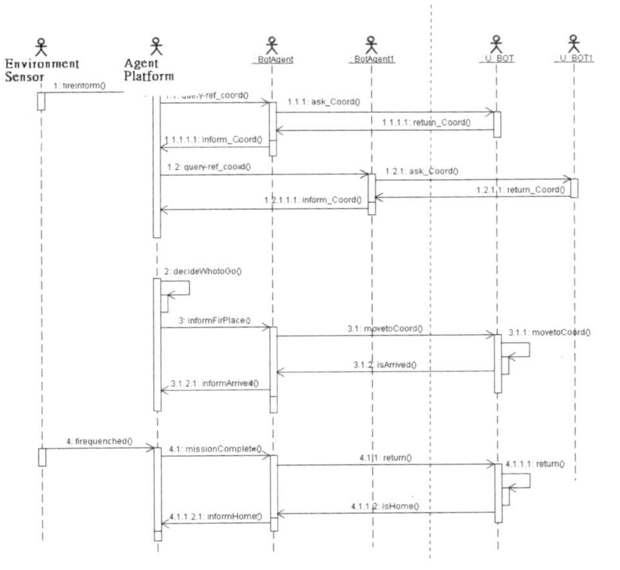

Figure 0.4. Illustration of Agent Communication

All three robots employed in this scenario possess the same navigation speed. Hence the calculation of time required for reaching a point is based on their path. The accurate distance between two points in this case cannot be approximated by a straight line since obstacles are present in the domain. For this reason, path planning is required to determine a more accurate robot path and distance between the robot's present locations and the fire point. Table 0.1 is a summary of the path planning results for all three fire points (*X*, *Y* & *Z*) and the origin and Table 0.1 gives the costs of each path (i.e. the length of each path).

Table 0.1. Summary of path information

	Path 1	Path 2	Path 3	Path 4
Cost/total distance (m)	7.8486	7.8361	5.7885	5.6971
Path Definition	Origin to Y	X to Y	Y to Z	X to Z

Initially, all three robots were located near the bottom left corner of the demonstration field (i.e., at the location x = 0, y = 0). This multi-robot fire rescue scenario is demonstrate in a site of dimension 10 m x 5 m with pre-allocated obstacles, as previously described, initiating the fire alarm at location *X*. At this point, the *MAP* receives the coordinates of location *X* and determines a suitable robot to send for assistance. This decision normally would be based on evaluating factors such as the capacity and efficiency of extinguishing the fire and the time required to reach the destination point. In this simulation, the only consideration is the time factor which is evaluated based on the distance between each robot's position and the *X* location. The chosen robot simply relocates horizontally from their initial position to reach the first fire point.

From Table 0.1, the distance of both path 1 & 2 are very close to each other with path 2 slightly shorter than path 1. *MAP* would choose to proceed with the option involving path 2. However, by implementing path 2, it means the robot which was originally at *X* location needs to be relocated to *Y* point. This implies that the *MAP* needs to send another robot from the origin to the *X* point for reinforcement. The distance the robot is required to travel horizontally from the origin to the *X* point is approximately 5 m, which isfeasible. Hence, when the fire alarm at the fire point is triggered, one robot is relocated from *X* point to *Y* point and another robot is sent from the origin to the *X* point.

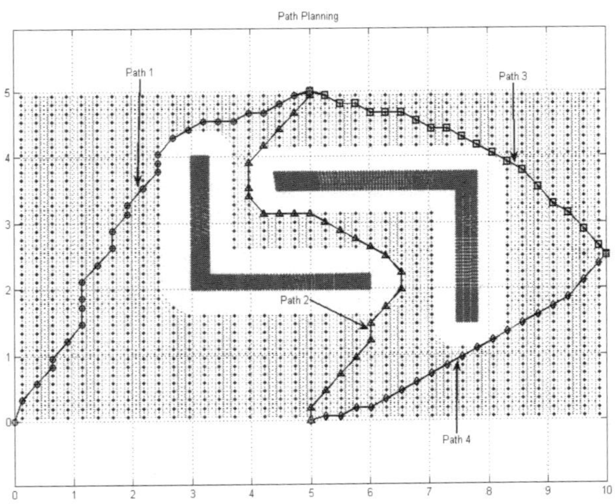

Figure 0.5. Robot path planning

After the last fire point (the *Z* point) is triggered, the *MAP* evaluates the feasibility of the three robots and relocates the robot at *X* point to *Z* point based on the shortest distance (see table 1.1 for comparison of path 3 and 4). At the same time, the only robot still at the origin is relocated to *X* point.

At this point, the fire signal from *X* point is switched off and the fire intensity level at *Y* is increased to medium. The observed behaviour is that the *MAP* relocates the robot at *X* location to *Y* point via path 2. When the fire alarm at *Y* location is switched off and the fire intensity at *Z* location is maximized,both robots at *Y* location are relocated to *Z* point via path 3.

Approximately two to four seconds of delay has been observed from the time when the *MAP* performs an evaluation and sending of commands to robots. This delay may be accumulated due to wireless network message exchange, the time required in computation andthe time the robots require to interpret the received commands.

7. Concluding Remarks and Future Work

This work addresses a multiagent system from an industrial point of view. A framework has been described and a multi-robot fire rescue simulation has been implemented. This work has focused on bridging the gap between an agent platform and robotic agent groups. Future work will address the integration of environment agent and user agent groups into the framework. It is also expected, in the future, to upgrade to a chosen environment such as home, factory, hospital, security systems and office via multi-agent based approach to provide higher quality of services anda wider range of services via selfish and cooperative agent platforms. This future objective requires bridging the communication gap between each agent both to the main agent platform and also to each individual agent group together with an effective control structure among the agents.

Acknowledgement

The financial support from the National Science Council (NSC) of Taiwan to the 35th International MATADOR conference is gratefully acknowledged.

References

[1] Sycara K.P. (1998) Multiagent systems, The American Association of Artificial Intelligence, 79-92.

[2] Kwon, D.S., Oh, J.H., Kwak, YK etal., (2001), A Novel Interactive Robot Soccer System, IEEE/ASME lntematinal Conference on Advanced Intelligent Mechatronics Proceedings, 1082-1086.

[3] Albus, J.S. and Meystel, A.M (1996), A reference model architecture for design and implementation of intelligent control in large and complex system, International journal of intelligent control and systems, 1(1), 15-30.

[4] Nguyen, D.T, Oh, S.R., and You, B.J. (2005), A framework for internet-based interaction of humans, robots, and responsive environments using agent technology, IEEE transactions on industrial electronics, 52(6),1521-1529.

[5] Wilson, L.F, Burroughs, D.J, Kumar, A, and Sucharitaves, J. (2001), A Framework for Linking Distributed Simulations Using Software Agents, Proceedings of the IEEE, 89(2), 186-200.

[6] Gerkey, B.P. and Mataric, M.J. (2002), Sold!: Auction Methods for Multirobot Coordination, IEEE transactions on robotics and automation, 18(5), 758-768.

[7] Moorehead, S., Simmons, R., and Whittaker, W., (2001), Autonomous exploration using multiple sources of information, in Proc. IEEE Int. Conf. Robot. Autom. (ICRA), 3, 3098–3103.

[8] Burgard, W., Moors, M, Stachniss, C. and Schneider, F.E. (2005), Coordinated Multi-Robot Exploration, IEEE transactions on robotics, 21(3), 376-386.

[9] Choset, H. (2001). Topological simultaneous localization and mapping (SLAM): toward exact localization without explicit localization, IEEE Trans. Robot. Autom., 17(2), 125–137.

[10] Stentz, A. (1994), Optimal and efficient path planning for partially-known environments, in Proc. IEEE Int. Conf. Robot. Autom. (ICRA), 4, 3310–3317.

[11] Billard, A., et al., (2000), A multi-robot system for adaptive exploration of a fast changing environment: probabilistic modeling and experimental study, Connection Science, 11(3/4), 357–377.

[12] Cohen, W., (1996), Adaptive mapping and navigation by teams of simple robots, J. Robot. Autonomous Syst., 18, 411–434.

[13] Koenig, S., Szymanski, B., and Liu, Y., et al., (2001) Efficient and inefficient ant coverage methods, Ann. Math. Artif. Intell., 31, 41–76.

[14] Endres, H. et al., (1998), Field test of a navigation system: autonomous cleaning in supermarkets," in Proc. IEEE Int. Conf. Robot. Autom. (ICRA), 1779–1781.

[15] Simoncelli, M. et al., (2000), Autonomous pool cleaning: self localization and autonomous navigation for cleaning, J. Autonomous Robot., 9(3), 261–270, 2000.

[16] Jäger, M. and Nebel, B., (2002), Dynamic decentralized area partitioning for cooperating cleaning robots, in Proc. IEEE Int. Conf. Robot. Autom. (ICRA), 3577–3582.

[17] Huang, H. et al., (1986), Automatic operation for a robot lawn mower, in SPIE Conf. Mobile Robots, 727, 344–354.

[18] Kuffner, J.J. and LaValle, S.M., (2000), RRT-Connect: An efficient approach to single-query path planning. In *Proc. IEEE Intl Conf. on Robotics and Automation (ICRA'2000)*, 995-1001.

[19] Blaer, P., (2005), Robot path planning using generalized Voronoi diagrams. http://www.cs.columbia.edu/~pblaer/projects/path_planner/voronoi.html

[20] Zelinsky A. et al., (1993), Planning paths of complete coverage of an unstructured environment by a mobile robot, in Proc. Int. Conf. Adv. Robot (ICAR), 533–538.

Motion Diagnosis of a Mobile Platform by Impedance Transduction

Y.L. Kuo[1], J.Y. Lin, Y.N. Lee, and Y.S. Sun

Mechanical and Systems Laboratories, Industrial Technology Research Institute, Hsinchu, Taiwan
[1]ylkuo@itri.org.tw

Abstract. This paper presents a new technology of impedance transduction applied to motion detection of an Omni-directional mobile platform without implementing conventional mechanical sensors. Since the experimental platform is driven by four DC motors, the electro-mechanical transduction of a DC motor is identified by both theoretical derivation and experimental evaluation. Also, the motion modeling of the mobile platform is presented. Based on the evaluation of mechanical impedance, an experimental example demonstrates that the platform has the function of road surface detection, which can be used to be a base of controlling the motion of the platform.

Keywords: electrical/mechanical impedance; transduction matrix; Omni-directional mobile platform.

1. Introduction

Impedance control, developed by Hogan [1] in 1985, is mainly applied to robots and manipulators. It aims at force control without tracking force trajectory but rather to regulate mechanical impedance. This augments the complexity and the implementation cost of entire systems. Anderson and Hagood [2] proposed an approach to obtain the deformation of a cantilever beam based on the electro-mechanical modeling of a piezoelectric actuator, which is coated on the beam. The simultaneous actuation and sensing has successfully predicted and controlled the vibration of the cantilever beam.

Lin *et al.* [3-4] measured the point impedance at the actuating point of a structure by utilizing a so-called piezoelectric inertial actuator (PIA) solely. This method successfully removed the loading effect caused by the shaker without solving the governing equations. Also, the authors [5-6] utilized a PZT-coated cantilever under a decoupling effect as an electro-mechanical transducer. They provided dynamic measurement based on simultaneous sensing and actuating.

A variety of mobile vehicle designs have been developed. For safety of mobile platforms interacting with environments, the implementation of controllers is commonly applied. Conventionally, various sensors to detect position, velocity, and force should be installed on mobile platforms. This paper presents a new technology to achieve the motion diagnosis of an Omni-directional mobile platform by impedance transduction.

The paper is organized as follows: electro-mechanical transduction, theoretical derivation and experimental evaluation of transduction matrix of DC motor, modeling of an Omni-directional mobile platform, motion experiment of mobile platform, and conclusions.

2. Electro-Mechanical Transduction

An actuator can be considered an electro-mechanical transducer. When the loads of mechanical systems change, the signals of electro-mechanical actuators vary. Since the characteristics of actuators are inalterable, it is feasible to evaluate mechanically physical quantities by detecting electrical signals of actuators. This process relies on electro-mechanical transduction of actuators. Based upon this concept, an actuator can simultaneously actuate and sense a mechanical system.

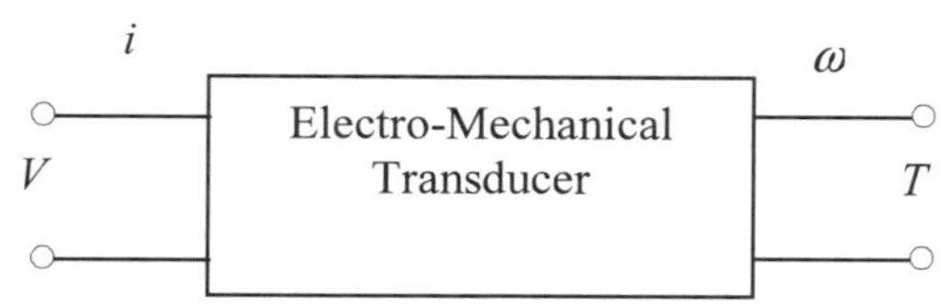

Figure 1.1. Two-port model of electro-mechanical transducer

An electro-mechanical transducer can be considered as a two-port model, if it is linear and reciprocal [7]. Fig. 1.1 illustrates a two-port model of an electro-mechanical transducer. Electrical port of an electrical motor includes voltage (V) and current (i), and its mechanical port composes torque (T) and rotating speed (ϖ). Thus, an electro-mechanical transduction based on frequency response functions (FRFs) can be expressed as

$$\begin{Bmatrix} V \\ i \end{Bmatrix} = [t]\begin{Bmatrix} -T \\ \omega \end{Bmatrix} \tag{1.1}$$

where T has the minus sign, because we define that the mechanical port applies the torque to the transducer; $[t]$ is called transduction matrix [3]:

$$[t] = \begin{bmatrix} t_{11} & t_{12} \\ t_{21} & t_{22} \end{bmatrix} = \begin{bmatrix} \left.\dfrac{V}{-T}\right|_{\omega=0} & \left.\dfrac{V}{\omega}\right|_{T=0} \\ \left.\dfrac{i}{-T}\right|_{\omega=0} & \left.\dfrac{i}{\omega}\right|_{T=0} \end{bmatrix} \tag{1.2}$$

Eq. (1.1) can also present electrical/mechanical impedances, V/i and T/ϖ, which imply characteristics of systems. Thus, impedance can be regarded as a performance index of an electro-mechanical system.

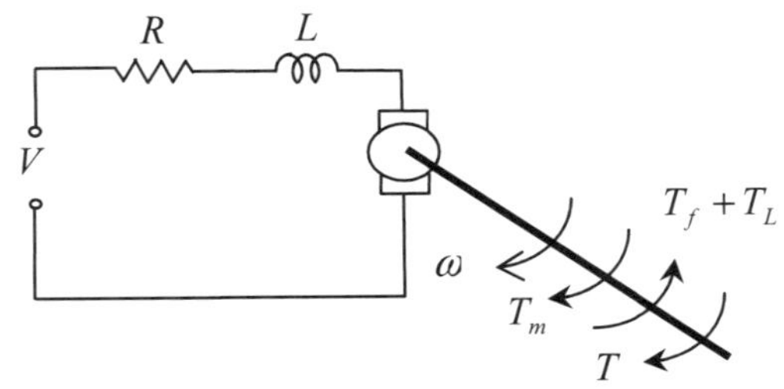

Figure 1.2. Equivalent circuit of DC motor

3. Theorectical Derivation of Transduction Matrix of DC Motor

The derivation is based on an equivalent circuit (Fig. 1.2) and a block diagram (Fig. 1.3) of a DC motor. According to Kirchhoff's and Newton's Laws, the electrical and mechanical governing equations are expressed as [8]

$$V = R \cdot i + L \cdot \frac{di}{dt} + V_{\text{emf}} \tag{1.3}$$

$$T_m = T_f + T_L + J \cdot \frac{d\omega}{dt} + B \cdot \omega \tag{1.4}$$

where L is an equivalent inductor of DC motor; T_m is motor's torque; T_f is static frictional torque; T_L is loading torque; J is moment of inertia of DC motor's rotor; B is dynamic damping coefficient; t is time.

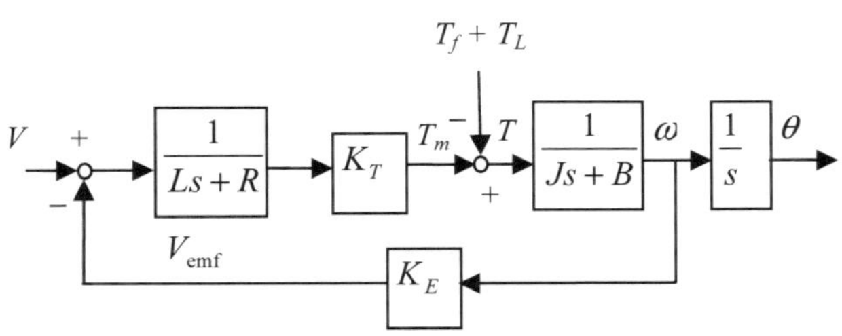

Figure 1.3. Block diagram of equivalent DC motor

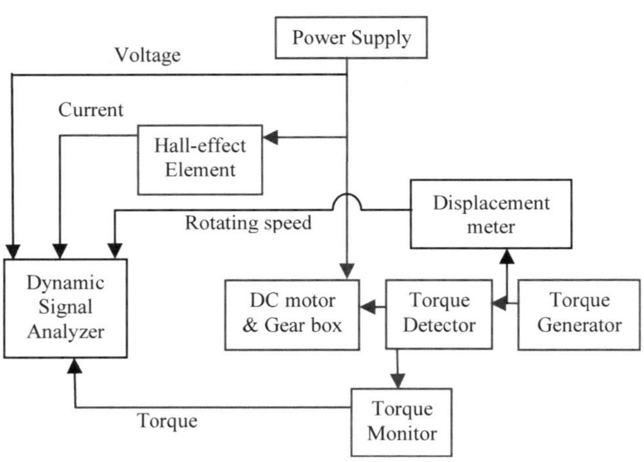

Figure 1.4. Schematic diagram of experimental setup [9]

Also, back emf voltage and motor torque are proportional to rotating speed of motor and current, respectively, as follows [8]

$$V_{\text{emf}} = K_E \cdot \omega \tag{1.5}$$

$$T_m = K_T \cdot i \tag{1.6}$$

where K_E and K_T are back emf constant and torque constant, respectively.

Based on Eqs. (1.1) to (1.5), the transduction matrix of a DC motor can be obtained as follows:

$$[t] = \begin{bmatrix} -\dfrac{Ls+R}{K_T} & K_E \\ -\dfrac{1}{K_T} & 0 \end{bmatrix} \tag{1.7}$$

where s represents the Laplace transform.

To obtain the transduction matrix, all parameters in the equivalent circuit of a DC motor are evaluated first, and then the transduction matrix can be expressed as [9]:

$$[t] = \begin{bmatrix} -0.2854 & 2.204 \\ -0.4537 & 0 \end{bmatrix} \begin{pmatrix} \text{Volt/(Nt}\cdot\text{m)} & \text{Volt}\cdot\text{sec/rad} \\ \text{Amp/(Nt}\cdot\text{m)} & \text{Amp}\cdot\text{sec/rad} \end{pmatrix} \tag{1.8}$$

where the determinant is calculated as

$$\det([t]) = 1.0000 \tag{1.9}$$

4. Experimental Evaluation of Transduction Matrix of DC Motor

The measurement of voltage, current, torque, and rotating speed is performed by the experimental setup illustrated in Fig. 1.4. A fixed voltage provided from a power supply input the DC motor, and a digital signal analyzer recorded all signals. Thus, the transduction matrix of a DC motor can be evaluated by using those experimental data and the least-square method.

According to the experimental results, the transduction matrix is determined as [9]

$$[t] = \begin{bmatrix} -0.4036 & 2.1737 \\ -0.4531 & 0.1348 \end{bmatrix} \begin{pmatrix} \text{Volt/(Nt}\cdot\text{m)} & \text{Volt}\cdot\text{sec/rad} \\ \text{Amp/(Nt}\cdot\text{m)} & \text{Amp}\cdot\text{sec/rad} \end{pmatrix} \tag{1.10}$$

where the determinant is calculated as

$$\det([t]) = 0.9304 \tag{1.11}$$

Comparing the transduction matrices obtained by the theoretical derivation and experimental evaluation, the error of the determinant is 6.96%. The major difference between both matrices is the component t_{22}.

Figure 1.5. Omni-directional mobile platform

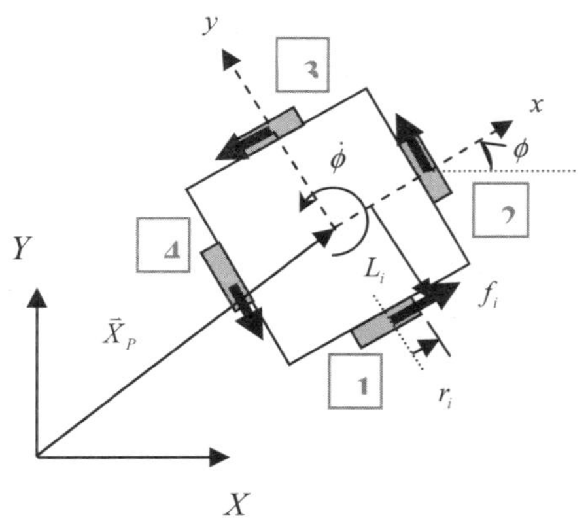

Figure 1.6. Coordinate systems of mobile platform

5. Modeling of an Omni-Directional Mobile Platform

The Omni-directional mobile platform (Fig. 1.5) has four Omni-directional wheels driven by a DC motor. With the translating velocity and the rotating speed at the central point of the platform, the rotating speed of each wheel can be evaluated and given via the driver of the motor and controller of the platform.

5.1 Translating Velocity and Rotating Speed at the Center of the Mobile Platform

The platform is considered as a rigid body with four wheels, as shown in Fig. 1.6. Thus, The translating velocity and rotating speed at the center of the platform are expressed as

$$\begin{Bmatrix} \dot{X}_P \\ \dot{Y}_P \\ \dot{\phi} \end{Bmatrix} = \begin{bmatrix} \cos\phi & -\sin\phi & 0 \\ \sin\phi & \cos\phi & 0 \\ 0 & 0 & 1 \end{bmatrix} \begin{bmatrix} \frac{1}{2} & 0 & -\frac{1}{2} & 0 \\ 0 & \frac{1}{2} & 0 & -\frac{1}{2} \\ \frac{1}{4L_1} & \frac{1}{4L_2} & \frac{1}{4L_3} & \frac{1}{4L_4} \end{bmatrix} \begin{Bmatrix} v_1 \\ v_2 \\ v_3 \\ v_4 \end{Bmatrix} \quad (1.12)$$

And,

$$v_i = \omega_i \cdot r_i \quad (1.13)$$

where L_i is the distance between the centers of the platform and the ith wheel and ϖ and r are rotating speed and radius of the wheel, respectively. The subscript represents the ith Omni-directional wheel.

5.2 Resultant Forces and Moment at the Center of the Mobile Platform

According to Eq. (1.4), the traction force of the ith wheel is expressed as

$$f_i = (T - T_f - J_i \cdot \frac{d\omega_i}{dt} - B_i \cdot \omega_i)/r_i \quad (1.14)$$

And, the resultant forces and moment at the center of the platform are expressed as

$$\begin{Bmatrix} F_X \\ F_Y \\ M \end{Bmatrix} = \begin{bmatrix} \cos\phi & -\sin\phi & 0 \\ \sin\phi & \cos\phi & 0 \\ 0 & 0 & 1 \end{bmatrix} \begin{bmatrix} 1 & 0 & -1 & 0 \\ 0 & 1 & 0 & -1 \\ L_1 & L_2 & L_3 & L_4 \end{bmatrix} \begin{Bmatrix} f_1 \\ f_2 \\ f_3 \\ f_4 \end{Bmatrix} \quad (1.15)$$

5.3 Governing Equations

Based on the velocities and the resultant forces shown in Eqs. (1.12) to (1.15), the governing equations of a rigid-body mobile platform are given as

$$\begin{cases} m\ddot{X}_P = F_X \\ m\ddot{Y}_P = F_Y \\ I\ddot{\phi} = M \end{cases} \quad (1.16)$$

6. Motion Experiment of an Omni-Directional Mobile Platform

In the motion experiment, there are four test cases for the mobile platform with a clockwise constant-speed rotation. The first test case is free rotation under hanging, and the other three cases are the platform rotating on three road surfaces: plastic floor, rubber mat, and carpet.

Based on the four test cases mentioned above, voltage and current signals were measured first, and mechanical impedance was calculated by the transduction matrix and shown in Fig. 1.7.

The results show the rank of mechanical impedance from the largest to the smallest for the four test cases – rubber mat, carpet, plastic floor, and free rotation. Free rotation under hanging can be considered as a case of unloading. It is understood that its mechanical impedance is the smallest. The rubber mat used in the experiment is very soft, so the platform is resistant to movement due to the wheels sunk into the rubber mat. This case has the largest mechanical impedance. This experiment shows that the Omni-directional mobile platform has the function of detecting road surface, and this function can be used a base to control the motion of the platform.

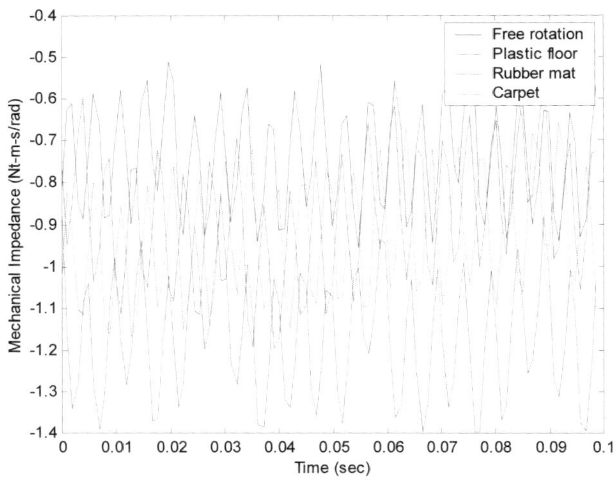

Figure 1.7. Mechanical impedance for four test cases

7. Conclusions

This paper proposed a new technology to evaluate motion of mobile platforms without implementing conventionally mechanical sensors. The technology utilizes the measurement of voltage and current of a DC motor to evaluate torque and rotating speed based on a transduction matrix, which is obtained by both theoretical derivation and experimental evaluation. Both results demonstrate the error of matrix determinant is 6.94%.

The technology was applied to an Omni-directional mobile platform driven by four DC motors, and made the platform having the function of detecting road surface. The idea is to evaluate mechanical impedance of various road surfaces, which can be used as a base of controlling the motion of the mobile platform. The experimental results demonstrate that the technology can accurately evaluate the rotating speeds of the platform and the DC motors.

Acknowledgement

The completion of the research project was sponsored by the Ministry of Economic Affairs, R.O.C. The technology transfer of electro-mechanical transducer from Professor Shih-Fu Ling, Nanyang Technological University, Singapore, was greatly acknowledged. Technical work supported by the researchers and the engineers of Industrial Technology Research Institute, Dr. S.K. Ho, M.C. Chieh, W.F. Hwang, Y.W. Chen, Y.S. Sun, Y.L. Chow, and J.H. Wu, was also thankful.

References

[1] N. Hogan, "Impedance control: An Approach to Manipulation: Parts I, II, and III", ASME Journal of Dynamic Systems, Measurement, Control, Vol. 107, pp. 1-24, 1985.

[2] E.H. Anderson and N.W. Hagood, "Simultaneous Piezoelectric Sensing/Actuation: Analysis and Application to Controlled Structures", Journal of Sound and Vibration, Vol. 174, pp. 617-639, 1994.

[3] S.-F. Ling and Y. Xie, "Monitoring Structural Integrity Using a Piezoelectric Inertial Actuator Cum Sensor", Journal of Sound and Vibration, Vol. 247, pp. 731-737, 2001.

[4] S.-F. Ling and Y. Xie, "Detecting Mechanical Impedance of Structures Using the Sensing Capability of a Piezoceramic Inertial Actuator", Sensors and Actuators A: Physical, Vol. 93, pp. 243-249, 2001.

[5] S.-F. Ling, D. Wang, and B. Lu, "Dynamics of a PZT-Coated Cantilever Utilized as a Transducer for Simultaneous Sensing and Actuating", Smart Materials and Structures, Vol. 14, pp. 1127-1132, 2005.

[6] S.-F. Ling, X. Hou, and Y. Xie, "Decoupling Loading Effect in Simultaneous Sensing and Actuating for Dynamic Measurement", Sensors and Actuators A: Physical, Vol. 120, pp. 257-265, 2005.

[7] E.O. Doebelin, System Modeling and Response: Theoretical and Experimental Approaches, Wiley, New York, NY, USA, 1980.

[8] S.J. Chapman, Electric Machinery Fundamentals, McGraw-Hill, New York, New York, USA, 2005.

[9] Y.L. Kuo, "Sensor-less Impedance Control: Verification of System Dynamic Characteristics", Technical Report, No. 53-3-95-0803, Industrial Technology Research Institute, Taiwan, Dec, 2006.

16

Sensor Technology

in situ Measuring of Temperature and Humidity within the Membrane Electrode Assembly by Micro-Sensors

C.Y. Lee*, S.J. Lee, G.W. Wu

Department of Mechanical Engineering, Fuel Cell Research Center, Yuan Ze University, Taoyuan, Taiwan, R.O.C.
*cylee@saturn.yzu.edu.tw

Abstract. This study investigated the micro array sensors which were fabricated by micro-electro-mechanical-systems (MEMS) fabrication technology for monitoring the temperature and humidity distributions within the membrane electrode assembly (MEA) of fuel cells. The micro array sensors were embedded in parylene substrate, the sensors were made from chromium and gold. The advantages of micro temperature and humidity sensors are their small volume, which enables them to be placed on MEA with high sensitivity and accuracy. The dimensions of the temperature and humidity sensors are 180 μm × 180 μm and 180μm × 220μm, respectively. The operating conditions involve temperatures from 30 to 100°C and the resistance of temperature sensor varied from 31.685 to 36.798 Ω. The measuring results reveal that the temperature is almost linearly related to the resistance and the accuracy and sensitivity are less than 0.3°C and 2.3×10^{-3}/°C. The humidity sensor showed that the capacitance changed from 12.14 to 15.03 pF, the relative humidity from 25 to 95 % RH, and the accuracy and sensitivity is less than 0.25 % RH and 0.04 pF/% RH.

Keywords: MEMS, MEA, Micro array sensors.

1. Introduction

Green energy has been globally promoted in recent years. Fuel cells are the most likely technology to replace the petrochemical fuel technology to become the origin of new energy in the future. However, they still have such problems as low power density, high cost, poor long-term stability and the difficulty of in-situ monitoring of the chemical and physical characteristics within the fuel cell. Water and heat management are key determinants of the performance of the fuel cell [1,2]. The temperature and humidity conditions of a membrane electrode assembly (MEA) determine the performance of fuel cells extremely [3]. MEA has the better performance since its ion conductivity and gas reaction rate increase with temperature. When the MEA is at low humidity, the protons cannot easily cross the MEA. Although the water generated at the cathode can wet MEA, it also blocks the feed gas distribution when water floods the flow channel, and the performance of the fuel cell substantially declines as the gas pressure falls.

The volume of traditional temperature and humidity sensors is too large to allow them to be used to measure the distribution of temperature and humidity in the MEA of fuel cells. Measurements cannot necessarily be made where required. They measure only the temperature and humidity distribution outside the fuel cells and yield results with errors that exceed those of measurements made in MEA. There are several researches in this area to monitor the temperature and humidity distribution of fuel cell [4-8]. Some of them used bigger commercial sensors [9] to monitor the outside of fuel cell and the others used the position of sensors which could only be placed at a gas inlet and outlet, which could not be placed at the position inside the fuel cell. To the best of our knowledge so far, there is no research to monitor both temperature and humidity at the same time in-situ.

In this research, we employed micro-electro-mechanical-systems (MEMS) fabrication technology to fabricate micro array temperature and humidity sensors on parylene substrate [10-12] to monitor the temperature and humidity distribution of fuel cell within MEA in-situ. This fabrication technique has the advantages of (1) mass production, (2) small size, and (3) measure position that could be placed precisely.

2. Methodology and experiments

2.1. The theory of temperature sensor

In this study, the temperature sensor was the resistance temperature detector (RTD). When the environmental temperature increases, the resistance of RTD will also increase. This is because the metal conductor has the physical property of positive temperature coefficient (PTC). If the temperature variation range of RTD is linear, the relationship between the measured resistance and temperature change can be expressed as

$$R_t = R_i(1+\alpha_T\ \Delta T) \tag{1}$$

$$\Delta T = t - t_i \tag{2}$$

where R_t and R_i is the resistance of RTD at t °C and i °C; is positive temperature coefficient of RTD; ΔT is temperature variation with respect to the reference temperature; t and t_i is the temperature of RTD at t °C and i °C. The Eq. (1) can be rearranged as

$$\alpha_T = \frac{R_t - R_i}{R_i(\Delta T)} \tag{3}$$

where α_T is the sensitivity of the temperature sensor (1/°C).

2.2. The theory of humidity sensor

In terms of the difference in the polymer material, the polymer humidity sensors could be divided into resistive-type and capacitive-type. Because the resistive-type sensors will be influenced by temperature, therefore we adopt the capacitive-type humidity sensors in this research, whose electrode structures are interdigitated electrodes. When the polymer material absorbs the water, the dielectric constant will raise with environmental humidity. The variation of capacitance of humidity sensor can be expressed as

$$C = 2\, n\, \varepsilon\, w\, (L-x)/d \tag{4}$$

where C is the capacitance of humidity sensor (F); n is the fingers number of interdigitated electrodes; ε is the dielectric constant of the sensing film; *w and L* is the thickness and length of interdigitated electrode (m); x and d are the distance between two electrodes at different directions (m). The structure of capacitive type humidity sensor with interdigitated electrodes is shown in Figure 1.1 and the sensitivity of the humidity sensor can expressed as Eq. (5)

$$\alpha_H = \frac{\Delta C}{\Delta \%RH} \tag{5}$$

where α_H is the sensitivity of the humidity sensor (*F/%RH*).

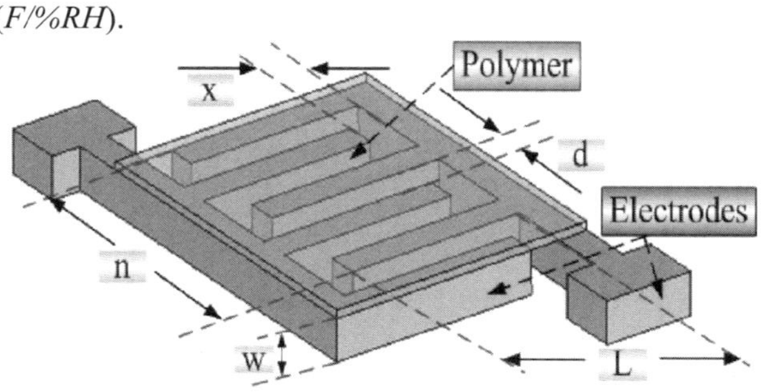

Figure 1.1. The structure of humidity sensor with interdigitated electrodes

3. Design and fabrication process

3.1 Schematic of the entire system

The entire system schematic is shown in Figure 1.2. When a proton passes through an MEA from anode to cathode, the chemical reaction will make the temperature of the MEA rise, and water will be produced at the cathode. For measuring the temperature and humidity distribution of MEA at different flow channel positions in-situ, the micro array temperature and humidity sensors are placed on MEA relative to the feed gas inlet, outlet and the position, which are to be measured precisely. The two poly-para-xylylene (parylene) thin film layers are used to protect the micro array sensors to resist stretching force from MEA.

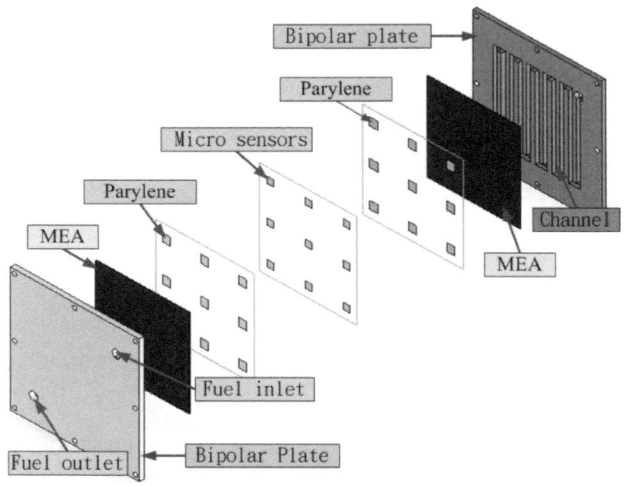

Figure 1.2. Schematic of the entire system

3.2. Fabrication process of micro sensors

Figure 1.3 presents the process flow for fabricating the temperature and humidity sensors on parylene-based. The process starts with a layer of photoresist which is spun on a silicon wafer as a sacrificial layer and deposited a thin film polymer parylene with 1μm thickness as the protective layer (Figure 1.3a). The first photolithograph is employed to define the pattern of the temperature and humidity sensors (Figure 1.3b). The Cr (100 Å) and Au (1000 Å) are deposited in turn on parylene substrate by using e-beam evaporator, the Cr is the adhesion layer between Au and parylene and lift-off technique is applied to form the temperature and humidity sensors (Figure 1.3c).

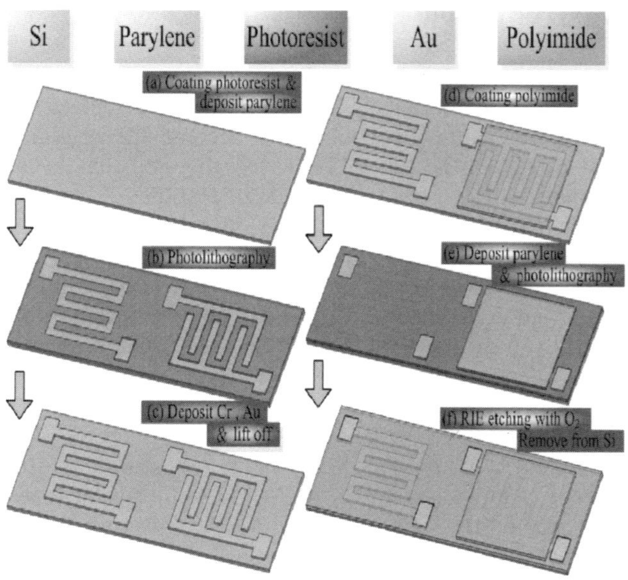

Figure 1.3. Fabrication process of micro sensors

The polyimide was coated on humidity sensors as the sensing film (Figure 1.3d). Another parylene thin film is deposited to embed the micro array sensors completely. The second photolithograph process is utilized to define the region of contact pads and sensing film of micro sensors (Figure 1.3e) for wire-bond with conducting lines and the parylene was etched by oxygen plasma reactive ion etching. Finally, the silicon substrate is segmented by dicing saw and then the micro sensors are peeled form the wafer in a water bath (Figure 1.3f).

4. Results and discussions

In this study the temperature sensor is the gold resistance temperature detector (RTD) and the humidity sensor is the capacitive-type sensor with interdigitated structure electrodes and their optical microscopy photo is shown in Figure 1.4 and Figure 1.5, respectively. The dimensions of the temperature and humidity sensors are 180μm × 180μm and 180 μm × 220 μm, respectively. At first, the micro temperature and humidity sensors are placed in the programmable temperature and humidity chamber to calibrate micro sensors by controlling the temperature and humidity in the chamber (Figure 1.6). The calibration curve of temperature and humidity sensor is shown in Figure 1.7 and Figure 1.8. The range of setting temperature and relative humidity changed from 30 to100°C and 25 to 95% RH, the resistance and capacitance varied from 31.685 to 36.798 Ω and 12.14 to 15.03 pF. The accuracy and sensitivity of temperature and humidity sensor is less than 0.3 °C, 2.3×10^{-3}/°C and 0.25% RH, 0.04 pF/% RH. The accuracy is defined by programmable temperature and humidity chamber.

The standard deviation of temperature-resistance and relative humidity-capacitance can be expressed as

$$S = \sqrt{\frac{1}{n-1}\sum(x_i - \bar{x})^2} \tag{6}$$

where S is the standard deviation, Σ is the sum, x_i is the individual score, $\bar{x}$ is the mean of all scores, and n is the sample size (number of scores). The standard deviation of temperature-resistance and relative humidity-capacitance are 22.22~6.94 and 22.36~0.44, respectively.

The correlation coefficient for temperature and resistance can be obtained with

$$r = \frac{1}{n-1}\sum_{i=1}^{n}\left(\frac{x_i - \bar{x}}{s_X}\right)\left(\frac{y_i - \bar{y}}{s_Y}\right) \tag{7}$$

where r is the correlation coefficient, Σ is the sum, x_i or y_i is the individual score, $\bar{x}$ or $\bar{y}$ is the mean of all scores, s_x or s_y is the standard deviation of x or y, and n is the sample size (number of scores) [13]. The correlation coefficient temperature-resistance and relative humidity-capacitance is 0.999324 and 0.965327 approximately 1 (linear). The experimental results reveal that temperature and relative humidity are almost linearly related to resistance and capacitance.

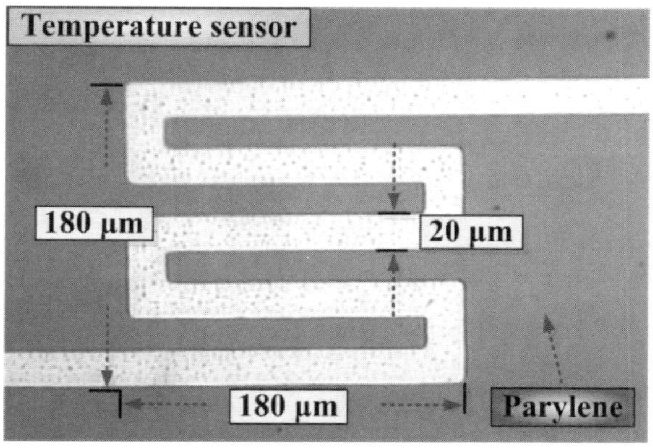

Figure 1.4. Optical microscopic photograph of micro temperature sensor

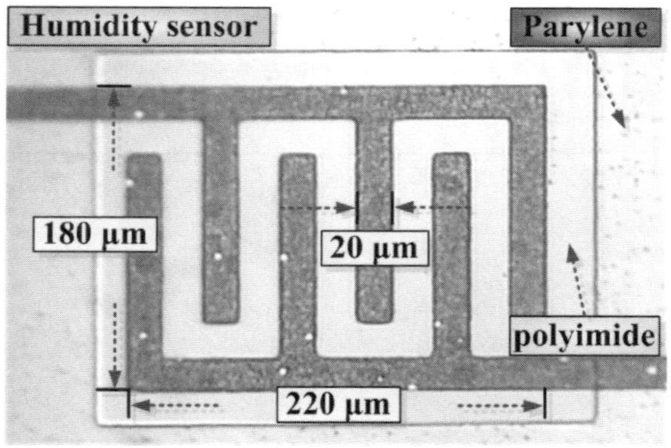

Figure 1.5. Optical microscopic photograph of micro humidity sensor

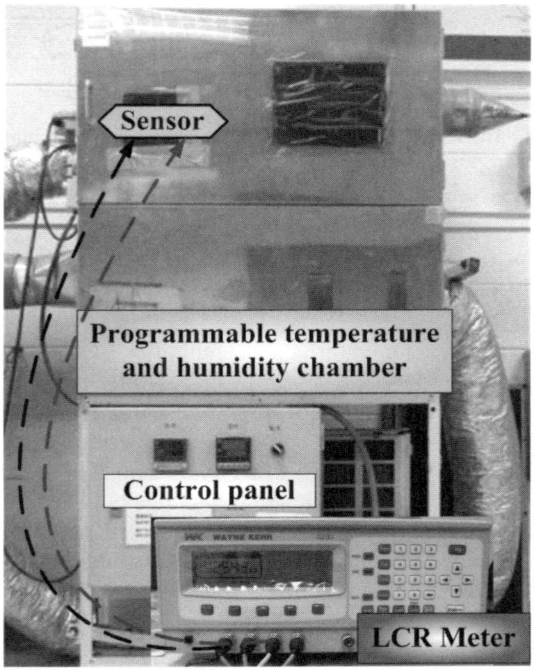

Figure 1.6. Experimental setup for the temperature and humidity sensor calibration

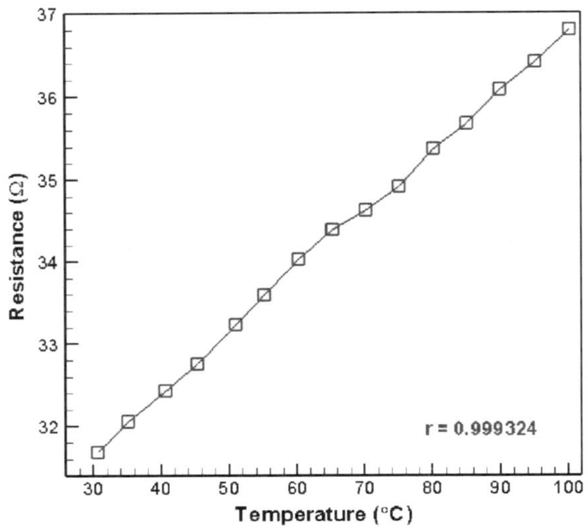

Figure 1.7. Temperature vs. resistance curve for temperature sensor

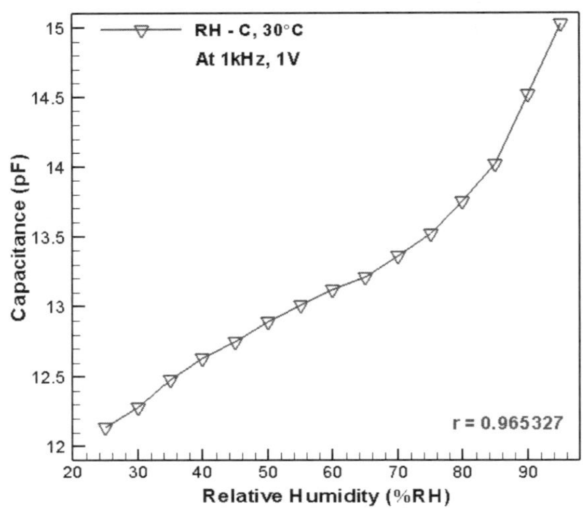

Figure 1.8. Relative humidity vs. capacitance curve for humidity sensor

5. Conclusion

The micro temperature and humidity sensors were fabricated on parylene substrate successfully in this work. We could use micro temperature and humidity sensors to monitor the temperature and humidity distribution of fuel cell within MEA in-situ in the future and this technique could provide optimal operation parameters to fuel cell as the reference to achieve the maximum cell performance, and be used in commercial products of fuel cells.

Acknowledgements

This work was accomplished with much needed support and the authors would like to thank the financial support of this research from the aim for the top university project of Ministry of Education of R. O. C. and YZU Fuel Cell Center through the grant No. 0950026846. The authors also like to thank Professors Lung-Jieh Yang, Shih Hung Chan, Ay Su, Fangbor Weng, Guo Bin Jung for their valuable advice and assistance in experiment. In addition, we would like to thank the TKU MEMS Lab, YZU Fuel Cell Center and NTU NMES Research Center for providing access to their research facilities.

References

[1] J. Larminie, A. Dicks, 2003, Fuel Cell Systems Explained, 2nd Edition, John Wiley and Sons, Ltd., Chichester, United Kingdom.
[2] Cho E.A., Ko J.J., et al: 2003, Characteristics of the PEMFC Repetitively Brought to Temperatures Below 0°C, Journal of Electrochemical Society, Vol.150, PA1667- A1670.
[3] A. Gasteiger, W. Gu, et al, 2003, Beginning-of-life MEA Performance-Efficiency Loss Contribution, in: Handbook of Fuel Cells, Vol.3, Wiley, P593-610.
[4] M.H. Wang, H. Guo, et al, 2003, The First International Conference on Fuel Cell Science, Engineering and Technology, Rochester, NY, USA.
[5] Vie P.J.S., Kjelstrup S.: 2004, Thermal Conductivities From Temperature Profiles in the Polymer Electrolyte Fuel Cell, Electrochim. Acta, Vol.49, P1069-1077.
[6] He S., Mench M.M., et al: 2006, Thin Film Temperature Sensor for Real-Time Measurement of Electrolyte Temperature in a Polymer Electrolyte Fuel Cell, Sensors and Actuators A, Vol.125, P170-177.
[7] M.M. Mench, D.J. Burford, et al, 2003, Proceedings of 2003 ASME International Mechanical Engineering Congress, Heat Transfer Division, Washington, DC, USA.
[8] Burford D.J.: 2004, Real-time Eelectrolyte Temperature Measurement in an Operating Polymer Electrolyte Membrane Fuel Cell, Master thesis, The Pennsylvania State University, University Park.
[9] Nishikawa H., Kurihara R., et al: 2006, Measurements of Humidity and Current Distribution in a PEFC, Journal of Power Sources, Vol.155, P213-218.
[10] Fan Z., Engel J. M., et al: 2004, Parylene Surface-Micromachined Membranes for Sensor Applications, Journal of Microelectromechniacl Systems, Vol.13, P484-490.
[11] Huang A., Lew J., et al: 2004, Microsensors and Actuators for Macrofluidic Control, IEEE Sens. J., Vol.4, P494-502.
[12] Satyanarayana S., McCormick D.T., et al: 2006, Parylene Micro Membrane Capacitive Array Sensor for Chemical and Biological Sensing, Sensors and Actuators B, Vol.115, P494-502.
[13] E. W. Weisstein, 2002, CRC Concise Encyclopedia of Mathematics, Second Edition, Chapman & Hall/CRC.

Application of Micro-sensors on Diagnosis of Micro-reformer

Chi-Yuan Lee[1*], Shuo-Jen Lee[1], Chin-Hua Wu[1], Chien-Heng Lin[1], Ming-Der Ger[2]

[1]Department of Mechanical Engineering, Fuel Cell Research Center, Yuan Ze University, Taoyuan, Taiwan, R.O.C., *cylee@saturn.yzu.edu.tw
[2]Department of Applied Chemistry & Materials Science, Chung Cheng Institute of Technology, National Defense University, Taoyuan, Taiwan, R.O.C.

Abstract. With advances in micro fuel cell development, the production of hydrogen for micro reformer has become increasingly important. However, some problems regarding the micro reformer are yet to be resolved. These include reducing the size, reducing the quantity of CO and combining the fuel cell, among others. Accordingly, in this investigation, array micro temperature sensors and heaters are combined within a silicon-based micro reformer to measure and control the temperature and thus improve performance and minimize the concentration of CO. In this work, micro-electro-mechanical-systems (MEMS) of the micro channel type are fabricated on a silicon substrate to enhance the methanol conversion ratio. Array micro temperature sensors and heaters are made of platinum and placed inside the micro reformer. Although the micro temperature sensor and heater have already been used to measure and control temperature in numerous fields, they have not been employed in micro reformer and commercial products. Therefore, this study presents a new approach for combining array micro temperature sensors and heaters within a silicon-based micro reformer to minimize the size and improve performance.

Keywords: Micro reformer, Array micro sensors.

1. Introduction

Proton exchange membrane fuel cells (PEMFC) use hydrogen and oxygen as fuel. The oxygen comes from the atmosphere. There are two kinds of methods for producing hydrogen. The first involves the refinement of fossil fuel by steam reforming, partial oxidation, coal gasification and so on. The second is the manufacture of hydrogen not from fossil fuels. The main examples include water splitting, photoelectrochemical water splitting and a few other hydrogen fabrication methods. Since obtaining fossil fuel is easy and the reforming reaction enables hydrogen to be produced, hydrogen can be used in fuel cells to produce electricity. The use of the reformer in fuel cells has led to its popularity in producing hydrogen.

The reaction between hydrocarbon and water is endothermic, meaning that the reactant must be heated for the reaction to proceed. The reaction produces more hydrogen as the temperature is increased. Some oxygen is typically fed to make the reaction exothermic but heat is still required to begin the reaction. Accordingly, a heater must be utilized to supply heat. In this investigation, micro heaters and temperature sensors are employed to control the temperature precisely to save energy and increase conversion efficiency.

Reformers with two structures are adopted of reformers herein. The first is a cylinder-type structure [1,2] and palladium (Pd) alloy is used to purify hydrogen directly. The second is a planar structure [3,4], with much reduced volume. Hence, the planar reformer design was selected in this study.

The silicon wafer is used to fabricate a flow field plate and the porous structure is combined with the Pd film [5,6]. The porous structure serves as both the gas diffusion layer and the supporting layer. The Pd film is deposited on a porous structure to ensure that the hydrogen is pure. Since 1956, hydrofluoric acid (HF) has been used as an etching liquid in electrochemical etching related research has been performed for several years. In particular, in 1990, Lehmann [7] described in detail the characteristics of porous silicon.

2. Theory and experiments

The main components of a reformer are the flow field plate, the porous silicon, the micro platinum temperature sensors, the micro heaters and the Pd film. The micro fuel cell will be integrated with the micro reformer to develop a light and portable power system, with the planar structure and which produces high hydrogen by the steam reforming method.

2.1. Fabrication of flow field plate and porous silicon

The serpentine flow field plate is used to form the gas channel. This optimal channel design outperforms parallel and mesh flow fields [8]. The flow field plate was of the parallel-type. To discuss the channel-type is related with reformer.

A p-doped, double-side polished (100) silicon wafer with a thickness of 525±25 μm was used initially. Low pressure chemical vapour deposition (LPCVD) was employed to deposit Si_3N_4 with a thickness of 5000 Å on one side of the silicon wafer as the flow field plate. Photolithography was utilized to define the pattern and by reactive ion etching (RIE) to transform the pattern. Then, a 45wt% KOH solution was used in the wet-etching fabrication process. In controlling the etching time, the etching temperature and the stirring of the solution, the depth and width of flow channel is 250 μm and 500 μm respectively after etching and the remaining 270 μm thickness of the silicon wafer: was regarded as the gas diffusion layer and the supporting layer.

The silicon wafer required an anodic potential to enable electrochemical etching to produce the porous silicon in the HF etching solution. The current density and the ion doping concentration in the silicon wafer greatly influence the size of the etched hole. In this work, porous silicon is used to etch nano-level hole, because of the need to support the Pd film.

Figure 1.1 presents the experimental equipment used to fabricate porous silicon. The dry-etching was employed to remove nitride from one side of the silicon wafer and then the silicon wafer was placed in the experimental equipment. The process was operated at a constant voltage to etch porous silicon to a depth of 50 μm. The etching solution was a mixture of HF, ethanol and water. HF acid was not used directly to produce the hole to prevent the leakage of dangerous HF after the etching solution is passed through the silicon.

Etching compensation was utilized to develop the passing through characteristics of the porous silicon and the channel. The holder, which functions as an etching mask, was adopted to protect the porous silicon that had been formed and prevent ongoing etching by KOH. Figure 1.2 displays the micro channel. The etching time of KOH had to be controlled precisely during etching compensation. Figure 1.3 shows an SEM photograph of the finally produced porous silicon. Porous silicon can act as the gas diffusion layer and a supporting layer in the Pd film, as shown in Figure 1.4.

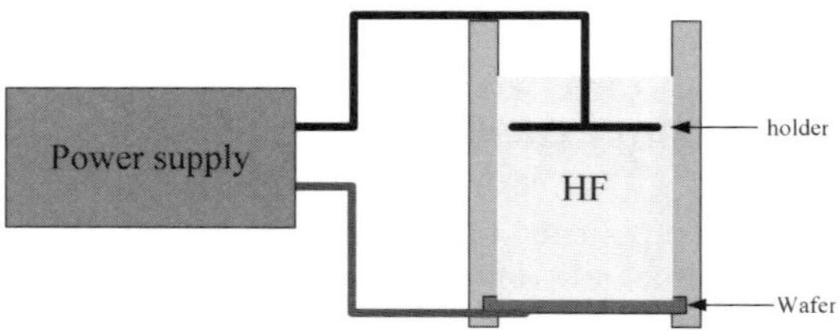

Figure 1.1. Experimental equipment used to etch porous silicon

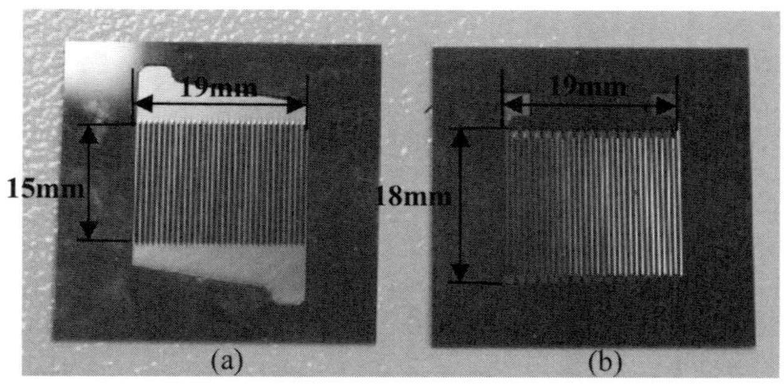

Figure 1.2. Silicon wafer etching of KOH. (a) parallel-type flow field (b) serpentine flow field

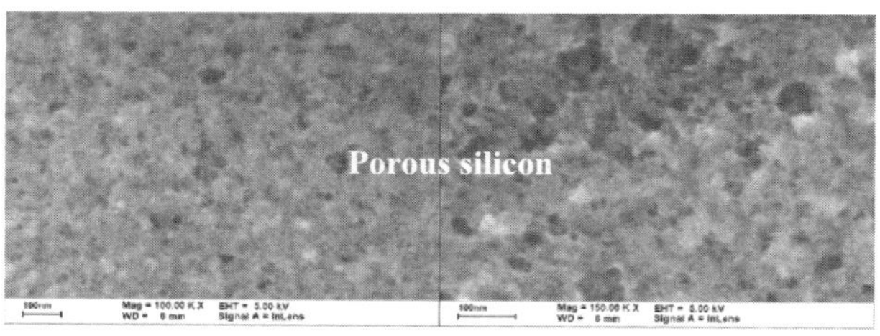

Figure 1.3. Top and section SEM photographs of porous silicon

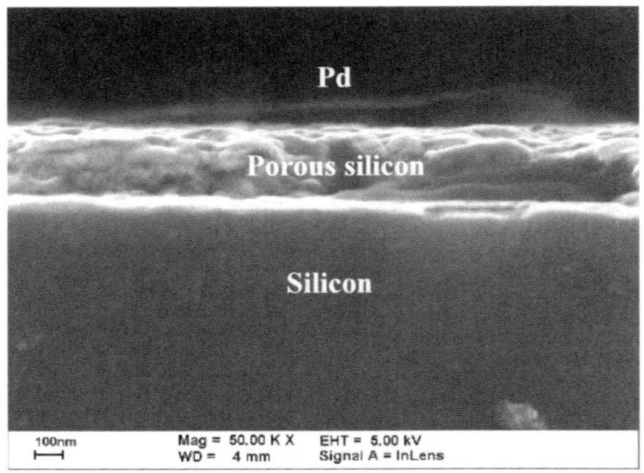

Figure 1.4. SEM photographs of Pd film

2.2. Theory of temperature sensor and heater

2.2.1. Temperature sensors and heaters

The general temperature sensor has four main parts: a platinum resistance temperature detector (PRTD), a thermal sensitive resistor (thermistor), a thermocouple, and a mercury-in-glass thermometer. The non-contact-type can be divided into a thermal detector and a quantum detector infrared receiver temperature sensor. Generally speaking, the volume of the traditional temperature sensor is relatively larger and unable to be applied in the microchannel. The measured position may not necessarily be at the place to be measured. The fluid flow would be influenced by the size of the sensor, which is relatively larger than the microchannel [9].

The micro heater consists of mainly the resistive heating wire and the conducting electrodes. Platinum and poly-silicon are typically selected as the resistance material. Platinum has been chosen as the conducting wire between the conducting electrodes [10]. Platinum was utilized herein to fabricate the PRTD and heater directly. The PRTD and heater are in common use. Accordingly, the micro-electro-mechanical-systems (MEMS) fabrication technique was applied herein to fabricate the micro platinum temperature sensors and heaters to measure and control the temperature within the micro reformer.

2.2.2. Characteristic of platinum

Platinum (Pt) is a soft and silvery-white metal, which is extremely malleable. Its resistance changes linearly in a wide range of temperature with resistance value and the measure range widely of temperature (-260 to 1000°C). Even though the surrounding environment temperature is over 1000 °C, it still keeps its steady characteristic and doesn't undergo any significant physical or chemical changes. The error range can get up to the high accuracy of ±0.06% (or ±0.15°C) at 0°C. Platinum cannot be etched in strong acid or alkali. Therefore, it is very suitable as the material of choice for the temperature sensor.

2.2.3. Theory of temperature sensor and heater

The resistance of a general metal is expressed as equation (1)

$$R = \rho \frac{L}{A} \quad (1)$$

where R is the resistance (Ω); ρ is the resistivity (Ω-m); L is the length of wire (m); A is the cross-section area (m^2). The resistivity of platinum is 1.042×10^{-9} Ω-m at room temperature. When it is used in the thin film temperature sensor, its temperature coefficient of resistance would differ with different thin film thicknesses and the value would range from 0.00375~0.00385. The resistance of a heater directly converts the energy associated with electrical current in the workpiece into heat [11].

The standard deviation of temperature and resistance can be expressed as equation (2)

$$S = \sqrt{\frac{1}{n-1}\sum(x_i - \bar{x})^2} \quad (2)$$

where S is the standard deviation, Σ is the sum, x_i is the individual score, $\bar{x}$ is the mean of all scores, and n is the sample size (number of scores).

The correlation coefficient for temperature and resistance can be obtained with equation (3)

$$r = \frac{1}{n-1}\sum_{i=1}^{n}\left(\frac{x_i - \bar{x}}{s_X}\right)\left(\frac{y_i - \bar{y}}{s_Y}\right) \quad (3)$$

where r is the correlation coefficient, Σ is the sum, x_i, y_i are the individual score, $\bar{x}$, $\bar{y}$ is the mean of all scores, S_x, Sy are the standard deviation of x and y, and n is sample size (number of scores).

2.3. Fabrication of micro temperature sensor and heater

The micro temperature sensor and heater were fabricated by the procedure that is illustrated in Figure 1.5. The micro temperature sensor and the heater both have a core platinum sensor. In both wet etching and lift-off methods, the insulating material must first be deposited as the insulating layer. Nano facility center of national chiao-tung university was entrusted to grow SiO_2 to a thickness of 1000 Å. An e-beam evaporator produces a high temperature and degrades the photoresist, causing the lift-off method to fail. Therefore, the aqua regia etching method was adopted herein to define the pattern.

An e-beam evaporator was employed to evaporate titanium (100 Å) as the adhesion layer and then evaporate platinum (700 Å). Photolithography was conducted to define the pattern of the micro temperature sensors and the heaters. The platinum was etched by aqua regia at an etching temperature of 75°C. It was then soaked in the acetone to remove the photoresist to finish the micro PRTD and the heater. Figure 1.6 displays the micro temperature sensors and the heaters, with dimensions 42 mm × 42 mm.

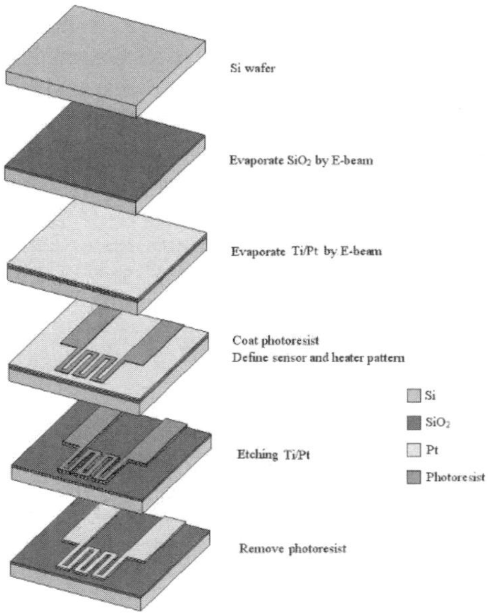

Figure 1.5. Fabrication process flowchart for micro temperature sensor and heater

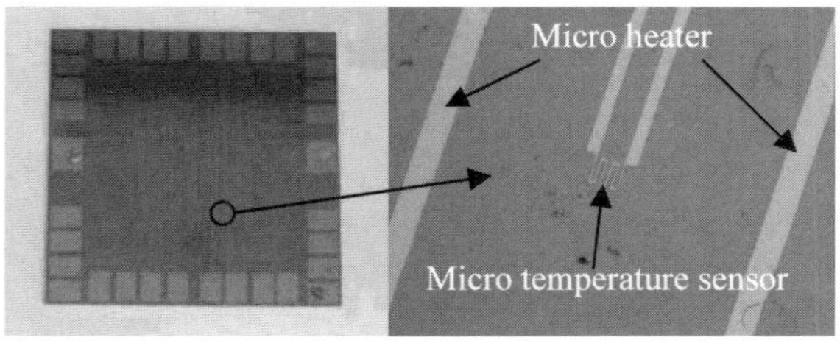

Figure 1.6. Micro temperature sensor and heater

2.4. Fabrication of Pd film and related theory

Hydrogen passes through the Pd film in four steps, described in the solution-diffusion model as follows.

(1) H_2 diffused to the surface of the Pd film to begin reversible dissociative chemisorption, allowing the H_2 molecules to dissociate into the H atom;
(2) Reversible dissolution of H atom into Pd film;
(3) Diffusion of H atom within Pd film;
(4) Recombination of H atom into H_2 on Pd film surface to start reversible fall off.

The Pd film was deposited on a porous structure to ensure proper functionality and required efficiency. Hence, physical vapor deposition (PVD) was utilized to deposit the Pd film (300 Å) on porous silicon to purify hydrogen.

3. Results and discussions

The MEMS technique was employed to fabricate a flow field plate and porous silicon on a silicon wafer. Porous silicon with nano-holes. The Pd film was evaporated on a porous silicon structure at low temperature with low stress using the e-beam evaporator. The electrodes of the micro temperature sensors and the heaters all used platinum to reduce cost and increase convenience. Figure 1.7 shows the package diagram. The micro temperature sensors and the heaters are integrated with the reformer to reduce the volume. The temperature sensor detected temperatures from 32.6°C to 79.9°C and the resistance varied from 8627.3 Ω to 9077.1 Ω. The correlation coefficient was 0.99676 approximately 1 (linear). The experimental results demonstrate that the relationship between temperature and resistance was linear (Figure 1.8) and the accuracy and sensitivity were less than 0.3°C and 1.102×10^{-3}/°C, respectively.

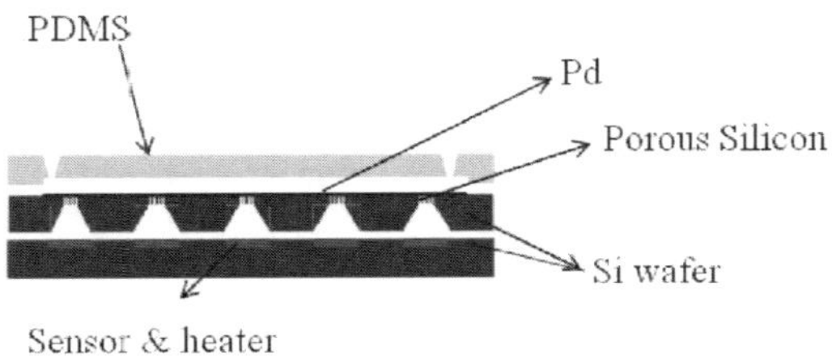

Figure 1.7. Package diagram

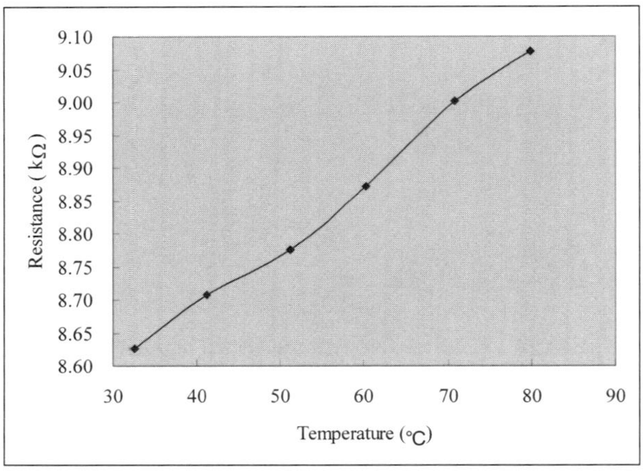

Figure 1.8. The temperature – resistance curve

4. Conclusions

Micro temperature sensors and heaters were successfully integrated with the micro methanol reformer. Using micro temperature sensors and heaters in a micro reformer will yield operating parameter as references and improve the performance. When a Pd film is deposited on porous silicon, the purity of the hydrogen that is produced by the micro methanol reformer is increased. The micro reformer will eventually be able to be directly combined with micro fuel cells.

Acknowledgements

This work was accomplished with much needed support and the authors would like to thank the financial support of this research from the aim for the top university project of Ministry of Education of R. O. C. and YZU Fuel Cell Center through the grant No. 0950026846. The authors also like to thank Professors Shih Hung Chan, Chuin Tih Yeh, Kuen Song Lin, Y. C. Liu, S. D. Lin, Ay Su, Fangbor Weng, Guo Bin Jung of the Yuan Ze University for their valuable advice and assistance in experiment. In addition, we would like to thank the YZU Fuel Cell Center and NTU NMES Research Center for providing access to their research facilities.

References

[1] Howard B.H., Killmeyer R.P.: 2004, Hydrogen Permeance of Palladium-copper Alloy Membrance Over a Wide Range of Temperature and Pressure, Journal of Membrane Science, Vol.241, P207-218.

[2] Han J., Kim I. S.: 2002, High Purity Hydrogen Generator for On-Site Hydrogen Production, International Journal of Hydrogen Energy, Vol.27, P1043-1047.

[3] Kawamura Y., Ogura N.: 2006, A Miniaturized Methanol Reformer with Si-based Microreactor for a Small PEMFC, Chemical Engineering Science, Vol.61, P1092-1101.

[4] D.E. Park, T.K. Kim, 2006, Micromachined Methanol Steam Reforming System Integrated with Catalytic Combustor Using Carbon Nanotubes as Catalyst Supports, The 4th International Conference on Fuel Cell Science Engineering and Technology, Irvine, CA, June.

[5] Ryi S.K., Park J.S.: 2006, Characterization of Pd-Cu-Ni Ternary Alloy Membrance Prepared by Magnetron Sputtering and Cu-reflow on Porous Nickel Support for Hydrogen Separation, Separation and Purification Technology, Vol.50, P82-91.

[6] S.Y. Ye, S. Tanaka, 2005, Thin Palladium Membrane Microreactor with Prous Silicon Support and Their Application in Hydrogenation Reaction, The 13th International Conference on Solid-State Sensors, Actuators and Microsystems, Seoul, Korea, June. P2078-2082.

[7] Lehmann V., Gösele U.: 1991, Porous Silicon Formation: A Quantum Wire Effect, Appl. Phys. Lett., Vol.58, P856-858.

[8] H. Gregor, 2003, Fuel Cell Technology Handbook, CRC Press, New York.

[9] Yang W.R.: 2001, The Modified Fabrication of the On-chip Measurement of Temperature for Microchannels, master's thesis, Tamkang University Department of Mechanical and Electro-Mechanical engineering.

[10] S.J. Chang, 2004, Design Fabrication and Characterization of Micro Probes Integrated with Micro Heaters and Their Applications on Nano-Machining, master's thesis, National Cheng Kung University Department of Engineering Science.

[11] C.F. Huang, 2004, Fabrications of Micro Heaters and Micro Sensors, master's thesis, National Sun Yat-Sen University Department of Mechanical and Electro-mechanical engineering.

16–3

The Development of a Diffraction Optoelectronic Inclinometer

Wenyuh Jywe[1,a], Chun-Jen Chen[2,b] Jian-Ming Lai[3,c] Hsin-Hung Jwo[1,d]

[1]Department of Automation Engineering, National Formosa University, Huwei, Taiwan ROC
[2]Department of Mechanical Engineering, National Cheng-Kung University, Tainan, Taiwan ROC
[3]Institute of Electro-Optical and Materials Science, National Formosa University, Huwei, Taiwan ROC
[a]jywe@nfu.edu.tw, [b]n1892130@ccmail.ncku.edu.tw, [c]mainboy77@hotmail.com, [d]schong@nfu. edu.tw

Abstract. Conventional levels are employed to measure just one dimensional inclinations each time, and are frequently used to measure inclination, straightness and flatness. There are two disadvantages when such levels are used to measure two dimensional inclinations (such as flatness): (1) two measurement steps are needed to measure two orthogonal directions in flatness; and (2) the two directions cannot be perfectly perpendicular to each other because of setting errors. This paper describes the development of a diffraction optoelectronic level which employs 1 pendulum, 1 1D reflection grating, 1 reflection mirror, 1 2D PSD and 1 laser diode. In this system, the laser light projected onto the reflection grating which is fixed on the pendulum, and the +1 order diffraction light ray of reflection grating was reflected to 2 dimensional position sensing detector (2D PSD) by the first surface mirror. The laser diode, first surface mirror, 2D PSD and pendulum are fixed on a fixture. Thus, the inclination measurement result can be obtained from the output of the PSD. A novel system was established and proven in experiments. The measurement range of rotation along x-axis and rotation along y-axis are ±4000 arc sec and ±4000 arc sec respectively. The accuracy of them are about ±1.2 arc sec and ±1.0 arc sec respectively.

Keywords: optoelectronic level, reflection grating, 2D inclination measurement

1. Introduction

Angle measurement is very important in the measurement of precision machines. The conventional instruments are the rotary encoder, laser interferometer, autocollimator and precision level. A rotary encoder [1] is commonly used in indexing measurement in a rotary machine, e.g. the spindle of machine tools and the indexing of a ball screw. However, the rotary encoder does not suit inclination measurement and geometric errors in machine tools. The laser interferometer [2] can be used to measure small angles and machine tools' geometric errors, but its setting up is difficult and it can measure only one kind of geometric error each time. It requires a great deal of time for all a machine tools geometric errors to be measured. The autocollimator [3] is frequently used to measure small angle and machine tools' geometric errors, and it can be applied to two dimensional angle measurement (pitch error and yaw error), but its disadvantage is that it is difficult to set up.

The conventional inclination measurement system uses a precision level [4], which can be used for inclination measurement and machine tool geometric error measurement. Conventional precision levels include precision vials, electrolyte-vials, inductive measuring systems, capacitance measuring systems and optoelectronic measurement systems. Capacitance measuring system is the commonest precision level in precision machines, and is based on the pendulum properties of a friction-free supported disc. A frequency is supplied to two ceramic- and copper- plated electrodes, which together with the pendulum disc supported in the shielded and dust proof gap between them, builds a differential capacitor made of 2 ceramic- and copper-plated electrodes, which delivers the angular signal. This unit is encapsulated and protected from outside electronic influence. The laser interferometer and autocollimator are frequently used to measure inclination, but they can only be used for relative inclination measurement. The precision level can be used to measure absolute inclination, because it is based on gravity.

The advantages of capacitance levels are high resolution and easy setup. The capacitance level can measure one inclination with each measurement. When using a precision level to measure flatness, the level must be rotated after each measurement. Rotating the level can easily cause errors, and the measurement directions of before rotating and after rotating will not be orthogonal. Therefore, a two dimensional inclination measurement system needs to be developed.

In recent research, angular measuring techniques have focused on interferometric techniques. In 1992, Huang et al. [5] developed a small angle measurement system which was based on the internal reflection effect in glass boundary and interferometric techniques. In Huang's system, the resolution was 0.2 arc sec and the measuring range was 3 arc sec. In 1996, Xiaoli et al. [6] established a two-dimensional small rotation angle measurement system by using two different parallel interference patterns (PIP) that were orthogonal to each other. The standard deviation of Xiaoli's system was 0.6 arc sec. In the following year Xiaoli et al. [7] improved their system so that its resolution was 0.2 arc sec and measuring range was ±5 arc- min. In 1997, Chiu et al.[8] established a modified angle measurement technique, with a resolution of 1.6 micro rad, and a measuring range of ±5.6 deg. At its optimum performance, the system's resolution was 8×10^{-5} deg. In 1998, Zhou and Cai [9] established an angle measurement technique which was based on the total-internal reflection effect and heterodyne interferometry. The system resolution was better than 0.3 arc sec, depending on the refractive index selected. In 1998, Huang et al.[10] established a method of angle measurement, based on the internal reflection effects, that used a single right-angle prism. They demonstrated that angle measurement with a range of ±500 arc min, a nonlinearity error of ±0.1%, and a resolution of 0.1 arc sec could be readily achieved. In 1999, Guo et al. [11] developed an optical method for small angle measurement based on surface-plasma resonance (SPR), and a measurement resolution of 0.2 arc sec was achieved experimentally. In 2003, Ge and Makeda [12] developed an angle-measurement technique based on fringe analysis for phase-measuring profilometry. The measurement range was ±2160 arc sec and that the deviation from linearity was better than ±0.02 arc sec. In 2004, Chiu et al. [13] developed an instrument for measuring small angles by use of multiple total internal reflections in heterodyne interferometry, and the angular resolution was better than 2.2×10^{-6} rad over the measurement range $-2.12° \leqq \theta \leqq 2.12°$ for 20 total-internal reflections.

Conventional angle measurement technique researches focuses on one-dimensional inclination measurement and interferometric angle measurement. Two-dimensional measurement also focuses on interferometric techniques, but interferometric systems are expensive and complex, it cannot be used extensively in industry.

This paper establishes a novel optoelectronic level based on total internal reflection in two reflection mirrors and using a friction-free pendulum. This system consists of a laser diode, reflection mirrors, position sensing detector and pendulum. This system can measure two dimensional inclinations at a time.

2. The measurement system

This system consists of 1 laser diode, 2 reflection mirrors, 1 pendulum, 1 2D PSD (UDT SC-10D), grating , 1 PSD signal processor (On-Trak OT-301), A/D card (Advantech PCI-1716 16-bit, maximum sampling frequency 250 kHz) and a PC. Figure 1.1 is the schematic diagram of our presented system. The laser diode projected the laser light onto reflection grating, which was fixed on a pendulum. The reflection grating produced +1 and -1 order diffraction light, and one of the diffraction lights was projected onto the reflection mirror, which reflected the diffraction light to the PSD. The laser diode, pendulum, mirror and PSD were fixed on one fixture. Therefore, the two dimensional inclination measurement results can be obtained from the output of the PSD.

The PSD output was a voltage-signal, which had to be transformed to an angle by calibration. Figure 1.2 is the photograph of the system calibration. In Figure 1.2, the metal plate was the bottom layer, the goniometer was the second layer, the plate was the third layer, and the measurement system and one capacitance level was the top layer. The goniometer was a manual adjustment device for two dimensional inclinations. Its adjustment range was ±20 deg, and its angle resolution was 1 arc min. The plate assisted in fixing the measuring system on the goniometer.

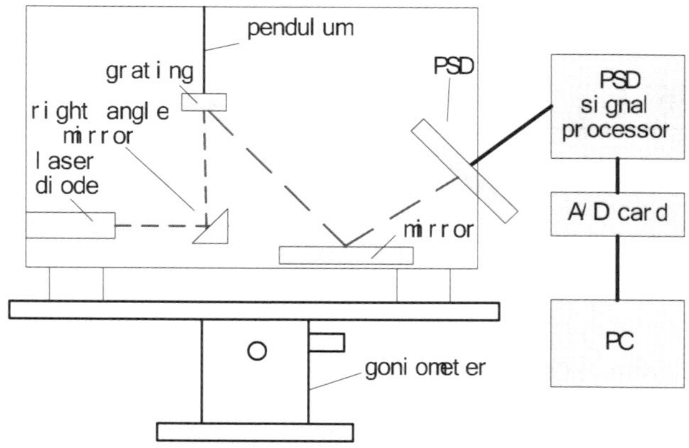

Figure 1.1. Schematic diagram of optoelectronic level

In the procedure of system calibration, the goniometer was mounted on a metal plate after resetting the goniometer. Another metal plate was placed above the goniometer and both the measuring system and the capacitance level were placed above the plate. After setting up the measuring system, the PSD signal wire was connected to the PSD signal processor, and the PSD signal processor output was connected to a PC. After attaching the signal wires, the laser light point on the 2D PSD was moved to the 2D PSD center by adjusting the measuring system fixture in order to ensure that the system had its maximum measuring range.

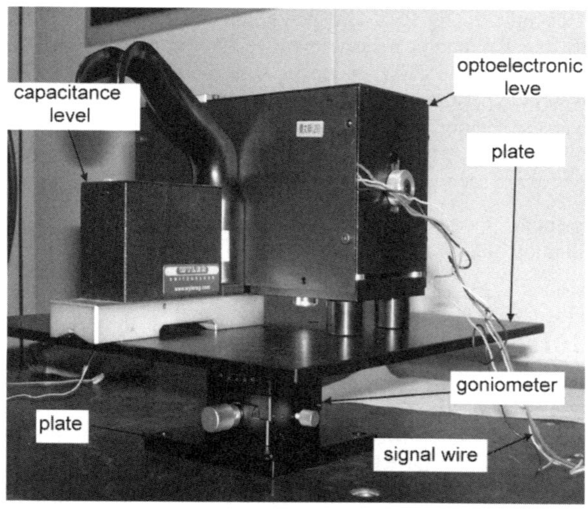

Figure 1.2. Photograph ofoptoelectronic level

After setting up the calibration platform and measuring system, the first calibration was implemented. First, the goniometer was adjusted along the +A-axis (rotated along the x-axis), and the PSD signal recorded every 20 arc min. The goniometer was adjusted along the – A-axis and the previous tasks were repeated until the laser point reached the PSD boundary. When the laser point reached the PSD boundary, the goniometer was reset. After the A-axis calibration, the goniometer was adjusted along the +B-axis (rotated along the y-axis), and the PSD signal was adjusted every 20 arc min. The goniometer was adjusted along the – B-axis and the previous tasks repeated until the laser point reached the PSD boundary. In order to check the repeatability of the system, the calibration was implemented three times.

3. Experimental results and discussion

The resolution of the A/D card described in this paper was 16 bit and the measuring range was ±10 V. The 2D PSD measurement area was 100 mm^2.

The calibration results of the A-axis are shown in Figure 1.3(a), and the standard deviation during A-axis calibration is shown as Figure 1.3(b). The results of the B-axis calibration are shown in Figure 1.3(c), and the standard deviation during B-axis calibration is shown in Figure 1.3(d). The measuring range of the A-axis was ±2000 arc sec, the resolution of the A-axis was 0.13 arc sec, and the accuracy was 1.5 arc sec. The measuring range of the B-axis was ±4000 arc sec, the resolution of the B-axis was 0.15 arc sec, and the accuracy was 3 arc sec.

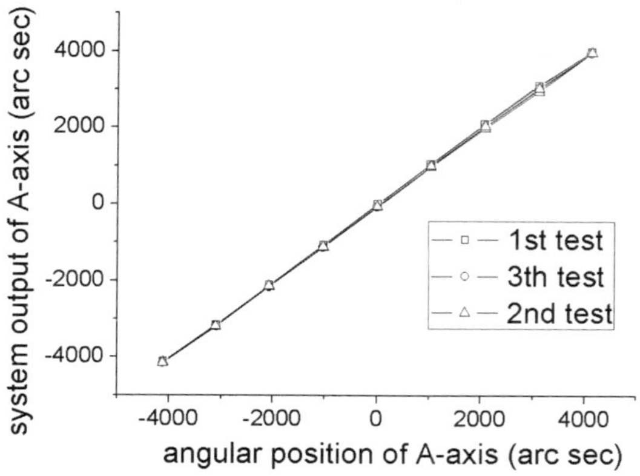

(a)

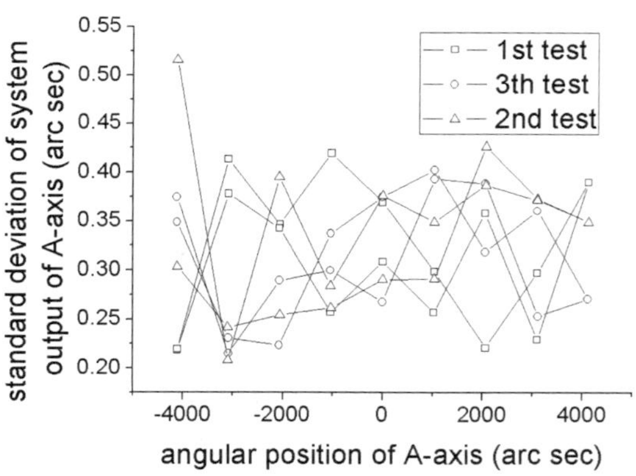

(b)

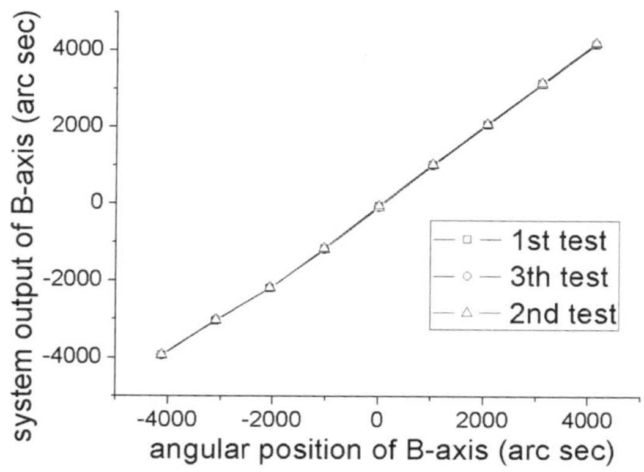

(c)

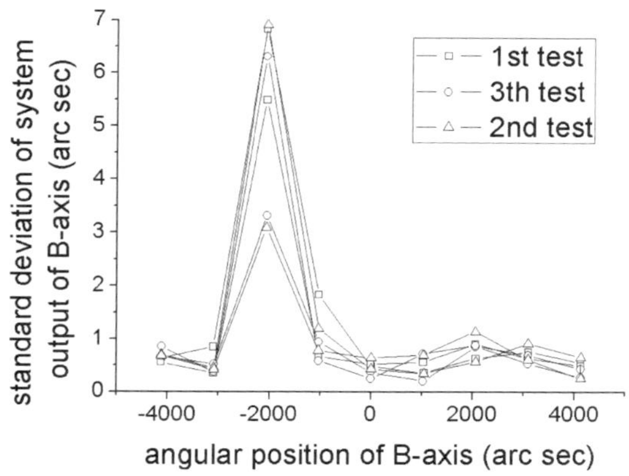

(d)

Figure 1.3. Results of system calibration (a) PSD signal of A-axis rotation, (b) standard deviation during A-axis rotation, (c) PSD signal of B-axis rotation, (d) standard deviation during B-axis rotation

4. Conclusion

This paper establishes a novel optoelectronic level, which is able to measure simultaneously two inclinations in two orthogonal directions. The system calibration and stability tests were completed. The measuring range of the A-axis was ±1.3 deg, and the measuring range of the B-axis was ±1.7 deg. This measuring range was larger than that of a laser interferometer and autocollimator. The resolution of the A-axis was 0.13 arc sec and the resolution of the B-axis was 0.15 arc sec. This system can be used in inclination measurement, and can be used to measure the straightness, parallelism and flatness of machine tools.

Acknowledgements

The work was supported by the National Science Council, Taiwan, Republic of China (NSC 93-2622-E-150-016-CC3)

References

[1] *Angle Encoder* (Heidenhain GmbH, 2004)

[2] Agilent 5529A Dynamic Calibrator (Agilent Technologies, 2000)

[3] LDS Electronic Autocollimator (NewPort Corporation, 1998)

[4] Inclination measurement with compendium yesterday today tomorrow (Wyler AG, 2001)

[5] Huang, P.S., Kiyono, S. and Kamada, O., Applied Optics, Vol.31, No.28, October 1992, pp.6047-6055

[6] Xiaoli, D., Osami S., John, E.G. and Takamasa S., Applied Optics, Vol.35, No.28, October 1996, pp.5657-5666

[7] Xiaoli, D., Osami S., John, E.G. and Takamasa S., Applied Optics, Vol.36, No.25, September 1997, pp.6190-6195

[8] Chiu, M.H. and Su, D.C., Applied Optics, Vol.36, No.28, October 1997, pp.7104-7106

[9] Zhou, W. and Cai, L., Applied Optics, Vol.37, No.25, September 1998, pp.5657-5666

[10] Huang, P.S. and Li, Y., Applied Optics, Vol.37, No.28, October 1998, pp.6636-6642

[11] Guo, J., Zhu, Z. and Deng, W., Applied Optics, Vol.38, No.31, November 1999, pp.6550-6555

[12] Ge, Z. and Takeda, M., Applied Optics, Vol.42, No.34, December 2003, pp.6859-6868

[13] Chiu, M.H., Wang, S.F. and Chang, R.S., Applied Optics, Vol.43, No.29, October 2004, pp.5438-5442.

A Novel Passive Pressure-relief Valve for Blood Pressure Monitor

T.-H. Chen, C.-M. Su, W.-T. Lin and Y.-C. Yang

Industrial Technology Research Institute, 30 Ta Hsueh Rd., Hsinchu, Taiwan 30080, ROC, tinghsuan@itri.org.tw
Industrial Technology Research Institute, 30 Ta Hsueh Rd., Hsinchu, Taiwan 30080, ROC, CMSu@itri.org.tw
Industrial Technology Research Institute, 30 Ta Hsueh Rd., Hsinchu, Taiwan 30080, ROC, WinTiLin@itri.org.tw
Industrial Technology Research Institute, 30 Ta Hsueh Rd., Hsinchu, Taiwan 30080, ROC, ctyang@itri.org.tw

Abstract. In the process of blood pressure measurement, the pressure of the cuff wrapped around human arms is slowly decreasing for blood pressure analysis. Therefore, a relief valve which renders constant pressure relief rate would improve precision of measurement. Here we demonstrate a novel pressure relief valve for blood pressure monitor. Based on the Bernoulli equation and ideal gas assumption, the relationship of the pressure relief rate and the pressure value is quadratic, leading to a reduction in variation of pressure relief rate. In contrast to the other valve designs, the functionality of this valve is passive and free from any moving parts. Besides, the fabrication of the devices incorporates MEMS technique for bulk fabrication, so the manufacturing cost can be substantially reduced. Thus the device was considered reliable and promising to be commercialized.

Two experiments were conducted to test the performance of the valve. First, the relationship between the pressure of cuff and its corresponding flow rate was determined. The results confirmed that their relationship was essentially quadratic. In another experiment, the pressure relief rate was obtained directly. The results demonstrated the relief rate lay between 2 to 4 mmHg/s, satisfying the suggested relief rate. Hence, the improvement of blood pressure measurement may be achieved for blood pressure monitors.

Keywords: Bleed, Relief, Valve, Blood Pressure

1. Introduction

Recently, due to the increase in cardiovascular disease, the demand of healthcare monitor for blood pressure measurement becomes more and more significant in our daily life. Among the methods of blood pressure measurement, the non-invasive one is more general and popular for the household instruments, such as mercury sphygmomanometer or electronic blood pressure monitor. For the type of the non-invasive methods, a cuff, which contains inflatable bladder, is wrapped around a human arm or wrist and provides an external pressure source. Generally, when the measurement is processing, the pressure of the cuff is rising beyond normal systolic pressure; then the gas is slowly bled from inflated cuff to gradually reduce the pressure. During the deflation, pressure oscillation generated from the partially-occluded artery is transmitted through the cuff, and the systolic pressure and diastolic pressure is analyzed by the type of oscillation. Therefore, the accuracy of measurement is affected by the pressure relief rate (mmHg/s). Typically, a relatively constant pressure relief rate, about 3 to 4 mmHg/sec, is desirable since the blood pressure measurement would then be more easily to execute and its result more reliable.

Generally, the pressure relief valve was simply designed as a pinhole or slit, so the performance of pressure relief is similar to the laminar pipe flow where the relationship between pressure relief rate and pressure value is linear [1]. Although it is the simplest way to manufacture a relief valve, a uniform flow rate is hard to achieve under a wide range of pressure conditions. Accordingly, literature and patents providing the valve design which involves moving part [2,3] to adjust the diameter of the valve's outlet. However, the structure is complex and expensive to manufacture. Besides, utilizing the electronic active control to modify the valve outlet was also demonstrated [4]. Indeed, it is the most reliable method to define the pressure relief rate, but the cost would limit the practical applications.

Upon such limitations, a passive valve design that contains non-moving parts is more suitable for industrial requirements. Here we demonstrate a novel passive design that is fabricated by MEMS (MicroElectroMechanical System) process. As shown in Figure 1.1, the valve is a circular plate where a through hole is formed in its center. The operation principle of the device is analogue to the concept of an orifice, as shown in Figure 1.2. Based on the Bernoulli equation, the relationship between pressure relief rate and pressure value is quadratic for orifice with low aspect ratio, therefore, in contrast to the laminar pipe flow, the variation in pressure relief rate could be reduced under wide pressure range. Furthermore, since the MEMS process allows bulk fabrication, the cost of manufacture could be reduced effectively. Consequently, the measurement accuracy of blood pressure monitors can be easily improved by the device without additional cost.

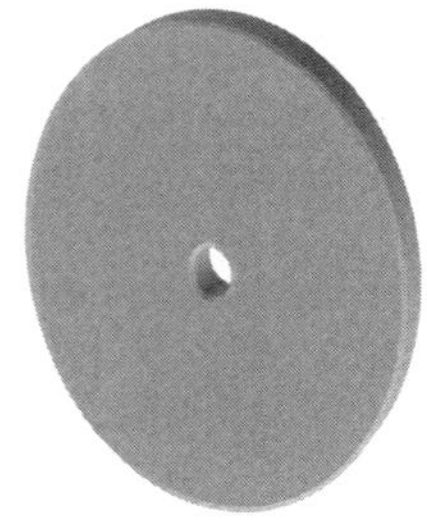

Figure 1.1. Scheme of passive pressure relief valve

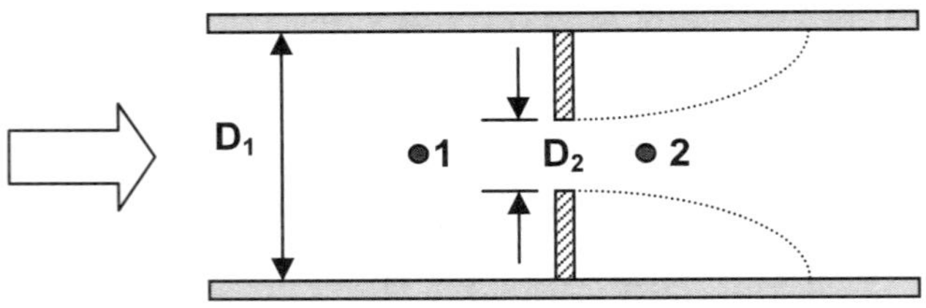

Figure 1.2. Operation principle of pressure relief valve. D_1 is the diameter of flow pipe, and D_2 is the diameter of valve orifice. Point 1 and 2 represent the location before and after the valve, respectively. Similar with the performance of orifice, the relationship between pressure relief rate and pressure value is quadratic in low aspect ratio orifice. Therefore, the variation in pressure relief rate could be reduced.

2. Theory

To better understand the operation principle, a theory was introduced through ISO 5167-3 [5]. Figure 1.2 shows the scheme of relief valve assembly where the valve was installed between a high-pressure cuff (left side) and ambient environment (right side) within a flow pipe. The diameters of flow pipe and orifice of relief valve were denoted as D_1 and D_2, respectively. The theory was developed based on the following assumptions:

- Steady flow
- Non-friction
- The gravity is negligible

As shown in Figure 1.2, point 1 means the location before passing the valve, and point 2 indicates the location after passing the valve. During the deflation of the cuff, a fully developed flow was prevailing the upstream of the pipe, and the presence of the valve made the flow to accelerate through it, thus the velocity is increased at point 2.

According to ISO-5167-3, the mass flow rate was determined as follows [5]:

$$Q_m = \frac{C}{\sqrt{1-\beta^4}} \varepsilon \frac{\pi}{4} D_2^{\,2} \sqrt{2(P_1 - P_2)\rho_1},$$
$$\beta = \frac{D_2}{D_1} \qquad (2.1)$$

where Q_m means the mass flow rate, ρ denotes the density of fluid, P is the pressure value, and subscript 1 and 2 symbolize the locations before and after the relief valve, respectively. In most cases, the coefficient of discharge, C, is a function of $Re(D)$ and β only. Therefore, we can deduce that C is independent of the pressure value.

In our case, air was employed as the working fluid, so the compressibility of the fluid is not negligible here. As included with Eq. (2.1), the compressibility is typically expressed by the expansibility factor, ε. In 1932, the general expression of ε had been derived by Buckinham [6], as shown in follows:

$$\varepsilon = 1 - (0.41 + 0.35\beta^4)\frac{(P_1 - P_2)}{1.4P_1} \qquad (2.2)$$

Notably, ε is relevant to the pressure value, so the relationship between the mass flow rate (and thus the pressure relief rate) and the pressure value might not be purely quadratic. However, in our case, D_2 was designed to be 60 μm or 115 μm, and D_1 of 2 mm was adopted. Accordingly, β equaled to 0.03 or 0.0575, and thus β^4 can be approximated as 0. Furthermore, for the application in which our valve was intended to be used, i.e. the blood pressure measurement, the term $(P_1 - P_2)/P_1$ was in general less than 0.2; hence ε can be calculated as:

$$0.94 \le \varepsilon \le 1,$$
$$\varepsilon \approx 1 \qquad (2.3)$$

As a result, the relation between mass flow rate and pressure difference $(P_1 - P_2)$ was approximately quadratic in this system. Besides, based on the assumption of ideal gas, the flow velocity is proportional to the pressure relief rate. Therefore, we can conclude that the pressure relief rate exhibits quadratic relationship with the pressure value; thus, the variation in pressure relief rate can be reduced during blood pressure measurement.

3. Fabrication

From the theory mentioned above, all designs containing low aspect ratio orifice would follow the rule based on Eq. (2.3). However, in order to satisfy the suggested pressure relief rate, 2 to 3 mmHg/sec, the diameter of the orifice should be constrained to microscale. So far, though, the technology for micro fabrication, such as laser beam machining or electroplating, is mature enough to fabricate the microscale orifice, it is still a challenge because the cost may not be suitable for manufacturing consideration. Here MEMS technology was introduced to fulfill the geometrical requirement since it allows bulk fabrication, so that lots of devices can be produced in a single fabrication process. On the other hand, recent MEMS technology incorporated polymer material to form micro devices; hence, cost from materials is reduced also, being more practical for industry.

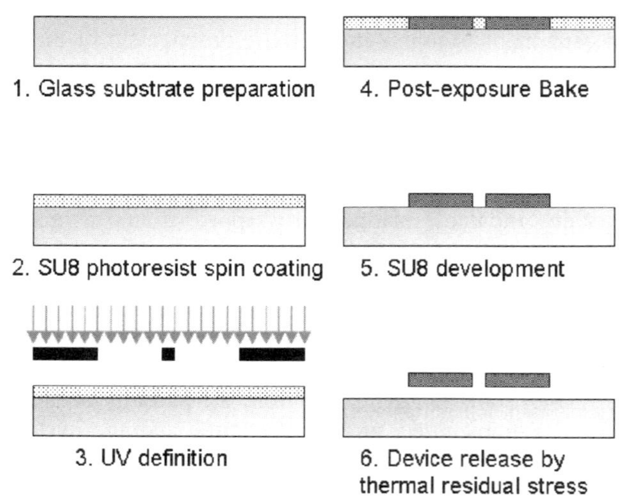

Figure 1.3. Fabrication process of pressure relief valve

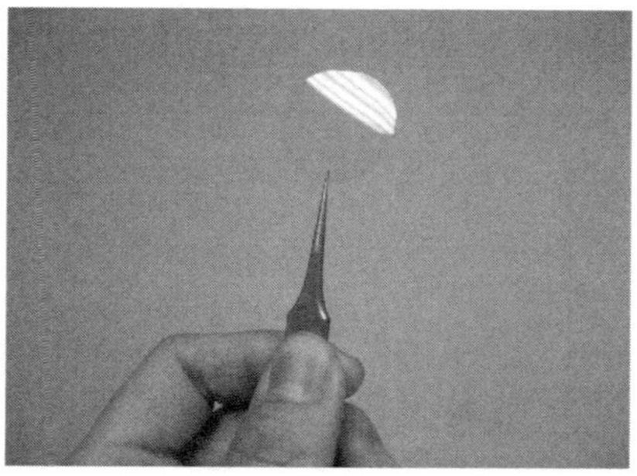

Figure 1.4. Macroscale photograph of pressure relief valve

The fabrication process is illustrated in Figure 1.3. First of all, a glass substrate was prepared by standard cleaning procedure (step 1). Then, negative UV photoresist, SU8 (SU8-2035, MicroChem Corp., USA), was spin-coated onto the glass substrate (step 2). The process parameters were adjusted to precisely yield the designed thickness, 100 μm. After soft baking, the valve formation was defined by UV exposure (step 3). Then the SU8 layer was cross-linked by post-exposure baking (step 4). Afterward, the non-exposured SU8 was developed by SU8 developer (PGMEA, MicroChem Corp., USA) (step 5). Finally, the valve was released from the glass substrate by thermal residual stress (step 6). Figure 1.4 represents the macro photograph of the valve body, and the diameter of the valve orifice was measured to be 60 μm and 115 μm.

4. Experimental Setup

4.1 Flow Rate Measurement

The scheme of experiment setting is illustrated in Figure 1.5. In this experiment, a controlled pressure parameter was used to simulate the pressure of a cuff, and the corresponding flow rate was obtained to determine their relationship. As shown in Figure 1.5, the experiment was set by installing the device between the high pressure resource (Gas Tank) and ambient environment (Outlet). When the experiment was started, the control valve was opened and adjusted to control the pressure, and the pressure value was read by the pressure meter set behind the control valve. Next, the pressure relief valve was installed within a clamping apparatus. The flow rate was measured by a flow meter installed upstream of the clamping apparatus. Consequently, the relationship between the pressure and flow rate was determined by adjusting and maintaining different inlet pressures of the clamping apparatus.

4.2 Pressure Relief Rate

On the other hand, another experiment was conducted to measure the pressure relief rate directly. The scheme of the experiment setting is represented in Figure 1.6.

In contrast to the previous experiment, an inflatable container of predetermined volume was employed to simulate the real cuff in blood measurement monitor. Simulated cuffs with different volume were adopted to simulate the wrist-type and arm-type devices. Both the pressure value and the pressure relief rate can be obtained directly by the readings of the pressure meter.

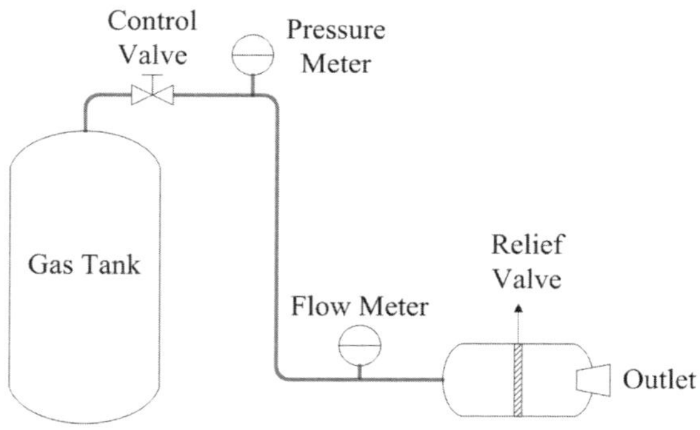

Figure 1.5. Scheme of experiment setting. The pressure relieve valve was installed between pressure resource and ambient environment. Pressure meter and flow meter are connected to monitor mass flow rate and pressure value. The control valve is used to start the experiment process and adjust the present pressure value.

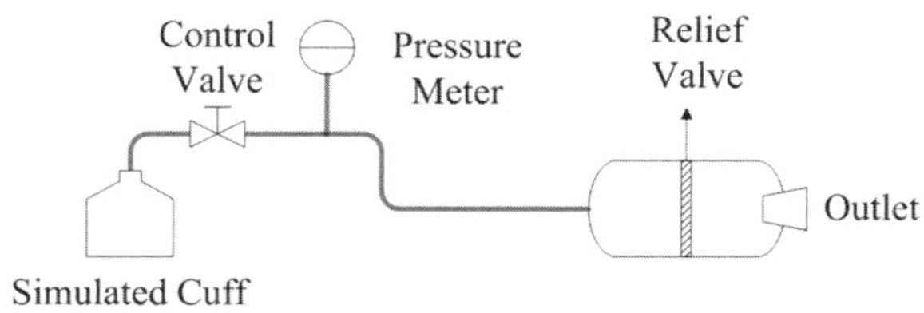

Figure 1.6. Scheme of the experiment incorporating simulated cuff. By reading the pressure value with a timer, the pressure relief rate is determined directly.

5. Results

5.1 Flow Rate Measurement

Based on the desired pressure relief rate, different diameters of orifice, 60 μm and 115 μm, were used for the wrist-type and arm-type measurement. Figure 1.7 shows the result of flow rate measurement using the experimental setup shown in Figure 1.5. The results for orifice with diameters of 115 μm and 60 μm are presented in Figure 1.7 (a) and (b), respectively. When the diameter of orifice was chosen as 115 μm, the variation of actual flow rate was between 30 ml/min and 80 ml/min while the pressure difference between upstream and downstream of the valve varied from about 9 kPa to 30 kPa. On the other hand, for the orifice of diameter 60 μm, the variation of actual flow rate was from 11 ml/min to 22.5 ml/min while the pressure difference varied from about 4 kPa to 27 kPa. These results demonstrated that the desirous quadratic relationship between pressure difference and flow rate is confirmed.

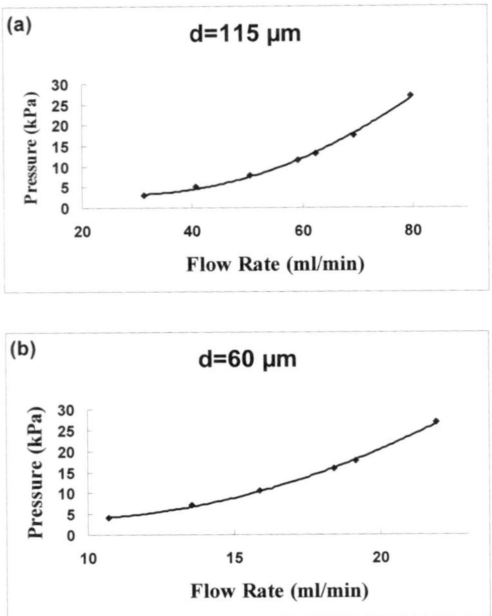

Figure 1.7. Experiment result of the flow rate measurement where the diameter of orifice was chosen as (a) 112.5 μm and (b) 60 μm. The results verify the concept is functional to perform the quadratic relationship.

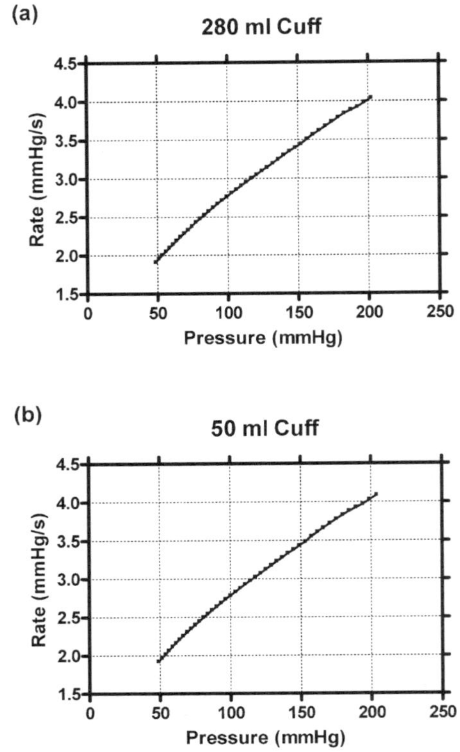

Figure 1.8 Experiment results of pressure relief rate. These results show the relief rate, 2 to 4 mmHg/s, fitting the suggested relief rate for accurate measurement.

5.2 Pressure Relief Rate

In this experiment, two types of inflatable containers, 280 ml and 50 ml, were used to simulate the wrist-type and arm-type measurement. The orifice diameters of 115 μm and 60 μm were set for containers of 280 ml and 50 ml, respectively, and their results are shown in Figure 1.8 (a) and (b), respectively. Both figures demonstrate that, for the range of pressure typically encountered in the application considered here, i.e. 60 mmHg to 180 mmHg, the pressure relief rate lies between 2 and 4 mmHg/s. The results were satisfactory considering its passive nature and the fact that the device was designed to be used in low cost products. In addition, the improvement of valve performance, as compared with traditional ones, was also confirmed.

6. Conclusion

In this paper, a passive pressure relief valve for blood pressure monitors was demonstrated. According to the Bernoulli equation, the relationship between pressure and pressure relief rate is quadratic in low aspect ratio orifice; hence, the valve was designed based on such theory to reduce the variation in pressure relief rate during the course of taking blood pressure. To satisfy the suggested flow rate and industrial consideration for mass production, the valve was fabricated by MEMS technology, providing a solution for low cost manufacture. In the experiments of valve performance testing, the results confirmed subtantially the quadratic relationship and yielded pressure relief rates lay between 2 mmHg/s and 4 mmHg/s. Therefore, it is deemed that blood pressure measurement can be improved for monitors make using this pressure relief device.

Acknowledgement

This work was sponsored by the Bureau of Standards, Metrology and Inspection, Taiwan R.O.C., under project code 96-1403-36-H01-00-00-0024. The authors would like to thank Prof. Fan-Gang Tseng for his support in the experimental facilities during the course of the study.

References

[1] Berg RF, (2005), Simple flow meter and viscometer of high accuracy for gases. Metrologia, Vol. 42, P11-23

[2] Matsuura M; Ichinomiya T and Abe H, (1988), Exhaust valve having a constant bleed rate. U. S. Patent, Re. 32587

[3] Ezekiel FD, Lexington, Aisenberg S and Natick, (1980), Automatic air deflation regulator for use in an instrument for measuring blood pressure. U. S. Patent, 4198031

[4] Ducloux O, Talbi A, Deblock Y, Gimeno L, Tiercelin N, Pernod P, Preobrazhensky V and Merlen A, (2006), Magnetically actuated microvalve for active flow control. J. Phys.: Conference Series, Vol. 34, P. 631-636.

[5] International Organization for Standardization, (2002), Measurement of fluid flow by means of pressure differential devices - Part 3: Nozzles and Venturi nozzles, ISO 5167-3. General International Organization for Standardization.

[6] Buckingham E, (1932), Notes on the orifice meter: the expansion factor for gases. J. Research, Research Paper No. 459, Vol. 9.

16–5

3D Data Acquisition and Configuration by Means of Binocular Stereopsis

H. Otsuka[1], Y. Murakami[2], Z. Qiu[3], H. Ojima[4], L. Zhou[5], J. Shimizu[6] and H. Eda[7]

[1]Ibaraki University, Japan, 3168511, 06nm908f@hcs.ibaraki.ac.jp
[2]Ibaraki University, Japan, 3168511, t63082r@hcs.ibaraki.ac.jp
[3]Ibaraki University, Japan, 3168511, qzj@mx.ibaraki.ac.jp
[4]Ibaraki University, Japan, 3168511, ojima@mx.ibaraki.ac.jp
[5]Ibaraki University, Japan, 3168511, lbzhou@mx.ibaraki.ac.jp
[6]Ibaraki University, Japan, 3168511, jshimizu@mx.ibaraki.ac.jp
[7]Ibaraki University, Japan, 3168511, eda@mx.ibaraki.ac.jp

Abstract. Image processing technology has now been widely applied into many areas such as electronics, manufacturing and security. However, most of those applications are based on two dimensional images. Recently, there are increasing demands of 3D information for new applications. The incoming visual information via a digital imaging device (ex. CCD) is digitized into pixels with color/gray scale. As each pixel contains 2D positional information, the CCD thus offers an 2D orthogonal coordinate, but loses the 3rd dimension in depth for objects in view. The aim of research is to develop an algorithm for acquisition and configuration of 3D profile by means of binocular stereopsis. Finally, the 3D data acquisition and configuration are performed on both static stereopair and live stereo vision.

Keywords: Image processing, Binocular stereopsis, Pattern matching, 3D data acquisition.

1. Introduction

With the cost-reduction in high-end CCD and CPU, the image processing technology has now been widely applied into many areas such as electronics, manufacturing, robotics and medical services for the purposes of inspection, automation, counting, measuring and visualization [1]. However, most of those applications are based on two dimensional images, due to the nature of the imaging device. Recently, there are increasing demands of 3D information in new applications like monitoring, positioning and ITS (intelligent transportation system).

As each pixel contains 2D positional information, the CCD thus offers an orthogonal coordinate, but loses the 3rd dimension in depth for objects in view. The 3D data acquisition and configuration are essential for those applications which require the 3D information. One of the fundamental methods to derive the 3D information of the object from 2D image is binocular stereopsis. The aim of research is to develop an algorithm of acquisition and configuration of 3D data by means of binocular stereopsis.

2. Principle of binocular stereopsis

Binocular stereopsis method obtains the 3rd dimension through stereopair or stereo vision via the stereo disparity [2]. Figures 1.1 and 1.2 show the concept of binocular steropsis method and the mesurement principle of the parallel optical axes arrangement.

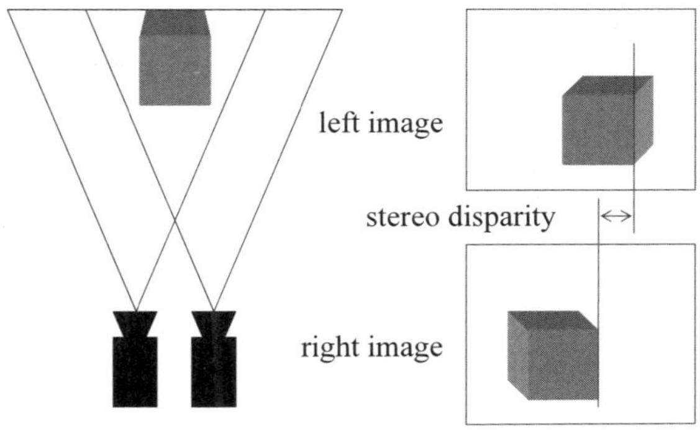

Figure 1.1. Concept of binocular stereopsis

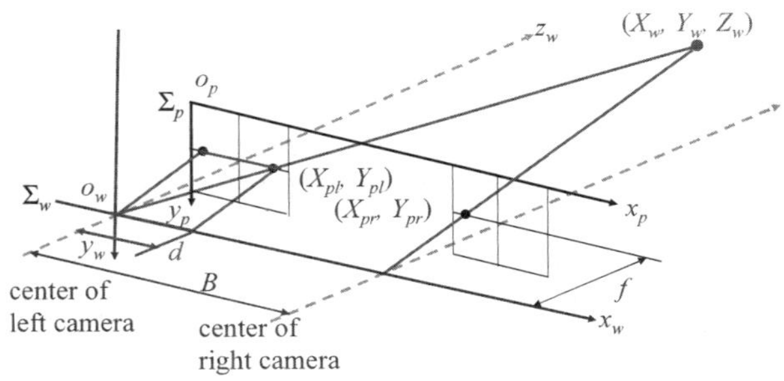

Figure 1.2. Mesurement principle of binocular steropsis

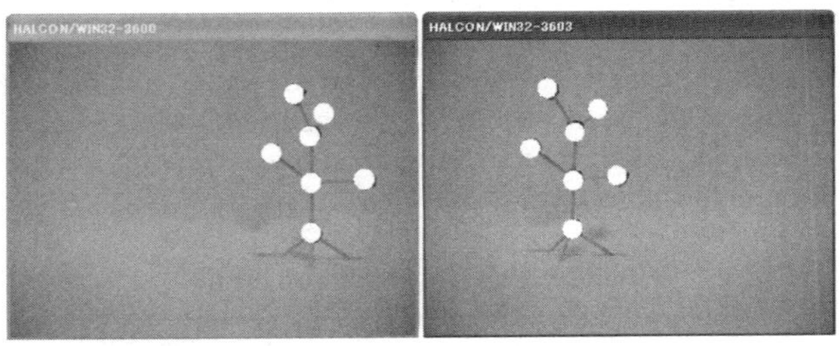

Figure 1.3. Feature based pattern matching

Let Σ_w be world coordinate system, Σ_p be pixel coordinate system, d be stereo disparity, B be the distance between cameras, f be focal length. Transformation from Σ_w to Σ_p is given by

$$\begin{Bmatrix} x_p \\ y_p \\ 0 \end{Bmatrix} = \frac{1}{T_w} \begin{bmatrix} 1 & 0 & 0 \\ 0 & 1 & 0 \\ 0 & 0 & 0 \end{bmatrix} \begin{Bmatrix} x_w \\ y_w \\ z_w \end{Bmatrix} \tag{1.1}$$

where T_w is scaling factor that is length per pixel. With the use of geometric similarity, T_w is given by

$$T_w = \frac{B}{d} \tag{1.2}$$

Consequently, the 3rd dimension Z_w is acquired by

$$Z_w = T_w f = \frac{Bf}{d} \tag{1.3}$$

3. Pattern matching

As indicated above, 3D data acquisition requires the disparity d of correspondent points in the stereopair. Unfortunately, difficulty in pairing the correspondent points of stereo image constitutes a barrier against applying this method in practical use. The technique to solve this problem is pattern matching [3]. Two types of pattern matching technique are developed for pairing the correspondent points; one is the gray scale based pattern matching which is designed to attain high accurate 3D data, while another is the feature based pattern matching which is oriented to make the process speed fast enough to respond to live or moving images.

In this research, feature based pattern matching is employed. Figure 1.3 shows the appearance of the feature based pattern matching for live stereo vision. This method is based on the edge data and its normal direction data. It is comprised of two steps.

The first step is to create the shape model for pattern matching. In this step, a feature in common shape for stereopair is pre-registered as the templete for successive pattern matching. For example, the templete simply are circles in Figure 1.3 The second step is to find features in both images matching with the templete. This procedure returns the coordinate values of matched feature points in pixel. As a result, the correspondent point can be found in stereopair and their disparity can be calculated.

4. 3D data acquisition and configuration

In this research, the feature based pattern matching algorithm has been developed and applied into two kinds of field test, namely, static stereopairs and moving images as listed in Table 1.1.

Table 1.1. Details of mesurement

	Object	Imaging device	Mesurement principle
static image	Microscopic topography	SEM	ref. Figure 95.4.
moving image	polyhedron models	CCD cameras	ref. Figure 95.2.

The procedures of measurement for the static stereopair are listed below;

1. Photo taken by SEM (scanning electron microscope)
2. Preparation
3. Feature point extraction
4. Pattern matching
5. Data processing
6. 3D display

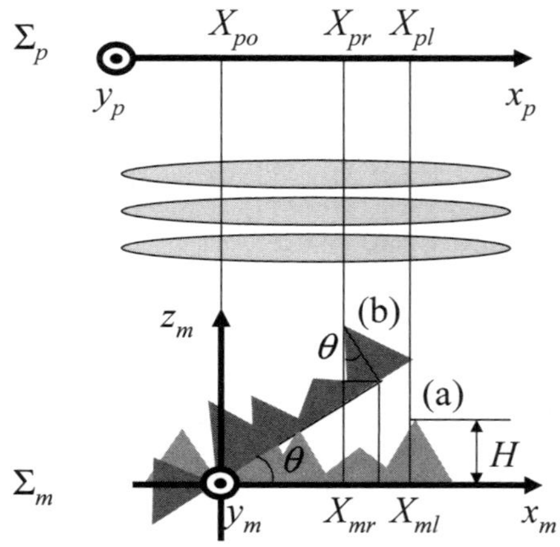

Figure 1.4. Camera model and sample arrangement for SEM photo taking

Figure 1.4 shows the camera and sample arrangement in SEM. Two photos are taken at a different tilting angle of θ degree, which compose of the stereopair (a) and (b), respectively. Let Σ_m be the model coordinate system, Σ_p be the pixel coordinate system, θ be the tilting angle. At a telecentric optical system, transformation from Σ_m to Σ_p is given by

$$\begin{Bmatrix} x_p \\ y_p \\ 0 \end{Bmatrix} = \frac{1}{T_m} \begin{bmatrix} 1 & 0 & 0 \\ 0 & 1 & 0 \\ 0 & 0 & 0 \end{bmatrix} \begin{Bmatrix} x_m \\ y_m \\ z_m \end{Bmatrix} \tag{1.4}$$

where T_m is scaling factor. The correspondent points in the left image X_{ml} and the right image X_{mr} are related by

$$X_{mr} = X_{ml} \cos\theta - H \sin\theta \tag{1.5}$$

If X_{ml} and X_{mr} are properly measured and matched, the correspondent height H at each point is given by

$$H = \frac{X_{ml}}{\tan\theta} - \frac{X_{mr}}{\sin\theta} = T_w \left(\frac{X_{pl}}{\tan\theta} - \frac{X_{pr}}{\sin\theta} \right) \tag{1.6}$$

Figure 1.5 shows a stereopair of SiC (Silicon carbide) sample image captured by SEM. Next, two steps of preparation are implemented. One is histogram equalization which enhances contrast and another is edge preserving smoothing which removes noises. After preparation, feature point is extracted. In this reserch, the feature points are the saddle points determined by partial differentiation of image brightness. Figure 1.6 (a) shows the scope of feature point extraction, whereas Figure 1.6 (b) shows extracted feature points.

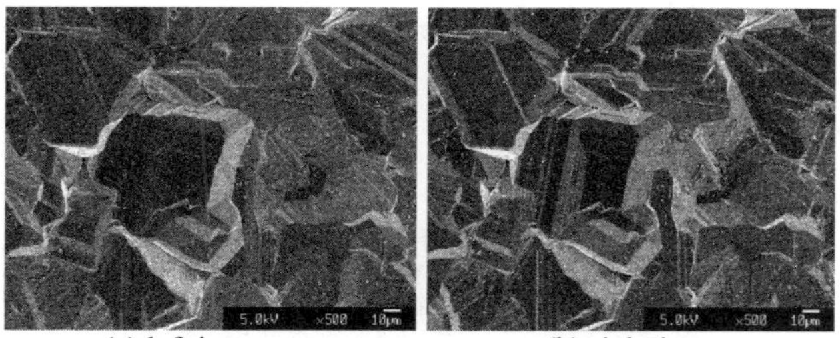

(a) left image (b) right image

Figure 1.5. SEM Stereopair of SiC surface

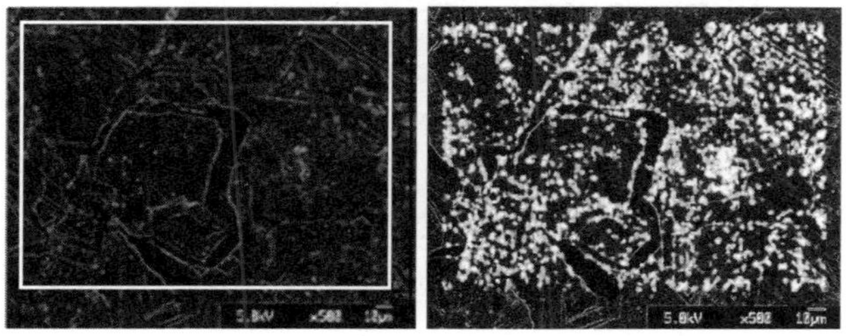

(a) Scope of feature point extraction (b) Extracted feature points

Figure 1.6. Feature point extraction

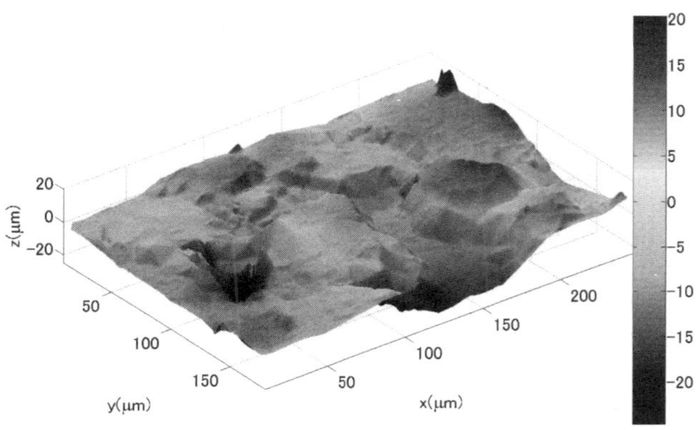

Figure 1.7. Mesurement result of SiC surface

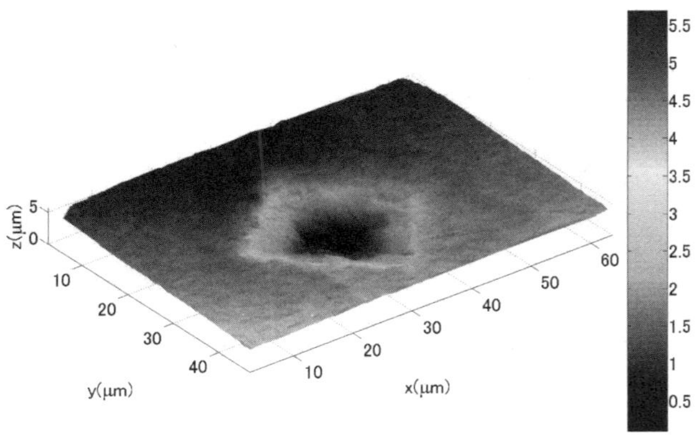

Figure 1.8. Measurement result of indentation on brass surface

Subsequently, pattern matching is implemented on the feature points. According to the matching result, the height of each feature point is calculated by Equation (1.6). Next, the outlier analysis has been done to eliminate the points mismatched from the calculated results. In this research, outliers are defined as those points whose heights exceed three times of standard deviation σ from the mean value *m* (or $m \pm 3\sigma$), as the result of try-and-error. It should be noted that the scope of outlier largely varies with the profile of objects and the quality of images. Figure 1.7 is the 3D result of the stereopair shown in Figure 1.5, which demonstrated a good agreement with the actual profile.

In order to quantitatively verify the accuracy of this method, an indented brass sample was purposely prepared. It was measured by both stereopsis and a COLOR 3D Laser Scanning Microscope (KEYENCE VK-9700). The result of stereopsis is shown in Figure 1.8. The comparison on its cross sectional profiles are given in Figure 1.9. Figure 1.9(a) shows the result by Laser Scanning Microscope, while (b) is the result by proposed stereopsis technique. The errors are 10% and 15% in horizontal direction and depth direction resepectively. As they are similar one to another, the errors are likely resulted from the inaccurate scale factor of T_m.

The procedures of measurement for the moving stereo vision are shown below;

1. Capturing a 3D object by two CCD cameras
2. Pattern matching
3. Data processing
4. 3D display

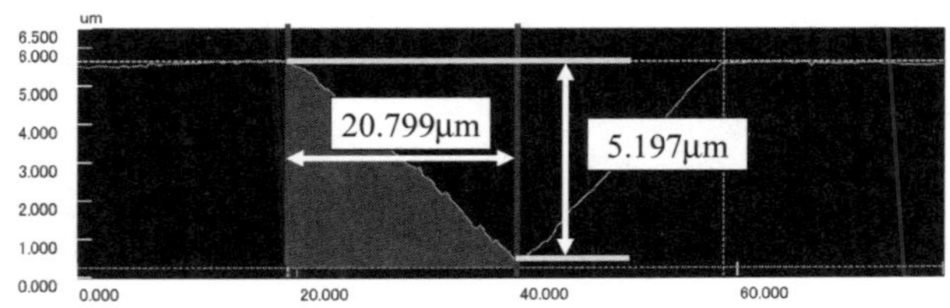

(a) Result by Laser Scanning Microscope

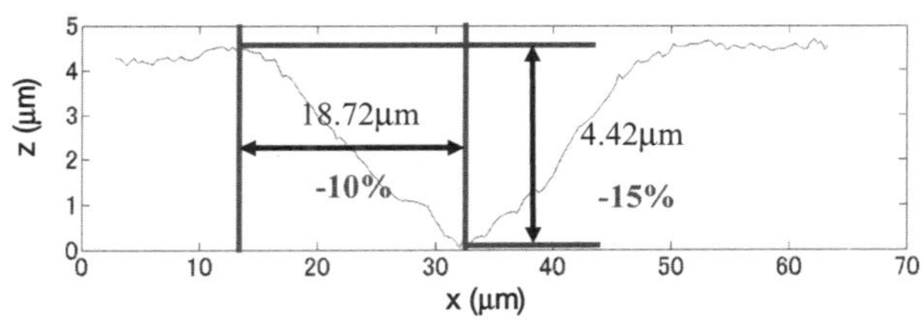

(b) Result by binocular stereopsis

Figure 1.9. Comparison of depth profile

Figure 1.10 shows a series images of a moving object captured by two CCD cameras. The optical axes of CCD cameras are configured perallelly in order to use the Equations (1.1) - (1.3). In Figure 1.10 (a) to (b), the object is moved along z direction from back side toward the camera (or front side), wheares from Figure 1.10 (b) to (c), vice versa. Thirteen continuous stereopairs are captured as the object is moving back and forward along z direction.

Subsequently, pattern matching is implemented. Seven circles are used as the templete in this case. As the object moves along optical axis, the size of the object and scaling factor are changed in a non-telecentric optical system. In this research, since processing is simplified to place priority on the development of algorithm, only the change in size is taken into account when matching of the correspondent points is performed. Then their disparities and 3D positions including the value in z-axis (or optical axis) are calculated by use of Equations (1.1) - (1.3).

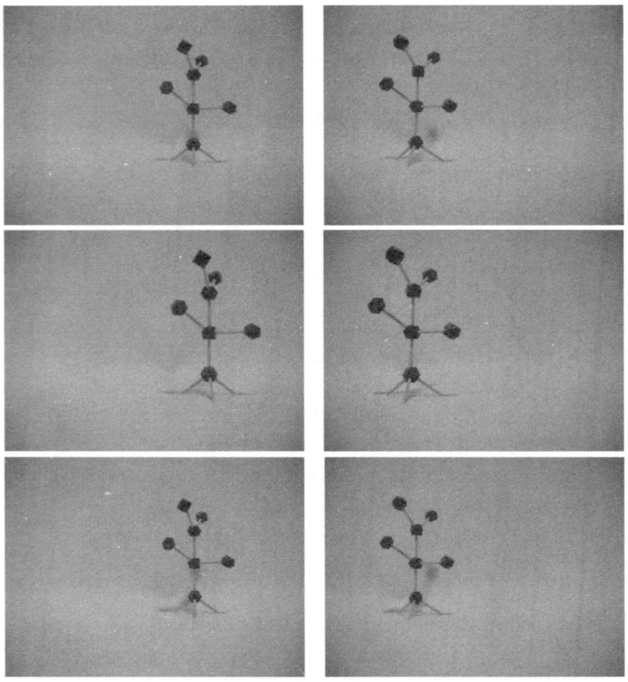

Figure 1.10. Moving stereo vision

Table 1.2. Specifications of hardware and prameters for fine tuning

Specifications of hardware				
OS	Windows XP			
CPU	Intel(R) Pentium(R) 4 3.4GHz			
Memory	1.5GB RAM			
Camera	SONY DCR-HC90 / 3.0Mpixels			
Prameters for fine tuning				
	AngleExtent (radian)	MinScore	Greediness	Time (sec.)
Before	0.78	0.50	0.90	4.0-4.6
After	0.0	0.45	1.0	0.70

According to the acquired 3D data, the object is re-constructed graphically by use of Open-GL. One sample result is demonstrated in Figure 1.11. By repeating the above image capturing, feature pattern matching, data processing and displaying, 3D data acquisition and configuration is continuously implemented for a moving object.

In order to improve the pattern matching speed, the parameters for feature based pattern matching are fine turned in aspect of rotation (AngleExtent), the minimum concordance (MinScore) and threshold of searching time (Greediness). The details are listed in Table 1.2 together with specifications of hardware used in this research. The processing time is able to be reduced from 4.0 - 4.6 s to 0.70s. It should be noted that improvement in processing speed is trade-off between processing time and accuracy.

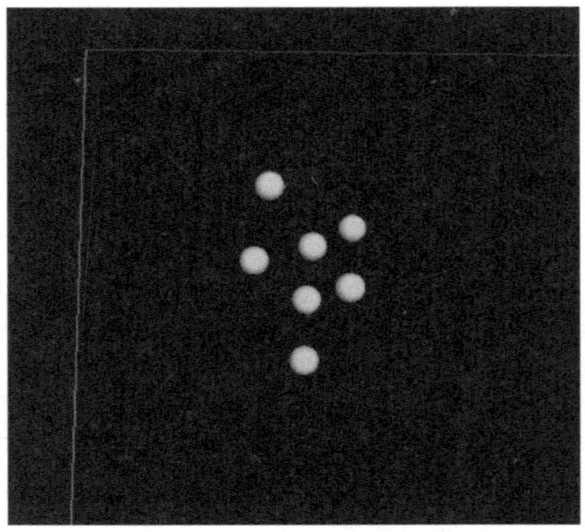

Figure 1.11. Graphic display of measurement result

5. Conclusion

In this research, an algorithm by means of binocular stereopsis has been developed for 3D acquisition and configuration, and applied into both static stereopairs and moving images.

In the case of static stereopairs, 3D topography is able to be acquired and configured at the level of micrometer. The accuracy is able to be further improved by proper calibration of scaling factor.

In the case of moving images, 3D position of a simple object is acquired and re-constructed graphically. Further improvement in processing speed is necessary to accommodate to the commercial motion capture systems in the video rate.

References

[1] The Japan Society for Precision Engineering, 2000, Image processing application system, Tokyo Denki University Press, (in Japanese), p5-7.

[2] Hideyuki TAMURA, 2002, Computer image processing, Ohmsha, (in Japanese), p227-230.

[3] LinX Corporation, 2004, HALCON Katsuyouhou, LinX Corporation, (in Japanese), p128-146.

[4] Z. Qiu, L. Zhou, K. Nishida, J. shimizu, H. Eda and T. Ishikawa, 2004, Development of Cell Manipulation System under Microscope, Proc. of the 1st Int. Conf. on Positioning Technology, pp.457-461.